INTERNATIONAL NEURAL NETWORK CONFERENCE

INNC 90 PARIS

Volume 2

INTERNATIONAL NEURAL NETWORK CONFERENCE

JULY 9-13, 1990
PALAIS DES CONGRES - PARIS - FRANCE

UNDER THE PATRONAGE OF THE COMMISSION OF THE EUROPEAN COMMUNITIES

THE INTERNATIONAL NEURAL NETWORK SOCIETY (INNS), THE IEEE NEURAL NETWORK COUNCIL COOPERATING SOCIETIES

KLUWER ACADEMIC PUBLISHERS
DORDRECHT / BOSTON / LONDON

ISBN-13:978-0-7923-0831-7 e-ISBN-13:978-94-009-0643-3
DOI: 10.1007/978-94-009-0643-3

IEEE Catalog Number 90-TH-0318-6

Published by Kluwer Academic Publishers,
P.O. Box 17, 3300 AA Dordrecht, The Netherlands.

Kluwer Academic Publishers incorporates
the publishing programmes of
D. Reidel, Martinus Nijhoff, Dr W. Junk and MTP Press.

Sold and distributed in the U.S.A. and Canada
by Kluwer Academic Publishers,
101 Philip Drive, Norwell, MA 02061, U.S.A.

In all other countries, sold and distributed
by Kluwer Academic Publishers Group,
P.O. Box 322, 3300 AH Dordrecht, The Netherlands.

Printed on acid-free paper

All Rights Reserved
© 1990 Kluwer Academic Publishers
No part of the material protected by this copyright notice may be reproduced or
utilized in any form or by any means, electronic or mechanical,
including photocopying, recording or by any information storage and
retrieval system, without written permission from the copyright owner.

THE ORGANISATION

The conference INNC-90-PARIS held at the Palais des Congrès from July 9 - 13, 1990 is divided into four main parts with 24 sessions:

- Applications (Image, Speech and Signal Processing, Robotics and Control, Optimization, Classification, Decision).
- Implementation (Neurobiological models, Cognitive science, Electronic and Optical neurocomputers, Implementations on Parallel Hardwares, Languages and Development environment).
- Theory (Supervised and Unsupervised Learning, Associative Memories, Architectures, Analysis of Network Dynamics, Statistics).
- Commercial (European and Government programs, Commercial Products, Company Strategy).

The trade show to display the latest innovations on neural networks will have approximately 40 exhibitors.

On July 9th, nine tutorials will be given by the world's leading specialists in the field.

453 papers were received from 35 countries (half from Europe):

. 221 papers were accepted for oral presentation.

. 164 papers were accepted for poster presentation.

Three-fourths of the papers were from scientists, one-fourth from industrialists.

We wish to thank for their help and their confidence: the INNS, the IEEE, the sponsors, the chairs, the speakers, the exhibitors, the reviewers, the participants and particularly the THOMSON, Bernard ANGENIOL and Bernard WIDROW who made this first large European Conference on Neural Networks supported by the INNS and the IEEE possible.

Nina THELLIER
INNC-90-PARIS Conference manager

INNC-90-PARIS CONFERENCE COMMITTEE

CONFERENCE CO-CHAIRS:

Bernard WIDROW: Professor at Stanford University, President INNS
Bernard ANGENIOL: THOMSON-CSF, Chairman of Pygmalion, Esprit Project on Neural Networks

PROGRAM CHAIR:

Teuvo KOHONEN: Professor at the Helsinki University of Technology

PROGRAM COMMITTEE:

Igor ALEKSANDER: Professor at the Imperial College London
Shun-Ichi AMARI: Professor at the University of Tokyo
Leon COOPER: Nobel Prize laureate, Professor at Brown University, co-founder of Nestor Inc.
Rolf ECKMILLER: Professor at the University of Dusseldorf
Françoise FOGELMAN: Professor at the University of PARIS XI
Stephen GROSSBERG: Professor at Boston University
Federico FAGGIN: President and founder of SYNAPTICS
Philip TRELEAVEN: Professor at the University College London
Fernando ALDANA: Vice-president Universidad Politechnica de Madrid

CONFERENCE MANAGEMENT:

Nina THELLIER, NTC: 19, rue de la Tour, 75116 PARIS

INNC-90-PARIS CONFERENCE SPONSORS

MAJOR SPONSOR OF THE CONFERENCE:
THOMSON

SPONSORED BY:
SUN
The British Computer Society
FIEE
FRAMENTEC
INRIA
NESTOR Inc.
PHILIPS
SIEMENS
SIREN
The DRET
The SPRINT Program of the Commission of the European Communities
01 INFORMATIQUE

FOREWORD

Neural Networks have been the theater of a dramatic increase of activities in the last five years.
The interest of mixing results from fields as different as neurobiology, physics (spin glass theory), mathematics (linear algebra, statistics...), computer science (software engineering, hardware architectures...) or psychology has attracted a large number of researchers to the field. The perspective of dramatic improvements in many applications has lead important companies to launch new neural network programs and start-ups have mushroomed to address this new market.

Throughout the world large programs are being set-up: in Japan the government has committed more than $18 million per year to its 20 year Human Frontier Science program; the DARPA and the US Navy have alloted more than $10 million per year each and other US government agencies are contributing to important but less ambitious programs. Neural networks are also a major research are in the supercomputing initiative.

Europe has from the beginning taken an active part in funding major projects in the new field with BRAIN, BRA, ANNIE and PYGMALION (Esprit).
Approximately $20 million has been invested to date since 1988 and new programs of nearly $30 million are being funded for the next 3 years. National projects in certain countries may globally double these amounts.

Neural network conferences are attracting larger audiences than ever before. Prior to 1987 attendance never surpassed 300.
The June 1989 IJCNN conference in Washington had over 2200 participants.
In Europe many workshops have taken place during the past 3 years but **INNC-90-PARIS** is the first large conference to take place under the patronage of the Commission of the European Communities, with the INNS and IEEE acting as cooperating societies. The marked interest of the Americans for developing scientific and commercial exchanges with their European counterparts is also significant.

We would like to thank and congratulate all those who have contributed to the Conference: the authors and reviewers of papers, the chairmen of the sessions, the speakers, panelists and trade show participants, all the members of the program committee, and especially Teuvo Kohonen, the program chair.
Finally we wish to thank the NTC team for the tremendous work they did in the practical organization.

Bernard Angeniol, Bernard Widrow
INNC-90-PARIS Conference co-chairs

TABLE OF CONTENTS

Volume 1

IMAGE PROCESSING I

- Chair: Kunihiko FUKUSHIMA

* ORAL PRESENTATION

* POSTER PRESENTATION

IMAGE PROCESSING II

- Chair: Errki OJA

* ORAL PRESENTATION

* POSTER PRESENTATION

SIGNAL PROCESSING

- Chair: David BOUNDS

* ORAL PRESENTATION

* POSTER PRESENTATION

SPEECH PROCESSING

- Chair: Hervé BOURLARD

* ORAL PRESENTATION

* POSTER PRESENTATION

ROBOTICS

- Chair: Jacob BARHEN

* ORAL PRESENTATION

* POSTER PRESENTATION

OPTIMIZATION

- Chair: Chen-Hang SUNG

* ORAL PRESENTATION

* POSTER PRESENTATION

CLASSIFICATION - DECISION - PREDICTION I

- Chair: Françoise FOGELMAN

* ORAL PRESENTATION

*POSTER PRESENTATION

CLASSIFICATION - DECISION - PREDICTION II

- Chair: Larry JACKEL

* ORAL PRESENTATION

* POSTER PRESENTATION

CONTROL AND EXPERT SYSTEMS APPLICATIONS

- Chair: Bernard WIDROW

* ORAL PRESENTATION

* POSTER PRESENTATION

NEUROBIOLOGICAL MODELS

- Chairs: Elie BIENENSTOCK and Vitaly KRYUKOV

* ORAL PRESENTATION

* POSTER PRESENTATION

COGNITIVE SCIENCE

- Chair: Rolf ECKMILLER

* ORAL PRESENTATION

* POSTER PRESENTATION

Volume 2

ELECTRONIC NEUROCOMPUTERS

- Chair: Federico FAGGIN

* ORAL PRESENTATION

* POSTER PRESENTATION

OPTICAL NEUROCOMPUTERS

- Chair: Demetri PSALTIS

* ORAL PRESENTATION

* POSTER PRESENTATION

IMPLEMENTATION OF NEURAL NETWORKS ALGORITHMS ON PARALLEL HARDWARE

- Chair: Igor ALEKSANDER

* ORAL PRESENTATION

* POSTER PRESENTATION

NETWORK DEFINITION LANGUAGES AND DEVELOPMENT ENVIRONMENT

- Chair: Philip TRELEAVEN

* ORAL PRESENTATION

* POSTER PRESENTATION

BENCHMARKING

- Chair: Philip TRELEAVEN

* ORAL PRESENTATION

SUPERVISED LEARNING

- Chairs: Yann LE CUN and Luis BORGES de ALMEIDA

* ORAL PRESENTATION

* POSTER PRESENTATION

UNSUPERVISED LEARNING

- Chair: Stephen GROSSBERG

* ORAL PRESENTATION

* POSTER PRESENTATION

ASSOCIATIVE MEMORIES

- Chair: James ANDERSON

* ORAL PRESENTATION

* POSTER PRESENTATION

ARCHITECTURES

- Chair: Robert HECHT-NIELSEN

* ORAL PRESENTATION

* POSTER PRESENTATION

ANALYSIS OF NETWORK DYNAMICS I

- Chair: David WALLACE

* ORAL PRESENTATION

* POSTER PRESENTATION

ANALYSIS OF NETWORK DYNAMICS II

- Chair: Shun-Ichi AMARI

* ORAL PRESENTATION

* POSTER PRESENTATION

STATISTICS AND PROBABILITIES

- Chair: Shun-Ichi AMARI

* ORAL PRESENTATION

* POSTER PRESENTATION

EEC AND GOVERNMENT SUBSIDIZED PROJECTS

- CHAIR: Jean-Jacques LAUTURE

* ORAL PRESENTATION

COMPANY STRATEGY

- CHAIR: Wolfram BUTTNER

* ORAL PRESENTATION

ELECTRONIC NEUROCOMPUTERS

Chair: Federico FAGGIN

LARGE SCALE OPTOELECTRONIC INTEGRATION OF ASYNCHRONOUS ANALOG NEURAL NETWORKS.

Gert Cauwenberghs, Charles F. Neugebauer, Aharon J. Agranat, Amnon Yariv
Departments of Applied Physics and Electrical Engineering
128-95 California Institute of Technology
Pasadena CA 91125, U.S.A.

Abstract *A simple circuit architecture in standard CMOS technology for the optoelectronic implementation of analog continuous-time neural networks (NN) is presented. The circuit enables the implementation of recurrent NN models with analog synapses and neurons, with continuous dynamics. The basic cell consists of a synapse coupled with a distributed neuron, where synaptic linear superposition and neuron nonlinear thresholding are combined together, using only five MOS transistors and one phototransistor per synapse. The synaptic interaction matrix is imaged continuously on the chip from a spatial light modulator, thus allowing fast reprogramming of the connections. The performance of the proposed system is illustrated by some measurements of synapse and neuron characteristics on a 16 neuron (256 synapse) prototype fabricated in MOSIS CMOS technology. The expected performance and limitations of a scaled up system are discussed.*

Careful examination of existing analog NN hardware reveals a tradeoff between the complexity of the implemented model and the size of the implemented network [1]. We henceforth present an optoelectronic NN implementation with fully analog neurons and synapses, using one phototransistor and five MOS transistors per synapse. The underlying principle is to image the synaptic interaction matrix from a spatial light modulator onto a VLSI circuit which performs the dynamics of the NN. Similar implementations were demonstrated earlier for binary neurons and analog synapses [2,3]. The system described here extends the previous implementations to analog neurons.

The general asynchronous analog NN model we seek to implement satisfies [4]:

$$\begin{aligned}(\tau_i \frac{\mathrm{d}}{\mathrm{d}t} + 1)\, u_i &= I_i + \sum_j T_{ij} V_j \\ V_i &= \sigma(u_i - u_i^{ref})\end{aligned} \tag{1}$$

where T_{ij} is the strength of the synaptic interconnection from neuron j to neuron i, I_i the external input i, and τ_i the time constant of neuron i. V_i is the output state of neuron i, related to the neuron potential u_i by the sigmoid thresholding function σ and threshold potential u_i^{ref}. The most commonly used sigmoid σ is the logistic function:

$$\sigma(u) = \frac{1}{1 + e^{-u}}. \tag{2}$$

A straightforward way to implement (1), found in most analog electronic NNs today, is to design a linear network synthesizing the sum of weighted outputs $\sum T_{ij} V_j$, followed by a thresholding device σ [1]. The linearity requirement on the weighted sum often necessitates a large number of transistors per synapse, as in [5] where four-quadrant multipliers are employed to generate the product $T_{ij}V_j$. Our approach here avoids the linearity requirement, by combining the thresholding $V_j = \sigma(u_j - u_j^{ref})$ and the synaptic contribution $T_{ij}V_j$ in a single device. Interesting enough, it is extremely simple to generate the quantity $T_{ij}\ \sigma(u_j - u_j^{ref})$ with just a few transistors. The device we need is simply a *differential pair* (DP): two common source MOS transistors [6].

The idea is illustrated in Figure 1(a), where an inhibitory synapse is shown. A current source provides the magnitude of the synaptic connection T_{ij}. The DP divides this current between i_1 and i_2. The fraction i_2 depends on the differential voltage $u_j - u_j^{ref}$, and when both transistors of the DP are biased into the subthreshold regime (realized in practice by limiting the currents to low levels), i_2 becomes

$$i_2 = T_{ij}\ \sigma(u_j - u_j^{ref}) = T_{ij}\ \frac{1}{1 + e^{-(u_j - u_j^{ref})}}. \tag{3}$$

(Here, voltages are expressed in units of an intrinsic voltage $kT\ /\ \kappa q$, where κ is the only process-depending parameter, accounting for the back-gate effect [6].) The current i_2 is then collected, ranging over all j's, on common node u_i. Hooking up a simple RC- circuit to that node, as shown in Fig. 1a, its voltage u_i then satisfies exactly eqn. (1), with the logistic sigmoid of eqn. (2) and where τ_i is given by the RC time constant.

Phototransistor current sources, which set the amplitudes of the synaptic weights, can be fabricated using a vertical bipolar *pnp* transistor in a $n-$well CMOS process. The synaptic weight matrix can be continuously loaded in parallel from an external optical device when we incorporate an array of phototransistors, each supplying a synapse current, into the circuit. This hybrid optoelectronic approach enables rapid modification of the connection matrix through the high bandwidth inherent in parallel optics, solving the classical electronic memory access bottleneck associated with densely interconnected systems.

To enable synapses of arbitrary polarity (inhibitory as well as excitatory), we choose to compose each synapse of two parts. One part is inhibitory and adjusted optically, supplied with the current of a local phototransistor I_{ph}. The other is excitatory and fixed, supplied by a current I_b through an MOS transistor with gate voltage V_b, common for all nodes. The fixed excitatory contribution I_b thus provides the necessary bias for optically adjustable synapses $T_{ij} = I_b - I_{ph}$ of both polarity. The excitatory portions of the synapses are made in a way similar to the one described above for the inhibitory parts, using a scheme complementary to that of Fig. 1(a). The final version of the basic circuit cell is shown in Figure1(b).

We designed, submitted for fabrication, and tested a prototype circuit, consisting of a 16×16 array of combined neuron and synapse units of Fig. 1(b). We used a MOSIS $2\mu m$ n-well process, with a $1600 \times 1600\ \mu m^2$ active area, of which each unit cell occupies about $100 \times 100\ \mu m^2$ (Most of this area is covered by the phototransistor). In the version reported here, the resistors and capacitors are to be connected externally. Results reported here originated from measurements on the output current at the neuron nodes u_i, at uniform illumination levels. These tests carry sufficient information to get an idea about the performance of the analog circuitry and the operational limit it implies when implemented on larger scales.

One important aspect of these measurements concerns the following: nonuniformities between the MOS transistors introduced at the fabrication stage affect the characteristics of the DPs (figure 2), thus introducing a nonuniformity of $\sigma(u)$ at different synapses from the same neuron. Careful analysis indicates that by using the MOS transistors in the subthreshold regime, we minimize the impact of these nonuniformities on the analog performance of the circuit. In the above-threshold regime, intrinsic differences between n-type and p-type transistors arise, as shown in Figure 3, which have a large impact on the neuron sigmoid and synapse characteristics (Figure 4).

As is clear from this analysis, we need to restrict the circuit operation to the subthreshold regime, which implies the use of rather small currents (in the order of nA in our design). This reduction of the synapse currents, however, necessitates the use of large output resistances, to obtain the same magnitude of synaptic connections. This slows down the time constants τ_i and hence the operating speed. In our tested system, τ_i is, under subthreshold conditions, limited to about $200\mu sec$. Although this is slow compared with discrete time hardware, the circuit is very attractive for the implementation of large scale networks with fully analog continuous time dynamics.

This research has been supported by the Defense Advanced Research Projects Agency and the Air Force Office of Scientific Research.

[1] H.P. Graf, L.D. Jackel,"*Analog electronic neural network circuits*," IEEE Circuits and Devices Mag., 5, 44 (1989).

[2] A.J. Agranat, C.F. Neugebauer, A. Yariv,"*Spatial light modulators as parallel memories for optoelectronic neural networks*," SPIE Proceedings **1150** (1989).

[3] E.A. Rietman, R.C. Frye, C.C. Wong, C.D. Kornfeld,"*Amorphous silicon photoconductive arrays for artificial neural networks*," Appl. Opt. **28**, 3474 (1989).

[4] J.J. Hopfield,"*Neurons with graded response have collective computational properties like those of two-state neurons*," Proc. Natl. Acad. Sci. U.S.A., **81**, 3088 (1984).

[5] S. Satyanarayana, Y. Tsividis, H.P. Graf,"*Analogue neural networks with distributed neurons*," Electr. Lett. **25**, 302 (1989).

[6] C.A. Mead,"*Analog VLSI and neural systems*," Addison Wesley (1989).

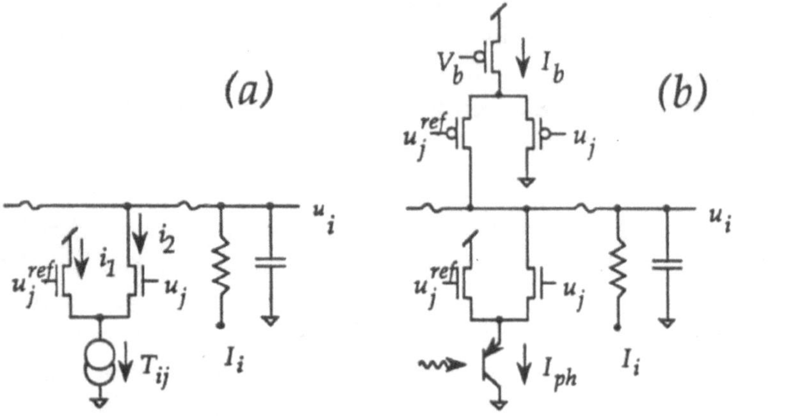

Figure 1: The basic circuit cell, combining neuron sigmoid and synapse. (a): inhibitory synapse; (b): synapse of arbitrary polarity.

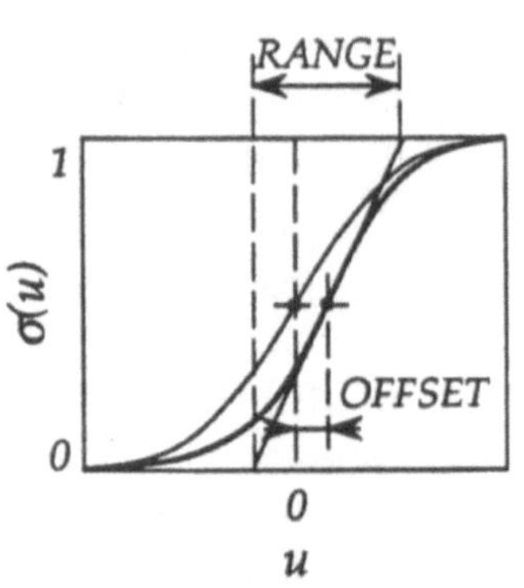

Figure 2: Definition of Offset and Range of differential pair sigmoid. Thin line: slope; thin curve: ideal sigmoid.

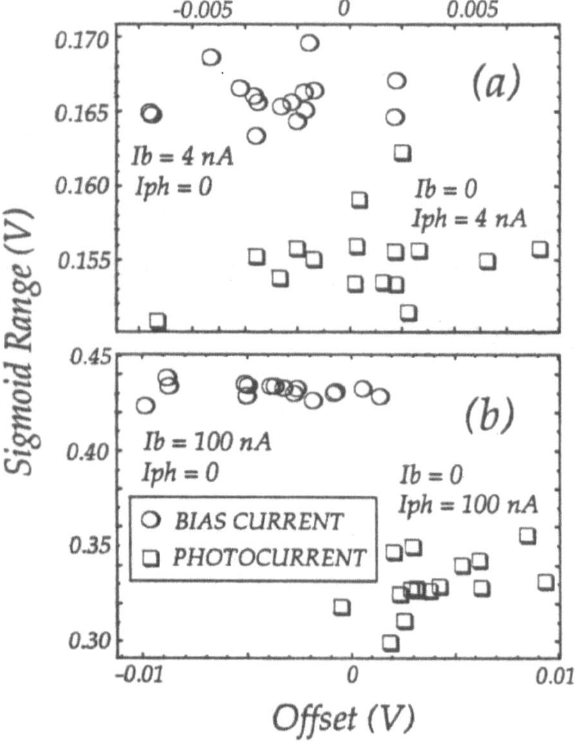

Figure 3: Experimental Offset and Range variations of n-type (box) and p-type (circle) differential pairs on one column of the synaptic array. (a): in subthreshold conditions; (b): in above-threshold conditions.

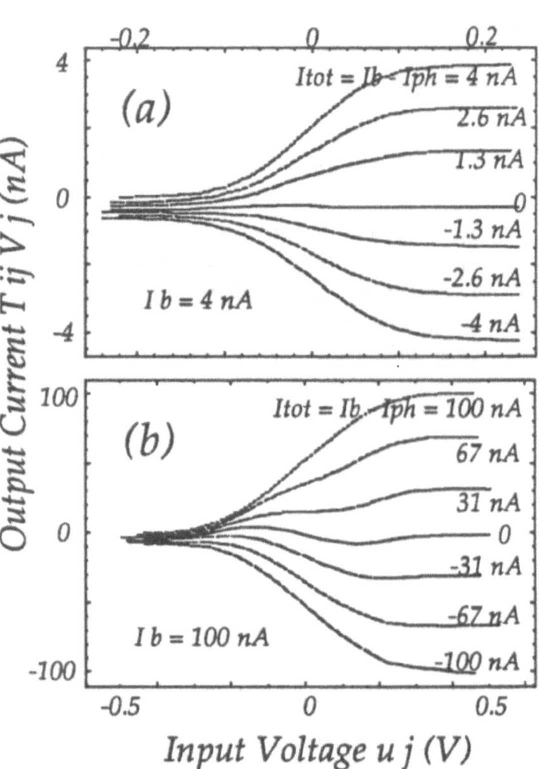

Figure 4: Experimental combined neuron sigmoid and synapse characteristics, measured on one synapse for different illumination levels. (a) and (b): same as Figure 4.

A HARDWARE EMULATOR FOR BINARY NEURAL NETWORKS

Marcin Skubiszewski

Digital Equipment Corporation Paris Research Laboratory (DEC PRL)
85, av. Victor Hugo, 92563 Rueil Malmaison Cedex
skubi@prl.dec.com

ABSTRACT

A neural network emulator is described in this paper. The device can implement neural networks of various architectures, provided that the inputs and outputs of neurons are binary.

The device is fully digital. It is mainly composed of 128 serial adders, and can be viewed as a 128-processor specialized SIMD machine. It can emulate neural layers of any size, only depending on the amount of RAM attached (a fully connected Boltzmann machine with 351 neurons has been implemented).

As an example of application, graph bisection using a Boltzmann machine is discussed on the basis of results obtained using the prototype.

1. Basic definitions; the task of the emulator

1.1 A neuron

A neuron is a small computing device with many inputs (called e_i) and one output (called s). It performs the operation

$$s = f\left(\sum e_i w_i\right) \qquad (1)$$

where f is the *threshold function* and w_i's are the neuron's *weights* (real numbers approximated in some arithmetic). The form of the threshold function is fixed for any given type of neural network, but the weights as well as the parameters of the threshold function are different for each neuron.

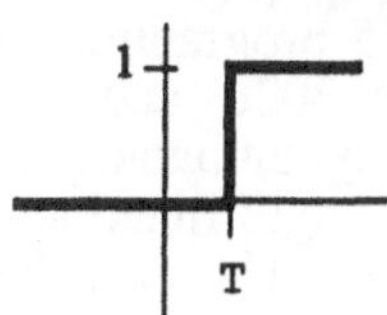

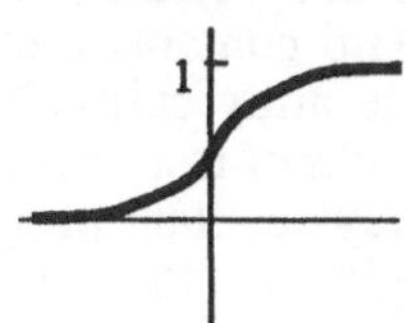

$$f(x) = \begin{cases} 1 & \text{if } x \ge T \\ 0 & \text{if } x < T \end{cases} \qquad f(x) = \frac{1}{1+\exp(-x)}$$

Binary threshold *implemented* — Smooth threshold *not implemented*

Fig. 1 *Usual threshold functions*

Fig. 1 represents the two usual forms of threshold functions. The emulator described here implements networks of binary neurons. Such neurons use the binary threshold function (fig.1, left) and have binary inputs and outputs.

1.2 Neural layers

A *neural layer* (fig. 2) is a group of neurons sharing a common set of inputs. A layer is the biggest component of a neural network whose structure does not depend on the network's architecture. So, to be general, the device implements neural layers rather than whole networks. These layers are viewed as atomic instructions by the software programs implementing particular neural networks.

1.3 The formulae of a neural layer

The following notations and formulae result from the remarks above. They describe the neural layer to be implemented by the emulator.

- n : number of inputs
- m : number of outputs (each output corresponds to one neuron)
- e_i : inputs $(0<i<n-1)$
- s_j : outputs $(0<j<m-1)$
- N_j : neuron computing s_j
- T_j : threshold of N_j
- w_{ij} : the weight attached to e_i in N_j.

The neuron N_j computes s_j:

$$s_j = \begin{cases} 1 \text{ if } \sum_i w_{ij} e_i \ge T_j \\ 0 \text{ otherwise} \end{cases} \qquad (2)$$

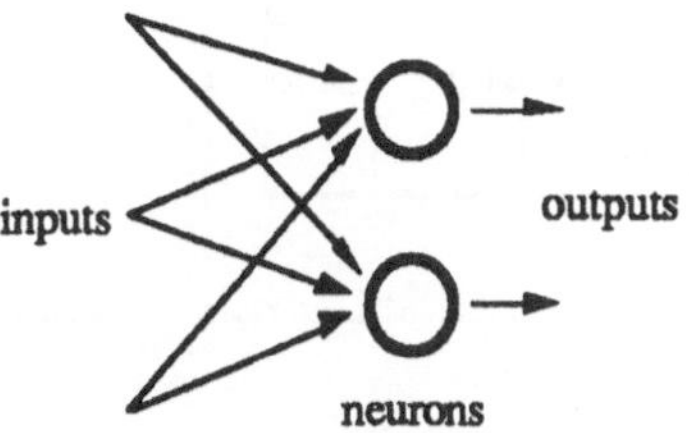

Fig. 2 *A neural layer*

2. The basic processor

The *basic processor* (fig. 3) is intended to implement one neuron at a time. It computes in fixed point two's complement arithmetic, which is also used to represent the weights; let's call b the precision (number of bits) of the arithmetic.

The weights are sent to the processor one by one and bit serially through the wire *weights*. The neuron's binary inputs are sent through the wire *inputs* one by one and in such a manner that during the transmission of the weight w_{ij}, the corresponding input e_i is present on the wire. Consequently, the AND gate delivers, also one by one and bit serially, all the products $e_i w_{ij}$. These products are added altogether by the serial adder composed of the 1-bit full adder labelled Σ, the *partial sum register* and the *carry register*.

To compute the neuron's threshold function, we add to the *partial sum register* the fictive input-weight product $e^* w^*_j$ with $e^* = 1$ and $w^*_j = -T_j$; then, the sign bit of the result (in the *partial sum register*) is the negation of s_j as defined by (2). This bit is sent outside as the processor's output.

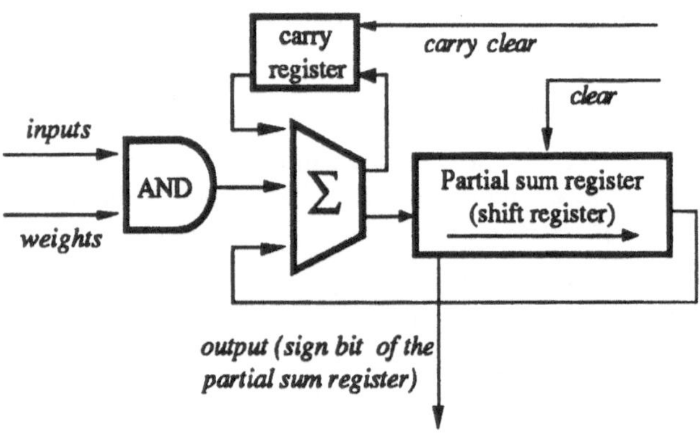

Fig. 3 *The basic processor*

3. The row of basic processors

32 processors are assembled to form a *row* (fig. 4). The row can implement a neural layer of up to 32 neurons or part of a larger layer.

During a given computation, each processor implements one neuron. The only individual properties of a neuron N_j are its weights $(w_{ij})_i$ and threshold $T_j = -w^*_j$. These data are delivered to the corresponding processor by its wire *weights*. This wire is fed by one of the 32 data bits of the *weight RAM*. At the same time, the inputs (common to the whole layer) are delivered to all the processors by the *data RAM* through the wire *inputs*. At the end of the computation, each processor sends to the data RAM, through its *output* wire, the output s_j of the neuron it implements. Thus, the row of processors can implement in one parallel computation a neural layer with any number of inputs and with 32 outputs.

Since a neural layer is nothing more than a group of neurons sharing a common set of inputs, we can split the neurons in a layer into several subsets and consider each subset as being a layer by itself. To implement a layer with more than 32 outputs, it is sufficient to follow this procedure to split it into several smaller layers; then, the outputs of the smaller layers are computed, one layer at a time, by the row of emulators, to be concatenated together.

As can be deduced from these remarks, a neural layer is implemented in $n*b*\lceil m/32 \rceil$ cycles.

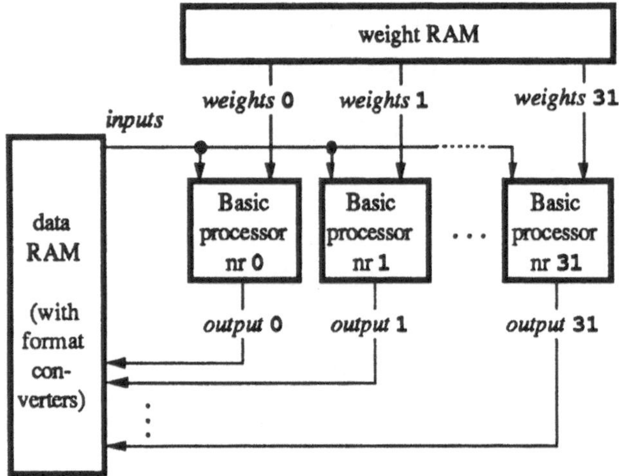

Fig. 4 *The row of basic processors*

4. A global view of the emulator

The neural network emulator is implemented on PRL's prototype *programmable active memory* board Perle-0 (see [1]) (fig. 5). The board contains a matrix of 25 programmable gate array chips (Xilinx LCA 3020) and two blocks of fast memory. The programmable gate arrays contain programmable logic functions, registers and routing resources. They can be configured from the host computer to turn Perle-0 into various special purpose digital devices, in this case our neural network emulator.

The number of processors in a row is 32 because the block of RAM used to store the weights has a 32 bit wide data bus, which cannot feed a larger row at full speed. Since available logic resources let us lay out 128 processors, we operate with four such rows. These rows always receive the same data from the weight RAM; they therefore simultaneously implement the same set of 32 neurons. Each row receives specific inputs from the data RAM and sends back specific outputs. Thus, the four rows of processors always implement the same neural layer, but with different inputs and outputs.

This form of parallelism is useful in many cases. In learning applications, the most time-consuming use of neural networks, we process many example data sets through the same network. In optimization with a Boltzmann machine (see below), we can apply the machine to four different random sets of initial inputs to obtain four different solution at the end of a run.

5. Precision of the arithmetic

The precision required in computations (called b) depends on the application. Since the computing time and the amount of RAM required by the weights are proportional to b, it is useful to finely adjust this value. For this purpose, b is a compilation parameter of the description of the device. We actually generate a different hardware for each value required by any particular application.

6. Speed

The emulator works with a clock period of 55 ns. With b=15, it computes 155 Mega synapses per second. This figure is not very relevant to the actual speed of any hardware+software application, but it provides raw data to compare this neural emulator with others.

7. Implementing Boltzmann machines

The reader of the following part of this paper is supposed to be familiar with Boltzmann machines. Ref. [2], chapters 8 and 9, provides a good introduction to the subject.

Several neural architectures have straightforward implementations on the emulator. The Boltzmann machines, which are a very interesting application of the emulator, are harder to implement because of their probabilistic neuron equations and the lack of inherent parallelism.

7.1 Implementing non deterministic neurons

The neurons in a Boltzmann machine obey the formula

$$s_j = \begin{cases} 1 \text{ with probability } \dfrac{1}{1+\exp\left(\dfrac{-\Delta E_j}{\theta}\right)} \\ 0 \text{ otherwise} \end{cases} \quad (3)$$

where $\Delta E_j = \sum_i w_{ij} e_j - T_j$

θ, the *temperature*, is a positive real number.

(3) is different from (2) and cannot be directly computed by a basic processor. Fortunately, (3) may be transformed into

$$s_j = \begin{cases} 1 \text{ if } \sum_i w_{ij} e_i \geq T_j - \theta * \ln\left(\frac{1}{A} - 1\right) \\ 0 \text{ otherwise} \end{cases} \quad (4)$$

where A is a random variable uniformly distributed on [0..1]. (4) describes the same neuron as (2),

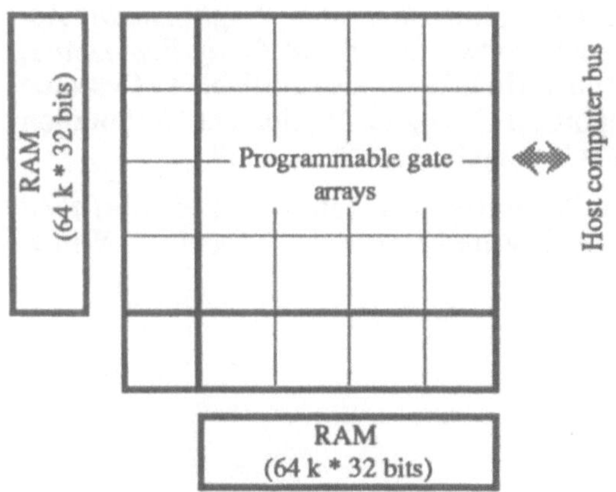

Fig. 5 *The board Perle-0*

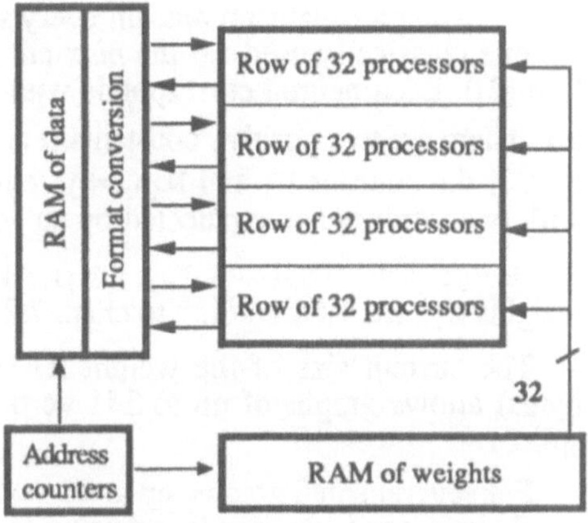

Fig. 6 *Overview of the neural emulator*

with the only particularity that in (4) the threshold $T'_j = T_j - \theta * \ln(\frac{1}{A} - 1)$ is a random variable. For each output to be computed, a new value of T'_j needs to be generated.

When the emulator has to compute the outputs of a layer of non deterministic neurons, the required values of T'_j are computed in the host and sent to the device to be used as thresholds. Then, the rows of basic processors compute the outputs as described in chap. 3.

7.2 Parallelism

In a correct implementation of a Boltzmann machine, only one neuron at a time can compute its output. This would lead to a sequential implementation and ban the notion of neural layer. The emulator implements Boltzmann machines in an approximate way: neurons are grouped by 32 into layers, and all neurons in a layer use a row of basic processors to simultaneously compute their outputs.

To compensate for the anomalies introduced by this implementation, we modify the Boltzmann machine. The weights w_{ii} (coupling each neuron with its own previous output) are, instead of being equal to zero, set to a positive constant called *persist*. This gives the neurons a tendency to keep the same value through many computing steps. Boltzmann machines modified in this way and implemented on the emulator appear to give good results in the examples described below.

8. An application: graph bisection

The graph bisection problem considered here consists in splitting the vertices of a non-oriented graph into two equal subsets in such a way that a minimal number of edges passes between the subsets. This problem has applications in electronic CAD and is well known to be NP-complete. We solve it using a Bolzmann machine very similar to the one classically used for the *max cut* problem (see [2]). Each neuron corresponds with one vertex. There are two positive constants α and β such that, if the neurons N_i and N_j $(i \neq j)$ correspond with two vertices non connected by an edge, then $w_{ij} = w_{ji} = -\alpha$; otherwise $w_{ij} = w_{ji} = \beta - \alpha$. The weights w_{ii} are set according to chap. 7.2.

The current size of the weight RAM (256 k bytes) allows graphs of up to 351 vertices to be split.

For several small graphs, up to 30 vertices, optimal bisections have been found using an exact algorithm, then compared with the results from the neural emulator. The emulator has also been run on simple graphs of various sizes, whose optimal bisections are known. In most cases, the device found optimal solutions, in others the best solutions found were within 1% from the optimum. For all other graphs tested, the device produced apparently good solutions; they are possibly optimal, although we have no way of checking this fact.

For a graph of 351 vertices, each run lasts 16.2s (2.9s to compute the weights and thresholds, 13.3s for the neural computation itself). Since the emulator always handles four sets of data at a time, the run delivers four different results.

During a run, the output of each neuron is computed 687 times for each of the four sets of data. The whole application executes thus 25.5 M synapses per second, which is 6 times slower than the raw speed of the underlying device. Most of the time is spent on generating the random weights T'_j in the host computer. The computer is a vintage Sun 3/160, and porting the application to a faster workstation should speed up the whole computation by a factor of 3 to 6.

Two other graph-related applications have been tested on the device: the *max cut* problem and the *independent set* problem, both described in ref. [2]. The results obtained were very similar to those of the graph bisection problem described here.

ACKNOWLEDGEMENT

The author wishes to thank Jean Vuillemin for his continued assistance throughout the project and both Jean Vuillemin and Alfred Permuy for defining, with the author, the guidelines of the hardware architecture.

REFERENCES

[1] Patrice Bertin, Didier Roncin and Jean Vuillemin, *Introduction to Programmable Active Memories*, in *Systolic Array Processors*, Prentice Hall, 1989; also available as Research Report Nr 2, Digital Equipment Corporation Paris Research Laboratory, 1989.

[2] Emile Aarts and Jan Korst, *Simulated Annealing and Boltzmann Machines*, Wiley, 1989.

AN INTEGRATED ARTIFICIAL NEURON BASED ON JUNCTION-CCD TECHNOLOGY

J.Hoekstra
Delft University of Technology
Department of Electrical Engineering
P.O.Box 5031, 2600 GA Delft
The Netherlands

Abstract

This paper briefly describes the design principles for an implementation of an artificial neuron in the form of a in silicon integrated circuit based on junction charge-coupled device technology. The significant features of the design are: (1) the possibility to have fixed as well as variable weights, with analog values; (2) analog output values; (3) suitable to support pulse-coded neural network algorithms; and (4) realized in a standard bipolar technology. As a simple example a McCulloch-Pitts neuron is fabricated and tested at a clock frequency of 40 MHz.

1 Introduction

In this paper some basic design principles behind a integrated circuit implementation of a artificial neuron are discussed. These principles, which can be used to implement artificial neural networks (ANNs), are illustrated by a simple test chip on which a McCulloch-Pitts neurons is realized. A complete neural network is not implemented, because the fan-out of the artificial neuron is not exactly known at this momemt; research on this point is going on. The design is based on the concept of junction charge-coupled devices (JCCDs), which is a bipolar version of the more conventional metal-oxide-semiconductor (MOS) charge-coupled device (CCD). Research into artificial neural network implementations based on MOS-CCDs has been done by groups in the U.S. [1,2]. The essential differences between MOS-CCDs and our bipolar CCD are (1) the use of the capacitance of a reverse-biased pn-diode, and (2) the possibility to have vertical charge transport, that is, transport of charge perpendicular to the charge transport in the plane above the silicon substrate. The charge packets represent

[1]This work has been supported by the Dutch Organization of Fundamental Research FOM

[2]The integrated circuit was fabricated at Philips Elcoma, Nijmegen, The Netherlands

analog information, and are transmitted to the artificial neuron, where the charge packets can be converted into a current representing the neurons analog output value. JCCD technology is adequately developed. It is used for solid-state imagers, filters, and logic circuits [3]. The technology is developed at the Delft University.

The design of the artificial neuron incorporates the following features:

- electrical programmability
- pulse-firing
- built-in learning capability
- analog storage
- analog output values

In the next section, first, the basic principles the Jucntion-CCD, needed to understand the operation of the artificial neuron, is described. As a simple example of the JCCD technology, the test-results of a Mcculloch-Pitts neuron are shown.

2 structure and operation

First, the overall physical realization is described, and then the details of its operation are covered. Basically, a JCCD consists of a p-type substrate and a n-type layer with diffused p-gates.

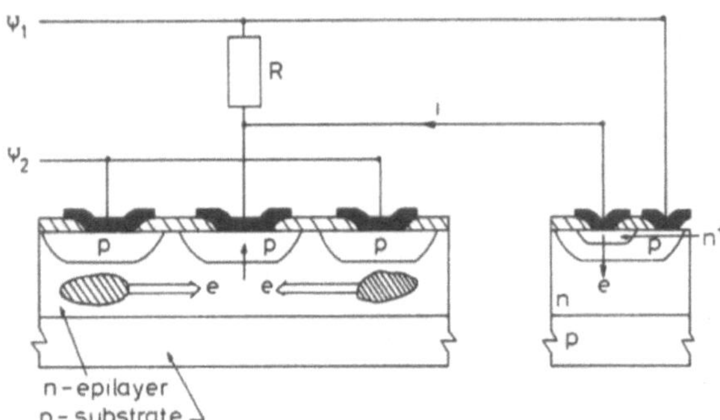

Figure 1: JCCD structure

If appropriate voltages are applied to gates, moving potential wells are created in which charges (electrons) can be stored. As is depicted in fig.1, charge packets can be moved to each other to be stored in a single potential well. If to this well more electrons are offered then can be contained in this well, a current results at the gate's contact. This current, which results if more charges are offered then a certain threshold value, can be injected elsewhere on the chip. This 'threshold' gate produces an artificial neuron's output. To obtain a structure that is capable to act as an artificial neuron, the input potential wells must be varied in size. The depth of the potential well can be changed by applying a changing voltage to the gate or by pre-charging the well itself.

In more detail, a JCCD consists of a lightly doped p-type substrate and a n-type epitaxial layer (referred to as epilayer) with diffused p-gates. The drain consists of a n+ diffusion. For

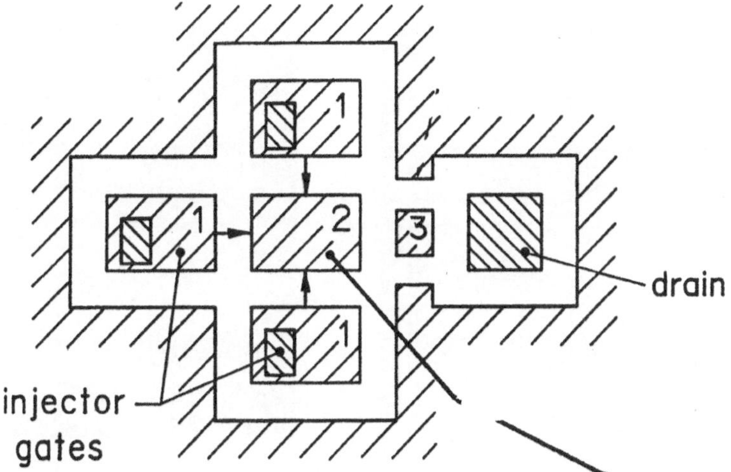

Figure 2: McCulloch-Pitts neuron layout 'threshold' gate

a given JCCD structure all electrons are removed from the epilayer by applying a sufficiently large positive voltage to the drain, while the gates and the substrate are kept at ground potential. In this state, the depletion layers extending from the gate-epilayer interface and from the substrate-epilayer interface touch. At this point the electrical potential in the semiconductor reaches a maximum. This is a global maximum; for an ideal JCCD it is constant throughout the epilayer. In operation, a local potential maximum in the epilayer, underneath a gate, is created by clocking a gate to a positive voltage. This local maximum serves as a well for electrons. The maximum potential in the device is referred to as the channel potential.

JCCD artificial neural circuits combine the properties of JCCDs with charge transport vertically through one of the p-n junction diodes, induced by surplus charge. The gate under which the surplus charge is created is called a threshold gate. The operation condition for obtaining vertical charge transport is that the gate voltage on the threshold gate is taken above the channel potential of a grounded gate. In this condition, the channel potential under the threshold gate—if the potential well is filled completely—is more positive than the channel potential under the neighboring gates, which have zero gate voltages. In addition, surplus charge will forward bias the p-n junction [4]. The surplus charge is created by combining different charge packets in a well that can only contain one charge packet. The structure will act as a pnp-transistor; the gate (emitter) will inject holes into the epilayer. This charge flow, called overflow current, will continue until all surplus electrons have been removed.

The overflow current is used to inject charge into another JCCD channel. This can be performed by an injector structure, which consists of an n+-diffusion placed in a p-gate. The overflow current is used to forward-bias this n+-p junction; and charge is injected into the JCCD channel, which acts as a collector of this vertical n-p-n transistor. Using vertical charge transport, artificial neurons can be realized [5].

3 McCulloch-Pitts neuron

Figure 2 shows a drawing of the tested device, which performs the function of a McCulloch-Pitts neuron with three input signals. Not shown are the input channels, which create the (weighted) charge packets.

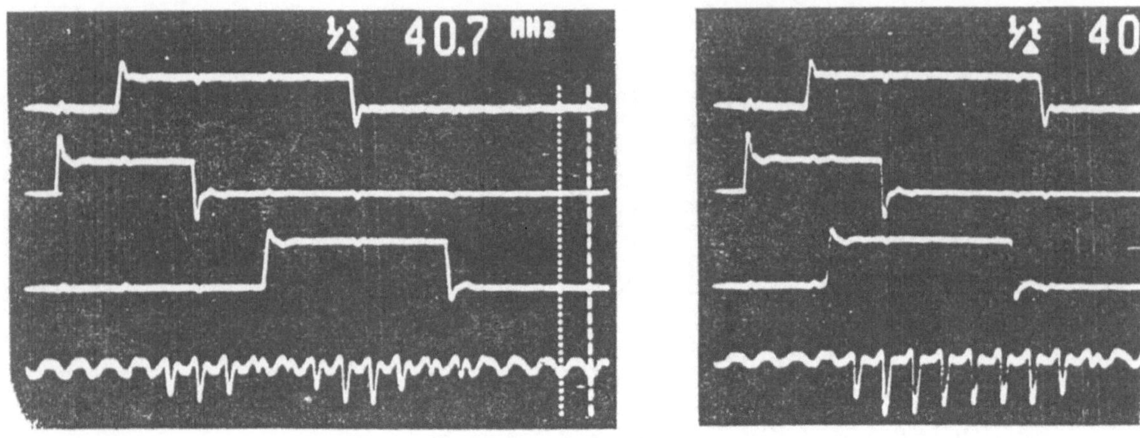

Figure 3: McCulloch-Pitts neuron test results

The charge packets under the input gates are transported towards the threshold gate where they are summed. Because the threshold gate can only contain one charge packet, the overflow current represents the threshold function, > 1 . The main parameters in this device are a 5 μm epilayer thickness and gate voltages between 0 - 7 Volts. The response of the neuron circuit at a clock frequency of 40 MHz is shown in Figs. 3a and 3b. The lower traces represent the output signal across 50 Ohm, 10 mV/div. The other traces represent the input signals: the high level voltage implies an input of charge, periodically sampled by one clock phase. The clock voltage is 7 V, the internal load resistors are 2.2 kOhm. The photographs show all permutations of the input signals.

References

[1] Psaltis, D. , Sage, J., 'Advanced implementation Technology, Part IV of the DARPA Neural Network Study', Fairfax: AFCEA International Press, 1988.

[2] Sage, J.P., Thompson, K., Withers, R.S., 'An Artificial Neural Network Intergrated Circuit Based on MNOS/CCD Principles', AIP Conf. Proc. 151, 381, 1986.

[3] Hoekstra, J., 'Simple JCCD Logic at 20 MHz', Electronics Letters, Vol. 23, 246, 1987.

[4] May, E.P., van der Klauw, C.L.M., Kleefstra, M., and Wolsheimer, E.A., 'Junction Charge-Coupled Logic (JCCL)', IEEE Journal of Solid-State Circuits, 1983, Vol. SC-18, pp 767-772.

[5] Hoekstra, J., Some Models and Implementations of Digital Logic Functions Using Junction Charge-Coupled Devices, PhD-thesis, Delft University of Technology, 1988.

Fast Generation of Neuro-ASICs

J. OUALI , G. SAUCIER
Institut National Polytechnique de Grenoble / CSI
46 Avenue Félix Viallet 38031 Grenoble Cedex FRANCE

Abstract:

This paper proposes a distributed, synchronous architecture for artificial neural networks. A basic processor is associated to a neuron and is able to perform autonomously all the steps of the learning and the relaxation phases. Data circulation is implemented by shifting techniques. Customization of the network is done by setting identification data in dedicated memory elements. The neuron has been implemented on silicon. It is shown that in a silicon compiler environment dedicated networks can be easily generated by cascading these elementary blocks.

Introduction

The goal is to provide an implementation frame for a large variety of architectures (Hopfield, layered network) including both learning and relaxation phases. Some other parameters such as precision are also taken into account as design parameters. A silicon compilation approach allows to provide architectures with a given number of bits for state and coefficient encoding without additional design efforts.

I. The target architecture [1 to 4]

The architecture is a distributed synchronous architecture. Each neural processor has autonomous control and is able to perform all the phases of neuro-computing (potential and activation function computations, learning and relaxation phases).

I.1 The neuron processor

A physical entity is associated with each neuron. It is an application specific processor, the controller of which implements both learning and relaxation phases. The neural processor (Figure 1) is made up of :

1) seven identification registers (ID_R):
- N1 containing the number of bits encoding the state
- N2 containing the number of rows
- N3 containing the number of coefficients stored in the memory
- N4 containing the shifting direction of the inputs in the layer.

for example : if N4=0 the shifting of the inputs is performed down
if N4=1 the shifting of the inputs is performed up

- N5, N6 ,N7 containing the numbers of shiftings required to format the result of each multiplication.

2) a local memory storing the coefficients {Cij} and the parameters used to approximate the activation function.
3) a data path able to perform mutiplications and additions/subtractions
4) a controller controlling the computation of the potential, the state of the neuron and the data transfer with the other neurons

5) an input register storing and shift the input value
6) an output register storing the state of the neuron

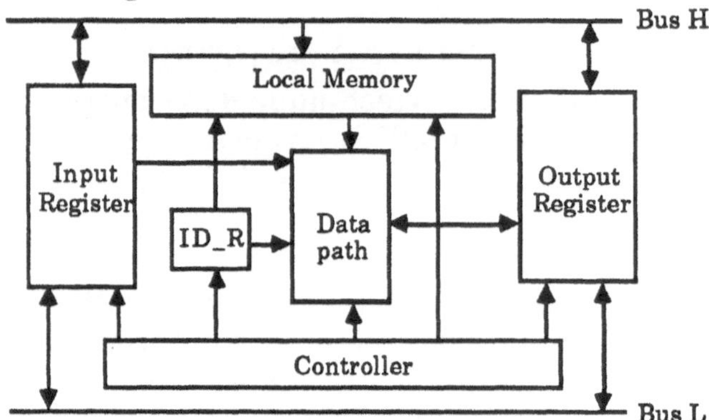

Figure 1 : Neural processor

I.2. Connections between neurons and global architecture

It has been pointed out frequently, that a large number of connections between the neurons is a severe drawback for their implementation on silicon. Practically, virtual connections are implemented to replace spatial connections. This means that if the processor cannot be connected to a large number of neighbors, the state of these neighbours will circulate in a shift register and the processor will pick up at the right instant the values of the state of its predecessors.

The processors are organized in rows between two buses. These buses are made up of bus segments (figure 2) which can be connected by soft switches (software programmable switches) when required during the phases of neuro-computing.

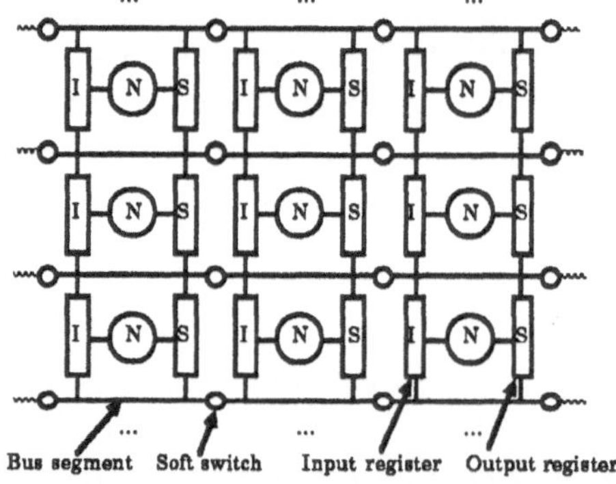

Figure 2 : Global architecture (partial view)

We will detail this global architecture using the example of 8 processors realized on a 4 x 2 array. Complete architectures will be built later on by assembling such basic blocks. First, let us show how the computation of potentials is performed ; the equations of potentials Vi are given in figure 3.

In a first computation step, bus segments are connected as shown in figure 3b. The switches of the second column are open. The state of neuron N1 is available on the bus segment of the second row for both neurons N1 and N2, the state of neuron N3 is available similarly on the next bus segment and the same for neurons N5, N7. The computations performed during this step are shown in figure 3a. In the second step of the computation, a circular shifting is performed in the first layer as indicated in figure 3a. The state of neuron N1 has been shifted on the next bus segment and is now available for both neurons N5 and N6.

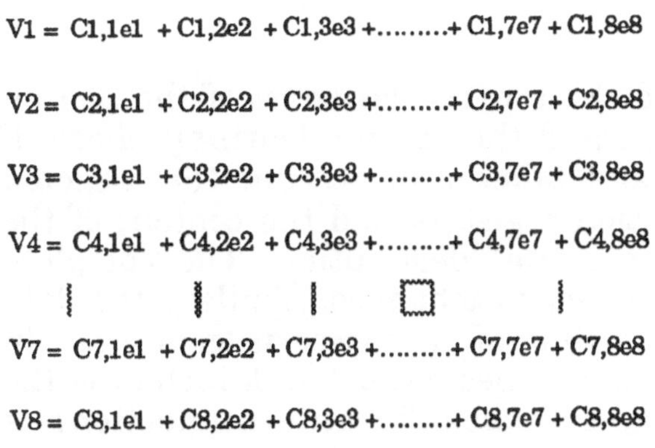

Figure 3: operations required for evaluating the potentials of the neurons

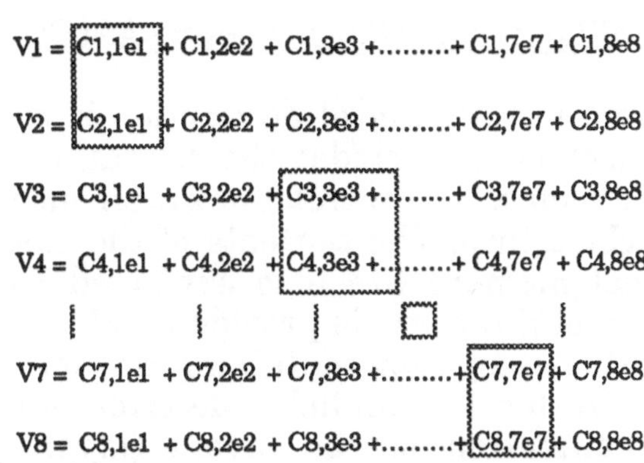

Figure 3a : operations performed in the first computation step

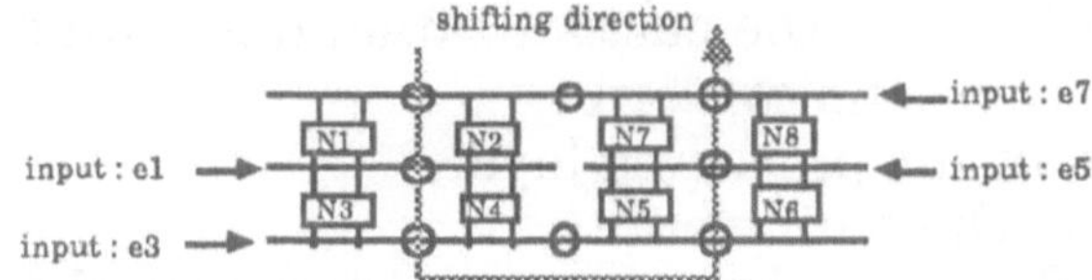

Figure 3b : A block of 8 neurons.

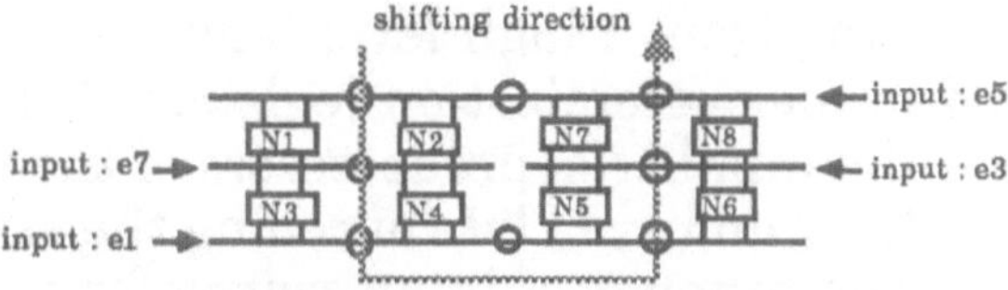

Figure 3c : Second computation step

It is clear that 4 elementary shiftings - a complete circular shifting - will be necessary to perform half of the complete computation. Two complete circular shiftings (8 elementary shiftings) will be required to end up the computation. Once this computation is over, the controller initializes the computation of the activation function. Generally speaking a row of N rows will be implemented on a (N/k x k) array. In any case N elementary shiftings are required and k/N circular shiftings.

The first prototype (figure 4) has been implemented in a 2µm CMOS technology using the block generation techniques of VLSI Technology and works perfectly well up to 20MHz.

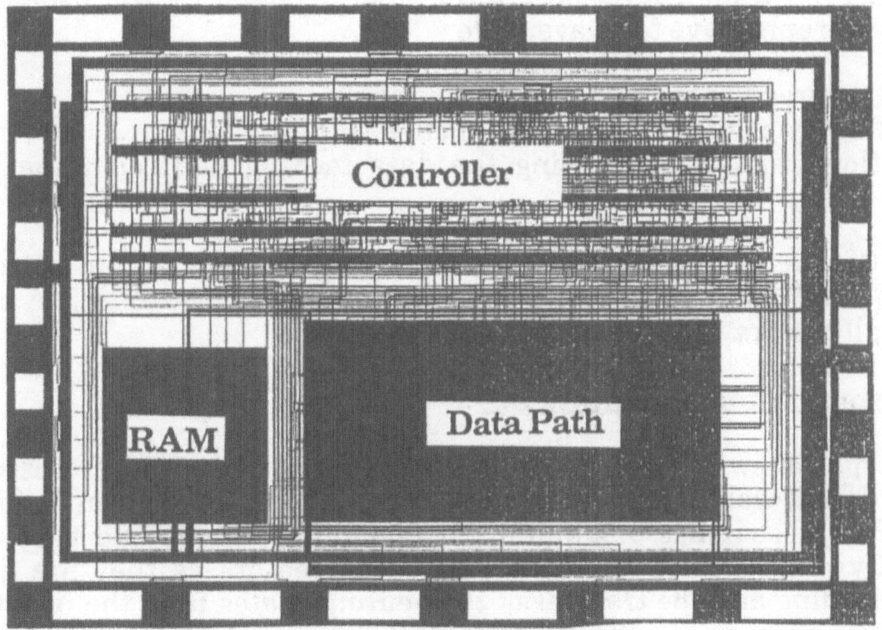

Figure 4 : Layout of the neuron

II. Silicon compiler for neuro-ASICs

From the high level description is extracted the control flowchart of the neuron processor. It includes the activation function and the specific learning phase, if necessary, as control flowgraph modules, and takes into account the targeted architecture. The contents of the customization registers and the content of the local memory are also extracted from this initial description. The complete control flowchart is provide in VHDL to allow easy exchanges. Similary the data path description containing an adequate amount of resources is generated in VHDL too. In parallel, a description of the interconnection network including the organization of the layers (layer folding, switches for interconnection) is generated.

II.1. Parameters from a high level description language :

Starting from a high level language description [5] , the customization data can be fed to the customization registers. These informations concern :
1 - the number of predecessors of the neuron in the preceding layer.
2 - the number of bits encoding the states and the coefficients.
3 - the number of input waves for a layer if the number of inputs is greater than the number of elements in the layer.
4 - the position of the neural processor in the physical and logical layer as a logical layer may be implemented in several physical layers.
5 - the shifting direction of the input values in the layer. For example, neuron from N1 to N4 (figure 3b and 3a) perform the shifting of the data down and from N5 to N8 perform the shifting data up
6 - the number of shiftings required to format the result of each multiplication.

II.2. Block generation and blocks interconnection

The local memory is generated by a ROM or a RAM if the relaxation phase only is implemented. The data path is automatically generated from an architectural description in terms of registers, ALU, bus connections and bit slice structure. Extracted from a high level description tool or given by the user, the following control flowgraphs have to be available

- control flowgraph of the potential computation
- control flowgraph of the activation function
- global flowgraph implementing the data feeding and including the previous ones.

A FSM (finite state machine) compiler generates then an automatic state assigment and a standard cell layout for the global controller. A chip compiler connects all the previous blocks and adds the pads.

IV.3. Customization of the switches :

Two types of switches implement the connection between processors. The first ones are definitevely fixed as they are always closed for a given target chip. They will be PROM switches or aluminium segments. The second type switches are electrically programmed from the neuron processors as they are closed during the data loading and the transfer of the neurons states from the first to the second layer steps, or open during the different steps of the potential and the state computations. We give an example of a 2 layers network, with 6 neurons in the first layer and 4 neurons in the second one.

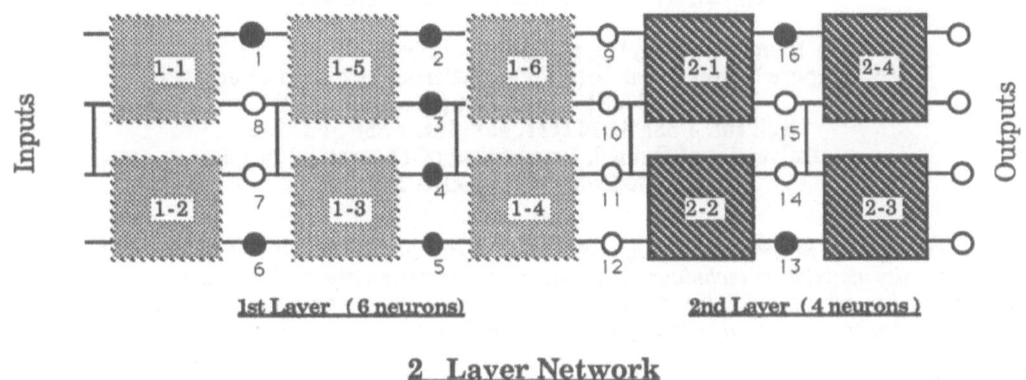

2 Layer Network

○ Electrically programmed

● Always closed

The switches from 9 to 12 are used to separate physically the two layers. They are always disconnected during the potential and state computations steps. All the switches are closed during the data loading and the transfer of the neurons states from the first to the second layer steps. The folding of the first and the second layers are respectively made by the switches 7, 8 and 14, 15

Conclusion

A synchronous distributed architecture based on a cascadable neuron processor has been demonstrated. The neuron processor is dedicated to a given application by using a silicon compiler techniques. This approach implements through a finite state machine compiler the desired activation function or any learning algorithm in the processor. The neuron processor may be cascaded into large chips until yield limitations are reached.

References:

[1] "A Flexible, Universal Wafer Scale Neural Network". J. Ouali, G. Saucier and J. Trilhe. 3rd International Workshop on Wafer Scale Integration, Come, Italy, June 1989.

[2] "A Flexible Wafer Scale for Neural Networks". J. Ouali, G. Saucier. International Conference on Computer Design (ICCD'89), Cambridge, USA, October, 1989.

[3] "A Distributed Architecture for Neural Networks Based on a Neural Processor". J. Ouali, G. Saucier. 2sd International Workshop Neural Networks and their Applications (Neuro-Nîmes 89), Nîmes, France, November 1989.

[4] "Customizable Neural Networks on silicon". J. Ouali, G. Saucier and J. Trilhe. IFIP Work. on Parallel Architectures on Silicon, Grenoble, France, December 1989

[5] "ESPRIT II - Pygmalion Conference". Brussels, Belgium. November 1989

VLSI ARCHITECTURE OF THE BOLTZMANN MACHINE ALGORITHM

Jouni Tomberg, Harri Raittinen[†] and Kimmo Kaski[†‡]
[†]Tampere University of Technology, Microelectronics Laboratory
P.O.Box 527, SF-33101 Tampere, FINLAND
Tel: Int. + 358 31 162111, Fax: Int. + 358 31 162620
[‡]University of Oxford, Department of Theoretical Physics
1 Kebble Road, Oxford OXI 3NP, UK

ABSTRACT - A new efficient programmable implementation of Boltzmann Machine algorithm will be presented. It is based on pulse-density modulation technique. Advantages of the design are simple structure of a synapse and thus small area, modularity and expandability. Furthermore, these structures can be used for various other neural network architectures. Applications for this type of networks can be found in the area of pattern recognition, image restauration and various optimization tasks.

1. Introduction

A major goal of neural network implementations on silicon is to have a large amount of simple computational elements "neurons" and "synapses" connected together via signal lines. Although the idea for these networks is based on the biological world, their models are considerable simpler. A neuron can be modelled by a nonlinear amplifier with a sigmoid shape transfer function. Neurons are connected to other neurons with synapses of variable strengths. These connections are in fact responsible for the information storage. From the data processing point of view a synapse can be considered as a multiplier. In the most simplified neural network model the connection weights are symmetric. Such a network is often called Hopfield type network /1/. Hence in a fully connected network with N neurons there are N x N synapses, one between each neuron pair. The state of this network can be fully described by the energy of the system. Thus a pattern recognition task, for example, proceeds by letting the network relax to the state of the lowest energy near the starting point. Some of the essential characteristics and advantages of the networks are asynchronous communication and massive parallelism. These features can be exploited in such applications as pattern recognition, image restauration, classifiers and combinatorial optimization, to mention few.

2. Algorithm and Implementation

The Boltzmann Machine algorithm is a kind of stochastic modification of the Hopfield type network /2/. In this case the state of the neuron is based on a statistical rather than a deterministic approach. The idea is to make the nets to find the global minima of the energy function instead of local minima as in case of Hopfield nets. The Boltzmann Machine is suitable for various optimization tasks although the algorithm is computationally very heavy because of the simulated annealing method requiring a very efficient implementation. Nevertheless, the structure of the Boltzmann Machine network differs somewhat from that of the Hopfield type network. In the Boltzmann Machine there are three different kind of neurons or nodes; input nodes, hidden nodes and output nodes. Input nodes are always clamped to a certain input value. The output nodes can be either clamped to a certain output value or they can change their states freely. The hidden nodes are internal nodes of the network and their states are freely running and not read out from the network. The state of a neuron is binary like in the Hopfield model. The output and hidden nodes are usually fully connected with feedback connections, but the input node layer forms just feed-forward connections to all of the output and hidden nodes (Fig. 1). Thus, the fully connected part tries to find a global minima from the energy space defined by the state of input nodes. The output state of a node or neuron depends statistically on the input values, i.e. a noise parameter is added to the nonlinearity function of the neuron. This parameter is called temperature of the network and it defines the probability of the network to be in either one of the binary states. If the temperature is zero the node operates just like a neuron in the Hopfield network. If the temperature is infinite the output value can be either one of the binary values with equal probability.

Training /2/ of such a network is based on collecting statistics of the network states during the learning and changing accordingly the values of the weights. The idea is to get the configuration such that the probability distribution of the I/O-node states is the same for clamped and free-running cases. The first step is to select randomly one of the training pairs and clamp it to the input and output nodes. The temperature of the network is set initially to some high value and then it is decreased slowly to a low temperature. During the annealing procedure the statistics are collected. After this the output nodes are unclamped and they can change their states freely. The annealing procedure is also done for this free output case and again the output values are collected. This sequence is repeated for all of the training pairs. The last step is to adjust the synaptic weight values according to the collected values. These steps are repeated until the network has learned the training set well enough. The weight adjustment depends on many different parameters and is also very case sensitive. When the Boltzmann Machine is performing an optimization task the input vector is clamped to the inputs and the network is annealed from a high to low temperature. The output nodes are in this case unclamped settling freely to the result values which can be read out. The advantage of this statistical algorithm is that it finds the global minimum of the energy function rather than the local one. This property is very desirable in many optimization tasks. The disadvantage, however, is that it is computationally very heavy; the annealing requires many additional computations compared to other algorithms. The parameters, e.g. the annealing rate, are also very case sensitive so it is difficult to find an optimal procedure for the network annealing. There exists also a number of variations to this algorithm which try to overcome this disadvantage /3/.

In the pulse-density arithmetic /4,5/ numbers are represented as streams of digital bits, 0 and 1. In contrast we interpret these values to be the sign bits of two's complement numbers. The value of a pulse-density number depends on the relation of 0's (+) and 1's (-) in a given window. Furthermore, we define that all the values in the pulse-density representation are fractional numbers between -1 and +1. Thus the value of a bit stream including N zeros and M ones is (N-M)/(N+M) and by a number including P bits we can represent P+1 values. The value of a pulse-density stream is continuous, i.e. we can take a sample of P bits any time from the stream and it represents an initial value. In the normal binary arithmetic we can represent 2^P values with a P-bit word. In that case the overall structure of the arithmetic is much more complicated and requires more area on a chip. In contrast the arithmetic of pulse-density numbers is quite easy to realize. We must, however, assume that the density of zeros and ones in a pulse-density stream is smoothly and randomly distributed inside a 'window' (length of the number). Thus all the initial values and results are rather statistical in nature than exact values of the signal. Nevertheless, this representation is suitable for the neural network algorithms, due to the **robustness** and **fuzzy** functions. The multiplication of two numbers can done by using simple exclusive-or function, as can be easily seen from the number representation. However, the addition is somewhat more complicated. An analog implementation of such an adder is quite straightforward when adding together currents or charges /4/. On the other hand a digital implementation of the pulse-density adder is found to be tricky /5/. This is because the pulse-density signal has a continuous value and thus its value inside the defined window can be only an approximation.

Pulse-density technique is appealing due to its small sensitivity to error and retain similarity of signal coding in biological neural nets. In artificial structures a synapse (Fig. 2a) multiplies the pulse-density stream coming from a neuron along the 'axon' line by the synaptic weight value, which is stored in a ring register also in the pulse-density form. The multiplication is performed by a simple XOR-gate and the result is added to the output values of the other synapses in the 'dendrite' line, which connects synapses of the same row together and to a neuron. The synaptic weight register is a simple shift register which can operate in two different modes. The registers of different synapses are chained together so that in the loading mode data is moving through the registers. When the data has moved to the right place in the register it is connected to the ring mode. In this mode data is moving round the register as a pulse-density weight value. The amount of artificial synapses and thus the network area, especially if it is fully connected, is proportional to the square of the number of neurons. Thus it is very important to minimize the area of one synapse when aiming for large networks. This can be done by simplifying the circuit of a synapse and by using minimal area dynamic structures. The dynamic implementation is very suitable for neural network architectures, because of the continuous nature of neural algorithms. The loading of the weights is done serially to minimize the complexity and the structure. The area of one synapse depends also on the length of the synaptic weight register, which in our design is 16 bits giving 17 different values between -1 and +1. By using the CCD technique for the implementation of shift-register the area could be further minimized and thus larger networks would be possible to implement on a single chip. In a neuron (Fig. 2b) the nonlinear function is performed on the pulse-density stream value coming from synapses along the dendrites in a window with the size equal to the length of the weight register, in this case 16. The output going to the synapses is again a bit stream of "-" and "+". The nonlinear function has ideally the shape of a sigmoid function, though approximations of it can also be used. Even a sharp step function gives good results. We have implemented the artificial synapse and neuron structures by using both analog switched-capacitor and fully digital implementations /4,5/. The advantage of the switched-capacitor structure is its small area and thus larger networks can be implemented on a chip. However, the digital structure offers better connectivity between single chips and thus very large networks can be obtained by connecting chips together. Both structures are very modular and expandable.

3. Architecture

The integrated circuit implementations have always restrictions compared to the ideal realizations of algorithms. The basic restriction is of course the accuracy of the number representation. In the digital weight implementations, as in this case, we have a certain amount of discrete levels for the weight representation. In the Boltzmann Machine algorithm the probability function of a neuron state is another property, which needs some kind of approximation to ensure an efficient implementation. We have used a linear probability function instead of the sigmoid one. Thus, we end up with two main *modifications* namely **linearizing** the probability functions and **quantizing** the synaptic weight values and the value representation in general. The effects of these modifications have been tested by simulation.

In the simulations we have studied the learning process of the Boltzmann Machine as a function of the number of learning steps. The network consists of an input layer and a fully connected hidden layer with one node being the output node. In the test cases we have used the parity problems of 2, 3 and 4 bit long binary numbers. The results of learning the XOR-problem (2-bit parity) are shown in Figure 3. The network consists of three input nodes and two to four hidden nodes. In Figure 3a the simulation results of the unmodified Boltzmann Machine algorithm are shown. It learns the XOR-problem within 1000 steps. In Figure 3b we show the results for the case of linearized probability function. The linearization has hardly affects the learning at all. In both of these cases two hidden nodes have been used. In the third case the algorithm is both linearized and quantized (Fig. 3c). The number of the quantization levels was in this case 17. As can be seen, the performance of the network decreases quite dramatically staying below 90%. In the last case we added two hidden nodes and increased slightly the amount of the quantization levels in order to get the performance back to the acceptable level (Fig. 3d). From these examples we can see that linearizing the probability functions has quite small effect on the performance. In contrast the quantization decreases the performance significantly. The reason is that the rough weight estimations does not allow the network to find an exact solution. By increasing the amount of hidden nodes we can decrease this effect, because the importance of a single node is smaller. Thus, we have two properties which has the same effect on the network performance, namely the number of weight levels and hidden nodes. Both increase the performance but naturally at the expense of increasing the area in implementations. The problem is to find an optimal solution, which seems to depend also on the network size and application.

The architecture of the circuit is shown in Figure 4. The circuit is designed to operate together with a host processor, which controls the learning and optimization operations. Thus, no control structures are implemented on chip in this first version. In reference /6/ an analog implementation of the Boltzmann Machine algorithm was introduced. It has 6 nodes and includes a control part. However, we wanted first to implement an efficient test structure for a larger network and include the control structures in a later designs. The network consists of the input nodes and fully connected block of hidden/output nodes. The operation is controlled via several registers, which are written and read by the host processor. The network configuration is selected via a special register which controls the I/O -nodes. During the learning the nodes which are selected to be output nodes can be clamped to desired values by writing the value in the I/O-register and selecting the clamped nodes via the control register. All the I/O-node values, both hidden and output nodes can be read via the I/O-register. The updating is performed by selecting the desired row via the row decoding register and updating the entire row by the value written in the update register. The simulated "temperature" parameter of the algorithm is controlled with a register. The 16-bit parallel Random Number Generator (RNG) /7/ block generates a new random value during each clock cycle. These random numbers can be considered as pulse-density coded white noise for 16 different channels. The random number generator is based on the shift-register structure with XOR feedback connections giving a very efficient implementation. The temperature parameter of the network can be changed by modifying the amplitude of the noise added on the dendrite lines. This is performed by feeding the noise via pulse-density synapses controlled by the temperature register. The noise is multiplied by the temperature value written in the register.

Both digital and switched-capacitor *implementations* of the structures based on the pulse-density modulation technique can be used. The switched-capacitor implementation leads to smaller area, but the digital implementation has the advantage of easier off-chip expandability. The active area of the circuit implemented by the 2.5 μm CMOS process is approximated to be 40 mm^2 with the switched-capacitor structures and 50 mm^2 with the fully digital structures.

4. Conclusions and Acknowledgements

We have designed an efficient VLSI implementation of Boltzmann Machine algorithm by using switched-capacitor structures and pulse-density modulation technique. These techniques offer very efficient, robust and flexible structures for the integrated circuit implementations. The design was targeted to 2.5 μm CMOS process, but can be implemented by any common CMOS process. However, the switched-capacitor structures require some kind of on-chip capacitor implementations. This work has been supported in part by Finsoft and Microelectronics programs of the Technology Development Centre in Finland (TEKES) and the Academy of Finland.

References

/1/ J.J.Hopfield: "Neural Networks and Physical Systems with Emergent Collective Computational Abilities", Proc. Nat. Acad. Sci. U.S., vol. 79, pp. 2554-2558, Apr. 1982.

/2/ H.Sompolinski, "Statistical Mechanics of Neural Networks", Physics Today, pp. 70-80, December 1988.

/3/ G.Barna and K.Kaski, "Variations on Boltzmann Machine", Report 2-89, Tampere University of Technology, Microelectronics Lab., Finland, 1989.

/4/ J.Tomberg, T.Ritoniemi, H.Tenhunen and K.Kaski: "VLSI Implementation of Pulse Density Modulated Neural Network Structure", in Proc. International Symposium on Circuits and Systems, Oregon/USA, pp. 2104-2107, May 9-11, 1989.

/5/ J.Tomberg, T.Ritoniemi, K.Kaski and H.Tenhunen, "Fully Digital Neural Network Implementation Based on Pulse-Density Modulation", in Proc. IEEE Custom Integrated Circuits Conf., pp.12.7.1-12.7.4, San Diego/USA, May 15-17, 1989.

/6/ J.Alspector, B.Gupta and R.B.Allen, "Performance of a Stochastic Learning Microchip", Advances in Neural Information Processing System, Denver, November, 1988.

/7/ R.C.Tausworthe, "Random Numbers Generated by Linear Recurence Modulo Two", Mat.Comput. 19, pp. 201-209, 1965.

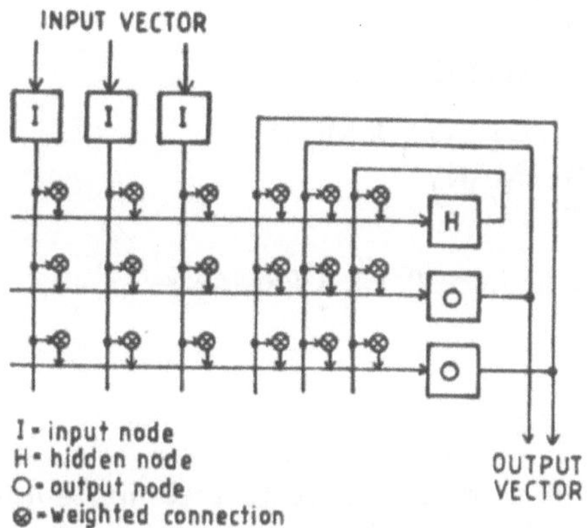

Figure 1. Structure of the Boltzmann Machine network.

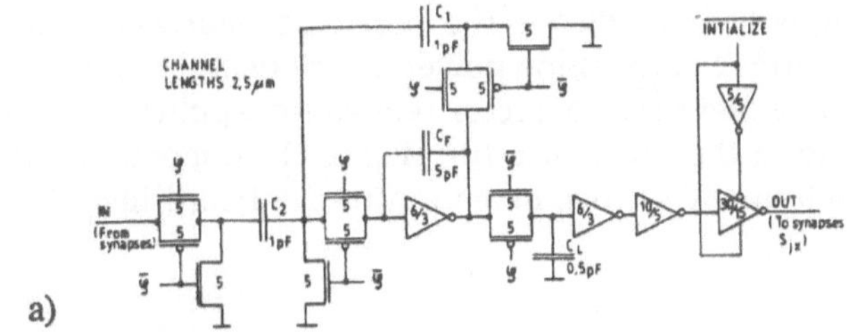

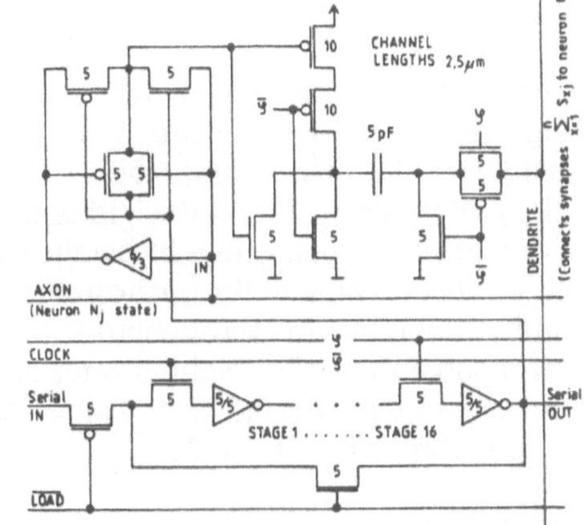

a) b)

Figure 2. Pulse-Density implementations of artificial a) synapse and b) neuron.

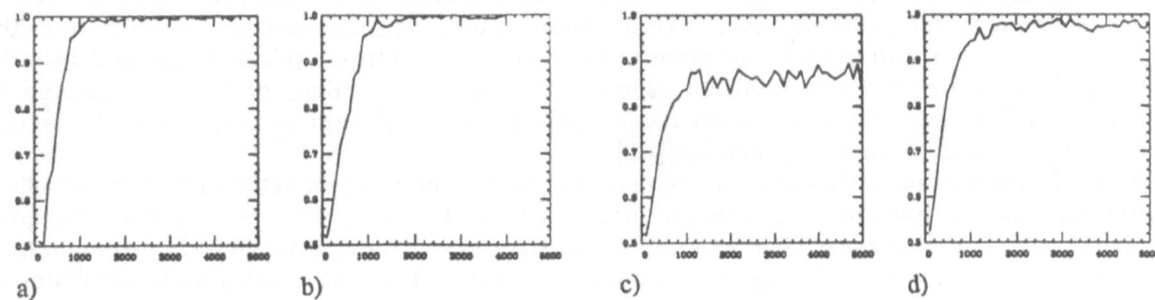

a) b) c) d)

Figure 3. Learning performance of modified Boltzmann Machine network.

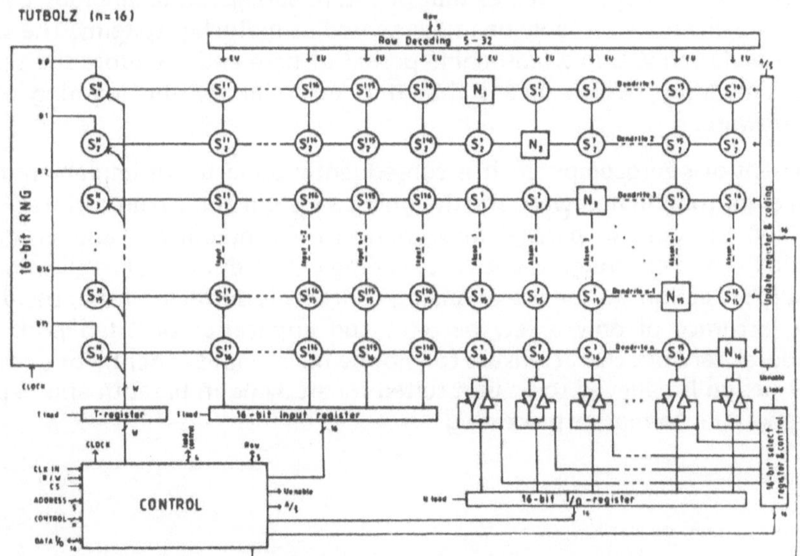

Figure 4. Architecture of the Boltzmann Machine circuit.

SYSTOLIC SYNTHESIS OF NEURAL NETWORKS

ULRICH RAMACHER, MATTHIAS WESSELING
SIEMENS AG , CORPORATE RESEARCH & DEVELOPMENT
D-8000 MUNICH 83, F.R. GERMANY
K. GOSER
UNIVERSITY OF DORTMUND, D-4600 DORTMUND 50, F.R. GERMANY

ABSTRACT

The analysis of today's neural paradigmas brings to light a set of elementary compute-intensive algorithmic strings which are shared by all neural models and, thus, make sense to be implemented in hardware. 2D arrays composed of a systolic neural signal processor module that integrates these elementary strings as hard-wired functional blocks present a favourable solution to the architectural problem of mapping neural parallelity and adaptivity into silicon. The proposed neurocomputer concept is sizeable to the applicational domain in terms of processing power, memory and flexibility, and is designed for throughput rates which enable the user to access real-world applications in reasonable time. Throughput rates at the chip site of the order of $5 \cdot 10^2$ MC/sec (1 Connection = 16 bit) are to be expected with 0.8μm CMOS technology. By systolic extension to the board level 10^5 MC/sec should be attainable.

INTRODUCTION

The response and the characteristics of present models of artificial neural nets are primarily investigated by simulation on vector computers, workstations, special coprocessors or transputer arrays. The fundamental drawback of such simulators is that the spatio-temporal parallelism in the processing of information that is inherent to the neural net is lost entirely or partly and that the computing time of the simulated net especially for large associations of neurons (tailored to application-relevant tasks) grows to such orders of magnitude that a speedy acquisition of "neural" know-how is hindered or made impossible[1,2].
An appreciable reduction in computing time for the simulation of neural nets and thus the handling of largish tasks or those that are to be executed in realtime become possible with specially designed neural hardware. A great number of technologies have been tried for neural net implementation and a dozen chips built [3]. For several reasons (see [4], pp.22) their use is restricted to simulation of simplified models of real-world scenaries.

Generally, neural nets for real-world pattern recognition overtax the single-chip integration potential of present technology as well as that of future submicron-technology by whole orders of magnitude. Such applications can only be implemented as multichip systems. The size of such a net will not permit simulation within a reasonable period of time and therefore the weights cannot be determined by simulation, which means that it is essential for the learning algorithms to be supported by hardware.

In the development of neurocomputers it is consequently a matter of implementing the compute-bound learning algorithms in hardware and designing a system and circuit architecture that supports in optimal fashion the massive parallel networking of the neural net and produces a sufficient measure of flexibility and expansion capacity for coping with a domain of applications (e.g. vision) or with general-purpose action. As the use of analog concepts is detracted from, besides others, by the low computing accuracy of only a few percent, and application-oriented problem analysis and modelling of neural network characteristics can not be made independently of circuit design, digital neurocomputer design is believed to be best suited for studying in breadth and depth the potential of neural networks for information processing.

ALGORITHMIC ANALYSIS OF NEURAL PARADIGMAS

A versatile neurocomputer architecture must absorb in silicon the elementary algorithmic strings shared by all known neural paradigma. In addition, a set of universal substrings has to be derived from those algorithmic strings which are not common to all the models. Elementary strings and substrings would compose the functional blocks, which have to be implemented in silicon, whereas concatenation of strings or substrings would be controlled by instructions. As the variety of models can be created to a large extent by the selection and order of these instructions, the design of a neurocomputer architecture splits into two parts: 1. design of the functional blocks, 2. design of control and communication blocks. This paper concentrates on the definition of the elementary neural strings and the problem of mapping neural parallelity and adaptivity into hardware.

Formula 1 displays the set of algorithms that define static neural nets with gradient-descent learning:

$$y_{ip} = f_i\left(\lambda \cdot \left(\sum_{j=0}^{N} W_{ij} y_{jp} + Y_i\right)\right) \quad , i = 1 \,..\, N,\ p \varepsilon P \ ; \tag{1a}$$

$$z_{jp} = D(f_j) \cdot \left[\sum_{i=0}^{N} W_{ij} z_{ip} + (t_{jp} - y_{jp}) \cdot \chi_j\right] \quad , j = 1 \,..\, N,\ p \varepsilon P \ ; \tag{1b}$$

$$\delta W_{ij} = \alpha \cdot \sum_{p \varepsilon P} z_{ip} \cdot y_{jp} \quad , i,j = 1 \,..\, N \tag{1c}$$

A neuron´s individual input is denoted by Y_i, the parameter λ controls the slope of the neuron´s discriminator function f_i or the temperature in case that a Boltzmann net is to be generated. $D(f_i)$ denotes the derivative of f_i, t_i the reference output in case of supervised learning, and χ_i the characteristic function of the output neurons. N designates the number of neurons and P the set of patterns.

It is quite remarkable that the second equation governs recurrent networks [5,6] and is structurally invariant under various choices of error functions. By concatenating the time-consuming matrix-vector or outer product operation with a few more "low cycle" operations (like adding "noise" to an updated weight) it is possible to create a large variety of neural models including for example Hopfield and Boltzmann type neural networks, Multi-layer Perceptrons and Kohonen maps (7,8,9).

ARCHITECTURAL ANALYSIS OF THE MAPPING PROBLEM

There exist various ways to implement formula (1) . Which one will be chosen depends on the applicational needs, technological constraints, the functional balance between chip and board architecture and the system environment [10,11]. Figure 1 shows the architectural scheme of a linear array of 8 multipliers and accumulators, respectively. Circles indicate registers, a number x in a circle a x-stage shift register. During the search phase the chain is fed by the sequence of weights W_{1j}, W_{2j}, ... , W_{8j} and by the sequence of inputs y_{j1}, y_{j2}, ..., y_{j8} , respectively; as mentionned above, y_{jp} denotes the j-th pixel of the p-th image. The i-th multiplier picks up W_{ij} and holds it for 8 cycles until the j-th pixel of all 8 images has been computed and separately stored, then proceeds by picking up $W_{i\,j+1}$ and computing the (j + 1)-th pixel of all 8 images and adding these separately to the previous ones, and so on. After completion of the weighting of the first 8 neurons, the accumulator registers are queued-up and read out. Then the procedure is repeated with the weights for the next 8 neurons in the network under consideration, etc..

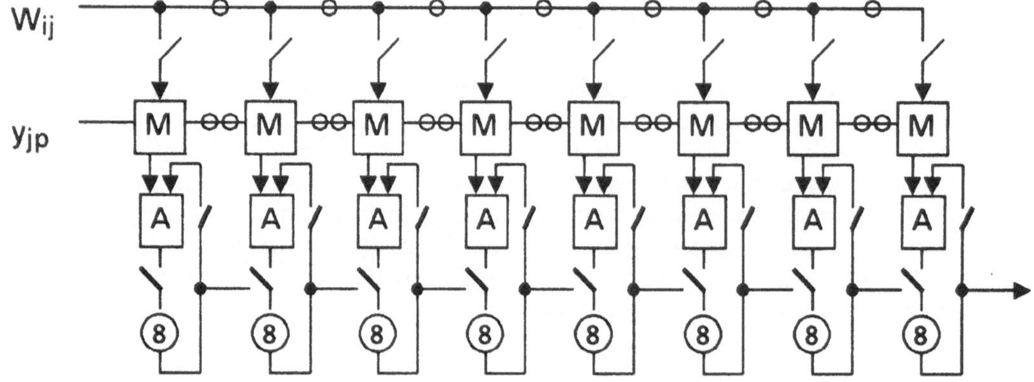

Figure 1 Systolic chain in the search mode

The chain can be used for learning, too, by re-arranging the formulas 1b and 1c, respectively:

$$\sum_{i=0}^{N} W_{ij} \cdot z_{ip} = \sum_{i=0}^{N} W^t_{ji} \cdot z_{ip} = \sum_{i=0}^{N} z^t_{pi} \cdot W_{ij} \; ; \tag{2a}$$

$$\sum_{p \varepsilon P} z_{ip} \cdot y_{jp} = \sum_{p \varepsilon P} y_{jp} \cdot z^t_{pi} = \sum_{p \varepsilon P} z_{ip} \cdot y^t_{pj} \; . \tag{2b}$$

It is left to the memory control and bus circuitry to transpose the weight matrix (middle expression of (2a)) or to change from row-wise to column-wise read-off (right expression of (2a)) , respectively. This is not a problem if on-chip storage of weights is possible or affordable. With off-chip storage of weights, however, the design of the memory control and bus circuitry gets complicated, especially for large memory banks (as needed for vision [2], for example). In this case it is desirable to use a fixed schedule for serving the weights and to add some extra circuitry to the chain.

Figure 2 shows how the search chain must be modified if the weights are fed into the chain in the

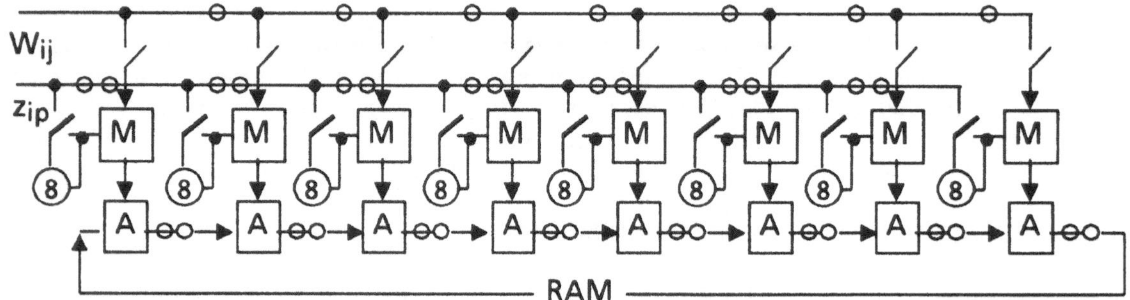

Figure 2 Systolic chain in the learning mode

same order as in figure 1. As the first weight index is distributed along the chain, the accumulation must be distributed in the same manner, according to (2a). The 8-stage shift register of the i-th multiplier is loaded with z_{ip}, p = 1..8, and repeatedly used for multiplying W_{ij}, j = 1..N. Note that there is a summation rippling along the chain for every pair of indices (j,p). Accordingly, the size of the (off-chip) RAM is at most proportional to the number of neurons and patterns. The content of the RAM is fed back to the chain in order to continue the accumulation along the next 8 rows of the weight matrix, together with the next block of 8 pixels of the same set of 8 patterns.

To calculate a weight update by means of the learning configuration as shown in figure 2 the following identity is to be observed:

$$\sum_{p \varepsilon P} z_{ip} \cdot y_{jp} = \sum_{p \varepsilon P} y^t_{pj} \cdot z^t_{pi} = \sum_{p \varepsilon P} z^t_{pi} \cdot y^t_{pj} . \qquad (2c)$$

As the right expression of (2c) is structurally equal to the left expression of (2a) it follows that the systolic chain shown in figure 2 can be used for weight updating as well.

The systolic structure of figure 1 and 2 may be absorbed by a configurable functional block FuncBloc which contains a multiplier, an adder, a shift register and some switches. This block is then repeatedly implemented. Note that the longer the chain is the more weights can be processed per second. It means also that more pixels of patterns need to be interleaved for the search mode because the number of interleaved patterns is equal to the lenght of the chain. The degree of interleaving can be reduced, thus, by parallelizing chains of reduced lenght. According to formula 1a this is done by serving either a parallel set of weights (W_{i_1j},W_{i_2j}), but one set of pixels y_{jp} of half the number of patterns (figure 3a)

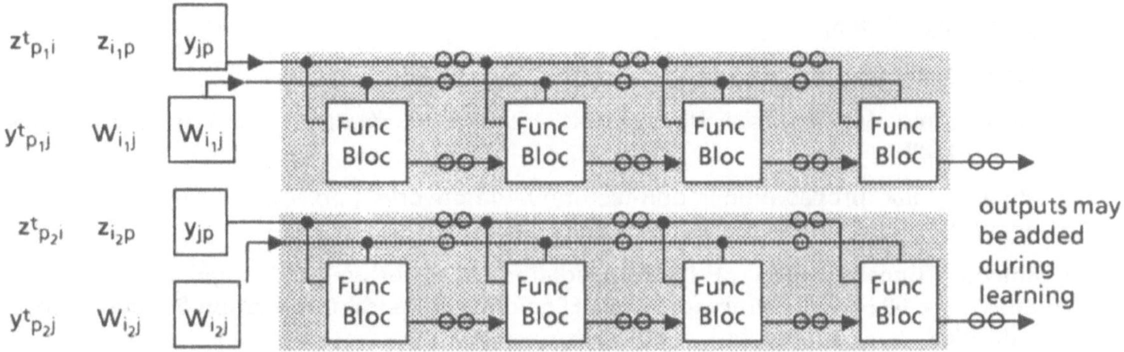

Figure 3a Parallelizing 2 chains in the row index

or a parallel set of weights (W_{ij_1},W_{ij_2}) and a parallel set of pixels (y_{j_1p}, y_{j_2p}) (figure 3b) . From formula (1b) follows that z_{ip} occurrs parallel if y_{jp} is served serially, and vice versa. Similiar reasoning holds for the update computation.

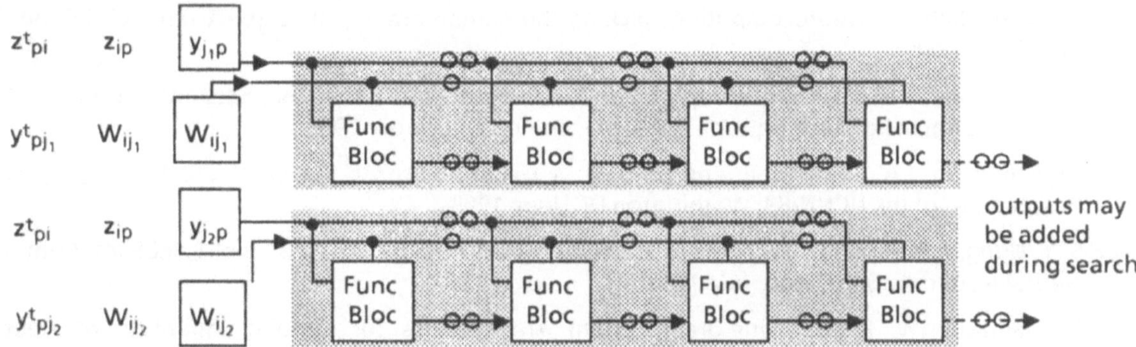

Figure 3b Parallelizing 2 chains in the column index

It depends on the application which configuration is selected; figure 3a offers serial input for the search mode and parallel input for the learning mode, whereas figure 3b offers parallel inputs in the search mode and serial input for the learning mode. With present technology 16 functional blocks configured as linear chain or parallel system of chains according to figures 1-3 may be implemented at chip site allowing for processing of about 640 MC/sec (1 Connection = 16 bit). To increase the processing power further, a 1- or 2-dimensional array of chips can easily be arranged for [13] .

CONCLUSION

The systolic approach as described in [12], for example, and in this paper is well suited to solve the neural net interconnection problem and cope with neural application areas like vision or speech. The systolic data flow structure is sizeable to the applicational domain in terms of memory, processor power and flexibility. At the chip site a processing power in the order of $5 \cdot 10^2$ MC/sec (1 Connection = 16 bit) is to be expected with 0.8μm CMOS technology. By systolic extension to the board level 10^5 MC/sec should be attainable.

REFERENCES

[1] DARPA Neural Network Study, pp.34 figure 2.14-15, AFCEA International Press, Nov. 1988

[2] ibid. pp.330 figure 28.5,

[3] ibid. pp.372 table 33.1-5

[4] "Microelectronics for Artificial Neural Nets", editors H. Klar, U. Ramacher, VDI-Verlag, Düsseldorf, 1989

[5] L.B. Almeida, "Backpropagation in non-feedforward networks", pp. 74, in : Neural Computing Architectures, edited by I. Aleksander, MIT Press 1989

[6] F.J. Pineda, "Generalization of Backpropagation to Recurrent and Higher Order Neural Networks", pp. 602-611 in: Proceedings of IEEE Conf. on Neural Information Systems, Denver, Colorado 1987

[7] R. P. Lippmann, "An introduction to computing with neural nets", IEEE ASSP Magazine, pp. 4-22, April 1987

[8] P.K. Simpson, "A Review of Artificial Neural Systems", part I&II, CRC Critical Reviews in Artificial Intelligence, 1988

[9] R. Hecht-Nielsen, "Neurocomputing: picking the human brain", IEEE Spectrum, vol. 25, no. 3, pp. 36-41, 1988

[10] Proceedings of the IFIP Workshop on Parallel Architectures on Silicon, Sessions 1-3, pp. 1-115, Institut National Polytechnique de Grenoble, France, December 1989

[11] M. Duranton, J.A. Sirat, "Learning on VLSI: A General-Purpose Digital Neurochip", pp. II-613, Proceedings of the IJCNN-89, Washington DC, June 1989

[12] S. Y. Kung, J. N. Hwang, "Parallel architectures for artificial neural nets", vol.2, IEEE Int. Conf. on Neural Networks, San Diego, July 1988

[13] U. Ramacher,W. Raab: "Fine-grain System Architectures for Systolic Emulation of Neural Algorithms", to appear in: Proceedings of the Int. Conf. on Application Specific Array Processors, Princeton NJ, Sept. 1990

SMART: HOW TO SIMULATE HUGE NETWORKS

Jean-Christophe LAWSON, Ali CHAMS & Jeanny HERAULT
INPG-ENSERG Labo TIRF, 46 Avenue Félix Viallet, 38 031 GRENOBLE CEDEX

Artificial Neural Networks often rhyme with massive parallelism. We need parallelism to simulate neural nets and fully benefit of their promises but we do need an efficient way to implement it. So we propose a dedicated approach, general enough to encompass a large class of models. It consists in performing arithmetics oriented algorithms. That leads to the joint development of hardware and software tools: the *SMART* machine. SMART stands for Sparse Matrix Adaptive Recursive Transforms, the general framework we use to discribe ANN. In order to handle networks with a sparse or dynamic connections topology we propose an original architecture. Using high speed CMOS technology it will deliver up to 300 MFlops in a deck side format, supporting up to one million connections: from 1000 fully interconnected neuronlikes to 10 000 sparsely interconnected ones. With the user-friendly UNIX Workstation host and integrated high speed VME channels we get a *Neurostation* exhibiting a power to price ratio many times greater than a supercomputer one. In this paper, we discuss first the SMART approach and its appeal. Second, we shortly present an hardware architecture for *SMART* sparse calculus. Third, we address the programming issue and display early results. Finally we summarize the advantages of the proposed machine with performance forecastings.

I/ THE SMART APPEAL :

A connections matrix is the most shared formalism in papers about ANNs. As a matter of fact it is the regular way of setting coupled equations, hence we can find very general models for which theoretical proofs are derived **[GRO88] [KOH84]**. But for further studies and practical implementations, this straightforward view seems computationally unmanageable. More and more, different networks are linked together: each of them is dedicated to a particular task such as visual perception, association, planning or motor control...**[KUP88][LAL88]** And it does not sound well to melt all these *neurons* in the same pot, with the same dynamic and any individual fully connected to the others. For sure there would be a lot of learning and stability problems provided we can afford the processing and memory size required by this connections mess. It is more appealling to have a dynamic topology, deleting unused connexion after a first learning, thus frozing basic concepts, then adding new connexions to relate these rough shapes while further learning.

As far as ANN software simulations are concerned, an oriented description for objects (actors...) is the answer of the moment **[DER89][KRA89]**. Any instance of a *neuron* object stores its weighted links and an activation program, enabling whatever hybrid network description we dream of. Then it is easy to derive one task per *neuron* and as we have a lot of *neurons*, we get the required number of tasks for a massive parallelism. But what about the implementation? The dynamic load balancing upon a large set of processors of such a number of asynchronous tasks is still an issue **[MUN89]**. Since we have very few a priori constraints on the *neurons* -different links numbers, adaptation rules, activation laws...- a static balancing is difficult as well. We get the feeling of a great overhead load for the sake of synchronizing a lot of a priori asynchronous tasks which perform more or less the same program. Obviously for any probable implementation, we will need only a small set of objects to refer. Most likely we shall let each neuron vary from the model through learning rather than have it customized.

The SMART approach infers matrices from the object oriented description. It simply groups *synapses* in clusters as large as possible, according to the various dynamics and adaptation rules. Hence we get a small set of connection matrices, presenting of course a sparse topology. Then *neurons* are grouped according to their activation laws. If we get too many matrices or groups, there is a design problem. Some reductions are to perform. For instance, in order to get different *neural* responses let us set a function with one extra argument to select the suitable response. Thus a set of a few very large tasks is defined. We might call them sequentially but each task is easy to parallelize since it consists in the same operation to perform many times on independent data.

However there are many ways to perform one operation, e.g. the information *weighting* through a link. The plainiest one is to multiply the *neuron* output with a weighting factor. It generally assumes multivalued responses which account for time and space averaging of spikes over a cluster of real neurons. SMART adopts this side since a useful use of the information coded by pulse sequences would require very complex links and might lead to the same average result **[HER80]**. As arithmetics become cheaper and the Amdhal law is still enforce**[AMD67]**, the point is to perform arithmetics rather than conditional operations and a lot of steps of a simple algorithm rather than few steps of a complex one which do the same.

This general view defined, we can devise an efficient architecture for a static load balancing, with no contention or synchronization problems. Sparse matrix handling will allow powerful implementations of complex networks:

1/ The dot product can be extended to a convolution with a matrix to compute each term, hence the weighting will become a recursive filtering.
2/ A few matrices with various dynamics will lead to a multi-synchronous network.
3/ Networks with dynamic topology will be legible at well, provided we support a quasi-static load balancing.

II/ AN EFFICIENT HARDWARE ARCHITECTURE :

Many techniques have been suggested for the efficient multiplication of sparse matrices by vectors; they involve often highly irregular storage schemes and manipulation algorithms for non zero elements in the matrix **[GEO81]**. Although these techniques lead to very powerful sequential implementations; they are, in general, not very suitable for parallel architectures. A few systolic architectures were implemented using some of these techniques **[MEL89]**, but they were not designed to support matrix updating. Meanwhile, *CRASY*, an other systolic architecture dedicated to ANN simulations exactly solved this problem **[GUE88]**. Unfortunately it did not deal with sparse matrix computing which impose an advance addressing to avoid the storage of zero elements and to limit the calculus cycle number.

Using *CRASY* architecture, we are designing *SMART* to carry on sparse matrix multiplication by vector as well as matrix updating. It is composed of a *SPARC®* micro-processor as control unit, and an array co-processor as processing unit (see Figure 1). A number of identical blocks, called slices, form the array co-processor. Each slice has its own local memory and two links to its neighbourhood, while broadcasting between the slices is performed by a common bus (see Figure 2).

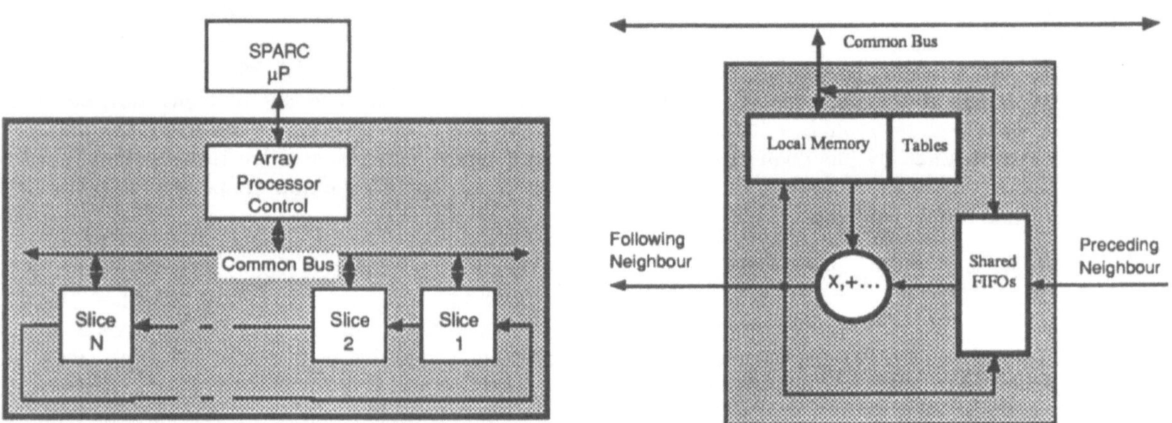

Figure 1 : SMART architecture

Figure 2: Slice architecture

The originality of this architecture comes from three major points:

1/ The way we control slices permits to achieve both parallel and systolic operating modes, but any mixing of the two modes is potential.
2/ In order to resynchronize the executed instructions, FIFOs are placed between slices. Therefore parallel and systolic instructions can share the same pipeline to get the best performance.
3/ The method to avoid multiplication by zero, when we multiply a sparse matrix by a vector, is to construct some tables that point to a vector component location for any given non-zero coefficient in the matrix. In other word these tables represent the result of compacting sparse matrices: zero coefficients which represent unused synapses are not stored. Part of the hardware is designed to support these tables with no extra load.

The early simulation for this architecture yields encouraging intermediary results, some of these are illustrated below:

1/ The maximum efficiency per slice can reach up to 20 MFLOPS affected by compacting sparse matrices performance.

2/ The number of connexions between neurons assigned per slice can reach up to 64000 connexions affected by compacting sparse matrices performance.
3/ The maximum number of neurons assigned per slice is about 1000 neurons according to the simulated network architecture.

Figure 3 displays simulations of the compacting performance mentioned above. Synapses are randomly distributed on a network of 1000 neurons for various average numbers of synapses per neuron. Then the corresponding matrices are squeezed and efficiency factor is deduced from slice allocation on an eight slices machine configuration.

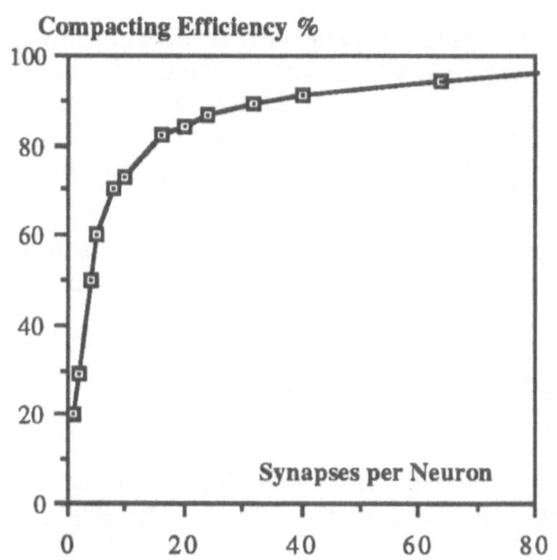

Figure 3 : Performance of the allocating algorithm

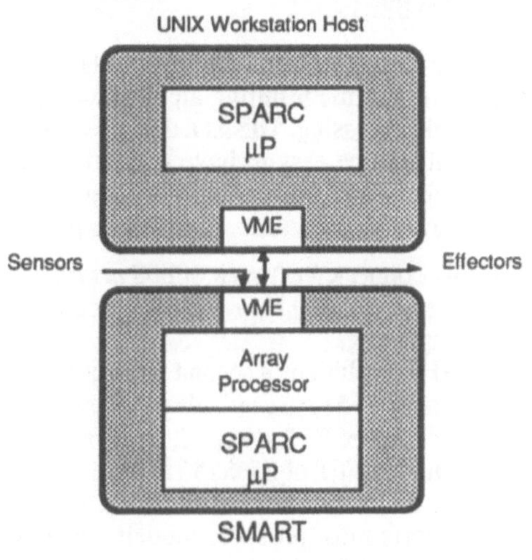

Figure 4 : System combination

III/ AN OPTIMIZING COMPILER :

The design of *SMART* hardware is coming with some software tools development, since such a machine cannot be efficient unless a suited software environment is provided. So, SMART will be linked to a workstation to get the benefit of a standard friendly environment (see Figure 4). The combination (workstation + *SMART*) forms a parallel system in which each machine is based on the same *SPARC®*processor model. This point of view allows to define a simple model for the combination.

A new language programming the combination appears necessary to get the best implementation of ANN simulations. This language should have certain features, as the following:

1/ The ability to implement other algorithms than the ANN ones, for instance signal processing, image processing, etc..
2/ The ability to manipulate matrices and vectors, in other word, the language should have few instructions to define them, and to apply all the possible operations between them. Sparse matrices will be treated as a special type.
3/ The ability to describe parallel tasks. In fact, we impose parallel notions on this language for tasks have to be shared between the two machines. *SMART* executes all vectorial instructions concerning matrices and vectors, whereas the host performs monitoring. Thus a data flow is estabilished between the machines. It is controlled by a special protocol introduced into the language. The user may assign the tasks for each machine in his program, therefore the synchronization is achieved using rendez-vous placed by either the user or the compiler.

A good choice is made for the language by extending C language to a vectorial one. This way, we use all the softwares available on the host by adding a simple interface to each. This language will be the first step toward an implementation of a high level specification language for ANN on *SMART* **[BES90]**.

An advanced compiler should be developed for the extended C according to *SMART* architecture. This means, the compiler should handle the following points:

1/ All the classical obstacles caused by the parallelism like communication, synchronization ... should be considered.

2/ The memory management for matrices, especially sparse ones, is an important task to perform. As we mentioned before, in describing hardware architecture, matrices should be stored in memory in the shape of lists which provide information about matrix structures. Then the compiler should deliver these lists for each matrix declared in a program, using some strategies to optimize memory allocation and load balancing upon the slices.

3/ The various number of slices allowed by SMART has to be transparent. A recompilation is required for programs when change in slices number takes place.

CONCLUSION :

We adopted a straightforward view to simulate neural networks which leads to the joint development of hardware and software tools. But the resulting architecture *SMART* is general enough to address other computer intensive tasks such as signal processing. This is a decisive advantage since ANNs are often related with low level processings and perception. It will be easy to have networks cooperating with classical pre-processing algorithms while the host performs higher level processing. Furthermore, *SMART* features all the advantages of a usual workstation with a very good power to price ratio. Eight slices provide 150 MFlops:

- a fully connected network of 800 *neurons* relaxes 120 times a second or 30 times with connections updating,
- a network of 5000 *neurons*, each neuron connecting to an hundred, relaxes 140 times a second or 40 times with updating,
- 75000 convolutions a second using a 1024 points kernel,
- 20 images of 512x512 pixels processed a second using a convoluting kernel of eight points,
- 8000 complex FFT a second, using a 256 points resolution,
- 15 complex FFT of 256x256 images per second...

But our scalable architecture accomodates up to sixteen slices delivering then 300 MFlops. Last but not least, *SMART* supports most probable models for networks of a significant size: sparse or dynamic topology, various networks connected together...

REFERENCES :

[AMD67] AMDHAL G., "The validity of the single processor approach to achieving large scale computing capabilities", AFIPS conf proc., Vol. 30, 1967, SJCC, pp.483-485

[BES90] BESSIERE P., CHAMS A., "Virtual model for ANN programming", submited to *INNC90*, Paris (FRANCE), July 9-13, 1990

[DER89] DERO T., ESCANDE P., MOULINOUX C., "A neuron-oriented programming environment", *NEURO-NIMES'89*, Nîmes (FRANCE), November 13-16, 1989, EC2 Editeur, pp183-200

[GEO81] GEORGE A.& LIU J., "Computer solutions of large sparse positive definite systems in computational mathematics", *Prentice-Hall*, Englewood Cliffs. NJ, 1981

[GRO88] GROSSBERG S., "Non-linear neural networks: principles, mechanism, and architectures", *Neural Networks*, Vol. 1, N°1, pp17-61,1988

[GUE88] GUERIN A., HERAULT J., "An efficient vector and scalar processing structure for adaptive algorithms", *Signal Processing*, IV, pp.331-334,1988

[HER80] "Le traitement de l'information dans les structure nerveuses. Etude par la simulation numérique et éléctrique. Application au traitement des signaux", thèse de doctoral ès sciences, I.N.P.Grenoble, 1980.

[KOH84] KOHONEN T., "Self organization and assiociative memory", Springer, 1984

[KRA89] KRAFT T.T., FROSTROM A., MAC RITCHIE B., ROGERS A.E., "The specification of a concurrent back-propagation network architecture using actors", *NEURO-NIMES'89*, Nîmes (FRANCE), November 13-16, 1989, EC2 Editeur, pp169-182

[KUP88] KUPERSTEIN M., "Neural model of adaptive hand-eye coordination for single postures", *Science*, Vol. 239, 11 March 1988, pp1308-1311

[LAL88] LALONDE W.R., GRAF D.H., "Neuroplanner: mechanisms for subcognitive control", *NEURO-NIMES'89*, Nîmes (FRANCE), November 13-16, 1989, EC2 Editeur, pp169-182

[MEL89] MELHEM R., "A systolic accelerator for the iterative solution os sparse linear systems", *IEEE Transaction on Computers*, Vol. 38, N°11, November 1989, pp1591-1595

[MUN89] MUNTEAN T., "Supernode : architecture parallèle et dynamiquement reconfigurable de Transputers", *11èmes Journées francophones sur l'informatique*, Nancy (FRANCE), janvier 1989, EC2 Editeur

AN ENHANCED PARALLEL PLANAR LATTICE ARCHITECTURE FOR LARGE SCALE NEURAL NETWORK SIMULATIONS

Yoshiji FUJIMOTO
Central Research Laboratories
Corporate Research and Development Group, SHARP Corporation
2613-1 Ichinomoto-cho, Tenri-shi, Nara 632, Japan

Abstract

In previous papers, the author proposed an enhanced parallel Toroidal Lattice Architecture (TLA) for simulations of large scale neural networks and implemented it onto a Transputer array. This paper proposes a Planar Lattice Architecture for Neural Network Simulations (PLANNS) as an improved version of TLA. The processor connections of this architecture are configured in a completely planar lattice structure, the most efficient structure for implementation using Wafer Scale Integration and increased parallel expandability. The performance of the PLANNS is almost proportional to the number of processors. Furthermore, this architecture exhibits great flexibility in simulating the various neural network architectures and a variety of neuron models.

1. Introduction

A neurocomputer is an essential tool for the research and development of neural networks in order to simulate neural networks with various configurations, parameters and a large number of learning data. A neurocomputer should have the following characteristics: ① Considerable computational power, ② Sufficient capacity for large scale neural networks, ③ Sufficient flexibility to simulate various neural network architectures and a variety of neuron models.

A spontaneous solution to the problem of computational power is parallel processing. The neural network simulators are implemented on general purpose parallel machines [3][4] and special purpose neurocomputers [5][6]. However, in simulations of large scale neural networks with a large number of connections, it was evident that the conspicuous performance degradation of parallel machines is caused by the data transmission bottleneck between processors[7]. No definite solution has ever been proposed for this connectivity problem in the parallel machines above mentioned.

In this paper, the author proposes an improved version of TLA[1][2], a Planar Lattice Architecture for Neural Network Simulations (PLANNS) which gives one solution to the connectivity problem and satisfies the three conditions above mentioned. First the general neuron model is defined and the PLANNS on Virtual Processors(VPs) is proposed. Then, the simulation of the Back Propagation on the PLANNS is described. Finally, the mapping from the VPs to physical Node Processors(NPs) with the same PLANNS is presented. This mapping is done by row and column partition. Before partitioning, the row and column permutations are carried out for NP load balancing.

2. General Neuron Model

A general model of artificial neurons can be presented by Eq.(1) and Eq.(2).

$$y_j = \Phi(\phi(v_1),\cdots,\phi(v_i),\cdots,\phi(v_{H+J})) = \Phi(\phi(v_i),\Phi(\phi(v_1),\cdots,\phi(v_{i-1}),\phi(v_{i+1}),\cdots,\phi(v_{H+J}))) \quad \cdots\cdots (1)$$

$$z_j = \Psi(y_j) \quad \cdots\cdots (2)$$

Where $\mathbf{z}=(z_1,\cdots,z_j,\cdots,z_J)$ is the output of neurons and $\mathbf{x}=(x_1,\cdots,x_h,\cdots,x_H)$ is the input to the neural network and $\mathbf{v}=(v_1,\cdots,v_h,\cdots,v_{H+J})=(x_1,\cdots,x_H,z_1,\cdots,z_J)$. Though this model is very simple, it represents the essential functions of a biological neuron. Eq.(1) which satisfies the associative rule, can be considered to correspond to the function of synapses and dendrites, as shown in Eqs.(3) and (10), where w_{ij} is the weight of a connection between neurons and inputs. Eq.(1) can also correspond to other multiply-accumulations as in Eqs.(8) and (16). Eq.(2) corresponds to the functions of a cell-body. The function $\Psi(y_j)$ is a sigmoid function Eq.(4) for the Perceptron or given by a differential equation Eq.(11) and a sigmoid function Eq.(12) for the Hopfield model.

The basic idea of the general neuron model is that a neuron is split into synapses and a cell body as a unit of simulation. Furthermore, Eq.(1) can be changed into a recursion formula Eq.(13) which corresponds to each synapse function.

【 Multi-Layer Perceptron 】

[Activation Forward Propagation]

$$y_j(n)=\sum_{i=1}^{H+J} w_{ij}(n)\cdot v_i(n) \qquad (3)$$

$$z_j(n)=f(y_j(n))=\frac{1}{1+e^{-y_j(n)/T}} \qquad (4)$$

[Error Back Propagation]

$$\Delta w_{ij}(n)=\alpha\cdot\delta_j(n)\cdot v_i(n) \qquad (5)$$

$$w_{ij}(n+1)=w_{ij}(n)+\Delta w_{ij}(n)+\beta\cdot\Delta w_{ij}(n) \qquad (6)$$

〔 Output Layer 〕

$$\delta_j(n)=z_j(n)\cdot(1-z_j(n))\cdot(t_j(n)-z_j(n)) \qquad (7)$$

〔 Hidden Layer 〕

$$S_j(n)=\sum_{k=1}^{J} w_{g(j),k}(n)\cdot\delta_k(n) \qquad (8)$$

$$\delta_j(n)=z_j(n)\cdot(1-z_j(n))\cdot S_j(n) \qquad (9)$$

【 Hopfield Model 】

$$y_j(t)=\sum_{i=1}^{N} w_{ij}(t)\cdot z_i(t)-x_j(t) \qquad (10)$$

$$u_j(t)-u_j(t-1)=-\gamma\cdot u_j(t)+y_j(t) \qquad (11)$$

$$z_j(t)=f(u_j(n))=\frac{1}{1+e^{-u_j(n)/T}} \qquad (12)$$

$$\Phi^i=\Phi(\Phi^{i-1},\phi(v_i)), \qquad where\ \Phi^i=\Phi(\phi(v_1),\cdot\cdot,\phi(v_i)) \qquad (13)$$

This recursion formula is applied to Eqs.(3),(10) and (8) as shown in Eq.(14) and (15),respectively.

$$y_j^i(n)=y_j^{i-1}(n)+w_{ij}(n)\cdot v_i(n) \quad or \quad y_j^i(n)=y_j^{i+1}(n)+w_{ij}(n)\cdot v_i(n) \qquad (14)$$

$$S_k^i(n)=S_k^{i-1}(n)+w_{k,i}(n)\cdot\delta_i(n) \quad or \quad S_k^i(n)=S_k^{i+1}(n)+w_{k,i}(n)\cdot\delta_i(n) \qquad (15)$$

3. Planar Lattice Architecture of Virtual Processors

A Planar Lattice Architecture of VPs which consist of a Synapse Processor(SP), a Cell Processor (CP), an Input Processor(IP), an Output Processor(OP) and an Input/Output Processor(IOP) is proposed as shown in Fig. 1. The SP has synapse functions which are the product-sum operations given in Eqs.(3),(8) and (10), and the updating operations of its weight given in Eqs.(5) and (6). The CP has cell body functions such as the threshold functions given in Eqs.(4) and (12), the neuro-dynamic functions given in Eq.(11) and the error evaluations given in Eqs.(7) and (9). The IP, OP and IOP have communication functions with the host computer through a bus line. VPs are arranged on the nodes of a planar lattice as shown in Fig. 1. CPs are placed on the diagonal of the J × J square region. The SPs and CPs are each connected to their four nearest neighbors with bidirectional channels. This configuration is called "PLANNS".

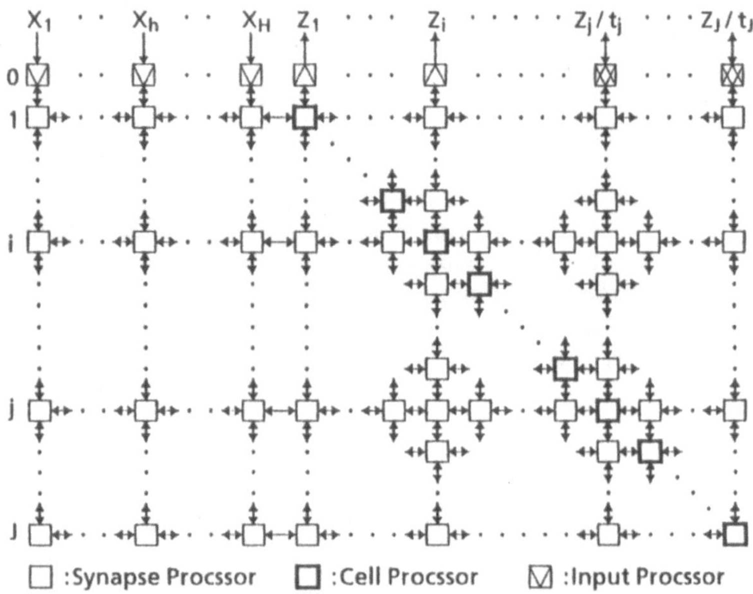

Figure 1 : Virtual Processors with the Planar Lattice Architecture

4. The Simulation of Multi-Layer Perceptron

The simulation of a Multi-Layer Perceptron(MLP) on the PLANNS is described. In this paper, the main Activation Forward Propagation(AFP) and Error Back-Propagation(EBP) procedures for simulation of a MLP are described.

[Activation Forward Propagation]

The product-sum operations given in Eq.(14) are executed on the SPs and a CP through the row shown in Fig.2. First, each CP delivers its output to all SPs in the same column. Each IP delivers training data to all SPs in the same column. In each SPs of right and left edges, the partial-accumulation is initialized at $y_j^1 = w_{1,j}(n) \cdot v_1(n)$ or $y_j^Q = w_{Q,j}(n) \cdot v_Q(n)$. Each SP on the left side of a CP receives the partial-accumulation from the left adjacent SP and adds it to the product of the training data or output and the weight value stored in its memory. The result of the product-sum operation is sent to the right adjacent SP. Each SP on the right side of the CP executes the same operations from right to left. The product-sum operations are completed when both of the partial-accumulations sent from the right and left sides, reach the CP. The CP then adds them of $y_j^k = y_j^{k-1} + y_j^{k+1}$ and executes the activation function given in Eq.(4) so as to get the output of a neuron. To propagate activations from the input to the final output layer, the above procedure is repeated the same number of times as the number of hidden and output layers.

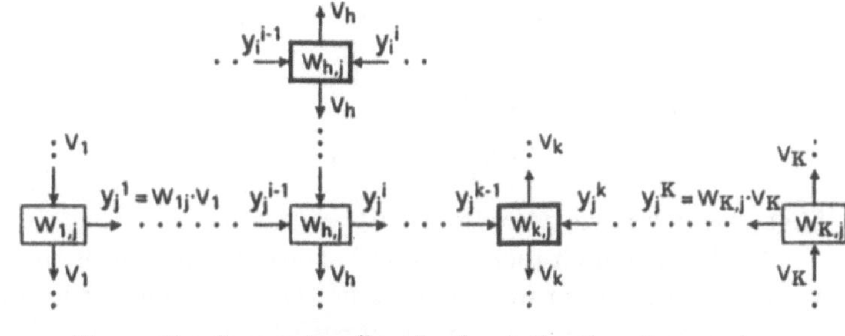

Figure 2 : Processing flow in the Activation Forward - Propagation where k = g (j), K = H + J

[Error Back-Propagation]

The product-sum operations of the EBP given in Eq.(15) is executed vertically on SPs in the same way as the AFP by exchanging row and column as shown in Fig.3. And then a CP executes the calculation given in Eq(9) to evaluates the error level.

One important reason for the above procedures is that the switchover between the horizontal and vertical product-sum operations are smoothly executed by the horizontal and vertical data deliveries from the diagonally arranged CPs.

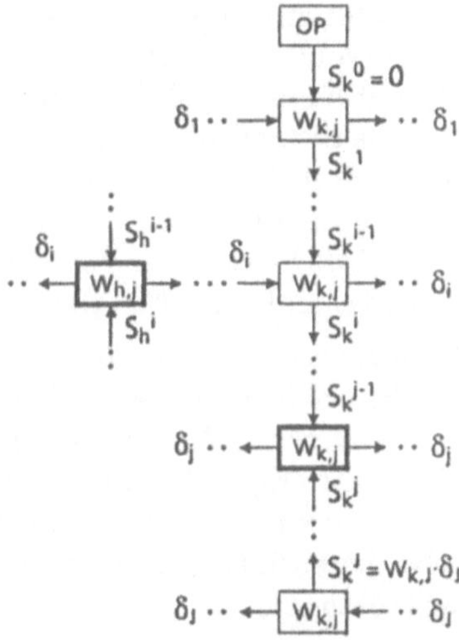

Figure 3 : Processing flow in the Error Back -Propagation where k = g(j), h = g(i)

5. Planar Lattice Architecture of Physical Node Processors

The number of VPs with the PLANNS is proportional to the square of the number of neurons and quite large for millions of neurons. Therefore, mapping the VPs onto a feasible number of physical processors is made. The rows of the Planar Lattice are partitioned into Q parts and the columns are partitioned into P parts. A physical processor called a 'Node Processor(NP) is assigned to the VPs in each rectangular region formed by the partitions and each NP is connected to its four nearest neighbors with bidirectional channels in the same manner as the VPs. Consequently, the P×Q NPs form a Planar Lattice as shown in Fig. 5. The NPs in the first row are connected to the host computer by a bus line. Each NP sequentially executes the functions of the VPs included in the assigned rectangular region. The product-sum operations of the SPs are integrated into one equation. Eqs.(3) and (8) are changed into Eqs.(16) and (17), (18) and (19), respectively. The partial-accumulations in Eqs.(16) and (18) are calculated on all NPs simultaneously, and the partial-accumulations in Eqs.(17) and (19) are calculated through the NPs in the same row and column sequentially.

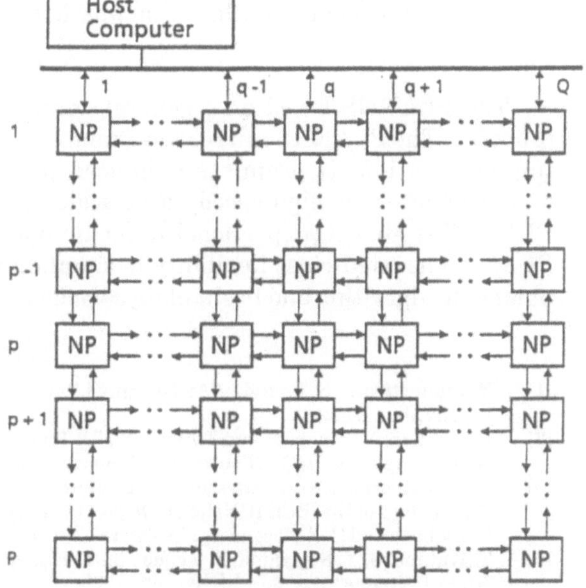

Figure 5 : Node processors (NPs) with the Planar Lattice Architecture

$$B_j^q(n)=\sum_{i=Ibq}^{Ieq} w_{i,j}(n)\cdot v_i(n) \quad \cdots\cdots\cdots\cdots (16)$$

$$y_j^q(n)=y_j^{q-1}(n)+S_j^q(n) \text{ or}$$

$$y_j^q(n)=y_j^{q+1}(n)+S_j^q(n) \quad \cdots\cdots\cdots\cdots (17)$$

$$D_j^p(n)=\sum_{k=Jbp}^{Jep} w_{g(j),k}(n)\cdot \delta_k(n) \quad \cdots\cdots\cdots (18)$$

$$E_j^p(n)=E_j^{p-1}(n)+D_j^p(n) \text{ or}$$

$$E_j^p(n)=E_j^{p+1}(n)+D_j^p(n) \quad \cdots\cdots\cdots (19)$$

[Activation Forward Propagation]

Each NP which includes a CP sends the output of the CP to the upper and lower adjacent NPs. Each NP in the top row which includes an IP sends the input data to the lower adjacent NPs. Each NP receives the data $v_i(n)$ from the upper or lower adjacent NP, stores it in the buffer memory and sends it to the lower or upper adjacent NP, respectively. All NPs execute the partial product-sum calculations in Eq.(16) simultaneously. Each NP in the left- and right-most columns initializes partial accumulations at $y_j^1(n)=B_j^1(n)$ or $y_j^Q(n)=B_j^Q(n)$ and sends to the right or left adjacent NPs, respectively. Each NP receives a partial accumulation $y_j^{q-1}(n)$ or $y_j^{q+1}(n)$ from the left or right adjacent NP and adds it to the partial product-sum of the j-th row $B_j^q(n)$ (Eq.(17)). The NP sends the result as the new partial accumulation $y_j^q(n)$ to the right or left adjacent NP respectively. When each NP which includes a CP in the assigned j-th row receives both $y_j^{q-1}(n)$ and $y_j^{q+1}(n)$ from the left and right adjacent NPs, the NP adds them to the partial product-sum of the j-th row, $y_j(n)=S_j^q(n)+y_j^{q-1}(n)+y_j^{q+1}(n)$, makes the result equal to the active potential and performs the sigmoid calculation (Eq.(4)) to determine the activation level.

[Error Back-Propagation]

The product-sum of the EBP calculations given in Eqs.(16) and (18) is executed vertically in the same way as the AFP calculations by exchanging row and column. Finally, an NP which contains a CP executes Eq.(9) to determine the error level $\delta_j(n)$.

In these procedures, one remarkable point is that the transmission of partial accumulations of active potentials and error levels can be executed simultaneously during calculations given in Eqs.(16) to (19).

6. Load Balancing of Node Processors

The load balancing of NPs is the most important problem for efficient parallel processing. In the previous paper[1][2], an algorithm which executes the row and column permutations of the VP Lattice before mapping VPs to NPs, was provided to solve this problem. It was proven by implementing the TLA on the 4×4 Transputer array that the performance of TLA is almost proportional to the number of processors. Though this algorithm was developed for the TLA, it is also applicable to the PLANNS.

7. Conclusions

The PLANNS based on a general neuron model as an improved version of the TLA has been proposed. The PLANNS can realize a completely planar architecture that is the most efficient for WSI implementation. It promises enhanced parallel processing and greater expandability of the neurocomputer for high speed, large scale simulations of neural networks. The performance of the PLANNS is almost proportional to the number of NPs as well as that of the TLA. Furthermore, the PLANNS has sufficient flexibility to simulate various neural network architectures by using a load balancing algorithm and to simulate a variety of neuron models by programing them on the NPs.

8. References

[1] Y. Fujimoto and N. Fukuda, "An Enhanced Parallel Toroidal Lattice Architecture for Large Scale Neural Networks," IJCNN, June 1989.

[2] N. Fukuda, Y. Fujimoto and T.Akabane, "A Transputer Implementation of Toroidal Lattice Architecture for Parallel Neurocomputing," IJCNN'90 Winter, January 1990.

[3] D. A. Pomerleau, G. L. Gusciora, D. S. Touretzky and H. T. Kung, "Neural Network Simulation at Warp Speed," Proceedings of the IEEE ICNN, San Diego, CA., July 1988, Vol.II, pp.143-150.

[4] G. Blelloch and C. R. Rosenberg, "Network Learning on the Connection Machine," Proc. of the IJCAI, August 1987.

[5] T. Watanabe, Y. Sugiyama, T. Kondo and Y. Kitayama, "Neural Network Simulation on a Massively Parallel Cellular Array Processor:AAP-2," IJCNN, June 1989.

[6] A. Iwata, Y. Yoshida, S. Matsuda, et al., "An Artificial Neural Network Accelerator using General Purpose 24 bits Floating Point Digital Signal Processors," IJCNN, June 1989.

[7] "DARPA Neural Network Study," AFCEA International Press, November 1989.

AN ANALOG CHIP SET FOR MULTI-LAYERED SYNCHRONOUS BOLTZMANN MACHINES

P.GARDA, E.BELHAIRE

I.E.F. - C.N.R.S. U.R.A. 22 BAT. 220 UNIVERSITE DE PARIS SUD 91405 ORSAY FRANCE

1. ABSTRACT

Many current neural nets experiments use multi-layered networks and the Generalised Delta Rule algorithm [1]. Whereas the experimentation speed is insufficient, no cellular circuits have been realised for this algorithm, because neither its implementation nor its parallelisation are obvious.

On the other hand, Boltzmann Machines [2] show a number of very attractive features : their asymptotic behavior is related to a mathematical model, the weight update rule is local and simple and the unit activation states are binary. Few software experiments are carried out - the simulations of the model are desperately slow -, but mixed analog/digital implementations have been described ([3], [4], [5]). However they are targeted towards fully connected networks.

In this paper, we describe the architecture of an analog chip set for multilayered Synchronous Boltzmann Machines, which improves the size of practically achievable systems and the effective use of silicon and board area in this context. Moreover, it uses analog circuits both for the computations and the storage, and this improves the relaxation speed and the density of computing units per chip achievable with today I.C. technologies..

2. SYNCHRONOUS BOLTZMANN MACHINES.

Actually, it is natural for a fully analog implementation to be fully synchronous, i.e. update simultaneously all the neuron states. However the already described implementations hardwire the weight update rule derived by Hinton and Sejnowski [2] under the assumption of an asynchronous neuron update, and this gives bad synchronous relaxation results, as pointed out in [5].

A Synchronous Boltzmann Machine model has been described by Prf. Robert Azencott in [6]. He succeeded in the derivation both of an asymptotic model of the network behaviour and of a weight update rule for networks using synchronous update. Designing circuits for this model results in considerable improvements in relaxation speed over circuits using asynchronous update, and in considerable improvements in learning and recognition results over circuits using synchronous update combined to Hinton and Sejnowski weight update rule.

The Synchronous Boltzmann Machine is operated as follows. Let $(u_i)_i$ be a set of units, x_i^n the activation state of u_i at instant n and w_{ij} the weight between units u_i and u_j. Let $V_i^n = \Sigma_j w_{ij} x_j$ be the potential of u_i after instant n. Then the activation state of u_i at discrete time step (n+1) is tossed at random with the probability : $P(x_i^{n+1} = 0) = 1/[1 + \exp(V_i^n/T)]$ (1)

Now let us consider the learning process. For this we have to choose input, output and hidden units. The learning process is repeatedly performed for all the pattern associations, and for each of them, it consists in two phases, during which a cooccurrence counter is associated to each weight, which computes the following expression : $p_{ij} = (1/N) [\Sigma_n x_i^{n-1} x_j^n + x_j^{n-1} x_i^n]$. (2)

During the clamped phase, a pattern is imposed both on the input and output units, and the cooccurrence p_{ij}^+ is computed, whereas during the free phase, the input pattern is presented to the input units while the output units are left free, and the cooccurrence p_{ij}^- is computed. After these clamped and free phases, the weights are updated with the rule : $\Delta w_{ij} = \text{sign}(p_{ij}^+ - p_{ij}^-)$ (3).

3. ANALOG STORAGE AND SYNAPTIC CELL.

A significant amount of chip area may be gained from replacing digital memories by analog ones. We state that the weight and cooccurence memories may successfully be carried out by capacitors at room temperature both in the learning and in the recognition phase. In the former the resulting slow weight decay favours the learning (cf [3]). In the later we use a suitable refreshing scheme : it is built out of an external digital memory (RAM), which is scanned in order to sequentially refresh the analog weight value. In this way we get a very small area on-chip storage without any specific technological developments such as circuit cooling or analog floating gates. Of course some external hardware is required, but it is accessed sequentially and thus no bottleneck is introduced.

A synaptic cell (figure 1) consists in an analog cooccurrence counter, which is incremented during the clamped learning phase and decremented during the free learning phase. At each relaxation iteration, the counter is sequentially modified twice, by considering a value which is firstly $x_i^{n-1}x_j^n$ and secondly $x_j^{n-1}x_i^n$. For that purpose two wires, x_i and x_j, reach the synaptic unit and are input to a nand gate whose output is connected to the cooccurrence counter. The output of the counter is fed into a comparator to determine the sign of $(p_{ij}^+ - p_{ij}^-)$. According to this sign the weight counter is incremented or decremented. Finally the weight capacitor drives a voltage-controlled current source whose output is connected to the net V_i in order for the synaptic unit to provide on V_i a current representing the contribution w_{ij} x_j. All computations are single rail and two clocks are required, one latching $x_i^{n-1}x_j^n$ or $x_j^{n-1}x_i^n$, the other latching Δw_{ij}.The same cell may be used to compute the asynchronous learning rule.

SYNAPTIC UNIT

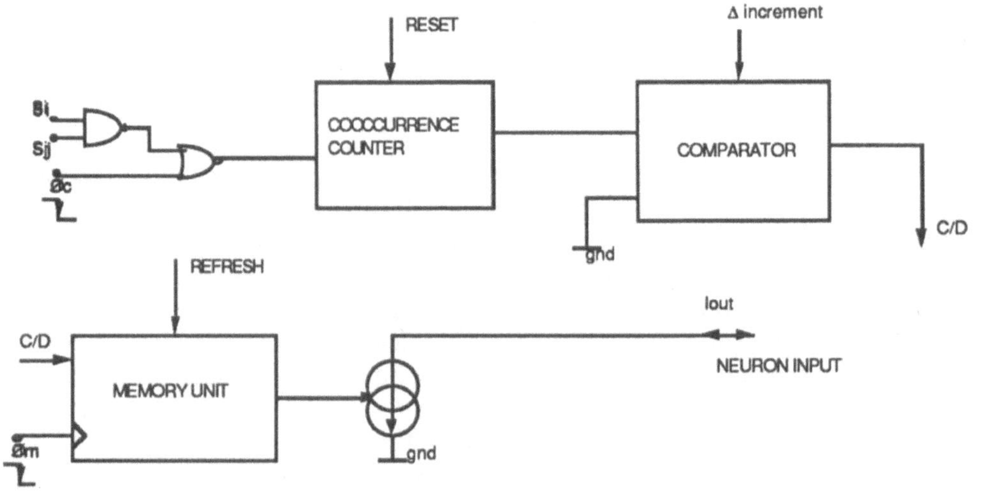

4. RANDOM NUMBER GENERATION AND NEURON CELL.

The random tossing required by the Boltzmann Machine results in some troubles : the use of resistor thermal noise (cf [3]) leads practically to correlated generators, as pointed out in [5] and confirmed experimentally by [7]. On the other hand, the use of digital pseudo-random generators uses a lot of surface and leads also to some problems (cf [4]).

We investigate a solution based on a uniform law random number generator : the binary value is resulting from the comparison of its output U_i to the sigmoïdal function $Z_i = 1/(1 + \exp(-V_i/T))$ of the potential V_i. This uniform law is built out of the integration of a sequence of unbiased independent binary values, which result from the sampling of an optical random physical system, the speckle of laser, which is made time independent by a suitable scheme described in [8].

Finally the neuron cell has two inputs, the binary random samples and the potential V_i, and one output, the activation state x_i. It consists (cf figure 2) in a current to voltage converter, sigmoïdal function, an integrator and a comparator.

5. Chip set.

Now let us describe the way the synaptic and neuron chips for arbitrary multi-layered networks. In order to keep on with the symetry of the weights which is required by the Boltzmann Machine model while using an analog storage, we have decided that each cooccurrence and weight counters should have a single physical representation. As each weight w_{ij} is used for the computation of the two contributions $w_{ij}\ x_i$ and $w_{ij}\ x_j$ to the potentials V_i and V_j, this means that the current representing these two contributions has to be computed in a single synaptic unit. Now in general this occurs between two fully interconnected successive layers. This means that p units $(u_i)_{1\leq i\leq p}$ of one layer are fully interconnected to q units $(v_j)_{1\leq j\leq q}$ of the other layer. The interaction between these two sets of units is described by a matrix $(w_{ij})_{1\leq i\leq p, 1\leq j\leq q}$, and this suggests a matrix floor plan for the synaptic cells matrix.

However, with the current design of the synaptic cell, at most 1000 synaptic units fit in a single CMOS 2μ chip. On the other hand, typical values of p and q are of the order of 100. If we group P by Q synaptic cells on a chip, with PxQ = 1000, then we need a matrix of [100/P]x[100/Q]

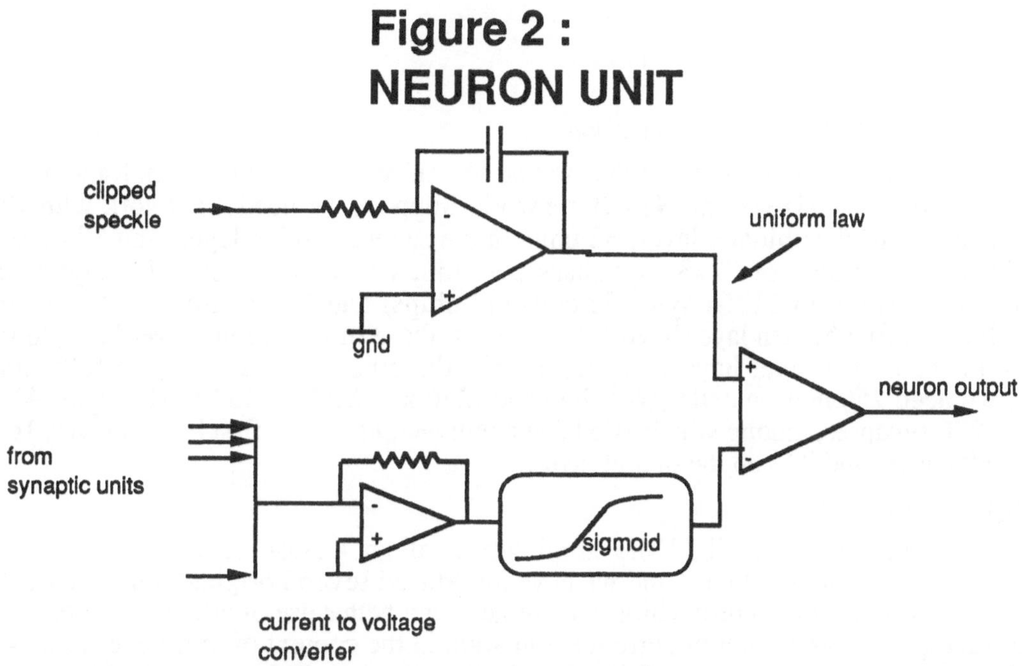

synaptic chips to fully interconnect two 100 units layers. This shows that the synaptic chips have to be cascadable in both directions. Now each off-chip connection will introduce some degradation in the computation of the current representing V_i, and thus their number should be minimised for all V_i nets. Thus a balanced choice is $P = Q \sim \text{sqrt}(1000) \sim 30$.

As the synaptic matrix have to be cascaded, they should not include any neuron cell, as this would lead to a large number of unused neuron cells in the system, and thus to a waste of chip area. Thus we need a separate neuron cells chip, which consists in a column of r neuron cells.

Now in the case of an input layer, the units activation state are always clamped by the environnment, and it is thus useless to compute their potential V_j. This leads to a synaptic chip whose units compute V_i only, and which has less analog output pads than the previous one : as the number of input units is large, this chip may allow the realization of larger system thank to a light extra design (the two chips share all their basic cells). However it increases the chips cost.

6. Multi-layered networks implementation.

Now a similar bidirectionnal synaptic chip has been used by Alspector to implement a multi-chip fully connected network [3]. Now let us explain how we use our chips to implement multi-layered network (figure 3). For each hidden layer, the contributions from the previous and the next layer are added through the connection of the corresponding potential nets. The unit activation states are propagated in the reverse way. Each layer is built out of several juxtaposed neuron chips, and each synaptic matrix is built out of a bidirectionnal cascade of synaptic chips.

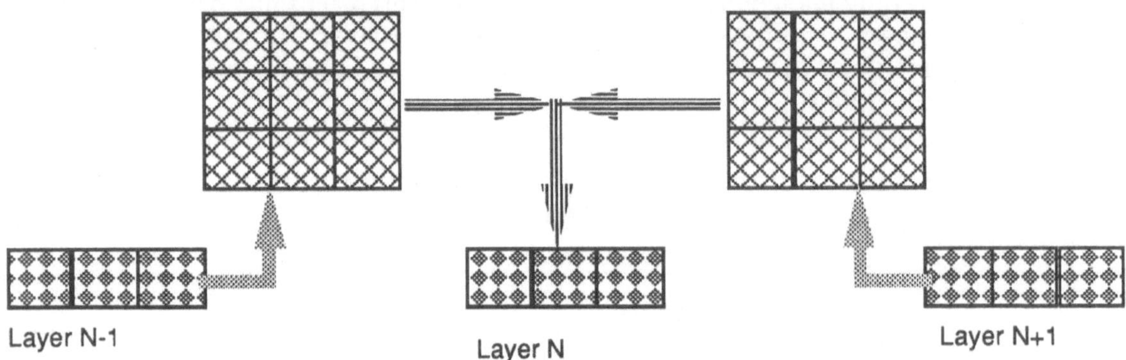

Figure 3 : hidden layer connexion.

Let us now compare the use of these circuits to that of a fully connected network for a toy example, which is roughly of the size of the Nettalk network. Suppose we need a network with 128 input units, 64 units in the first hidden layer, 32 units in the second hidden layer, and 32 output units, and we have 32 neurons or 32x32 synapses per chip, where two successive layers are fully connected. Then we need 11264 synaptic cells (11 chips) and 128 neuron cells (8 chips). The neuron cell of the first hidden layer have 150 synapses, those of the second layer have 96 synapses and those of the output layer have 32 synapses. On the other hand if we use a fully connected network, we need 256 neuron cells (16 chips) and (256 x 255)/2 = 32640 (31 chips). All neuron units have 256 synapses, among which 106 have a null weight for the first hidden layer, 160 on the second hidden layer and 224 on the output layer.

7. Conclusion.

In this paper, we have described the architecture of an analog chip set for multilayered Synchronous Boltzmann Machines, and we have introduced several original points. By separating and cascading synaptic and neuron chips, we are getting a better use of silicon and board area and we are lowering the degradation of current summation in the context of multilayer networks, thus paving the way to larger systems. By following the Synchronous Boltzmann Machine model from D.I.A.M.E.N.S., we are getting considerable improvements in relaxation speed and in learning and recognition results. By using analog circuits both for the computations and the storage, we are improving the relaxation speed and the number of computing units per chip. Finally by using an optical setup from I.O.T.A. we are getting random numbers free of correlation at a very high rate.

8. Acknowledgments.

This work has been supported by D.R.E.T. under Contract 87/292 and by C.N.R.S. under P.R.C.-A.M.N. project RA. We thank Robert Azencott, from D.I.A.M.E.N.S. in Paris, Pierre Chavel and and Philippe Lalanne, from I.O.T.A. at Orsay, Francis Devos and Kurosh Madani from I.E.F. for fruitful discussions.

9. References.

[1] G.Hinton,& al. : Generalised Delta Rule, C.M.U. T.R. CMU-CS-84-119, C.M.U. 1984
[2] G.Hinton,& al. : Boltzmann Machines, C.M.U. T.R. CMU-CS-84-119, C.M.U. 1984
[3] J.Alspector & al. A neuromorphic V.L.S.I. learning system. 1987 Stanford Conf. on V.L.S.I., M.I.T. Press.
[4] J.Alspector & al. Stochastic learning network and their electronic implementation. Procs N.I.P.S. 87 A.I.P.
[5] I.Kreuzer & al. : A modified model of Boltzmann Machines for W.S.I. realization. Signal Processing IV, 1988.
[[6] R.Azencott : Synchronous Boltzmann Machines and their learning algorithms. NATO ARW. Les Arcs, 02 1989.
[7] J.Alspector & al. : public & private communication, 22nd Asilomar Conf., Pacific Grove, Nov. 1988.
[8] P.Lalanne : Journées d'Electronique de Lausanne, Nov. 1989.

DNNA: A DIGITAL NEURAL NETWORK ARCHITECTURE

Max Stanford Tomlinson Jr., Ph.D., Dennis J. Walker
Neural Semiconductor, Inc.
5973 Avenida Encinas
Carlsbad, CA 92008 USA
Voice: (619) 931-7600, FAX: (619) 931-7825

ABSTRACT

A novel method for performing neural network computations is discussed. Because the architecture is fully digital it permits the fabrication of neural network integrated circuits using a variety of standard semiconductor production techniques. However, instead of representing neuron activations and synaptic weight values using fixed or floating point formats, a stochastic "pulse train" is employed. This representational scheme allows simple logical operators to be used to perform network computations. High densities (number of weights per area of silicon) comparable to that of analog implementations have been achieved. Furthermore, DNNA integrated circuits can be used as "building-blocks" to construct large neural networks with many chips in various topologies while maintaining constant execution speed.

INTRODUCTION

Figure 1 illustrates a small section of a "typical" neural network. Input activations arrive either externally or from the outputs of a previous layer of neurons. The activation of neuron *i*, denoted as o_i, is bounded, usually in the range from 0 to 1. These activations are modified locally by the synaptic weights. The weight to neuron *j* from neuron *i* shall be denoted as w_{ji}.

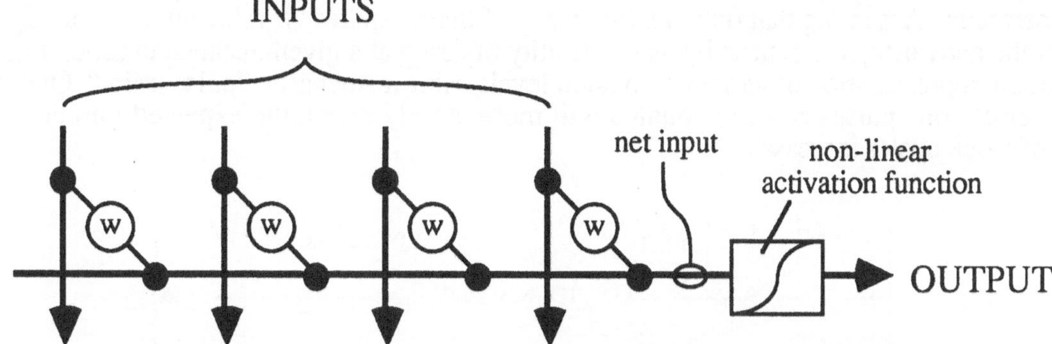

Figure 1: "Typical" Neural Network Computation

Often the weighted inputs are summed linearly to produce a "net input" value to a neuron. Finally, this potentially large net input value is then "squashed" to the activation range of the neurons by a non-linear activation function, *f*. For the most part, the characteristic equations (1) and (2) summarize (non-adaptive) neural network computation [1]:

$$o_j = f(n_j), \tag{1}$$

$$n_j = \left(\sum_i w_{ji} o_i\right) + b_j, \tag{2}$$

where b_j is a bias weight, n_j is the net input, and $f(n_j)$ is a non-linear activation function.

IMPLEMENTATION REALITIES

While these equations are convenient for simulating neural networks with standard von Neumann computers, they pose a serious challenge for implementation in silicon. A directly implemented, fully parallel (DIFP) neural network integrated circuit must not only implement every weight and neuron with separate circuitry, it should permit the easy expansion of such circuitry for the construction of large networks. Often the toughest challenge is in calculating and representing the (possibly infinitely) large net input term, n_j.

An analog implementation has the potential for high densities and fast operation. However, analog circuitry poses many inherent difficulties. It is susceptible to noise and changes in temperature. Chip-to-chip variations make it difficult to produce integrated circuits that are functionally identical. Combining multiple IC's together is difficult. Furthermore, storing weights for long periods often requires special fabrication techniques.

A digital implementation of neural network integrated circuits eliminates a majority of the problems associated with analog circuitry. Digital integrated circuits are easy to manufacture and are functionally identical. Weight storage is straightforward and can be realized using volatile (e.g. static or dynamic RAM), non-volatile (e.g. EPROM), or fixed (e.g. ROM) techniques. Unfortunately, the representations often used for storing digital values, fixed and floating point formats, require significant silicon area to implement the multiplication and addition operations. The use of several bits to represent a single value also impedes the parallel transmission of multiple values between IC's.

INSPIRATION FROM REAL NEURONS

Although the underlying computational method of the brain is inherently of analog nature, neurons effectively communicate their activations over long distances by using a "digital-like" transmission of action potentials. The more active a neuron is, the more frequently it produces action potentials. Assuming that only the frequency of these "spikes" is relevant, one can digitally represent the activation of a neuron by its probability of firing at a given instance in time. Figure 2 illustrates the representation of various activation levels using a stochastic "pulse train." Of course, the number of "on" pulses actually counted will more closely match the expected number as the number of clock cycles increases.

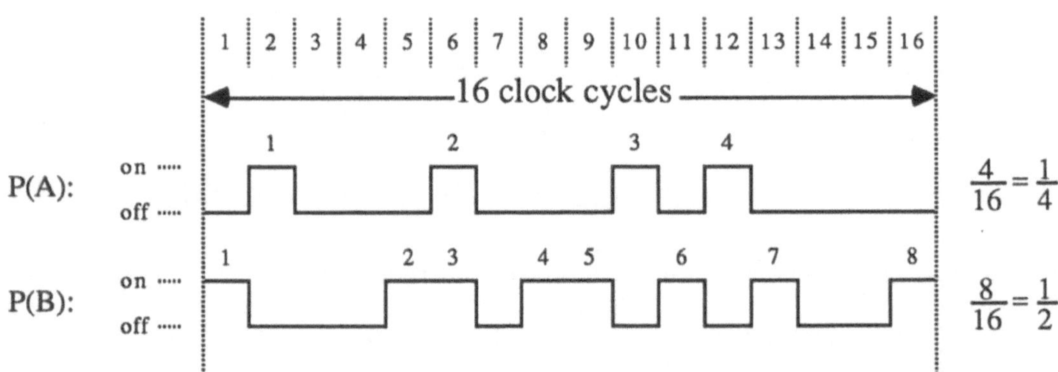

Figure 2: Stochastic Pulse Train Representation

If activations are represented by firing frequency, then the function of a synaptic weight is to alter this firing frequency. By representing a weight as a stochastic pulse train, the multiplication $w_{ji}o_i$ can be performed with a single AND gate (Figure 3). From simple probability theory, P(A AND B) = P(A)P(B), assuming P(A) and P(B) are statistically independent. Interestingly, this digital multiplication using a single AND gate requires only two transistors – a savings of 1/3 over the number required for a simple 6-transistor analog Gilbert multiplier.

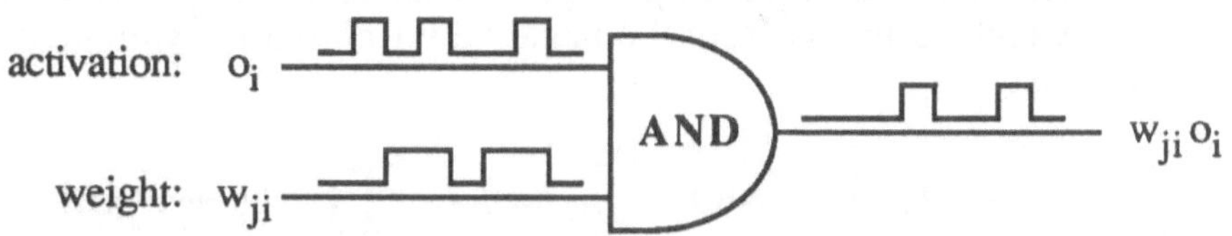

Figure 3: Multiplication of Pulse Trains

NON-LINEAR SUMMATION OF WEIGHTED ACTIVATIONS

The Digital Neural Network Architecture employs a simple AND gate to compute the weighted-activation terms $w_{ji}o_i$. In a similar vein, the DNNA performs a unique non-linear summation by employing a simple OR gate. As Figures 4 and 5 illustrate, a logical OR function both "sums" the weighted input activations and "squashes" the result to the range $0 \leq n_j \leq 1$. Again from simple probability theory, P(A OR B) = 1 - (1 - P(A))(1 - P(B)), assuming P(A) and P(B) are statistically independent.

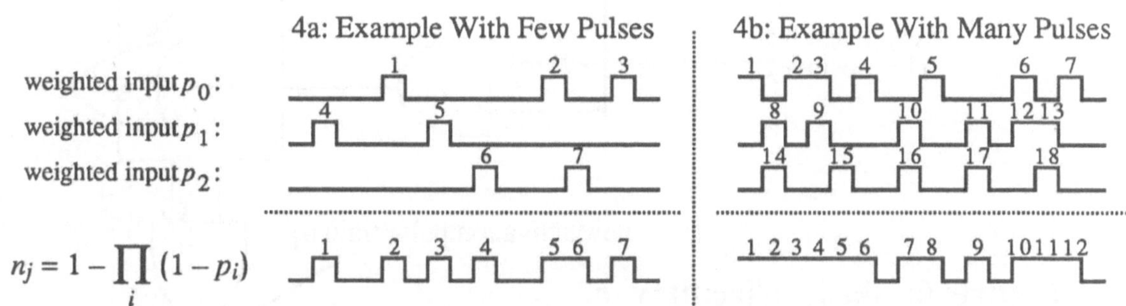

Figure 4: OR'ing Weighted Activation Pulse Trains

In Figure 4a, the weighted input pulse trains from three neurons are OR'ed together to produce a net input pulse train, n_j. Because the pulses do not occur very frequently, no pulses collide and a linear summation is performed (7 pulses in, 7 pulses out). In Figure 4b, several pulses collide since the pulse trains occur more frequently. The result is that some pulses are lost during the OR'ing (18 pulses in, 12 pulses out).

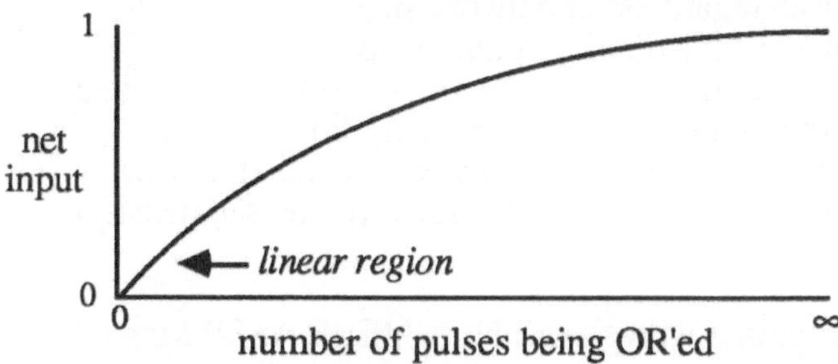

Figure 5: Non-linear "Squashing" Effect as Pulses Increase

Squashing...
As the number of weighted input pulses increase, the net input term approaches its limit at 1. Figure 5 illustrates this non-linear net input summation as a function of the number of input pulses. This function takes the form of the "upper-half" of the sigmoid function often used for a non-linear activation function.

NEGATIVE WEIGHTS

The Digital Neural Network Architecture handles negative weights in a unique manner. Rather than including all weighted inputs in a single net input term, n_j, DNNA separates the positively (excitatory) weighted inputs from the negatively (inhibitory) weighted inputs. This creates two net input terms:

$$n_j^+ = 1 - \prod_{i:\, w_{ji} \geq 0} \left(1 - w_{ji} o_i\right), \quad and \quad n_j^- = 1 - \prod_{i:\, w_{ji} < 0} \left(1 + w_{ji} o_i\right). \tag{3}$$

$$o_j = n_j^+ \left(1 - n_j^-\right). \tag{4}$$

The positive and negative net inputs (3) are combined to compute the activation, o_j, as shown in (4). Figure 6 illustrates the circuitry needed to implement (3) and (4).

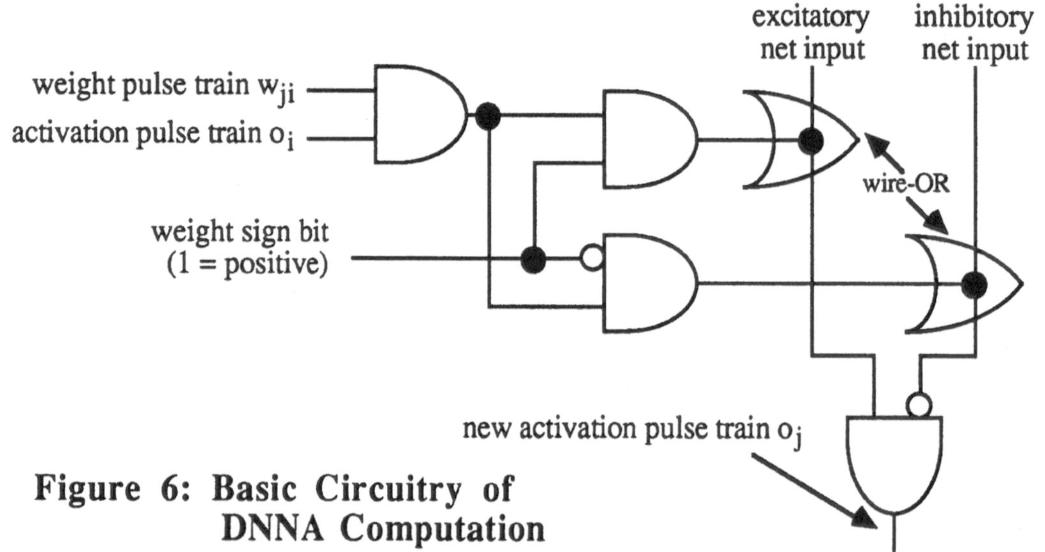

Figure 6: Basic Circuitry of DNNA Computation

SUMMARY

A novel method for performing neural network computations has been discussed. The following points summarize the significant features that make DNNA a practical solution for the direct implementation of fully parallel neural network integrated circuits.

- fully digital architecture results in predictable and reliable components
- fully parallel guarantees constant speed regardless of network size
- low transistor count per weight provides high density components
- flexible wire-OR chip interconnection permits construction of networks of unlimited dimensions and various topologies (e.g. multi-layered, recurrent, time-delayed)
- speed vs. accuracy tradeoff: 100,000 pattern classifications per second at 25Mhz
- variety of weight storage techniques: volatile, non-volatile, fixed (increased density)

REFERENCES

[1] R.P. Lippman, "An Introduction to Computing with Neural Nets," IEEE ASSP Magazine, Vol. 4, Number 2, April 1987

[2] M.S. Tomlinson, "Implementing Neural Networks," Dissertation Thesis, University of California, San Diego, June 1988

DNNA™ and DIFP™ are trademarks of Neural Semiconductor, Inc.

THE LNEURO-CHIP: A DIGITAL VLSI WITH ON-CHIP LEARNING MECHANISM

J.B. Theeten, M. Duranton, N.Mauduit and J.A. Sirat
Laboratoires d'Electronique Philips
B.P. 15 - 3, avenue Descartes
94451 Limeil-Brevannes Cedex
France

ABSTRACT

Neural network simulations are often limited because of the time required for both the learning and the evaluation phase of the simulation. Our parallel digital LNeuro circuit drastically reduces these times by updating synaptic coefficients related to one neuron in parallel. Contributions of 'input' neurons to one output neuron are also computed in parallel.

LNeuro-chips can easily be associated using Transputer microprocessors as controllers. Boards communicate through the reconfigurable links provided by a SuperNode architecture. This allows to simulate large size-networks or structured network architectures like Multi-Layer Perceptrons.

We report demonstrations of a parallel system built with several LNeuro-chips, which include a local learning rule (on a 'real-time' application), and the famous Backpropagation algorithm.

INTRODUCTION

LNeuro is in 1.5 μm CMOS digital technology for easy interfacing, known precision, well-defined states and full control of parameters. It is designed to be highly efficient in matrix operations as well as in implementing most learning rules. Special attention was paid to make the device as flexible and cascadable as possible, so that networks of virtually any structure and any size can be built with associations of chips working in parallel.

The description of the architecture of LNeuro and references to other hardware implementations are given in [1] [2] [3] [4] .

The core of the chip is the synaptic RAM (*fig.* 1). It contains 1024 'synapses' (for 16 input and 64 output neurons). LNeuro implement a parallelism of 16 allowing to simulate layers of 32 input and 32 output 8 bit neurons with 16 bit synaptic coefficients (with asynchronous update of neurons). It could also implement layers of 256 binary input neurons and 4 output neurons. The VLSI performs two generic vectorial operations needed in neural computation:

- The *scalar product* $\sum_{j=1}^{N} W_{ij} V_j$,
- The *Hebbian Learning Step* which modifies all the synaptic weights W_{ij} converging to a given output neuron i in a single time step according to: $W_{ij}(t+1) = W_{ij}(t) + \Delta_i V_j$, where Δ_i is an integer (scalar) 'increment' that depends only on the output neuron i. V_j is the state of the input neuron j which may be replaced by some other quantity related to input neuron j in special cases.

Various learning rules are implemented only by modifying the microcode controlling the general learning unit, and by doing several elementary steps. Detailed programming can be found in [2] [3] [4] .

SYSTEM ARCHITECTURE

Parallel oriented processors (INMOS Transputers, for instance) are well suited to design systems making use of these circuits, as they are fast enough to perform part of the communication protocol required when using several neuromimetic chips.

The frame of this machine is Supernode, a general purpose parallel computer based on INMOS Transputers. The architecture of Supernode is modular, and so well suited to our prototype : it can gather up to 1024 processors and their associated memory. Its parallel architecture allows fast communication between the processors, through four asynchronous links. The communication path between all processors can be dynamically reconfigured with a crossbar interconnection network. Four Transputers take place on every motherboard; each of these controls four LNeuros.

The different blocks of a neurochip can be addressed by a control Transputer as a part of its memory.

DEMONSTRATION WITH A LOCAL LEARNING RULE

In many learning rules, modification of W_{ij} involves only the input neuron state V_j and the output neuron state V_i . (V_j and V_i might be replaced by other quantities related to neurons j and i respectively, e.g. desired states). We call them local as the procedure needs no information from neurons that are not connected by the considered W_{ij} . The implementation of these rules is rather straightforward in LNeuro: loading the microcode of the rule, determining the right variables (Δ_i and V_j), and executing several elementary steps.

In particular, neural algorithms for PCA (Principal Component Analysis) that are proved to be more efficient than standard covariance matrix diagonalization. This is due to their local aspect which simplifies computation (see e.g. [6] and references therein). We use such a scheme for an application which require real-time learning (TV-image compression, see [5] [7]).

DEMONSTRATION WITH THE BACK-PROPAGATION ALGORITHM

In contrast to the previous learning rule, it is non-local. In fact modification of synaptic weight W_{ij} requires information located in other units than i or j. Thus, hardware implementation of this approach is not straightforward.

We already have proposed a special configuration which yields maximum computing speed and minimum inter-chip communication [2] [3] [4] . *Fig.* 2 displays the arrangement of circuits for implementing the error back-propagation algorithm on a 2-layer Perceptron. Two subsystems perform forward propagation of neural states (3 chips on the left), and back-propagation of errors (1 chip on the right), respectively. It also illustrates cascadability ability of the circuit.

As for the first demonstration, we apply a neural scheme to image compression. The architecture of the net is : N input neurons, n hidden neurons, N output neurons ($n < N$). When training e.g. on blocks of pixels extracted from a TV image, the net is able to restitute the whole image with good quality. In the 'standard' case it is related to principal component analysis, but some variants allow to improve the performances (see [5] [7]).

CONCLUSION

The arrangement of several LNeuro-chips in a Transputer based computer allows in principle to emulate most of the formal neural nets (Hopfield, Kohonen, Multilayer Perceptrons) with the associated learning schemes, including the error back-propagation. As the architecture is fully cascadable, any size and architecture of neural nets is also possible.

We have demonstrated the capabilities of our system in two relevant cases:

- for a 'real-time' application (with a local learning rule)
- with the Error Back-Propagation algorithm.

ACKNOWLEDGEMENTS

We would particularly like to thank C.Dufour, J.R. Viala and J.L.Zorer for software simulations, L. Davidovic for his contribution to the VLSI design, A. Aglan and D. Zwierski for Transputer programmation.

[1] M. DURANTON, J. GOBERT and J.A. SIRAT, *Digital VLSI Module for Neural Networks.* Proc. of 'nEuro', Paris, June 6-8 1988

[2] M. DURANTON, J.A. SIRAT, *Learning on VLSI: a General Purpose Digital Neurochip* IJCNN conference on Neural Networks, Washington DC, June 18-25,1989

[3] M. DURANTON, N. MAUDUIT *A General Purpose Digital Architecture for Neural Network Simulation* First IEE conf. on Artificial Neural Networks. Londres, October 16-18 1989.

[4] M. DURANTON, J.A. SIRAT *Learning on VLSI: a General Purpose Digital Neurochip* Mini et Micros, N°323 , 29 Mai 1989

[5] J.A. SIRAT, J.R. VIALA, C. REMUS *Image Compression with Competing Multilayer Perceptrons* First IEE conf. on Artificial Neural Networks. Londres, October 16-18 1989.

[6] T.D. SANGER *Optimal Unsupervised Learning in Single Layer Feed Forward Neural Network* Neural Networks, vol.2, N°6, pp 459-473, 1989

[7] G.W. COTTRELL, P.W. MUNRO, D. ZIPSER *Image compression by Error Backpropagation: A Demonstration of Extentional Programming* Advances in Cognitive Science, vol. 2, suppl.1, 399.

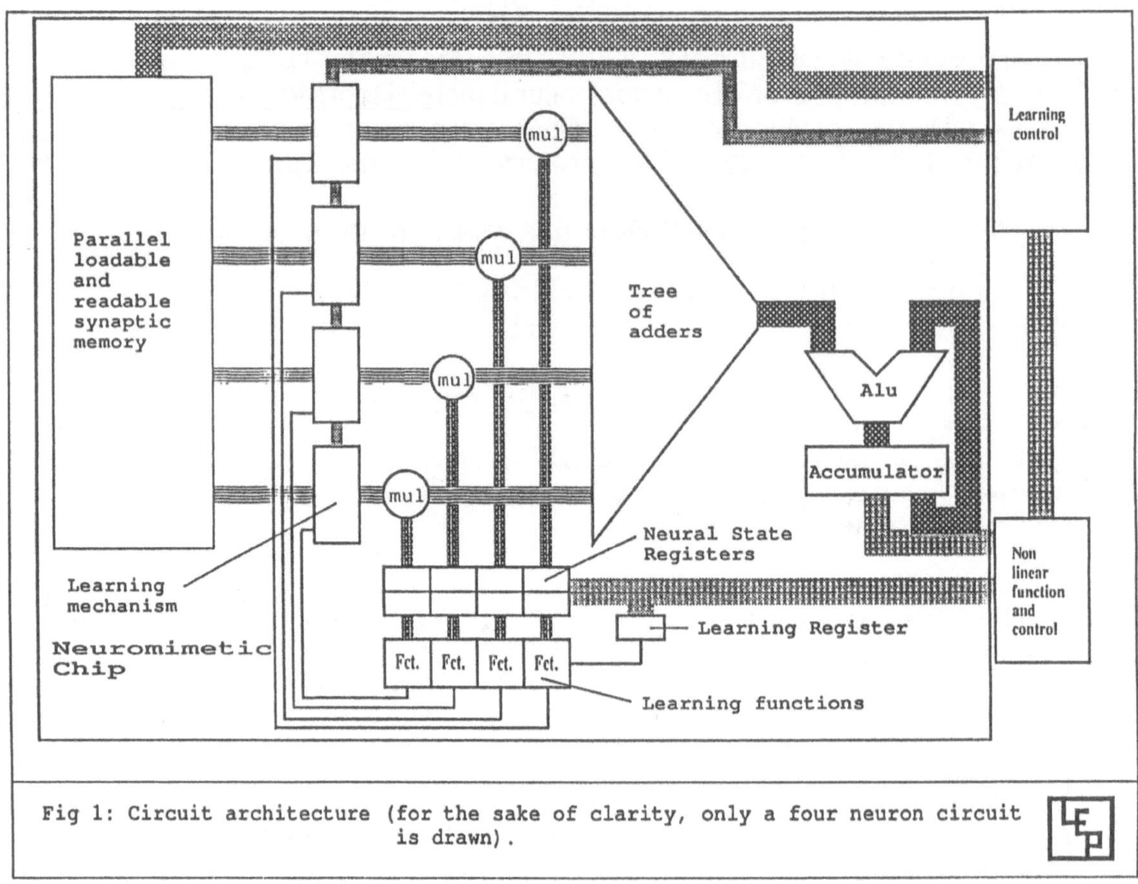

Fig 1: Circuit architecture (for the sake of clarity, only a four neuron circuit is drawn).

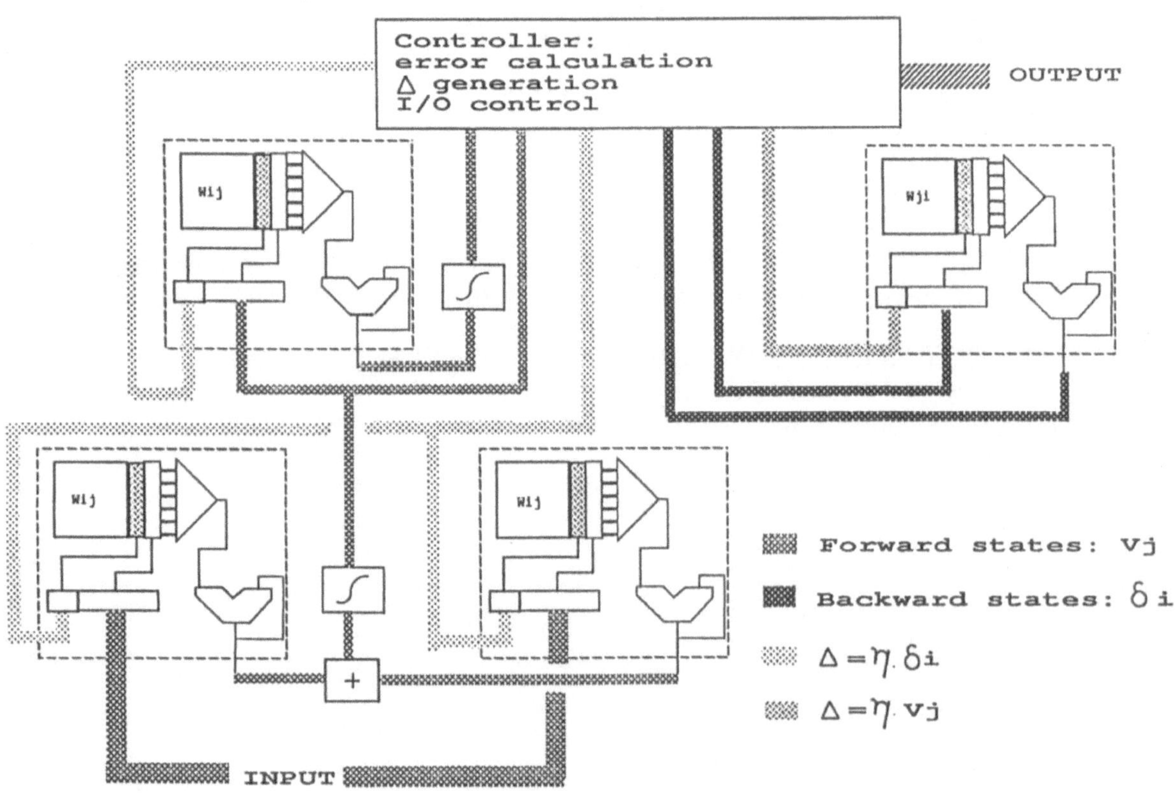

Fig. 2 : MLP with Error back-propagation learning rule.

IMPLEMENTATION OF A VLSI FEEDBACK NEURAL NETWORK CHIP WITH INTERNAL AUTOMATIC IDENTIFICATION OF SUCCESSFUL PROTOTYPE RETRIEVAL

Jean-Dominique Gascuel, Antoine de Maricourt, Pierre-Yves Alla, Joël Roman, Michel Weinfeld
Laboratoire d'Informatique, Ecole Polytechnique, 91128 Palaiseau Cedex, France

Abstract : Most neural networks, especially feedforward and feedback, do always produce an output for any input stimulation, the relevance of which is being assessed by the supervisor, who marks a succes in case of recognition or correct classification, and a failure if it is not the case. It is particuarly interesting, in view of multi-networks architectures, that any network may be made autonomous, capable of self-identification of success or failure.

We have devised a digital feedback network, including 64 binary neurons, capable of internal learning, and also having the capability of a rather crude but efficient annealing mechanism for improving the overall attractivity of the stored prototypes. We use a small part of each stored vectors for labeling purposes, the label being calculated by a classical cyclic error correcting code.

Whe show that, after relaxation, the coherence between the label field and the information field in the attractor (in the sense of the cyclic code) is a strong indication that this attractor is a learnt prototype, and is a spurious state in the opposite case. This identification may be as high as 98% accurate if the label field is long enough, for instance six bits for a total vector of 64 bits.

This feature, easily implemented in silicon, allows to envision various architectures involving the cooperation of several networks with different learnt patterns.

TFT TECHNOLOGY FOR VARIABLE-SYNAPSE ELECTRONIC NEURAL NETWORKS

P.GUYADER laboratoire de recherche sur les composants et systèmes de visualisation I.U.T de Lannion BP 150 22302 Lannion
S.SALAUN , J.GUERIN C.N.E.T Lannion LAB/OCM/TEP Route de tregastel 22302 Lannion

Abstract: An analog memory device based on thin film technology has been prepared. The Ids current of a MNN'S (N for nitride , N' for graded nitride and S for amorphous silicon) can be increased or decreased incrementally by negative or positive pulses applied on the gate . The modification of Ids can be kept for a long time . This component is applicable to the storage of weighting values in thin film implementations of learning neural networks . The first experimental results are reported .
Keywords : neural networks , TFT , memory , analog device .

Analog VLSI for Connectionist Learning

D. B. Schwartz and V. K. Samalam
GTE Laboratories, Inc.
40 Sylvan Rd.
Waltham, MA 02254 USA

Abstract

We have built analog VLSI circuits to implement connectionist networks with on chip learning. The weights are stored as charge on MOS capacitors and the weight changes are computed locally to each weight by a pair of simple circuits.

Since few of the learning algorithms now known are well suited for VLSI implementation we have provided generic facilities suggested by examination of several different learning rules. To facilitate the development of learning algorithms for these circuits the results of our measurents will be reduced to behavioral models in an object oriented computer language that can be linked to conventional connectionist simulators.

A single weight occupies a $50\mu \times 250\mu$ area in 2μ CMOS suggesting a maximum of 8000 weights/cm^2. More than half of this area is taken up by the ciruits required to support learning.

OSCILLATORY STATES IN COUPLED NEURAL OSCILLATORS

T. Gencic and G. Dangelmayr

Institute for Information Sciences, University of Tübingen
Köstlinstr. 6, 7400 Tübingen, FR Germany

Abstract

Two different kinds of analogous model neurons have been investigated so far, namely, the Hopfield type neuron characterized by a sigmoid function, and Hoppensteadt's voltage controlled oscillator (VCO). Both neurons can be implemented by means of electronic circuits composed of operational amplifiers and linear elements. Whereas Hopfield nets have been mainly applied to cognitive tasks such as content addressable memory, and more recently also to the storage of cycles of states, VCO neurons are used to simulate endogenous bursting neurons (EBN), i. e., they produce bursts of output peaks with a constant frequency. In order to provide a link between Hopfield and VCO neurons we construct a nonlinear oscillator by coupling a pair of Hopfield neurons asymmetrically. This oscillator can be regarded as one EBN as it is mimiced by a VCO, where its square wave output is interpreted as the envelope of an EBN's burst. We have implemented a network of four cyclically coupled oscillators (eight neurons) electronically and compared the observed wave forms with those predicted by the Hopf bifurcation theory with symmetry. The experiments show that all generic oscillatory states of the theory can occur in the network.

THE MINCHINTON CELL - ANALOG INPUT TO THE N-TUPLE NET

P.R.Minchinton, J.M.Bishop & R.J.Mitchell
The Neural Network Research Group
Cybernetics Department, Reading University, Whiteknights
READING, Berkshire, Great Britain, RG6 2AL.

The Minchinton Cell is a pre-processing element, enabling efficient use of multi-level analog input with digital n-Tuple RAM cells. A significant application of the cell is to make n-Tuple vision classification schemes more tolerant to changes in lighting levels.

DIGITAL NEUROCOMPUTER VLSI-SYSTEMS WITH PARALLEL ARCHITECTURE

Kalyayev Anatoly Vasilyevich - associate member of the Academy of Sciences of the SU, doctor of technical sciences, professor. The head of the Research institute for Multiprocessor Computer Systems (RIMCS)
Galuyev Gennady Anatolyevich - candidate of technical sciences, chief scientist. The head of the laboratory of neuronlike systems at RIMCS
Official address: 347928 Taganrog, ul.Chekhova, 2, RIMCS

The construction principles and the element base of digital neurocomputers with parallel architecture oriented to a VLSI-based realization and representing the new class of neurocomputer systems are proposed here.

The posibilities of use of the given approach for the construction of neurocomputer VLSI-systems of visual processing are presented.

A SYSTOLIC IMPLEMENTATION OF THE SELF ORGANIZATION ALGORITHM

François Blayo and Christian Lehmann
Laboratoire de Microinformatique
Ecole Polytechnique Fédérale de Lausanne
Ecublens, CH-1015 Lausanne, Switzerland

Abstract

The VLSI implementation of neural networks is a source of complex problems generated by the large number of interconnections associated with massive parallelism. In this paper, we propose an operational and structural analysis which underline the common computation requirements between several connectionist algorithms. Then, a systolic implementation is described implementing both the Hopfield and Kohonen maps in the same chip design. An architecture of the elementary cell is presented and gives a first evaluation of the final silicon area.

DIGITAL NEUROCOMPUTERS ORIENTED TO A VLSI -REALIZATION.

Kalyayev Anatoly Vasilyevich - associate member of the Academy of Sciences of the SU, doctor of technical sciences, professor. The head of the Research Institute for Multiprocessor Computer Systems. Bozhic Vladimir Ivanovich - candidate of technical sciences, assistent professor of the Department of Computer Engineering.

Official address: USSR, 347928, Taganrog, Chekhov st.,2,RIMCS

The architecture of digital neurocomputers as the distributed systems with multilevel hierarchy satisfying modern countrains of a VLSI -technology are presented in this paper. A neurocomputer architecture is considered in terms of physical and functional structures. It's shown that the use of a cubic interface allows to design the neurocomputer for simulation of neural networks of random configuration including fully connected neural structures with local circuits of information processing. The features of the development of digital neurocomputer with architecture of hypercubic type are presented here. The given approach will allow to extend the application of neurocomputers at the expence of functional possibilities of dentride circuits of a dynamic neuroprocessors.

OPTICAL NEUROCOMPUTERS

Chair: David PSALTIS

Network Analysis of an Optically Implemented Connectionist Architecture

Michael G.Robinson, Kristina M.Johnson & Lin Zhang.

Center for Optoelectronic Computing Systems, Engineering Building, University of Colorado, Boulder, Colorado 80309.

Abstract

We investigate the system error tolerance of optical connectionist architectures, which utilize liquid crystal spatial light modulators to represent neurons and weights in a neural network configuration. Experimental and computer simulation results show training on one layer networks compensates for errors in the optical and electronic hardware. The mathematical description of the least mean square algorithm used to train these networks shows that this is to be expected. This tolerance, however, is not shown in multiple layer networks implemented with similar architectures. To overcome problems associated with linear dependence between training patterns and maintain the error tolerance of the one layer machines (without going to multiple layer networks), an architecture with higher level interconnections is proposed. This fully interconnected second order architecture requires only a single optical pass before updating the weights. It also uses a very similar training algorithm to that of the single layer network.

Introduction - The inherent parallelism of optics lends itself to the massively interconnected architectures required for neural network and linear algebraic computation. The accuracy however of such optical processors is poor compared with that of a conventional digital computer [1]. In this paper we address the extent to which the processing accuracy effects the training and final error in an optical connectionist architecture utilizing liquid crystal spatial light modulators (SLM's) as input units and interconnection weights. Of the two optical connectionist machines to be considered in this paper, the main difference between them is the use of nematic liquid crystal SLM's in one and ferroelectric liquid crystal (FLC) SLM's in the other. By using FLC's the addressing of the separate SLM's can be three orders of magnitude faster than that of SLM's fabricated from NLC's. The surface stabilized FLC modulators are however binary devices in contrast to the analog twisted nematic SLM's. In order to obtain grey levels in the interconnection weights with the FLC SLM's we use spatially encoded weights.

The complete neural network architecture can be considered as a hybrid electro-optic system consisting of an all optical matrix/vector multiplier (MVM) interfaced to a controlling personal computer (PC). The PC is responsible for updating the weight values and carrying out the sigmoidal non-linearities. Using this architecture it is possible to perform both single and multi-layer neural connectionist algorithms; the extension to multi-layers being made by multiple passes through the MVM.

Experimental - A schematic of the architecture is shown in Figure 1 below.

Figure 1 - Optical connectionist architecture

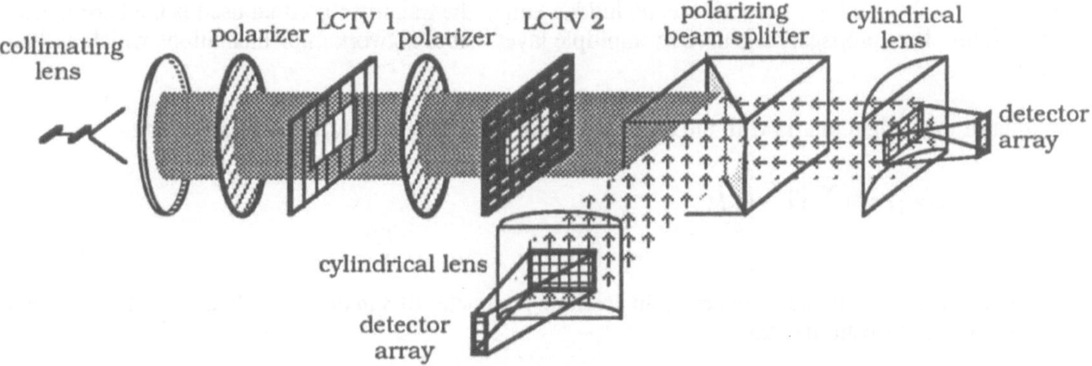

The optical MVM module has been reported previously [2] and incorporates polarization based logic, which allows negative weight values to be realized. The binary input vector components are represented by vertical stripes coded onto the beam using the input spatial light modulator. These stripes are then passed through a second weight matrix which consists of a checkerboard of pixels, with each pixel representing a weight value. In the case of the FLC weight matrix SLM, grey levels are obtained by spatially multiplexing the weight value, with each pixel representing a weight value consists of a 4x4 square array of binary coded mini pixels. The effect of the weight matrix is to rotate the polarization of the optical beam between vertical and horizontal such that the subtraction of the individual polarization intensities yields the correct multiplication. The intensities of the individual polarization components for the different matrix multiplication products are split by a polarizing beam splitter, summed with cylindrical lenses and are detected by two linear detector arrays. The spatial filtering optics used to remove diffracting light between the successive stages of the MVM was not used in the FLC based OCM making the architecture more compact despite it having an increased input vector dimensionality (32 bit as compared with 16 for the NLC machine).

For each of the machines the MVM results were poor, typically errors in the output vector components were 20% and 10% for the NLC and the FLC machines respectively. This low accuracy was directly due to the non-ideal performance of the systems. A list of the system error terms effecting the MVM together with the size of such terms for the two machines is given in Table 1 below.

Table 1 - System errors

System error	NLC machine	FLC machine
Contrast ratio of input SLM	15:1	25:1
Optical beam Gaussian profile Int. input bit 0 / max intensity	0.6	0.55
Detector mismatch	≈ 3%	≈ 3%
Cross talk between adjacent channels (input to weight SLM/ at detector)	(0.5% / 2%)	(1% / 10%)
Random noise (Electronic)	≈2%	≈3%
Non-linearities	Large	Small

The main source of noise in the optical system are the non-linearities associated with the liquid crystals. In the case of the NLC implementation, the analog rotation of the optical polarization is not linearly mapped onto the desired weight value (w_{ij}) is given instead by the following non-linear mapping:

$$\theta_{ij}(\text{rad}) = 0.717 + 0.472.w_{ij} + 0.154.w_{ij}^2 \qquad (1)$$

This, coupled with the non-linearity due to clipping of the weights i.e. restricting weight values between the values of -1 and 1, and further non-linearities associated with the polarization based logic, makes the NLC based MVM highly non-linear. In contrast, the spatially multiplexed FLC weight values, although discrete, are essentially linear with desired weight value.

Without compensating in either the hardware or the controlling software, the two MVM's are operated as part of a neural network. In the one layer network (i.e. no hidden units), the training algorithm used is the least mean square (LMS) algorithm. Backpropagation is used for multiple layer neural network implementations which will be discussed later.

The updating of the weight matrix with the LMS algorithm is given by the following expression:

$$\Delta w_{ij} = \eta \sum_{p} \left(T_i^p - O_i^p\right) I_j^p \qquad (2)$$

where i and j are vector component indices, η the learning parameter (0.3 in our case), I^p and O^p the input and output vectors and p the pattern index.

By using this algorithm it is possible to train a single layer net to associate input pattern vectors with desired output vectors. This can be seen in the Figure 2, where 8 random patterns are associated with 8 random targets. Identical pattern/target files were used for the two machines indicating that the use of FLC SLM's improves the performance of the single layer network.

Figure 2 - Training of FLC and NLC based single layer neural networks

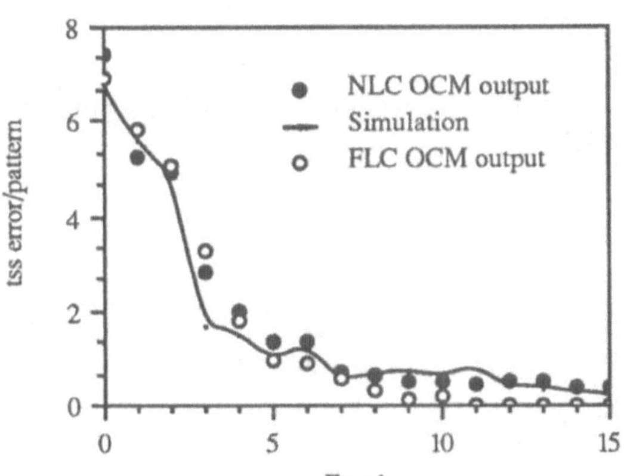

Figure 3 - Effect of system error terms on training

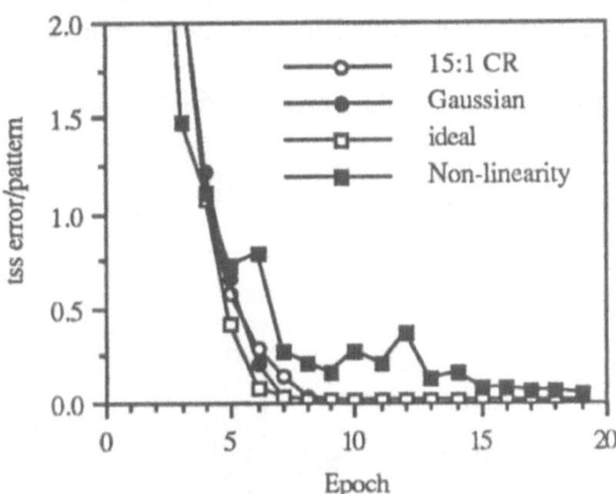

Also shown in Figure 2 are results from a computer simulation of the NLC machine, which incorporates the various system errors. This simulation has been used to investigate the effect of such terms on the training performance as is discussed in the next section.

Network analysis - Using the computer simulation it is possible to analyze the effect of the different system error terms on the training performance of the machine. This is best illustrated in Figure 3, in which the total sum squared error per pattern between the output vectors and the target vectors is shown for the ideal machine (i.e. no system errors other than clipping of the weights) and when various errors are introduced. This figure only shows the NLC machine simulation but similar results are found for the FLC machine. Using quantized weight values (>5 levels) in this machine does not seem to significantly affect the machine's performance from an ideal analog machine [3]. For clarity, the effects of detector mismatch and random noise are not shown in this figure. Both these terms effect the total sum squared (tss) error in a similar way to the contrast ratio and gaussian profile.

This result can be expected from the training algorithm given in (2). Ideally for a case which has imperfect or distorted input vectors caused by the optical encoding the ideal change in the weight value is given by:

$$\Delta w_{ij} = \eta \sum_{p} \left(T_i^p - O_i^p\right) I^{*p}_j \qquad (3)$$

Where I^*_j are the distorted input vector components. From this expression it can be seen that such errors as contrast ratio and gaussian profile effect the components of this vector. For a finite contrast ratio, the actual weight update (eqn. 2) does not compensate for I^*_j components in the input vector that are ideally zero but actually allow light through. However, this results in a larger error in the output component O_i which then leads to compensation by subsequent non-zero updates of the weight matrix. Similar arguments can be made for the other linear error terms. The effect of the non-linearities are however more difficult to determine. Qualitatively, these non-linear terms bend the decision hyperplanes and translate them from the origin in the space defined by the input vectors. These effects increase the training error considerably, and is the main reason why the FLC machine with its linear weight mapping performs better than its NLC counterpart. It is interesting to note that the influence of the contrast ratio on the learning rate still remains small even in machines with large numbers of processing units. Simulations show that even a 10:1 contrast ratio in a 256 bit optical connectionist machine does not severely effect its training capability. However, a vector matrix multiplier of the same size with the same low contrast SLM's would not accurately perform the desired processing.

Multi-layer and higher order networks - The architecture described above has been used to implement two layer networks (one hidden layer) using the optical MVM to perform the matrix multiplications between the input and hidden layers, and then between the hidden and output units. In such an implementation, backpropagation was used to update the weights. The results indicate that the the FLC based machine learns successfully the XOR problem but the NLC does not. The final tss error per pattern however, for the FLC was only half that obtained by the single layer net (which cannot learn such an association). This indicates that the backpropagation algorithm is not best implemented using an optical MVM. In order therefore to learn XOR associations or more complex 'real' problems with associations, a higher order net may be more suitable [4]. With such a net there would be only one pass through the weight matrix and the updating would be similar to that used in the single layer net. An architecture for a fully interconnected second order net is shown in Figure 4.

Figure 4 - Fully interconnected, second order optically interconnected optical connectionist module

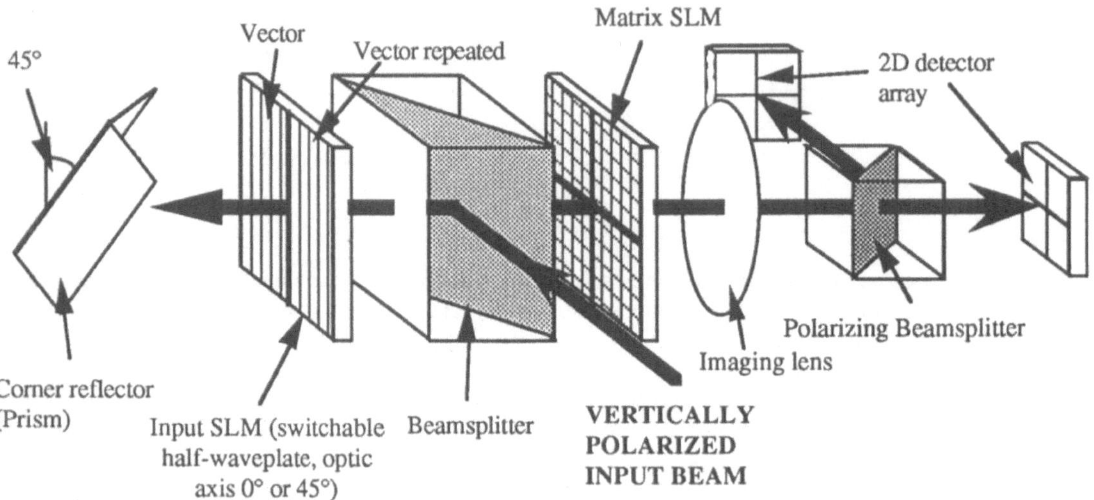

Here the input vector (coded with 1's and -1's) is physically rotated and passed through the input SLM a second time, carrying out the required multiplications. The higher order terms then pass through an interconnection matrix as before and are detected using polarization-based logic, which allows the representation of negative weight values. In this paper we will present results of an optical implementation of a higher order net.

Conclusions - We have shown that single layer neural network algorithms can be successfully implemented on a hybrid elecro-optic connectionist machine. Using the simple least mean square algorithm, system errors associated with the all-optical matrix vector multiplication can be successfully compensated. This architecture, however, when used for two layer neural network implementations, is much less successful at overcoming these errors. This is probably to low accuracy in the weight values used to backpropagate the error. In order, therefore, to implement networks capable of complex associations and which are error tolerant, we propose having to go to higher order networks. We have presented a fully interconnected second order architecture which is being investigated experimentally at the present time.

References

1. D.Psaltis and R.A.Athale, "High accuracy computation with linear analog optical systems: a critical study", Appl. Opt. **25** p.3071 (1986).

2. M.Kranzdorf, K.M.Johnson, B.J.Bigner & L.Zhang, "An Optical Connectionist Machine with Polarization-Based Bipolar Weights", Opt. Eng. **28** p. 844 (1989).

3. M.G.Robinson, K.M.Johnson, L.Zhang & B.J.Bigner, "Optical neural networks using smectic liquid crystals", OE LASE meeting in Los Angeles, 14-19 Jan. 1990. SPIE Proc. 1215.

4. A.P.Ittycheriah, J.F.Walkup, T.F.Krile & S.L.Lee, "Outer product processor using polarization encoding", Appl. Opt. **29** p.275 (1990)

OPTICAL DISK BASED PROCESSOR FOR HANDWRITTEN CHARACTER RECOGNITION

M. A. Neifeld, S. Kobayashi, A. A. Yamamura and D. Psaltis
Department of Electrical Engineering 116-81
California Institute of Technology
Pasadena, California 91125

ABSTRACT

We describe an optical disk based system for the recognition of handwritten numerals. The recognition scheme is based on a K nearest neighbor strategy using a template library of 650 exemplars. The optical system compares an unknown input against the template library at a demonstrated rate of 26,000 comparisons/sec.

INTRODUCTION

In this paper we describe an optical disk based implementation of a handwritten character recognition system. Handwritten character recognition is a pattern classification task of considerable practical value. Although shift, rotation and scale invariances may be effectively eliminated through normalization procedures,[1] author dependent distortions are often dealt with using statistical techniques based on a large training set of exemplar patterns. Such approaches are typically based on computationally intensive algorithms requiring a great deal of both time and memory.[2] Whereas such computationally intensive algorithms are difficult to realize on conventional computers, parallel access optical storage technologies such as the optical disk, provide an efficient implementation. The optical disk is an optically addressable medium capable of storing $\approx 10^{10}$ bits of information. Optical parallel access to information stored on the disk provides a mechanism for high data retrieval rates and concomitantly large processing speeds.[3] The combination of parallel access and large storage capacity makes optical disk based architectures well suited to the efficient implementation of several of the traditional pattern recognition paradigms as well as the more contemporary neural network models. In this paper we will experimentally demonstrate an optical system which performs handwritten numeral recognition using the K nearest neighbor algorithm (KNN).

OPTICAL IMPLEMENTATION AND RESULTS

Given a set of M training vectors which we will refer to as templates, the KNN algorithm classifies an unseen vector according to the class most strongly represented by it's K nearest neighbors in the template set.[4] In order to realize this algorithm we must compute the M distances $|\underline{x} - \underline{x}^i|$ for $i = 1, ..., M$ where $\underline{x}$ is the unknown input vector and the $\underline{x}^i$s are the stored templates. The success of this algorithm depends on how well the template set represents the underlying probability distributions of the problem at hand. As M becomes very large, the probability of error for the KNN algorithm is known to approach the optimum value, but the computational requirements become impractical for conventional computers. The optical system described here however, is capable of performing $\approx$ 40 Million such comparisons per second and it is possible to store $\approx 10^6$ templates of 10^4 bits each on a single disk.

A database of 950 16X16 handwritten numerals (95 per class) was used to construct the training and testing sets for our experiments. 65 vectors were chosen randomly from each of ten classes to generate the 650 element training set. The remaining 300 vectors were used as a testing set. It was found that for a fixed template set size of 650 vectors, the recognition performance of the KNN algorithm was improved dramatically if the 16X16 binary input vectors were first normalized for position and scale. Accordingly, a preprocessing step consisting of a centering operation followed by a scaling of the centered 16X16 character to a 10X10 window was performed. The 10X10 templates were unravelled and stored as 100 dimensional vectors on the optical disk. Along with each binary template $\underline{x}$, we generate it's complement $\overline{\underline{x}}$ (ie. $\overline{\underline{x}}+\underline{x} = (1, 1, ..., 1)$) and store it also on the disk. Thus, we can implement bipolar valued templates by subtracting the inner products between the same input and the two stored vectors. The 650 templates and their complements were recorded as 1300 radial lines on a Sony "Write-Once" optical disk which served as a parallel access template library in our experiment. The remaining 300 characters were preprocessed in the same way and recorded on transparencies to serve as testing inputs to our optical system.

The optical system is shown in Figure 1. An image of the input vector located on an input Spatial

Light Modulator (SLM) is formed on the optical disk along a radial line as shown. The light diffracted from the disk is collected on the photodetector and represents the inner product between the input vector and the vector recorded along the illuminated line. As the disk rotates the input is compared against all of the stored templates and an electrical signal representing the result of these comparisons is analyzed by postprocessing electronics to determine the correct classification of the unknown vector. In our experiment, the postprocessing electronics were responsible for sampling the output of the photodetector and using the inner product data to calculate the KNN of the input vector. The inner product based distance metric used is given by :

$$y = \frac{\underline{x} \cdot (\underline{x}^i - \overline{\underline{x}^i})}{|\underline{x}^i|^2}$$

where the $|\underline{x}^i|^2$s were computed using the optical system output when presented with the input vector having all its components equal to one. The disk rotation rate in our system was 20Hz therefore, the number of binary inner products being computed per second is 26,000. The resolution of the optical disk will allow storage of up to 10^6 templates of dimension 100X100. At a 20Hz rotation rate, this corresponds to a computing rate of $20X10^{10}$ binary operations per second.

When tested on the 300 remaining vectors, the optical system achieved a recognition rate of 71% using a K=5 KNN algorithm. It should be noted here that the 300 testing vectors had not been seen by the system prior to testing. This performance should be compared with a simulation result of 83% correct classification, and with a model of the optical system which predicts a 73% rate. This model included various error sources such as beam nonuniformity, electrical noise in the postprocessing electronics, sampling phase jitter and quantization error but we found that the most critical parameter is the input image contrast. These results are summarized in Table 1. A plot of recognition rate vs. input image contrast is shown in Figure 2. The data plotted includes experimental values of the error sources mentioned above in addition to finite contrast. It can be seen from the graph that for our system operating at a measured contrast of $< 20:1$, an expected rate of $< 75\%$ is obtained in agreement with experiment. It is clear that noise sources only account for $\approx 2\%$ additional recognition error and upon improvement of the input contrast, our optical system should perform near the simulation rate of 83%.

CONCLUSION

We have demonstrated an optical system capable of comparing an input vector against a large number of stored templates at MHz rates. Our system realized a processing rate of 26KHz. This system is attractive from the perspective of data reduction in image recognition oriented tasks. Our optical system can be envisioned to be an efficient preprocessor for high dimensional input data effectively reducing the dimensionality by projecting the input image onto stored templates or feature vectors stored on the optical disk. The reduced dimensionality output of the optical system is well suited to not only the KNN recognition algorithm demonstrated here, but also to many other pattern recognition and neural network schemes such as Parzen windows, multilayer networks and associative memories, as well as hypersurface reconstruction networks using radial basis functions.[5] In general, we can envision hybrid pattern recognition systems utilizing high speed optical preprocessors followed by more conventional electronic computing elements which together will be capable of realizing many different algorithms and networks in a flexible and efficient fashion.

REFERENCES

[1] Y. S. Abu-Mostafa and D. Psaltis, Image Normalization by Complex Moments, IEEE Transactions on Pattern Analysis and Machine Intelligence, Vol. PAMI-7, No. 1, Jan. 1985.

[2] Y. Le Cun, et.al., Handwritten Digit Recognition: Applications of Neural Network Chips and Automatic Learning, IEEE Communications Magazine, Nov. 1989, pp 41-46.

[3] D. Psaltis, A. A. Yamamura, M. A. Neifeld, S. Kobayashi, Parallel Readout of Optical Disks, Proceedings of the Third Topical Meeting on Optical Computing, Salt Lake City, Utah, Feb. 1989.

[4] R. Duda and P. Hart, Pattern Classification and Scene Analysis, John Wiley and Sons, 1973.

[5] T. Poggio and F. Girosi, A Theory of Networks for Approximation and Learning. MIT AI Laboratory and Center for Biological Information Processing, Whitaker College, AI Memo No. 1140, CBIP paper No. 31, July 1989.

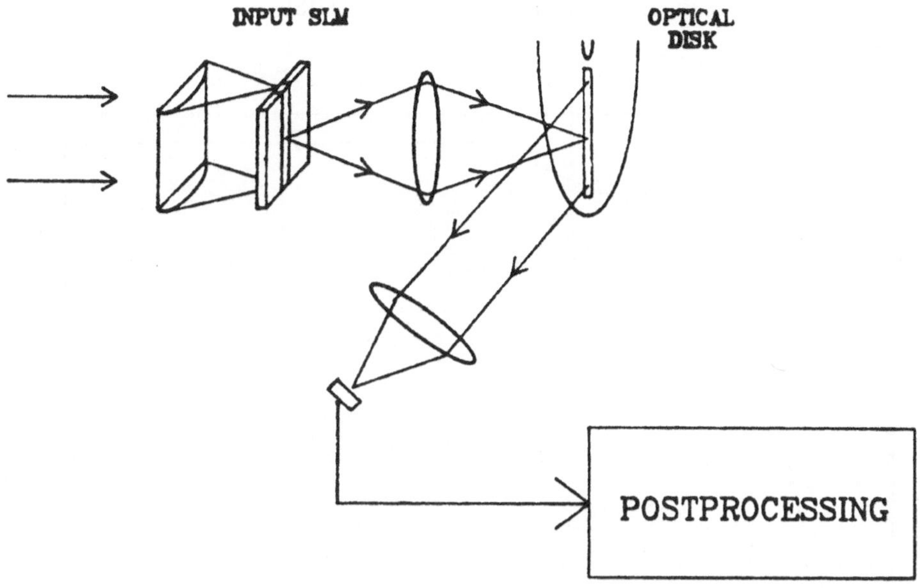

Figure 1 : Schematic of optical handwritten character recognition system

CLASS	SIMULATION	EXPERIMENT	MODEL
0	28	28	28
1	27	19	17
2	24	28	24
3	23	9	15
4	28	24	26
5	24	26	21
6	22	23	20
7	24	26	17
8	23	21	26
9	26	9	24
	83%	71%	73%

Table 1 : Results of optical character recognition system for K=5.
Entries indicate number of correct classifications out of 30.

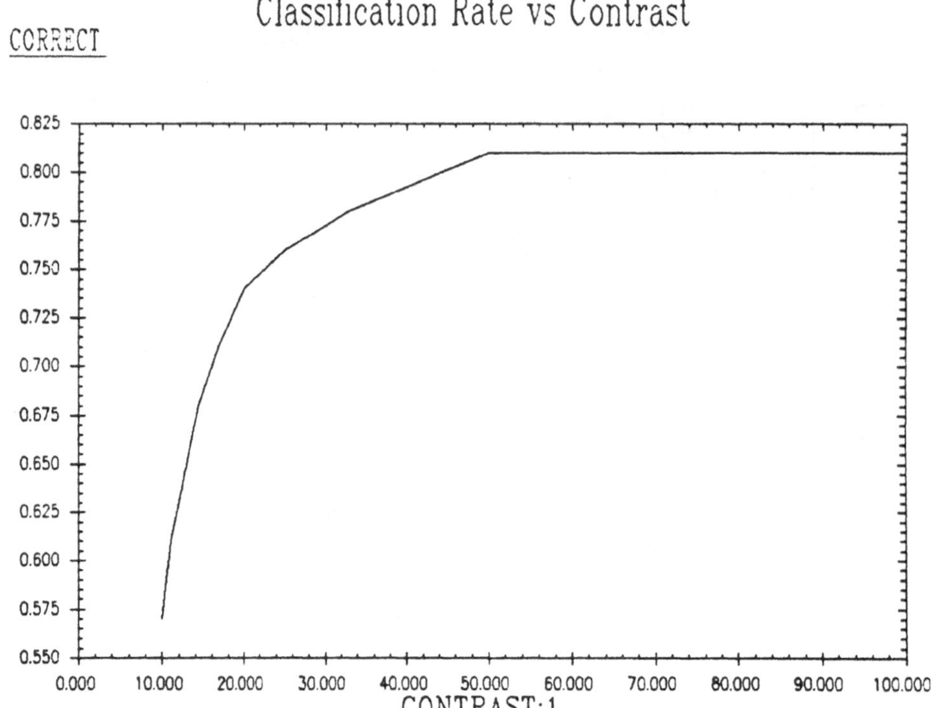

Figure 2 : Predicted classification rate vs contrast based on a computer model

OPTICAL IMPLEMENTATION OF NEURAL NETWORKS WITH FIXED GLOBAL INTERCONNECTION AND LOCAL ADAPTIVE GAIN-CONTROL

Bo-Yun Koh, Yung-Wan Kwon, Hyuek-Jae Lee, Soo-Young Lee, and Sang-Yung Shin
Department of Electrical Engineering
Korea Advanced Institute of Science and Technology
P.O.Box 150, Cheongryangni, Seoul, Korea

A new adaptive learning algorithm has been developed for optical implementation of large-sized neural networks. In this model N^4 interconnections for $N \times N$ 2-dimensional neurons are composed of two different types, i.e. global fixed N^4 interconnections and N^2 adaptive gain control. The former may be achieved by a multifacet(N^2 facets) hologram, where each holographic facet stores N^2 interconnections. The latter may be done by spatial light modulators (SLMs) of N^2 elements. This model allows us to implement neural networks of $N \times N$ neurons with SLMs of only $N \times N$ elements. The adaptive learning algorithm is based on gradient descent and error back-propagation, and easily extendable to multilayer structures. Performance of this model has been investigated, and its electro-optical implementation will be proposed.

INTRODUCTION

Neural networks have been widely recognized as having good potential to solve complicated classification and adaptive control problems[1]. While adaptive trainability by simple learning algorithms provides flexibility to perform complex tasks, special hardwares are required to take advantage of massive parallelism and analog asynchronous operation. Therefore, many researchers have been looking for advanced implementation techniques for artificial neural networks. The well-developed semiconductor VLSI technology may not be the best choice to implement enormous interconnections required for artificial neural networks with large number of neurons. Optical technology has theoretical limits much larger in terms of storage (interconnections) and speed (interconnections/sec) [1], but requires high performance spatial light modulators(SLMs) beyond current availability. For autoassociative memory model of $N \times N$ two-dimensional images N^4 interconnections are required, and have been implemented by volume holograms [2] or multifacet holograms[3-6]. For the latter N^4 fixed interconnection are achieved by film holograms, but real-time interconnection changes for adaptive learning need SLMs of N^4 elements and set practical limitation on achievable number of neurons. In this paper a new adaptive learning algorithm is presented to best utilize high interconnection density of optical holograms, while circumventing needs of N^4 element SLMs. Performance of the proposed model has been investigated by computer simulation, and its optical implementation will be discussed.

NETWORK ARCHTECTURE

In this model the N^4 interconnections are composed of global fixed interconnections and local adaptive gain-controls. The former may be achieved by a multifacet (N^2 facets) holograms, where each holographic facet stores N^2 interconnections [3]. The latter may be done by SLMs of N^2 pixels. Fig. 1 shows this architecture in a simple form. In mathematical notations each output y_{ij} is represented as

$$y_{ij} = S(\sum_{k,l} v_{ij}\ T_{ijkl}\ w_{kl}\ x_{kl}), \tag{1}$$

where x_{kl} is input, $v_{ij}T_{ijkl}w_{kl}$ is interconnection between input neuron (k,l) and output neuron (i,j), and S $(\cdot)$ is a Sigmoid function. The interconnection consists of fixed global interconnection T_{ijkl} and adaptive local gain-control v_{ij} and w_{ij}. In Ref. [7] we had shown that adaptive learning of bit-significance w_{ij}'s greatly increases storage capacity and error-correction performance for the Hopfield model. Unlike the previous model T_{ijkl}'s are pre-determined in this new model. They may be randomly generated, or obtained from any learning algorithm for standard input data. It is worth mentioning that the Hopfield interconnections T_{ijkl} look like random numbers for a huge set of independent input data. For handwritten character recognition applications one may obtain T_{ijkl}'s for typed characters, and use this new learning algorithm of w_{ij}'s and v_{ij}'s for handwritten characters.

This combination is a way to compromise between global interconnections and adaptability. Provided the multifacet holograms had enough information storage capacity for fixed N^4 interconnections, the number of implemented neurons is mainly limited by resolution of available SLMs, and this model allows us to implement artificial neural networks of $N \times N$ neurons, instead of $N^{1/2} \times N^{1/2}$ neurons in Ref. [4], with SLMs of $N \times N$ elements. In practice multifacet hologram is time consuming to make for large facets. However, once it is made, it is easy and also very cheap to copy, and is applicable to large wide of applications with this model.

ADAPTIVE LEARNING ALGORITHM

We have adopted gradient-based least-square error minimization algorithm for the adaptive learning. Following Ref.[8], the total error E is defined as

$$E = \frac{1}{2} \sum_{s} \sum_{i,j} (y_{ij}^{s} - t_{ij}^{s})^2, \tag{2}$$

where s is an index over classes (input-output pairs), (i,j) is an index over output units, y is the actual state of an output unit, and t is its desired state. To minimize E by the steepest descent method it is necessary to compute the partial derivative of E with respect to each adaptive elements, w_{ij} and v_{kl}. By applying chain rule one obtains

$$\frac{\partial E}{\partial v_{kl}} = \sum_{s} \gamma_{kl}^{s} x_{kl}^{s} \,/\, v_{kl}, \tag{3}$$

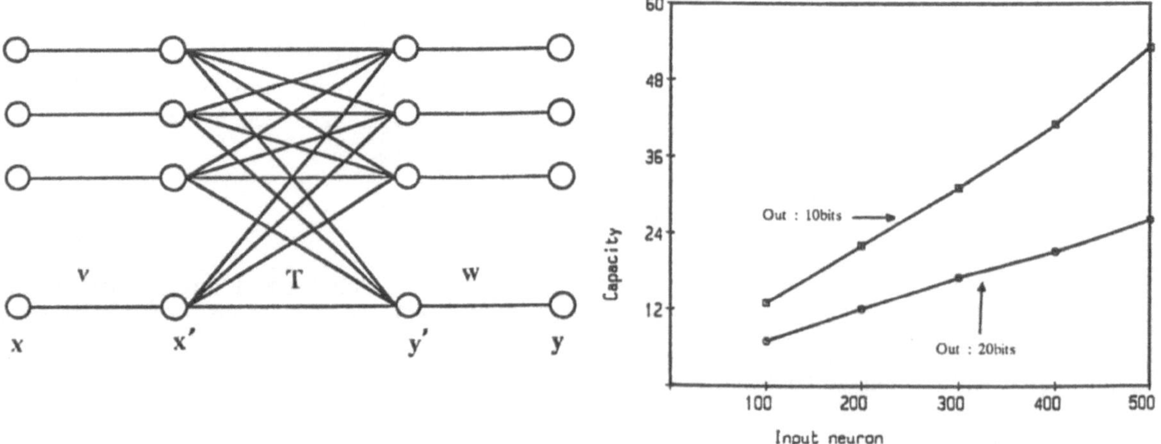

Fig. 1 Neural network architecture with fixed global interconnection(T) and local gain-control(v and w)

Fig. 2 Storage capacity vs. number of input neurons for hetero-associative memory

$$\frac{\partial E}{\partial w_{kj}} = \sum_{s} \delta^s_{ij} \hat{y}^s_{ij} \;/\; w_{ij}, \tag{4}$$

where $\hat{y}^s_{ij}$ is the argument of the Sigmoid function in Eq. (1) with input x^s_{ij}, δ^s_{ij} and γ^s_{ij} are output and input errors, respectively, and defined as

$$\delta^s_{ij} = (\, y^s_{ij} - t^s_{ij} \,) S'(\, \hat{y}^s_{ij} \,), \tag{5}$$

$$\gamma^s_{kl} = \sum_{i,j} \delta^s_{ij} w_{ij} T_{ijkl} v_{kl}\,. \tag{6}$$

It is worth noting that the input error γ^s_{kl} may be calculated by back-propagation of output error, δ^s_{kl}. This error back-propagation allows us to extend this model to multi-layer structure, which is quite similar to multi-layer perceptron. However, unlike the multil-layer perceptron, gradient calculation of this model does not involve any vector-matrix type multiplication. Only point-to-point scalar multiplication is enough.

Storage capacity of this model for heteroassociative memory is presented in Fig. 2. Maximum number of stored image pairs $(\mathbf{x}^s, \mathbf{y}^s)$ versus number of input neurons are plotted for given number of output neurons. As the number of input neurons increases, the storage capacity increases. However, as the number of output neurons increases, the ratio of adaptive-to-fixed interconnections decreases and so does the storage capacity. In brief the storage capacity is roughly estimated as number of input neurons divided by number of output neurons. Other simulation results show that this model requires less number of adaptive elements than perceptron for classification of given number of classes.

OPTICAL IMPLEMENTATION

The proposed model is actually designed for optical implementation of large-sized artificial neural networks. Resolution of available SLMs has been one of the most critical limitations on achievable number of adaptive interconnections. In globally connected neural network models such as perceptron and Hopfield model, it directly limits achievable input neuron numbers multiplied by output neuron numbers. However, in our model, only sum of input neuron and output neuron numbers are limited by SLM resolutions.

Fig. 3 shows schematic illustration of optical implementation for the proposed model. It has two paths both controlled by a Personal Computer(P.C.). At recall stage only the upper path works. The lower path is designed for error back-propagation at adaptive learning stage. The local gain-controls v_{kl} for foward path and w_{ij} for backward path are combined with input x_{kl} and output error δ_{ij}, respectively, and implemented by 2-D SLMs with gray levels. The output gain-controls w_{ij} and v_{kl} for forward and backward paths are implemented by the P.C. Multifacet holograms store the fixed global interconnections. It is worthy noting that the forward and backward paths use different N^4 interconnection schemes in order to utilize same multifacet hologram for both paths[9]. Calculation of error gradients requires only scalar multiplication, and may easily be done in the P.C. One may also put 2-D SLMs in front of the detectors for these calculations.

CONCLUSION

We have developed a new adaptive learning algorithm for optical implementation of large-sized artificial neural networks. This model requires less adaptive elements for same classification performance than the perceptron, and may easily be implemented by multifacet holographic interconnection techniques with efficient utilization of SLM resolution.

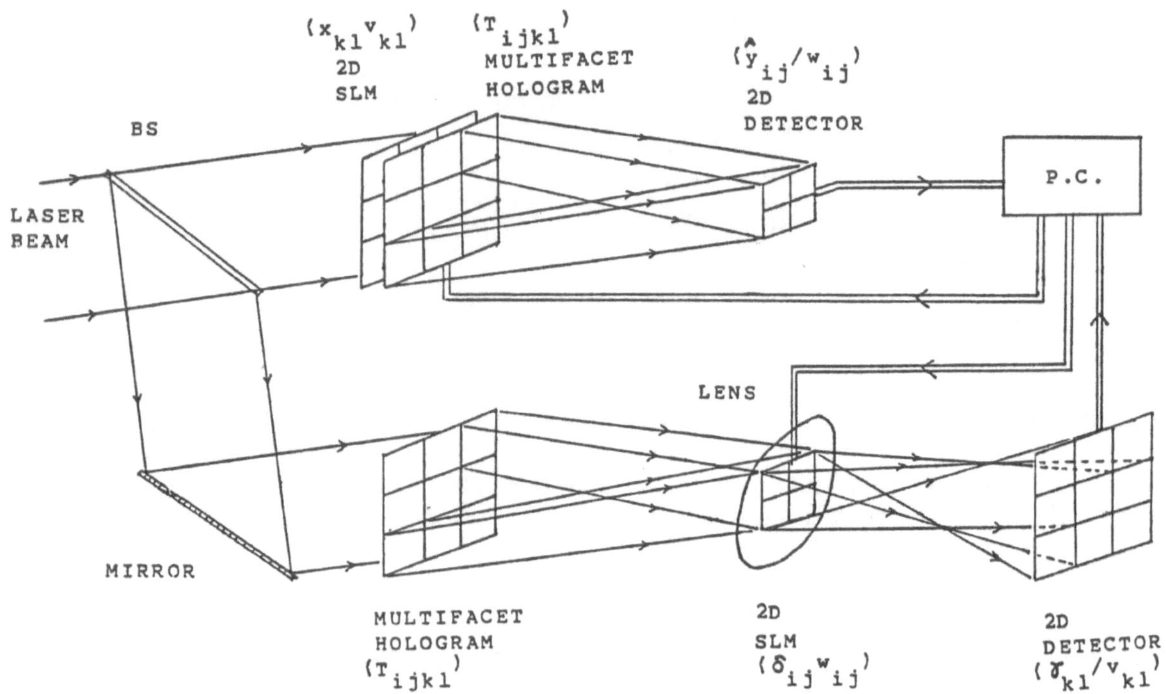

Fig. 3 Schematic illustration for electro-optical implementation

References

[1] *DARPA Neural Network Study*, AFCEA International Press, 1988.

[2] D. Brady, X.-G. Gu, and D. Psaltis, "Photorefractive crystals in optical neural computers," Pro. SPIE 882 *Neural Network Models* for *Optical Computing*, pp. 132-136,1988

[3] J.S. Jang, S.W. Jung, S.Y. Lee, and S.Y. Shin, "Optical implementation of the Hopfield model for two-dimensional associative memory," Opt. Lett. Vol. 13, pp. 248-250, 1988.

[4] J.S. Jang, S.Y. Shin, and S.Y. Lee, "Programmable quadratic associative memory using holographic lenslet arrays," Opt. Lett. Vol. 14, pp. 838-840, 1989.

[5] H.J. Caulfield, "Parallel N^4 weighted optical interconnections," Appl. Opt. Vol. 26, pp. 4039-4040, 1987.

[6] H.J. White and W.A. Wright, "Holographic Implementation of a Hopfield model with discrete weights," Appl. Opt. Vol. 27, pp. 331-338, 1988.

[7] S.Y. Lee, J.S. Jang, J.S. Park, S.Y. Shin, and C.S. Shin, "Modification of the Hopfield model and its optical implementation for correlated images," Proc. SPIE 963 Optical Computing, pp. 504-511, 1989.

[8] D.E. Rumelhart, G.E. Hinton, and R.J. Williams, "Learning representations by back-propagation errors," Nature Vol. 323, pp. 533-536, 1986.

[9] J.S. Jang, S.Y. Shin, and S.Y. Lee, "Parallel N^4 weighted optical interconnections : comments," Appl. Opt. Vol. 27, pp. 4364, 1988.

OPTICAL NEURAL NETWORKS USING ELECTRON TRAPPING MATERIALS

S.Jutamulia, G.M.Storti, J.Lindmayer
Quantex Corporation, 2 Research Court, Rockville, MD 20850

ABSTRACT

We demonstrate that novel electron trapping (ET) materials are capable of performing optical outer-product, inner-product, as well as the formation of a reconfigurable interconnection matrix. 2D and 1D associative memories based on inner- and outer-products employing ET materials have been demonstrated.

1. ELECTRON TRAPPING (ET) MATERIALS

Unlike common spatial light modulators (SLMs) which modulate the phase and amplitude of a wavefront passing through it, the electron trapping (ET) materials can emit different output photons which correlate spatially in intensity with input photons. The ET materials are IIa-VIb compounds (alkaline-earth chalcogenides) with two or three rare-earth dopants added. ET devices consisting of layers of $Ca_xSr_{1-x}S$:Eu,Ce,Sm have been employed in optical experiments. The mechanism for the light emission of ET materials is described as follows. Both ground and excited states of each impurity exist within the band gap of the wide-band-gap (4.5 eV) host material. Short wavelength visible (e.g., blue) light excites electrons from the ground state to an excited state of one of the dopants from whence the electrons transfer over to the second dopant. The electrons remain in the ground state of the second dopant for very long times. However, subsequent exposure to IR light excites the trapped electrons to the excited states of the second dopant, the electrons transfer to the excited states of the first dopant and return to the ground state of the first dopant with the emission of orange light. The ET materials can be used to store optical information as trapped electrons.

In addition to storage, the ET materials are a subject of interest as hardware devices capable of performing multiplication, addition, and subtraction within a dynamic range covering 4 orders of magnitude. The orange emission intensity is proportional to the product of the blue write-in intensity and the IR read-out intensity. The addition and subtraction are performed by increasing and decreasing the number of trapped electrons. These operations are physically carried out by exposing the ET thin film to blue and IR light, respectively.

The ET thin films can be fabricated using standard thin film deposition technologies. Consequently, the fabrication process is potentially much more cost effective than that of other types of SLMs. We have e-beam deposited ET thin films of 4 micrometer thickness on sapphire substrates. The resolution of the ET thin films is currently about 40 lp/mm as shown in Fig.1. The resolution will be improved to the neighborhood of 500 lp/mm when the ET thin films are deposited employing a sputtering technique. It has been observed that the response time of ET materials is on the order of several tens of nanoseconds.

2. OPTICAL FORMATION OF INTERCONNECTION MATRIX FOR AN OUTER-PRODUCT ASSOCIATIVE MEMORY

The Hopfield neural network model assumes that an associative memory retrieval process is equivalent to a matrix vector outer-product expressed as follows.

$$V_i \rightarrow 1 \qquad \text{if} \quad \sum_{j=1}^{N} T_{ij} V_j \geq 0$$
$$V_i \rightarrow 0 \qquad\qquad\qquad < 0 \quad , \qquad (1)$$

where V_i is the output, and V_j is the input. The interconnection matrix T_{ij} is defined by the following prescribed learning.

$$T_{ij} = \sum_{m=1}^{M} (2V_i^m - 1)(2V_j^m - 1) \quad , \text{ for } i \neq j$$
$$= 0 \quad , \text{ for } i=j \quad , \qquad (2)$$

where V_i^m and V_j^m are ith and jth elements of the mth memory vector V^m.

The ET materials can be applied to a neural network as follows. The interconnection weight function of the matrix is stored as an optical memory in the form of a density pattern of trapped electrons in a higher energy level as a consequence of the blue light absorption. The matrix vector multiplication expressed by Eq.(1) can be optically performed using a fan-out and a fan-in cylindrical lens. The emission intensity from the ET thin film is proportional to the product between the intensities of the write-in blue matrix and the read-out IR vector.

The terms of $(2V_i^m -1)(2V_j^m -1)$ in Eq.(2) will be either +1 or -1. Thus, a matrix element is a summation of a series of +1s and -1s. The matrix elements of +1 are written with blue light into the ET thin film to increase the trapped electrons. Similarly, the elements of -1 are written with IR light into the ET thin film to decrease the trapped electrons. The contributions of +1 and -1 are determined through the XNOR and XOR between vector elements V_i^m and V_j^m respectively. A bias level is required to avoid the need for negative numbers of trapped electrons (a physical impossibility) in the ET thin film. After the matrix has been formulated, an input vector carried by IR light is projected onto the ET thin film. Consequently, an emission representing the matrix vector outer-product or a 1D associative memory will be obtained.

An experiment has been conducted to demonstrate the optical formation of an interconnection matrix. An input vector (1,0,0,1,1) was stored in a matrix form. Figure 2 shows the emission from the ET device with an uniform IR light stimulation verifying that the matrix was correctly formed. Three distinctive gray levels show matrix elements of -1, 0, and +1. The optically formed matrix shown in Fig.2 was then used for retrieving the associative memory.

3. OPTICAL INNER-PRODUCT FOR 2D ASSOCIATIVE MEMORY

We have also demonstrated an algorithm of a one-loop 2D inner-product association which can be implemented with ET thin films. There are two operations to obtain the associative memory. These are: (1) compute the inner-product of the input and each memory, and (2) if the inner-product equals or exceeds the threshold, then the output is the corresponding memory. The alternative is as follows: (1) compute the inner-product of the Fourier power spectra of the input and each memory, and (2) do the same as that of the previous approach.

We have experimentally demonstrated algorithms mentioned above using an ET thin film. A beam splitter was employed to combine the blue and IR images onto the ET thin film. A photomultiplier was used to detect the integrated intensity emitted from the ET thin film as shown in Fig.3. The detected inner-product of Fig.4(a) and 4(b) was about 9 V, while the auto-inner-product of Fig.4(a) was about 11 V. A thresholding circuit with a threshold set at 10 V was then employed to recall the associative memory.

4. CONCLUDING REMARKS

We have described two optical methods to implement neural networks using ET materials based on outer- and inner-product for 1D and 2D associative memories. Experimental demonstrations have been performed. In addition to the capability of performing analog multiplication, addition, and subtraction as shown in this paper, the ET materials have some inherent advantages over currently available SLMs such as: (1) capability of memory storage, (2) erasability, (3) blue write-in energy < 10 mJ/cm2 and IR read-out energy < 0.03 mJ/cm2, (4) several tens of nanosecond response time, (5) high resolution (currently about 40 lp/mm), and (6) cost effectiveness. The resolution can be substantially improved as thin film deposition techniques are refined. The erasability of the ET materials certainly can be utilized for the synthesis of a learning machine. Applications of the ET based optical neural networks to solving real world problems are under study.

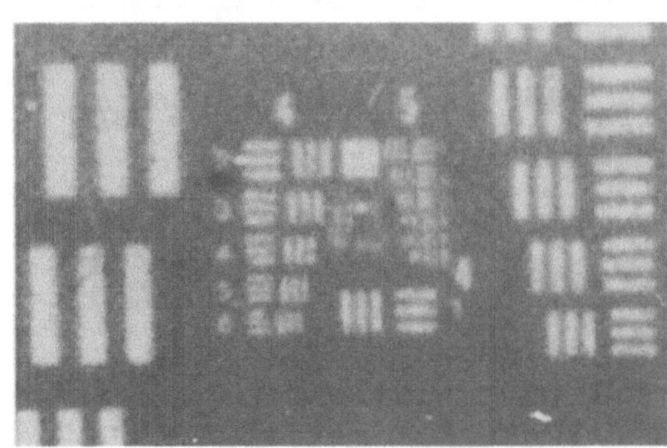

Figure 1

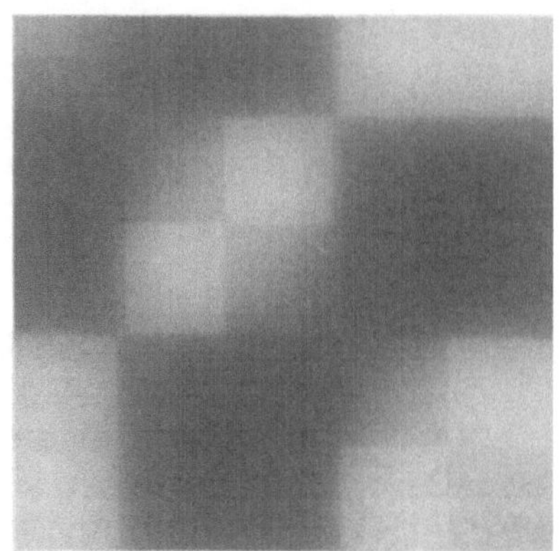

Figure 2

Figure 3

Figure 4(a)

Figure 4(b)

Learning experiment using erasable optically stimulable phosphor in optical neural network

Ken-ichi Kitayama, Fumihiko Itoh ,and Yasuaki Tamura*
NTT Transmission Systems Laboratories
1-2356 Take, Yokosuka, Kanagawa, 238-03 Japan
*NTT Opto-Electronics Laboratories
162 Tokai, Ibaraki, 311 Japan

Abstract A novel application of erasable optically stimulable phospher(OSP) to memory device for modifiable synaptic weights in learnable optical neural network is experimentally demonstrated. OSP has many attractive features for mass storage of synaptic weight that is analog, accessible in parallel, and erasable and rewritable as well as high in resolution and fast in response.

This paper describes that a novel application of erasable and rewritable optically stimulable phospher(OSP), which is a class of electron trapping material, to memory device for modifiable synaptic weights in learnable optical neural network is exprimentally demonstrated. To the authors' knowledge, only an experiment of associative memory has been reported by using similar OSP device.[1] OSP provides many attractive features as a memory device for synaptic weight that is analog and accessible in parallel as well as high in resolution and fast in response.

According to delta rule learning algorithm, update of synaptic weight is given by $\Delta w_{ij} \propto \delta_i v_j$ where δ_i is an error signal, and v_j is an input vector.[2] In the case of autoassociation δ_i becomes $(v_i - u_i)$ being u_i the output vector.

Figure 1 shows the experimental setup. The writing beam is visible light(Argon laser: λ=515 nm). The readout emission of w_{ij} is obtained near λ=680 nm by uniformly illuminating light(He-Ne laser: λ=1150 nm). The system output is obtained optoelectronically. Readout synaptic weights is detected and amplified by the image intensifier. The computations of $\sum_j w_{ij} v_j$ followed by thresholding and error signal are executed by using a computer. The weight change is generated optically by using two SLMs. In the case with $\Delta w_{ij} > 0$, argon laser beam passing through a specific pixel position (i, j) of SLMs overwrites Δw_{ij} on previous w_{ij} in OSP. The negative weight change can be made by taking erasure process with the illumination of IR beam.

OSP film prepared for the experiment was fabricated from a mixture having the composition: 100 g CaS, 6 g CaF_2, 0.04 g Eu_2O_3, and 0.06 g Sm_2O_3. The particle size is 1 to 10 μm. The film thickness is approximately 20 μm. In Fig. 2, the intensity of luminescence is plotted as a function of writing energy of visible light. This linearity is preferable for the present analog storage.

In the experiment, the following three 16-bit long vectors are used: $v^{(1)}$ = (1111111100000000), $v^{(2)}$ = (1111000011110000), and

$v^{(3)}$=(1010101010101010). In Fig.3, the readout emissions of OSP observed in the weight change process are shown in order. The output vector $u^{(p)}$ (p=1,2,3) for each input $v^{(p)}$ is illustrated. The blurred pattern is due to insufficient spatial resolutions of OSP and image intensifier. After twenty learning cycles, the weight matrix and output converge as shown in (c). Here, $u^{(3)}$ does not become equal to $v^{(3)}$ because three vectors exceed the memory capacity for sixteen neurons. By comparing it with the theoretical w_{ij} in (d), a good agreement is found in general.

In conclusion, a novel application of erasable optically stimulable phospher to memory device for the modifable synaptic weight in learnable optical neural network has been experimentally demonstrated. By developing high-resolution thin film crystalline OSP, a scale-up of the implementation, equivalent of increase of neuron number, will presented at the conference. Due to the fast response in nanoseconds with promising 10^2lp/mm in resolution,[1] the system might attain ultimately tera OPS/cm^2 processing capacity.

The authors would like to thank Dr.K.Ono for invaluable discussions. They would also like to thank Dr.S.Shimada, Dr.H.Kimura, Dr.H.Murase, and Dr.T.Matsumoto.

References

1. A.D.McAulay et al., IJCNN'89, II-483(Washington D.C. July '89).
2. D.E.Rumelhart et al., Parallel Distributed Processing vol.1, The MIT Press, Cambridge(1986).

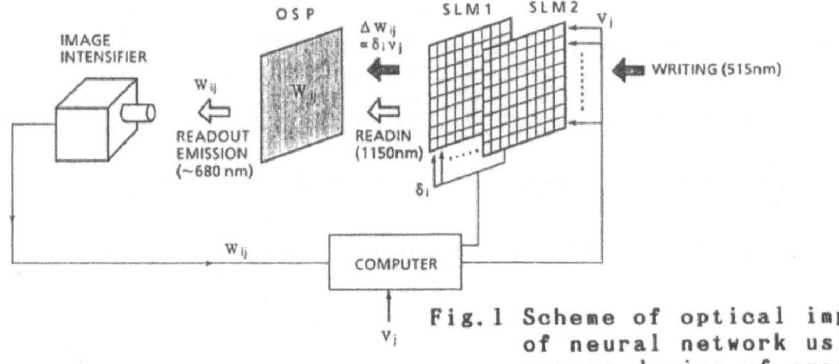

Fig. 1 Scheme of optical implementaion of neural network using OSP as memory device of synaptic weights.

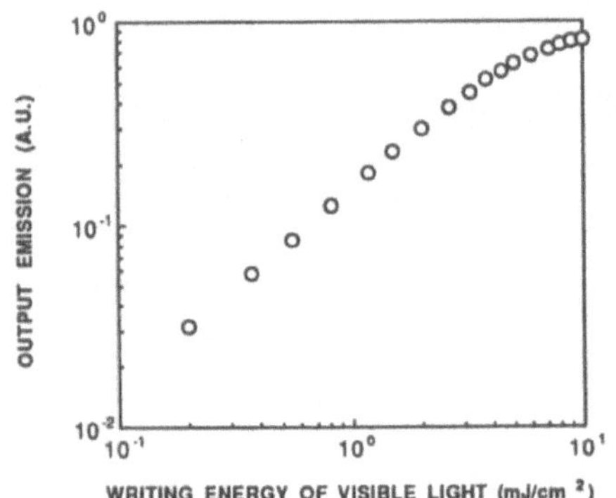

Fig. 2 Intensity of luminescence at λ=680nm plotted vs. writing energy of visible light at λ=515nm.

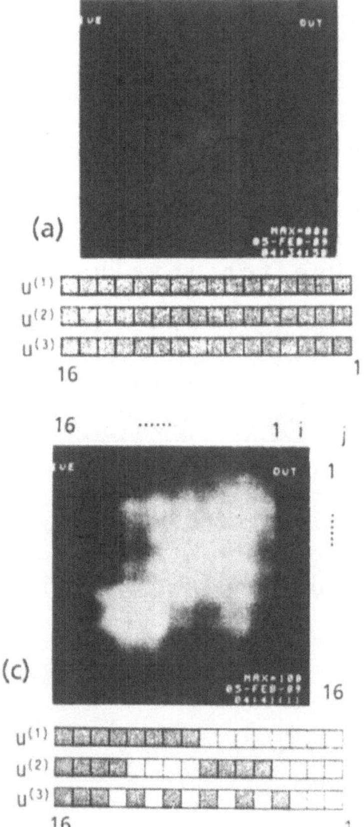

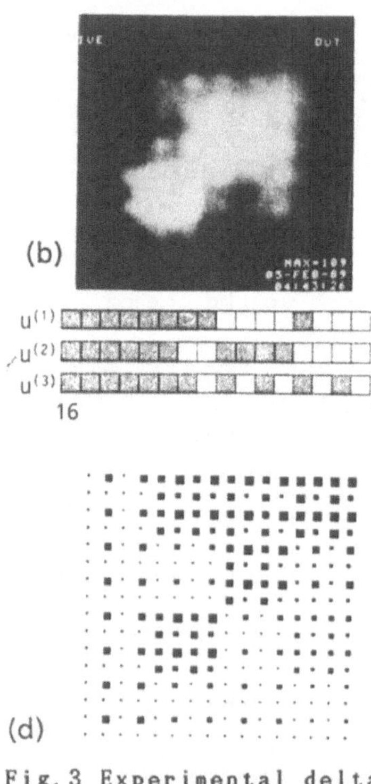

Fig. 3 Experimental delta rule learning processes shown in order from (a) to (c). (d) Theoretical synaptic weight after completion of learning.

OPERATING CHARACTERISTICS OF A SECOND-ORDER NEURAL NETWORK CLASSIFIER SYSTEM

P. Horan, D. Uecker* and A. Arimoto,
Central Research Laboratories, Hitachi Ltd.,
Kokubunji, Tokyo 185, Japan.
*now at Uni. of California at Santa Barbara

Abstract. The operational characteristics of a simple model of a translation invariant second-order discriminator, such as fault and noise tolerance, are investigated. A particular optical implementation, which exploits the features of the model suited to optics, is detailed and the operation of the system described.

SIMULATION OF AN OPTOELECTRONICALLY IMPLEMENTED NEURAL NETWORK FOR EARLY VISUAL PROCESSING[1]

Courosh Mehanian
MIT Lincoln Laboratory
244 Wood Street
Lexington, MA 01273 USA

Abstract

Simulations of the operation of a neural network architecture for boundary segmentation, an important stage of early visual processing, are presented. The network is based on the CORT-X model, a multiple spatial-scale, feedforward architecture for boundary segmentation of noisy binary images. The CORT-X architecture can be made to accommodate, with slight modifications, existing optoelectronic hardware capabilities. Network performance is evaluated with respect to deviations from ideal response along two dimensions: (1) contrast ratio and (2) nonuniformity. The effect of these parameters on network performance as a function of the input image signal-to-noise ratio (SNR) is evaluated. With ideal response (perfect uniformity and infinite contrast) the modified CORT-X architecture performs as well as the original model. Finite contrast does not significantly degrade network performance as long as there is some reasonable contrast. On the other hand, nonuniformities as small as 10% degrade performance even for high SNR.

[1] This work was sponsored by the Defense Advanced Research Projects Agency.

IMPLEMENTATION OF NEURAL NETWORK ALGORITHMS ON PARALLEL HARDWARE

Chair: Igor ALEKSANDER

THE COGNITIVE CHALLANGE FOR NEURAL ARCHITECTURES

Igor Aleksander
Professor of Neural Systems Engineering
Imperial College, London, SW7 2BT, UK

Helen Morton
Lecturer in Human Sciences
Brunel University, Uxbridge, 8UB 3PH, UK

Abstract
The very fact that neural computing is attracting a considerable amount of research attention makes it important to develop a long-term perspective by identifying some targets which are ambitious but worth achieving. Here we shall concentrate on just one target which is in the domain of general computing: the design of machinery which , in cognitive (i.e knowledge-based speech, vision and natural language) applications, is more competent than that which is normally available. Cognitive tasks have been the target for much of artificial intelligence work based on pre-programming, but this has resulted in only limited successes. Do neural nets hold the promise of some further achievement? In this paper we suggest a general hybrid architecture which is the basis of a major project currently in place at Imperial College: the RUNES project (Real Understanding of Neurally Engineered Systems). This is based on *weightless* neural designs (Aleksander and Morton 1990, a). We provide some examples that are both a design challange and guide for such an architecture.

1. A General Neural Architecture

During the time that a concept is in the early stages of development it is fashionable to suggest that the idea is so powerful that it will replace all that went before it. So it was with artificial intelligence and expert systems, and so it seems to be with neural computing. The view taken here is that the neural will not overshadow the conventional. Instead, neural computing is likely to act as a powerful addition to the armoury of techniques currently available in conventional computing. We believe that the two methodologies working together hold the greatest promise, and it is this notion which leads to the speculative architecture shown in fig. 1.

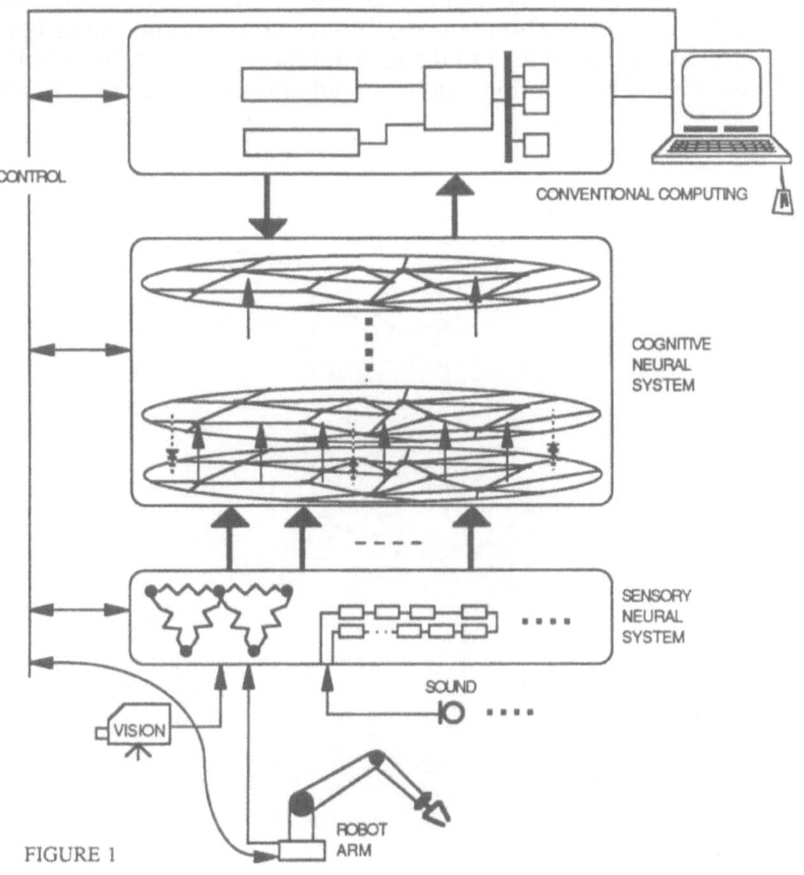

FIGURE 1

There are three major parts to this architecture: a sensory neural level, a cognitive neural level, and a conventional computing level. Peripheral connections are made to appropriate video cameras, microphones and other electronic sensors on the input side. On the output side the system may have the ability to connect to robot arms, voice synthesizers, graphics screens and so on. The conventional computing level of the system would have provision for conventional communication with the user: keyboard, mouse, display screen and the like.

There is a need to distinguish between the sensory neural level and the cognitive level. The main difference between these two lies in the fact that the sensory level contains fixed processing equipment **that does not learn** but is helpful in distinguishing between different forms of input. No more needs to be said about this but for the fact that the work of Carver Mead (1989) is precisely that which underpins this section of the machine. The cognitive level has loosely coupled layers, each layer being autoassociative and (in our work) is being designed using *weightless* methodologies (Aleksander and Morton, 1990, a).

2. The Cognitive Processing Challange: Language/Vision

The major challenge for neural computer designers is the need to coordinate various neural processes each of which is known to be feasible. We illustrate this with some specific tasks (which are described as challenges rather than recipes for a solution).

Is it a box?

Clearly, the cogntive layer would be responsible for taking (say) the line features produced by a sensory layer and exclaiming (say): "it's a box". As shown in fig. 2, part of the cognitive net needs to perform a pattern completion exercise. This could occur in the manner of **schemata** discussed by Sharkey (1989). The features of the image activate a box-like schema at some layer L in the cognitive neural system (fig. 2(a)). Other arrangements of features may activate a pyramid-like schema in the same layer (fig 2 (b)). The stable "box" schema in (a) leads to a localized identification at some B-neuron in a higher layer (M) while another recognition (of the pyramid, say) may lead to the activation of some P-neuron in layer M. It now remains to be said how such a recognition can be turned into an answer to the spoken phrase "What do you see?".

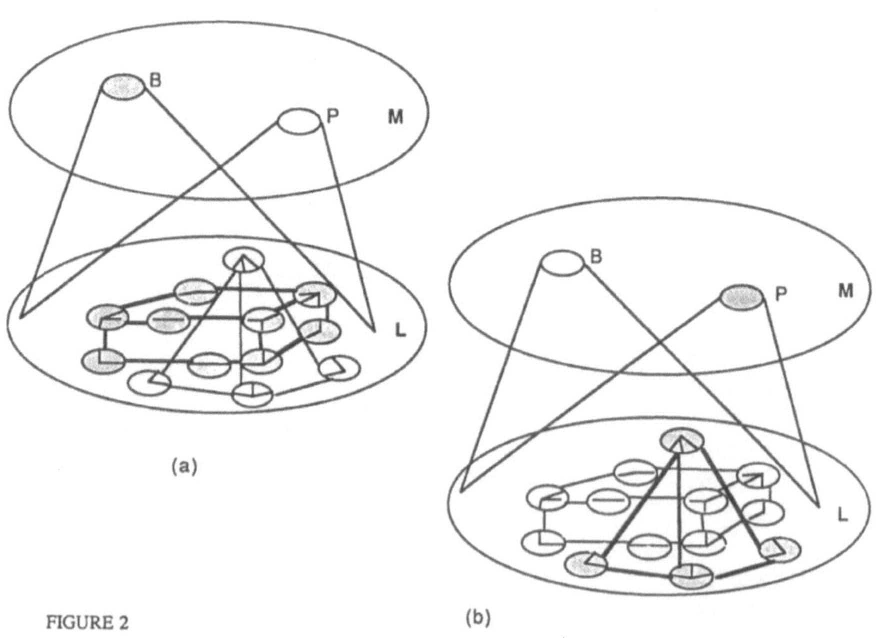

FIGURE 2

At the lower levels of cognitive processing, it is clear that several recognition activities must be going on in parallel. Such activities are related to the different senses: while part of the neural circuitry decodes what the eye "sees" as in fig. 2, another must be devoted to the extraction of meaning from what the ear "hears". Calling again on the method of schemata, but now on the speech understanding side, the elements "what", "you" and "see" could be thought of as forming a stable schema at one layer which, at a higher layer, is interpreted by the activation of one (or a few) units as "introspective description (voiced sequence) required on the vision side of the net". In such a layer within the net, other localized representations may be: "introspective description required on the auditory (tactile, ..) side", "movement required of the robot arm to coordinates x,y, z", "voice utterance required in response to state of vision side of the net" etc. ... .

In a living organism, this process of localized interpretation in a higher net of a stable schema in a lower net must eventually lead to the appropriate actions in the organism. (Note: some of the "actions" could be internal, such as "thinking about an event".) In a man-made machine, however, there comes a point at which conventional processes of computation could be triggered off by the state of the neural net. For example, the state "introspective description (voiced sequence) required on the vision side of the net" is a clear instruction which could be interpreted by a conventional computer and a voice synthesizer. The machine would "address" the upper vision layer, discover that "box" is activated, interpret "introspective description required" as an instruction to output, via the voice synthesizer, the phrase "it's a <content of vision layer>".

3. The Cognitive Processing Challange: Language and Temporal Knowledge

Sharkey (1989) suggests generating an appropriate description of a sequence of actions under the control of a stable schema (e.g. the actions required by the schema of catching a train). This is an example of "temporal neural processing": it requires the recall of sequences of events and the actions that would change such sequences. This aspect of neural computing is still very much in its infancy, and is a fruitful area for further research.

"To get to Covent Garden, change at Green Park"

Temporal processing includes the ability of living neural systems in living organisms to predict sequences of events on the basis of past experience. A central and general question is how sequences of actions leading to given outcomes (like" If you change to the eastbound Piccadilly line at Green Park you will first get to Piccadilly Circus and then Covent Garden") are learnt and recalled by humans. The accurate prediction of events is helpful not only in order to find one's way around the London Underground when there are no maps to be seen, but , more importantly, it underlies some of the ability of individuals to survive. To illustrate what may be needed, we continue with the example of the underground journey. Consider part of the London Underground Map shown in fig.3.

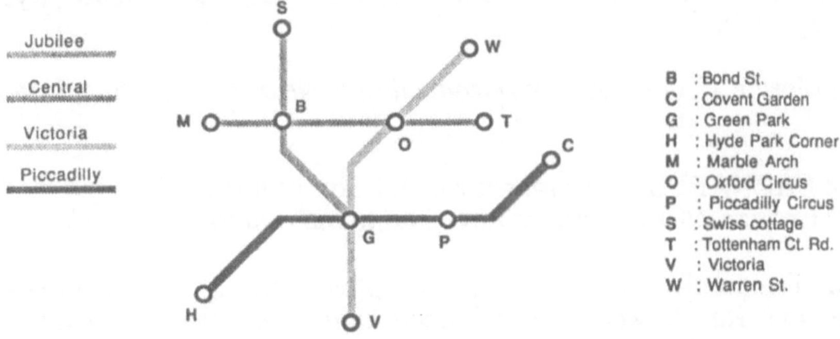

FIGURE 3

Say that a fully interconnected layer in a cognitive net is capable of learning a sequence of states by "being " in a state, and being trained to change to the next. These states could be names of underground stations. So if the state of the net is P, it could learn that the next state will be C, in the sense that if the net is clamped to P and released, it will change to C. However, a characteristic of such a layer is that one state can only lead to another specific state. Therefore merely labelling states as stations would lead to confusion as there are many states that can follow G in our example - something else is also needed.

Clearly, the net has to store context as a "state" (e.g. "I am on the Victoria line at O and I am trying to get to C") to determine the next state which should be "I am at G changing to the Piccadilly line towards P trying to get to C". The way in which these sequences are **learnt by exploration** rather than the memorization of a map is one of the open questions in neural computing.
That is, the problem is not how such knowledge may be represented in neural nets (Sharkey and others provide sensible suggestions for that), but how this knowledge could get there, particularly through exploration. The challenge therefore is for a neural net to gather sufficient experience to be able to answer questions such as " How do I get from Victoria to Covent Garden?"

Although many examples of temporal processing may be found in the literature (too many to quote individually), they rarely address the difficult problem of learning to predict in environments that have language - like automata-theoretic structures in which control contexts have to be learnt at a different level from that of the surface sequences.

4. Comment

In this paper we have outlined some challanges which face us in the design of a major system which is to be more competent at cognitive tasks than conventional systems. Should this succeed, we feel that such an architecture would not only be seen as a "neural" architecture, but may become a model for general purpose computational devices of the future.

5. Notes

More information on the background and philosophy of the highly compressed descriptions in this paper may be found in a published text (Aleksander and Morton, 1990b) and interal reports by members of the Imperial College team: Myers (1989) on learning in temporal environments and Fulcher (1989) on sequential knowledge acquisition. This work is supported under a UK Information Engineering Directorate grant No. 3/1/1005.

References

Aleksander I. and Morton H. (1990,a): An overview of Weightless Neural Systems, *IJCNN Washington,* January 1990.

Aleksander I. and Morton H. (1990,b): *An Introduction to Neural Computing,* London, Chapman and Hall .

Fulcher, E. (1989): BARNABUS: A neural net which learns sequences by a bidirectional chaining of events. *Imperial College, Neural Systems Engineering Internal Report,* EF1/89.

Myers, C. (1989): Temporal credit assignment: adaptive learning when results are delayed or interleaved in time. *Imperial College, Neural Systems Engineering Internal Report,* CM2/89.

Sharkey, N. E. (1989): A PDP approach to natural language understanding. In Aleksander (Ed.) *Neural Computing Architectures*, Boston, MIT Press, pp 92-116,

IMPLEMENTATION OF BACK-PROPAGATION ON A VLSI ASYNCHRONOUS CELLULAR ARCHITECTURE

Bernard FAURE and Guy MAZARE
IMAG-LGI Groupe Circuits
46 avenue Félix Viallet
38031 Grenoble cedex, France

The inherent parallelism of any neural network structure can be efficiently taken into account by massively parallel architectures. However, a communication problem remains since the neurons are highly interconnected. A communication system, based on message transfers and without need for allocating a physical link for each connection seems to be a solution for any parallel machine but is very hard to implement efficiently on hypercubes. This paper present a VLSI architecture of asynchronous cells arranged in a two-dimensional array with a hardware-based array-wide message transmission mechanism, the choices induced by the VLSI implementation constraints and its simulated performances. Our first implementation of the back-propagation algorithm is approximately twice faster at evaluating the NETtalk text-to-speech network than the Warp, a 20 processor systolic array, which ought to be the fastest back-propagation simulator reported in the literature. We can take more advantage of the architecture by allowing the pipe-lining of multiple recalls and then increase the performance by 3 to 5 times. Our results indicate that two-dimensional arrays can be good candidates for neural processing, providing they can handle high communication rates.

INTRODUCTION

The application of neural networks to everyday problems that involve misunderstood faculties of the brain requires a huge amount of data processing and thus high computing power. Neural networks typically learn to perform these tasks from the iterative presentation of an important set of examples. Since the learning is achieved by minimizing the error detected from the network output of each presented example, the total error-minimization time must be as short as possible. One possibility to increase the computing power is to use a parallel machine but, in such architectures, the bottleneck is the interprocessor communications. In large grain architectures, subsets of the neural network are processed by a single processor. The network mapping, which must minimize interprocessor data transfers and take advantage of the parallelism to be efficient, is a hard work. The one-neuron-per-processor association, which first comes in mind, can be used with fine grain architectures. Such an architecture can be composed of a large number of processors which are reduced to simple automata performing the algorithms of a neuron. In fine grain multi-dimensional hypercube machines, with respect to the network topology, the communications are short but difficult to embed in the 3-dimensional space because of the high number of links a single processor must handle. Such a machine needs connection switches to be configured to a particular application. A communication system based on message transmission between processors can avoid the handling of a huge number of physical links.

We propose a message transmission mechanism distributed in each processor. The messages are passed from a processor to one of its neighbours until they reach their destinations ; their paths are set dynamically by each passed-through processor. This communication mechanism is hardware-implemented to favour the transmission speed and then meet the high communication rates required by neural networks applications. Because each processor can communicate with potentially any other, this communication system can better take advantage of the plane than of the 3-dimensional space. The proposed architecture is derived from the communication system we have decided to use : it consists of a two-dimensional array of asynchronous processors (called cells because they are devoted to the simple algorithms of a neuron and to the message transmission), each of which being physically connected to its four immediate neighbours. The description of back-propagation networks [7] as a set of neurons arranged in layers of different sizes (with only interlayer connections most of the time) leads to a disorganized parallelism that can be efficiently handled by this cellular array.

THE CELLULAR ARCHITECTURE

Each cell performs a simple local function and is physically connected to its four immediate neighbours through eight unidirectional buffers, one for each way of the four directions. A flip-flop based mechanism included in each buffer prevents multi-access form the two cells sharing it. Another important feature of this array is that it is dedicated to fit a particular application : the basic cell can have the minimum complexity for a given task.

Our research team has already studied and designed cells and small arrays dedicated to logical simulation [2] and image reconstruction [6]. We are now studying data flow processing [8] and neural computing. Figure 1 shows the topology of the cellular array and the structure of the message.

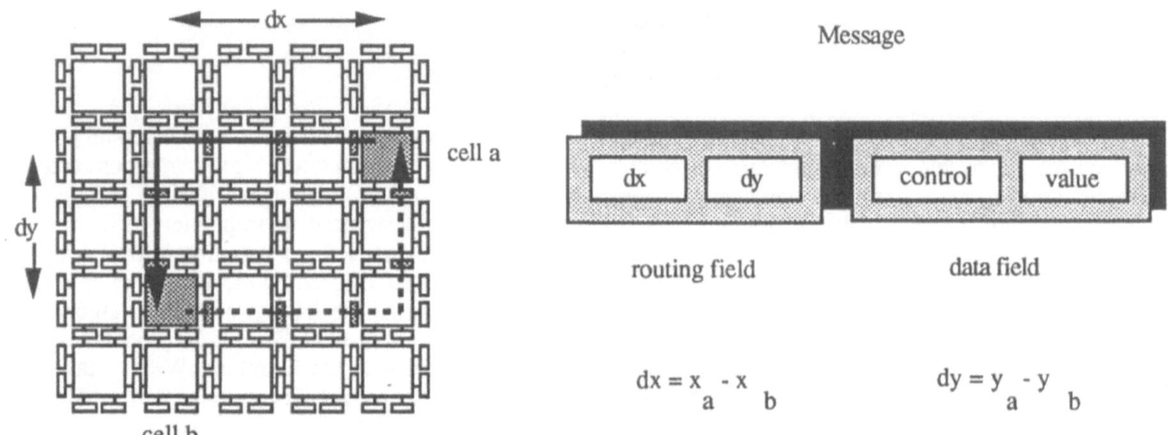

FIGURE 1 : Message transmission

To be handled by the array, a message must holds a data field containing the information to be processed by the application mapped on the array and a routing field containing some information needed by the array for its own use : the relative displacement dx and dy from the current to the addressee cell. This relative displacement fully describes the path the message will follow : dx and dy are updated in each passed-through cell. This allows us to use a non trivial routing policy that can take advantage of a WSI technology [1]. The cell surface is large enough to permit the use of parallel buffers and then parallel transmission of the messages. The cell performs two concurrent tasks and thus is divided in two asynchronous parts : *the routing part*, dedicated to the message transmission, which reads a message from an input buffer, selects an output buffer according to the routing policy, updates the message routing part and stores the new message in the selected output buffer if empty, otherwise, it holds the message in the input buffer and selects another possible input ; *the processing part* which emulates the neuron, that is processes both the recall and the back-propagation algorithms.

THE ALGORITHMS IMPLEMENTED

In such a fine grain architecture, the compromise surface versus time is critical. To match this, and due to the high number of computations that have to be done, we must simplify the functions used,limit the data path width and decrease the memory requirements as much as possible. First, we have decided to implement only one input and one output buffer for the communication between the processing and routing parts of a cell. The processing part performs the computations in a different order from the standard series : sum the weighted incoming inputs, compute its activation and output a weighted activation on each downstream path during the recall phase as update the weights of the downstream links with the incoming gradient, weight and sum the incoming gradients, compute and output the local gradient on each upstream path during the learning phase. This is easily done by associating the weights inside a cell to the downstream links instead of the upstream links in the standard neuron model. We use a linearly interpolated transfer function. The connection weights and neuron activations are scaled between 1 and -1 and the gradients between 2 and -2. The absolute values of all the data handled by a cell range from 2^1 to $2^{-(n-2)}$, n being the resolution of a cell ($n \geq 16$ bits). The final mapped network can be allowed a wider data path that takes round-off errors into account. We limit the connectivity (fan-in and fan-out) of all cells in the array by adding sublayers to the initial neural network topology and then map the increased network on the array.

This architecture has been simulated, operating on different neural networks : the exclusive-or (a 2-2-1 units, 3 layers network), the Little Red Riding Hood [4] (a 6-3-7 units, 3 layers network), the numeral and character pattern classification (respectively 45-10 and 45-26 units, 2 layers networks). The cellular array can memorize all the presented patterns but needs many more learning steps than with unrestricted computations using 32 bits floating-point operations and a sigmoid transfer function [3]. The overhead depends on the initial neural network topology :

wide layers demand more learning steps. A primary version of the cell, with a limited fan-in and fan-out of 8, has been designed in 1988. The number of transistors ranges between 11 and 19000 for a cell that includes the routing part, four external buffers and a connectivity between 2 and 8 and increases until 35 to 48000 for a connectivity of 64. As the cellular array is a square matrix, its size, which depends on the cell connectivity ranging from 2 to 64, is : 4 and 3 for the exclusive-or, 10, 6 and 5 for the Little Red Riding Hood, 32, 19 and 15 for the numeral pattern classification, 50, 30, 21 and 19 for the character pattern classification, 80, 47, 32, 24, 15 and 14 for a typical example (a 25-60-25 units, 3 layers network), 167, 97, 65 and 59 for the NETtalk [10] application (a 203-60-26 units, 3 layers network). Its speed, which is directly derived from the number of active cells is impressive. It appears that we do not need basic cells with a connectivity higher than 4 or 8 to perform neural network processing. The use of low connectivity cells demands a wider data path (30, and 19 bits for the NETtalk application with a cell connectivity of 2 and 4) but 16 bits for higher cell connectivities.

THE ARCHITECTURE SPEED

The simulations have shown that there is no need for an efficient placement. The performance between a good and a bad placement differ only by a few percent : from a 3 % time increase for a very small network (xor and lrrh) to 0.2 % for pattern classification and 0.04 % for NETtalk. The time increase is due to the message retention along the most used paths that are less in a bad mapping, the neurons being more spread over the array. The cell connectivity must raise to at least 32 and the mapping very bad (let say less than 25 % of active cells in the array induced by the mapping of io cells only on the array periphery) to see paths with very high connection rates seen only for the pattern classification and NETtalk problems. The activity can be increased by allowing the use of the two peripheral layers of the cellular array for io cells, but the recall and learning times remain about the same. The speed of the cellular array was evaluated for a cell having a 16 bits operative part, running at a 20 MHz clock speed and able to transmit a message in 40 ns from any input buffer to any output buffer [5]. During the simulations, we have assumed that the messages are treated by the outside world at the array speed. We can compare the simulated performances achieved for the NETtalk simulation, measured in MCUPS (millions of connection updates per second), with those of other machines we found in [9]. Table 1 shows that our 65 x 65 array of 8-connectivity cells, 4175 of which are neural units is approximately twice faster than the most powerful non dedicated machine : the 20 nodes Warp, four times faster than the 64K connection machine 1 and more than 1000 times faster than a Vax 780. We could not find any reported speed evaluation for other dedicated VLSI circuits running one of the tested applications.

machine	MCUPS	relative speed
Ridge 32 (Vax equivalent)	0.05	1
Convex C-1	1.8	36
16K CM-1	2.6	52
64K CM-1	13.0	260
10 Warp	17.0	340
20 Warp	32.0	640
65 x 65 array (connectivity 8)	51.4	1028

TABLE 1 : Speed comparisons for various machines

The activity of the cellular array is low : less than 25 % for the numeral pattern classification, 26 % for the typical example and 10 % for NETtalk. The low activity during the learning phase is due to the fact that the weight update can only be performed when the array has output the activation of all the neurons in the output layer. This activity can be lightly raised by forcing the array interface to process an output message as soon as it is present in a peripheral buffer of the array. With this amelioration, the array processing time decreases a little bit but the activity raise is not noticeable. The activity during the recall phase can be increased by forcing each cell to send acknowledge messages to its connected cells in the upstream layer. This allows the possible pipe-lining of multiple pattern recalls. This local resynchronization done for each recall even during the recall preceding the weight update does not influence the total learning time. The introduction of acknowledge messages can induce a drop in the recall time down to a factor 2 for the xor, 2.5 for the lrrh, 3 for the numeral pattern classification and 4.5 for the typical example while the overhead introduced in the learning phase is not perceptible most of the time (a maximum 8 % increase for the numeral pattern classification and 10 % for NETtalk in some particular configurations). We hope to raise the performances by at least a factor 2 for NETtalk but we have not completed the simulations yet.

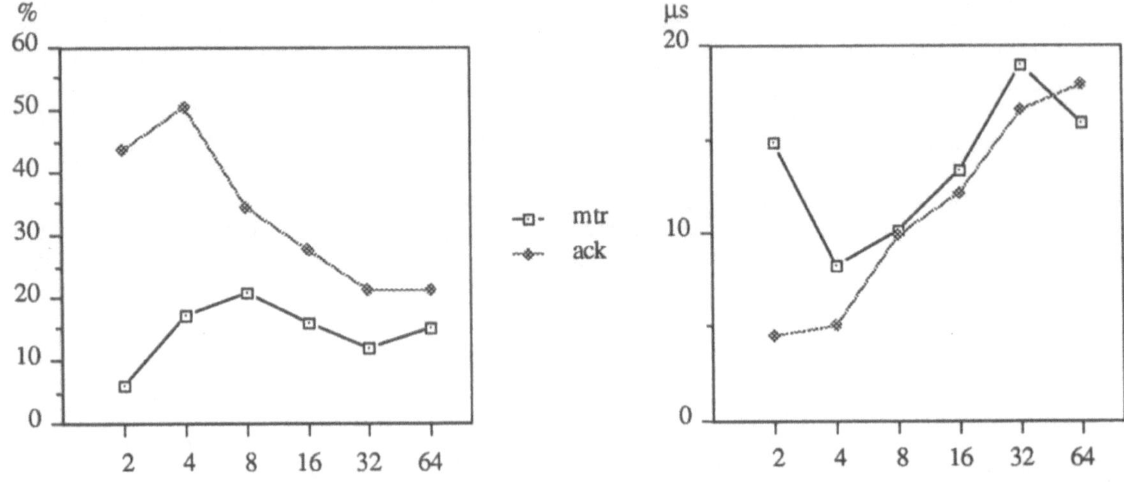

FIGURE 2 : Average number of working cells

FIGURE 3 : Time to recall a pattern

Figures 2 and 3 show the activity and time performances of the cellular array during the recall phase as a function of the maximum connectivity of a cell for the numeral pattern classification problem. Two series are shown : the recall of one pattern with acknowledge messages, and the recall of one pattern without acknowledge messages.

CONCLUSION

There are three options to increase the performances of neural network simulations : keep the Von Neumann concept with a more advanced processor able to perform complex operations, take advantage of multiprocessor architectures where the work is distributed over a limited number of processors, but this approach demands a non trivial mapping of the network to fit the architecture constraints, or map directly the neural net over a large number of simple processors. Most of all, the use of fine grain architectures can help in keeping the initial topology with its underlying massive parallelism and robustness features as a neural network is well defined by the distributed algorithms performed by a neuron and by the interactions between neurons.

REFERENCES

[1] Ansade, Y., Cornu-Emieux, R., Faure, B., Mazaré, G., WSI asynchronous cell network, in : proceedings of the IFIP WG 10.5 Workshop on Wafer Scale Integration, Grenoble, France, march 1986, pp.77-88.

[2] Cornu-Emieux, R., Mazaré, G., Objois, P., An integrated highly parallel architecture to accelerate logical simulation, in : proceedings of the IEEE ISELDECS 87, December 1987.

[3] Faure, B., Mazaré, G., A VLSI asynchronous cellular architecture dedicated to multilayered neural networks, in : proceedings of the nEuro'88 Conference, Paris, France, June 1988, pp. 710-719.

[4] Jones, W. P., Hoskins, J., Back-propagation : a generalized delta learning rule, in : BYTE Magazine, October 1987, pp. 155-162.

[5] Karabernou, M., Etude et réalisation d'un mécanisme d'acheminement de messages dans un réseau cellulaire, DEA de microélectronique, Université de Grenoble, June 1988.

[6] Lattard, D., Mazaré, G., Image reconstruction using an original asynchronous cellular array, in : proceedings of the IEEE ISSCS89, May 1989, pp. 13-16.

[7] Le Cun, Y., Modèles connexionistes de l'apprentissage, Thèse d'informatique, Université de Paris 6, 1987.

[8] Payan, E., Mazaré, G., Programable highly parallel architecture : functional definition and performance evaluation, accepted at ICNC, Dusseldorf, March 1990.

[9] Pomerleau, D. A., Gusciora, G. L., Touretzsky, D. S., Kung, H. T., Neural network simulation at Warp speed : how we got 17 million connections per second, in : proceedings of the IEEE International Conference on Neural Networks, San Diego, Ca, July 1988.

[10] Sejnowski, T. J., Rosenberg, C. R., Parallel networks that learn to pronounce English text, in : Complex Systems, n° 1, 1987, pp. 145-168.

VLSI Implementation of a Neural Associative Memory and its Application to Vector Quantization

Rodney M. Goodman and Tzi-Dar Chiueh
Department of Electrical Engineering (116-81)
California Institute of Technology
Pasadena, CA 91125, U.S.A.

Abstract

In previous papers we have proposed a new high capacity associative memory which we call the exponential correlation associative memory (ECAM). In this paper we describe the VLSI design of a programmable ECAM which has been implemented and tested in 3 micron CMOS. The prototype chip is capable of storing 32 memory vectors of 24 bits each. The high capacity of the ECAM is partly due to the use of special exponentiation neurons, which are implemented via sub-threshold MOS transistors in this design. The prototype chip is capable of performing associative recall in 3 μs, and we demonstrate its capabilities for real time processing using binary vector quantization as an example application.

Architecture

Previously [1,2], we have proposed a general model for correlation-based associative memories, which includes a variation of the Hopfield memory and the high-order correlation memories as special cases. This model is based on an architecture consisting of binary connection weights, simple hard-limiter neurons, and specialized nonlinear circuits as shown in Figure 1. The evolution equation of this general model is

$$\mathbf{x}' = sgn\left\{ \sum_{k=1}^{M} f(<\mathbf{u}^{(k)}, \mathbf{x}>)\, \mathbf{u}^{(k)} \right\}, \tag{1}$$

where $\mathbf{u}^{(1)}, \mathbf{u}^{(2)}, \cdots, \mathbf{u}^{(M)}$ are the M memory patterns. $\mathbf{x}$ and $\mathbf{x}'$ are the current and the next state patterns of the system respectively, and sgn is the threshold function, which takes on the value $+1$ if its argument is nonnegative, and -1 otherwise.

We addressed, in particular, the case where $f()$ is of the exponentiation form, namely, when the evolution equation is

$$\mathbf{x}' = sgn\left\{ \sum_{k=1}^{M} a^{<\mathbf{u}^{(k)}, \mathbf{x}>}\, \mathbf{u}^{(k)} \right\}, \tag{2}$$

and a is a constant greater than unity. This new exponential correlation associative memory (ECAM) possesses a very large storage capacity, which scales ***exponentially*** with the length of memory patterns. Furthermore, it has been shown that the ECAM is asymptotically stable in both synchronous and asynchronous updating modes [1,2].

The ECAM chip we have designed is ***programmable***, that is, one can change the set of memory patterns to be stored at will. To perform an associative recall, one first loads a set of memory patterns into the chip. The chip is then switched to the associative recall mode, an input pattern is presented to the ECAM chip, and the ECAM chip then computes the next state pattern according to Equation (2). The components of the next state pattern appear at the output in parallel after the internal circuits have settled. Feedback is easily incorporated by connecting the output port to the input port. In this case the chip will cycle until a fixed point is reached.

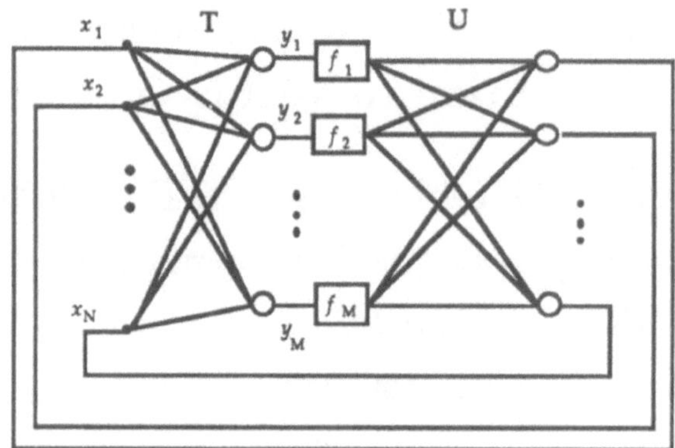

Figure 1: Architecture of the general correlation-based associative memory

According to the evolution equation of ECAM, we notice that there are essentially three circuits that need to be designed in order to build an ECAM chip. They are :

- $<\mathbf{u}^{(k)}, \mathbf{x}>$, the correlation computation circuit.
- $\sum_{k=1}^{M} a^{<\mathbf{u}^{(k)}, \mathbf{x}>} \mathbf{u}^{(k)}$, exponentiation, multiplication and summing circuit.
- $sgn(\cdot)$, the threshold circuit.

In our VLSI design, the functions of correlation computation, exponentiation, multiplication, and summation are all incorporated into an ECAM block. This block, together with a standard static RAM cell, makes up the final ECAM cell as illustrated in Figure 2. The final design of an exponential correlation associative memory that holds M N-bit memory patterns can then be obtained by replicating the basic ECAM cell in the horizontal direction M times and vertical direction N times.

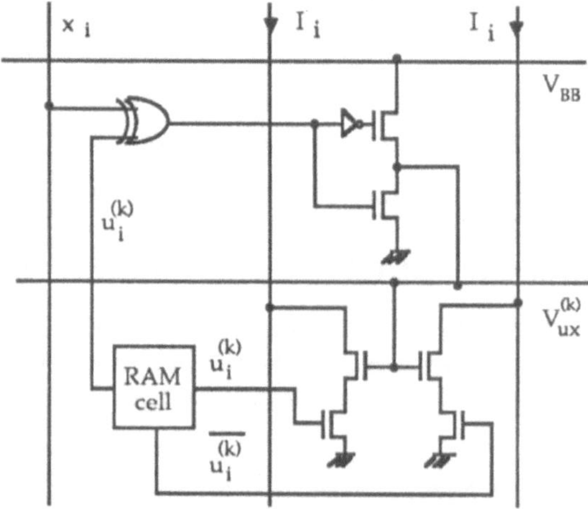

Figure 2: Circuit diagram of the basic ECAM cell

The correlation circuit is controlled by **XOR** gates and has the form of a voltage-divider type circuit consisting of NMOS transistors working as controlled resistors (linear resistors or open circuits) which compute the correlation between the input pattern $\mathbf{x}$ and a memory pattern $\mathbf{u}^{(k)}$. The correlation is accumulated as a voltage on the horizontal wire shown, and has a maximum value which is controlled by an externally supplied bias voltage Vbb. The exponentiation function is implemented by an NMOS transistor whose gate voltage is this correlation voltage $\mathbf{V}_{\mathrm{ux}}^{(\mathrm{k})}$. Since the bias voltage Vbb (the maximum value that $\mathbf{V}_{\mathrm{ux}}^{(\mathrm{k})}$ can assume) is normally set in our design to be lower than the threshold voltage (Vth), the NMOS transistor is in the subthreshold region, where its drain to source current depends exponentially on its gate to source voltage [3]. Since the multiplier $u_i^{(k)}$ assumes either $+1$ or -1, the multiplication can easily be done by forming two branches, each made up of a transmission gate transistor in series with an exponentiation transistor whose gate voltage is $\mathbf{V}_{\mathrm{ux}}^{(\mathrm{k})}$. Summation of M terms in the evolution equation is done by current summing on the vertical wires shown. The final results are two currents $\mathbf{I}_i^+$ and $\mathbf{I}_i^-$, which need to be compared by a threshold (differential amplifier) circuit to determine the sign of the ith bit of the next state pattern x_i'.

We have implemented a prototype ECAM chip, which is made up of 32 × 24 basic cells, read/write circuits, sense amplifiers, address decoders, and I/O multiplexers. The chip can store 32 memory patterns, each 24 bits wide.

Vector Quantization

We have tested the speed of the ECAM chip using binary image vector quantization as an example problem. Vector quantization is a means of data compression (source coding) on information to be transmitted or stored, e. g., speech waveforms, images, etc. [4]. In principle, a vector quantizer should, given a set of codewords and an input, find the nearest codeword to the input. Then only the index of the nearest codeword is transmitted or stored instead of the information itself. Usually, the number of possible codewords is much less than that of possible information patterns, hence vector quantization can reduce the bandwidth (number of bits) needed.

The ECAM chip performs binary image vector quantization on 4 × 4 blocks of pixels. A set of 32 codewords corresponding to all white, all black, horizontal edge, vertical edge, and diagonal edge blocks are chosen and programmed into the ECAM chip. The codewords used were not particularly optimized, and are shown in Figure 3.

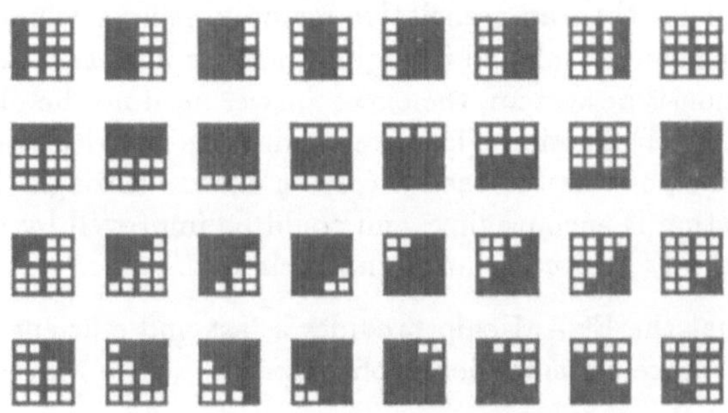

Figure 3: Vector quantization templates

The quantization process is performed by associatively recalling the nearest codeword to an input 4 × 4 pixel block. The index of the codeword is then transmitted, achieving a compression ratio of 16/5. The reconstructed image is formed by replacing each block by its corresponding codeword. However, there are times when the output pattern of the ECAM chip is not a codeword (remember there are many spurious stable states in any associative memory), in which case an all white block is generated instead. Figure 4 shows the effect of vector quantization on a typical image. It can be seen that the quality of the reconstructed binary images is somewhat poorer than the originals, yet this is the price paid for reduced transmission or storage bandwidth.

The speed at which the ECAM chip can vector-quantize these binary images is of interest. We find experimentally that the ECAM chip is capable of doing one associative recall operation in less than 3 μs. This projects to approximately 49 ms for a 512 $\times$ 512 binary image, or more than 20 images per second. For larger images, more ECAM chips can work together since each block is quantized independently.

Figure 4: Comparison of (right) the original girl image and (left) the reconstructed girl image after vector quantization by the ECAM chip

Conclusions

The ECAM chip is more robust than an associative memory using a winner-take-all function. In the latter system, if the decision of the winner-take-all is wrong, then the final answer will be erroneous. For the ECAM used as a feedback evolution-type system, the exact answer need not be obtained at once. Since even if several out of the N binary (+1 or -1) decisions (for determining the polarities of N components of the next state pattern) are wrong, it is still possible to reach a correct answer through iteration. The performance of the ECAM for vector quantization is encouraging, and could be improved by template optimization, and the use of analog correlation at the ECAM correlation input circuits.

In conclusion, we believe that the ECAM chip provides a fast and efficient way for solving associative recall problems and minimum distance classification problems.

References

[1] T. D. Chiueh and R.M. Goodman, "VLSI Implementation of a High-Capacity Neural Associative Memory," in *Advances in Neural Information Processing Systems II*, Ed. David S. Touretzky, Morgan Kaufmann, 1990.

[2] T. D. Chiueh and R. M. Goodman, "High Capacity Exponential Associative Memory," in *Proc. of IEEE ICNN*, Vol. I, pp. 153–160, 1988.

[3] C. A. Mead, *Andlog VLSI and Neural Systems.* Reading, MA : Addison-Wesley, 1989.

[4] R. M. Gray, "Vector Quantization," *IEEE ASSP Magazine,* Vol. 1, pp. 4-29, 1984.

IMPLEMENTING SEMANTIC NETWORKS IN AN ELECTRONIC NEURAL NETWORK

†Yuzo HIRAI and ‡Qing MA
†Institute of Information Sciences and Electronics, University of Tsukuba
1-1-1 Tennodai, Tsukuba, Ibaraki, 305 JAPAN
‡Doctoral Program in Engineering, University of Tsukuba
1-1-1 Tennodai, Tsukuba, Ibaraki, 305 JAPAN

Abstract

A semantic network based on an *is-a* hierarchy with exceptions has been implemented in an electronic neural network system which we have developed. The system consists of a control computer and a neural network hardware, which is composed of digital neuro-chips and contains fifty-four neurons fully interconnected via both excitatory and inhibitory six bit modifiable synaptic connections. The output of each neuron is encoded by impulse frequency, as are biological neurons. The behaviour of each neuron follows a nonlinear first order differential equation. By simply connecting the neuro-chips, neural networks of any size can be constructed.

The structure of the semantic network implemented in the neural network system is assembled within the framework of HASP, which is an associative memory model proposed by one of the authors. Each node is represented by a neuron, and each link is represented by a synaptic connection between neurons. *Is-a* link is implicitly represented by connecting a subordinate node to its parent node. Since the activity of a subordinate neuron spontaneously spreads to the superordinate ones, traversals in the *is-a* hierarchy can be achieved automatically. Each label on a property link is represented by a neuron, and the label can be specified by activating the corresponding neuron. By property inheritance given by the *is-a* hierarchy with exceptions, the semantic network can answer *what* type and *yes-no* type queries in a few milliseconds.

1. Introduction

The implementation of semantic networks in parallel hardware or artificial neural networks has been proposed by several authors [1][2], but to our knowledge there has been no actual implementation. In this paper we describe the implementation of a semantic network in the electronic neural network system which we have developed [3].

2. Semantic Network

Figure 1 shows a small part of a somewhat artificial semantic network. Ovals represent concepts, labels on links represent properties, and rectangles at the arrowheads represent property values. The central idea of the semantic network is the property inheritance by which the property of a superordinate concept is inherited by its subordinates via *is-a* links. For example, the question *"What is the nutritive source of John?"* can be answered by following the *is-a* link emanating from John to person and from person to mammal. In Falman's NETL, this function can be performed by the marker passing mechanism [1]. Our network is similar to Anderson's spreading activation model [4], but the underlying structure differs from his model.

There are two types of queries which can be answered by the semantic network shown in the figure: *what* queries and *yes-no* queries. An example of a *what* query is *"What is the colour of Bill?"* In the figure we can see that *"The colour of Bill is pink"* which is an exception

to the general colouring of people, which may be white, yellow or black. The treatment of exceptions in semantic networks has been addressed in recent studies [5]-[8]. In our model, the colour *pink* can be retrieved by providing *Bill* and *colour* simultaneously. The retrieval is performed by the parallel *set intersection* operation, which is the central feature of the associative memory model HASP proposed by one of the authors [9]. By activating *Bill*, all property values directly assciated with him will be activated. By activating *colour*, all *colours* will be activated. *Pink* is in the intersection of the sets whose members are activated by *Bill* and *colour*, respectively. However, since *Bill* will activate *person* node *person* will activate *white*, *yellow* and *black*, all colour values will eventually be activated. In our network it is assumed that all property value nodes are mutually inhibited. And since it takes time to spread activations, *pink*, which is activated at first, will win the mutual competition over the other colour values. Therefore the exception will always be retrieved by suppressing the others which are associated with the superordinate nodes.
An example of a *yes-no* query is *"Is John clever?"* By activating *John* and *clever* simultaneously, the *clever* node receives further activation from the *John* node. Therefore if the activation of the property value node exceeds the direct activation level, the answer must be yes. If the *clever* node does not receive additional activation, as in the case of *"Is Bill clever?"*, the answer must be no, which is guaranteed by the closed world assumption [6]. The semantic network we have implemented in hardware can perform these two kinds of query-answering by exploiting the parallel processing capabilities of neural networks.

3. Implementing Semantic Network in a Neural Network System

The semantic network described in the previous section has been implemented in a previously described electronic neural network system [3]. The system consists of neural network hardware and a control computer. The hardware consists of fifty-four neurons fully interconnected through both excitatory and inhibitory synaptic connections. It was built with neuro-chips relying sololy on digital circuitry. A chip contains six neurons and each neuron has seven excitatory and seven inhibitory synaptic connections, so that there are eighty-four synapses in a single chip. By simply connecting these chips, neural networks of any size can easily be made. Each neuron outputs an impulse train whose frequency is proportional to the positive internal potential. When the internal potential is negative, the output frequency becomes zero, so that nonlinear analog threshold function is realized. The behaviour of each neuron follows a nonlinear first order differential equation.
The synapse transforms the input impulse frequency to a frequency proportional to the synaptic weight. This transformation is carried out by a 6 bit rate multiplier. The weight can be set by the control computer. The output impulses from excitatory synapses are summed by OR gates and are fed to the up input of the 12 bit up/down counter in the cell body circuit. The outputs from inhibitory synapses are fed to the down input of the counter. The contents of the up/down counter represent the internal potential of a neuron, which can be monitored and set by the control computer.
The structure of the implemented semantic network is shown in Figure 2. There are three network compoments: an *Is-a* network, a *porperty-value* network and a *winner-take-all* network. Each neuron in the *is-a* network, denoted by AN(y), represents a concept node. An *is-a* relation is stored by strengthening the positive feedback connection from a subordinate node to the parent node, denoted by $W_{SA}(y_2,y_1)$, from zero to some positive constant. By supplying input to one of the nodes, *Bill*, for example, all the superordinate nodes, *person* and *animal*, are activated through the feedback. Each neuron in *the property-value* network, denoted by S(n), represents a property value. The relationships between a concept, *Bill*, for example, and its property values, *pink* and *small*, are stored by strengthening the connections denoted by $W_{SN}(n,y)$ from zero to a positive constant W_{SN}, so that a set of property values

can be specified by the concept. The relationships between a property, *colour*, for example, and its values, *pink*, *white*, *yellow* and *black*, are stored by strengthening the connections denoted by W_{SP}(n,m) from zero to a positive constant W_{SP}, so that a set of property values can be specified by the property input. The hatched neuron denoted by I_{KP} is inhibitory and has fixed excitatory connections from property inputs, denoted by W_K, and it inhibits all *property-value* neurons. By making $W_K=W_{SP}=W_{SN}$, property values in the intersection of the sets, which is specified by the conceptual inputs and the property input, can be retrieved.
To answer the *what* queries, *"What is the colour of Bill?"*, for example, activate *Bill* and provide *colour* input, then the value *pink* will be activated at first, and then *white*, *yellow* and *black* will become activated as the activation propagates up through the *is-a* hierarchy. The *winner-take-all* network selects the one which is activated at first, so that *pink* will be retrieved as the answer.
To answer the *yes-no* queries, *"Is John clever?"*, for example, activate *John* via a T_N(y) input and activate *clever* via a T_P(n) input, then the *clever* neuron will become active if it receives activation from *John* or its superordinate nodes. But if it does not receive activation from concept nodes, it will not become active because of the inhibition through I_{KV}, as in the case of I_{KP} inserted in the property input line.
To show the performance of the neural network system running the semantic network, some video demonstrations will be presented at the conference.

4. Conclusion

A semantic network has been implemented in a parallel neural network hardware system. The network can answer two types of queries, *what* queries and *yes-no* queries. We have more recently completed implementation of another type of semantic network which can solve *recognition* problems, in which a concept satisfying a set of input *property-values* is retrieved by a closest-match mechanism.

References

[1] Fahlman,S.E.:Representing implicit knowledge. in Parallel Models of Associative Memory. Hinton,G.E. and Anderson,J.A. eds. Lawrence Erlbaum Associates, Inc. 1981
[2] Hinton,G.E.:Implementing semantic networks in parallel hardware. ibid.
[3] Hirai,Y. et al.:A digital neuro-chip with unlimited connectability for large scale neural networks. Proceedings of IJCNN'89, Vol.2, 163-169, 1989
[4] Anderson,J.R.:The Architecture of Cognition. Harvard University Press, 1983
[5] Winograd,T.:Extended inference modes in reasoning by computer systems. Artificial Intelligence, Vol.13, No.1,2, 5-26, 1980
[6] Reiter,R.:A logic for default reasoning. ibid, 81-132, 1980
[7] Touretzky,D.S.:The Mathematics of Inheritance Systems. Pitman, 1986
[8] Etherington,D.W.:Reasoning with Incomplete Information. Pitman, 1988
[9] Hirai,Y.:A model of Human Associative Processor (HASP), IEEE Trans. on SMC, Vol.SMC-13, No.5, 851-857, 1983

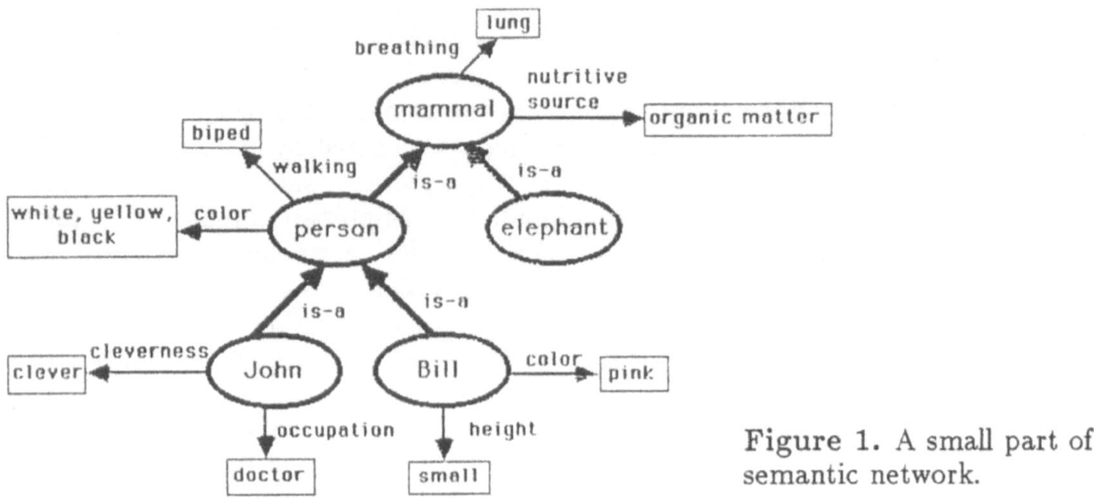

Figure 1. A small part of semantic network.

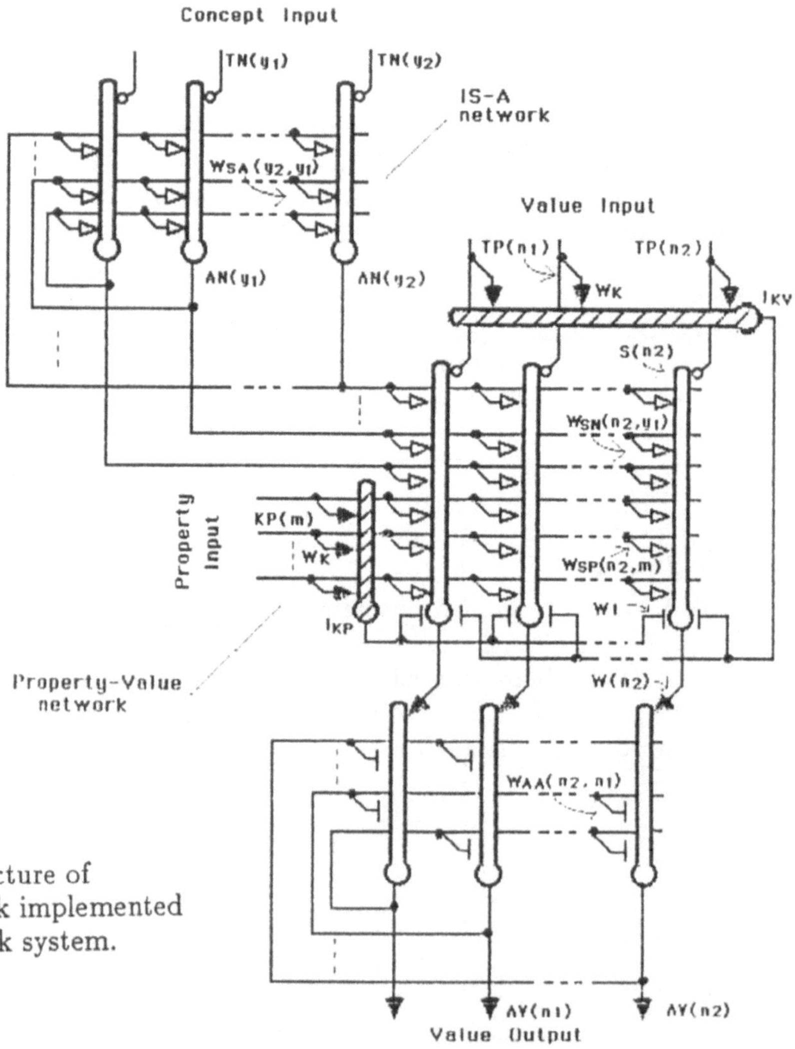

Figure 2. The structure of the semantic network implemented in the neural network system.

KOHONEN NETWORKS ON TRANSPUTERS: IMPLEMENTATION AND ANIMATION

H.P.Siemon, A.Ultsch
Department of Computer Science
University of Dortmund
D-4600 Dortmund
PO Box 500 500
Federal Republic of Germany

Abstract

Self organizing feature maps have been introduced in 1982 by the Finish phycicist Kohonen [KOHO 82]. Since then they have been used for a variety of applications. Implementations of the algorithm on conventional hardware are rather slow for big problems; direct VLSI or special purpose hardware implementations are rather expensive.

In this paper we describe an implementation of the algorithm on a network of transputer. The network makes efficient use of the algorithm´s inherent parallelism. The computational power of the net can easily be extended to almost any desired range by adding more processors; the ratio of price to performance is very good as only off-the-shelf components are used. The implementation allows flexible reconfiguration and adaption to all network and vector sizes.

The network offers a speed of up to 2.7 Mega CUPS. This allows to train even fairly big nets of more than 10,000 units within less than 30 minutes. These good performance characteristics give the possibility to animate the training process in real time. The resulting pictures are not only aesthetic in their own right but give some insight into the algorithm´s behaviour at the same time.

1. Introduction

We assume that the reader is familiar with the theoretical background of the Kohonen network algorithm and we will just review (informally) the essential ideas.

Given a set of input vectors **I**= { $x_1, x_2, ..., x_m : x_i \in R^n$} and a (usually) rectangular, two dimensional array **U** of units, $u_{ij} \in R^n$. We define a distance **d**(x_i, u_{ij}); in our case **d** is the Euclidean distance. For every input vector x_i we can determine the unit u_{ij} which is next to it. We than change all units within a given environment of size **e** so that they become more ´similar´ to x_i (one training step). The learning stops after **T** steps. The size of **e** decreases in a linear way when **T** becomes bigger.

From the above we can see that there are three basic operations: (1) calculate the Euclidean distance of a given vector to all units in the array **U**; (2) determine the global minimum distance; (3) update all units within the current change environment. To get a rough estimate of the total number of operations needed for one training step we make the following simplifying assumptions : the number of units is N^2; additions, subtractions and comparisons take the same time, multiplications take twice as long as additions; every vector has a dimension of **n**; the average size of **e** is $^1/_4$ **N**. We then have for one single training step:

$$O_{train} = (n\,N^2)_{add} + (n\,N^2)_{sub} + (n\,N^2)_{mult} + (N^2)_{comp} + (0.25\,n\,N^2)_{add} + (0.25\,n\,N^2)_{sub} + (0.25\,n\,N^2)_{mult} \approx (5\,n\,N^2)_{ops}$$

Even for very moderate sizes of **N** (100), **n** (10) and **T** (50,000) the value for O_{train} will be rather high: $(25 * 10^9)_{ops}$.

This is one of the reasons why most software implementations only work with very small networks or very small vector dimensions. To overcome these limitation VLSI implementations have been proposed which promise very high performance rates [MAGO 87].

2. Implementation

The hardware implementations proposed make use of the very high degree of parallelism which is inherent to the algorithm. This parallelism could also be exploited if one used a general purpose multiprocessor MIMD system. We have access to a transputer network (2 Parsytec Multicluster, 17 and 16 T800 transputers, 1 MB local memory each [INMO 89]) and implemented the Kohonen algorithm on it (see Fig. 1) using the **occam** 2 programming language [INMO 88b,JOGO 88].

Transputers can be linked together to form almost any topology; processes have to communicate through messages they send over one of their four links. We used a simple ring network for our implementation. This topology is easy to wire and program, test runs can be performed on sub nets without disturbing any other user and the computational power can be extended by either simply adding more TPs to the same ring or by attaching more rings to the host marked 'Ho' (Fig.1). The communication overhead for this kind of structure is usually less than 5 % of the total run time.

Smaller overheads could be achieved if one used a 2- or 3-tree network, or if the input data vectors could be held directly on the network TPs. The first is possible if we were able to reconfigure the network electronically (which is possible with a special device we have not got yet); the latter can always be done if the amount of input data is small and static.

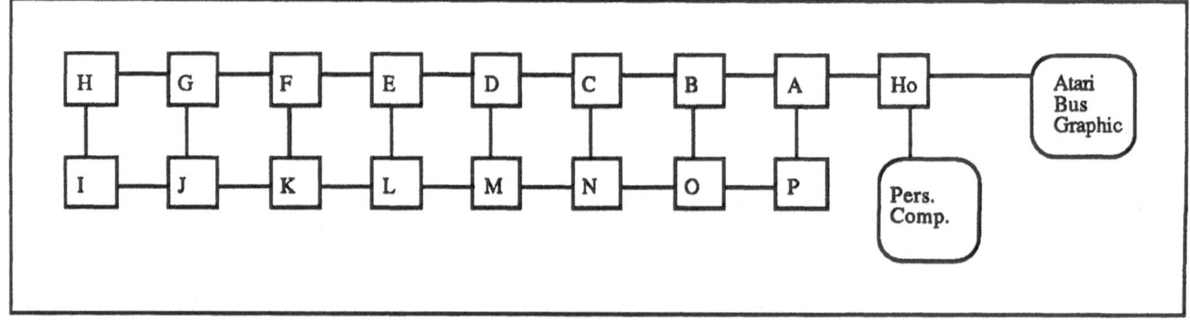

Fig. 1

Every TP hosts two processes which run in parallel: (1) a communication process passing messages through the ring and (2) a work process which does the actual calculation. The host contains a general ring control process and the animation software. Units have to be mapped to TPs in such a way that a maximum degree of parallelism is achieved for all of the 5 operations described below. Therefore mapping is done in an interleaved fashion: u_1, u_{k+1}, u_{2k+1}, ... are mapped to TP_1; u_2, u_{k+2}, ... are mapped to TP_2 and so on (k being the number of TPs).

There are five basic steps which have to be executed: (1) propagate the data vector to all TPs; (2) calculate the local minimum for each TP; (3) find the global minimum distance value; (4) propagate the coordinates of the global minimum value; (5) update units which are within a given distance of it.

Step one and two can be performed in parallel. The host sends data vector x_i to the TP_1. On receipt of x_i the communication process of TP_j sends x_i to its work process first and then to TP_{j+1}. As soon as the work process of TP_j receives the data it can start to calculate the minimum distance value for all units which were mapped to TP_j. So the min values for all k TPs are calculated in parallel. The operation is initiated by the flow of data through the ring and not by any external beat or clock. The feasibility of this data flow model can be shown when one adds more TPs to the ring: doubling the number of TPs cuts the run time to half of the former value.

Step three is a sequential step as all TP local min values have to be gathered and compared to find the *global* minimum (remember: one TP contains only part of all units). Again this step is driven by a token being send round the ring; at the end of its journey it contains (as its data part) the desired value. Step four and five again can be performed similar to steps (1) and (2).

With this mode of operation we ensure a maximum degree of parallellism. For a net of 16 TPs and values N = 128, n = 17, T = 25,000 the training time is 2546 seconds which is equivalent to 2.7 Mega CUPS. The higher the workload per TP (i.e. the number and size of units per TP) the better the performance characteristics. More test run data can be found in [ULSI 89].

3. Animation

Animation of algorithms plays an important role in the understanding of their behaviour and in the detection of otherwise missed features [BROW 88].

We have access to a (fairly) high resolution graphics device and plot package [BOSI 90] which can be linked to any TP via one of the TP's links. The device is an Atari with a purpose built bus adapter. The device is slow compared with other 'real' graphic devices; it allows about 5 complete pictures to be transmitted and displayed per second which is enough for our puposes.

Two sorts of animation display are supported: a distribution plot and a simplified U-matrix display [SIUL 89]. The first plot (see Fig.2) simply shows the current position of each input data vector which is applied to the net (i.e. the coordinates of the unit with min distance to this vector). The display is updated whenever the min position of any vector changes. Fig. 2 shows the initial distribution of vectors for one of our test data sets; see [ULSI 89] for details.

The simplified U-matrix display (Fig. 3 and 4) shows the length of each unit vector as a grey scale value. The length of a vector is defined as $\sum | x_i |$, i= 1..n. The display is updated after every learning epoch, i.e. after one complete application of the input data vectors.

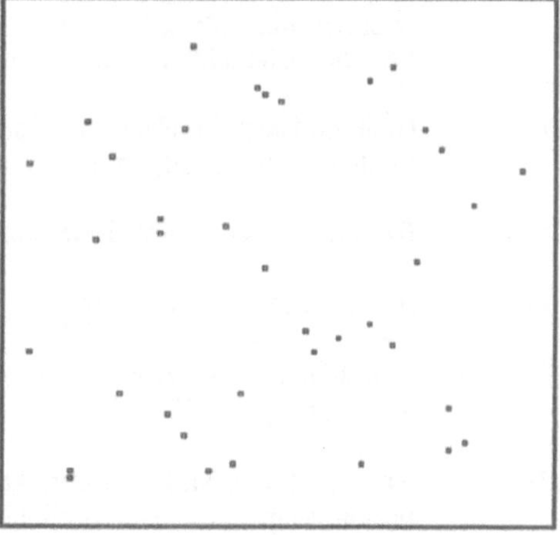

Fig. 2

This kind of animation is rather expensive as it involves a number of operations in the order of $(nN^2)_{ops}$.

The simplified U-matrix display gives a first indication of data clusters [DETR 85]. Fig. 3 and 4 show the displays for blood data samples from [DETR 85]. The first figure shows the initial state of the net, nothing has been learned yet. Fig. 4 shows the net after 25,000 training steps; the clustering obtained is comparable to [DETR 85]. Vectors which are of the same kind are situated in regions which have the same height. So a change in grey value indicates a change in the kind of data, i.e. indicates the clustering of the input data.The expressive power is clearly not as high as that of the U-matrix method [ULSI 90] but this sort of display is much easier to obtain.

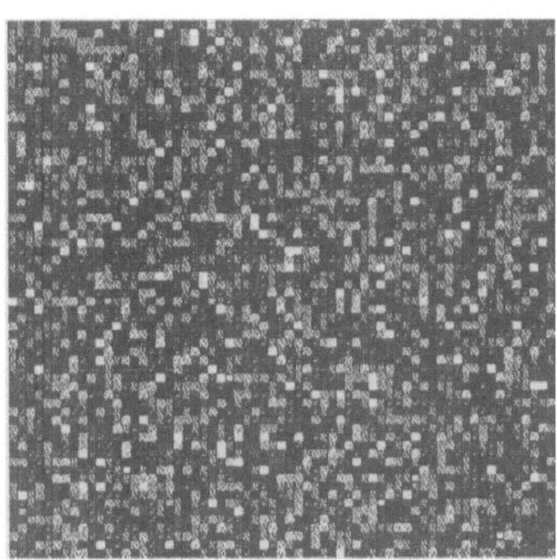

Fig. 3

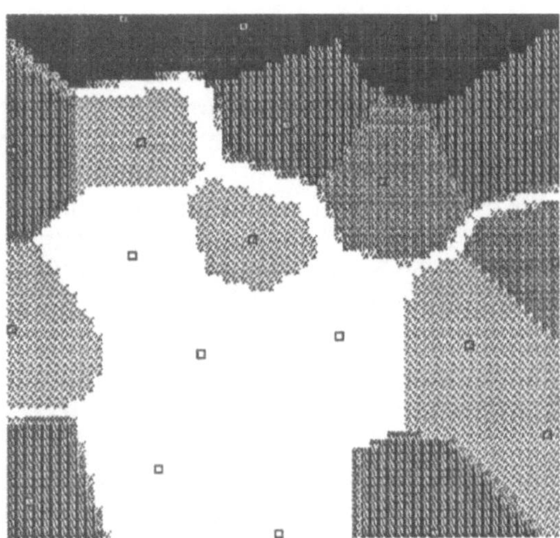

Fig. 4

[BROW 88] Marc H. Brown, Algorithm Animation, MIT Press, 1988

[BOSI 90] Max Boehm, H.Peter Siemon, A Poor Man´s Graphic Device for the Transputer, Forschungsberichte Infomatik Nr. 330, Universität Dortmund, 1990

[DETR 85] Guntram Deichsel, Hans Joachim Trampisch, Clusteranalyse und Diskriminanzanalyse, Gustav Fischer Verlag, Stuttgart, New York 1985

[INMO 88a] INMOS Ltd, Transputer Reference Manual, Prentice Hall International, 1988

[INMO 88b] INMOS Ltd, occam 2 Reference Manual, Prentice Hall International, 1988

[KOHO 82] T. Kohonen, Self Organized Formation of Topologicaly Correct Feature Maps, Biol. Cybern. 43, 1982, pp. 59-69

[MAGO 87] Karl M. Marks, Karl F. Goser, Analysis of VLSI Process Data Based on Self-organizing Feature Maps, Proc. Neuro-Nimes 88, pp.337-348

[ULSI 89] Alfred Ultsch, H.Peter Siemon, Exploratory Data Analysis: Using Kohonen Networks on Transputers, Forschungsberichte Infomatik Nr. 329, Universität Dortmund, 1989

[ULSI 90] Alfred Ultsch, H.Peter Siemon, Kohonen´s Self Organizing Feature maps for Exploratory Data Analysis, submitted for publication INNC 1990

A PARALLEL BOLTZMANN MACHINE SIMULATOR FOR DISTRIBUTED MEMORY MULTIPROCESSOR SYSTEMS

A. Ferscha
Institut für Statistik und Informatik, Universität Wien
Lenaugasse 2/8, A-1080 Wien

Abstract

A Boltzmann Machine (BM) is set of probabilistic binary processors (*units*) connected to each other by a set of bidirectional connections of variable strength. Each processor is either in state 1 ('*on*'), expressing the acceptance of some domain specific hypotheses, or in state 0 ('*off*'), signaling rejection of that hypothesis. The amount of confirmity of all the underlying hypotheses is expressed in terms of the activation vector (set of all unit states), and every unit tries to change state in order to achieve maximum overall confirmity.

An asynchronous parallel simulation model for BM's allowing units to change state concurrently and independently of each other is proposed, introducing a new source of exploitable parallelism compared to approaches where a single unit is chosen at random to calculate the next state of the BM. For the production phase of the BM, i.e. calculating an activation vector such that overall conformity reaches maximum, a distributed cooling algorithm for execution on a loosely coupled multiprocessor system with medium-grained local memory and fixed neighborhood is given. The connection strength matrix is partitioned and equally distributed among the processing elements (PE's), each of them working on a dedicated region of the BM. The interconnection topology of the PE's is chosen so as to facilitate optimum data broadcasting by applying generalized chordal rings, supporting simple, regular, scalable and self-synchronising routing mechanism. For a target hardware with node degree $d = 4$ ideal interconnection topologies are given by example.

1 An asynchronous Boltzmann Machine

Formally a Boltzmann Machine is a triple $BM = \{\mathcal{G}, \mathcal{W}, s\}$ where $\mathcal{G} = (U, E)$ is an undirected graph with N units forming the set of nodes $U = \{u_0, u_1, \ldots, u_{N-1}\}$, and $E = \bigcup_{i,j \in \{0,1,\ldots,N-1\}}(u_i, u_j)$ is the set of all *unit connections*. $\mathcal{W}$ is a $N \times N$ weight matrix with entries $w_{i,j} \in I\!R$ representing the *connection strenghts* of edges (u_i, u_j). If there is no connection (u_i, u_j) for some i, j, then $w_{i,j} = 0$. $s = \{a_0, a_1, \ldots, a_{N-1}\} \in S$ is the *state (vector)* of BM charcterized by the *activation values* $a_i \in \{0, 1\}$ of all the units u_i. $S = 2^N$ hence is the set of all possible states of BM. We denote a BM being in state $s \in S$ by BM^s, the activation value of unit u_i in state s by a_i^s.

An edge $e_{i,j} = (u_i, u_j) \in E$ is said to be *active* if $a_i = a_j = 1$. The connection weights $w_{i,j}$ represent a measure of the *desirability* [Aart 88] of $e_{i,j}$ being active: $w_{i,j} \gg 0$ expresses that the units u_i, u_j attract each other to both being 'on', whereas negative weights $w_{i,j}$ state the undesirability of an active edge $e_{i,j}$. The overall *conformity* (consensus [Aart 89]) $C(BM^s)$ of a BM in state s is defined upon the set of active edges:

$$C(BM^s) = \sum_{e_{i,j} \text{is } active} w_{i,j} = \sum_{e_{i,j} \in E} a_i^s a_j^s w_{i,j} \qquad (1)$$

Let $\{s(0), s(1), \ldots s(N-1)\}$ be the set of neighboring states of state $s = \{a_0, a_1, \ldots, a_{N-1}\}$ where $s(i)$ is derived from s by changing the activation value a_i of u_i to the complement of its current activation ($a_i^{s(i)} = 1 - a_i^s$), then the *conformity neighborhood* is given by the vector ($C(BM^{s(0)})$, $C(BM^{s(1)})$, ..., $C(BM^{s(N-1)})$). Given that $C(BM^s)$ is known, then $C(BM^{s(i)})$ can be calculated by

$$C(BM^{s(i)}) = C(BM^s) + (\sum_{j|e_{i,j} \in E} (a_i^{s(i)} - a_i^s)\, a_j^s w_{i,j}). \tag{2}$$

From (2) it is easy to see, that the conformity neighborhood of some state s can be calculated in parallel for every component $C(BM^{s(i)})$ on the basis of only the weights assigned to edges incident to u_i and the states of all units u_j adjacent to u_i (*locality*).

Let $\delta^{s \to s(i)} = (a_i^{s(i)} - a_i^s) \sum_{j|e_{i,j} \in E} a_j^s w_{i,j}$ denote the difference in overall conformity when changing from state s to $s(i)$ and the value of the binary variable b_i be 1 if $\delta^{s \to s(i)} > 0$ (0 otherwise), respectively. In order to maximize overall conformity (1) every unit u_i has to behave so as to make its utmost contribution to C, i.e. u_i changes state if $b_i = 1$ and remains in its current state otherwise. Hence the vector $B^s = (b_0, b_1, \ldots, b_{N-1})$ gives the optimum state changes for all units. Note that again every component of B^s can be calculated locally in parallel. In traditional definitions of the BM one unit u_i is chosen at random to change its state ([Ackl 88], [Aart 88]) so as to bring the machine into state $B^{s(i)}$. We allow all units to change states according to B^s concurrently, being aware of the possible error that might appear (temporarilly) due to state changes of neighboring units. On the other hand this approach is able to improve runtime performance of finding the maximum conformity by introducing a kind of parallelism exploitable for distributed memory parallel machines.

Cooling the Boltzmann Machine For maximization of the conformity state changes are randomized using a *cooling algorithm* [Aart 86] based on simulated annealing [Kirk 83]. In our network every unit u_i determines its next state $\overline{a_i}$ at random time, asynchronously and independently of all other units according to the probabilistic decision rule

$$\overline{a_i} = \begin{cases} 1 & \text{with probability} \quad \frac{1}{1+e^{-\frac{\delta^{s \to s(i)}}{T}}} \\ 0 & \text{with probability} \quad 1 - \frac{1}{1+e^{-\frac{\delta^{s \to s(i)}}{T}}} \end{cases} \tag{3}$$

where $T \in I\!R^+$ is a predetermined global parameter usually called the *temperature*. Maximization of (1) starts with a randomly chosen state s and initially 'high' temperature T, and performs by iteratively making *local* decisions based on (3) and simultaneously decreasing T (*cooling*). The stochastic decision scheme in (3) is devised to also allow locally wrong decisions ('down-hill climbing') in order to atain global advantage later, i.e. the probabilities for locally wrong choices depend on the ratio of $\frac{\delta^{s \to s(i)}}{T}$ and decrease with $T \to 0$. This mechanism allows to avoid 'local maximum traps' if the schedule for decreasing T, along with the starting temperature and the stop criterion (*cooling scheme*) is appropriate. Finding such a scheme that guarantees asymptotic convergence to global optimum in short time is problem dependent and cannot be treated in general [Aart 89], [Aart 86]. Especially for the proposed asynchronous update strategy the asymptotic convergence properties of the simulated annealing algorithm can no longer be assumed to hold, but as by rule (3) the probabilities for making erroneously state changes decreases (with T) downto 0, one can neglect this source of error.

2 Implementation

Simulation Model Assume a loosely coupled multiprocessor system with $M \ll N$ PE's $P_0, \ldots, P_{M-1}$, comprising local processing facilities, local memory and fixed neighborhood (consider for example transputer-based systems). Because of $M \ll N$ U has to be partioned into disjoint sets of equal size, such that (approx.) $\frac{N}{M}$ units are simulated on a single PE by time-multiplexing their activies. Interconnections among units are multiplexed over physical hardware links because of the fixed (node) degree of PE's. Let $U = \bigcup_{i=0}^{M-1}\{u_{\pi(0*i)}, u_{\pi(1*i)}, \ldots, u_{\pi((\frac{N}{M}-1)*i)}\}$ be such a partition (for simplicity: N *modulo* $M = 0$), where π denotes a permutation over the indices of elements of U. We assign sets of rows of $\mathcal{W}$ to PE's such that P_i holds all connection weigths $w_{k,l}$ with $k \in \{\pi(0*i), \ldots, \pi((\frac{N}{M}-1)*i)\}$, $l \in \{\pi(0), \ldots, \pi(N-1)\}$ to achieve equal distribution of data among all the PE's in the network. It is easy to see, that if the units $\{u_{\pi(0*i)}, u_{\pi(1*i)}, \ldots, u_{\pi((\frac{N}{M}-1)*i)}\}$ are (virtually) simulated on P_i it is sufficient to have this row-block of $\mathcal{W}$ on P_i to calculate the conformity neighborhood of these units, because it contains weights to all neighbors of all these units and because of the locality of (2). Furtheron the binary vector s defining the current state of BM is assigned to every P_i. Note that it is not necessary to know $C(BM^s)$ in every P_i, as the decision rule for state changes in (3) requires only $\delta^{s \to s(i)}$ and the current T.

Simulation of cooling to find maximum overall conformity (1) is then performed as follows: In a first step a host processor distributes a random state vector s along with a cooling scheme (starting/stopping temperature T_0 and T_{Stop}, temperature decrements $\eta(T)$ and the number of iterations at proper temperature $\mu(T)$). Then every P_i behaves according to the following procedure: Given state s then first choose a random binary vector $r = (r_0, r_1, \ldots r_{\frac{N}{M}-1})$, determine $\delta^{s \to s(\pi(i*j))}$ and apply (3) for every component $r_j = 1$ in parallel. (This parallelism often cannot be fully exploited, but is realized in true parallelism for at least M units on the M PE's, and in pseudo parallelism for all other units.) According to the probabilistic decision in (3) the corresponding components in B^s are set and s is updated by $s = s$ XOR B^s locally in every P_i. This procedure is iterated in every P_i $\mu(T)$ times, influencing only components a_j, $j \in \{\pi(0*i), \ldots, \pi((\frac{N}{M}-1)*i)\}$ of s. Finally this *section* j of s is broadcasted to all other PE's by concurrently receiving 'local' sections of s from all other PE's. We call $\mu(T)$ updates of s along with the broadcast operation one *step*. After each step T is lowered by $\eta(T)$; steps are repeated until the stop criterion holds.

```
procedure asynchronous-parallel-cooling
 par for all P_i
     while T > 0
          for i = 1 to μ(T)
               set binary random vector R
               par for all r_j = 1
                    evaluate δ^{s→s(π(i*j))}
                    set b_π(i*j) in B^s according to (3) .
          s = s XOR B^s
          par for all j ≠ i
               broadcast (a_π(0*i), ..., a_π((N/M-1)*i)) to P_j
               collect (a_π(0*j), ..., a_π((N/M-1)*j)) from all P_j
          T = T − η(T)
```

Processor Topology For the simulation of the broadcast-operation we have to consider several criteria when choosing the interconnection topology. As data has to be broadcasted from every PE very frequently, very *short distance* is desirable between any two PE's. The topology should be *regular* in the sense that routing algorithms can easily be derived (supporting straightforward or even automated implementation), and *scalable* in that routing strategies could be adopted to larger M with no or minor changes. Buffering of message packets in PE's (store-and-forward routing) should be avoided to prevent the burden of additional memory and processing requirements. The broadcast mechanism has to self-synchronising among all PE's (and deadlock-free across all PE's), as inhomogeneous execution times for single steps in the PE's are to be expected according to the random selection of units for simultaneous activation updates. Finally for our dedicated target hardware (transputer-based) a maximum node degree $d = 4$ has to be considered.

To cope with all these requirements we propose processor topologies based on *generalized chordal rings* (CR) [Berm 86]. A CR connects M nodes (PE's) numbered $0, 1, \ldots, (M-1)$ in a way that if node i is connected to node j then node $i \oplus q$ has to be connected to $j \oplus q$ where q is a divisor of M and $\oplus$ is the addition *modulo* M. For given M there are of course several possibilities of

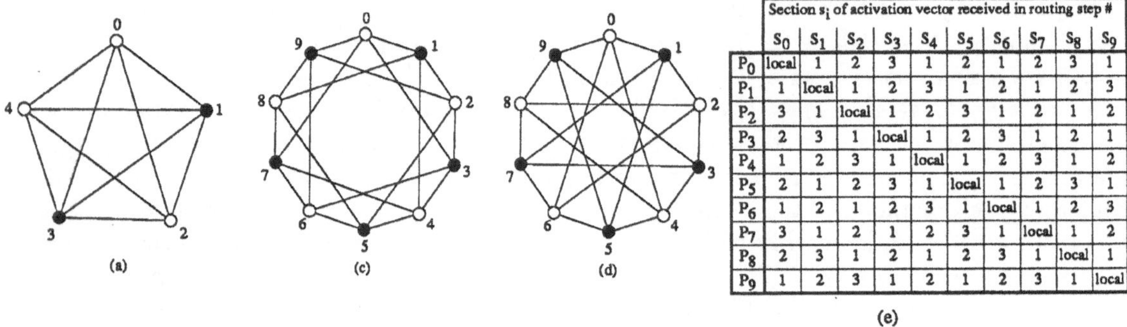

	Section s_i of activation vector received in routing step #									
	s_0	s_1	s_2	s_3	s_4	s_5	s_6	s_7	s_8	s_9
P_0	local	1	2	3	1	2	1	2	3	1
P_1	1	local	1	2	3	1	2	1	2	3
P_2	3	1	local	1	2	3	1	2	1	2
P_3	2	3	1	local	1	2	3	1	2	1
P_4	1	2	3	1	local	1	2	3	1	2
P_5	2	1	2	3	1	local	1	2	3	1
P_6	1	2	1	2	3	1	local	1	2	3
P_7	3	1	2	1	2	3	1	local	1	2
P_8	2	3	1	2	1	2	3	1	local	1
P_9	1	2	3	1	2	1	2	3	1	local

(e)

Figure1: Chordal ring topologies with routing mechanisms.

constructing CR's, but we are only interested in those ones with node degree $d = 4$ and 'small' *diameter* Δ (maximum of minimum distances between any pair of nodes), but at the same time we want to have regularity to support efficient routing. From the Moore-bound we know that a graph with degree d and diameter Δ can have at most $2\Delta + 1$ nodes for $d = 2$ and at most $\frac{d(d-1)^{\Delta}-2}{d-2}$ for $d > 2$. For $d = 4$ and desired $\Delta = 1$ we can use only five PE's and find the ideal interconnection of figure 1 (a). Using this topology the broadcasting and collection step of the cooling algorithm can be done in a single message exchange step, because the transputer provides four bidirectional links able to send and receive (i.e. handling eight message transfers) concurrently. The routing is straightforward: every PE sends its section of the state vector s to all of his four neighbors while simultaneously receiving the four other sections of s necessary to build the new activation vector s'.

A graph with $d = 4$ and $\Delta = 2$ can comprise at most 17 nodes. In [Erdo 80] it has been proven that such a graph cannot have even 16 nodes. In [Berm 86] an optimal ($d = 4, \Delta = 2$)-graph connecting 15 nodes is given, but it is not regular in the CR-sense and would cause sophisticated routing. So we could trie to find a CR with 14 nodes and $d = 4, \Delta = 2$ etc. In fact it is not necessary to have minimum diameter for optimum broadcasting as we wish to illustrate with the two CR's in figure 1 (c) and (d), (c) having $\Delta = 3$ and (d) having $\Delta = 2$. Complete message exchange between all PE's can be performed in (c) as well as in (d) within 3 routing steps: In the first step (CR (d)) every P_i sends its local section of s to all of its neighbors $P_{(i\oplus 1)}$, $P_{(i\ominus 1)}$, $P_{(i\oplus 4)}$ and $P_{(i\ominus 4)}$, by concurrently receiving sections $i \oplus 1$, $i \ominus 1$, $i \oplus 4$ and $i \ominus 4$ of s. In the second step P_i broadcasts section $i \oplus 1$ and in the third step section $i \ominus 1$ of s (both have been received in step 1). (See the table in figure 1 that every P_i has a complete new state vector s after 3 routing steps.) For broadcasting among 10 PE's both topologies are ideal in that no interconnection can be found to support full message exchange in less than 3 steps. This result can easily be extended to larger M (if M is even).

References

[Aart 86] E. H. Aarts, F. de Bont, E. Habers, and P. J. van Laarhoven. "Parallel Implementations of the Statistical Cooling Algorithm". *INTEGRATION, the VLSI journal*, No. 4, pp. 209 – 238, 1986.

[Aart 88] E. H. Aarts and J. H. Korst. "Boltzmann Machines and Their Applications". In: J. W. de Bakker, A. J. Nijman, and P. C. Treleaven, Eds., *PARLE. Parallel Architectures and Languages Europe, Vol I.: Parallel Architectures. Eindhoven, June 15-19, 1987.*, pp. 34 – 50, Springer Verlag, 1988.

[Aart 89] E. H. Aarts and J. H. Korst. "Combinatorial Optimization on a Boltzmann Machine". *Journal of Parallel and Distributed Computing*, Vol. 6, No. 2, pp. 331 – 357, Apr. 1989.

[Ackl 88] D. H. Ackley, G. E. Hinton, and T. J. Sejnowski. *A Learning Algorithm for Boltzmann Machines*, Chap. 10, pp. 285 – 307. Ablex Publ., 1988.

[Berm 86] J.-C. Bermond, C. Delorme, and J.-J. Quisquater. "Strategies for Interconnection Networks: Some Methods from Graph Theory". *Journal of Parallel and Distributed Computing*, Vol. 3, No. 4, pp. 433–449, Dec. 1986.

[Erdo 80] P. Erdos, S. Fajtlowicz, and A. J. Hoffmann. "Maximum degree graphs of diameter 2". *Networks*, Vol. 10, pp. 87–90, 1980.

[Kirk 83] S. Kirkpatrick, C. D. Gelatt, Jr., and M. P. Vecchi. "Optimization by Simulated Annealing". *Science*, Vol. 220, No. 4598, pp. 671–680, May 1983.

Optical Character Recognition and Neural-Net Chips.

Y. Le Cun, L. D. Jackel, H. P. Graf, B. Boser, J. S. Denker,
I. Guyon, D. Henderson, R. E. Howard, W. Hubbard, and S. A. Solla
AT&T Bell Laboratories, Holmdel, NJ 07733, USA

Abstract

Neural Network research has always interested hardware designers, theoreticians, and application engineers. But until recently, the common ground between these groups was limited: the neural-net chips were too small to implement any full-size application, and the algorithms were too complicated (or the applications not interesting enough) to be implemented on a chip. The merging of these efforts is now made possible by the simultaneous emergence of powerful chips and successful, real-world applications of neural networks. Here, we discuss how the compute-intensive part of a handwritten digit recognizer, based on a highly structured backpropagation network, can be implemented on a general purpose neural-network chip containing 32k binary synapses. Using techniques based on the second-order properties of the error function, we show that very little accuracy on the weights and states is required in the first layers of the network. Interestingly, the best digit-recognition network is also the easiest to implement on a chip.

1 Introduction

An ongoing problem faced by neural-net hardware designers is that with currently available technology, it is impossible to design versatile chips, both general enough to accommodate many architectures, and large enough to handle real-world applications. Most neural-net chips designed so far assume fully-connected layers of neurons with as many neurons and connections as will fit in the system. This is clearly of limited value as the number of connections grows quadratically with the number of neurons in the layers. Even in a small image-recognition problem, where the input field is around 16x16, a single fully-connected layer of processing easily requires 100,000 connections. The largest neural-net chip made so far [2] has 32,000 1-bit connections, so it apparently falls short of the needs. Moreover, single-bit weights are likely to be insufficient, and several 1-bit weights may have to be combined to provide the required analog depth.

The prospects for the near-term availability of really useful hardware are not as bleak as the previous paragraph might suggest. Several investigations have shown that, on perception tasks, locally-connected networks with replicated weights dramatically improve the classification accuracy [3], and result in practical solutions to real-world applications [4]. These network architectures also have the tremendous advantage of being efficiently implementable on time-multiplexed hardware, where the same processing element is used in sequence for several parts of the network. We will describe a network architecture for handwritten character recognition, and then discuss how it can be partially implemented on our new (and fully tested) neural-net chip.

2 Handwritten Character Recognition

We have studied the task of identifying handwritten characters from a pixel image. In particular, we have developed several systems that recognize handwritten zipcode digits collected by US Postal Service contractors from real envelopes. Figure 1 shows a few typical examples among the 9300 digits in the database after segmentation and size-normalization (7300 were used for training and 2000 for testing). Notice that many of the digits are poorly formed and challenge the recognition skills of humans. Two additional sets of 2500 and 700 machine printed digits were added to the training set and the test set respectively.

Figure 1: Examples of normalized digits from the testing set.

A naive approach to this task would use a large, fully-connected backpropagation net. This approach has severe, fatal deficiencies. Either the number of weights is kept reasonable, and the network is not even able to learn the training set accurately, or it is made big enough, and the excessive number of parameters causes overfitting of the training data with devastating effect on the generalization performance. Even if it were good enough, the huge number of weights would make it too big to fit on current chips. It is, however, possible to build a large back-prop net that has excellent generalization performance, and that can also be efficiently evaluated in a time-multiplexed manner on an available neural-net chip, as described in the next section.

3 A Constrained Backpropagation Network

One alternative to a fully connected net is of course a locally connected net. The design of the connection pattern must be guided by our knowledge about shape recognition. Because there are well-known advantages to performing shape recognition by detecting and combining local features, our network has only *local connections* in all but the last layer. Furthermore, salient features of a distorted character might be displaced slightly from their position in a typical character, or the same feature can appear at different locations in different characters. Therefore a feature detector that is useful on one part of the image, is likely to be useful on other parts of the image as well. This can be implemented by forcing a set of units, located at different places on the image, to have identical weight vectors. The outputs of such a set of neurons constitute a *feature map*. A sequential implementation of this would be to scan the input image with a single neuron that has a local receptive field, and store the states of this neuron at corresponding locations in the feature map. This operation is equivalent to a convolution with a small size kernel, followed by a squashing function. The process can be performed in parallel by implementing a plane of neurons that *share* a single weight vector [1]. That is, units in a feature map are constrained to perform the same operation on different parts of the image.

An interesting side-effect of this *weight sharing* technique, already described in [6], is to reduce greatly the number of free parameters, since a large number of units share the same weights. In addition, this builds a certain level of shift invariance into the system. In practice, multiple feature maps, extracting different features from the same image, are needed.

The idea of local, convolutional feature maps can be applied to subsequent hidden layers as well, to extract features of increasing complexity and abstraction. Interestingly, higher level features require less precise coding of their location. Reduced precision is actually advantageous, since a slight distortion or translation of the input will have reduced effect on the representation. Thus, each feature extraction in our network is followed by an additional layer which performs a local averaging and a subsampling, reducing the resolution of the feature map. This layer introduces a certain level of invariance to distortions and translations. A functional module of our network consists of a layer of shared-weight feature maps followed by an averaging/subsampling layer. This is reminiscent of the Neocognitron architecture [1], with the notable

[1] A better name for a feature map would be "regiment" since all units are controlled by the same kernel.

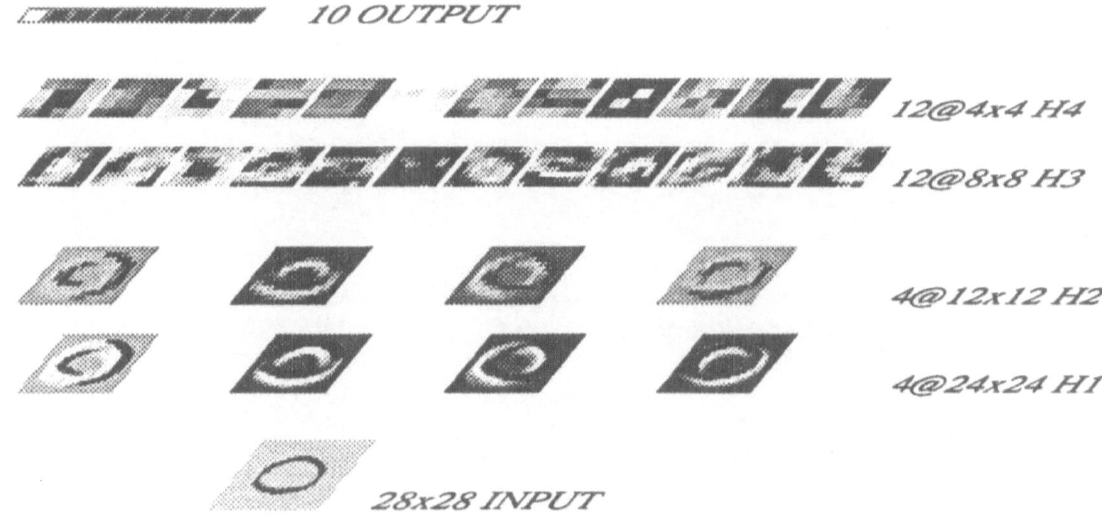

Figure 2: Network Architecture with 5 layers of fully-adaptive connections.

difference that we use backprop (rather than unsupervised learning) which we feel is more appropriate to this sort of classification problem.

The input of the network is a 28x28 image plane in which a 16x16 size-normalized character is centered (the extra units are added to avoid boundary effects). The network has 4 hidden layers labeled H1 to H4. H1 is composed of four 24x24 feature maps, in which each unit receives input from a 5x5 receptive field in the image plane. Units within a feature map are constrained to have identical weight vectors, constituting a convolution kernel. Layer H2 contains four 12x12 maps. Units in an H2 map receive inputs from non-overlapping 2x2 neighborhoods on the corresponding map in H1. All 4 input weights to a unit are equal, and all units in a map have identical weights. H3 and H4 play the same role as H1 and H2 except that each map in H3 receives input from one or two maps in H2. H3 has twelve 8x8 maps, and H4 has twelve 4x4 maps. The output layer has 10 units (one for each digit) and is fully connected to H4. The network has 4635 units, 98442 connections, but only 2578 independent parameters. On the test set, the network correctly classified 90% of the digits, misclassified 1%, and rejected 9% as unclassifiable. Such performance on this kind of difficult data is at the current state-of-the-art. When implemented on a AT&T DSP32C Digital Signal Processor, the network evaluation is performed in 30ms. The overall system throughput is 10 characters per second, with the size normalization taking about 60ms.

4 Network Hardware Requirements

Although the network is too big to fit on any present day chip in parallel, the convolutional nature of the first few layers facilitates time-multiplexing. We can break down the creation of the 12x12 maps in H2 to a serial process in which hardware is time-multiplexed, doing all the processing necessary to produce one or a few elements in the map in one step. Parallel hardware to evaluate these two layers in one step would require about 63,000 connections.

To produce one element in an H2 map, the net must perform four 5x5 convolutions centered on the four adjacent locations that stride a 2x2 region on H1. To do this in one step, the chip must be presented with a 6x6 region of the 28x28 input field, so there must be 36 gray-scale inputs. For each of the 5x5 kernels, four 6x6 kernels are generated, representing the original kernel with 0s appended to a row and column, effectively shifting the original kernel to all possible locations in a 6x6 array. Thus, to process the first layer we need

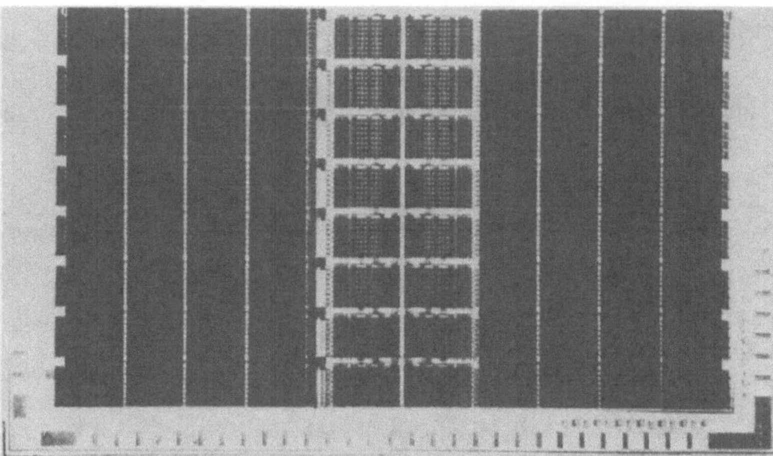

Figure 3: 32,000 synapse neural network chip.

4 (one for each feature) x 4 (number of ways a 5x5 kernel fits in a 6x6 field) = 16 neurons, all sharing the same 36-pixel input vector. We also need 4 (one for each feature) x 4 ((number of ways a 5x5 kernel fits in a 6x6 field) x 36 (each 6x6 kernel) = 16 x 37 = 576 connections, plus a few biases. To compress the four 24x24 maps to four 12x12 maps, we need one more neuron for each feature type, for a total of 20, and four 2x2 kernels, requiring an additional 16 connections for a total of 592. To compute the first two layers outlined above requires stepping a 6x6 region through 24x24 centers over the 28x28 field in 2x2 strides for 144 time steps. If we can accommodate 60 inputs (plus a bias) per neuron, then using similar arguments we can take 6x2 strides and evaluate these layers in 48 time steps (5x5 receptive fields arranged in 2 columns of 6 rows on a 10x6 input field). The first layer needs 2880 connections and 48 neurons (and biases). These units share the same 60-pixel input vector, but only 25 pixels are used by each of them. The second layer needs 12 neurons (and biases) and 48 connections. It is crucial that many units can share the same input vector, since this maximizes the amount computation per data transfer to the chip.

The chip designed in our group [2] has 32,000 reconfigurable binary synapses, partitioned in 256 neurons with 128 binary inputs each. All 256 neurons can be updated in parallel in a single 100ns clock cycle. The linear output of several neurons can be combined with power-of-two coefficients to provide up to 4 bits of analog depth on the weights. Analog depth on the state variables can be obtained by duplicating each input bit and using a "thermometer code". For example, a 64-input neuron with 3-level inputs and 4-bit weights can be implemented with four 128-input binary neurons. The first layer in the above example requires 192 binary neurons (48x4) with their 128 binary inputs (60 2-bit pixels). With that level of parallelism, it and will take 48 processing cycles (4.8 microseconds) to compute the entire layer. Similar analysis can be applied to subsequent layers in the net. Loading the 120-bit input vector through the 16-bit bus takes 8 clock cycles (multiplied by 48 for the whole image). Over 80% of the time would be spent communicating data to and from the chip, during which the processing power would be wasted: the system speed is limited by data exchange rather than by computation. External communications can be reduced drastically by providing sufficient on-chip storage, high speed data paths, and data shuffling units such as shifters, for storing and formating intermediate results. Fortunately, our chip has such capabilities, and the state of intermediate layers do not need to go off-chip. In fact, since most of the silicon area is used for digital storage, the chip should be considered mainly as a storage device with an *extremely fast* communication channel to and from the on-chip processor. The entire network can be computed in less than 2000 clock cycles (200 microseconds), including communications.

Two important questions remain: is the accuracy provided by the chip sufficient for our application? If it is, how can we optimize the precision needed at each layer in order to minimize the amount of resources used on the chip. Experiments with a previous architecture showed that only 2 to 3 bits on the states and 3 to 4 bits on the weights are needed in the feature extraction layers. The last layer needs more accuracy, but since it contains only a small portion of the connections, its computation can be performed on a slower, more accurate processor such as a DSP. A theoretical tool based on second order properties of the error functions was developed [5] to estimate the precision required for the network variables. It is shown that on this task, the precision increases with the layer index, the first layers require just a few bits, whereas the last layer requires around 8 bits.

5 Conclusion

These experiments showed that a practical-size network can be implemented on todays neural-net chips with very high performance. However, one might think that using such a system for recognizing *isolated* characters is a waste of resources, since the preprocessing necessary to segment and size normalize the individual characters is orders-of-magnitude slower than the network update. This conclusion would be premature since the chip performance makes "brute force" solutions possible. Instead of normalizing the size of the character, we can afford to have several identical networks that look at the character at various scales. Instead of segmenting the image, we can create a *fabric* of networks by replicating the original architecture over the entire region of interest, and using it as a character spotter. This is reminiscent of biological visual systems. Preliminary results indicate that replicated networks *can* function as object spotters.

Current silicon technology forces us to use time multiplexing, and therefore requires the use of external storage. Fast storage can only be implemented digitally, ruling out fully analog implementations. With these constraints, highly structured networks with shared weights are efficiently implemented because they reuse the same data and weights several times, thus limiting the amount of communications and storage required.

Acknowledgments

We thank the Postal Service contractors for providing the handwritten digit database, and Henri Baird of AT&T Bell Laboratories for providing the printed digit database. Network simulations were performed with SN2 by Léon-Yves Bottou and Yann Le Cun.

References

[1] K. Fukushima and S. Miyake. Neocognitron: A new algorithm for pattern recognition tolerant of deformations and shifts in position. *Pattern Recognition*, 15:455–469, 1982.

[2] H. P. Graf and D. Henderson. A reconfigurable CMOS neural network. In *ISSCC Dig. Tech. Papers.* IEEE Int. Solid-State Circuits Conference, 1990.

[3] Y. Le Cun. Generalization and network design strategies. In R. Pfeifer, Z. Schreter, F. Fogelman, and L. Steels, editors, *Connectionism in Perspective*, Zurich, Switzerland, 1989. Elsevier.

[4] Y. Le Cun, B. Boser, J. S. Denker, D. Henderson, R. E. Howard, W. Hubbard, and L. D. Jackel. Handwritten digit recognition with a back-propagation network. In David Touretzky, editor, *Neural Information Processing Systems*, volume 2, Denver, 1989, 1990. Morgan Kaufman.

[5] Yann Le Cun, J. S. Denker, and S. Solla. Optimal brain damage. In David Touretzky, editor, *Neural Information Processing Systems*, volume 2, Denver, 1989, 1990. Morgan Kaufman.

[6] D. E. Rumelhart, G. E. Hinton, and R. J. Williams. Learning internal representations by error propagation. In *Parallel distributed processing: Explorations in the microstructure of cognition*, volume I, pages 318–362. Bradford Books, Cambridge, MA, 1986.

Exploiting the Inherent Parallelism of Artificial Neural Networks to Achieve 1300 Million Interconnects per Second

Alexander Singer
Thinking Machines Corporation
245 First St.
Cambridge, MA 02142 USA

An artificial neural network implementation on the Connection Machine is presented which performs 1300 million interconnects per second. This implementation exploits training set parallelism and is discussed within the framework provided by the inherent parallelism of ANNs.

I. Introduction

The artificial neural network (ANN) paradigm is inherently parallel. Furthermore, ANN algorithms are known to scale very poorly [e.g. 1, 9]. These facts reveal the ease with which real–world ANN problems will overwhelm even the fastest serial supercomputer. Though parallel implementations of ANN algorithms will not reduce the amount of computation and communication required by ANNs, they provide an approach which, through a better match between algorithm and implementation, can make large scale ANN problems amenable to practical solution.

The contrast between the parallelism of neural networks and serial machines is in sharp contrast with the natural match between the Connection Machine (CM) and artificial neural networks. The CM is constructed from up to 65536 one–bit processors with local neighborhood communications capabilities and general router ability to send messages along arbitrarily connected networks. The design of the CM embodies precisely the same ideas which inspired the connectionist school: large numbers of relatively weak processing elements cooperating to solve a problem. As a result of this match, a variety of ANN implementations are practical on the CM [8]. This paper will describe the fastest of those implementations.

II. Training Set Parallelism within the ANN Paradigm

The parallelism inherent in the ANN paradigm itself provides a framework within which parallel implementations of ANN algorithms can be organized. In order to make this discussion concrete, let us assume that backpropagation [7] is, in some sense, typical of ANN algorithms, and let us accept it as a standard without passing judgement on its efficacy as an artificial neural network algorithm or on its general desirability as an approach to classification, signal processing, pattern matching, or function approximation problems. This will provide a foundation upon which we can base an analysis of inherent parallel dimensions of ANNs.

There are two obvious degrees of parallelism in backpropagation. First, there is the parallel processing performed by the many nodes of each layer. Second, there is the parallel processing of the many training examples. The latter derives from the fact that backpropagation and similar algorithms provide for the linear combination of the individual contributions made by each training pattern to the adjustment of the network's weights. This linearity means that the patterns can all be processed independently and hence, simultaneously.

A third, but less obvious, parallel aspect in backpropagation stems from the fact that the forward and backward passes of different training patterns can be processed in parallel, i.e. while the delta vector for a training pattern is being passed from the output layer back to a hidden layer, the input layer can be passing the next training pattern through the set of weights linking the input layer to the first hidden layer (see Figure 1). Which, if any, of these degrees of parallelism is exploited in a

Figure 1

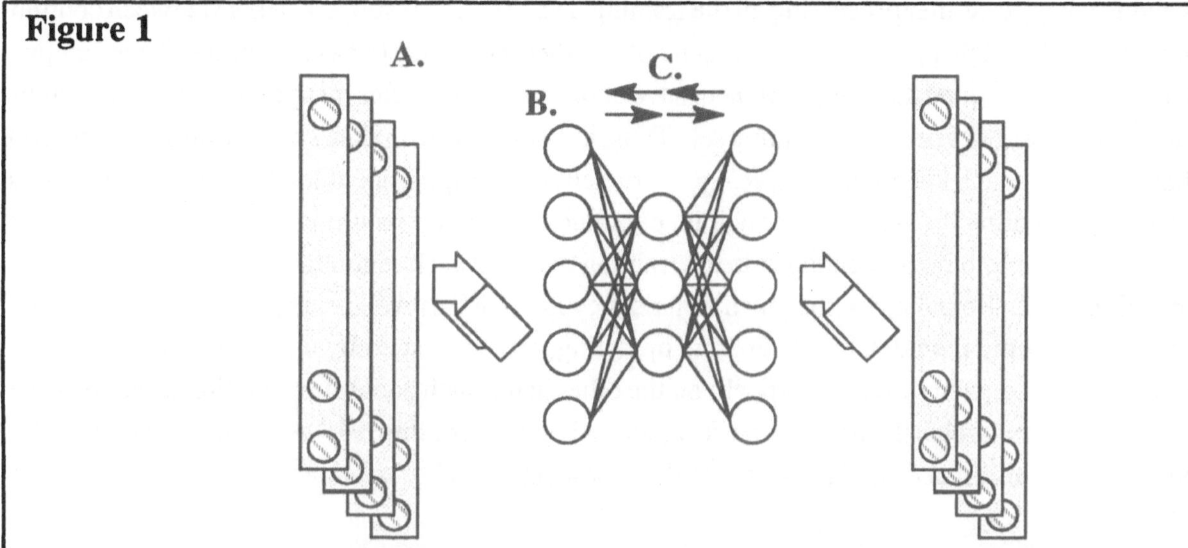

Figure 1. Three degrees of parallelism inherent in the artificial neural network paradigm provide a framework in which parallel implementations of ANNs can be organized: multiple training patterns can be processed in parallel (A); multiple nodes can be processed in parallel (B); and forward and backward passes can be pipelined.

particular implementation will depend on the constraints imposed by the hardware platform and by the constraints of the particular problem being solved.

Of the three parallel dimensions, the simultaneous processing of many training patterns is a particularly inviting option for ANN algorithms like backpropagation since there are commonly many more training patterns than nodes. An implementation which exploits this dimension, however, imposes two constraints on the ANN researcher seeking maximal performance. First, this implementation requires a training set large enough to merit parallel implementation. Second, this implementation requires that weights be updated after all the training patterns being processed in parallel have been seen, so–called "batch" updating.

With respect to the first issue, the degrees of freedom embodied by the explosive number of weights in a network combine with the "curse of dimensionality" to impose enormous data requirements on ANNs. Furthermore, these requirements persist in any network with pretensions of statistical validity. Thus, large training sets appear to be an inherent part of the ANN paradigm. Nonetheless, sparse sampling, advantageous posterior probability distributions, and even empirical claims that networks perform well in spite of discouraging *a priori* statistical analyses force at least some mention of this issue.

The second issue is more controversial, with contention between researchers who update network weights continuously, i.e. after each training pattern is presented and those who update weights only after some subset, or often after the entire set, of training patterns has been presented to the network [2]. To best exploit the training set parallelism mentioned above, batch updating after the entire training set has been processed, i.e. after each epoch, is to be preferred.

In the context of parallel ANN implementations, however, this should not present as divisive an issue. Backpropagation is known only to be performing gradient descent if the update of weights takes place after processing all the training data [7]. As a result, the debate between continuous and batch weight updating must be seen not as a theoretical one but as an empirical one: empirically, it has been found that it is *possible* to save expensive CPU cycles by updating network weights more often than once per entire training set. Thus, it is desirable to update connection strengths more than once per epoch in light of the already overwhelming computational load a serial machine must bear implementing the ANN. In principle, continuous updating would be expected to provide an advantage by exploiting redundancy in the training data; but such redundancy is by no means guaranteed to exist. Both the risk that such redundancy not exist, as well as any possible advantage in convergence time provided by continuous uptdating, are eliminated by a parallel implementation which updates weights after each epoch; on the other hand, such an implementation might be sufficiently fast to outweigh the possible performance advantages of this risk. A training set parallel implementation also makes the theory of gradient descent provably applicable.

III. A Training Set Parallel CM Implementation of Backpropagation

Though training set parallelism has been exploited by several researchers using relatively small systolic arrays of processors [5, 6], it is particularly efficient within an implementation on the CM where the number of processors can match the amounts of training data real–world ANNs require.

A very simple CM implementation can exploit training set parallelism by placing a complete copy of the nodes of the network along with a single training pattern (or single input/output training pair) in each CM processor [3]. In this implementation the connection weights are stored on the front end computer which feeds instructions to the CM. The implementation proceeds by broadcasting each weight to all of the CM processors and having each processor simultaneously accumulate, in turn, the activity of the nodes in the next layer for all training patterns. After the activities for all nodes in a layer have been computed, the sigmoid can be applied to each node in turn, but to all copies of the network simultaneously. The procedure can then continue on to the next layer. The only communication required by this implementation comes when it is time to update the weights. At this point, a set of summations over all processors, one for each interconnection weight, is required in order to collect the adjustments for each link weight. Backpropagation proceeds similarly except that the order in which layers and weights are accessed is reversed.

The implementation is serial in terms of the network weights, and hence any sparsity in the network can be exploited. Furthermore, this serial approach to the network itself allows the implementation to remain immune to the details of the network architecture and makes the implementation particularly easy to program for those accustomed to working on serial machines. These advantages of a serial approach to ANNs complement the training set parallelism.

Though the risk of "creative benchmarking" is great in a field where performance standards are only slowly beginning to evolve, this implementation is extremely fast by any measure. The speed of each of the serial steps combines with the massive savings provided by the simultaneous processing of all training patterns to provide orders of magnitude improvement over typical ANN speeds. Using, for example, interconnects per second, where this is defined as the number of training patterns multiplied by the total number of network connections divided by the time to do a forward pass (as might be inferred from [4]), we find this implementation running at 1300 million IPS. This time includes the entire forward pass through a two layer network, including the time required to retrieve individual weights from the front end computer, multiply all activation values by their corresponding weight, add the product into the activity of the next layer, and apply the sigmoidal nonlinearity; the time includes the manipulation of both sets of weights and both hidden and output layers. Further, this timing was generated for a fully connected network with 128 nodes in each of three layers and 65536 training patterns. The hardware configuration represented by the timing is a 64K CM–2 with 32–bit floating point hardware. For the sake of comparison with other units of measure, the time for full forward and backward passes plus weight update for this implementation is between 3 and 4 times greater than the time for the forward pass alone.

The 1300 million IPS appears as an asymptotal speed for this implementation where networks with fewer nodes are less efficient. The implementation has either constant time behavior or linear time dependence with respect to the number of training patterns, depending on the size of the Connection Machine used. On the other hand, a variety of known improvements in the low–level programming details promise a factor of between five and ten speedup.

IV. Conclusion

The exploitation of training set parallelism allowed by the linear fashion in which many ANN algorithms handle their training data, has led to a Connection Machine implementation of backpropagation which runs at 1300 million interconnects per second. Training set parallelism is only one of three parallel aspects inherent in the ANN paradigm and the implementation described here provides only one example of how those inherent parallel dimensions can be mapped onto parallel hardware. Hardware on which ANNs can be implemented as well as ANN algorithms are continually evolving; nonetheless, the implementation above, and the framework into which it fits, emphasize the importance of the interaction between parallel programming and knowledge of the parallelism within the ANN paradigm itself. Attention to this interaction leads to matches between implementation and algorithm that are more exact and more efficient; performance like that achieved here is the direct result of this interaction.

V. References

1. Blum, Avrim and Rivest, Ronald. (1988). "Training a 3–node neural network is NP–Complete." *Proceedings of the 1988 IEEE Conference on Neural Information Processing Systems — Natural and Synthetic.*
2. Fahlman, Scott E. (1988). "Faster learning variations on backpropagation: An empirical study." *Proceedings of the 1988 Connectionist Models Summer School,* pp. 38–50.
3. Farber, Robert M. (1989). personal communication.

4. MIT Lincoln Laboratory. (1988). *DARPA Neural Network Study — Final Report.*

5. Millán, J. and Bofill. (1989). "Learning by back–propagation: a systolic algorithm and its transputer implementation." Universitat Politècnica de Catalunya, Report LSI–89–15.

6. Pomerleau, Dean A., Gusciora, George L., Touretzky, David S., and Kung, H.T. (1988). "Neural network simulation at warp speed: How we got 17 million connections per second" in *Proceedings of the IEEE International Conference on Neural Networks, San Diego 1988.*

7. Rumelhart, D.E., Hinton, G.E., and Williams, R.J. (1986). "Learning internal representations by error propagation" in *Parallel Distributed Processing,* eds. James McClelland, David Rumelhart, and the PDP Research Group. Vol. 1, pp. 318–362. Cambridge:MIT Press.

8. Singer, Alexander. (1990) "Implementations of Artificial Neural Networks on the Connection Machine." Thinking Machines Corporation, forthcoming technical report.

9. Tesauro, Gerald. (1987). "Scaling relationships in backpropagation learning: dependence on training set size." *Complex Systems,* **1**:367–372.

BACHUS: A VLSI ARCHITECTURE FOR A LARGE BINARY ASSOCIATIVE MEMORY

Martin Huch, Werner Poechmueller, Manfred Glesner
Technische Hochschule Darmstadt
Institut fuer Datentechnik
Schlossgartenstr. 8
D-6100 Darmstadt
FRG

In recent years many attempts have been made to realise neural networks using VLSI technology. Thereby, major difficulties are how to implement several hundreds or, if possible, thousands of highly interconnected neurons with their synapses on the area of one chip or to develop cascadeable architectures. Due to the large number of conceivable architectures many, more or less convincing approaches have been made. Some of them are trying to take advantage of analog VLSI circuits, whereby one has to take into account many problems concerning parameters depending on fabrication and temperature. Other researchers developed wholly digital circuits, causing problems with chip area, synchronisation, and information exchange in highly interconnected networks. With this paper we want to present a digital implementation of a binary network using VLSI technology. Due to the lack of on-chip learning support and the use of industrial standard RAMs, we are able to offer an extremely dense storage of binary synaptic weights. Therefore, it is feasible to realise network structures with several thousand completely connected neurons. Such a network may be used for applications in the areas of speech and image recognition.

The BACHUS Approach: A Binary Neural Network System in Parallel Hardware

Current attempts to implement neural networks in hardware concentrate on one hand on dedicated accelerator boards for standard computers which allow a more efficient simulation of neural networks and on the other hand on a way to integrate neural networks on VLSI-Chips, often using analogue circuit techniques.

Many non-stochastic implementations of neural networks are based on a common model which calculates the output of a neuron Nj as follows:

$$N_j = f\left(\sum_{i=1}^{n} T_{ij} * N_i + \sum_{i=n+1}^{n+e} * E_i \right)$$

The Networks only differ in some parameters, as the wordlength and representation of the synaptic weights T_{ij} and neuron states N_j or different areas of the synaptic weight matrix which are filled with "0"

The Technische Hochschule Darmstadt presents an approach, where the model of a neural network is simplified so far, that it is suitable for a purely digital VLSI implementation. The model uses

binary values for the representation of the neuron states and the synaptic weights.
As a tradeoff for this simplification, all input and output vectors for the network must be sparsely coded bit strings. Due to this simplification, the main operation of neural network hardware, the accumulation of weighted inputs, is transformed to a counting procedure, where all locations in the synaptic weight matrix are counted if the matrix bit and the correspondig input vector bit are both set to one. This operation is pointed out in the following figure:

```
           IN
Ramadr 1    0       0 0 0 0 0 0 0 0
       2    1 ->    0 1 0 0 0 1 0 0
       3    0       0 0 0 0 1 1 0 0
       4    0       1 1 0 0 0 0 1 1
       5    0       0 1 0 0 0 0 1 0
       6    1 ->    0 1 0 0 0 1 1 1
       7    1 ->    0 1 0 0 0 1 1 0
       8    0       0 0 0 0 0 0 0 0
                      |       | | |

                    0 3 0 0 0 3 2 1   out
                      |       | |

                    0 1 0 0 0 1 1 0   Thresholded out
```

The operation of this network may be regarded as an optimisation process, where the network finds the stored bit vector which is closest to the bit vector presented at the inputs.

A standard cell chip has been developed by TH Darmstadt which contains the counter logic for 32 neurons plus some address generator logic. The synaptic weight matrix of the chip is not stored inside the chip but in external standard RAM chips. This chip is fully cascadable and allows to build up neural networks of nearly any size.
A neural network system consisting of 16000 fully connected neurons is planned, using 256 of these chips.
This hardware may be used for applications like speech recognition, obstacle avoidance of an autonomous vehicle, and other supervised classification tasks.

Chip architecture

The standard cell chip called "BACHUS" comprises 32 counter cells plus some control and address generation logic (see fig.1). Counters are used to realise the functionality of neurons (summation of input data and threshold detection). One counter on the chip with its registers and control logic realises one neuron. Address generation logic is necessary for reading the final result, available on the neuron outputs.

Improvement of the BACHUS Design

The BACHUS chip has been designed and fabricated by use of the VENUS 3 standard cell system and successfully been tested.
Actually, a prototype system with 512 neurons is under construction.
The BACHUS architecture causes some problems if large systems shall be built:

- The chip has separate busses for access of the internal registers and for the hit address busses. The total number of pins is thus about 110. The chip is packaged in a 120 pin PGA package which is very expensive.
- The storage of new patterns in the weight matrix is rather circumstantial

Therefore a redesign called BACHUS II has been done and is actually fabricated.
This redesign has a reduced pin count and fits into a 68 pin PLCC package. The main saving in pins has been done by taking the chip address registers off the chip and by using the chip data bus also for outputting the remaining 5 bits of the hit address.

System Architecture

The entire Neural Network system consists of a host computer (a VME-bus system), a controller board and a number of BACHUS boards (see fig. 2).
One BACHUS board will hold 8 BACHUS chips with their 4 RAM chips each thus giving 256 neurons per board.
A complete rack may hold 32 boards, the controller and the host computer giving a network size of 8000 neurons.
The controller contains a state machine, which generates the appropriate command sequences for the BACHUS chips. A FIFO register on the controller holds the input vector, the output vector and intermediate results.

Literature

G. Palm
On Associative Memory
Biol. Cybernetics 36, pp. 19-31, 1980

G. Palm, T. Bonhoeffer
Parallel Processing for Associative and Neural Networks
Biol. Cybernetics 51, pp. 201-204, 1984

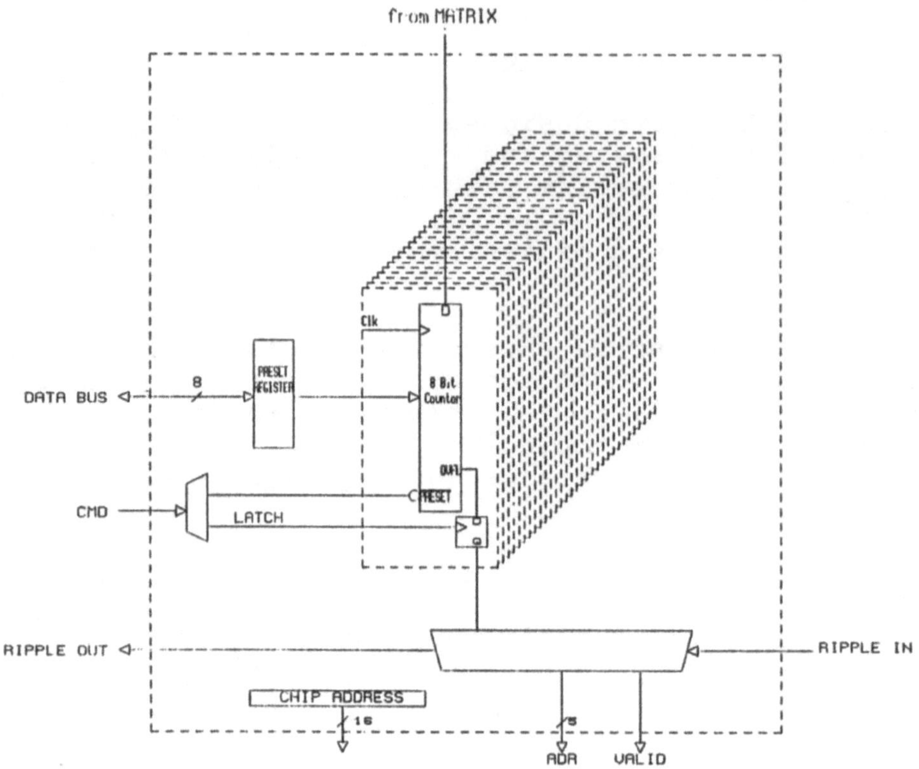

Fig.1: BACHUS Chip Architecture

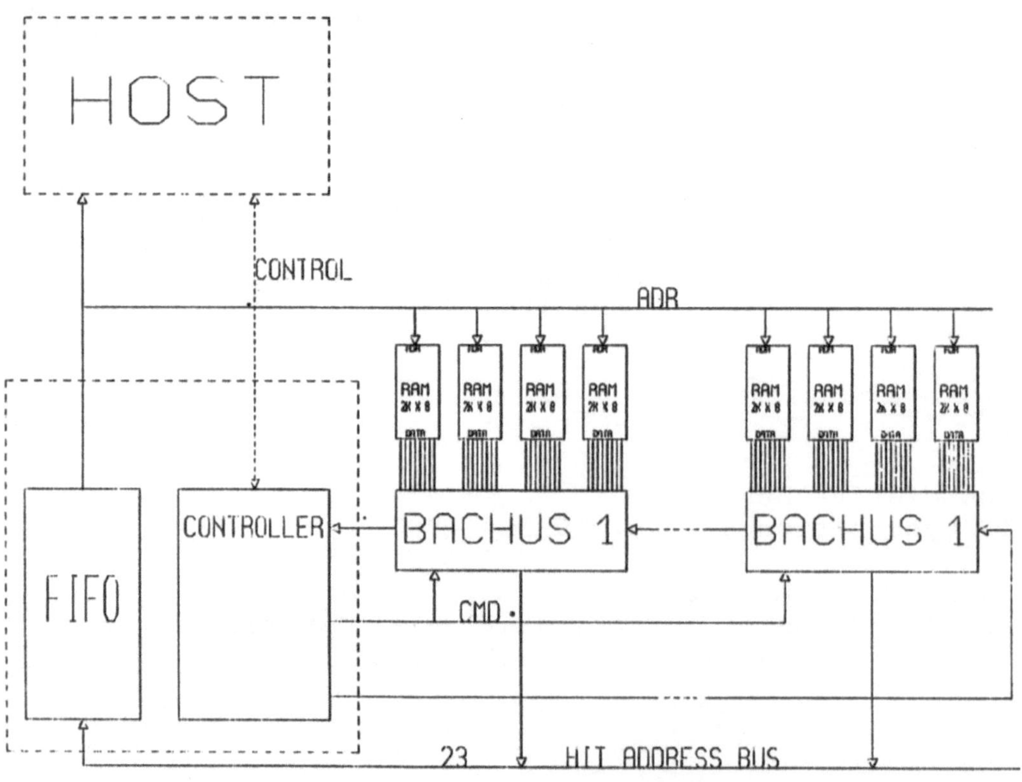

Fig.2: BACHUS System Architecture

A RECONFIGURABLE ARCHITECTURE FOR A VLSI IMPLEMENTATION OF ARTIFICIAL NEURAL NETWORKS ——— A VLSI Design of a Basic Neural Unit

Bo Jin
Department of Electrical Engineering
The Pennsylvania State University
University Park, PA 16802
U.S.A.

Belgacem Raggad
Division of Business Administration
The Pennsylvania State University
Middletown, PA 17057
U.S.A

Abstract

In this paper, a VLSI design of a Basic Neural Unit (BNU) for artificial neural networks (ANNs) is proposed. The BNU is used as a basic processing element in constructing an ANN. An important issue to address in the design of BNU is to reduce the interconnection complexity of ANNs. When the size of an ANN grows, the number of synaptic weights connected to each neuron increases rapidly. From the hardware implementation point of view, an expandable and reconfigurable architecture design for ANNs is in demand. A new approach has been used in designing a BNU. This BNU design is the foundation of a reconfigurable architecture for constructing an ANN with a very strong expandability.

Weight-centered approach, in contrast with the traditional neuron-centered view, is the key point to this new architecture design. The weight-centered idea intends to transfer the processing load from a neuron to a number of synaptic weight branches, thus decentralizing the computation task. Adding more neurodes, when enlarging the size of an ANN, is accomplished by merely appending more weighted branches. In VLSI implementation, this is achieved by plugging in one or more BNU chips. The weight-centered design approach enhances the reconfigurability and expandability of an ANN. This design has since been fabricated on a silicon chip by the IBM corporation.

1. Introduction

Artificial neural networks are large-scale adaptive learning systems. They attempt to achieve good performance via dense interconnection of simple computation elements. An ANN stores, represents, retrieves and manipulates data on a purpose of achieving human-like performance. The brain performs its processing feat through massive parallelism, using 10 billion neurons and more than 1000 times that many interconnections. To simulate massive parallelism, an ANN consists of setting up a network of processing elements, the electronic analogy to neurons.

In an ANN model suggested by Hopfield, each processing element is connected to every other neuron, which requires $O(N^2)$ interconnections where N is the number of processing elements in the system. This will complicate the layout tremendously when the system is implemented on a VLSI silicon chip, when N is real big in a large ANN. What design approach can reduce this complexity? From an application point of view, reconfigurability of an ANN is necessary due to the fact that the application domain covers a dynamic problem solving space. Being used in an application environment, an ANN is often required being reconfigurabled. What design approach can be used to achieve this?

2. A Basic Neural Unit (BNU) — Chip-level Design

Artificial neural networks are composed of processing units(PEs) and synaptic weight interconnections. This is shown in Figure 1. Once a PE is implemented on a silicon chip, it is hard to expand the size of the network with the increased number of inputs. We consider this as a neuron-centered design approach. A new approach, weight-centered, is proposed. A basic neural unit is designed by using this approach. Figure 2 shows a 4-input, 1-output BNU which can be used to construct a larger size neural network. Figure 3 shows a 1-layer 8-input, 8-output ANN which can be easily expanded from the BNU of Figure 2. Using this approach, an ANN has a very strong expandability. In VLSI design of a basic neural unit (BNU), we maximized the number of weight cells that can be put on a silicon chip. The BNU is implemented and fabricated on an IBM 4.2 mm image silicon chip. Figure 4 shows the circuit schematic of a BNU. It consists of 10 inputs, 3 outputs and 30 synaptic weights. Figure 5 is the photomicrogragh of a BNU chip. It is packaged in a 120-pin package.

3. Using BNU Chips to Construct an ANN

The BNU chip is designed for easy expansion. A number of BNU chips may be used to construct an ANN. The idea of using BNU chips offers a great expandability and reconstructivity to the network. Depending on the application problem, three features can be reconfigured or expanded: the number of neurons of the input layer, the number of neurons on the hidden layers and the number of neurons on the output layer. To add some number of neurons on each layer is simply to plug in certain number of BNU chips. In a 3-layer ANN, those BNU chips representing the weight matrix from input layer to hidden layer are built on one network board. The weight matrix from hidden layer to output layer is represented by another group of BNU chips and can be constructed on one network board. Figure 6 shows a 3-layer ANN constructed by two network boards, one with 12 BNUs and another with 8 BNUs. This ANN has 30 input neurons, 12 hidden neurons and 24 output neurons. It is not difficult to see that expanding more input neurons can be achieved by adding some number of BNU rows on the input layer-to-hidden layer board, whereas plugging in more BNU columns is to add more number of hidden neurons. Output neurons can be expanded through output board. Most of the ANNs used in problem solving are 3-layer networks. Nevertheless, using BNU chips, more hidden layers can be expanded into an ANN. Figure 7 illustrates how a 4-layer ANN is constructed. It has two hidden layers, the first has 21 hidden neurons and the second has 9.

Conclusions

In conclusion, we have proposed a new approach in constructing an ANN with a very strong expandability and reconfigurability. It is very convenient to use BNU chips to construct an ANN with a flexible network configuration. Such a network may also be reconfigured. All the network boards are built in a ready-to-plug mode. We believe that the design of BNU chip is a very good approach in design for reconfigurability.

References

[1] S.Y.Kung. Parallel Architecture for Artificial Neural Nets. In Proceedings, IEEE Intl' Conf. in Systolic Arrays, pp163-174, Feb, 1988.

[2] S.Y.Kung and J.N.Huang. Digital VLSI architectures for artificial neural nets. In Proceedings, Neural Networks for Computing, Snowbird, Utah, April, 1988.

[3] R.P.Lippmann. An introduction to computing with neural nets. IEEE ASSP magazine, 4: pp 4-22, April, 1987.

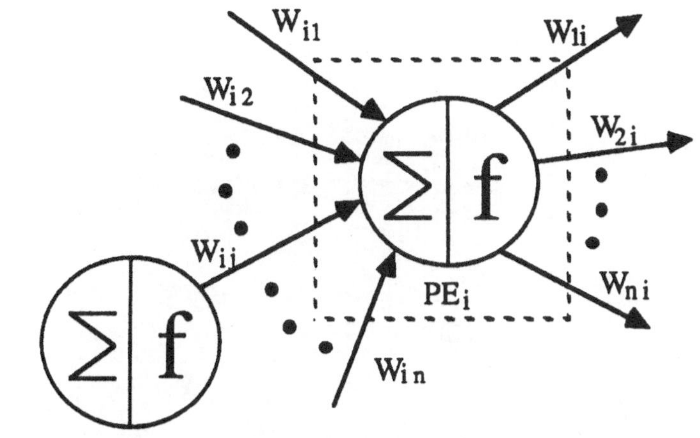

Figure 1. A neural-centered processing element

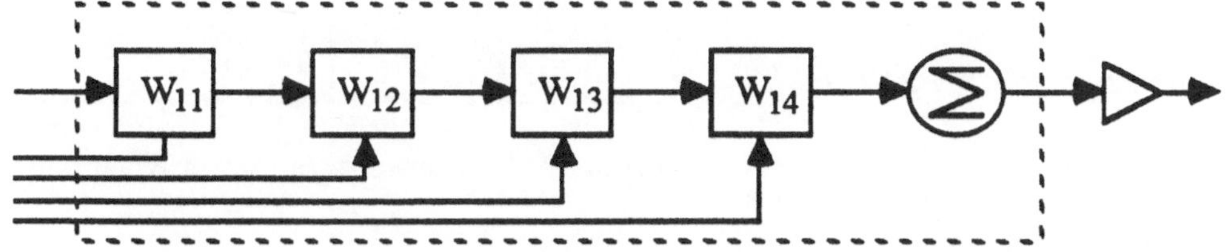

Figure 2. A weight-centered processing element (4-input, 1-output)

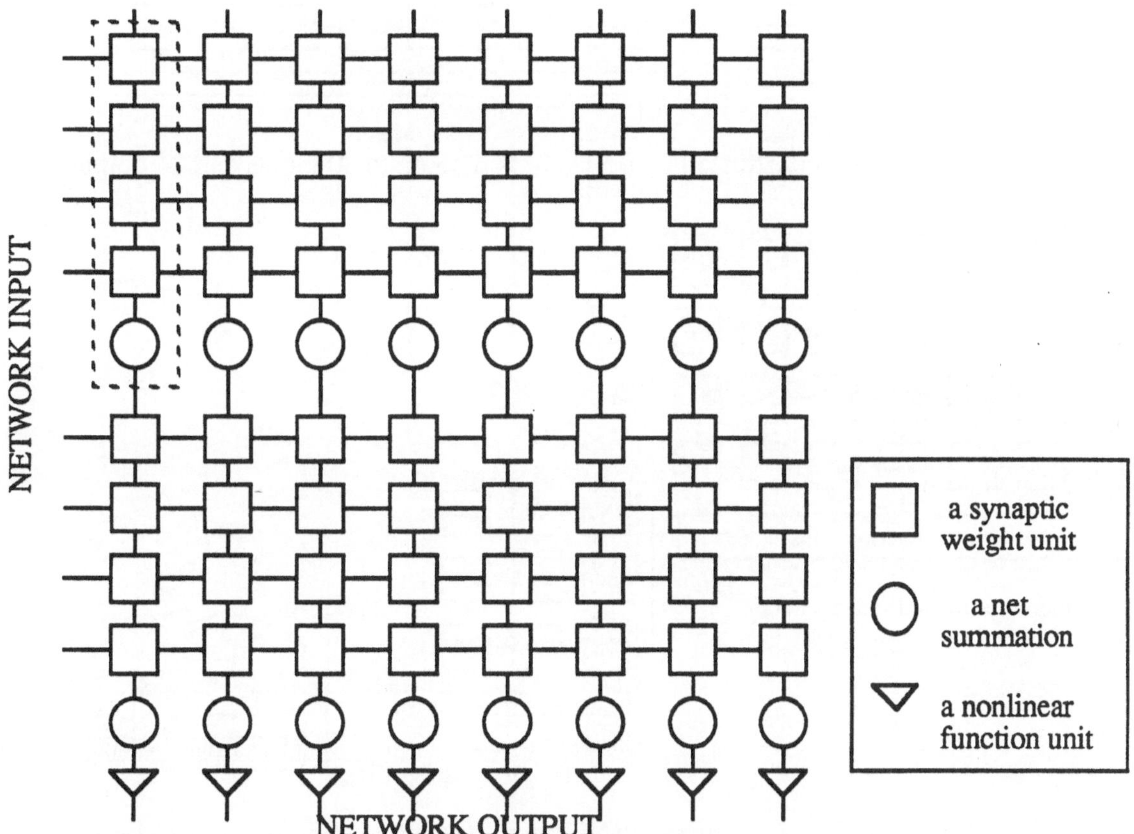

Figure 3. An example of constructing an ANN (8-input, 8-output)

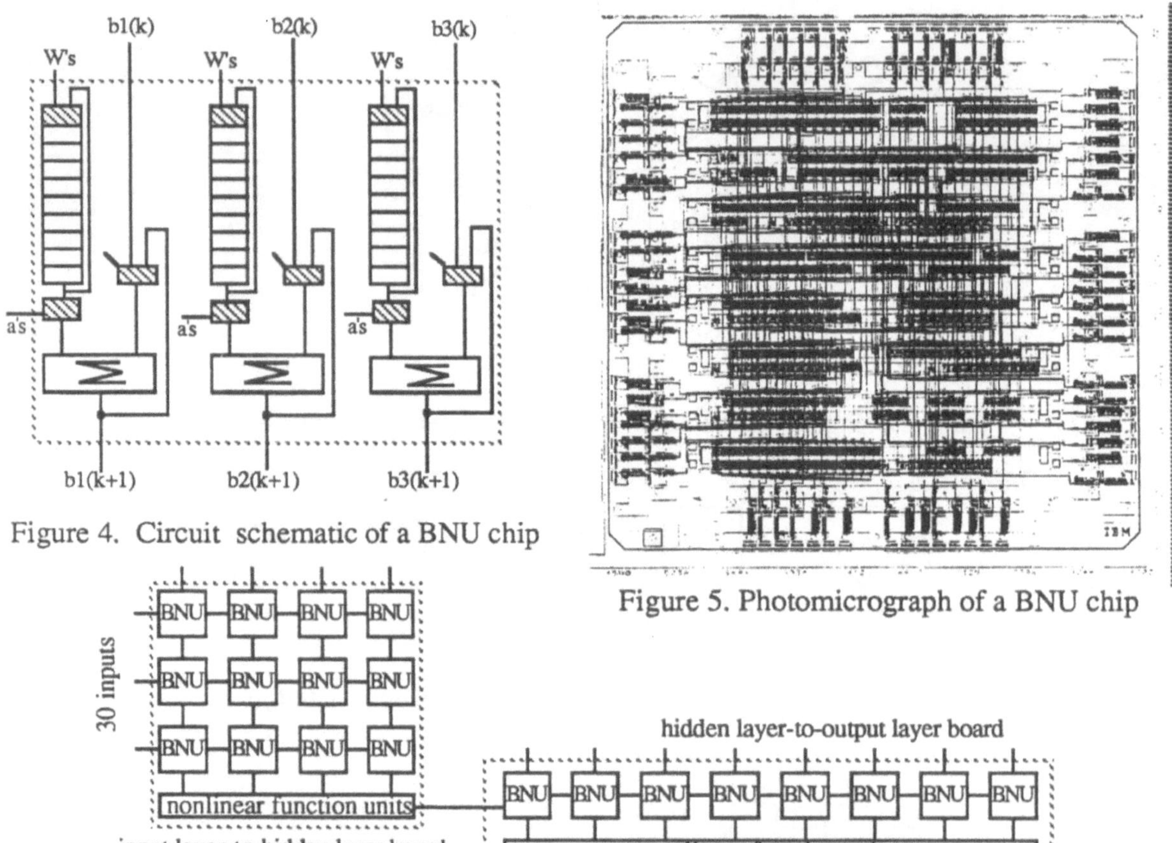

Figure 4. Circuit schematic of a BNU chip

Figure 5. Photomicrograph of a BNU chip

Figure 6. A 3-layer ANN constructed by BNU chips with up to 30 inputs and 24 outputs

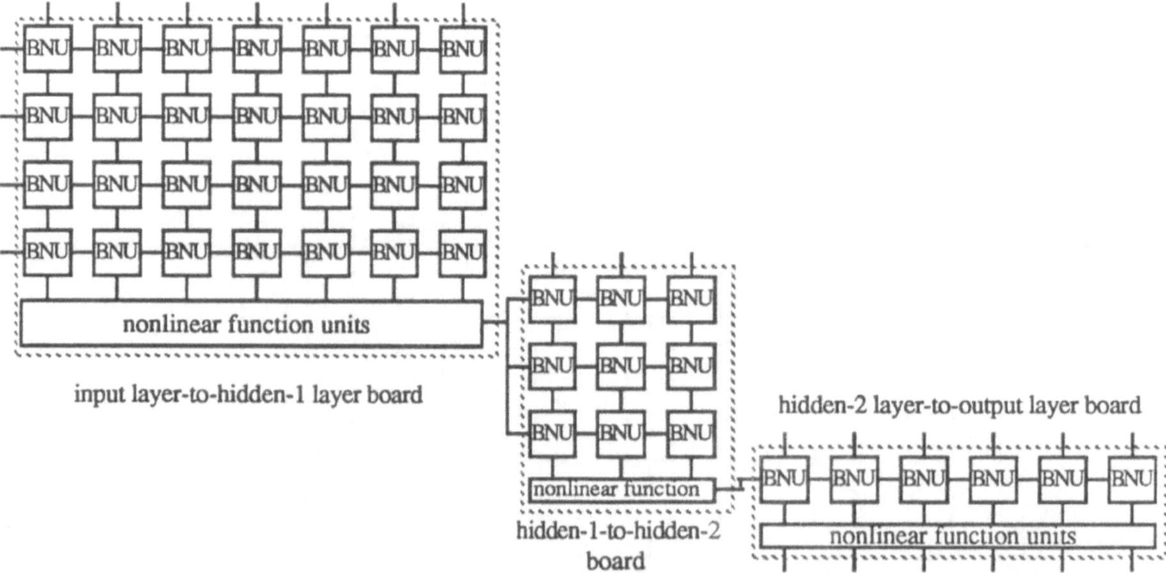

Figure 7. A 4-layer ANN constructed by BNU chips with up to 40 inputs and 18 outputs

MULTILAYER NEURAL NETWORKS ON DISTRIBUTED-MEMORY MULTIPROCESSORS

Hyunsoo Yoon and Jong H. Nang

Department of Computer Science
Korea Advanced Institute of Science and Technology
P.O.Box 150 Cheongryang, Seoul 130-650, Korea
e-mail : hyoon@sorak.kaist.ac.kr

Abstract: In this paper, it is investigated simulating/implementing a fully connected multilayered feedforward neural network using the backpropagation learning algorithm on a distributed-memory multiprocessor system. Each layer is partitioned into p disjoint sets and each set is mapped on a processor of a p-processor system. A fully distributed backpropagation algorithm, necessary communication among the processors, and its time complexity are investigated. The p-processor speed-up of the backpropagation algorithm over a single processor is analyzed theoretically for some popular processor interconnection topologies, which can be used as a basis in determining the most cost-effective or optimal number of processors.

Introduction

Simulating a large artificial neural network to solve the real-world application problems such as pattern recognition, vision and speech recognition can require huge amounts of computational resources. There have been some research efforts of making fast simulator by using a parallel computer or parallel processing technique. For example, the Warp [Pome88], the Connection Machine [Depr89], the MPP [Hick88], and the BBN Butterfly [Feld88] have been used to implement fast neural network simulators at various research institutions.

In this paper, we investigate the neural network simulation on the distributed-memory, message-passing multiprocessor (DMM) system such as the Intel Hypercube, Ncube, or the Inmos Transputer system which are more widely available than the aforementioned parallel computers. In particular, we investigate the fully connected multilayered feedforward network using the backpropagation learning algorithm, the most common and popular neural network [Darp88], on the DMM.

We partition the multilayered feedforward network into p subnetworks, map each subnetwork on a processor of the p-processor DMM, and derive the fully distributed backpropagation algorithm. By computing the time complexity of the p-processor distributed backpropagation learning algorithm, and comparing it with the single processor's case, we show the p-processor speed-up bound, $S(p)$, over the single processor. This speed-up ratio is studied for some popular processor interconnection topologies such as the ring, hypercube, and mesh.

A Fully Connected Multilayered Feedforward Network

A fully connected multilayered feedforward network consists of L+1 layers. The first layer at l=0 is the input layer and contains n_0 units. Subsequent layers are labeled $1 \leq l \leq L$; the l-th layer contains n_l units. Each unit on a layer is connected to all nodes on the next layer. Associated with each unit i on layer l is an activation value $a_i(l)$, and attached to each connection, connecting unit i on layer l to unit j on layer l+1, is a weight $w_{ji}(l,l+1)$.

The Backpropagation Learning Algorithm

The backpropagation learning algorithm [Rume87] consists of three passes; forward execution, backpropagation of the error, and weight update. In the forward execution pass, the network, given an input pattern presented at the input layer, is executed by the following feedforward equations,

$$a_i(l) = f\left(\sum_{j=1}^{n_{l-1}} w_{ij}(l-1,l)\cdot a_j(l-1)\right),\ l=1,...,L \text{ and } i=1,...,n_l \tag{1}$$

where f is a nonlinear sigmoid function of the form of $f(x) = (1+e^{-x})^{-1}$. The second pass involves the comparison between the actual output pattern and the desired one, and the propagation of the error, which is governed by the following equations,

$$\delta_i(l) = \begin{cases} [t_i - a_i(L)]\cdot[a_i(L)(1-a_i(L)], & l = L \\ [\sum_{k=1}^{n_{l+1}} \delta_k(l+1)\cdot w_{ki}(l,l+1)]\cdot[a_i(l)(1-a_i(l)], & l = L-1,...,1 \end{cases} \tag{2}$$

where $\delta_i(l)$ is the error value of unit i on layer l and t_i is the desired value of unit i in the output layer. Weight update is performed in the third pass according to the following equations,

$$\Delta w_{ij}(l-1,l) = \eta\cdot\delta_i(l)\cdot a_j(l-1) \tag{3}$$

where η is a learning rate. The second and the third pass may be combined into one pass.

Distributed Memory Multiprocessor Simulation

A DMM in our model consists of p processors, has no shared memory, and communicates only by message passing. Each communication link interconnecting two processors is a point-to-point bidirectional link. Each layer of n_l units is partioned into p parts, and each part of n_l/p units is assigned to each processor. Each processor maintains in its local memory the activation values, the error values, and the input weight vectors of the assigned units. In this way, all the values of activations, errors, and weighs are completely partitioned into p disjoint sets, reducing by p times the total memory requirement of a processor of a p-processor DMM system over a single processor system.

A DMM executes the backpropagation algorithm following the three passes expressed in Eqs.(1)-(3). Each pass is conceptually subdivided into two phases, communication and computation phases, though in practice they can be interleaved. In the communication phase of the first pass, every processor needs to know the activation values of all other processors, which can be accomplished by *all-to-all broadcasting* [John89]. All-to-all broadcasting is a process whereby a set of messages, with a distinct message initially residing at each processor, is disseminated so that eventually a copy of each message comes to reside at each processor. Once all-to-all broadcasting is completed, every processor is informed of all the necessary activations values of all other units in a layer, and can execute Eq.(1) for the assigned units. Note that weight values need not be exchanged since each processor keeps locally the input weight vector of each assigned unit. In the second pass, we need more complex communication patterns. To compute an error value $\delta_i(l)$ of a unit i in layer l, a weight value $w_{ik}(l,l+1)$ multiplied by an error value $\delta_k(l+1)$ of a unit k in layer l+1 is necessary for all units in layer l+1. In other words, each unit in layer l+1 needs to send n distinct values (weight multiplied by the error value) to the n units in layer l. This kind of information exchange can be accomplished by *all-to-all personalized communication*, by which it is meant that each processor sends a unique message to every other processors. Once all-to-all personalized communication is completed, each processor can execute Eq.(2) for the assigned units independently. The third pass, the weight update pass is similar to the first pass, requiring all-to-all broadcasting to exchange the activation values of units in layer l-1. Note that all-to-all personal communication is required in the second pass since each processor keeps the input weight vector of the assigned units. If the output weight vector is stored instead of the input weight vector, all-to-all personalized communication is required in the first and the second passes.

The Time Complexity and the Speed-up Ratio

The time required for a single processor to execute the backpropagation algorithm for a layer of n units is approximately given by $T_1 = t_1 + t_2 + t_3$, where t_i is the time to execute the i-th pass , *i.e.*, Eq. (i); t_1 = n(n·M_a + F), t_2 = n(n· M_a), t_3 = n(n·M_a), where M_a is a multiply-and-add time of two floating-point numbers and F is the time to evaluate the sigmoid function. Without loss of generality, it is assumed in evaluating the time complexity that $n_l = n$ for l=0,...,L.

The time complexity of a distributed backpropagation learning algorithm running on a p-processor DMM with a point-to-point communication can be given by $T_p = t_1' + t_2' + t_3'$; $t_1' = (n/p)\cdot AAB(p) + (n/p)\cdot(n\cdot M_a + F)$, $t_2' = (n/p)^2\cdot AAPC(p) + (n/p)\cdot(n\cdot M_a)$, $t_3' = (n/p)\cdot AAB(p) + (n/p)\cdot(n\cdot M_a)$, where AAB($p$) is the time to complete all-to-all broadcasting, and AAPC(p) to complete all-to-all personalized communication on the p-processor DMM. The factor (n/p) accounts for the fact that every processor is assigned n/p units per layer.

The AAB(p) and the AAPC(p) are analyzed for some popular symmetric interconnection topologies, and are summerized in Table 1. The mathematical proofs for the time complexity and the corresponding routing algorithms are omitted due to the space limit. (The hypercube case was obtained independently in [John89].)

It is assumed in evaluating the AAB(p) and the AAPC(p) that during each period of unit time, a processor can send and receive a unit of message on one of its ports. The port on which a processor sends and receives can be different. If a processor can either send or receive a message during each period of unit time, the time complexity given in Table 1 increases by at most a factor of two. The communication time unit for a processor sends and/or receives a unit of message is defined as C which includes the overhead of setting up the necessary communication mechanism and the actual message transfer time. The message unit is a word representing the floating-point number of the activation value or the weight.

Table 1. The time complexity of AAB(p) and AAPC(p).

	Ring	(Illiac) Mesh	Hypercube	Complete Connection
AAB(p)	p - 1	p - 1	p - 1	p - 1
AAPC(p)	$\lfloor p^2/4 \rfloor$	$1/2\cdot\sqrt{p}\cdot(p-1)$	$p/2\cdot\log_2 p$	p - 1

The most important parameter is the communication/computation ratio which is defined as $\Delta = C/M_a$ [Anna89], and our testbed, a Transputer-based system, has approximately the value of Δ=1.3. With the values in Table 1, and the definition of $\Delta = C/M_a$, and letting $F = \theta\cdot M_a$, the p-processor speed-up ratio, S(p)=T_1/T_p, can be obtained, which is shown graphically in Figure 1.(a), where n=20000 units/layer, Δ=1.3 and θ=40 were taken.

Once the learning is complete for a specific application, the application can be hard-wired, and the network may execute only the forward execution pass. Figure 1.(b) shows the p-processor speed-up, $S'(p)=t_1/t_1'$, for this case.

Summary and Conclusion

The backpropagation learning can take huge amount of time for a practical application. One natural way to overcome the time and space limit is to use parallel computers. We studied the backpropagation on the p-processor DMM. We derived the fully distributed backpropagation algorithm, identified the necessary communication mode (all-to-all broadcasting and all-to-all personalized communication), derived the time complexity of communication and computation, and thereby obtained the theoretical upperbound of speed-up. The most important parameters for the speed-up are the communication/computation ratio and the processor interconnection topology. The speed-up in the ring and the mesh topology does not increase monotonically as the

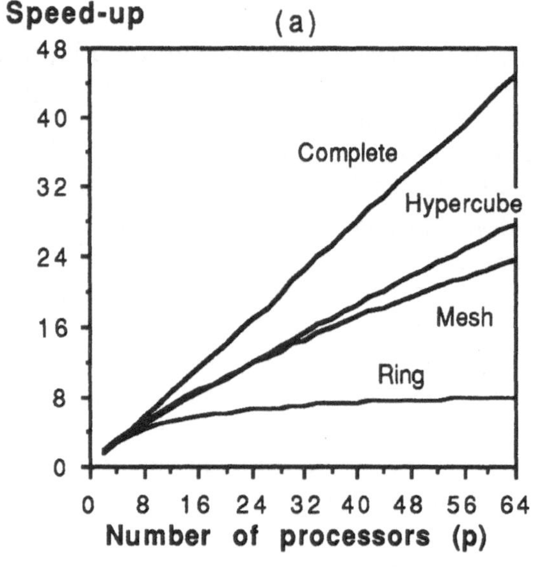

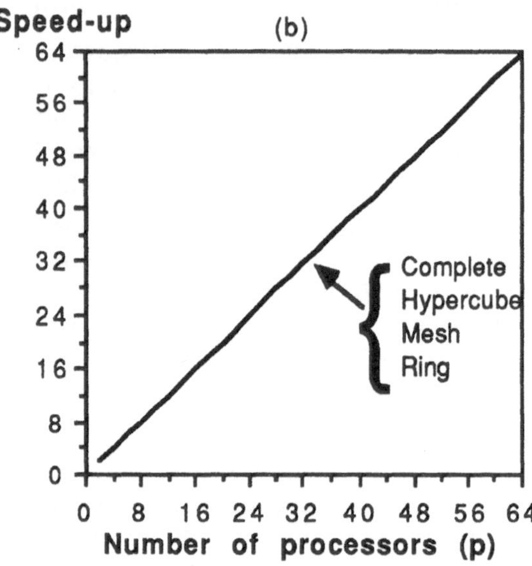

Figure 1. (a) The p-processor speed-up of the backpropagation algorithm. (b) The p-processor speed-up of the forward pass of the backpropagation algorithm.

number of processor increases, due to the excessive communication overhead. In the forward execution pass, the ring topology achieves the lower bound of communication complexity, and thus no more powerful topology than the ring need to be used. Our model and equation for the speed-up can be very useful for studying the effect of the processor interconnection topology, the number of units, the communication/computation ratio, the sigmoid function evaluation time, and the number of processors on the backpropagation algorithm speed-up. We implemented our model on the Inmos B008 System, a Transputer(T800)-based DMM, and our experimental results confirmed the model and analysis.

References

[Anna89] M. Annaratone, C. Pommerell, and R. Ruhl, "Interprocessor Communication Speed and Performance in Distributed-Memory Parallel Processors," *Proc. of the 16th Annual Int'l Symp. on Computer Architecture*, pp.315-324, 1989.

[Darp88] *DARPA Neural Network Study*, AFCEA International Press, 1988.

[Depr89] E. Deprit, "Implementing Recurrent Back-Propagation on the Connection Machine," *Neural Networks*, Vol.2, pp.295-314, 1989.

[Feld88] J. A. Feldman and et al., "Computing with Structured Connectionist Networks," *Comm. of ACM*, Vol.31, No.2, pp.170-187, 1988.

[Hick88] J. Hicklin and H. Demuth, "Modeling Neural Networks on the MPP," *Proc. of the 2nd Symp. on the Frontiers of Massively Parallel Computation*, pp.39-42, 1988.

[John89] S. L. Johnsson and C. T. Ho, "Optimum Broadcasting and Personalized Communication in Hypercubes," *IEEE Trans. on Computers*, Vol.38, No.9, pp.1249-1268, 1989.

[Pome88] D. A. Pomerleau and et al., "Neural Network Simulation at Warp Speed: How We Got 17 Million Connections Per Second", *Proc. of IEEE 2nd Int'l Conf. on Neural Networks*, Vol. II, pp. 143-150, 1988.

[Rume87] D. E. Rumelhart, G. E. Hinton, and R. J. Williams, "Learning Internal Representations by Error Propagation," In D. E. Rumelhart and J. L. McClelland (Eds.), *Parallel Distributed Processing: Explorations in the Microstructure of Cognition*, Vol.1, pp.318-362, MIT Press, 1987.

Neural Network Simulation on the MasPar MP-1 Massively Parallel Processor

K. A. Grajski, G. Chinn*, C. Chen*, C. Kuszmaul** and S. Tomboulian***

*Ford Aerospace
Advanced Development Department
San Jose, CA 95161-9041
kamil@wdl1.fac.ford.com

**MasPar Computer Corporation
749 North Mary Avenue
Sunnyvale, CA 94086
argosy!sherry@decwrl.dec.com

ABSTRACT

Progress in neural network R&D and technology transfer to real-world applications depends critically upon the availability of high-speed, massively parallel computing. The MasPar computer is a SIMD architecture well-suited for neural network implementations. We present an implementation of multi-layer perceptron learning using the back-propagation algorithm. The algorithm design is based on a virtual processor approach. An initial implementation of an unoptimized, general design yields approximately 306K connection updates per second (CUPS). This is 75% of previously reported results for an IBM 3090 (listed in Watanabe, et al., 1989, Table II) and 17-fold greater than our own Sun 3/80. We expect significant improvement to several 10's of MCUPS as we begin to address load-balancing issues, better usage of the (1K) register space available to each processor and development of virtual processor depth reduction techniques.

ARTIFICIAL NEURAL NETWORK ON A MASSIVELY PARALLEL ASSOCIATIVE ARCHITECTURE

A. Krikelis
Aspex Microsystems Ltd.
Brunel University
Uxbridge UB8 3PH, United Kingdom

ABSTRACT

An implementation of a fully connected artificial neural network using the multi-layered perceptron model is described. The neural network is implemented on the ASP (Associative String Processor). The ASP is a fine-grain massively parallel SIMD architecture, emerging from research at Brunel University and being developed by Aspex Microsystems Ltd., based on associative processing. Neural networks readily map onto the ASP architecture. The neural network described in this paper is a multi-layered perceptron model which uses the back-propagation learning paradigm. The network has 137 nodes in three layers and is trained to recognise letters of the alphabet. The work is the basis for the implementation of a massive artificial network environment (with 10^5 nodes and 10^8 connections) on the ASP using the back propagation and alternative learning techniques.

Using Xputers as Universal Accelerators for Neuro Network Simulation and its Applications

by

R. W. Hartenstein, A. G. Hirschbiel, M. Weber
Universitaet Kaiserslautern, Bau 12,
Postfach 3049,
D - 6750 Kaiserslautern, F. R. G.

Extended Abstract

category: B. Implementation

Abstract.
It is the goal of the paper to draw the attention of the neuro network scene to results which mostly have been published somewhere else. It focuses on the use of xputers as universal accelerators, which promise extraordinarily high throughput also in neuro network simulation - on a surprisingly simple hardware. Acceleration factors of >100 are expected for neuro networks generally. The machine principles of xputers will be illustrated briefly. It will be shown, that xputers are as universal as computers. Because of xputer basic high performance machine principles acceleration factors between 100 and more than 2000 have been obtained for a number of important applications - although using a hardware, the non-programmable parts of which are more simple than that of a single RISC microprocessor. That's why xputers are universal accelerators. Some performance figures will be given which have been obtained experimentally on the MoM xputer environment. Because of such high acceleration factors in many applications specialized accelerators for neuro network simulation are not needed. Cheap mass production ASICs (gate arrays) may be derived from an xputer program without any extra design effort, since xputers also may run in stand-alone mode.

NES: a Neuron-like net for a diagnostic Expert System

BURATTINI ERNESTO
Istituto di Cibernetica - CNR
Via Toiano 6 - Arco Felice - 80072 - Naples - ITALY
E-mail: ernb%arco@inria.uucp

ABSTRACT.
In the light of the most recent results obtained in the design and construction of parallel machines, like connectionist ones, we propose here a diagnostic expert system shell based on a neuron-like net.

The causal reasoning, pursued by an expert when diagnosing some events starting from observed findings, is typically a parallel process. Roughly speaking, we represent each finding and event by a single neuron, a mathematical formalization of which will be given in the paper, and the relationship between findings and events by links among neurons. The expert's knowledge, i.e. euristics and strategies diagnosing the happened events, are embedded in the values of coupling coefficients, theresholds, excitation decay laws, characterizing each neuron.

A prototipical model of this net has been implemented in PROLOG, awaiting a parallel machine, and tested and analyzed to verify the net structure and the handling level of the knowledge representation.

The results obtained have been so encouraging as to persuade us, for the next step of the research, to try to apply such shell to a medical knowledge base, already built for a more classical system, and to compare both diagnoses.

HARDWARE DESIGN CONCEPTS FOR A CALM NEURAL NETWORK USING 400 SIMPLE PROCESSORS: SYSTEM ARCHITECTURE

J. Hoekstra[1], J.N.H. Heemskerk[2], A.J. Klaassen[1], R.H. Phaf[2], P. Knoppers[1], P.T.W. Hudson[2]

Dept. Electrical Engineering
Delft University of Technology
PO Box 5031
2600 GA Delft, The Netherlands

January 18, 1990

Abstract

This paper discusses the system architecture of a network consisting of 400 simple microprocessors for simulating neural networks with 400 nodes (neurons). In particular, the multiprocessor network should be able to simulate the CALM Neural Network approach. CALM employs a modular structure, which is used to simplify the hardware design. A system architecture is developed, which is based on (1) parallel units that consist of several processors, and (2) a special network controller that arranges the inter-unit communication.

[1]Authors are at Delft University of Technology, Dept. Electrical Engineering, PO Box 5031, 2600GA Delft, The Netherlands.

[2]Authors are at Leiden University, Dept. Psychology, PO Box 9555, 2300RB Leiden, The Netherlands.

Transputer based simulation of a general purpose, fault tolerant neural network

Prof. F. E. Lauria*
Istituto di Fisica Teorica - Università degli Studi di Napoli
Pad. 19, Mostra d'Oltremare - I-80125 Napoli, ITALY
G-Mail: lauria@vaxna.infnet.it

Dott. M. Sette
Istituto per la Ricerca dei Sistemi Informatici Paralleli - CNR Napoli
G-Mail: sette@vaxna.infnet.it

January 15, 1990

Abstract

Research on neural networks is to a large extent dependent upon the use of computer simulations. The availability of the Transputer hardware [1] and its accompanying OCCAM [2, 3] software together with the highly parallel nature of the neural algorithms has led us to develop a Transputer based simulation of a neural network.

The simulated model has deterministic activation function and syncronous updating (according to the classification of neural models proposed in [4]), and its processing elements are linear threshold units.

We look at the network as a parallel general purpose computer which can be programmed assigning the connection weights. To this end it has been developed an assembly programming language which we use to configure the network and to make it partially fault tolerant.

*Supported in part through the MPI 40% fund and a contract with the CNR "Progetto finalizzato Informatica" and one with the "Progetto finalizzato Robotica".

RESOURCE-ALLOCATION IN A FAULT TOLERANT NEURAL NETWORK

Dott. P. De Pinto
Istituto per la Ricerca dei Sistemi Informatici Paralleli - CNR Napoli.
Prof. F.E.Lauria
Dipartimento di Scienze Fisiche dell'Università di Napoli *
Mostra d'oltremare Padiglione 19 Napoli, I-80125 Italy

Abstract

A problem of the neural network theory is the physical resource allocation. In our network model, differing from the McCulloch and Pitts one because the instructions are stored as weights associated to the coupling between the nodes, the coupling coefficients values correspond to the semantic interpretation of the well-formed strings of its associated language. We present here a simple assembler and the language associated with the network. In order to obtain a higher network reliability and precision we increase the number of network nodes, however the number of coupling coefficients does not grow. The assembler has been designed to be implemented in the OCCAM simulation of our neural network.

*This research was supported in part through the MPI 40 % fund and a contract with the CNR "Progetto finalizzato Informatica" and one with the "Progetto finalizzato Robotica".

A DATA-FLOW LINEAR ARRAY IMPLEMENTING NEURAL NETWORK ARCHITECTURES

M. Marchesi, G. Orlandi , F. Piazza, A. Uncini*

Dip. di Elettronica ed Automatica, Università di Ancona
Via Brecce Bianche - 60131 Ancona, Italy

* Dip. INFOCOM, Università di Roma "La Sapienza"
via Eudossiana, 18 - 00183 Roma, Italy

ABSTRACT

A linear digital data-flow architecture for VLSI implementation of neural networks is proposed. It is characterized by true local connections, full expandibility and reconfigurability. It is composed of a linear array of processing elements (PE), each simulating a single neuron or a set of neurons, depending on the granularity of the actual implementation. Every PE is connected only to its two nearest neighbours through three buses. The proposed architecture is proved able to run many different neural network models. Among them, the multi-layer perceptron with Back-Propagation learning algorithm and the Counter Propagation network. Processed data are pipelined across the architecture and the computational efficiency can reach up to 100% efficiency in favourable cases.

SELF-ORGANIZING NONSYMMETRICAL ALGORITHM AND NONLINEAR SYNAPTIC MODEL OF NEURAL NETWORKS

Algis Garliauskas

Institute of Mathematics and Cybernetics
Lithuanian Academy of Sciences
Akademijos str.4, 232021 Vilnius, Lithuania, USSR

ABSTRACT

In case to improve information technology by using neural networks the spurious states in nonsymmetrical networks were investigated, and the self-organizing algorithm, in which spurious states represent the learning criterion, has been created. The modified symmetrical, nonsymmetrical, and self-organizing algorithms are realized in software. The modelling of neural network with nonlinear synapses is considered. The synapses have N-shaped characteristics with two stable states. The analysis of nonlinear feedbacks in neural network opens the perspective to search for the methods of numerical calculation in network and information memory on the level of dissipative structures.

ON HUMAN ASSOCIATIVE ABILITY (ASSOCIATIVE MEMORY)

Simeon J. MRCHEV (ul. Jordan Mishev 21A; 8600 JAMBOL, Bulgaria)

Modern conceptions classify associations on the basis of the following aspects: *FIRST: Psycho-Linguistic Aspect:* According to their application in cognitive processes associations can be divided into: 1. Memory associations; 2. Thought and speech associations; B: According to their psycho-linguistic complexity: 1. Simple association by proximity/contiguity; 2. Simple association by contrast; 3. Simple association by similarity; 4. Simple second-signal associations; 5. Complex verbal second-signal associations; C: According to the stage of the association formation in the cognitive system: 1. Local association; 2. Personal system association; 3. Intra-system associations; 4. Inter-system associations (inter-object); D: According to the type of relation between stimulus and reaction in verbal memory: Predicative associations of syntactic relations denoting and ACTION (action, agent, condition, instrument, place, modality, object, purpose, quality, addressee, time, truth) and an OBJECT (indefinite, definite: material, immaterial (name, notion, situation, time, space)); *SECOND: Anatomo-physiological Aspect:* Depending on what the engramme associative networks comprise: 1. Network formation on synaptic contact modifications (stochastic wave character, fuzzy matrix character of accumulation and reproduction, etc.); 2. Network formation on the basis of inner feed back and lateral inhibition (time sequences, iterational processing, orthogonal filtres, antagonistic coding, etc.); 3. Network formation on local (neuron, dendrite, synaptic, etc.) and global (the connections between the parts and the whole vertical hierarchy of the brain) plastic abilities and ontogenetic development (outgrowths, micromovements, and other self-organizing connections) in the presence of novelty filtration, adaptive formation of certain quality detectors, etc.; 4. Network formation on unit-columns in which each neuron is an element functionally related to the other elements of the cólumn; 5. The model analogy the network is built upon. G. Hopfield (1982) has drawn a formal analogy between the functioning of simple neuron networks and "spin glass" behaviour (the total spin behaviour in magnetic systems with exceptional magnetic characteristics is similar to the behaviour of atoms in structureless substances like glass). All spin configurations in spin glass are stable. Therefore, spin glass just like memory has in store a number of spatial patterns.

REFERENCES

1. S.J. Mrchev, Holographic Associative Semantic Model of Human Long Term Memory, IEEE/Engineering in Medicine and Biology Society Ninth Annual Conference, 13–16 November 1987, Boston, USA
2. G.V. Primov and S.J. Mrchev, A Holographic Semantic-Associative Human Memory Model and Some Ax Consequences, International Symposium on Optical Memory 1987, 16–18 September 1987, Tokyo, Japan
3. S.J. Mrchev, Modelling and Application of Distributive Model of Human Memory as Distributed Parameter System, IMACS/IFAC International Symposium on Modelling and Simulation of Distributed Parameter Systems, 6–9 October 1987, Hiroshima, Japan
4. J.P. Marinov and S.J. Mrchev, Podhod dlya algoritmicheskogo opisanie-modeli assotsiativnosti chelovecheskoi pamyati, I-ya mejdunarodnaya konferentsiya po osnovnim problemam bioniki "BIONIKA-75", 17–21 sentemvri 1975, Varna, Bulgaria (in russian)

THE STATIC RECONFIGURATION OF A SERIAL SUPERCHIP SYSTEM BASED ON NEURAL WEIGHTS

Y. Bissessur and R.N. Gorgui-Naguib, BSc,MSc,PhD,DIC,CEng,MIEE,MIEEE

Department of Electric and Electronic Engineering
University of Newcastle upon Tyne
Newcastle upon Tyne, NE1 7RU, England.

Abstract

The superchip architecture presented in [1] is aimed at achieving both static and dynamic reconfigurability. With the advent of VLSI technology, it has become cost-effective to include a large number of general purpose processing elements (PE) in a multiprocessing system. This, combined with the varying states of the transmission gates (TX) causes the optimisation of the routing algorithm to develop into an increasingly complex problem. Since the latter then has an immense number of variables, a search for the mathematical optimum by tree search can be of considerable combinatorial difficulty and hence time consuming.

A neural network approach has been suggested in [2] for the implementation of the serial superchip structure. Based on that approach, this paper develops a novel algorithm for finding a near optimum solution to the routing problem posed by the presence of static faults within such structure. The key idea here is to assign a weight value to each of the TXs depending upon their individual current states, and to set up links to these cells with weights equal to their respective values. Once a certain threshold has been defined, it is then possible to discriminate between links to fully working cells and those to faulty ones. The latter links, having weight less than the threshold value, are deleted during one simulation step leaving the structure with only valid links to be considered. This greatly helps in choosing a preferential direction in the route finding process.

PARALLEL IMPLEMENTATION OF THE KOHONEN SELF-ORGANIZATION ALGORITHM

Bernadette Dorizzi and Jean-Marie Auger
Institut National des Télécommunications
9 rue Charles Fourier
91011 EVRY
FRANCE

email : DORIZZI@FRINT51

Abstact :

We study the implementation of Kohonen's self-organization algorithm on a loosely coupled architecture consisting of sixteen T800 Inmos Transputers structured as an hypercube. We show that both the general algorithm and its application to the Traveling Salesman Problem are accelerated by a factor of about 13, which demonstrates the efficiency of our parallelization .

ARCHITECTURES FOR UNCERTAIN REASONING

James Austin, Tom Jackson, Alan Wood

Department of Computer Science,
University of York, York, YO1 5DD UK.

ABSTRACT

This paper presents the basis of a neurocomputer architecture for uncertain reasoning. The arguments for applying neural networks to this domain are discussed together with the broad architecture of the system. The work considers the boarder range aspects of a neural computer architecture, not just the implementation of the neural network. It is shown how a two stage system comprising of a front end SIMD processor and a tightly coupled back end neural processor is to be used in initial investigations. A neural associative architecture is considered for the back end processor that is based upon the ADAM system, a high speed extensible network developed at York. The major issues that relate to the full development of the system are presented, along with an indication of the applications of such a system.

NETWORK DEFINITION LANGUAGES AND DEVELOPMENT ENVIRONMENT

Chair: Philip TRELEAVEN

ORTHANC - AN N-TUPLE NETWORK DEVELOPMENT ENVIROMENT

J.M.Bishop, P.R.Minchinton, R.J.Mitchell.
The Neural Network Research Group,
Cybernetics Department, Reading University, Whiteknights,
READING, Berkshire, Great Britain. RG6 2AL.

The Neural Network Research Group of the Department of Cybernetics at the University of Reading is interested in various topologies of n-Tuple pattern classification networks. These include novel single-layer networks as well as multiple layer networks with new learning strategies. These topologies can be investigated using software simulations for which many features are similar in the different networks. So that these features do not have to be rewritten for each investigation, a flexible network definition environment has been developed which can be used by various members in the group. These definitions are given in a very general manner so as to allow the investigator to configure any form of n-Tuple network. This paper describes the facilities provided.

Introduction

The object of the enviroment is to allow the user to configure any n-Tuple network. Therefore it must be flexible. Typical networks include the WISARD type standard multi discriminator network [1,2], simple feedback pattern association networks [3], the authors' pattern separation technique [4], or multiple layer networks, for example using Probabilistic Logic Neurons (PLNs) [5]. Each of these networks can be configured using a set of standard structures, which are described below.

Environment Description

The basic block in an n-tuple network is a Cell, as shown in figure 1, which provides a very simple model of a neuron. Each cell computes a micro function on its k element input vector [I]. This function is described by a vector of 2^k values [V]. The cell output, given an input vector [i] is thus [Vi]. Hence the n-tuple cell, unlike the generalised McCulloch/Pitts cell, can compute all 2^k functions of its input. Each cell can be used in one of two ways: either information can be stored in the Cell, or data can be read from the Cell.

The major data structure of the system is the Discriminator, shown in figure 2. This consists of an input vector, a block of M Cells and an M element output vector. This is also used in one of two modes: input patterns are learnt or analysed. In each case, for each of the M cells a k bit tuple is formed by suitably sampling the input vector. In learn mode, an appropriate micro function on this tuple is stored in the mth Cell, as described above. In analyse mode, the response of the mth cell to the tuple is used to determine the mth element of the output vector. The response of the discriminator is usually determined by examining its output vector.

A third fundamental structure is the Vector, consisting of a list of data elements.

A complete Network consists of a matrix of interconnected discriminators. The output vector of one discriminator can be passed to the input of the same discriminator (to provide feedback) or to the input of another discriminator in a multi-layer network. In pattern associative networks, the pattern to be learnt can be stored in the output buffer.

The basic Discriminator learn function consists of the following:

```
FOR each Cell DO
   Form tuple by processing the input vector;
   Learn suitable data in to Cell at the tuple address
END
```

The basic analyse routine is:

```
FOR each Cell DO
   Form tuple by processing the input vector;
   Read from Cell at the tuple address and write into
       appropriate location in output vector
END
```

And the response of the network is:

```
Process each element of the output vector
```

The enviroment is implemented by four basic modules, written in Modula 2, which define Networks, Discriminators, Cells and Vectors. An object orientated approach was adopted for their implementation, so each module is self contained, having both the appropriate data structures and the associated procedures. This ensures that programming errors are minimal and easy to find. Also provided in the enviroment are facilities for conducting tests and logging the results.

Customisation

The user is able to configure the network by specifying the number of Discriminators, the sizes of the Vectors and Cells, and such factors as the size of a tuple. Default procedures for learning, etc., are provided, but the user can customise the operation of the network by redefining any appropriate procedures.

Implementation of a Standard n-tuple network

In a standard n-tuple pattern classification network, all elements in each cell are set to ZERO before any data are learnt. Then the learn function consists of writing a ONE into each addressed element in a Cell, where the addresses are usually generated by random sampling of the input vector. On analysis, the contents of each addressed Cell are copied into the appropriate location in the output vector, and the response of the network is the number of ONEs in the output vector.

The Cells and the output vectors contain binary data, and for binary input images, the input vector contains binary data.

Thus the configurable procedures for such a network are those which form the next bit of the tuple, learn a ONE into the addressed element in a Cell, check whether such an element has a ONE stored in it (in analyse mode), and, for finding the response, reporting if a ONE has been stored in the appropriate element in the output vector.

Other Networks

Different requirements are needed for other networks. For example, in the authors' Pattern Separtion technique, the Cells must contain data which can take four values, not the usual two binary values. Also, the learn process requires that an element of the class identifer and the contents of an addressed Cell be both processed to determine the value to be stored in the Cell. In the analysis phase, both the addressed Cell and the input vector must be processed to determine the output. These can all be achieved by suitable configuration of the development enviroment and writing a few extra procedures.

Conclusion

The ORTHANC system has enabled the quick and compact generation of Modula 2 programs for the analysis of various n-tuple networks. By careful use of the library of routines provided it is possible to go from idea to the logging of test results in under an hour.

References

[1] Bledsoe, W.W, & Browning, I:
"Pattern recognition and reading by machine".
Proc Eastern Jnt Comp. Conf, 1959, pp252-232.

[2] Aleksander, I, Thomas, W.V & Bowden, P.A.
"WISARD: a Radical Step Forward in Image Recognition"
Sensor Review. 4(3),120-124(1984)

[3] Aleksander, I. & Mandani, E.H:
"Microcircuit learning nets: improved recognition by means of pattern feedback".
Electron. Lett. 1968, 4, (20), pp425-426.

[4] Aitken, D., Bishop, J.M., Mitchell, R.J., & Pepper, S.
"Pattern Separation in digital learning nets".
Electron. Lett. 1989, 25, (11), pp685-686.

5] Aleksander, I.
"The logic of connectionist systems"
Neural Computing Architectures. Ed. Aleksander, I.
North Oxford Academic. 1989, Ch. 8. p133-155.

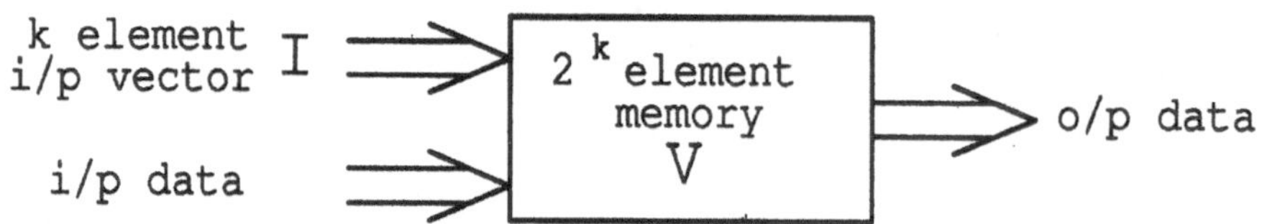

Figure 1. Block diagram of a Cell

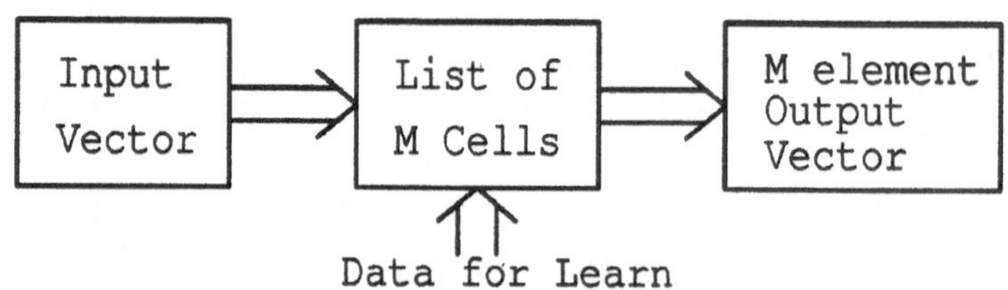

Figure 2. Block diagram of a Discriminator

A VIRTUAL MACHINE MODEL FOR ARTIFICIAL NEURAL NETWORK PROGRAMMING

Pierre Bessière[1], Ali Chams[2] & Traian Muntean[3]

ABSTRACT

This paper introduces the model of a virtual machine for A.N.N. (Artificial Neural Networks).

The context of this work is a collaborative project to study new V.L.S.I. implementations and new architectures for neuronal machines[4]. The work consists in the specification and a prototype implementation of a description language for A.N.N., of the associated virtual machine, of the compiler between them and of the compilers mapping the virtual machine on different highly parallel computers.

In this short paper we present the virtual machine model which combines the features of various parallel programming paradigms. Our model allows, in particular, to have the same A.N.N. program running on both synchronous or asynchronous type of machines. In this framework a parallel architecture (S.M.A.R.T.) and a dynamically reconfigurable parallel machine of Transputers® (SuperNode) are considered as target machines.

---

[1]L.G.I./I.M.A.G. : Laboratoire de Génie Informatique / institut d'Informatique et de Mathématique Appliquée de Grenoble. Adress : B.P. 53X, F-38041 Grenoble, FRANCE. Phone: 33-76.51.45.72. E-mail: bessiere@imag.imag.fr.

[2]L.T.I.R.F. : Laboratoire de Traitement d'Image et de Reconnaissance des Formes / institut d'Informatique et de Mathématique Appliquée de Grenoble. Adress : 46 ave Felix Viallet, F-38031 Grenoble cedex, FRANCE. Phone: 33-76.57.45.50.

[3]L.G.I./I.M.A.G. : Laboratoire de Génie Informatique / institut d'Informatique et de Mathématique Appliquée de Grenoble. Adress : B.P. 53X, F-38041 Grenoble, FRANCE. Phone: 33-76.51.48.64 E-mail: muntean@imag.imag.fr.

[4]The work described in this paper is part of an ESPRIT-B.R.A. (Basic Research Action) project named NERVES (NEurocomputing Research and Vlsi for Enhanced Systems).

I. INTRODUCTION

We will enhance mainly the concepts of a virtual machine which combine various programming paradigms encountered in ANN.

The introduction of a virtual machine has several advantages:

- the definition of the virtual machine instructions obliges to clarify the fundamental concepts and their semantic properties;
- the programming language for A.N.N. is independent of the actual target machines;
- only the compiler between the virtual machine and the target machine should be written when a new target machine is considered.

In a first phase two target machines have been selected for experiments: S.M.A.R.T. (Sparse Matrix Adaptative Recursive Transforms) and SuperNodes of Transputers®.

SMART is a parallel machine dedicated to A.N.N., which is under development at L.T.I.R.F. (Laboratoire de Traitement de l'Image et de Reconnaissance des Formes). It has been designed to be especially efficient for sparse matrix operations which are some of the basic computations needed for A.N.N. For further descriptions see: [Jutten90] & [Lawson90].

SuperNode is a general purpose, dynamically reconfigurable, parallel machine based on Transputer®. The underlying programming paradigm is C.S.P. [Hoare85]. SuperNode machines have been developed in the ESPRIT project 1085 between 1986 and 1988. For detail descriptions see: [Harp86] & [Muntean89].

II. MOTIVATIONS AND MAIN DIFFICULTIES

II.1. Motivations

A.N.N. are intrinsically highly parallel. However, most existing implementations are still sequential. Numerous works are under way to build parallel machines dedicated to A.N.N..

Such parallel computers need a programming environment. They need especially, user friendly interfaces and powerful programming languages.

The existence of such a programming environment is essential for a practical use of parallel computers to emulate A.N.N.

II.2. Main difficulties

Two main difficulties have been encountered:

- The design of a programming language should be considered independently of the features of the different computers. Classical sequential architectures, even if they lead to different computers, they share the same underlying programming paradigm. In parallel computing the situation is quite different, there are several very different parallel programming models due to various possible control structures of concurrency; synchrony or asynchrony of process behavior is to be considered, nondeterminism due to multiple execution paths can arise. A hard problem for a parallel programming language for A.N.N. is to make a synthesis between those different parallel programming paradigms.

- Usually, A.N.N. have simple treatment processes and complex interconnection schemes between them. Consequently, a programming language for A.N.N. has to propose powerful features to describe this aspect. The second difficulty has been precisely to find a right and efficient way to specify the network architecture and connectivity since it is not a common feature of usual programming languages.

This second difficulty will not be addressed in such a short paper.

III. APPROACH

III.1. Basic concepts of the virtual machine

The implementation principle is a classical virtual machine approach (see figure 1 below).

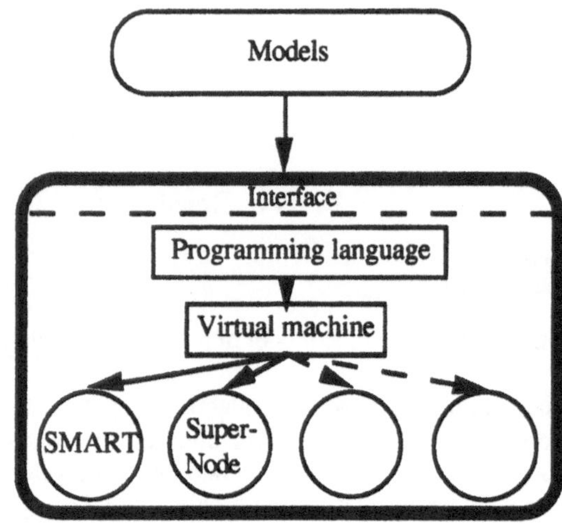

figure 1

This approach has some difficulties, especially, the clarification of the fundamental concepts when defining the virtual machine. This is one of the advanced research topics.

The synthesis between the different parallel programming paradigms is based on the following concepts introduced for the description of any A.N.N. at the virtual machine level:

- cell;
- aggregate;
- connector.

A network could therefore be represented graphically (figure 2) by using circles for cells, sets of cells for aggregates, and lozenges for connectors.

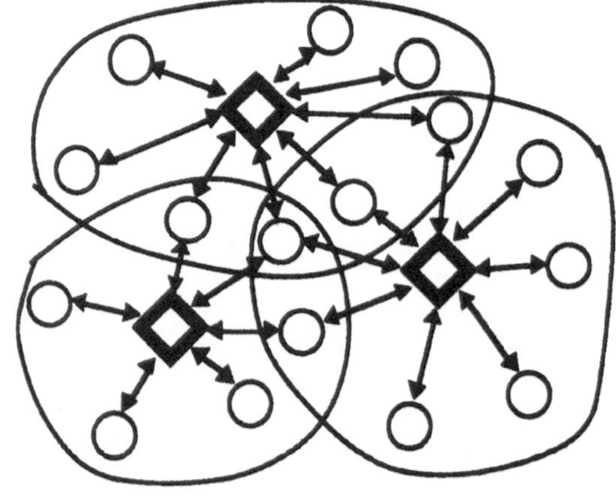

figure 2

III.1.1. Cell

A cell is a process executing a program as soon as its inputs are available. Consequently, there is no way to control explicitly the execution of cells' programs since they are triggered by the availability of their inputs from other cells. As cells behavior is only function of their interaction and their description program, control structures of the language should reflect the kind of synchronizations attached to any interaction. This function will be devoted to connectors.

III.1.2. Aggregate and connectors

An aggregate is a set of potentially fully connected cells, whose activities are temporarily correlated.

A cell can be a member of more than one aggregate. Any cell in an aggregate can exchange information with any other cell in the same aggregate. This exchange of information can be synchronous or asynchronous. In both cases, this correlation is expressed by a protocol. The connectors are communication objects defined to implement those protocols.

The principal information, stored in a connector, is a connection matrix W. If an aggregate connects N cells, the matrix size is N^2. In most cases the connection matrix is sparse, and it reflects the topological structure of the aggregate. In fact, the connection matrix represents more than communication capabilities, it contains entries (real or integer) usually called synaptic weights. These weights are used as parameters for a composition function C and an integration function I which are used to transform information exchanged by the cells; the informations should go through the same function IoC before being broadcasted to cells of an aggregate. Typically, C is a multiplication function and I is a summation one. A learning function A is associated to each aggregate. For any given connection matrix and some parameters reflecting the past activity of the network, A produces a new connection matrix.

Examples:

- For "game of life" [Gardner70] a possible modelization is made of one aggregate for the all population of cells. W is a very sparse matrix which encodes the 8-nearest-neighbors neighborhood. All its non zero entries are fixed to one. C is multiplication and I is summation. All the cells have the same program consisting in the usual test on the number of alive neighbors. A is obviously identity because no learning occurs in this model.
- For "competitive learning" models [Rumelhart86], two kinds of aggregates exist. One kind groups the cells of two connected levels, the other kind corresponds to the set of cells of a cluster. For the first kind, W encodes the full forward connection from level n to level n+1, weights are initialize at random values, C is multiplication and I is summation, A is the standard competitive learning function. For the second kind of aggregate, W encodes full connection inside a cluster, weights are all set to -1 but the entries of the diagonal of W which are set to n-1 if there are n cells in the cluster, C is multiplication and I is summation and each cell returns the value 1 if the result of IoC is positive and 0 otherwise.
- Some realistic asynchronous simulations of parts of brain could be encoded using one aggregate for each sets of neurons possibly interesting from a biological point of view.

A basic operation of the virtual machine is the activation of an aggregate. Roughly speaking, this activation consists in broadcasting the informations delivered by every cell of the aggregate to the others, through the transformation IoC. This process means the automatic execution of the programs associated to all the cells of an aggregate, as soon as their inputs are available. In fact, the differences between synchronous and asynchronous models appear in this operation.

III.2. Activation of synchronous aggregate

If we suppose that each cell delivers an output, so each synchronous aggregate has a vector V representing the output values for all the cells; then v_j (jth component of V) gives the output of the jth cell in the aggregate.

The composition function C can be applied for each couple (w_{ij}, v_j) to obtain c_{ij}, then the integration function I produces for all N-tuples of c_{ij} the result vector R. In the case where C is multiplication and I is addition, the whole operation can be written using a matrix operation $R=W*V$.

Each r_i can be an input for the jth cell, whenever the inputs are available, the jth cell re-execute automatically its own program to produce the new v_j, for another iteration.

Examples:

- The synchronous activation of aggregate for game of life is the usual scheme. If V is generation n, then R is generation n+1.
- For competitive learning we have first synchronous activation of the aggregate corresponding to the transition from layer 1 to layer 2, the activation of the aggregates corresponding to the clusters of layer 2, then activation of the aggregate for the transition from layer 2 to layer 3, and so on until output layer is reached.

III.3 Activation of asynchronous aggregate

The asynchronous model has "time" as an extra information to be exchanged between cells. Three values are delivered by an asynchronous cell after execution of its program:

- the outputs;
- the time when the values are delivered;
- the expected time for the next exchange of information between the connector and the cell.

The temporal informations are controlled by the connector which will wake up the cells at the right time. Asynchronous models use the data structures described previously for synchronous models, with the additional items:

- a vector ***H*** containing the arrival time of the last messages coming from each cell (h_i for the *i*th cell);
- a vector ***U*** containing the next activation time for each cell;
- a counter ***m*** representing the last activation time for a cell in the same aggregate.

After the initialization phase of ***W***, ***V***, ***H***, ***U***, and ***m*** (***m***=0), the activation of an aggregate is described by the following algorithm:

- search the next cell(s) to wake up and update m;
- send to all such cells the vector ***V*** after application of the composition function ***C***;
- send to these cells the vector ***H***;
- all cells execute the function ***I*** and their own programs as soon as ***V*** and ***H*** are available;
- each cell i sends to the connector its output values, the present time (h_i), and its next activation time (u_i).

Note that the synchronous model can be emulated in the asynchronous mode by waking up the cells at each clock step.

Examples:

- As function ***I*** is executed within each cell, its expression may be much more complex. In particular, ***I*** can take into account information on the state of the cell and on the delays (***H***), and may be used to simulate behaviors like attenuation of action potentials, refractory periods or hysteresis, etc.... This is, of course, of great importance for simulation of realistic biological model.

IV. CONCLUSION

Compared to the other works having similar purposes, our approach has two specificities:

- the choice of the virtual machine methodology to construct a programming environment for A.N.N.;
- the choice of massively parallel architectures as hardware support for this environment.

The basic concepts of this virtual machine for A.N.N. have been validated writing programs for several classical examples of neural networks models. Unfortunately, there is no room in this paper to describe those examples.

A first specification of the programming language has been produced. Implementation of the three needed compilers has not been done yet. This work will take place during the coming last two years of the NE.R.V.E.S. project.

We expect through this approach to come out with a tool usable on a wide range of machines and offering full programs portability.

We also hope that work about the virtual machine will have significantly contributed to clarify the A.N.N. paradigm.

V. BIBLIOGRAPHY

[Gardner70] Martin Gardner
The fantastic combinations of John Conway's new solitaire game of life
Scientific American, N° 223, 1970

[Harp86] J.G. Harp, C.R. Jesshope, T. Muntean & C. Whitby-Strevens
SuperNode project P1085: Developpement & application of a low cost high performance multiprocessor machine
ESPRIT 86, Bruxelles, 1986

[Hoare85] C.A.R. Hoare
Communicating Sequential Processes
Prentice Hall International, 1985

[Jutten90] C. Jutten, A. Guerin & J. Herault
Simulation machine and integrated implementation of Neural Networks
to appear Proceedings of EURASIP, Portugal, 1990

[Lawson90] J.C. Lawson, A. Chams & J. Herault
SMART: how to simulate huge networks
Submited to this conference

[Muntean89] T. Muntean
SuperNode: achitecture parallele et dynamiquement reconfigurable de transputers
in Proceedings of "Onzième journées francophones sur l'informatique"
EC2 éditeur, 1989

[Rumelhart86]D.E. Rumelhart & D. Zisper
Feature discovery by competitive learning
in *Parallel Distributed Processing*
M.I.T. Press, 1986

Optimisation of Network Structure using Genetic Techniques

Nigel Dodd*
Research Initiative in Pattern Recognition,
Royal Signals and Radar Establishment,
Malvern, WR14 3PS, UK
nigel@uk.mod.hermes

Abstract

We present arguments that it is necessary to use *structured* neural networks for the solution of certain problem types. Structure is imposed on connectivity, activation functions and other parameters of the network to simultaneously optimise generalisation ability and compactness of the network. An analogy is made between the development of biological nervous systems from their genetic coding and the generation of artificial neural networks from a parametric description. Experiments are described which use genetic techniques to optimise network structure for a specific class of problem. Results are given which demonstrate the effectiveness of genetic optimisation of network specifications in comparison with other optimisation techniques. The parallel asynchronous implementation of genetic algorithms on a network of Sun's is briefly described.

1 Introduction

An important property of the multi-layer perceptron (MLP) is *generalisation*. Without this ability an MLP may be considered to be simply a look-up table. Many examples from the literature demonstrate that, when the structure of a network addresses the underlying symmetries and invariances of the data classification problem, the network is able to generalise well, learn quickly and is often of a compact form. These examples include

Parity A conventional fully connected, layered MLP with logistic activation functions fails to learn parity of more than 4 bits. When the network is structured appropriately, the solution for an arbitrary number of bits is always found quickly using error backpropagation and the network is extremely compact [3].

Sequential Context Much work, [6], has been done using structured MLPs for speech recognition. [2] describes networks with delay-line and recurrent sub-structures which perform to the Baum-Welch limit in state-sequence classification.

Transform Invariance Recognition of handwritten characters [1] requires invariance to translation. Networks predisposed to these and other invariances can be built and are termed "spread networks".

*Supported by SMITHS INDUSTRIES Aerospace & Defence Systems, Bishops Cleeve, Cheltenham, GL52 4HZ, UK

For problem types where the mechanism of data production is well understood, human engineering of a suitable network structure is often possible. Where we do not have this knowledge it is desirable to make use of an automatic method for network optimisation. The formulation of such a method is the aim of this research.

The manner in which the genes map onto a network architecture is of crucial importance. If the mapping is such that a gene controls the presence or absence of a node or of a link, then it is possible for the generation process to create networks without a path existing between input and output or networks with unreachable subnetworks or other congenital defects [4]. Ideally we would like a genetic space of network structure which does not contain any unviable network genotypes, but which spans the space of all potentially useful network genotypes. A parametric description of a specific network type is such a space, and we used this level of genetic description for our experiments, choosing two examples whose invariance required a "spread" network. Results are presented for automatic generation of network structure for two example problems, one toy (with 100 connections) consisting of a line orientation problem, and one real (with 5000 connections) consisting of the identification of dolphin sounds. Both problems exhibit spatial invariance along one or two axes and so are amenable to the general class of spread networks. Optimisation is done over the height and width of the replicated weight pattern, the horizontal and vertical increment of this sub-structure and learning rate. Fitness is taken to be inversely proportional to residual error and network complexity.

2 Criteria for Choice of Optimisation Technique

In general the parameters of a network specification are qualitatively different from one another. It is desirable to use an optimisation method that is insensitive to correlations between the network parameters. Different network types will have quite different descriptive parameters and a general optimisation technique will cater for all network types.

It is desirable to explore several different regions of search in parallel since the parameters may have no correlation and the search space may be irregular. It may also be productive to swap the values of the parameters between the various search points.

Runs of the same network with different starting weights can yield quite different figures of performance. The optimisation technique must be tolerant of such noise present on estimates of fitness.

These and other requirements lead to the use of genetic algorithms [5] to find near optimal combinations of the network generation parameters. Two, more conventional, optimisation techniques were implemented for comparison.

3 Detail of Optimisation Techniques

Pure crossover was implemented as follows. A starting generation was produced from a flat distribution over the genetic space. Parents F and M were selected from the starting generation with a probability equal to their normalised fitness. From two parents an offspring genotype was created by randomly selecting a splice point within or at the extremes of the chromosome. To the left of the splice point genes of parent F were inherited, and to the right of the splice point genes of parent M were inherited. A population equal in number to the starting population was built up in this way, and the fitness of each member of the population was evaluated. Subsequent populations were similarly generated. Pseudo-code is given in figure 1

```
for(number of populations)
for(number in population)
 {
 select parents from population with
 probability proportional to fitness

 mate
 create 1 offspring
 calculate fitness
 }
```

Figure 1: Pseudo-code for Crossover

```
select a random genome G1
evaluate fitness, f1m, of G1
for(number of evaluations)
    {
    mutate G1, call the mutation G2
    evaluate fitness, f2, of G2
    f2m = mean(any previous
     fitness valuations of G2)
    if G1 = G2 then reevaluate f1m,
     the new mean
    if f2m > f1m then
     G1 = G2, f1m = f2m
    }
```

Figure 2: Pseudo-code for PGD

The two conventional optimisation techniques tested on the "lines" data were firstly a random search, and secondly a pseudo-gradient descent (PGD). The crossover technique finished with a generation of fit genotypes. A generation size of 100 individuals was selected as a satisfactory compromise. The comparison techniques were also implemented to finish with a set of 100 fit individuals. Estimates of fitness are taken as the mean over each set of 100 individuals.

1. For the random search, a number of evaluations equal to the total number of evaluations made in the genetic crossover approach were made. The values of the genes were uniformly distributed over the allowable genetic range.
2. The PGD pseudo-code is listed in figure 1. It is only possible to evaluate exactly a phenotype. The fitness of a genotype, of which the phenotype is an expression, is plagued with noise. By repeated evaluations of genotypes near the optima in gene space, the PGD method reduces the noise of genotype evaluation in the interesting regions of gene space.

The three algorithms described in this paper were all capable of parallel implementation and many of the experimental results were obtained using an asynchronous parallel implementation on a network of Sun 3's. A semaphore protocol was employed to avoid file access contention.

4 Comparison of the Optimisation Techniques

A comparison of the three optimisation techniques is shown in the following table.

	random	PGD	Crossover
optimum	10.2	19.3	17.0
re-evalutaion	6.1	8.5	17.0

It can be seen that the pseudo-gradient descent algorithm finds the fittest phenotypes, whereas crossover finds the fittest genotypes. We are interested in finding the fit genotypes since we want a network description which, given any weight start, will perform well. The phenotype in this context is the network plus a set of random weight starts which the PGD finds effectively. These apparently fit individuals, however, do not perform well when re-evaluated with another set of random weight

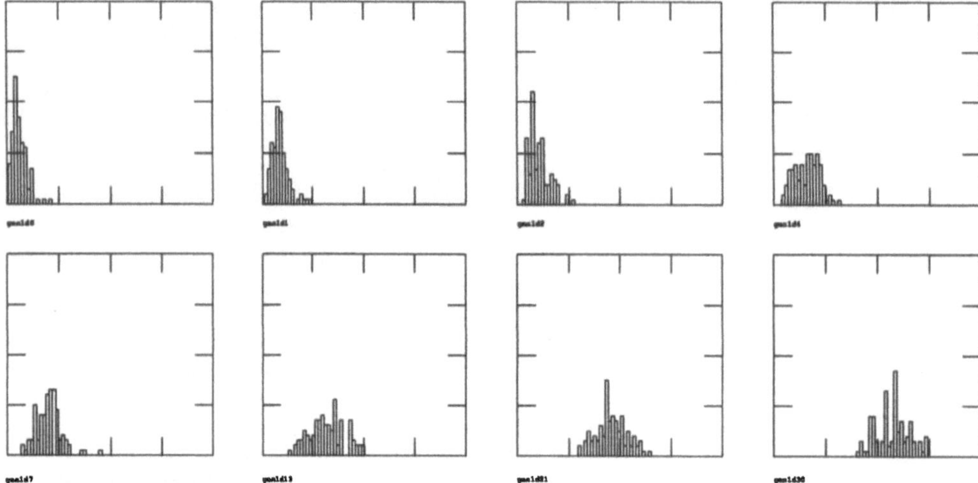

Figure 3: Frequency histograms (frequency of occurrence vs. fitness) for the "dolphin" data. The last number in the label corresponds to the generation number.

starts. The crossover technique is, by comparison, very effective in avoiding individuals which evaluate well with a special starting weight set. As expected for the crossover algorithm, the mean fitness of a re-evaluation of a generation is approximately identical to the original evaluation of fitness of the generation.

As a demonstration of the increased fitness of successive generations, histograms are given in figure 3. This shows the generations of dolphin sound recognition networks increasing by a factor of three, or so, over 30 generations. Even with the toy problem, it is found that the genetically derived optimal networks differ from, and are more efficient than, the networks engineered by humans.

References

[1] Y. Le Cun et al. Hand written digit recognition with a back-propagation network. In *Proceedings of Neural Information Processing Systems—Natural and Synthetic*, 1989.

[2] Nigel Dodd. *The GMLP program used as a state sequence classifier.* Technical Report RIPRREP/1000/42/89, Research Initiative in Pattern Recognition, 1989.

[3] Nigel Dodd. *Parity implemented on an MLP with a sinusoidal activation function.* Technical Report RIPRREP/1000/46/89, Research Initiative in Pattern Recognition, 1989.

[4] Steven A Harp, Tariq Samad, and Aloke Guha. Designing applications-specific neural networks using the genetic algorithm. In *Proceedings of Neural Information Processing Systems—Natural and Synthetic*, 1989.

[5] John H Holland. *Adaptation in Natural and Artificial Systems.* Ann Arbor: University of Michigan Press, 1975.

[6] Richard P. Lippmann. Review of neural networks for speech recognition. *Neural Computation*, 1(1), 1989.

NEURAL INTERACTIVE PARADIGM AS A MORE EFFECTIVE COMPUTER SYSTEM ENVIRONMENT

Yuri Shestov

Department of Computer Science
MET College, Boston University
Boston, MA 02115

TU, Inc.
8460 Tyco Rd.
Vienna, VA 22180

Abstract.

Implementation of neural networks as a backbone of memory and user interface management systems allows greatly improve user-system interaction, as well as increased computer system performance. It does not violate the native structure of the system, but enables adaptive, situation-sensitive capability through presentation and positioning of the memory items according to the most likely utilization at a given computing situation. That prediction capability comes through on-the-job learning experience of previous interactions. User interaction, caching mechanism, parallel or distributed processing may all utilize such information on the most likely computational flow. Appropriate algorithms for modification of synaptic strength in connections of memory items and attention span filtering scheme may fine tune the adaptation of the system for various users and tasks performed. This self-organizing adaptive environment, through normal use of the computer system, gains corresponding expertise, which is utilized in consequent interaction with the system.

Organization of memory.

Neural Interactive Paradigm (NIP) on one side utilizes the idea of hypertext or hypermedia-type connections of logically or procedurally related memory items.[1] The other side utilizes ideas of neural networks.[2] Effectiveness of the hypermedia scheme for development and use of many applications is well known, but has several drawbacks. Among those, two are of particular interest here. Numerous yet desirable connections are hard to navigate, they clutter the screen, and a connection important to one user or particular situation may be irrelevant to a different user or situation. Another typical restriction of the hypermedia memory organization is that it typically creates additional and specific memory structures of directly connected memory items, frequently duplicating information and forcing a specific format or size for connected items. Neural Interactive Paradigm resolves both problems and provides a greater number of advantages to main-stream computing through the adaptive and situation-sensitive organization of memory.

In NIP the later restriction is removed, not through the connection of the memory items themselves, but rather the connection of their small and easily manipulated representatives, called memory descriptors or information nodes. A memory descriptor provides the pointer to and other information on the corresponding memory item and information on connections to other memory descriptors. Because additional pertinent information may be stored in a memory descriptor, it carries the additional advantage of being a descriptor of information on a generic situation like Minsky's frame, which may be utilized as a part of a knowledge base.[3] With such a structure, NIP does not violate or segregate native file systems, connecting different type and size memory items into more uniform organization of the memory descriptors. Those become principle objects of manipulation up to the moment of the needed invocation of underlying memory.

Memory descriptors in NIP incorporate information on the neural state of the environment, including threshold, excitation, and normalized weights (strength of associations to other connected memory descriptors). A node fires (is active) when the excitation of this node is greater than a threshold (greater than zero). Neural excitation in such organized paradigm propagates (leaks) from firing (active) nodes through local connections to the neighboring ones. A special user node representing a current user of the system fires through certain user activity (e.g., via keyboard or mouse).

The number of nodes excited in the current situation may be considerable (e.g., a word processor, called by user, may excite a large number of files that could be edited with the word processor). Thus, to narrow the attention of the system or the user on a related subset (e.g., a preferred subset of files recently or frequently accessed by the user), it is appropriate to introduce a filter. An adjustable attention span serves such a purpose: to filter a mass of excited nodes to manageable size. Parameters of the attention span could be a number of nodes not to be exceeded, or a cut-off level of excitation for nodes to be filtered out. That mechanism alleviates principal drawbacks of traditional hypermedia environments. The same filter may play a role in mechanism of propagation of excitations. Another algorithm of auto-transition is introduced for automatic invocation of memory items and a corresponding change of the attention span, with a default time window to interrupt the process.

The total collection of all interconnected memory descriptors, with a specific excitation propagation algorithm, is called a memory manifold. At this point it is still a relatively simple structure, but with potentially great capabilities. One may experiment in this adaptive environment by choosing different excitation propagation schemes. Additionally, it is well suited to implement appropriate logical inference mechanisms - as with a power to explain actions of neural network decisions, since the later operates in a feature domain.[4] In NIP the border between knowledge base environment and neural paradigm need not be rigid - adaptive learning may make a conditioned expert rule, and a rule of the knowledge base may be scrapped or modified in some situations through adaptive learning. The two mechanisms are complementary to each other.

To summarize, the NIP structure is an arbitrary directed graph with labeled nodes and links, consisting of two functional layers. One layer is a collection of neurally interconnected memory descriptors with a second underlying layer of executables and data memory items (typically a file or a part of one). Those are pointed to from the corresponding memory descriptors. The memory item may be invoked from the corresponding controlling descriptor. The control of the interaction with the system and execution occurs with the help of the propagation of excitation on the top neural layer. The latter presents user or system with filtered, more probable in the situation, choices to continue the interaction. The NIP operation may proceed as follows: from a particular situation, reflected in memory manifold excitation pattern, attention span filters in an appropriate manageable subset; excitations are recomputed on that base; and, if not preampted, system transits into new situation. Several nodes of the system may trigger some distributed actions and other interacting systems may change the NIP excitations, which in turn may trigger new excitation flows. The resulting excitation pattern makes a new stage for the next cycle of interaction.

Learning in the system occurs through interaction with it. Weights can be modified when desired or actual transition differs from the one suggested by the system. This disparity is easily detected and the learning subsystem is invoked. That is, when the user or system forces transition to a particular situation different than that suggested by NIP, and the transition is satisfactory, the weights for the appropriately computed short path are modified (e.g., increased), and weights of the pertinent links are renormalized. Even connection architecture may be modified in some situations.

Implementation.

The implementation of such paradigm is completed in DOS as a Hyper Information Manager (HIM), which provides self-contained (with corresponding user interface), dynamically adaptable, interactive environment. A Markov chain is adopted as a restricted, but easily implemented, model for excitation propagation. Excitations and weights are real numbers from 0 to 1. A particular computer system situation is represented as a pattern of activity of the system memory - a distribution of excitations of memory descriptors. It is an n-dimensional vector of (e_i) excitations of memory nodes, which can be interpreted as a probability of each memory item n (corresponding to the component of the vector) to be invoked. Weighted connections are represented by n*n matrix (w_{ij}), correspondingly interpreted as probabilities of the transition of activity from node i to node j. One step of propagation in this model is equivalent to multiplication of a state vector of excitations by the transition matrix. The dynamic behavior of such system has been studied.[5] The system becomes fuzzy when propagation is computed from the filtered and normalized attention span mask.

HIM is a rather simple yet effective demonstration of the potential of NIP, inviting further exploration. A UNIX version of HIM is being developed with more interesting variations on neural dynamics in memory manifold. The mathematical foundation for this kind of memory organization is being investigated. The Neural Interactive Paradigm is wide open for different implementations and experiments.

Conclusion.

Proposed memory organization has the following theoretical and practical implications:

* A more streamlined interaction with complex systems, using experience of experts, is achieved through learning information, not interrogation, on typical computational flow and actions of experts actually using the system and suggesting those choices to less experienced users.
* An emergence of intelligent behavior in NIP through a reasonable anticipation and response from the knowledge of some of the most likely computational behavior patterns for given situations, recorded in the learned weights from previous experiences.
* An efficient cache memory procedure can be created by bringing into the cache those pieces of memory that are more likely to be used next, through look-ahead or learned anticipation for those from previous experience in similar situations.
* NIP is a natural environment for parallel and distributed computation, with automatic invocation of processes, based on the appropriate excitation of "ready to go" memory in the environment of those processes and information on most likely future for such distributed computational flow.
* Natural integration of neurocomputer architecture into existing and well established base of computing machines.

NIP offers an effective and natural environment to interact with, and manage, great volumes of information by compressing access to the information and making it handy - filtered and localized - according to situation-dependent demand, learned on similar situations. This is timely and important development in light of the rising information management bottleneck.[6] In the future, it may help to handle on-line gigabytes of interrelated information.

ACKNOWLEDGMENT

The research and development was done under a contract with TU, Inc.

REFERENCES

1. **Conklin, E.J.**, *Hypertext: An Introduction and a Survey.* IEEE Computer 2, 9, 17-41, Sept. 1987.
2. **Shestov, Yuri**, *A Neural Network Enhancement to von Neuman Computer Architecture.* Journal of Neural Network Computing, 41-45, Winter 1989.
3. **Minsky, Marvin**, *A Framework for Representing Knowledge.* In Mind Design, 95-128, edited by J. Haugeland, The MIT Press, 1981.
4. **Shestov, Yuri**, *A Rule Extraction from NNs Operating in a Feature Domains.* Draft.
5. **Feller, William**, *An Introduction to Probability Theory and Its Applications.* Vol. 1, Ch. XVff, John Wiley & Sons, 1957.
6. **Straub, D.W. and Wetherbe, J.C.**, *Information Technologies for the 1990s: An Organizational Impact Perspecitve*, Communications of ACM, 1328-1329, November 1989.

FUNCTIONAL DESCRIPTIONS OF NEURAL NETWORKS[1]

P.W.M. Koopman, L.M.W.J. Rutten, M.C.J.D. van Eekelen and M.J. Plasmeijer.
Department of Computer Science, University of Nijmegen,
Toernooiveld 1, 6525 ED Nijmegen, The Netherlands.

Abstract

In this paper the use of functional programming languages is proposed for the formal specification of neural networks. A formal specification written in a functional programming language has many advantages. First of all, a very high-level, compact, mathematically based description of the functional behaviour of a network and its components can be given. This is due to the presence of higher-order functions and the possibility to define high-level tools to manipulate functions. The compactness is illustrated by the fact that this paper contains a complete description of an xor network. Another advantage of such a specification is that it is directly executable. Hence it can serve as a prototype implementation of the network. Partial correctness of the specification (such as type consistency) is checked automatically by the compiler of the functional programming language. A specification in a functional language inherently makes it possible to combine classical computations with neural computations.

Introduction

In order to detect which aspects of neural networks can be applied with success in computer science, it is necessary to have a good understanding of the paradigm used in neural network computations. This can be done by studying each network separately or by developing a uniform description for the various neural networks. A high-level uniform description of the neural networks will reveal the essential aspects in which they differ and in which they are essentially the same. It is very important to develop such a uniform description since many existing networks contain components that were merely introduced for a particular application. In this paper the use of functional programming languages is advocated to obtain such a uniform specification.

Functional programming languages enable a high-level compact specification of neural networks since the mathematical concept of a function is incorporated in the language in a uniform manner. Functions are 'first class citizens': they can be passed as arguments to other functions and they can be the result of the evaluation of an expression. In the description of neural networks given below this is heavily used e.g. to parametrize neurons with their synapse and output generator functions.

The description of a neural network in a functional programming language

To illustrate the descriptive power of functional programming languages, a simple neural network is described. Only a small subset of the functional programming language Miranda[2] [Turner 86] is used. Language constructs are explained wherever they are used for the first time. There are several good textbooks introducing functional programming [e.g. Field 88]. Miranda program fragments in this paper will be written in `courier`.

First of all it must be decided at what level we want to describe the neural network. There are many possible descriptions for each biological network. These descriptions range at least from the chemical processes inside and between the neurons up to statistical descriptions of ensembles of neurons. Artificial neural networks are always an approximation of biological neural networks. A recent overview of artificial neural networks and neurocomputers is given by Treleaven [Treleaven 89]. We present a description of the neurons as computing elements comparable with the digital logic level in an ordinary computer.

A major simplification used in many descriptions is the use of discrete time. The output of each neuron changes only at discrete equally spaced time intervals. Also the information travelling over

---

[1]This work is partly sponsored by the Dutch ministry of Science & Education via the Neural Network Foundation.
[2]Miranda is a trademark of Research Software Ltd.

the axons changes only in this way. This implies that axons can be adequately modelled by their values at the sampling points. In a functional language these sampling points are collected in a list of values, in Miranda denoted as `[value]`. Miranda is a lazy functional language, which implies that the elements of a list are not computed until their value is required. Each neuron takes some axons as input and has one output axon. This output axon can be used as input by as many neurons as required. To represent the inputs, a list type will be used again.

```
axon      == [value]               || The type of an axon is a list of values.
neuron    == [axon] -> axon        || A neuron is a function from a list of axons to an axon.
```

The symbol '||' denotes in Miranda that the remainder of the line contains a comment.

Different delays on the axons can be modelled adequately by inserting some additional values on the axon. Each neuron can have its own output axon as input provided that a delay of at least one time unit is inserted. Delays are not considered in the remainder of this paper.

Each neuron consists of a synapse which produces an internal potential, and a generator which produces an output value based on this potential. This kind of element is introduced by McCulloch and Pitts [McCulloch 43] and has become well known through the work of Hopfield [Hopfield 82]. Elements of this kind are also used in Perceptrons [Rosenblatt 61].

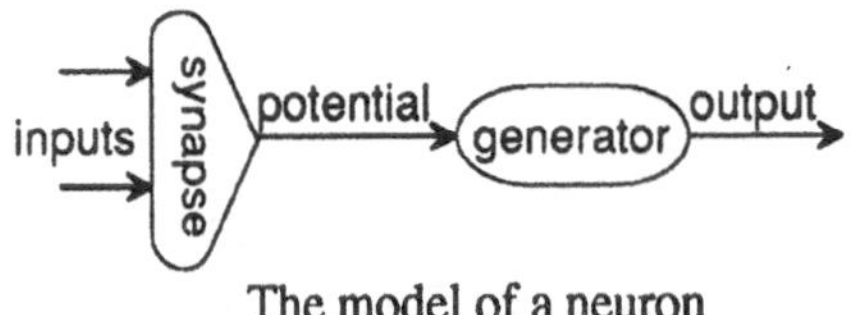

The model of a neuron

In the most simple model the potential is only determined by the current input values and the output value is a deterministic function of the potential. The current input values are called state. The states are obtained by taking the values corresponding to each sample time from the input axons. In our sample model the types are:

```
state     == [value]               || The input state is represented by the list of values.
potential == num                   || The potential is a (real) number.
synapse   == state -> potential    || A synapse is a function from state to potential.
generator == potential -> value    || A generator yields an value when a potential is given.
```

Each neuron in the network is an element determined by its synapse and generator. Such an element is described by a function that takes the synapse and generator functions as arguments to become a neuron (a function from a list of input axons to one output axon).

```
element :: synapse -> generator -> neuron    || The type of the function element.
element synapse generator                    || The function element takes 2 arguments
  = (map compute_value).transpose            || and is defined by this expression.
    where                                    || The following sub-definition holds:
    compute_value input_state                || this is a locally defined function,
      = generator potential                  || defined as apply generator to potential.
        where                                || The following sub-definition holds:
        potential = synapse input_state      || synapse applied to state yields potential.
```

The dot denotes function composition: `(f.g) x ≡ f (g x)`

The function `map`, predefined in Miranda, is a tool that takes a function and a list as arguments and applies this function to each element of the list. Here, the locally defined function `compute_value` is applied to each state. The predefined function `transpose` takes a list of lists as arguments, it delivers a list of lists of corresponding elements. It is possible to extend the description to handle more complex neurons. For instance, we may add a history for the potential or a stochastic generator for the output values.

The most common model for the synapse is a component that takes a weighted sum of the inputs. This synapse is modelled by a function that takes a list of weights as an additional parameter. In the rest of this paper we will employ numbers to represent the values on the axons.

```
value == num                          || The values on the axons are represented by (real) numbers.

wsum :: [num] -> synapse
wsum weights = sum.(listproduct weights)        || potential = ∑_{i=1}^N weight_i × input_i

listproduct :: [num] -> [num] -> [num]
listproduct (a:x) (b:y) = a*b:listproduct x y|| Multiply elements; continue with tail.
listproduct []    []    = []                 || No elements in lists: yield an empty list.
```

The predefined function `sum` computes the sum of the elements of a list.

Some well known generator functions are:

```
sigmoid :: num -> generator
sigmoid s potential = 1 / (1 + exp (-1 * s * potential))

clip :: num -> num -> generator
clip lower_bound upper_bound potential
  = lower_bound, if potential   <= lower_bound
  = potential  , if lower_bound <  potential < upper_bound
  = upper_bound, if upper_bound <= potential

delta :: num -> generator
delta bound potential = 1, if potential > bound
                      = 0, otherwise
```

A particular neuron can be specified by supplying an `element` with an appropriate synapse and generator. An `additive_element` is an element with a synapse that takes a weighted sum of its inputs and a delta function with a bound of 0.5 as generator. An individual additive element is specified by the weights of its synapse.

```
additive_element :: [num] -> neuron
additive_element weights = element (wsum weights) (delta 0.5)
```

A network is described by specifying the individual neurons and their interdependencies. In this functional description a network has a list of input axons as argument and yields a list of output axons. For small networks it is possible to list all neurons explicitly. For large networks other specifications will be used.

A very small neural network computing the exclusive OR of two inputs will be described as example. One hidden unit is used to compute the logical AND of the inputs. The output unit computes the logical OR of both inputs provided that the hidden unit is not active.

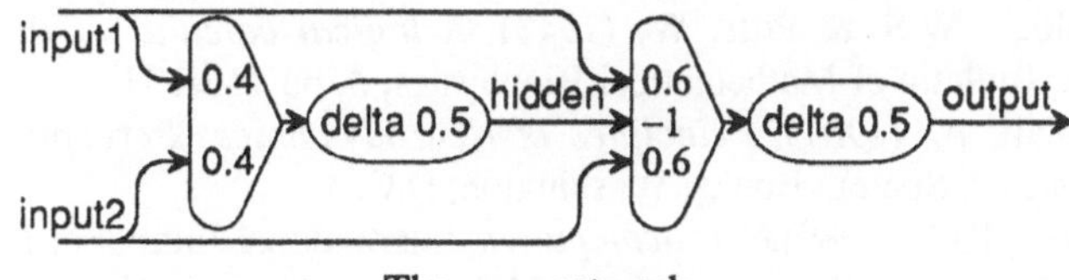

The `xor` network

```
network == [axon] -> [axon]

xor :: network
xor [input1,input2]
  = [output]
    where
    hidden = additive_element [0.4,0.4]     [input1,input2]
    output = additive_element [0.6,0.6,-1] [input1,input2,hidden]
```

This specification can be partially checked by the compiler for the description language. The compiler checks the type of all expressions and definitions. Furthermore undefined identifiers are indicated. Moreover, the specification given above is a complete implementation of the network. By supplying values for the input axons, the output of the network can be computed by evaluating the expression `xor [axon1,axon2]`. Expressions are evaluated by applying the defined functions. Suppose `axon1 = [0,1,0,1]` and `axon2 = [0,0,1,1]`, then the result of the computation is `[[0,1,1,0]]`.

Discussion and future work

As shown above, functional programming languages are very suited to specify neural networks. The Miranda fragments listed in this paper form a complete, high-level, compact, correct and executable specification of a neural network.

The specification method given above will be used in the near future to specify a large class of neural networks and learning algorithms. Learning in neural networks can be modelled by a function that adapts the weights supplied to the additive elements. To be able to express the stochastic behaviour found in many neural networks, functional programming languages have to be extended with a random generator. A random generator is not a 'pure' function since its result is not deterministically determined. In a purely functional language only pseudo random generators can be specified.

The absence of side effects in functional programming languages makes them very suited for parallel evaluation. By adding special annotations to the specification it can be made suited for execution on a parallel machine architecture [Eekelen 89].

Such formal descriptions will be used in future research to specify and compare variants of neural networks in order to achieve a unification of the essential computational aspects of neural networks with the functional paradigm.

Acknowledgement

We thank C. Gielen of the deparment of medical and biophysics of University of Nijmegen for proof reading this paper.

References

Eekelen 89 — Eekelen, M.C.J.D. van, Nocker E.G.J.M.H., Plasmeijer, M.J. and Smetsers, J.E.W. (1989) *Concurrent Clean*, Technical Report no. 89-18, University of Nijmegen, The Netherlands.

Field 88 — Field, A.J., Harrison, P.G. (1988) *Functional Programming*, Addison-Wesley Publisers Ltd.

Hopfield 82 — Hopfield, J.J., (1982) *Neural networks and physical systems with emergent collective computational abilities*, Proc. Natl. Acad. Sci. USA. **79** pp 2551-2558.

McCulloch 43 — McCulloch, W.S. & Pitts, W. (1943). *A logical calculus of the ideas immanent in nervous activity*, Bulletin of Mathematical Biophysics, **5**, pp 115-133.

Rosenblatt 61 — Rosenblatt, F., (1961), *Principles of Neurodynamics: Perceptrons and the Theory of Brain Mechanisms*, Spartan Books, Washington, D.C.

Treleaven 89 — Treleaven, P.C., (1989),*Neurocomputers*, International Journal of Neurocomputing, **1**, 1, pp 4-31.

Turner 86 — Turner. D.A., (1986), *An overview of Miranda*, SIGPLAN Notices, **21**, 12, pp 158-166.

A GRAPHICAL DEVELOPMENT LANGUAGE FOR INTEGRATED KNOWLEDGE SYSTEMS

Lee E. Plansky[1]
University of Idaho - Department of Metallurgical and Mining Engineering
Moskow, Idaho 83843

Abstract

An integrated knowledge system is one made up of the three computing technologies: neural computing networks; expert systems; and, procedural systems. This paper presents a simple, graphical approach to the organization, analysis, and development of such systems. Its focus is insight into the alternative organizations possible when developing integrated systems applications. The systematic presented herein is a prototype graphical system design language. It can serve as an aid to understanding, conceptualization, design and communications as well as a tool in the development of integrated system models of arbitrary size and complexity. Such models can serve a useful function as prototypes for larger systems or as research models, given technology limitations.

Introduction

Two of the computational reasoning technologies of knowledge systems, namely neural computing networks and expert systems, can be integrated with classical procedural computing. Such system solutions are finding their way into application and the demand for them is growing. Situations arise where one technology may be difficult or inefficient to use and more appropriate solutions may be found when two or more of the other technologies are used together. Integrated systems have been developed for such diverse applications as the detection of explosives in airport baggage inspections, and chemical analysis. A systematic follows which simplifies the structure, design and development of systems when using these technologies together.

Development

Three types of processing modules are considered: neural (N); expert (E); and, procedural (P). Figure-1 shows the three technologies functioning together as one. As one they can be looked at as a "molecule" of modules or components, where the components of a molecule are N, E or P routines. Insights can be obtained into how an integrated system can be structured and built by examining Figure-1. It leads to a structure, where the modules N, E or P and their functional objectives and requirements are the starting point for integration. The symbols of Figure-2 have been adopted to extend the graphic symbols used in Figure-1. A general example of how these symbols may be used in application is given in Figure-3.

A molecule can be defined as consisting of two or more components of type N, E or P, where a given type is present only once. Each component would perform a specific function as would the constructed molecule. There are nine possible module interfaces which must be considered in the construction of a molecule: N:N; N:E; N:P; E:P; E:E; E:P; P:N; P:E; and, P:P. How this can be done will be determined more by the state-of-the-art and capabilities of available software and hardware technology than by a particular set of design objectives; the same is true for data communications. This structure provides its own set theory and algebra to aid in analysis - if the hardware or interface requirements for P:E and N:P are known, for example, then the requirements for N:P:E are also known. These relations are not commutative in general, that is, the interface requirements for the interface P:E are not necessarily the same as those for the interface E:P.

---

[1] Mailing Address: POB 1625 E. G. & G. Idaho Inc. Idaho Falls, Id. 83415

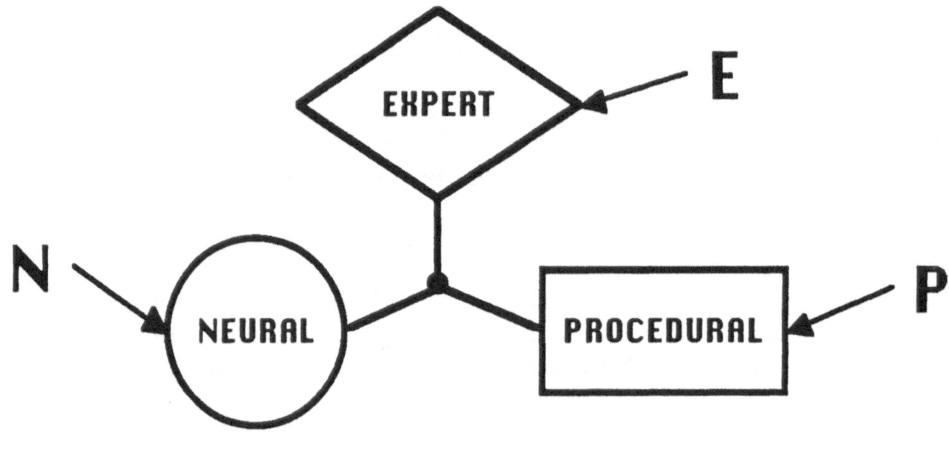

Figure - 1 --- A Molecule

An andral model can be defined as a model consisting of one or more molecules. The function of an andral model would be determined by the functions of its individual molecules. Two or more andral models would also be an andral model. A system could be built of one or more andral models with external, real-time, electro-mechanical components and sensors. A general andral model structure would be denoted by the number of constituent molecules and andral models: A^I_J, where I and J are the number of constituent molecules and andral models. The example of Figure-3 represents such a model at its highest level. Additional application examples are given in (Plansky and Glass, 1990). Such models could form the basis for generalized, synthetic systems as noted in (Dress, 1989; Abelson, 1988). Related advancements are: co-operating expert systems have been introduced (Bailin et al, 1989); systems made up of expert systems and procedural programs are in development and use (Boarnet et al, 1987); and, a system called "The Theoretician" is being studied which envisages integrated systems which can analyze scientific data and hypothesize the processes and causes of the results (Tenorio, 1989).

Summary and Conclusions

The systematic of this paper promotes an improved understanding of the complex task of integrating different computing technologies. Integrated computing would find use in applications where a single computer technology would be difficult or inefficient to apply and where two or more of the other technologies would work jointly - some types of applications may be easier to develop in this way. An integrated application is a severe challenge to knowledge, technology and the intellect. Applications of integrated computing should be started with full commitment once they have been identified and gained approvals. The technologies needed may require a long lead time to learn and apply. Developments could be lengthy and require basic research involving many disciplines where present day knowledge and technology are limited. Team development would be the norm. Technology surveillance is also necessary. Good software practice must be followed in system construction. As technology advances, parallel processing or object oriented, event driven systems may be used and the functional modules of the three technologies could be relegated to routine libraries for further simplification.

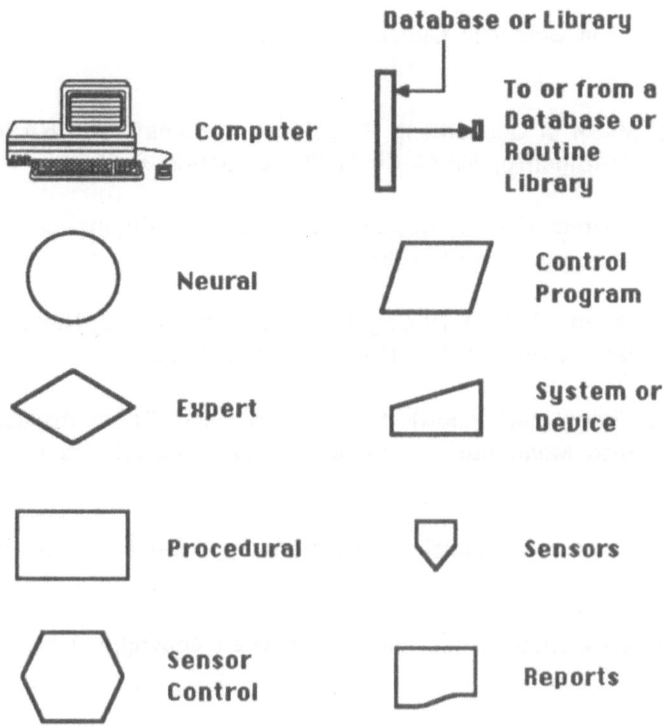

Figure -2 --- Andral Language Symbols

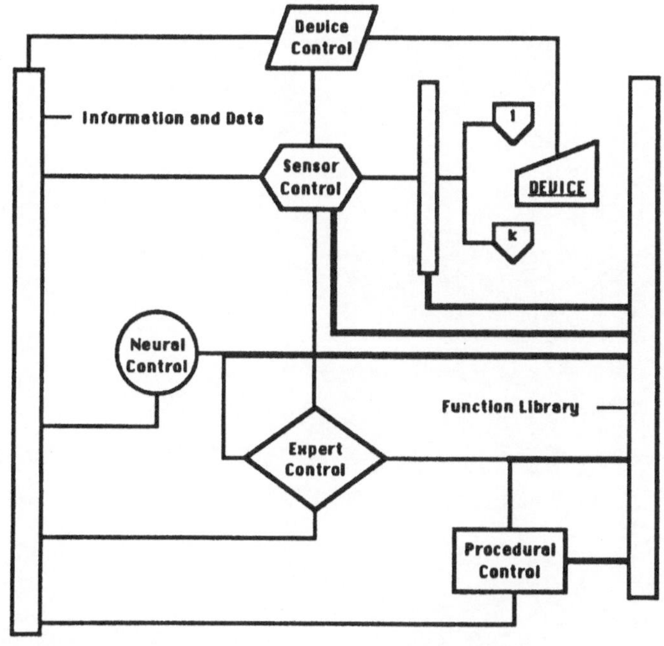

Figure - 3 --- Application Example

References:

Abelson, H. and others, "Intelligence in Scientific Computing", Communications of the ACM, Vol. 32 (5), pp.546-562, 1989.

Bailin, S.C. and others, "A Logical Model of Cooperating Rule-Based Systems" 1989 NASA Conference on Space Applications of Artificial Intelligence, NASA-CP-3033, pp. 319-333, 1989.

Boarnet, M.G., and others, "Using Hybrid Expert System Approaches for Engineering Applications", Engineering with Computers, Vol. 2 (2), pp. 95-110, 1987.

Dress, W. B., "Synthetic Organisms and Self Designing Systems", Oak Ridge National Laboratory Research Report, DOE Accession No. CONF-8905130-1/DE89-010318, 1989.

Plansky, L.E. and Glass, C.E., "Integrated Knowledge Systems and Their Application to the Geosciences", Oxford University Press Monograph of the International Association for Mathematical Geology, in publication, 1990.

Tenario, M.F., "The Natural Environment of Artificial Laboratories", Scientific Computing, pp. 7-8, Sept. 1989.

Note: The concepts and terms developed in this paper have been copyrighted by L. E. Plansky, 1989.

PYGMALION
Neural Network Programming Environment

M. Azema-Barac, M. Hewetson, M. Recce, J. Taylor, P. Treleaven, M. Vellasco

University College London

ABSTRACT

The PYGMALION Project[1], funded by the European Community ESPRIT Programme, is one of the major European neural computing projects. The objectives of PYGMALION are: firstly to demonstrate to European industry the potential of neural networks for various applications; and secondly to develop a European "standard" neural network programming environment.

The PYGMALION neural programming environment provides a rudimentary "platform", for neural network applications, that can be easily extended by the application builder. (It also allows trained networks to be easily moved between machines.) The environment provides a Graphic Monitor (using ***X-windows***); an algorithm library (containing back-propagation, boltzmann, hopfield, etc.); a high level neural programming language, called *N* (based on *C++*); and an intermediate level, neural network specification language (a subset of *C*). This paper introduces the PYGMALION neural programming environment, and describes its Graphical Monitor and its neural network specification language, called ***nC***.

1. PYGMALION ENVIRONMENT

The ESPRIT II PYGMALION project[1] is intended to provide a focus for neural computing research within the European Community. PYGMALION aims to promote the application of neural networks by European industry, and to develop European "standard" computational tools for programming and simulation of neural networks.

The design philosophy of the PYGMALION neural programming environment is twofold. Firstly, to provide an "open" programming environment - a rudimentary "platform" - that can be easily extended and interfaced to other tools. For this reason the core of the environment is ***X-windows***, ***C*** and ***C++***; running on a colour workstation. Secondly, to provide "portable" neural network applications, so that trained and partially trained networks can be easily moved from machine to machine. For this reason the (partially) trained neural network applications are specified in a subset of ***C***; essentially a ***C*** data structure.

The environment comprises 5 major parts:

- **Graphic Monitor**, the graphical software environment for controlling the execution and monitoring of a neural network application simulation. This includes a simulation command language for setting up a simulation, monitoring its execution, interactively changing values, and saving a trained network.
- **Algorithm Library**, the parameterised library of common neural networks, written in the high level language and providing the user with a number of validated modules for constructing applications.
- **High Level Language** *N*, the object-oriented programming language for defining, in conjunction with the algorithm library, a neural network algorithm and application, by describing the network topology and its dynamics.
- **Intermediate Level Language** ***nC***, the low level machine independent network specification language for representing the partially trained or trained neural network applications, a format analogous to P-code for PASCAL systems.
- **Compilers** to the target UNIX-based workstations and parallel Transputer-based machines.

This structure of the neural programming environment is illustrated by Figure 1.

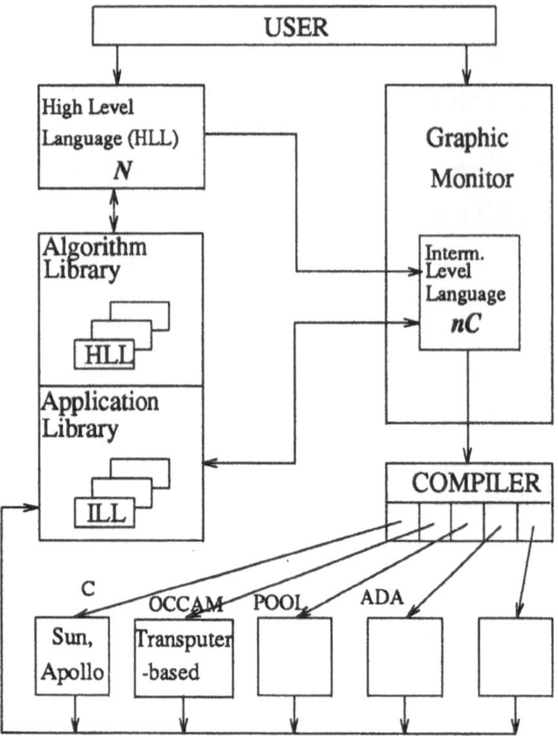

Figure 1: PYGMALION Neural Programming Environment

From existing commercial graphic monitors for neural networks, it is evident that users view networks in a hierarchical manner; net, layer, cluster/slab, neuron etc. Therefore the PYGMALION environment reflects this view, in a consistent way throughout the Graphic Monitor and the languages. For example, the Graphic Monitor displays a hierarchy of windows and ***nC*** embodies a hierarchical data structure encompassing: system, network, layer, cluster, plus neuron and synapse, as shown in Figure 2.

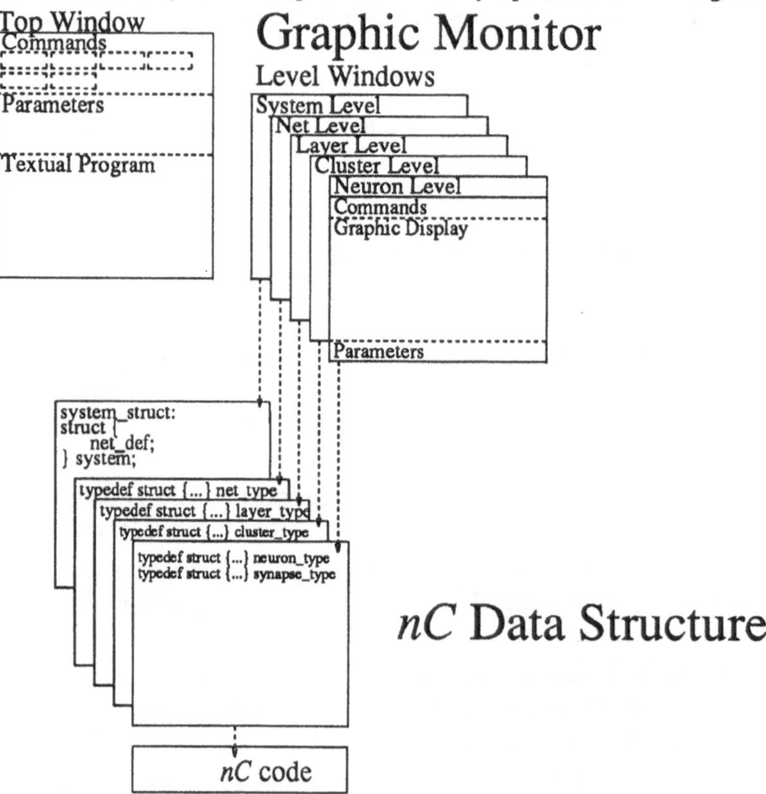

Figure 2: PYGMALION Graphic Monitor and *nC* Specification

2. GRAPHIC MONITOR

The Graphic Monitor sits on top of the *nC* data structure and displays its contents, with a window corresponding to each level in the data structure. There are two types of window:

- **Top Window**, providing facilities for controlling the simulation and displaying status information plus the program text of the neural network.
- **Level Windows**, providing control facilities, and displaying status information and a graphic representation, for each specific level.

The Top window consists of three areas: commands, parameters and program text, as shown in Figure 3a. The command area comprises a series of button boxes (i.e. labels), each associated with a command or a "pull-down" menu. The boxes are: **HLL, TARGET, TRAINING SET, RUN, PAUSE, INIT**(IALISE), **LOAD, SAVE** and **QUIT**. Next, the parameter area, defines the files and computers being used in the simulation. This is needed for the commands to operate. Lastly, the program text area displays the high level language (HLL) program text specifying the neural network application. An application can be created or modified using a standard text editor, and then compiled into the intermediate level language (ILL) *nC* specification for execution.

The Level windows also consist of three areas: commands, graphics, and parameters, as shown in Figure 3b. At the top is a series of button boxes. Examples of buttons are: learn, recall etc. Each button box corresponds to an *nC* rule specified in the associated level. Clicking the button causes the rule to be executed. Next, the graphic area gives a graphical display of the associated level and its pattern of intra-connectivity. (This display relates to the *nC* topological structure and substructure). Lastly, the parameter area displays status information in a textual form, that complements the graphic display. For example at the neuron/synapse level, this information comprises state, weights vector etc.

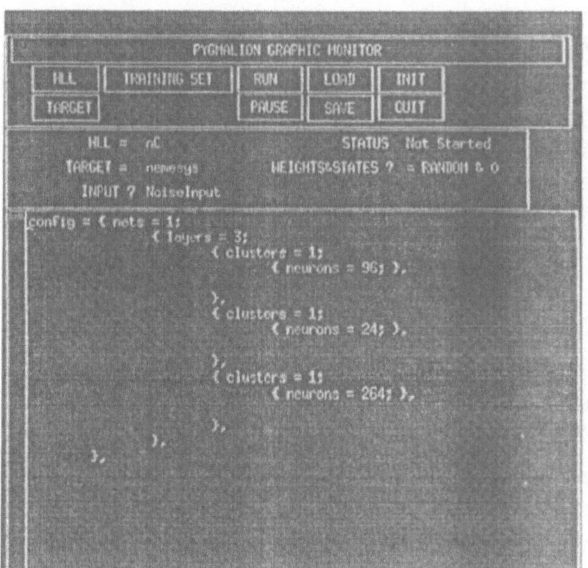

a) Top Window Layout

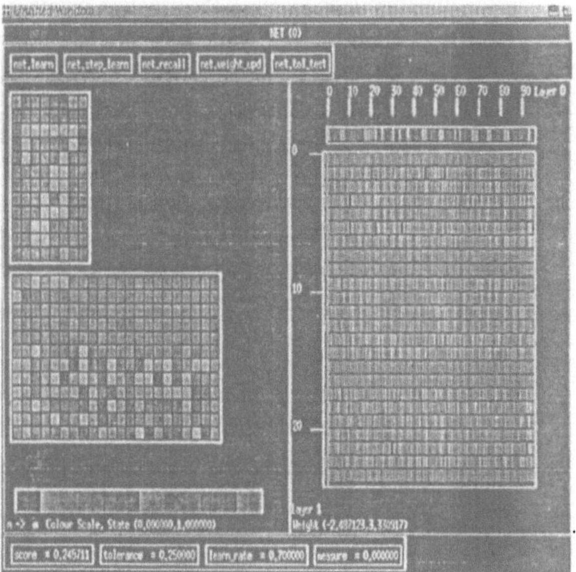

b) Level Windows Layout

Figure 3: Window Layouts

When the PYGMALION environment is loaded, initially the Top window is displayed. By clicking on the **HLL** button the user can choose a high level neural language and then enter its sub-window. This sub-window supports the programming and compilation of an application in the HLL. Alternatively, by selecting the **RUN** button, a user can run an already trained network.

Control of the windows is based upon a corresponding hierarchy of menus:

- **Global Menus** provide a set of window management functions, that can be applied to any window. Examples include Open, Close etc.
- **Local Menus** provide a set of functions specific to a window.

3. NETWORK SPECIFICATION LANGUAGE

The intermediate level network specification language ***nC*** was developed as a target language to accommodate the diverse requirements of neural network programming. For pragmatic reasons, ***nC*** is based on a small subset of the ***C*** language.

The ***nC*** language divides the neural network information into 4 different domains:

- the network *topology*, that describes the central hierarchical configuration as well as the connectivity of a neural network;
- the *data* of the system, which comprises the neuron's status and synaptic weights;
- the *functions* defining the processing in the network; and
- the *control* of the network activities.

These four domains are basically described by one special data structure: ***system***, as illustrated by Figure 2. The ***system*** structure defines the *topology* by giving the explicit location of each synapse and neuron inside the network. The ***system*** is composed of networks that have layers, that are composed of clusters, that comprise neurons, that contain synapses. Alternatively, synapses can be allocated as a cluster sub-level to cope, for instance, with models that use shared weights. Each level has the same underlying structure and semantic:

- **Rules**, which provide the control for the neural network operations;
- **Parameters**, which provides the neural network's parameters and their values;
- **Substructures**, which provide the topology of the neural network.

Rules correspond to commands, parameters to parameters, and substructure to graphics. For instance, each button box in a window is associated with an ***nC*** rule in the data structure of the level concerned. This hierarchy is very general, allowing the development of heterogeneous system consisted, for instance, of a Hopfield network and a Back Propagation network.

The *system* structure also encloses the *data* of the complete system, which comprise the neuron's status and the synaptic weights. The *functions* performed by each element in the system and the overall system *control* are defined by the concept of ***rules*** and ***meta-rules***, respectively. ***Rules*** and ***meta-rules*** are defined using the same data structure - **RULE** -, but are conceptually different. The **RULE** structure is basically composed of a *name*, a *function pointer* and a list of *parameter pointers*. This means that the function pointer points to a basic procedure that will be applied, when the rule is executed, to the rule's parameter pointers. What differentiates a *rule* from a *meta-rule* is the type of parameter pointers and, consequently, the type of function to be executed. At the *rule* level, the parameters are actual numbers that will be used by the function to calculate a certain value. On the other hand, at the *meta-rule* level, the parameters are pointers to the rules that should be executed to perform a higher level function, and the function pointer determines how these rules should be executed. There are basically two ways of executing rules: either in sequence, using the ***sexec*** function; or in parallel, utilising the ***pexec*** function. These two functions are built-in functions provided by the ***nC*** intermediate level language.

Finally, to overcome the inadequacy of the pure ***C*** syntax, we have introduced in the ***nC*** syntax an explicit control for parallel execution. This was accomplished by introducing a new control operator, namely **PAR**. The **PAR** control operator executes in parallel all instructions within the following open-and-close brackets. Furthermore, the **PAR** control operator also serves as an operation modifier when applied to a ***for*** statement. It converts an iterative ***for*** loop into a replication operation. This explicit control of execution provides complete information to guide the ***nC*** compilers for optimising the implementation of ***nC*** programs onto various parallel processing machines.

References

1. B. Angeniol and P. Treleaven, ''PYGMALION Neural Network Programming & Applications,'' *ESPRIT Conference (1989)..*

BRAINTRACER: A Software Package for the Simulation of Complex Neural Networks

A.Bartoli, G.Tononi, G.Buttazzo, P.Dario
"Scuola Superiore S.Anna"
Via Carducci 40 , Pisa Italy

ABSTRACT

BRAINTRACER is a software package dedicated to the development of neural networks with complex structure, inspired by the anatomy of real brains and to the simulation of their behavior on general purpose digital computers.

Unlike other connectionist software products, BRAINTRACER offers the opportunity of creating, modifying and evaluating the performance of different network structures rather than of different activation or learning algorithms.

A powerful command interpreter, enables the user to easily create and modify complex networks consisting of several classes of neurons, ordered and interconnected topologically, and to simulate and examine their behavior.

Conversely, there is only one general type of activation and learning rule; however, this is made very powerful and flexible by the use of several settable parameters.

Special features include the possibility of defining neurons with either specific or diffuse projection, of anisotropically ordering the values of some parameters within each class of neurons, and of using a variable temporal window and a reinforcement factor in the learning algorithm.

The program is fully interactive, except for the simulation of the specific environment chosen by the user, for which a simple self-contained C source is required. This allows for the easy reproduction and evaluation of many interesting neurobiological problems.

INTRODUCTION

This paper presents a software tool (BRAINTRACER) for the development and the simulation of neural networks with complex structure on general purpose digital computers.

BRAINTRACER offers the possibility of creating, modifying and evaluating the performance of different network structures rather than concentrating on activation and learning algorithms.

The underlying assumption, which originates from the study of biology, is that structure is the most important factor in determining what a given network can learn and perform when it interacts with the environment.

We briefly describe how BRAINTRACER deals with the creation of simulated networks of whatever structure, then with activation and learning rules and finally with the environment. At the end, we give a short example to illustrate the power of BRAINTRACER commands.

THE MODEL

Structure: The user begins by defining classes of neurons. A class consists of neurons which are topologically arranged in a bidimensional layer and which have similar projective and receptive fields as well as similar parameters controlling the activation and learning rules. Each class has three main attributes: *interface* (receptive / motor / internal), *projection mode* (diffuse / specific) and *type* (excitatory / inhibitory). The 'interface' attribute describes the relation between the class and external world: *receptive* means that the activation of its neurons is clamped by the environment, *motor* means that those activations have influence on the state of the environment, while *internal* means that the class have no direct interaction with it; the 'projection mode' attribute specifies whether each neuron within the class projects toward all neurons of the target class, or only toward a subset of them, and whether its activation level acts as a reinforcement signal controlling learning in the target neurons; the 'type' attribute specifies whether the neurons of the class excite or inhibit their targets.

Connections between neurons are specified by defining mapping between classes, that is by declaring a set of (source_class, target_class) pairs. The definition of each mapping automatically establishes, according to rules described below, the synapses between all neurons belonging to the specified classes. Each source neuron (SN) of the source_class is associated with a target neuron (TN) of the target_class on a topological basis. Each SN projects synapses toward a square of target_class centered on its TN, and the initial values for the weights are specified by a set of numbers W_i which must be declared together with the corresponding pair of classes. The values of the weights can be distributed either isotropically or anisotropically among all neurons of the source_class; in the case of anisotropic distribution, each value vary linearly with the ordinal position (along row or column) of each source neuron.

Activation: The activations of non-receptive neurons are computed at a rate associated with the class, according to the following law:

$$V_j(t) = K_{dec} V_j(t\text{-}1) + S_{exc}\Big(\sum_{j\,exc.} W_{ij} V_j(t)\Big)\, S_{in}\Big(\sum_{j\,inhib.} W_{ij} V_j(t)\Big)$$

where the two functions $S_{exc}()$ and $S_{in}()$ act as a 'filter' for the inputs from excitatory and inhibitory neurons, respectively.

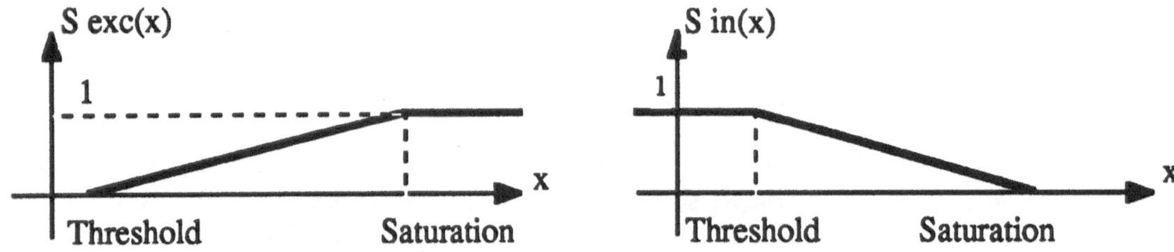

Fig. 1: Filter functions

Each neuron has a 'refractory period' T_r: when its activation reaches some predefined high value, during the following T_r instants it can only decay and ignore any incoming input.

The activation parameters are the following: a decay constant K_{dec}, a refractory period T_r, a threshold and a saturation value associated with the filter functions $S_{exc}()$ and $S_{in}()$. Although the activation parameters are similar within a class, the value of each parameter can be distributed either isotropically or anisotropically.

Learning: The synapses of a mapping can be fixed as well as modifiable at a rate associated with the mapping itself. the learning rule is:

$$W_{ij}(t) = K_d\,(W_{ij}(t\text{-}1) + E_{ij}(t) R_j(t))$$

where W_{ij} is the weight of the synapse from neuron 'j' toward neuron 'i', K_d is a decay constant and E_{ij} (eligibility cfr [1]), is associated with the synapse and is given by:

$$E_{ij}(t) = G E_{ij}(t\text{-}1) + (1\text{-}G)(K_{hebb} V_i(t) V_j(t) + K_{pre} V_j(t) + K_{post} V_i(t))$$

Each mapping have its own eligibility parameters G, K_{hebb}, K_{pre}, K_{post}. Because of the recursive definition of $E_{ij}(t)$, the user can choose, by appropriately tuning the value of G, a 'temporal window' over which the activation of the pre- and post-synaptic elements may influence learning. K_{hebb} is a hebbian term, while K_{pre} and K_{post} allow the user to separately enhance or reduce the pre- and post-synaptic components of learning (cfr.[2]).

The $R_j(t)$ (reinforcement signal) is defined as the weighed difference between the effects on neuron 'j' of diffuse projection classes, grouped in excitatory and inhibitory. Since it is associated with the target neuron, its effect is the same for all synapses incoming to it. The signal is given by:

$$R_j(t) = C_j^+\Big(\sum_{h\,edp} W_{jh}V_h(t)\Big) - C_j^-\Big(\sum_{h\,idp} W_{jh}V_h(t)\Big)$$

In this formula, the index 'h edp' refers to neurons 'h' belonging to excitatory diffuse projetion classes, while 'j idp' refers to inhibitory ones. Reinforcement curves C+() and C-() are specified through the following parameters: threshold, saturation, tonic and phasic levels. For a given mapping all learning parameters are identical except threshold and saturation levels, which can be distributed either isotropically or anisotropically among all neurons of the target class.

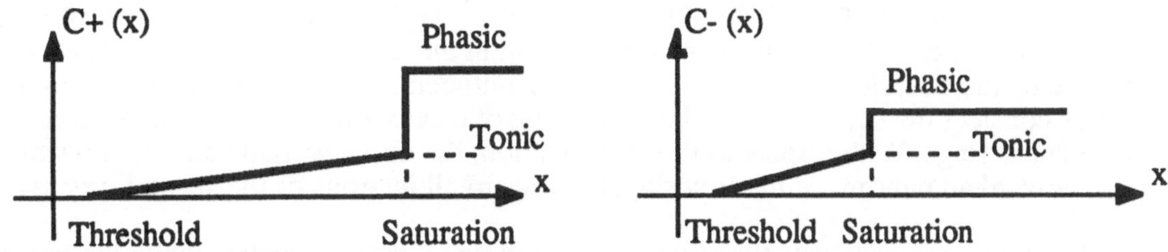

Fig. 2: Reinforcement curves

Environment: Of course, it is not possible to define interactively all the environments and stimuli which the user may wish to connect to a network. However, a network can interact with whatever simulated environment the user has developed simply by defining a suitable set of receptive and motor classes, and by preparing a file containing a C source of three simple functions. The definition of these functions requires no knowledge at all of the internal organization of the program: get_att() reads activity of motor neurons, world() computes the new state of the environment and set_att() writes new activity values for receptive neurons.

Run-time controls guarantee the consistency of the network with the simulated environment.

THE PROGRAM

The program is written in C and porting on various systems is available (UNIX™, VMS™, MSDOS™, OS/2™, MSDOS™, X/OS™).

The package consists of two executable modules; an *interpreter* module which creates files containing a description of the network, and a *simulator* module which simulates its evolution. The interpreter runs interactively, providing many commands for the creation, modification and analysis of a network as well as help and syntax checking facilities. The simulator module creates two files: the first contains a trace of the simulation and the second contains a description of the network as modified by the learning session (this file can be used again by either module).

An Example: The following simple example illustrates some of the interpreter commands. For instance, the user may create a small network in a few keystrokes simply by typing:

```
create class_1 e 10/17
create class_2 i 9/8
activ class_1 dec rows 0.85/0.35
```

The first two commands define two classes of neurons with symbolic names 'class_1' and 'class_2', dimensions (row, columns) 10/17 and 9/8, attributes excitatory (`e`) and inhibitory (`i`), respectively. By default, activation parameters are isotropically distributed with values defined at compilation time. The third command, however, changes the decay constant (`dec`) for neurons of 'class_1'. This activation parameter is now anisotropically distributed: its value vary linearly starting from 0.85 (neurons of the first row of class_1) and ending with 0.35 (neurons of the last row). Other possible commands are:

```
map class_1 to class_2 iso 0.9 0.7 0.4 0.2
learn m class_1 to class_2 post 0.22
set cycles = 900
define class_2 mot
```

First, the user defines a mapping where 'class_1' is the source class and the 'class_2' is the target one, the distribution of the synapses being in this case isotropic (`iso`). The values for the weights are described by the set of the four numbers. To understand the meaning of these, let us consider the square centered on the target of each source neuron as composed by a set of concentric rings. With respect to this organization, the initial weights are: 0.9 toward the target, 0.7 toward all neurons of the first ring, 0.4 toward all neurons of the second ring and so on.

The third command sets the end of the simulation after 900 cycles, while the last one defines 'class_2' as composed of motor neurons (`mot`). Other commands specify the rates of activation and learning, modify the attributes and the reinforcement parameters for classes or the learning parameters for mapping, define the simulation parameters, like the trace commands, and so on.

```
write net_pre
quit
simula.exe net_pre net_post
```

With these commands the user may finally save the network in a disk file (`net_pre`), quit the interpreter module and call the simulator (`simula.exe`) to start the simulation. The file '`net_post`' will contain the network as modified by the learning session.

CONCLUSIONS

By offering the possibility of simulating networks of great structural complexity, endowed with some crucial properties of real brains, BRAINTRACER aims at helping the scientist to approach difficult neurobiological problems in ways that are not yet generally feasible.

REFERENCES

[1] Barto A.G., Sutton R.S., Anderson C.W., "Neuronlike adaptive elements that can solve difficult learning controls problems", IEEE Transactions on Systems, Man and Cybernetics, vol.SMC-13, pg.834-846

[2] Linsker R., "Self-organization in a perceptual network", Computer March 1988, pg.105-117

THE MetaNet NETWORK ENVIRONMENT FOR THE DEVELOPMENT OF MODULAR NEURAL NETWORKS

Jacob M.J. Murre and Steven E. Kleynenberg[1]

Leiden University
Unit of Experimental and Theoretical Psychology
P.O. Box 9555, 2300 RB Leiden
The Netherlands

Abstract The MetaNet network environment allows users to build and examine modular neural networks, and to specify and run complex simulations. It consists of a graphical editor, a network compiler and a graphical (de)compiler, a network specification language (*MetaNet*), and hardware drivers. Its requirements are based on experiences with a text based network environment, which has been in use at our department since early 1988. Using the environment requires minimal programming experience. Currently, the system is implemented only on PCs. Off-loading of calculation processes to other machines is achieved through hardware drivers. It is possible to convert MetaNet code into ANSI C for direct compilation to stand-alone applications on any machine.

Writing code for the implementation of large neural networks can be a tedious and time consuming process. The *MetaNet network environment* emerged from our efforts to provide a general environment for the development of modular neural networks. The primary aim was that users with few programming skills be able to easily specify, run, and evaluate complicated simulations. First, we shall describe MetaNet's predecessor. Then, the objectives of the current environment will be described, which were largely based on experience with the older system. Finally, we shall give a brief description of the different components of the environment.

MetaNet's predecessor: the CALM development system

The MetaNet network environment is based on almost two years of experience with a system that was developed in early 1988: the *CALM development system.* (CALM stands for Categorizing And Learning Module). This system was build primarily for the simulation of CALM networks. CALM networks are learning neural networks that consist of CALM modules (for a general description, see Murre, Phaf, and Wolters, 1989; Murre, Phaf, and Wolters, 1990; Murre, Phaf, and Wolters, submitted; for theoretical and practical applications, see Happel, Phaf, and Murre, 1990; Phaf, Postma, and Wolters, submitted; Phaf, Postma, and Wolters, 1990). A CALM module can categorize and learn an arbitrary activation pattern. To explain the requirements of the system we shall briefly review the network architecture of CALM. A CALM module consists of two rows of n nodes, R(epresentation)-nodes and V(eto)-nodes, and two single nodes, the A(rousal)-node and the E(xternal)-node. We call n the size of a CALM module. All of these nodes are connected by non-modifiable connections according to a certain wiring scheme, as a result of which learning and categorization in CALM are sensitive to the novelty of the input activation pattern. Interconnecting two CALM modules F and T means connecting all R-nodes in F to all R-nodes in T. These connections are always learning.

Designing CALM networks invites thinking in terms of modules. One of the main objectives of the CALM development system, therefore, was to enable users to generate networks on the basis of high level (modular) specifications. For this purpose an interactive component was developed with a textbased user interface. Some of the

---

[1] The second author is also a partner in the Kleynenberg & Kleynenberg Company.

features of the interactive component of the CALM development system were:

- Generating a CALM network on the basis of a high level specification
- Saving and loading a network, including modified connections and activations
- Inspecting a network: activations, connections, modular structure, parameters, node status
- Changing a network: activations, parameters, node status
- Initializing a network: activations, connections, parameters
- Running a network: one cycle at a time, *i* cycles, until convergence

The interactive component was adequate for designing and testing CALM networks. It did not suffice, however, for extensive simulations of, for instance, psychological experiments. Therefore, the system was extended with a non-interactive, batch type component. Files with often used commands could be prepared in advance using a few simple syntactic rules. These command files could then be run from the interactive component. The results could be observed directly, or the file could be processed in a background process and be evaluated after the job had ended. In addition to this list of features, the non-interactive component also supported the specification of patterns (stimuli) in terms of node activations and status, as well as a variety of output formats. A minimal control structure was available for repetition of blocks of commands. The CALM development system has been used by numerous users during a two-year period. It has been implemented on VAXs (VMS), on PCs (MS-DOS), on Amiga computers (Amiga-DOS), and on T800 Transputer networks.

The MetaNet system

The CALM development system had a number of limitations: (1) It lacked graphical facilities. (2) The control structure of the non-interactive facility was very limited. (3) It could only run CALM type networks and it did not allow experimentation with other networks. (4) The system itself was very large, so that only relatively small networks could be run on a PC. (5) It was not possible to control processing in various kinds of hardware by a single system. The need for this was prompted, among others, by the development of a 400-processor network implementation of CALM (see Hoekstra, Heemskerk, Klaassen, Phaf, Knoppers, and Hudson, 1990). To overcome these limitations, the MetaNet network environment provides the following. (1) A graphical interface, which includes a graphical editor. (2) A graphical compiler and a network compiler. (3) MetaNet, a higher programming language for the specification of networks and simulations. (4) The option to convert MetaNet code to ANSI C code for compilation to stand-alone applications. (5) Hardware drivers to buffer the communication in the various processes, possibly on different machines. The MetaNet system is schematically drawn in the figure. A brief description of each component follows below.

Graphical editor. The current implementation of the graphical editor is aimed at PCs running MS-DOS. We judged X-Windows to be too slow on a PC for our purposes. Instead, we have tried to keep the system portable by using a graphical shell with functions found in most other graphical programming environments. The editor is event driven, and it is operated more or less as a 'paint' program. It enables the user to graphically edit a network. Nodes, modules, and other objects may be placed on the screen, and (generalized) connections formed. Objects are manipulated by means of a mouse. Activations, weights, thresholds, etc., can easily be adjusted. Activation, learning, and transfer functions can be chosen from a menu. Experienced users may define their own functions. Hooks with graphs or other status reports may be placed on objects, such as nodes and connections. A special feature is the possibility to automatically position objects using an 'charged-particle algorithm', in which the objects mimic the motions of charged particles.

The MetaNet Network Environment

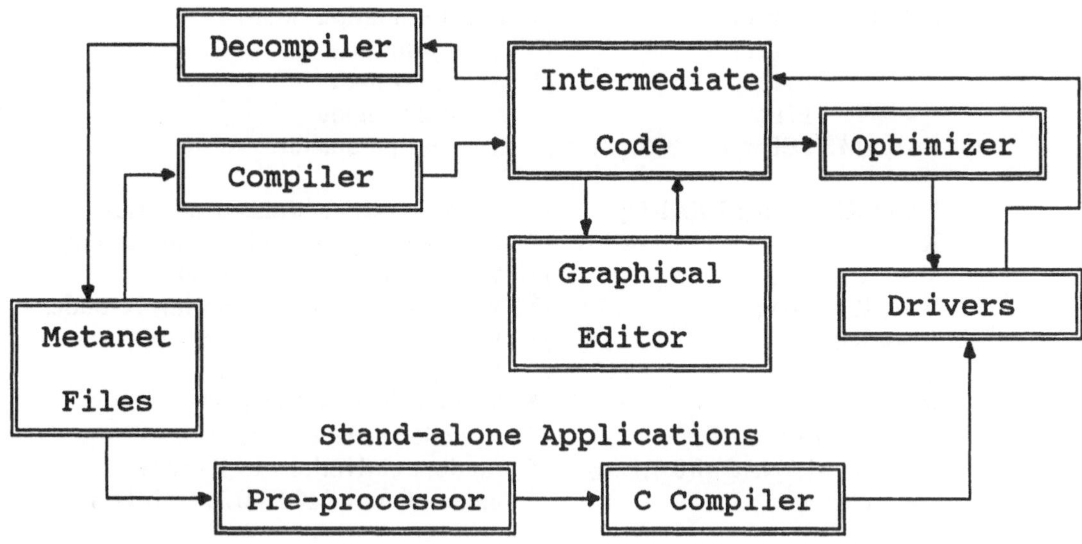

Compiler/Decompiler. Networks and complete simulations may be specified in MetaNet, a high level network language (see below). Inside the MetaNet environment networks and control flow are represented in *intermediate code*. This code is processed by an optimizer to increase performance, and is subsequently send to the hardware drivers, which actually run the networks. To convert a MetaNet file into intermediate code a graphical (de)compiler has been developed. Graphical compilers have been built, among others, for the purpose to directly convert logical structures into object code. An example is the GRASPL programming system, which generates code for micro controller chips based on direct compilation of state diagrams drawn by the user (Zieleman, 1988). In a similar fashion, the graphical decompiler enables the user to graphically edit networks and convert them to MetaNet files. It is also possible to go the other way. A network compiler takes MetaNet files as input and generates intermediate code. This code can directed to the graphical editor, where it can be changed (and possibly decompiled to MetaNet again), or it can be send directly to the optimizer and drivers. The results of the ongoing calculations can be directed to the graphical interface without enabling the editor. In this way, demos can easily be built. A version of the network compiler is implemented separately as a preprocessor, which converts MetaNet to ANSI C for compilation to stand-alone applications.

MetaNet language. MetaNet is a simple network specification language. It is a subset of C extended with network specific syntax, which has an object-oriented flavor. Networks can be defined by using the following MetaNet objects: *NODE*, *GROUP* (of nodes), *MODULE*, and *MODEL*. Each of these types has specific properties. They form a strict hierarchy (a *MODEL* consists of *MODULEs*, which consist of *GROUPs*, which consist of *NODEs*). *GROUPs* roughly correspond to layers, such as used in back-propagation and other layered networks. MetaNet supports a recursive structure as well, the *NETWORK*, which may consist of *NODES* or other *NETWORKs*. MetaNet's network specific syntax focusses more on structures than on processes. The idea is that for calculations (activation and learning rules, etc.) and sequential control (iterations and conditionals) a small subset of C is more appropriate. An example of MetaNet, defining the internal wiring scheme of the CALM module, is given below. Generalized connections between object (*GROUPS*, *MODULES*, etc.) are expressed using the symbols ->, -+, and -\. In this syntax, A -> B : 1.0 means: A is connected to B in an all-to-all fashion with weights 1.0. For instance, if A and B are *GROUPs*, all nodes in A are connected to all nodes in B. The -+ operator indicates one-to-one connections and the -\ operator all-to-all-but-one, the corresponding nodes are not connected.

```
typedef MODULE (int size)            /* define a new MODULE (sub)type       */
{
     GROUP  R[size],                 /* R(epresentation nodes               */
            V[size],                 /* V(eto) nodes                        */
            A[1],                    /* A(rousal) node                      */
            E[1];                    /* E(xternal) node                     */
     PARAMETER mu : .005;            /* learning parameter                  */

     EXTERN -> R : LEARN;            /* learning intermodular connection    */
     R -+ V : 1.0;                   /* R-nodes feed V-nodes (one-to-one)   */
     V -\ V : -1.0;                  /* V-nodes inhibit each other          */
     V -\ R : -10.0;                 /* V-nodes inhibit neighboring R-nodes */
     V -+ R : -1.0;                  /* V-nodes inhibit R-nodes             */
     V -> A : -.6;                   /* V-nodes inhibit A-node              */
     R -> A : .4;                    /* R-nodes excite A-node               */
     A -+ E : 1.0;                   /* A-node feeds E-node                 */
     E -> R : RANDOM(0.0,1.0);       /* random activations to R-nodes       */
} CALM_TYPE;                         /* name of new MODULE (sub)type        */
...
GROUP          input[5];             /* create GROUP                        */
CALM_TYPE little[10];                /* create MODULE of CALM_TYPE          */
...
input -> little : .6;                /* connect with initial weight .6      */
```

Network drivers. To increase machine independence hardware drivers are used. This also saves memory, because only the relevant drivers need to be loaded. The drivers buffer network calculation processes, possibly running on remote machines. Drivers may also do calculations themselves. The communication protocols with the other components of the MetaNet environment are handled by means of intermediate code used inside the MetaNet system. Networks may also be dumped directly (full dump) in a file using intermediate code.

The different constituent parts work together to provide an optimal performance environment, both for inexperienced and advanced users. The MetaNet environment still has to pass several testing cycles, so that we cannot yet reach a conclusion about its range of applicability. We expect that modular, or similarly structured, neural networks will benefit most from the approach taken in MetaNet.

References

Happel, B.L.M, R.H. Phaf, and J.M.J. Murre (1990) Categorization in multi-module CALM networks: recognition of handwritten digits. *INNC-90*, Paris, July 9-13.

Murre, J.M.J., R.H. Phaf, en G. Wolters (1989) CALM networks: a modular approach to supervised and unsupervised learning. *IJCNN-89*, Washington D.C., June 18-22, 1, 649-656.

Murre, Phaf, and Wolters (1990) Novelty dependent categorization and learning in CALM modules. *INNC-90*, Paris, July 9-13.

Murre, J.M.J., R.H. Phaf, and G. Wolters (submitted) CALM: Categorizing And Learning Module.

Phaf, R.H., E.O. Postma, and G. Wolters (submitted) ELAN-1: a connectionist model for implicit and explicit memory tasks.

Phaf, R.H., E.O. Postma, and G. Wolters (1990) Connectionist simulations of the dissociation between implicit and explicit memory. *INNC-90*, Paris, July 9-13.

Zieleman, J.G. (1988) A graphical editor and compiler for the GRASPL programming system. Master's thesis, University of Texas, Austin, Texas.

An Actor Language for Connectionism based on Cellular Automata

Thierry Cornu[1 & 2], Jean-Paul Haton[1]

(1) CRIN/INRIA
Université de Nancy I
BP 239
54506 Vandœuvre lès Nancy
France
tel:83 91 20 00

(2) Ecole Supérieure d'Electricité
Service Informatique
2 rue Edouard Belin
57078 Metz Cedex 3
France
tel:87 74 99 38

Abstract : We describe a parallel computer language designed for neural networks implementation. The underlying computational model is somewhat inspired from the one used in actor languages, and yet it is slightly different, in order to fit the connectionist concepts of computation. We hope this model to be able to give a valuable unifying framework to most of present neural network algorithms. We first present our computational model, comparing it with previous works like the actor model and the cellular automata models. We then describe a first version of a compiled language based on this model. We finally discuss some crucial points of a parallel implementation of such a system.

ARCHITECTURE OF THE NEURAL NETWORK SIMULATION ACCELERATOR NEUROSIM/L

Toshiya Nakajima
Scientific Systems Dept, Fujitsu Ltd.
1-17-25 Shinkamata, Ota-ku, Tokyo 144, Japan

Abstract:

This paper introduces architecture of hardware and software of the high-speed, window-based neural network simulator NEUROSIM/L which runs on personal computers. The simulator consists of hardware acceleration board with a 13.3MHz digital signal processor(DSP) and 4MB on-board memory and window-based software with network simulation program and man-machine interface(MMI). The simulator is for layered networks and provides four learning algorithms; the back propagation learning, the pseudo impedance learning, the supplementary learning and the incremental learning; the latter three algorithms are newly developed by Fujitsu to reduce probability to be trapped by local minima, and also to improve convergence speed of learning patterns.

The learning and recognition speed of the hardware is 1.5 mega connection per second(MCPS) at back propagation learning, 3.9MCPS at forward operation (recognition); both are more than five or six hundred times faster than pure software simulation (with a 20MHz 80386 CPU) and almost as fast as mainframe machines.

The software is not only learning algorithm implementation but also window (MS-Windows)-based graphical MMI which is easy-to-use to construct networks and learning patterns, and to monitor network status while learning such as total or each pattern's learning error vs. time, realtime display of network weight updating.

1. Network simulators and NEUROSIM/L

Several versions of simulators has been implemented for mainframe machines, super computers, workstations and personal computers at Fujitsu. Each simulator for each type of machine has its own purpose to use. The mainframe or super computer version written in Fortran is for extremely high-speed, large network simulation under batch operation. The workstation version written in C is for R&D to unite neurocomputing with other computation paradigms especially symbolic computation such as expert systems. The personal computer version, written in C with acceleration hardware, has effective man-machine interface to represent especially learning process of network and is for researchers who are interested to analyze network behavior.

NEUROSIM/L (NEURO SIMulator/Layered model) described in this paper is the personal computer version with high-speed, easy-to-use simulation hardware and software environment for layerd type neural networks.

The acceleration hardware uses a digital signal processor(DSP), which runs at 13.3MHz(75nsec) clock cycle, provides high-speed simultaneous floating-point multiplication/addition and resulted simulation speed is 1.5 mega connection per second (MCPS) for BP learning, 3.9MCPS for forward operation of networks. The simulation software provides four types of learning algorithms: the back propagation(BP), the pseudo impedance learning(PI), the supplemantary learning (SL), and the incremental learning(IL). The latter three are new, mainly to improve convergence speed of learning patterns.

2. Implemented learning algorithms

This section briefly describes three new learning algorithms[1)] implemented in the simulator.

2.1 The pseudo impedance learning(PI)

This algorithm is derived to cope with local minima problem of steepest descent method. Consider a situation that a ball rolling down from top of bumpy hill, the ball will pass through shallow minima if its inertia and speed is high enough. PI models this situation as a mechanical vibration system of mass-spring -dashpot and external force (eqn.1).

$$M \cdot \Delta \ddot{W} + D \cdot \Delta \dot{W} + K \cdot \Delta W = -\partial E / \partial W \tag{1}$$

where W is network weight, ΔW is weight update value, E is squared error, and M, D, K are coefficients which correspond mass, damping rate, spring constant respectively. Discretizing eqn.1 it can be written as eqn.2.

$$\Delta W_t = -\varepsilon \cdot \partial E / \partial W + \alpha \cdot \Delta W_{t-1} + \beta \cdot \Delta W_{t-2} \tag{2}$$

where $\varepsilon = (M+D+K)^{-1}$, $\alpha = (2M+D)(M+D+K)^{-1}$, $\beta = -M(M+D+K)^{-1}$.

If $\beta = 0$ eqn.2 becomes BP therefore PI is an extended algorithm of BP.

2.2 The supplementary learning(SL)

This algorithm is aimed to reduce learning time by excluding converged patterns from unconverged patterns while learning. If a pattern is converged and others are not, no further learning computation for the converged pattern is continued until the error for the pattern again exceeds allowance as a result of progression of other patterns' learning. Fig.1 shows comparison of learning time for SL is on and off. SL is used together with BP or PI.

2.3 The incremental learning(IL)

This algorithm is for efficient learning when new learning patterns are added to already learned network. IL requires learning patterns to be divided into groups beforehand and learning is proceeded groupwise with re-learning phase after one group has learned. For example, when patterns are divided into three groups such as A, B and C; network learns them as follows:(1)learns all patterns in group A,(2)learns all patterns in group B starting with the weight after learning of group A, (3)re-learns all patterns in group A and B, (4)learns all patterns in group C,(5)re-learns all patterns in group A,B and C. Fig.2 shows this process with correspondence of the numbers above.

Patterns in same group should be "similar" to each other because learning is efficient(completes fast) for similar patterns. In many cases, re-learning completes faster because for example after learning group B, network's weight is somewhat biased to the one after group A learning. So network can "remember" group A and does not "forget" group B by re-learning. IL is used together with BP or PI.

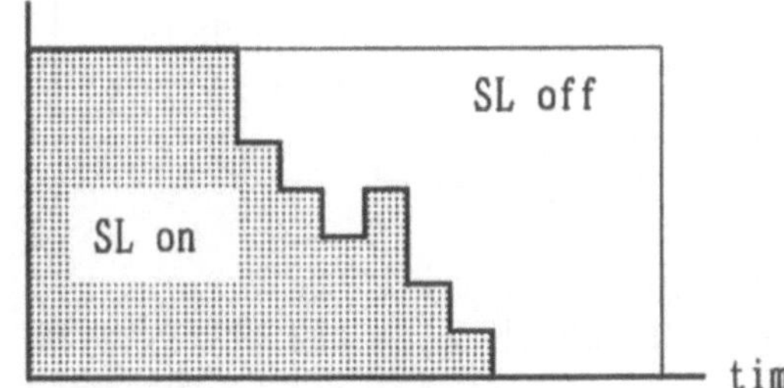

Fig.1 Effect of SL

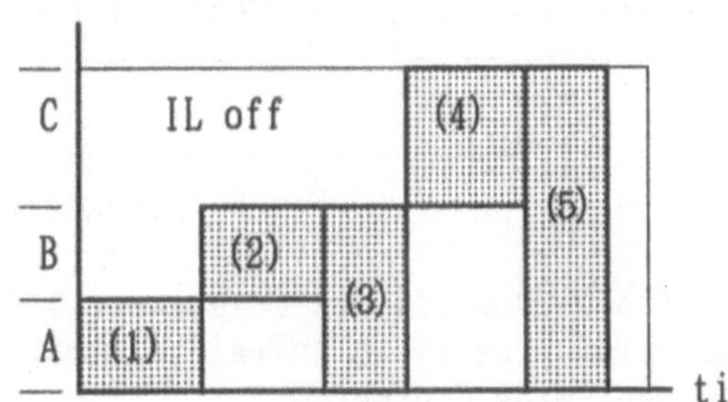

Fig.2 Effect of IL

3. Hardware

The acceleration board ("the Neuroboard") is a high-speed number cruncher with 4MB onboard memory to compute learning algorithms by simultaneous addition and multiplication by using a digital signal processor(DSP). Specification of the DSP is shown in Table 1.

Table 1 Specification of the DSP

name	process	machine cycle	floating point	external data memory
MB86232	1.2μm CMOS	75ns(13.3MHz)	24E8(IEEE)	64KWx32bitx16page

Although computation speed of the Neuroboard depends on network scale, the peak value is 1.5MCPS at BP learning and 3.9MCPS at forward operation; both are as fast as mainframe computer. Table 2 shows computation speed vs. network scale at learning and recognition. Software cases used 20MHz 80386 CPU.

Table 2 Speed of the Neuroboard

network scale(units/layer)	BP learning	recognition	recognition(software)
50 x 50 x 50	1.09 MCPS	1.14 MCPS	0.00368 MCPS
128 x 128 x 128	1.28	2.78	0.00405
240 x 240 x 240	1.54	3.60	0.00435
128 x 128 x 128 x 128	1.22	3.07	0.00393
240 x 240 x 240 x 240	0.84	3.93	——

The Neuroboard is an I/O peripheral of the Fujitsu's personal computer FMR. To operate the board, FMR first transmits firmware program of the DSP to the onboard memory, then writes start signal to I/O register. To read data from the board, FMR halts the DSP and writes start address of the data to the register. Fig.3 shows block diagram of the Neuroboard.

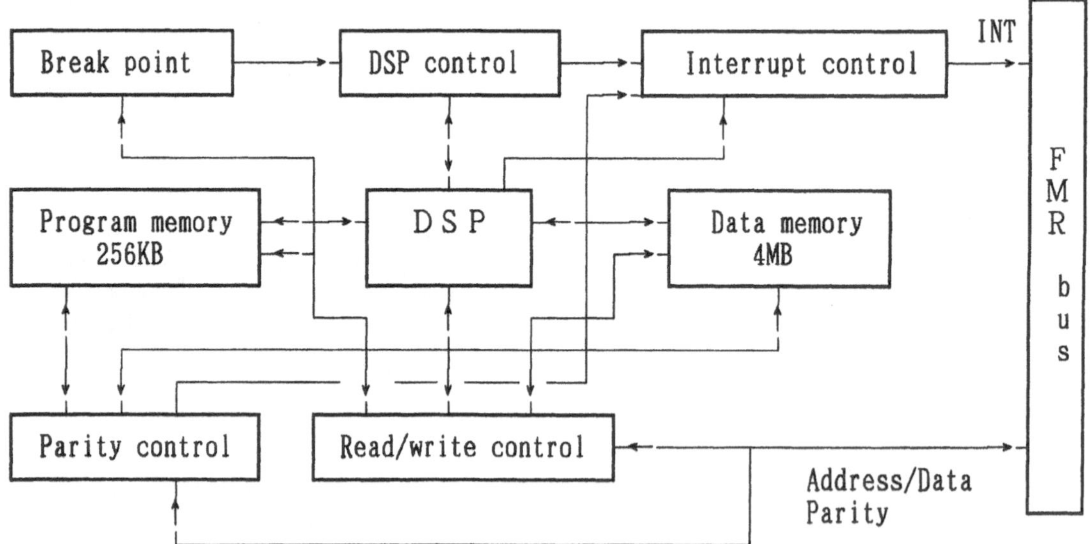

Fig.3 Block diagram of the Neuroboard

The Neuroboard has C language interface in the firmware where network simulation program is installed, therefore it is possible to use the program from user-made C program.

4. Software

This section describes man-machine interface (MMI). The MMI is MS-Windows based and supports many facilities needed to simulate neural networks. The number fields shown in Fig.4.1 are for specifying number of units of input, hidden1, hidden2(if necessary) and output layers. The left pattern shown in Fig. 4.2 is learning input to the network, the right is the desired output; both are made by dragging the mouse. Fig.4.3 is for input of learning parameters. Fig.4.4 shows total squared error while learning in the left window, and each pattern's squared error in the right window. After learning, a pattern shown in the left of Fig.4.5 input by dragging the mouse is recognized as the right pattern. Fig. 4.6 shows weight matrix between two layers which is displayed in gradated red (positive weight) or blue(negative weight).

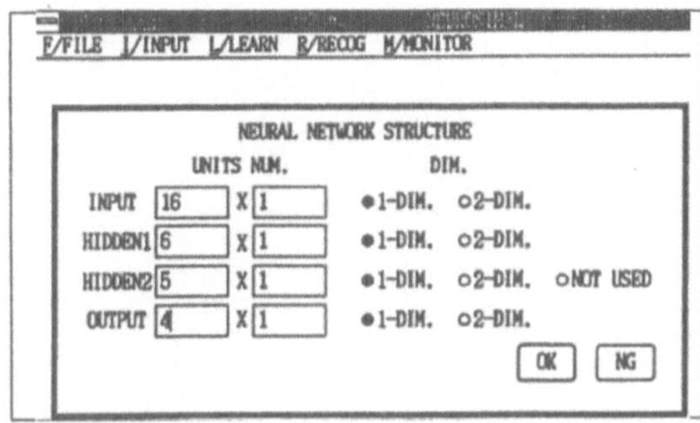

Fig.4.1 Network construction

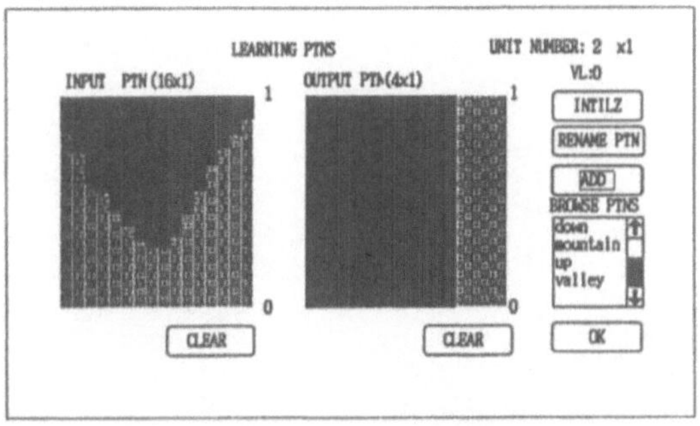

Fig.4.2 Learning pattern

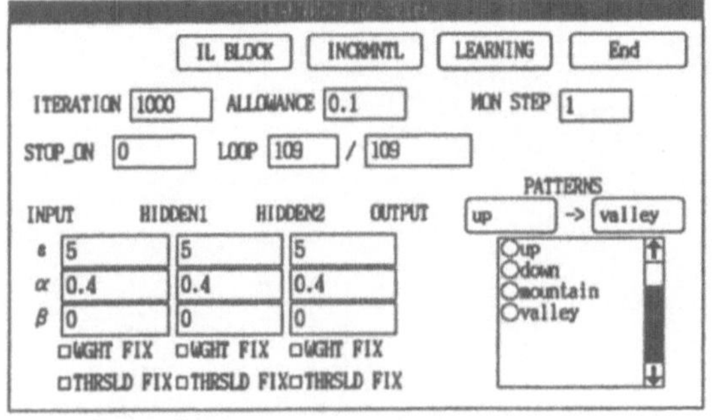

Fig.4.3 Learning parameters

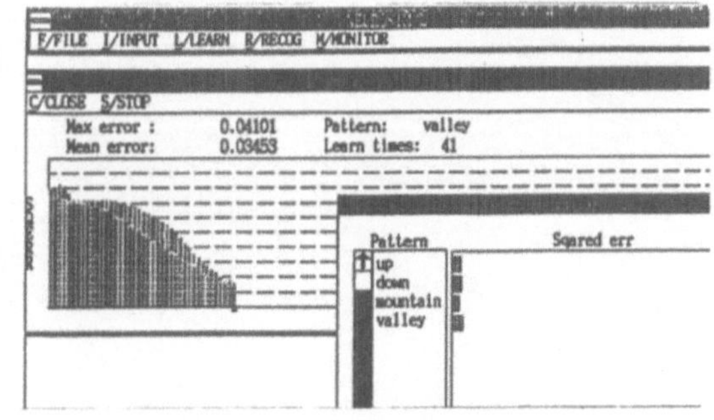

Fig.4.4 Error display

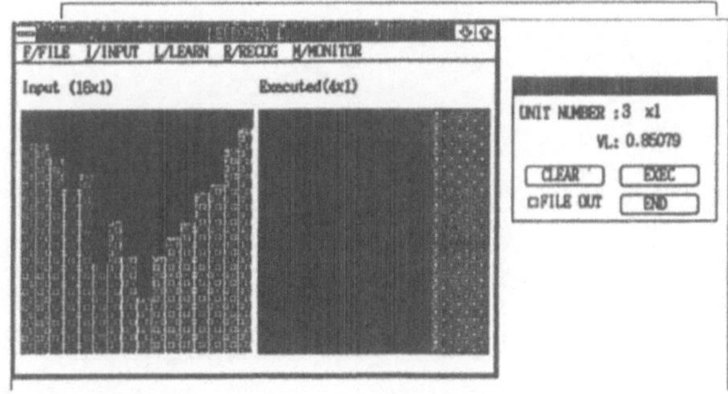

Fig.4.5 Recognition

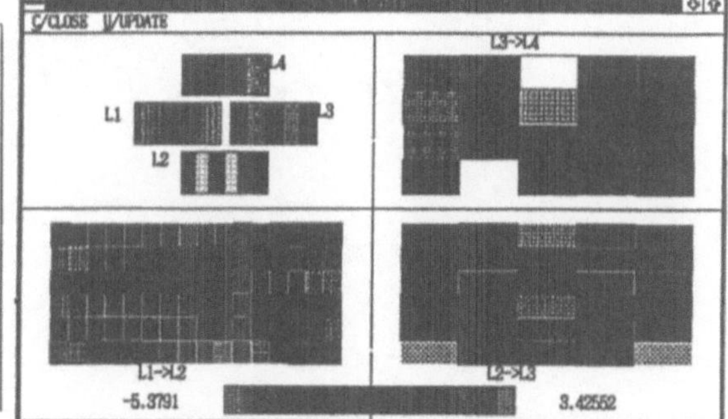

Fig.4.6 Weight matrix

5. Summary

NEUROSIM/L achieved both high-speed network simulation and easy-to-use MMI and provides new learning algorithms. It may be used by various kinds of users.

Reference: 1) Nagata,et. al., Control of Mobil Robots with Neural Networks, Neural Networks, Vol.1 Supplement 1, pp.349, 1988.

BENCHMARKING

Chair: Philip TRELEAVEN

Stability Study of Learning Vector Quantization

Ari Visa
Helsinki University of Technology
Laboratory of Information and Computer Science
Rakentajanaukio 2 C, SF-02150 Espoo
Finland

January 19, 1990

Abstract

In this paper the stability of Learning Vector Quantization is studied. The analysis is based on theoretical considerations and simulations. The Learning Vector Quantization (LVQ) is found to be quite insensitive to disturbances. The localization of the closest reference vector on the map is the most sensitive part LVQ method. Systematic errors that influence on a feature vector, the expected values of stochastic processes, have an effect on classification. The method is not sensitive to error in labelling during the supervised learning phase.

1 Introduction

In pattern recognition the classification is based on reduced pattern vectors, features. So is the case with the neural network methods. The features belonging to a same class build a cluster in a feature space, this means features contained by a class resembles each other. The way how the similarity is measured depends on the problem. It is common that features are considered as prototypes. This is very difficult to argue for, especially, considering real data. It is better to consider that a feature vector x is a random variable ($x = x_1, x_2, ..., x_n$) consisting of n stochastic processes. Each of them has an expected value E_i and a standard deviation σ_i, Figure 1 . This is particularly true when the feature vector is formed by concatenation of measured values. The standard deviation σ_i of each element x_i varies from element to element. This implies that a classification and learning process should have a certain amount of stability against disturbances. In this paper the stability of the Learning Vector Quantization (LVQ) [Ko 86] is studied.

2 Stability Analysis

The LVQ method consists of two parts: First the closest reference vector m_c is localized. Secondly the fine tuning of the map is taken place. The localization of the closest reference vector m_c is done by nearest neighbour method [Ba 74]. It is assumed that there is a map consisting of N elements. The topology of the map doesn't matter. N distances $D_j = ||x - m_j||$ are calculated to localize m_c to a feature vector x. The m_c is minimum of $D_j = ||x - m_j||$. The distance or the similarity can be calculated in many way. Different distance metrics can be found in references by Fu and Kohonen [Fu 76,Ko 83]. During the fine tuning phase labelled samples are assumed besides a map of N elements . The exact form of the fine tuning part is:

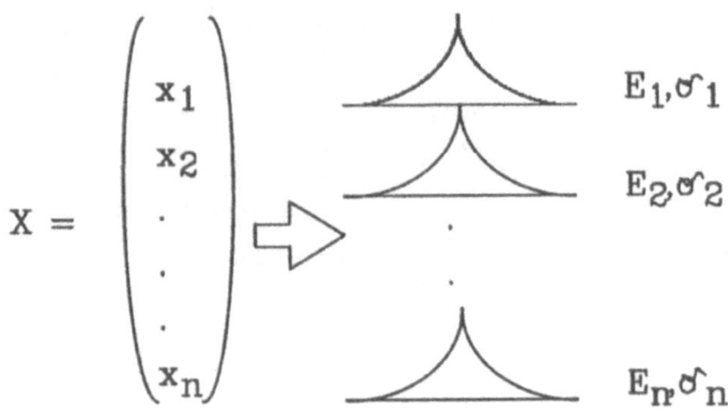

Figure 1: A feature vector x consisting of n random processes.

$m_c(t+1) = m_c(t) + \alpha(t) * (x(t) - m_c(t))$
if $x(t)$ and the closest unit belong to the same class,

$m_c(t+1) = m_c(t) - \alpha(t) * (x(t) - m_c(t))$
if $x(t)$ and the closest unit belong to different class,

$m_i(t+1) = m_i(t)$ for $i \neq c$,
where function $alpha(t)$ is decreasing function with properties:

$\sum_{t=-\infty}^{\infty} \alpha(t) = \infty, \sum_{t=-\infty}^{\infty} \alpha(t)^2 < \infty.$

The stability analysis will consider each part alone. First the map initialized by another algorithm

is stored in memory. This initiation can be made, for instance, by self-organization [Ko 83]. The stored map can be assumed to be invariable. The labelled learning samples might be drifted compared with those samples that created the map. Using Euclidean distance metric an expression to sensitivity of distance variation of features can be stated.

$$D_j = \sqrt{\sum_{i=0}^{n} (m_{i,j} - x_i)^2}$$

The error can be stated by deriving and differentiating.

$$\frac{\partial D_j}{\partial x_i} = \frac{(\sum_{i=0}^{n} m_{ij} - x_i)}{\sqrt{\sum_{i=0}^{n} (m_{i,j} - x_i)^2}}$$

The sensitivity can be studied by differentiation.

$$error_j = \frac{\sum_{i=0}^{n} |\Delta_i|}{\sqrt{\sum_{i=0}^{n} (m_{i,j} - x_i)^2}} \cdot$$

The Euclidean distance measure is quite insensitive to disturbances. Similar error expressions to other distance metrics are possible but they are more complicated. It is however published studies of instability to city block and square distance measures [Ba 74]. Batchelor [Ba 74] has also reported of instability of nearest neighbour classifier particularly associated with the use of non Euclidean distances. The LVQ method belongs to nearest neighbour methods [Ko 88]. The LVQ method should be sensitive to disturbances because of this heritage. The LVQ might be sensitive to slow changes between feature vectors and class reference vectors.

In fine tuning part the instability can only occur by $\alpha(t)$ or by erroneous labelling. Function

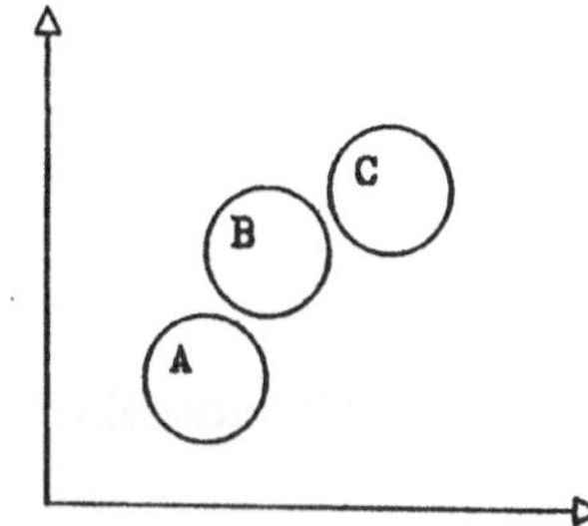

Figure 2: An illustration of the placement of classes A,B and C.

alpha(t) is decreasing function with given properties. This means that the potential disturbances will converge to zero. The only error source is wrong labelling. The effect of stochastic labelling errors in moderate numbers is limited due to the convergence properties of $\alpha(t)$.

3 Tests

The results of the analysis have been tested with simulations. A $8 * 8$ topological feature map has been used to tests. The map size has not had any effect on the results. The topological feature map has been taught by 2100 learning samples. This means that α has decreased from 0.01 to 0 over 2100 steps. The samples have been consisted of three classes.

The three taught classes have been stochastic processes with expected values (E_{11}, E_{12}, E_{13}), (E_{21}, E_{22}, E_{23}) and (E_{31}, E_{32}, E_{33}) and with standard deviations $(\sigma_{11}, \sigma_{12}, \sigma_{13})$, $(\sigma_{21}, \sigma_{22}, \sigma_{23})$ and $(\sigma_{31}, \sigma_{32}, \sigma_{33})$. The classes were close to each other but separate, Figure 2.
Four cases have been studied:

First LVQ has been tested with three feature vectors, stochastic processes, with expected values (E_{11}, E_{12}, E_{13}), (E_{21}, E_{22}, E_{23}) and (E_{31}, E_{32}, E_{33}) and with standard deviations $(\sigma_{11}, \sigma_{12}, \sigma_{13})$, $(\sigma_{21}, \sigma_{22}, \sigma_{23})$ and $(\sigma_{31}, \sigma_{32}, \sigma_{33})$. This answer made further comparisons possible.

Secondly LVQ has been tested with three modified feature vectors: The expected values have been (E_{11}, E_{12}, E_{13}), (E_{21}, E_{22}, E_{23}) and (E_{31}, E_{32}, E_{33}) and the standard deviations have been $(2 * \sigma_{11}, 2 * \sigma_{12}, 2 * \sigma_{13})$, $(2 * \sigma_{21}, 2 * \sigma_{22}, 2 * \sigma_{23})$ and $(2 * \sigma_{31}, 2 * \sigma_{32}, 2 * \sigma_{33})$. This test told about the behaviour of LVQ when the learning samples have not been representative.

Thirdly the LVQ has tested again with three feature vectors. This time the expected values have been $(E_{11} + a, E_{12} + a, E_{13} + a)$, $(E_{21} + b, E_{22} + b, E_{23} + b)$ and $(E_{31} + c, E_{32} + c, E_{33} + c)$ and the standard deviations have been $(\sigma_{11}, \sigma_{12}, \sigma_{13})$, $(\sigma_{21}, \sigma_{22}, \sigma_{23})$ and $(\sigma_{31}, \sigma_{32}, \sigma_{33})$. This test described the reaction of LVQ on a systematic error, a bias. Constants a, b and c were 2the expected values of the actual stochastic processes. The expected values come closer each other due to these biases.

The last test of LVQ has been with following data: the feature vectors with expected values (E_{11}, E_{12}, E_{13}), (E_{21}, E_{22}, E_{23}) and (E_{31}, E_{32}, E_{33}) and the standard deviations $(2 * \sigma_{11}, \sigma_{12}, \sigma_{13})$, $(\sigma_{21}, 2 * \sigma_{22}, \sigma_{23})$ and $(\sigma_{31}, \sigma_{32}, 2 * \sigma_{33})$. This was a test to study how nonlinear disturbances influence on LVQ.

For each class 1000 feature vectors have been generated. A set of 3000 feature vectors, samples have been used to each test. The achieved results have been represented in Table 1. The classification

classes	*test1*	*test2*	*test3*	*test4*
class A	6.4	19.0	9.9	13.1
class B	9.3	33.0	27.3	21.2
class C	8.5	13.5	12.2	3.1

Table 1: The missclassification percents are given to the described tests.

4 Conclusions

In some cases the stochastic nature of feature vectors, is disturbing the learning process or the pattern recognition. This is due to the real world constrains of data acquisition. Neural network methods have been discussed lately but hardly any results concerning the stability of the neural network methods have been represented.

The LVQ method is insensitive to disturbances. However systematic errors, for instance a drift, increase the missclassification rate. This can be avoided by careful design of the equipment. The choice of distance measure has also a stabilizing effect. Nonlinearities in data acquisition process increase also the missclassification. Because these errors influence on only some elements x_i of a feature vector x the total effect is less than with systematic errors. Of course, the learning data should be representative during the learning process. This is the greatest error source to LVQ but the situation is the same to other methods.

As a conclusion can be stated that LVQ is a practical and effective learning and pattern recognition method if the feature extraction process is relative well under control.

Acknowledgement The author wishes to thank Professors T. Kohonen and O. Simula of support.

References

[Ba 74] Batchelor, B,G., Practical Approach to Pattern Recognition, Plenum Press, London, New York, 1974.

[Fu 76] Fu, K.S.,(editor),Digital Pattern Recognition, Springer-Verlag, Berlin, Heidelberg,New York, 1976.

[Ko 83] Kohonen, T., Self-Organization and Associative Memory, Springer-Verlag, Berlin, Heidelberg, New York, Tokio, 1983.

[Ko 86] Kohonen, T., Learning Vector Quantization for Pattern Recognition, Helsinki University of Technology, Report TKK-F-A601, 1986.

[Ko 88] Kohonen, T., Barna, G., Chrisley, R., Statistical Pattern Recognition with Neural Networks: Benchmarking Studies, IEEE International Conference on Neural Networks, Sheraton Harbor Island, San Diego, California, July 24-27, 1988, 61-68.

AN EXPERIMENT WITH 3-D SURFACE MAPS TO ILLUSTRATE NEURAL NETWORK PERFORMANCE

Peter G. Raeth
Wright Research and Development Center
Avionics Laboratory; WRDC/AAWP-1
Electronic Support Measures Research Group
Wright-Patterson AFB, OH 45433-6543

Abstract

This paper suggests a possible means for evaluating the performance of neural networks from a global perspective in parameter-space. Traditional evaluations tend to focus on performance in weight-space or on overall output error during one training session. However, a global perspective of performance in parameter-space may be of primary importance during the initial stages of problem solution. During these stages, the researcher is typically trying to determine a network configuration and suitable values for its training equation parameters. Instead of a hit-or-miss approach, this paper describes an organized experimental method that identifies network configuration and parameter value choices which are not sensitive to minor variations for a standard training metric. The technique is illustrated for the network used by Hopfield and Tank to solve a traveling salesman problem.

Introduction

The application of neural networks to new and complex problems would be greatly aided by a global view, in parameter-space, of neural network performance. It is the author's experience that researchers tend to offer combinations of training equation parameter values and network configurations without explanation or apparent systematic choice. For instance, in traditional backpropagation, the values chosen for the Gain and Momentum parameters are typically not explained. When a new problem is tried, the original values may or may not permit the network to learn the new mappings even if they are of the same class as the original. The researcher then has to try many variations of parameter values and network configurations or do detailed studies in error or weight-space in order to get the network to learn the new mappings. It would be better if there were some systematic method to show ranges of parameter values and network configurations that would work well for a given class of mappings. This desire has led to this paper and a longer term research effort aimed at neural network performance evaluation.

Before proceeding, it is necessary to define two terms as used in this paper.

Performance: The number of training cycles the network needs to carry out its intended task. For input/output vector mapping networks: the number of random exposures to the training vector pairs the network needs in order to learn the input/output mappings represented by the vectors. For energy minimization networks used to solve optimization problems (such as Hopfield and Tank's): the number of node updates needed before the network settles into its minimum energy state within a given tolerance.

Global/Local Perspective: Instead of looking at performance during one training session (local perspective), the global perspective looks at performance over many training sessions.

Convergence Maps

Convergence maps are N-dimensional plots which show the ability of a neural network to converge on (learn) a given training metric. The traveling salesman optimization problem is a classic metric for testing energy minimization networks. This is the metric discussed in this paper.

Two dimensional convergence maps have been used in the past to illustrate the performance of neural networks. Among the recent papers are Cherkassky and Vassilas, Perugini and Engeler, and Levine. Once such global measures are taken in parameter space, plots of the error surface or of weight space can be developed during a specific training run. These latter plots are useful for observing the behavior of a network at the local level. Based on global observations, specific changes may be indicated in the values of training equation parameters or in network configuration. Modifications to the training method and/or training equation can be made based on local observations. Either global and/or local performance measures can be taken again to judge the results of the changes. In this way, an organized experimental approach to the selection of training equation parameter values, network architecture, or the training method/equation for the problem class represented by the training metric could develop.

Maps of Hopfield and Tank's Traveling Salesman Neural Network

Hopfield and Tank's neural network for solving the traveling salesman problem presents an opportunity for trying out the ideas behind convergence maps. The thought here is to take a global look at the network's performance relative to its goal of arriving at valid tours. In their summary, Wilson and Pawley state, "Our simulations indicate that Hopfield and Tank were very fortunate in the limited number of TSP simulations they attempted. .... their basic method is unreliable ...." If we could take a global look at the performance of Hopfield and Tank's network, we could see for ourselves whether or not the training of the network is reliable. (In this sense "reliable" means "convergence on valid tours is not sensitive to variations in parameter values".) The following paragraphs show convergence maps which give us a start at getting this global look.

The training equation of the Hopfield and Tank network as used in this paper is shown below. The form presented here is due to Little's analysis of Hopfield and Tank's original mathematics. To initialize the network, the

Uxi's are set to small random values. These are the values present in the network when t=DeltaT. Time is allowed to advance in steps of DeltaT until there are no further changes in the Uxi's (within a given tolerance). At this point, minimum energy has been achieved and a valid route should have resulted. As you will see, a valid tour does not always result. The frequency with which this happens led to Wilson and Pawley's remark.

$$U_{xi}(t+\Delta t) = U_{xi}(t) - \Delta t \Big\{ A \sum_{j \neq i} V_{xj} + B \sum_{y \neq x} V_{yi} + C\Big[\sum_{y}\sum_{j} V_{yj} - N\Big] + D \sum_{y} d_{xy}\left[V_{y,i+1} + V_{y,i-1}\right] + \left[U_{xi}(t)/\tau\right] \Big\}$$

The parameters in the training equation described above were set as follows except in the cases where they were varied to produce a given plot:

```
A      = 500
B      = 500
C      = 200
D      = 500
dxy    = Distance between cities x and y
DeltaT = 0.0001 (change in time t)
N      = 15
t      = time
Tau    = So big that  (Uxi(t) / Tau)  could be assumed = 0
U0     = 0.02
Uinit  = -0.5 * U0 * Ln(#CITIES - 1)
Vxi    = 0.5 * (1.0 + TANH[Uxi(t)/U0])
x,y    = City number
i,j    = Tour position
Region of random selection for Uxi initialization =
                                        U(-0.1Uinit,+0.1Uinit)
Node update method          = Synchronous
Training cutoff             = At valid tour, limit of 12000 node updates
```

It is possible to plot the number of node updates the network needed to converge on a valid tour. The plot could be based on 2500 training sessions where 50 value variations of one equation parameter are made for each of 50 variations of another parameter. The axes for such a plot are shown in Figure 1. Figures 2, 3, and 4 show plots for variations of the D & N, A & B, and Tau & DeltaT parameters respectively. By observing these plots it is possible to take an organized look at the Hopfield and Tank network from a global perspective. A similar method could be used to study the performance of other network types (see the author's other two papers on this subject). Some details on Figures 2, 3, and 4 are given below.

Figure 2. Choosing the above values for the network's training equation parameters and Hopfield and Tank's city locations as determined by Wilson and Pawley, the convergence map given in Figure 2 results if D = 0 - 600 and N = 0 - 30. As is readily evident, the map is generally a high table with a narrowing trench and one fairly wide pit where valid tours are reliably achieved within 12000 iterations of the equation for Uxi(t). This range of D and N was chosen because Hopfield and Tank suggest the above set of values and then say that some variation about those values may be necessary. The surface shows that between N=10.41 and N=19.59 over the entire range of D, there is plenty of opportunity for converging on a valid tour.

Figure 3. Fixing D at 12.24 and N at 11.63, the map in Figure 3 results under variations of A = 0 - 600 and B = 0 - 600. Observe the many opportunities for reliable convergence available when the values of A and B range over 220.41 - 600.0.

Figure 4. Setting A = 367.35, B = 428.47, D = 12.24, and N = 11.63; variations of DeltaT and Tau (0.0 - 0.7 and 0.25 - 2.0) produce the map in Figure 4. Notice that the surface is essentially low and flat except where DeltaT = 0.0 and for very low values of DeltaT coupled with very high values of Tau.

Acknowledgments

The author wishes to thank Dr Gordon Little for his analysis of the Hopfield and Tank network and Dr Steve Gustafson for his encouragement of this project. Both professors are with the University of Dayton Research Institute. They both reviewed this paper and offered many helpful comments. Their Jan 89 graduate course in neural networks saw the beginnings of this research project. Thanks are also due to Debbie Ables for her administrative support.

Bibliography

Cherkassky, Vladimir and Nikolaos Vassilas; "Performance of Back Propagation Networks for Associative Database Retrieval"; Proc: International Joint Conference on Neural Networks; pI-77; 1989

Hopfield, J.J. and D.W. Tank; "Neural Computation of Decisions in Optimization Problems"; Biological Cybernetics, 52, 141-152, 1985

Levine, R.Y.; "Neural Network Performance on the Stochastic Exclusive-Or Problem"; DTIC/NTIS # AD-A197-789; Jul 88

Little, Gordon; Class notes on Hopfield and Tank TSP; University of Dayton Research Institute; Apr 89

Perugini, N.K. and W.E. Engeler; "Neural Network Learning Time: Effects of Network and Training Set Size"; Proc: International Joint Conference on Neural Networks; pII-395; 1989

Raeth, Peter G.; "3-D Surface Maps for Neural Network Performance Evaluation"; Proc: ACM Dayton SIGART Aerospace Applications of Artificial Intelligence Conference; Oct 89

Raeth, Peter G.; "Event-Train Restoration Via Backpropagation Neural Networks (Jan-Jun 89)"; National Technical Information Service & Defense Technical Information Center, Cameron Station, Virginia (USA); TBP

Wilson, G.V. and G.S. Pawley; "On the Stability of the Travelling Salesman Problem Algorithm of Hopfield and Tank"; Biological Cybernetics, 58, p63, 88

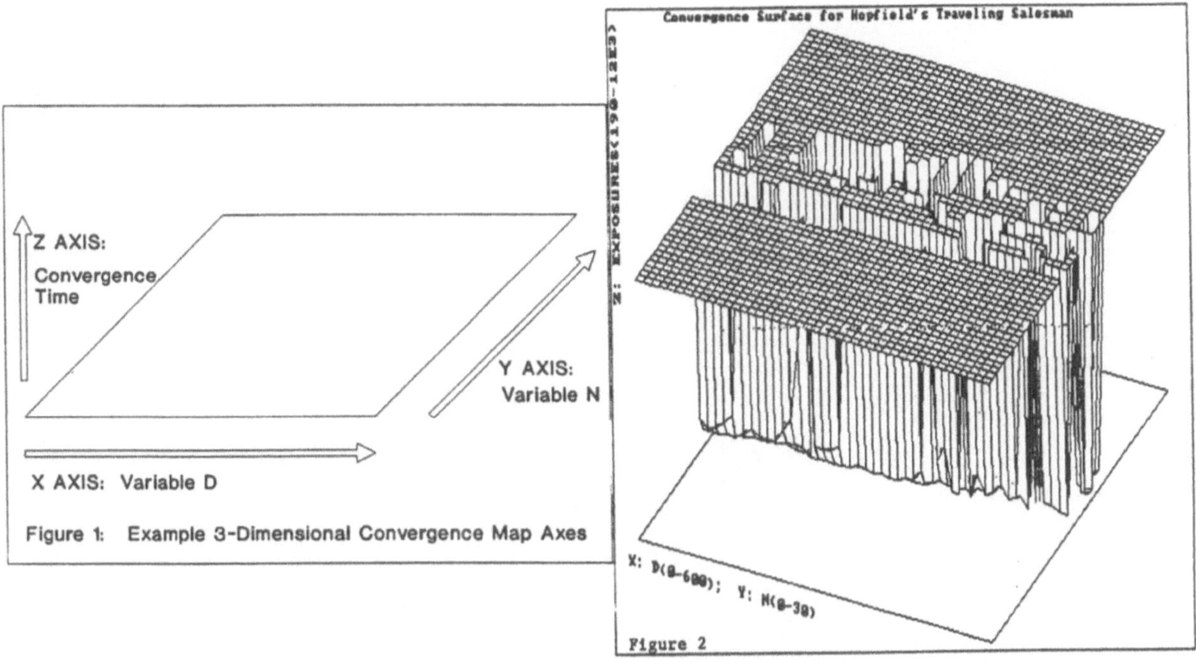

Figure 1: Example 3-Dimensional Convergence Map Axes

Figure 2

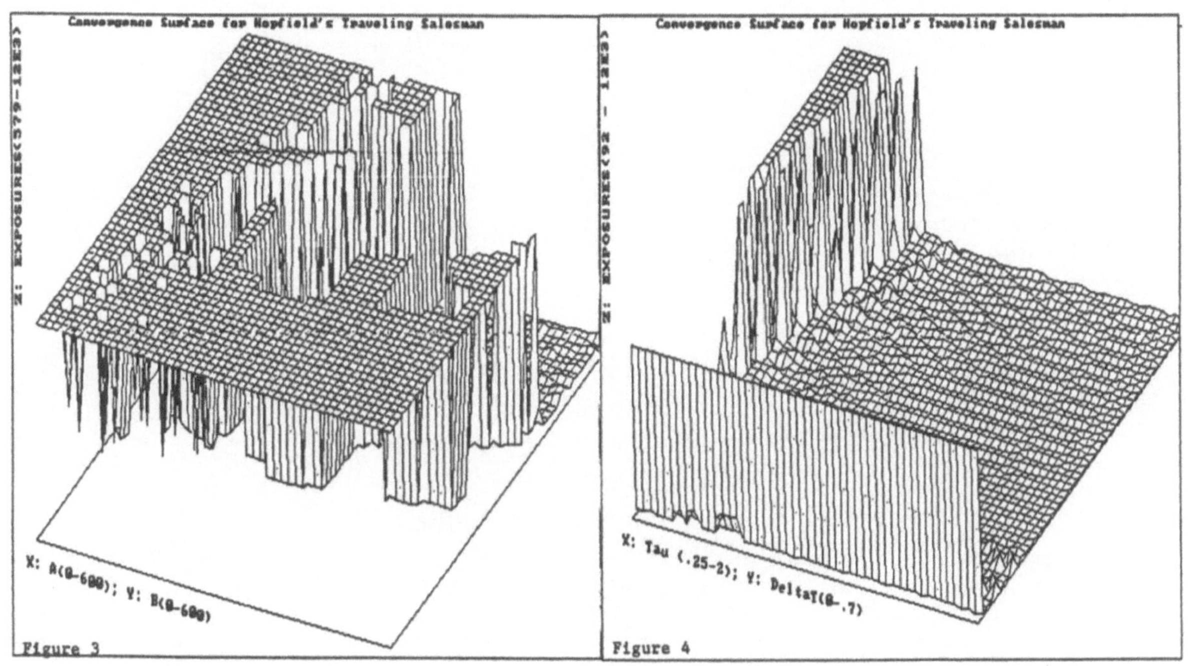

Figure 3

Figure 4

SUPERVISED LEARNING

Chairs: Yann LE CUN, Luis BORGES DE ALMEIDA

On Backpropagation Learning of Edited Data Sets

Martin A. Kraaijveld and Robert P.W. Duin

Pattern Recognition Group
Faculty of Applied Physics
Delft University of Technology
P.O. Box 5046
2600 GA Delft
The Netherlands

e-mail: martin@duttnph.uucp

Abstract

By the application of a technique from the statistical pattern recognition literature, the editing algorithm [Devijver 1982], a learning set can be transformed to a data set in which overlap between classes is effectively removed. Because it can be proven that an edited data set is close to Bayes-optimal for the nearest-neighbour classifier, it is very likely that a multi-layer network which classifies *all* samples in the edited learning set correctly, is also close to Bayes-optimal. In this paper we investigate the performance of the backpropagation algorithm on edited data sets. This leads to an optimal criterion to stop the learning phase and to a moderate improvement in learning speed.

Introduction

The backpropagation algorithm [Rumelhart 1986] is a well known method for training multi-layer feedforward networks. Practical applications (e.g. [Bounds 1988], [Gorman 1988], [Waibel 1989], etc.) as well as benchmarking studies (e.g. [Kohonen 1988]) have shown that classifiers based on the backpropagation algorithm have a good performance. Multi-layer networks therefore seem to offer a reasonable alternative for various parametric and non-parametric techniques of the statistical pattern recognition literature.

However, from a theoretical viewpoint, there are still some drawbacks:

The algorithm is essentially based on a gradient descent in an error space which contains local minima. Therefore, the learning phase can get stuck into a local minimum, resulting in a suboptimal performance of the classifier. It is important here to distinguish the cases of recognition problems with overlapping and non-overlapping classes:
When there is no overlap between the classes in the learning set, it is desirable that the network, for the learning set as well as for any test set, finally correctly classifies 100 % of the samples. Getting stuck in a local minimum implies that the performance is less than 100 %, and is therefore suboptimal.
When the classes in the learning set do overlap, however, the situation is completely different. When the learning set has a Bayes error (i.e. an intrinsic overlap) of e %, no classifier will ever achieve a performance higher than (100 - e) %. Therefore, when a large multi-layer network has achieved a performance of 100 % on the learning set, it must have made many small decision boundaries around samples in the overlapping part of the learning set and will certainly have a suboptimal performance on a test set. As the backpropagation algorithm tries to minimize the mean squared mapping error, it does in fact tend to a situation with decision boundaries around outliers, because this corresponds to a lower point in error space.
From this we conclude that the optimal classifier for a learning set with overlapping classes is surprisingly found in a *local* minimum of the error space. For overlapping classes it is therefore highly desirable that the learning phase gets stuck in a local minimum! (N.B. notice that for a learning set which is not sufficiently representative for the underlying distributions of the classes, the optimal classifier might not even correspond to a local minimum in error space).

A second problem is that there is no insight in the learning process. Unless there is some explicit knowledge of the underlying distributions of the classes in the decision problem, it is very difficult to decide in which state the learning phase remains. Is the current state the global error minimum? Is the current state close to a ravine in error space? Is the network currently learning an outlier of the classes in the learning set?, etc.
Furthermore, it is not clear if learning should continue until all samples are classified correctly (in the case of separable classes), or until another stopping criterion is reached (in the case of intrinsic overlap of the classes). Due to the lack of a stopping criterion that takes into account the state of the network in error space, ad hoc criteria are applied. Examples of these are: continue the learning phase until the mean squared error is sufficiently low, until the time-derivative of the mean squared error is sufficiently low, or until a sufficient percentage of the learning set is classified correctly.

The problems that are sketched above, bear a close parallel to problems that are related to the nearest neighbor classifier. For the nearest neighbor classifier, outliers of a different class will generate regions in which samples will be assigned to the wrong class. This deteriorates the performance of the resulting classifier. For a multi-layer network, the learning procedure will try to separate these small regions around the outliers, because this decreases the mapping error.
A solution for the problem of the outliers with the nearest-neighbour rule is described by Devijver and Kittler [Devijver 1982], and is called "editing". The editing algorithm effectively removes the overlap in case of overlapping classes, i.e. it effectively separates the classes. This results in a learning set from which all outliers are removed. The main contribution of this paper is to investigate the behavior of the backpropagation algorithm on edited data sets.

The Editing Algorithm.

The editing technique [Devijver 1982] is an algorithm that is used to improve the performance of the nearest neighbor classifier as well as to reduce the number of samples in the learning set. The algorithm removes the intrinsic overlap (if any) between the classes in the learning set in such a way that it hardly affects the position of the optimal decision boundary. However, erroneously classified samples are removed (see fig. 1). In the context of nearest neighbor pattern classification this largely improves the performance of the resulting nearest neighbor classifier. In fact it can be proven that, provided that the learning set is sufficiently large, the performance of the 1-nearest neighbor rule on an edited data set is very close to Bayes-optimal.

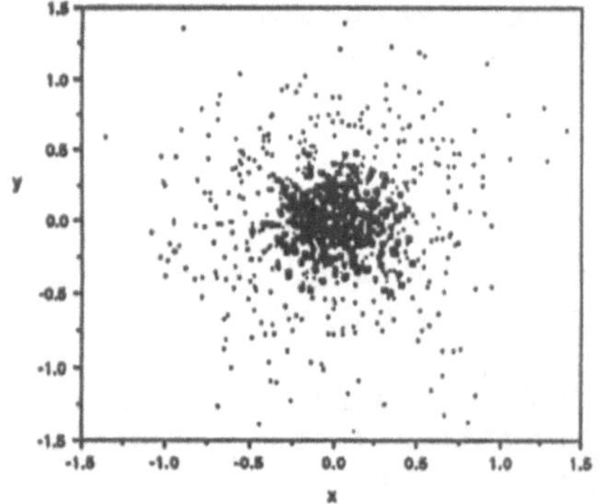

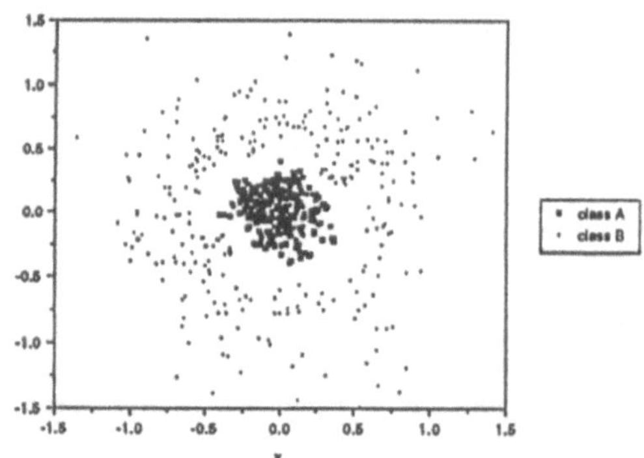

Figure 1: Data set 1 (left), and the data set after editing.

For the experiments described in this paper, a special version of the editing algorithm was used, which is called the multi-edit algorithm. It consist of 5 steps:

1. <u>Diffusion</u>: Make a random partition of the learning set S into N subsets S_1,........,S_N, $N > 2$.
2. <u>Classification</u>: Classify the samples in S_i using the 1-nearest neighbor rule with $S_{(i+1) \text{ Mod } N}$ as a training set.
3. <u>Editing</u>: Discard all samples that were misclassified at step 2.
4. <u>Diffusion</u>: Pool all the remaining data to constitute a new set S.
5. <u>Termination</u>: If the last I iterations produced no editing then exit with the final set S, else go to step 1.

Using an edited data set instead of non-edited data set for the training of a multi-layer network is expected to have the following consequences:

- By editing the data set, a very deep minimum is in the error space is introduced. Because the learning set is now completely separable, this error minimum is also a global minimum and has a error value of zero. The stopping criterion for the learning phase has now become trivial: as long as not all samples are classified correctly the learning phase should be continued.
- Because all outliers are removed from the learning set, there is no danger that the performance of the classifier will deteriorate by learning outliers, even when the network has a large number of hidden units.
- Although it is somewhat difficult to measure, it appears that learning an edited data set is faster than learning a non-edited data set. The difficulties with the measurements are caused by the fact that the learning time of an non-edited data set is based on a certain stopping criterion. Any figure for the learning time can be produced, however, when the parameters of the stopping criterion are changed. Learning is therefore certainly faster in the sense that we can make an earlier decision to stop the learning phase. Notice that this improvement in speed is derived by an operation on the data set instead of an adaptation of the learning procedure.

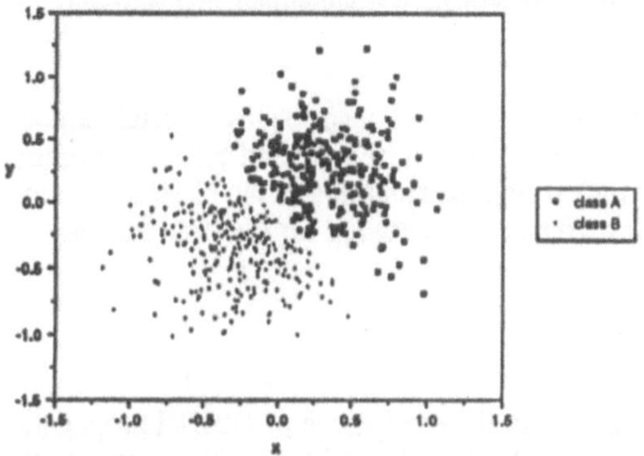

Figure 2: Data set 2 *after* editing.

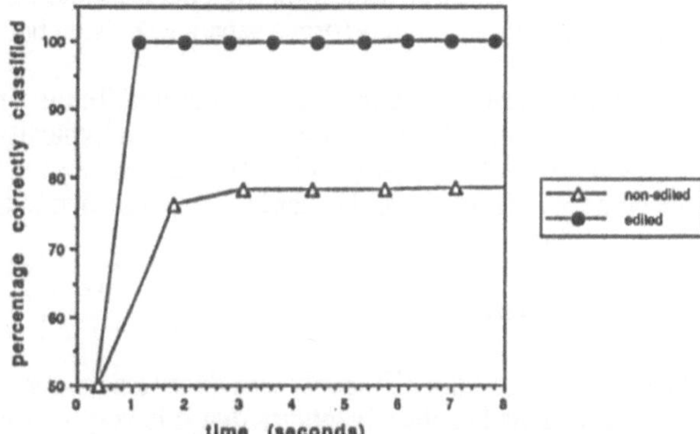

Figure 3: The *average* performance of a network with 10 hidden units on the learning set during the learning phase. The simulations were performed on a SUN 4/280.

Experiments.

To investigate the behavior of the backpropagation algorithm on edited data sets, some experiments were performed on two heavily overlapping circular symmetric Gaussian distributions. Data set 1 (figure 1) has both means on the origin (0, 0) and standard deviations 1.645 and 0.6076. Data set 2 (figure 2) consists of two classes with means (0.2, 0.2) and (- 0.2, - 0.2) and equal standard deviations 0.333. The Bayes error for both learning sets is 20 %. Each set consists of 1000 samples.

For both learning sets, the original set and its edited version were used as a training set for a number of multi-layer feedforward networks. The networks consist of 2 input units, 10, 20, 50 or 100 hidden units and 1 output unit. The parameters of the backpropagation algorithm were: learning rate $\eta = 0.1$ and momentum term $\alpha = 0.9$. A sample was considered to be correct when the difference between the actual output and the desired output was smaller than 0.5. The learning phase for the edited data sets was terminated when all samples were classified correctly.

Table 1: The average performance of 50 trained networks on a test set of 25,000 samples.

average performance	data set 1		data set 2	
	non-edited	edited	non-edited	edited
10 hidden units	80.7 %	79.1 %	80.1 %	80.0 %
20 hidden units	80.3 %	79.0 %	80.1 %	80.1 %
50 hidden units	80.2 %	79.0 %	80.1 %	80.0 %
100 hidden units	80.1 %	78.7 %	80.0 %	80.0 %
nearest neighbor classifier	74.7 %	78.9 %	74.0 %	79.8 %

Table 1 shows the performance of a trained network on a large test set. Figure 3 shows a plot of the average performance on the learning set as a function of the learning time. All simulations were performed on a SUN 4/280 system.

Discussion.

From the experiments that were performed, it appears that the backpropagation algorithm with an edited data set is a factor three faster when compared to a non-edited data set, *provided that we had a good criterion to stop the learning phase for a non-edited data set.* In practical situations therefore, the optimal stopping criterion results in a considerable improvement in speed. What we also gained with editing is the relative certainty that a reasonable classifier is found, due to the introduction of a large near-Bayes global minimum in the error space.
However, it is remarkable that the performance of a multi-layer network is not seriously deteriorated by the overlap in the learning set, as is the case with the nearest neighbor classifier. It is clear that the algorithm does not form small decision boundaries around outliers, as can be expected from a procedure that is based on the minimization of the mapping error. The search for a global minimum is apparently constrained by a mechanism that prevents that too small groups of outliers form a separate decision boundary.

Finally some remarks on the editing algorithm. In the first place, the editing step does also require a certain amount of computational effort; for the two data sets typically 20 seconds CPU time. In the second place, the editing algorithm requires that the data is sufficiently representative for the underlying distributions of the classes; i.e. when the data set is divided over N subsets, all these subsets must on their own be representative for the distributions.

Conclusion.

The editing algorithm effectively introduces a near-Bayes global minimum in the error space of the backpropagation algorithm. This has the advantages that it is very likely that the learning phase will end in this minimum and that an optimal stopping criterion is found.
Experiments indicate that the search for the global minimum in the backpropagation algorithm is essentially constrained by a mechanism that prevents the formation of decision boundaries around small groups of outliers. This results in a very good performance for overlapping classes.

Acknowledgements.

This work was sponsored by the Dutch Government as a part of the SPIN/FLAIR-DIAC project, and by the Foundation of Computer Science in the Netherlands (SION) with financial support from the Dutch Organization for Scientific Research (NWO).

Literature.

[Bounds 1988] Bounds, D.G., and Lloyd, P.J., "A Multi-Layer Perceptron Network for the Diagnosis of Low-Back Pain", proc. of the SGAICO conference 1988, University of Zürich, Oct. 1988.

[Devijver 1982] Devijver, P.A. and Kittler, J., "Pattern Recognition, A Statistical Approach", Prentice Hall, 1982.

[Gorman 1988] Gorman, R.P., and Sejnowski, T., Neural Networks 1, 75-89, 1988.

[Kohonen 1988] Kohonen, T., Barna, G., and Chrisley, R., "Statistical Pattern Recognition with Neural Networks: Benchmarking Studies", proceedings of the Neuro '88 conference, Paris, June 1988.

[Rumelhart 1986] Rumelhart, D.E., and McClelland, J.L., "Parallel Distributed Processing: Explorations in the micro structure of cognition", chapter 8, MIT press 1986.

[Waibel 1989] Waibel, A., Hanazawa, T., Hinton, G., Shikano, K., and Lang, K., IEEE Tr. on ASSP - 37, March 1989.

A STUDY OF LEARNING AND GENERALIZATION BY EXHAUSTIVE ANALYSIS

V. K. Samalam and D. B. Schwartz
GTE Laboratories
40 Sylvan Road
Waltham, Massachusetts 02254
USA.

Abstract

We have explored learning and generalization in layered networks with a specific example involving binary weights. The example chosen for analysis is the contiguity problem where the architecture is asked to classify binary input vectors according to the number of contiguous 1's present. The basic idea we have explored is that learning can be described as the gradual elimination of networks from an initial ensemble. The example is analyzed numerically by characterizing both the architecture and the learning process in an exhaustive fashion. The numerical results on the distribution of networks and learning are then analyzed theoretically. Our main conclusion is that the learning curve can be predicted from the initial distribution by means of a very simple formula. The theoretical analysis illustrates a method that can be followed in analyzing other examples of learning in layered networks and has the virtue that it is independent of the vagaries of any particular algorithm.

Introduction

One of the challenges in neural networks research is to understand what determines the generalization ability of a network architecture. In the past this question had been explored in the context of particular learning algorithms. It is only recently that attempts have been made to explore this question theoretically independent of any particular algorithm [1,2,3,4]. Our analysis is based on the idea that the learning process can be thought of as the gradual elimination of networks. By 'network' we mean a unique combination of values of all the parameters of the architecture such as weights and biases. We have explored this idea by exhaustively characterizing all possible networks for an architecture with binary weights. The architecture was designed to solve the contiguity problem where binary input vectors are classified according to the number of contiguous 1's present. Learning was simulated by systematically removing networks that gave incorrect results for test examples and then computing the average generalization score for the remaining networks. Our main conclusion is that the generalization or the error curve as a function of the number of training examples can be predicted completely from the initial distribution alone by means of a very simple formula [5].

Numerical Experiments

The architecture consists of m input units that pass the input values unchanged. The input vectors are all of length m and consist of 0s and 1s. The hidden layer has m units which have a bias of -0.5 and their outputs are either 1 or 0. Each input unit is coupled to a hidden unit by two weights so that there are a total of 2m weights. Finally the output unit is a simple thresholding unit which takes on the values of 1 or 0 with a bias of -2.5. The only parameter of the network that is not fixed is the value of the couplings between the input and the hidden layer which were restricted to be 1 or -1. The first part of the experiment consisted of grouping all the networks according to their g score where g is defined as the ratio of the number of input vectors the network correctly classifies to the total number of input vectors. The points in Figure 1a are the fraction of the networks which have a particular score of g. The second part of the experiment consisted of simulating learning in an exhaustive manner. First a random sequence of input vectors was generated and all the networks were tested against the first input vector. Only those networks that classified the vector correctly were kept. The average generalization score of these remaining networks was then computed from a sample that did not belong to the training set. These networks were then tested against the second input vector and again only those networks that classified it correctly were kept and their average generalization score computed. This process was repeated for all the vectors in the training set. To obtain good statistics the entire process was repeated with a number of different training sequences. The points in Figure1b are the average error as a function of the number of training examples [5].

Theory

The experimental distribution can be understood by noting that the networks are distributed over a set of microscopic states which for this problem is the set of all Boolean functions possible for the input vectors. The microscopic states are grouped by the g score and the picture is analogous to statistical mechanics where the microscopic states are grouped according to their energy. The distribution can be calculated if the microscopic probability, which is the probability that a network belongs to a state, is known. We have shown that if the microscopic probability p_i is of the form [6]

$$p_i = (\overline{g})^{I_c} \times (1-\overline{g})^{I-I_c}, \tag{1}$$

where $\overline{g}$ is the average probability for getting a single vector correct and I_c and I are the number of correct input vectors and total input vectors respectively, then the distribution is given by

$$P(g) = \frac{N(g)}{N^T} \approx \frac{1}{\sqrt{2\pi}\sigma} \exp \frac{-(g-\overline{g})^2}{2\sigma^2}. \tag{2}$$

In the above equation $N(g)$ is the number of networks with a g score of g, N^T is the total number of networks and $\sigma = \sqrt{\overline{g}/I}$. The solid line in Figure 1a is the function given above with $\overline{g} \approx .69$. While the form of the distribution near the peak can be derived, it is very difficult to derive from first principles the value of $\overline{g}$ since the process of choosing the architecture skews the microscopic probabilities in unpredictable ways.

We have also shown that the experimental learning curve can be explained by a simple formula which depends only on the initial distribution [5]. The average error of the architecture after training with n examples is

$$\varepsilon_n = 1 - \overline{g}_n = 1 - \frac{\int_0^1 g^{n+1} P(g)\, dg}{\int_0^1 g^n P(g)\, dg}. \tag{3}$$

The above formula has to be modified slightly because of errors introduced by averaging over small sample sizes. The solid line in Figure 1b is basically the theoretical formula given above modified slightly to take into account sampling errors and as can be seen the agreement is good. For small values of n, ($n < \tilde{n} = \sqrt{I\bar{g}}/2$), the distribution retains its form while the peak shifts slightly to higher values of g. The initial plateau in the error curve corresponds to this region. For $n > \tilde{n}$ the peak in the distribution disappears and the error curve is determined solely by the tail of the distribution. In this region the error depends exponentially on the number of training examples.

Conclusion

The basic idea that we have explored in this work numerically and theoretically is that learning can be described in terms of a probability distribution of networks defined over a function space. The initial distribution can be described in terms of the microscopic probabilities and the error as a function of the number of training examples can be derived from the particulars of the initial distribution alone. The analysis was done for an architecture involving binary weights and the challenge is to see if this analysis can be extended to continuous valued weights.

References

[1] John Denker, Daniel Schwartz, Ben Wittner, Sara Solla, Richard Howard, Lawrence Jackel, and John Hopfield.
Automatic learning, rule extraction, and generalization.
Complex Systems **1** (1987) 877-922.

[2] Eric B. Baum and David Haussler.
What size net gives valid generalization?
Neural Computation **1** (1989) 151-160.

[3] Naftali Tishby, Esther Levin, and Sara A. Solla.
Consistent inference of probabilities in layered networks: predictions and generalization.
IJCNN International Joint Conference on Neural Networks, IEEE, New York (1989) Vol. II, 403-409.

[4] Sara A. Solla.
Learning and generalization in layered neural networks: the contiguity problem.
Neural Networks: from Models to Applications, ed. by L. Personnaz and G. Dreyfus, I. D. S. E. T. , Paris (1989) 168-177.

[5] D. B. Schwartz, V. K. Samalam, S. A. Solla and J. Denker.
Exhaustive Learning
Neural Computation (To be published) (1989).

[6] V. K. Samalam and D. B. Schwartz.
A study of learning and generalization by exhaustive analysis.
GTE Laboratories Technical Report TM-0224-12-89-401 (1989).

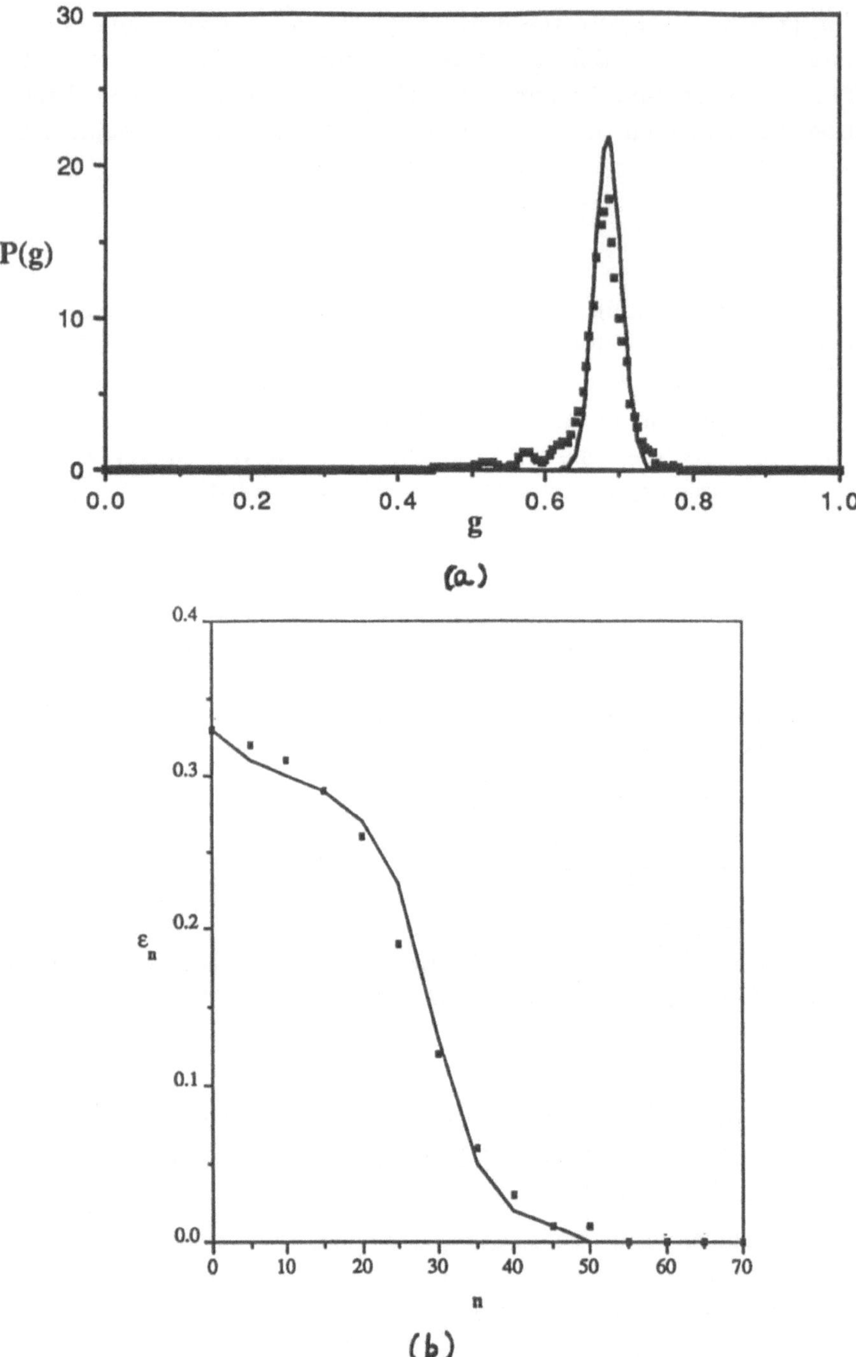

Figure 1: a) The distribution for 11 bit vectors. The points were obtained numerically and the solid line is from theory. b) The average error as a function of the number of examples. The points were obtained numerically and the solid line is from theory.

Curvature-Driven Smoothing in Backpropagation Neural Networks

C. M. Bishop

Theory and Computing Division
Culham Laboratory
AEA Technology
Abingdon, OX14 3DB. U.K.
(Euratom/UKAEA Fusion Association)

Abstract

A central problem in the theory of feed-forward neural networks is the determination of the number of hidden neurons. With too few neurons the network may be unable to achieve the desired accuracy, while too many neurons leads to over-fitting. In this paper we propose a modification to the standard mean square error function which will reduce the tendency to over-fit, and which can be controlled by a single scalar parameter. A new learning algorithm, which can be used to minimise this error function, is also described.

1 Curvature-Driven Smoothing

Feed-forward neural networks, with hidden neurons and non-linear activation functions, can generate a large class of non-linear multivariate mappings. There exists no theoretical basis, however, for determining the number of hidden neurons required for a specific application. If the network has too few hidden neurons the class of functions which it can generate is too restricted and it is unable to achieve the desired accuracy. With too many neurons, however, the network over-fits the data (sometimes called over-generalisation) and this reduces the useful predictive capability of the network. In this paper we propose the use of a modified error function designed to allow the use of many hidden neurons, whilst avoiding the problem of over-fitting. A generalisation of the backpropagation learning algorithm, which can be used to minimise this error function, is also described.

To begin with, consider continuous mappings of a single variable x to a single variable y, so that the network has one input and one output neuron. The network has a feed-forward architecture in which each neuron generates a non-linear function of the weighted sum of its inputs:

$$z_i = f(\sum_j w_{ij} z_j) \,, \qquad f(x) \equiv \frac{1}{1+e^{-x}} \tag{1}$$

where z_j is the activation of the $j^{\rm th}$ neuron, w_{ij} is the connection weight between neurons i and j, and the function f is taken to be the standard sigmoid. There also exist thresholds for each neuron. Since, however, these are equivalent to weights from an extra neuron whose output is permanently set to +1, they are contained in the formalism of Eq.(1) and we need consider them no further. By construction, the function $y(x)$ will be continuous, single-valued and differentiable.

Suppose we have a set of data points $\{x_p, t_p\}$, $p = 1, ..., N$, where t_p is the target value for y corresponding to $x = x_p$. The standard learning algorithm minimises the error function $E^{\rm S} = \frac{1}{2}\sum_{\{p\}}(y_p - t_p)^2$ where $y_p = y(x_p)$. This choice of error function can lead to a tendency to over-fit the data [1], as shown schematically in Figure 1. We now seek to modify the error function so as to generate smoother functions $y(x)$, as indicated in Figure 2.

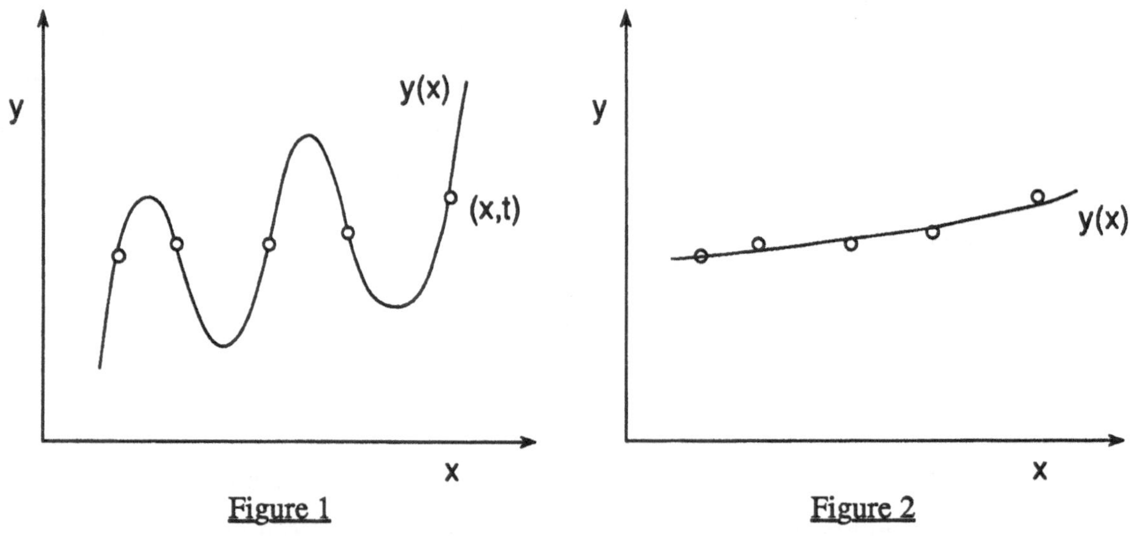

Figure 1 Figure 2

To achieve this we define a new error function

$$E = \frac{1}{2}\sum_{\{p\}}(y_p - t_p)^2 + \lambda E^{\rm C}, \tag{2}$$

$$E^{\rm C} = \frac{1}{2}\int_a^b \kappa^2 dx = \frac{1}{2}\int_a^b \frac{(y'')^2}{[1+(y')^2]^3}dx \simeq \frac{1}{2}\sum_{\{n\}}\frac{(y''_n)^2}{[1+(y'_n)^2]^3}\Delta x_n \tag{3}$$

where κ is the curvature of the line $y = y(x)$, primes denote d/dx, and $\Delta x_n = x_n - x_{n-1}$. The interval (a, b) spans the range of values of $\{x_p\}$. It will often be convenient to choose the points $\{x_n\}$ to coincide with the $\{x_p\}$ since this will make use of values for the neuron activations which need to be calculated anyway for the minimisation of E^{S}. If, however, the data points are too sparse in some regions for sufficient accuracy to be obtained, it is straightforward to include extra values of x. Equally, it may be acceptable to exclude a proportion of the data points from $\{x_n\}$ and so reduce the training time.

For a network with N input and M output neurons, the curvature term in the error measure can be generalised as follows:

$$E^{\mathrm{C}} = \frac{1}{2}\sum_{i=1}^{N}\sum_{j=1}^{M}\int_{a_i}^{b_i} \kappa_{ij}^2 dx_i \ , \qquad \kappa_{ij}^2 = \frac{(\partial^2 y_j/\partial x_i^2)^2}{\left[1+(\partial y_j/\partial x_i)^2\right]^3} \ . \tag{4}$$

2 Learning Algorithm

As usual, the network is trained by gradient descent so that

$$\Delta w_{ij}(m) = \ -\eta\frac{\partial E}{\partial w_{ij}} \ + \ \mu \Delta w_{ij}(m-1) \tag{5}$$

where m denotes the training step number, η is the learning rate, and μ is the momentum. The derivative $\partial E^{\mathrm{S}}/\partial w_{ij}$ is calculated using the standard backpropagation algorithm [2]. We now seek a procedure for calculating $\partial E^{\mathrm{C}}/\partial w_{ij}$. From Eq.(3) we have

$$\frac{\partial E^{\mathrm{C}}}{\partial w_{ij}} = \sum_{\{n\}}\left\{\frac{(y_n'')}{[1+(y_n')^2]^3}\frac{\partial y_n''}{\partial w_{ij}} - \frac{3(y_n'')^2 y'}{[1+(y_n')^2]^4}\frac{\partial y_n'}{\partial w_{ij}}\right\}\Delta x_n \ . \tag{6}$$

To calculate the partial derivatives in Eq.(6), we note that w_{ij} and x_n are independent variables, and so we can interchange the order of the derivatives:

$$\frac{\partial}{\partial w_{ij}}\left(\frac{dy}{dx}\right) = \frac{d}{dx}\left(\frac{\partial y}{\partial w_{ij}}\right) . \tag{7}$$

We now introduce an 'error' term σ_i for each neuron:

$$\frac{\partial y}{\partial w_{ij}} = \sigma_i z_j \ , \qquad \sigma_i = \frac{\partial y}{\partial z_i} z_i(1-z_i) \tag{8}$$

where we have used Eq.(1) together with the relation $f' = f(1-f)$. From the chain rule for partial derivatives it follows that

$$\sigma_i = z_i(1-z_i)\sum_k w_{ki}\sigma_k \ . \tag{9}$$

Using Eqs.(7) and (8) we can write

$$\frac{\partial y'}{\partial w_{ij}} = z_j \frac{d\sigma_i}{dx} + \sigma_i \frac{dz_j}{dx} \tag{10}$$

where

$$\frac{d\sigma_i}{dx} = z_i(1 - z_i) \sum_k w_{ki} \frac{d\sigma_k}{dx} + (1 - 2z_i) \frac{dz_i}{dx} \sum_k w_{ki}\sigma_k \; . \tag{11}$$

The partial derivatives z_i' can be calculated during the forward propagation phase using

$$\frac{dz_i}{dx} = z_i(1 - z_i) \sum_j w_{ij} \frac{dz_j}{dx} \; . \tag{12}$$

For the output neuron we have

$$\sigma_o = y(1 - y) \; , \qquad \frac{d\sigma_o}{dx} = (1 - 2y) \frac{dy}{dx} \; . \tag{13}$$

We can now summarise the learning algorithm as follows. First apply inputs $\{x_n\}$ and forward propagate to generate, layer by layer, the neuron activations z_{jn}, y_n using Eq.(1), and the derivatives dz_{jn}/dx_n using Eq.(12). Next, compute the error for the output neuron using Eq.(13) and backpropagate the errors using Eqs.(9) and (11). Finally, update the weights using Eqs.(5), (6), and (10). Analogous results for the second derivative term $\partial y''/\partial w_{ij}$ are easily obtained.

In this paper we have proposed a new error measure for feed-forward neural networks which is intended to bias the network in favour of smooth solutions and thereby avoid the problem of over-generalisation. A learning algorithm for the minimisation of this error measure has also been described. This algorithm is applicable to a wide class of error functions, and readily generalises to multivariate input and output data.

A major outstanding problem with feed-forward networks is the determination of the number of hidden neurons. With the technique described here it should be possible to use a network with a large number of neurons and to control the properties of the solution by varying a single scalar parameter λ instead of having to change the network architecture.

References

[1] A N Tikhonov and V Y Arsenin, *Solutions of Ill-Posed Problems*, Wiley (1977).

[2] D E Rumelhart, J L McClelland and the PDP Research Group, *Parallel Distributed Processing: Explorations in the Microstructure of Cognition*, Vol. 1: Foundations, MIT Press (1986).

Experimental Analysis of Performance of Temporal Supervised Learning Algorithm, Applied to a Long and Complex Sequence

Ryotaro Kamimura
Information Science Laboratory
Tokai University
1117 Kitakaname Hiratsuka
Kanagawa 259-12, JAPAN

April 19, 1990

Abstract

In the present paper, we evaluate the performance of temporal supervised learning algorithm(TSLA), developed by R.J. Williams and D. Zipser, and propose the computational methods which accelerate and stabilise the learning process of recurrent neural network with TSLA, when it is applied to long and complex sequences.

TSLA represents consecutive events in architecture itself, which enables the network to deal with long and complex time-changing phenomena without increasing the number of units. However, TSLA shows extreme instability, when it is applied to long and complex time-changing phenomena. Moreover, it tends to take a long time to finish the learning. It is absolutely necessary to evaluate the performance of TSLA and to develop some computational methods which improve the instability and reduce considerably the learning time. We attempt to remove the instability by using the variable learning rate, which means that the learning rate can vary, according to the progress of learning. The Minkowski-r power metrics are used to accelerate the learning time.

From the experiments, it was confirmed that the instability was removed by the variable learning rate and some other minor adjustment, and that the network with Minkowski-r power metric learned a relatively long sequence (English sentences) more than three times faster than that with ordinary error function.

1 Introduction

This paper is concerned with the evaluation of performance of temporal supervised algorithm(TSLA), developed by R.J. Williams and D. Zipser[5], and the computational methods which enable TSLA to cope with long and complex sequences. TSLA has used explicit representation of consecutive events. This means that TSLA deals with the natural temporal order through time-changing inputs or outputs. It is expected, thus, that it can deal with long time-changing phenomena by means of relatively small units, which is impossible for ordinary static network architecture. The network architecture must be unfolded so as to represent the time sequence as fully discussed by D.E. Rumelhart *et al.*[4].

TSLA has been evaluated, however, upon the experimental results of extremely short sequences, or sequences with explicit regularity. Thus, it is uncertain for the learning algorithm to be applied to actual time-changing phenomena, which show complex regularity, or whose characteristics can not be found in short sequences. For example, it has been well-known that natural languages have a long-distance correlation, which may span over the entire text. One of the reasons is that the structural dependency should be one of their characteristics.

TSLA shows extreme instability, when applied to long and complex sequences as suggested by F.J. Pineda[3]. The time of learning is also long. Thus, we attempt, in the following sections, to develop the computational methods, which improve the instability and reduce the learning time. The instability is removed by means of the variable learning rate and the learning time is greatly reduced by using Minkowski-r power metrics, introduced by S.J. Hanson and D.J. Burr[2].

2 Theory and Computational Methods

Let $x_i(t)$ denote the activity of ith unit at time t, and w_{ij} represent the connection strength from jth unit to ith unit, then the activity of ith unit at time $t+1$ is given by

$$x_i(t+1) = f(s_i), \tag{1}$$

where f is the logistic function, and

$$s_i(t) = \sum_j w_{ij} x_j(t). \tag{2}$$

Some of the units can be taken as a set of output units Ω, which should have their own target values. The network is forced to learn the target values by minimizing the difference between target values and actual values. Consider the difference at time t,

$$e_k(t) = \begin{cases} d_k(t) - x_k(t), & \text{if} \quad k \in \Omega, \\ 0 \quad , & \text{otherwise,} \end{cases} \tag{3}$$

where $d_k(t)$ represents a target value. The overall network error at time t is defined as,

$$J(t) = \frac{1}{r} \sum_k \mid e_k(t) \mid^r, \tag{4}$$

which is referred to as "Minkowski-r power metrics". The metrics are used in this paper only to reduce the learning time. The weight change for any particular weight at time t can be given by

$$\begin{aligned} \Delta w_{ij}(t) &= -\eta \frac{\partial J(t)}{\partial w_{ij}} \\ &= \eta \sum_k \mid e_k(t) \mid^{r-1} \left[\frac{\partial x_k(t)}{\partial w_{ij}}\right] sgn(e_k(t)) \end{aligned} \tag{5}$$

where η is a learning rate. The derivative $\partial x_k(t)/\partial w_{ij}$ can be computed by the equation

$$\frac{\partial x_k(t)}{\partial w_{ij}} = f'(s_k(t-1)) \Big[\sum_l w_{kl} \frac{\partial x_l(t-1)}{\partial w_{ij}} + \delta_{ik} x_j(t-1)\Big] \tag{6}$$

with initial conditions

$$\frac{\partial x_k(t_0)}{\partial w_{ij}} = 0. \tag{7}$$

Finally, the overall weight change is simply represented by

$$\Delta w_{ij} = \sum_t \Delta w_{ij}(t). \tag{8}$$

TSLA, formulated just above, is powerful when applied to relatively short sequences or sequences with explicit structure. TSLA has shown extreme instability, when applied to long and complex structures. This means that any weight tends to explode while learning sequences. We have tried to use multiple learning methods which prevent the explosion in learning. One of the most successful methods is ,without details, formulated as follows. The weight is updated by the formula:

$$\Delta w_{ij}^{(n)} = \eta\Big[\alpha \Delta w_{ij}^{(n-1)} - (1-\alpha)\sum_t \frac{\partial J(t)}{\partial w_{ij}}\Big], \tag{9}$$

where n is the number of iteration of learning. While α can remain a constant, the factor η should vary according to the degree of increase of mean square error between target values and actual values at the output units, so as to evade extremely high acceleration of learning. While computing the equation(6), it varies extremely, which causes the explosion of weight matrix. The equation should be multiplied by a small constant (ϵ) between 0.3 and 1.0, which greatly contributes to the extinction of explosion. Finally, it should be stressed that even if the learning rate is sufficiently small, the learning eventually fails unless all the parameters are appropriately given, according to the progress of learning.

In addition to the variable learning rate, it is necessary to use the teacher-forced learning algorithm[5] and employ several computational methods discussed by S.E. Fahlman[1], in order to accelerate the learning.

3 Results and Discussion

In our experiments, the parameter α was set to 0.9 for all cases. The uniform random numbers ranging between -0.15 and 0.15 were given as initial values. The sequence, used in the experiments, consisted of English sentences, whose maximum length was 100 letters. 26 alphabet letters and blank were represented in the binary numbering system with 5 bits. The network was fully connected and there were no input units. The number of output units were 5. Finally, all the computations were performed on SX-1 with FORTRAN77.

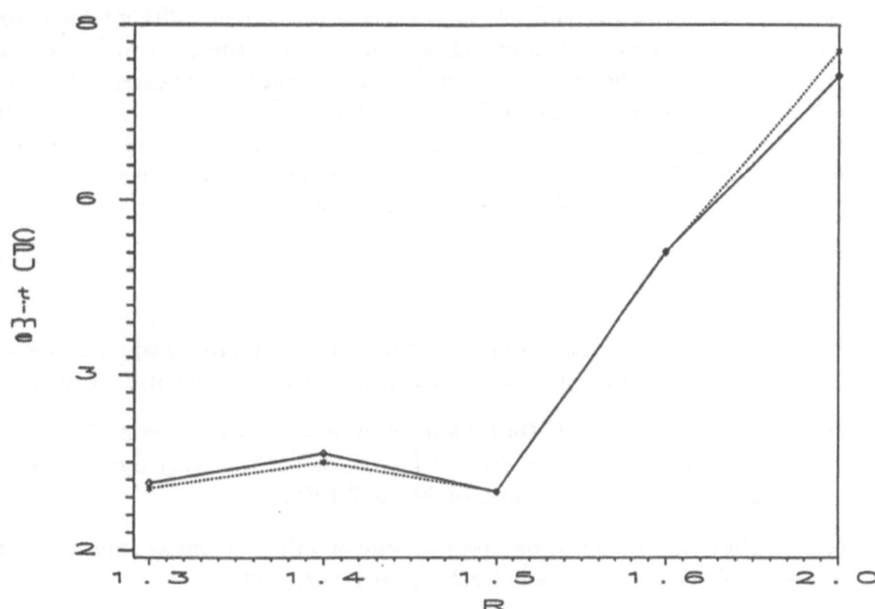

Figure 1: CPU time as a function of Minkowski-r

For Minkowski-r, we took 1.5 as a value of r, which gave the most rapid convergence of learning. As can be seen from Figure 1, the best performance was obtained, when r was 1.5. In this case, the network with $r = 1.5$, converged approximately more than three times faster than that with $r = 2.0$, which has been extensively used for the error function. With other values of r, which are not included in the figure, the learning failed. The network used in the figure, had 6 hidden units, dealing with a sequence whose length was 30. CPU time was measured in minutes.

We reproduce here sequences generated by the network at some stages of learning, to show how the network learns the target natural language. The recurrent network we used had 26 hidden units and 5 output units and ϵ was set to 0.3. After 888 learning cycles, we had:

MISS EKS EJS EJS EJS EJS EJS EJS EJS EJS EJW EJW E
*KW E S E S*EKS*EKS EIS EK S EKS EKS EKS EKS EKS EKS*

1833 learning cycles gave the sequence:

MISS MARTHA MEACHAM KEPT THE LITTLE BAKDPT THE OIU
*PW*ICSZEGRCEISSMKDTT THE LITTLE BAKDPT THE OIUPW*I*

Finally , the sequence obtained after 2374 learning cycles was given by

MISS MARTHA MEACHAM KEPT THE LITTLE BAKERY ON THE
CORNER THE ONE WHERE YOU GO UP THERE STEPS AND THE

which is a perfect original sequence. To reach this final point, it took about 2 hours.

The network began to learn naturally the sequence from the beginning. The word "Miss" was learned at the early stage of learning. The second point to be noted is that the network learned the letters or words with large frequency at the early stages. The letter "e", and the word "the" are the typical examples. Finally, the weight matrix obtained, showed rather vague regularity. To extract a certain kind of structure in the connection of units, the restriction should be imposed upon the network.

4 Conclusion

In this paper, we have examined the performance of TSLA and the computational methods which should improve its performance. TSLA can represent the time-changing sequence overtly. Thus, it is expected that TSLA can cope with the multiple time-varying phenomena without increasing the number of units in the network. TSLA is good at learning short sequences or sequences with explicit regularity. However, it shows extreme instability, when applied to the long and complex sequences. Our computational methods, in which the learning rate can vary, according to the progress of learning, improved the instability greatly, and the time of learning. Moreover, with Minkowski-r, the time of learning was further accelerated. Thus, it can be safely said that TSLA will be used for relatively long and complex sequences, and it will contribute to the understanding of complex time-changing phenomena.

References

[1] S. E. Fahlman, "Faster-learning variations on back-propagation: an empirical study," in *Proceedings of the 1988 Connectionist Models* , Carnegie Mellon University, pp. 38-51, 1988.

[2] S. J. Hanson and D. J. Burr "Minkowski-r back-propagation: learning in connectionist models with non-Euclidian signals", in *Neural Information Processing Systems*. D. Z. Anderson, Ed. New York: American Institute of Physics, pp.348-357,1989.

[3] F. J. Pineda, "Recurrent backpropagation and the dynamical approach to adaptive neural computation", *Neural Computation*, Vol.1, No.2, pp. 161-172, 1989.

[4] D. E. Rumelhart, G. E. Hinton and R. J. Williams, "Learning internal representations by error propagation," in *Parallel distributed processing* . D. E. Rumelhart, J. L. McClelland and the PDP research group, Ed. Cambridge, Massachusetts: The MIT Press, Vol.1, pp.318-362, 1986.

[5] R. J. Williams and D. Zipser, "Experimental analysis of the Real-time Recurrent Learning algorithm," *Connection Science*, Vol.1, No.1, pp.87-111,1989

BFGS Optimization for Faster and Automated Supervised Learning

Roberto Battiti and Francesco Masulli
Dipartimento di Fisica, Universita' di Genova
Via Dodecaneso 33, 16146 Genova, Italy
Computer address: battiti%genova.infn.it@iboinfn.bitnet

Abstract

Standard back-propagation learning (BP) is known to have slow convergence properties. Furthermore no general prescription is given for selecting the appropriate learning rate, so success is dependent on a trial and error process. In this work a well known optimization technique (the *Broyden-Fletcher-Goldfarb-Shanno* memoryless quasi-Newton method) is employed to speed up convergence and to select parameters. The strict locality requirement is relaxed but parallelism of computation is maintained, allowing efficient use of concurrent computation. While requiring only limited changes to BP, this method yields a speed-up factor of 100 – 500 for the medium-size networks considered.

Comparisons are done with a version of BP employing learning rate adaptation. This last method is in itself interesting, since it converges in a number of iterations close to that of optimized BP, with no need for parameter optimization.

1 Introduction

Back-propagation learning [8] is a gradient-descent method: in a given iteration the *search direction* is given by the negative gradient of the energy, while the step along this direction is given by a constant (*learning rate*) chosen by the user.

Now, it is well known from the optimization literature that pure gradient-descent methods tend to be very inefficient [3]. A case in which this happens is when "the search space contains long ravines that are characterized by sharp curvature across the ravine and a gently sloping floor" [8].

A recent overview of other heuristics employed to accelerate back-propagation (like the *momentum term*) has been presented in [5]. Unfortunately up to now there are no good general prescriptions for selecting the parameters defining the optimization strategy. It is usually left to the user to find a good or optimal combination of these parameters that leads to avoidance of local minima and fast convergence times. This process of *meta-optimization* (optimization of the optimization method) leads in general to a sizeable waste of computational resources.

The focus of this work has been on transferring some *meta-optimization* techniques to the learning algorithm itself. Since this involves measuring optimization performance and correcting some parameters while the optimization algorithm is running, some *global* information is required. *Parallelism* of computation is nonetheless maintained, resulting in efficiency close to 100% when the algorithm runs on a parallel computer.

In all cases the "standard" back-propagation algorithm is used to find the value of the negative gradient for a given configuration. The differences are in the definition of the *search direction* and/or in the selection of a *step size* along the selected direction.

In the "bold driver" method the search direction remains equal to the negative gradient but the step size is adapted during the computation. In the BFGS method both the search direction and the step size are changed in a very efficient way.

In both cases the network is updated only after the entire set of patterns to be learned has been presented to it.

2 The "Bold Driver" Method

This strategy has been suggested independently in [10,6,1]. The proposed heuristic is to start with a given learning rate and to monitor the value of the energy function $E(\mathbf{w}_n)$ after each learning cycle. If E decreases, the learning rate is increased by a factor ρ. Vice versa if E increases, this is taken as an indication that the step made was too long, the learning rate is decreased by a factor σ, the last change is cancelled and a new trial is done. The process of reduction is repeated until a step that decreases the energy value is found.

Heuristically, ρ has to be close to unity in order to avoid frequent "accidents", because the computation done in the last back-propagation step is wasted in these cases. We choose $\rho = 1.1$ and $\sigma = 0.5$.

The performance of this apparently "quick and dirty" method is close to and usually better than that obtainable by optimizing a learning rate that is to remain fixed during the procedure.

3 The BFGS Memoryless Quasi-Newton Method

Shanno [9] reviews several variations of the conjugate gradient methods for function minimization and suggests one method that "substantially outperforms known conjugate gradient methods on a wide class of problems". In the "one-step Broyden-Fletcher-Goldfarb-Shanno memoryless quasi-Newton" method the search direction for the n'th iteration is defined as

$$\mathbf{d}_n = -\mathbf{g}_n + A_n\mathbf{p}_n + B_n\mathbf{y}_n \tag{1}$$

where $\mathbf{g}_n = \nabla E(\mathbf{w}_n)$ is the gradient, the coefficients A_n and B_n are defined as follows:

$$A_n = -\left(1 + \frac{\mathbf{y}_n \cdot \mathbf{y}_n}{\mathbf{p}_n \cdot \mathbf{y}_n}\right)\frac{\mathbf{p}_n \cdot \mathbf{g}_n}{\mathbf{p}_n \cdot \mathbf{y}_n} + \frac{\mathbf{y}_n \cdot \mathbf{g}_n}{\mathbf{p}_n \cdot \mathbf{y}_n}\quad , \quad B_n = \frac{\mathbf{p}_n \cdot \mathbf{g}_n}{\mathbf{p}_n \cdot \mathbf{y}_n} \tag{2}$$

and the vectors used are differences between subsequent weights and gradients: $\mathbf{p}_n = \mathbf{w}_n - \mathbf{w}_{n-1}$ and $\mathbf{y}_n = \mathbf{g}_n - \mathbf{g}_{n-1}$. Every N steps (N being the number of weights in the network) the search is restarted in the direction of the negative gradient .

Successive approximations $\mathbf{w}_n$ to the minimizer $\mathbf{w}^*$ of $E(\mathbf{w})$ are generated using a fast one-dimensional minimization along the search direction:

$$\epsilon_n = \min_{\epsilon} E(\mathbf{w}_n + \epsilon\ \mathbf{d}_n) \tag{3}$$

$$\mathbf{w}_{n+1} = \mathbf{w}_n + \epsilon_n\ \mathbf{d}_n \tag{4}$$

The one-dimensional minimization used in this work is based on quadratic interpolation and tuned to back-propagation where in a single step both the energy value and the negative gradient can be efficiently obtained. A small number of energy evaluations is sufficient [1].

This method has shown a consistent superior performance on different test problems. The results for two examples are presented in the following sections[1].

[1] A similar optimization approach, using Polak-Ribiere optimization, is presented in [4]. Now, a major difficulty with the Polak-Ribiere algorithm is that the search directions obtained are not necessarily *descent* directions, and numerical instability can result.

4 Test: the Dichotomy Problem

This problem consists in classifying a set of randomly generated patterns in two classes. It has been demonstrated in [2] that an arbitrary dichotomy for any set of N points in general position in d dimensions *can* be implemented with a network with one hidden layer containing $\lceil N/d \rceil$ neurons. In this test the pattern coordinates are random values belonging to the [0-1] interval.

A dichotomy problem is defined by the number of patterns generated. The dimension of the space and the number of inputs is two, the number of middle-layer units is $\lceil N/2 \rceil$ by the above criterion and one output unit is responsible for the classification.

Simulation runs have been made starting from small random weights, with maximum size r equal to 0.1. Correct performance is defined as coming within a margin of 0.1 of the correct answer.

Average results for different test runs (the random number seed is changed) are given in Fig. 1. Cases for correct solutions or local minima are shown separately. The speed-up factor increases for bigger networks (the largest value obtained is 116).

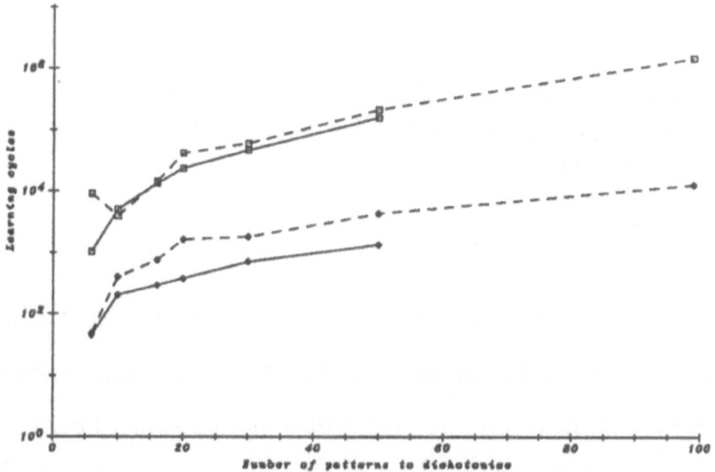

Figure 1: Performance comparison: "bold driver" method (□) versus BFGS optimization (◇). Continuous lines show the average number of cycles for convergence to correct solution, dashed lines for convergence to local minimum.

5 Test: Predicting the Logistic Map

Back-propagation has been applied to temporal series prediction in [7], where the logistic map is used for some tests. The logistic map is a discrete-time, nonlinear dynamical system. The value of a variable x_n is mapped to a new one x_{n+1} according to the recurrence relation

$$x_{n+1} = r x_n (1 - x_n) \tag{5}$$

where $0 < x_n < 1$. When $r = 4$, given an initial value $0 < x_n < 1$, the generated sequence is ergodic and chaotic. Although simple, this map captures the essence of a whole class of real-world phenomena. In the biological context, the parameter r is related to the fertility, while the factor $(1 - x_n)$ is viewed as a natural limitation of resources.

A 3-layer feed-forward neural network as described in [7] is used in order to make the comparison. It consists of one input, five hidden and one output unit.

BD		BFGS		speedup (BD/BFGS)
cycles (st.dev.)	generalization score	cycles (st.dev.)	generalization score	
1310680 (1564720)	0.0794	**2638** (2242)	0.0346	496

Table 1: Results for logistic map prediction.

A sequence of 10 example pairs (x_n, x_{n+1}) generated by the logistic map is used as the training set. After convergence (when all the desired outputs are reproduced with a maximum error of 0.01) the network is tested for generalization by presenting to it a new sequence of $N = 500$ example pairs and calculating the root-mean-square prediction error ("generalization score").

The average convergence and generalization results for 10 different networks are shown in table 1. The speed-up factor is almost 500 [2], while the generalization score obtained with the BD and BFGS methods are comparable (actually a better score is obtained with BFGS). No significant difference has been observed between the representations obtained with the two teaching methods.

Acknowledgments

Part of this work was completed while one of the authors (R. Battiti) was at the California Institute of Technology and was supported by DOE grant DE-FG-03-85ER25009, the National Science Foundation with grant IST-8700064 and by IBM.

References

[1] R. Battiti, "Accelerated Back-propagation Learning: Two Optimization Methods," *Complex Systems*, in press.

[2] E. B. Baum, "On the Capabilities of Multilayer Perceptrons," *Journal of Complexity* **4**, 193–215 (1988).

[3] P. E. Gill, W. Murray and M. H. Wright, *Practical Optimization* (Academic Press,1981).

[4] A. H. Kramer, A. Sangiovanni-Vicentelli, "Efficient Parallel Learning Algorithms For Neural Networks," *Advances in Neural Information Processing Systems* Vol. **1**, 75–89 (Morgan Kaufmann, CA, 1988).

[5] R. A. Jacobs, "Increased Rates of Convergence Through Learning Rate Adaptation," *Neural Networks* **1**, 295–307 (1988).

[6] A. Lapedes and R. Farber, "A self-optimizing, nonsymmetrical neural net for content addressable memory and pattern recognition," *Physica* **22 D**, 247–259 (1986).

[7] A. Lapedes and R. Farber, "Nonlinear signal processing using neural networks: Prediction and system modeling," Los-Alamos Preprint LA-UR-87-1662.

[8] D. E. Rumelhart and J. L. McClelland (eds.), *Parallel Distributed Processing: Explorations in the Microstructure of Cognition. Vol. 1: Foundations*, (MIT Press,1986).

[9] D. F. Shanno, "Conjugate gradient methods with inexact searches," *Mathematics of Operations Research* **3 - 3**, 244–256 (1978).

[10] T.P. Vogl, J.K. Mangis, A.K. Rigler, W. T. Zink and D. L. Alkon, "Accelerating the Convergence of the Back-Propagation Method," *Biological Cybernetics* **59**, 257-263 (1988).

[2] The actual computing time for BFGS is some minutes on a Sun workstation.

Grow-and-Learn: An Incremental Method for Category Learning

Ethem Alpaydın

Laboratoire de microinformatique
Ecole Polytechnique Fédérale de Lausanne
IN-F 1015 Lausanne Switzerland

Abstract. Learning by changing connection weights *only* is time-consuming and does not always work. Freedom to modify network structure is also needed. Grow-and-Learn (GAL) is a new algorithm that is able to quantize vectors as members of categories in an incremental fashion. When a new vector is encountered, it is tested as in nearest neighbor search and if it is not already quantized correctly, unit and links are added to accommodate this additional requirement. Thus network when learning, grows if and when necessary. As the structure of the resulting network in such a learning phase is dependent on the order of encountering the vectors, a second phase is added to eliminate old, no-longer necessary associations. In this phase, the network is closed to the environment and the input patterns are generated by the network itself during which relevance of units are computed and those who are not vital are removed. Simulation results when applied to character recognition is promising. Physiological plausibility and how the idea may be extended to unsupervised learning is discussed.

1. Introduction

Learning as opposed to adaptation requires structural changes additional to parametric ones. Small changes in the environment can be compensated for by modifying some parameter values, i.e. connection weights, in a neural network. However in initial learning phase when a certain network is to be created from scratch, the structure of the network, i.e. number of layers, units, connectivity, should also be determined. In algorithms like back-propagation this is done by the network designer most of the time by trial-and-error. It was noted before (Baum, 1989) that learning algorithms where units and links may also be added may lead to shorter learning times and better generalization. This paper explains one such algorithm for supervised learning of categories.

2. Network Structure and Learning

The network has the structure seen in fig. 1. The first layer is the layer of input units which may be binary or analog depending on the input representation. The second layer is the layer of exemplar units and the third is the layer of class units. A class may have more than one exemplar. W_{ie} and T_{ec} denote the connections from input unit i to exemplar unit e and from exemplar unit e to class unit c respectively. W_{ie} values depend on the input representation. T_{ec} are 1 or 0 depending on whether e is an exemplar of class c or not. When P is the input vector, the activation of an exemplar unit A_e is computed and then a winner-take-all form of non-linearity is applied:

$$A_e = \sum_i P_i * W_{ie}. \tag{1}$$

$$E_e = \begin{cases} 1, & \text{if } A_e = \max_i (A_i); \\ 0, & \text{otherwise.} \end{cases} \tag{2}$$

The unit whose weight vector is closest to the input vector will be the only active unit. The input and weight vectors should be normalized for the dot product to become a sensible distance criterion. For the class layer the response of a class unit is calculated as in (3) after which there will be exactly one active class unit:

$$C_c = \sum_e E_e * T_{ec}. \tag{3}$$

When a new input P is given as member of a certain class c, the system first checks if class c already exists. If not, a new class unit is created and labelled c. An exemplar unit e is created which currently is the only exemplar of that class. Connections are set as follows:

$$W_e = P.$$

$$T_{eo} = \begin{cases} 1, & \text{if } o = c; \\ 0, & \text{otherwise.} \end{cases} \quad (4)$$

If class c already exists, then P is given as input to the network and responses of exemplar and class units are calculated. If the class found by the network is the same as the class desired, no modification is done. If it is not, a new exemplar unit e is created and the W and T connections are set as in equation (4).

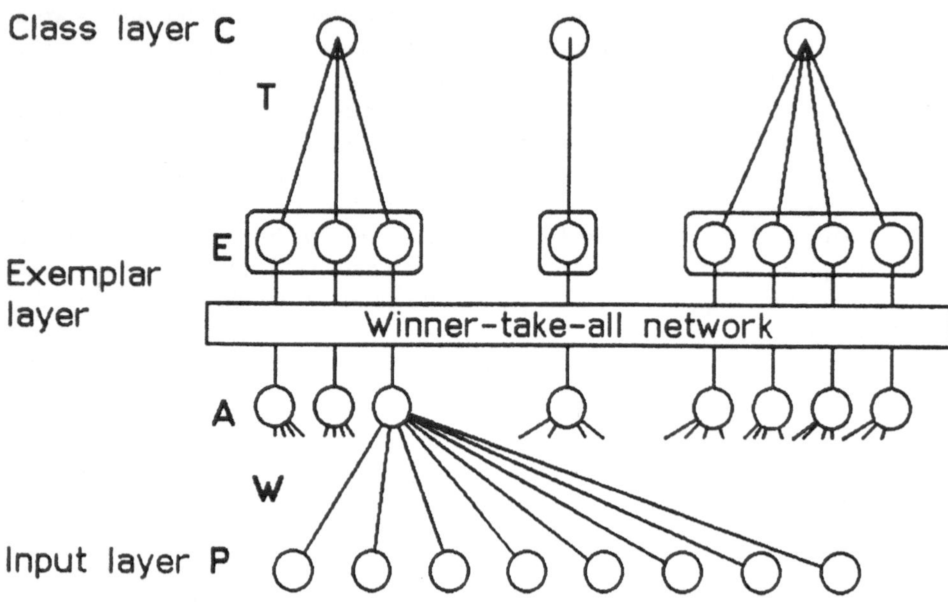

Fig. 1. GAL network structure.

In figure 2, how the number of exemplar units and error depend on the size of the training set for an example problem are depicted. In this problem, some classes have concave shapes, implying that to be able to get error less than 200 with back-propagation, at least three-layers are needed.

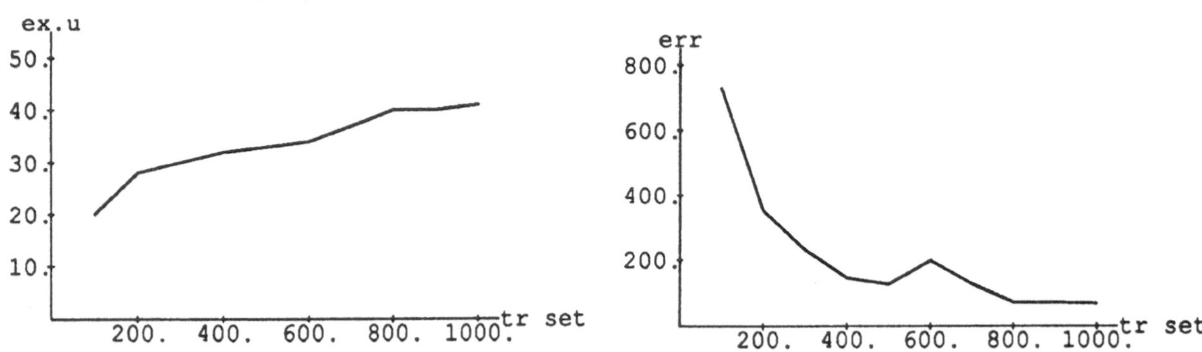

Fig. 2. Number of exemplar units and error vs. training set size for a given problem.

The idea of storing multiple exemplars incrementally when need arises exists also in the model proposed by Reilly, Cooper, and Elbaum (1982). In their model, decision for the winner is done by a variable thresholding mechanism which defines a hypersphere of domination region around exemplar units, i.e. Parzen windows. When the threshold is decreased due to an error in one dimension, the region of domination is effectively reduced in all dimensions. In a winner-take-all network, the regions dominated by units are separated by hyperplanes and these regions need not be of equal distances to exemplar coordinates in all dimensions. The price to pay is that winner-take-all is more difficult to compute than a set of threshold-type non-linearities. Work is under way in our lab to build specialized hardware to implement winner-take-all type non-linearity which is also used in Kohonen's self-organization map (Blayo & Lehmann, in review).

3. Forgetting

The algorithm tends to store those input patterns that are closest to the borders for finer separation of classes. In such a learning scheme, the actual exemplars stored depend on the order of encountering the vectors. The units previously stored which are further from the boundary when an exemplar that is nearer to boundary is stored become useless. Such exemplar units as now they are in the domination region of another unit of the same class, are useless and may be eliminated to decrease network size.

To accomplish this, in a so-called *sleep* mode, the system is closed to the environment and the exemplar patterns' stored W are given as input to the system in a random order and tested while that exemplar unit is disabled. If the system gives still the same class code as output, it can be deduced that that exemplar does not contribute to the success and is eliminated. When a different class code is given, the unit should be kept as its elimination will cause a wrong shift of the class boundary (fig. 3). However, such elimination of units may cause error to increase which may later be compensated for in an *awake* mode when the system is open once more to the environment. After successive *awake* and *sleep* sweeps over the training set, GAL settles to a set of exemplar units with 100% success where no further additions or deletions are possible. The drawback is that the particular subset of vectors chosen out of the training set and thus the success on the previously unseen test set is dependent on the order of encountering the vectors.

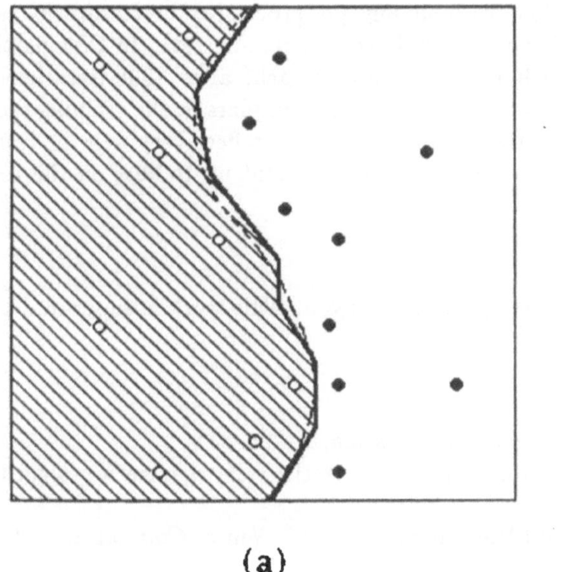

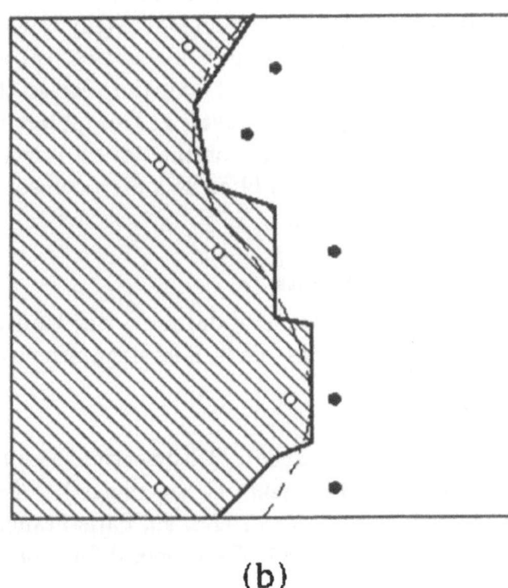

Fig. 3. Exemplar units (a) before, (b) after sleep.

The idea of decreasing network complexity and increasing the quality of generalization by eliminating irrelevant units was previously used with back-propagation (Mozer & Smolensky, 1989).

It was proposed by Crick and Mitchison (1983) that the REM sleep in mammals when dreaming takes place is for elimination of unwanted modes of behaviour due to growth of brain by accumulated experience. In this mode, when the system is closed to the environment, inputs are generated by the system and an active process of unlearning occurs. It is said that "we dream in order to forget."

Neuronal death as it involves thousands of cells every day and due to its relative speed (order of a few hours) is generally believed to play a positive rather than negative role in development and memory (Dawkins, 1971). It is reported (Hubel, 1988) that just by changing the visual stimulus, i.e. selective visual deprivation, significant physiological and morphological changes in the visual cortex take place. The is explained by the fact that neurons to stay "alive," need to receive a nourishing "trophic" substance which is taken up by the terminals and carried by fast retrograde axonal transport to the cell body. This may be a sort of positive feedback by which neurons are informed when they affect system output. It may be conjectured that starting with a random and highly redundant structure, during a period of "critical development," one may qualify units on the basis of their utility, i.e. firing correlation of presynaptic and postsynaptic neurons, and may then decide to eliminate those who did not much fire on the pretext that they do not contribute significantly to system response.

5. Conclusion

When compared with back-propagation or iterative algorithms generally, GAL takes less time and guarantees 100% on the training set. The risk of having a decoder-like structure at the end exists but finding a good representation to form clusters, is left to unsupervised learning (Barlow, 1989). As iterative algorithms require several sweeps, the training set should be stored throughout learning which is not the case in GAL. When the training set changes, e.g. a new association is added, in back-propagation the whole learning needs to be repeated, one cannot just "add" a new association because all associations share the same connections and modifying them in favor of one only will derange others. In GAL as the vectors are stored locally, adding a new unit or removing one does not disturb other stored vectors.

The method proposed may be described as a variant of nearest-neighbor classification where instead of all, only a subset of the exemplars are stored. Alternating modes of *awake* and *sleep* help the system to store a smaller number of exemplars, thus to have a smaller network. The system when tested on normalized bitmaps of handwritten numerals (Guyon et al., 1989) without any pre-processing or feature extraction gave 92.8%on the previously unseen test set (with of course 100% on the training set). In that particular run where the training set and the test set both include 600 exemplars each, after 1308 iterations (600 awake+108 sleep+600 awake), 112 exemplars were stored. (When all 600 exemplars in the training set were stored, success was 93.2% on the test set implying that the training set did not reflect very well the test set.) By adding a Gaussian filter as a preprocessor, with the same data success went up to 96% on the test set using 103 exemplars.

Acknowledgements

The handwritten numeral database was kindly supplied by Isabelle Guyon. This work is supported by the Fonds National Suisse de la Recherche Scientifique.

References

Barlow, H.B. (1989). "Unsupervised learning," *Neural Computation*, **1**, 295–311.

Blayo, F., Lehmann, C. (in review). "A systolic implementation of the self-organization algorithm," submitted for review to *INNC 90*, Paris-France.

Baum, E.B. (1989). "A proposal for more powerful learning algorithms," *Neural Computation*, **1**, 201–207.

Crick, F., Mitchison, G. (1983). "The function of dream sleep," *Nature*, **304–14**, 111–114.

Dawkins, R. (1971). "Selective neurone death as a possible memory mechanism," *Nature*, **229**, 118–119.

Guyon, I., Poujaud, I., Personnaz, L., Dreyfus, G., Denker, J., and Le Cun, Y. (1989). "Comparing different neural network architectures for classifying handwritten digits," *Proc. IJCNN 89*, Washington-USA.

Hubel, D.H. (1988). Eye, Brain, and Vision, New York: Scientific American Library.

Kandel, E.R., Schwartz, J.H. (1985). Principles of Neural Science, 2nd edition, New York: Elsevier.

Mozer, M.C., Smolensky, P. (1989). "Skeletonization: A technique for trimming the fat from a network via relevance assessment," *Connection Science*, **1**, 3–26.

Reilly, D.L., Cooper, L.N., and Elbaum, C. (1982). "A neural model for category learning," *Biological Cybernetics*, **45**, 35–41.

AN APPROACH TO GENERALIZATION PROBLEM IN BACK-PROPAGATION LEARNING

Kiyotoshi MATSUOKA

Division of Control Engineering
Kyushu Institute of Technology
Sensui 1-1, Tobata, Kitakyushu, 804 Japan

Abstract - Back-propagation learning is one of the most popular methods in neural networks and has successfully been applied to many practical problems. However, although the method might allow networks to acquire desired mappings for taught patterns, it does not necessarily give desirable results for untaught patterns. In order to deal with this problem, some criterion is necessary as to how the network should behave for untaught patterns. This paper proposes a method that minimizes not only the error between the actual output of the network and the desired output but also the gradient of the resultant mapping of the network.

1. INTRODUCTION

Back-propagation is one of the most popular learning algorithms used in artificial neural networks and has successfully been applied to many practical problems. This is the method by which, for a given set of pairs of input and output, the weights of connections among neural elements are iteratively modified so that the error between the actual output of the network and the desired output be minimized. However, even if the error becomes small enough for the presented patterns, there is no guarantee that the obtained network provides a desirable mapping for untaught (unknown) patterns. In fact, it is known that, is some cases, it is better to stop the learning process before the error function becomes very small (the problem of over-learning).

In order to deal with this problem, one must consider some criterion as to how the network should behave for untaught patterns. This paper describes an approach to this problem, in which not only the standard error function but also the gradient of the mapping are evaluated.

2. THE PROBLEM

A typical structure of feedforward network treated by the back-propagation learning is as follows. It consists of (L+1) layers (0-th to L-th), the k-th layer having N_k elements (neurons). The output pattern at each layer is given by the following equations.

$$y_0 = s,$$
$$x_k = W_k y_{k-1}, \quad y_k = G_k(x_k) \quad (k=1,..,L), \tag{1}$$
$$y = y_L.$$

Here, s and y are the input and output patterns of the overall network. x_k is an N_k-dimensional vector, the i-th component of which, $x_{k,i}$, is a total weighted sum of the signals from the (k-1)-th layer. W_k is the weight matrix representing the weights of the connections from the (k-1)-th to the k-th layer. y_k is the output of the k-th layer which is obtained by applying nonlinear function $g_{k,i}$ to each $x_{k,i}$ (G_k=diag[$g_{k,1},..,g_{k,Nk}$]). For the nonlinear functions the logistic function $g_{k,i}(z)=1/\{1+\exp(-z)\}$ is usually employed except an element in each layer (say the N_K-th element) used for a bias ($g_{k,Nk}(z)=1$).

Let a set of pairs of input and output patterns, $s^{(m)}$ and $y^{(m)}$ ($m=1,..,M$), be given with probabilities p_m. The problem is to determine the connection matrices W_k so that they minimize the standard error function defined by

$$Q'(W_1,..,W_L) = (1/2)\langle|y^{(m)}-F(s^{(m)})|^2\rangle \tag{2}$$

where $F(s)$ is the mapping embodied by the network, $|\bullet|$ denotes a Euclidean norm of a vector or a matrix, and $\langle\rangle$ represents a statistical expectation with respect to $s^{(m)}$.

In order to make clear the key point of the approach described in the next section, we first consider a two-layer network ($L=1$) where all elements are linear ones; $g_{1,i}(z)=z$ or $F(s)=W_1 s$. According to linear algebra, we find W_1 minimizing (2) as

$$W_1 = \langle y^{(m)}s^{(m)T}\rangle\langle s^{(m)}s^{(m)T}\rangle^+ + Z(I_{N0}-\langle s^{(m)}s^{(m)T}\rangle\langle s^{(m)}s^{(m)T}\rangle^+), \tag{3}$$

where Z is an arbitrary $N_1\times N_0$ matrix, I_{N0} is the N_0-dimensional identity matrix, and $()^+$ represents the pseudoinverse of $()$. Thus, W_1 cannot uniquely be determined in general. In order to determine W_1 uniquely, we choose W_1 to take a minimum Euclidean matrix norm among every W_1 given by (3), then we have

$$W_1 = \langle y^{(m)}s^{(m)T}\rangle\langle s^{(m)}s^{(m)T}\rangle^+. \tag{4}$$

This is nothing but the optimal associative mapping proposed by Kohonen.

Defining a new error function as

$$Q(W_1) = (1/2)\langle|y^{(m)}-W_1 s^{(m)}|^2\rangle+(1/2)\varepsilon|W_1|^2 \tag{5}$$

where ε is a small positive constant, (4) is achieved in an approximate sense by the following learning process.

$$dW_1/dt = -\partial Q(W_1)/\partial W_1 = -W_1(\langle s^{(m)}s^{(m)T}\rangle+\varepsilon I_{N0}) + \langle y^{(m)}s^{(m)T}\rangle. \tag{6}$$

W_1 stably approaches $W_1=\langle y^{(m)}s^{(m)T}\rangle(\langle s^{(m)}s^{(m)T}\rangle+\varepsilon I_{N0})^{-1}$ with time, and it converges to (4) with $\varepsilon\to 0$.

It may be a rare case that W_1 that minimizes (2) is not unique although it typically occurs when the number of the patterns to be learned, M, is smaller than the dimension of the input space, N_0. However, it is by no means rare that the patterns lie in a neighborhood of a subspace (a manifold in general) of the input space because the patterns to be learned are similar to each other in many cases. In such cases, the standard error function (2) becomes a long ravine as shown in Fig. 1. In this situation, even though W_1 minimizing (2), indicated by $\bar{W}$ in Fig. 1, can uniquely be determined, its value is very sensitive to the patterns that happen to be used for the learning.

This problem is nothing but the one which often occurs in linear regression analysis, namely, the problem of collinearity. An easy solution to this problem is to minimize (5) rather than (2) even when W_1 minimizing (2) can be determined uniquely. This prescription is the Ridge estimation well known in linear regression analysis. If (5) is used, the optimal mapping is obtained as a nearest point to the origin in the ravine ($\hat{W}$ in Fig. 1). In the next section we will implement this idea to the back-propagation learning of multilayer networks with nonlinear elements.

3. PROPOSED MODEL

The basic idea mentioned in the last section is to minimize $|y^{(m)}-W_1s^{(m)}|^2$ and $|W_1|^2$ simultaneously. $|W_1|$ represents the magnitude of the gradient of the mapping embodied by the two-layer linear network. In the case of general nonlinear networks the gradient of the mapping is represented by $\partial F(s)/\partial s$. So, it may be reasonable to find W_k such that

$$\text{(a)}\ \langle|y^{(m)}-F(s^{(m)})|^2\rangle,\ \text{and}\qquad \text{(b)}\ |\partial F(s)/\partial s|^2$$

take as small values as possible simultaneously. As for the minimization of (b), since $\partial F(s)/\partial s$ depends on s, two approaches are conceivable.

<u>Method 1</u> Minimize $|\partial F(s)/\partial s|^2$ for presented patterns, or $|\partial F(s^{(m)})/\partial s|$ (m=1,..,M), only.

In this case the error function to be minimized is

$$Q(W_1,..,W_L)=(1/2)\langle|y^{(m)}-F(s^{(m)})|^2\rangle+(1/2)\varepsilon\langle|\partial F(s^{(m)})/\partial s|^2\rangle. \tag{7}$$

Introducing a N_0-dimensional random vector, d, which is independent of $s^{(m)}$ with $\langle d\rangle=0$ and $\langle dd^T\rangle=\varepsilon I_{No}$, (7) is approximated by

$$Q(W_1,..,W_L) \simeq (1/2)\langle|y^{(m)}-F(s^{(m)}+d)|^2\rangle. \tag{8}$$

Here, $\langle\rangle$ denotes a statistical expectation with respect to both $s^{(m)}$ and the random variable. By considering a dynamics (locally) minimizing this function, we can derive the following learning algorithm.

$$\begin{aligned} &y_0 = s^{(m)} + d,\quad x_k = W_k(n)y_{k-1},\quad y_k = G_k(x_k)\quad (k=1,..,L), \\ &v_L = G_L'(x_L)(y^{(m)}-y_L),\quad v_k = G_k'(x_k)W_{k+1}(n)^Tv_{k+1}\quad (k=L-1,..,1), \\ &W_k(n+1) = W_k(n) + \eta v_k y_{k-1}^T, \end{aligned} \tag{9}$$

where $G_k'(x_k)=\mathrm{diag}[dg_{k,1}(x_{k,1})/dz,..,dg_{k,Nk}(x_{k,Nk})/dz]$. V_k are the well-known errors propagated backward through the network, and η is a small positive constant. (9) is almost the same as the standard back-propagation algorithm except that the random noise is superimposed to the input patterns.

<u>Method 2</u> Minimize $|\partial F(s)/\partial s|^2$ for every pattern in the input space.

In this case, we may evaluate $\mathrm{Max}_s|\partial F(s)/\partial s|^2$ for example, but it cannot be expressed in an explicit form. However, since $|\partial F(s)/\partial s|^2\leqq\Pi_k|W'_k|^2$ holds if the logistic function is used except the bias elements, it may be reasonable to minimize $\Pi_k|W'_k|^2$ instead of $\mathrm{Max}_s|\partial F(s)/\partial s|^2$. Here, W'_k represents the connection weights from the elements other than the bias element at the (k-1)-th layer to the elements at the k-th layer. This leads to the following error function to be minimized:

$$Q(W_1,..,W_L)=(1/2)\langle|y^{(m)}-F(s^{(m)})|^2\rangle+(1/2)\varepsilon\Pi_k|W'_k|^2. \tag{10}$$

For W'_k, the algorithm to achieve the minimization of (10) is,

$$W'_k(n+1) = W'_k(n) + \eta v_k y'^T_{k-1} + \delta \, \Pi_{\alpha\neq k}|W'_\alpha(n)|^2 W'_k(n) \quad (k=1,..,L), \tag{11}$$

where y'_k is the output of the neurons other than the bias element and δ is a small positive constant. For the connection weights (vector) from the bias element in the (k-1) layer to the elements in the k-th layer, w_k, we obtain

$$w_k(n+1) = w_k(n) + \eta v_k. \tag{12}$$

Calculations of x_k, y'_k and v_k are exactly the same as those in (9) except that noise is not used in this case.

Now, we shall show an example. The following three pairs of patterns

$$s^{(1)}=[0,0]^T,\ y^{(1)}=0, \qquad s^{(2)}=[.5,.5]^T,\ y^{(2)}=.5, \qquad s^{(3)}=[1,1]^T,\ y^{(3)}=0$$

are supposed to be learned by a three-layer network with two (+ a bias) elements for the input layer, five (+ a bias) elements for the middle layer, and one element for the output layer. The mapping of the network obtained by the standard algorithm (a), Method 1 (b), and Method 2 (c) are shown in Fig. 2 with contour mapps. Although the learning are successful for the presented patterns in every method, the conventional method usually generates distorted mappings as shown, while Methods 1 and 2 generate apparently simpler mappings. In the case of Method 1, the slope at $s=s^{(m)}$ is almost flat, but for some other input, for example $s=[.25,.25]^T$, the mapping shows a steep gradient. This is because the gradient of the mapping at the presented patterns are only evaluated. On the other hand, Method 2 gives a gentle slope for the whole input space.

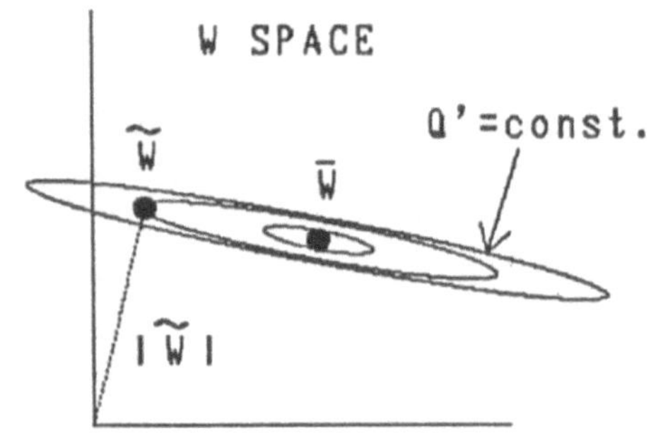

Fig. 1

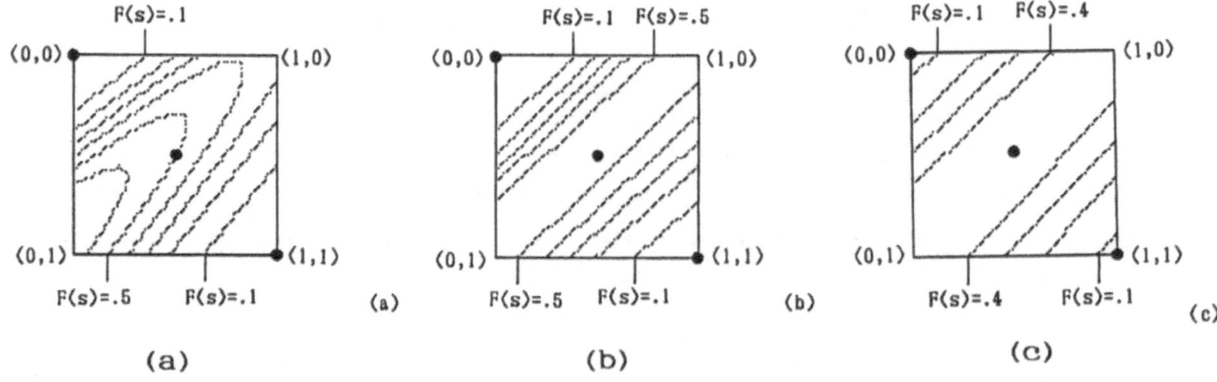

(a) (b) (c)

Fig. 2

A TRAINING ALGORITHM FOR A PIECEWISE LINEAR NEURAL NETWORK

Yong Hae Kong and Andrew S. Noetzel
Department of Electrical Engineering and Computer Science
Polytechnic University, 333 Jay Street, Brooklyn, New York 11201

Abstract

A neural network implementing the Piecewise Linear Machine is developed, and its operation is demonstrated. The weight vectors of the cells in the network's middle layer represent prototype vectors for the various classes. The network classifies patterns according the minimum Euclidean distance to the prototype vectors. The training rule is controlled by parameters ε and δ. If there is no cell of the correct class within distance ε of a training pattern, a new prototype cell is created. And all prototype cells within distance δ of the training pattern have their weights adjusted, either towards or away from the training vector, depending upon their class. Cells of the same class that are moved close together are merged. The parameters ε and δ can be varied during training, to achieve a low classification error rate with a minimum number of prototype cells.

Results of training and recognition experiments with this new neural network are presented, and compared with those of a backpropagation network of similar size trained with the same data. The classification regions used in the experiments were various forms of nonconvex regions. The new network performed better than the backpropagation network in all cases, with an error rate close to the theoretical minimum.

1. Introduction

A classifier whose discriminant functions for C classes have the form $g_i(X) = W_i \cdot X + w_{i_0}$ for $i = 1,\ldots,C$, is called a linear classifier. Its decision regions are convex. A generalization is the classifier called the Piecewise Linear Machine (PLM), defined by Nilson [4]. The PLM has discriminant functions

$$g_i(X) = \max_j g_{ij}(X)\,,\; i = 1,\ldots,C\,, \tag{1}$$

where the *subsidiary* discriminant functions g_{ij} are defined by

$$g_{ij}(X) = W_{ij} \cdot X + w_{ij_0}\,,\; j = 1,\ldots,n_i\,. \tag{2}$$

The decision regions of the PLM can be nonconvex, through the union of the convex regions of the subsidiary functions.

A most useful special case of the PLM obtains if the weight vectors W_{ij} of the subsidiary discriminant functions are considered prototype vectors for the class, and $g_{ij}(X) = g_i(X)$ when W_{ij} is the prototype vector of minimum distance from X; i.e., when $\|X - W_{ij}\| = \min_k \|X - W_{ik}\|$ where $1 \le k \le n_i$. The minimum distance decision is the same as the decision for minimum squared distance: prototype j is selected when $\|X - W_{ij}\|^2 = \min_k (X \cdot X - 2W_{ik} \cdot X + W_{ik} \cdot W_{ik})$. Then it is seen that $g_{ij}(X) = W_{ij} \cdot X - 1/2\, W_{ik} \cdot W_{ik}$ is a subsidiary linear discriminant function in the form of (2), that performs nearest neighbor classification.

There have been several attempts to develop training rules for PLMs. The models of Duda [1] and Kohonen [3] require that the number of prototypes be defined before training, either arbitrarily or based on the priori probabilities. Since the number of prototypes required depends entirely on the shapes of the classification regions, which can not be assumed beforehand, a systematic training rule controlling the number of prototypes is necessary.

Here, the Piecewise Linear Neural Network, or PLNN, is presented as a pattern classifying machine without assuming any probability distribution or number of prototypes. Its generalized learning rules are based on the following operations,

(a) birth and death of prototypes,
(b) selective adjustment of the cells within a coarseness threshold δ of the training pattern,
(c) Euclidean distance as measure of similarity.

The parameter δ is similar to a quenching threshold of Grossberg STM equation [2] with a sigmoid activation function. It allows one or more cells to become active. Reward and punishment feedback from the output classifying layer is applied to all cells above the δ threshold, resulting in more dynamic learning than winner-take-all learning.

2. The Piecewise Linear Neural Network.

The structure of the PLNN is shown in Figure 1. Layer L1 is the input layer, which accepts the pattern X from the environment. A cell k in layer L2 either represents a prototype vector for some class i, or else it is unassigned. Let E be the set of active, or assigned, L2 cells. For $k \in E$, let $\gamma(k)$ be the class, and $\sigma(k)$ the prototype vector number within the class. If $k \in E$, k is connected to L1 through the weight vector W_{ij}, where $i = \gamma(k)$ and $j = \sigma(k)$.

Each cell i in L3 represents a class i recognition, for $i = 1,..,C$. The L3 cell i is connected to L2 through the weights a_{ki}, each of which is one if $\gamma(k) = i$ and zero otherwise. The output of each L3 cell is the maximum of all its inputs from L2.

3. PLNN Training.

The PLNN is trained with supervision. Upon the presentation of a training vector X, each L2 cell $k \in E$ computes the distance $d(k,X) = \| X - W_{\gamma(k)\sigma(k)} \|$ between X and its prototype vector. Cell k then compares $d(k,X)$ with two global parameters ε and δ.

$$\text{Let} \quad S = \{ k \mid d(k,X) < \varepsilon \}, \tag{3}$$

$$\text{and let} \quad R = \{ k \mid d(k,X) < \delta \}. \tag{4}$$

Suppose the input is of class i, denoted $X^{(i)}$. If there are no class i cells in S, a new L2 cell J is assigned to class i,

$$E = E \cup \{J\}, \quad a_{Ji} = 1, \quad W_{iJ} = X^{(i)}. \tag{5}$$

If no new cells are created, all cells $k \in R$ are modified by

$$W_{\gamma(k)\sigma(k)} = W_{\gamma(k)\sigma(k)} \pm \alpha \, (X^{(i)} - W_{\gamma(k)\sigma(k)}), \tag{6}$$

where α is a parameter whose magnitude controls the learning rate, and whose sign is taken to be positive if $\gamma(k) = i$ and negative otherwise. If $R = \Phi$, the nearest prototype, regardless of class, is chosen for modification by (6).

A cell that is modified might possibly be merged with another of the same class. If $k \in R$ and $\| W_{\gamma(k)\sigma(k)} - W_{\gamma(l)\sigma(l)} \| < \eta$, for $l \neq k$ and $\gamma(l) = \gamma(k)$, then

$$W_{\gamma(k)\sigma(k)} = (W_{\gamma(k)\sigma(k)} + W_{\gamma(l)\sigma(l)})/2, \tag{7}$$

and cell l is removed from E. Here, $\eta \ll \delta$ is a global parameter that controls merging.

The cell creation parameter ε is provided to generate a new prototype if no prototype of class i exists within distance ε of the training pattern. Unlike other classifiers which place prototypes in arbitrary places before training, the generation procedure of (5) constructs prototypes wherever required. To prevent excessive prototype generation during learning, ε increases as learning proceeds, after starting at a small value.

More than one cell can be adjusted during a training trial. This mass interaction causes prototypes of a class to gathered in several clusters, while those clusters repel each other within the region. Merging of redundant prototypes of a class occurs by (7). That is, (a) within a cluster, same class prototypes are gathered and eventually merged into a single prototype, and (b) among clusters of a class, scattering places them as far apart as possible, which results in the best approximation of a connected class region with the given δ.

4. Results of Experiments.

Two-dimensional pattern vectors of two classes were generated and used to train the PLNN and backpropagation network[5]. The distributions of patterns are shown in figures 2a, 3a, and 4a. The small dots are the patterns of one class and the large dots are those of the other class. The remaining figures show the classifications (learned responses) of the PLNN and backpropagation network.

These classifications are again represented by small and large dots. The squares and triangles represent the PLNN prototypes for the two classes.

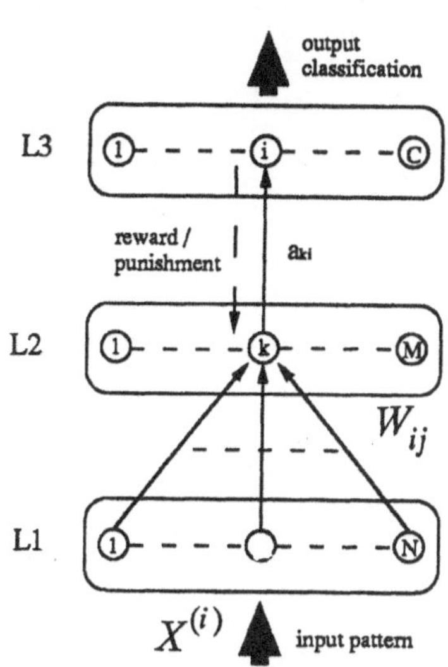

Fig. 1. PLNN structure

In figure 2a, the initial patterns are uniformly distributed over the unit square and divided into two classes. First, we show the results of training the PLNN with the merge operation (7) disabled. Figures 2b and 2c show the results in various stages of training, in which the prototypes of each class are seen to become clustered in several locations. Then we shows the results of training with a fixed number of prototypes, and the birth operation (5) disabled. Figures 2d and 2e exhibit the scattering of prototypes in a class regions. This results in the best approximation of the class region with a given number of prototypes. With both the merge and generation, the learned response of the PLNN is shown in figure 2f.

In another experiment, the patterns were again uniformly distributed over the unit square, but one class consisted of the patterns in a ring and the other the patterns in the disjoint regions inside and outside of the ring. Figure 3a shows this distribution. Figures 3b and 3c are the trained results of the PLNN and backpropagation network respectively. The PLNN approximates the nonconvex class regions successfully, while as shown in figure 3c, backpropagation network fails to recognize the region inside the ring.

Two dimensional patterns were generated with normal distribution with different means and variances for each class, as shown in figure 4a. The Bayes theoretical error rate for this pattern set is 9.23 percent. Figure 4b shows that the PLNN closely approaches the Bayes classifying boundary with five prototypes. The PLNN error rate is 9.55 percent. Figure 5c is the backpropagation network response, which has an error rate of 10.72 percent. Note that its classifying boundary is more like a line compared to the curved boundary of the PLNN.

5. Conclusions.

The multiple-winner learning rule for the middle layer, together with the generation and merging operations, allowed the PLNN to learn to recognize classes in nonconvex regions in the various experiments, with few prototypes. The PLNN approaches the theoretical optimum classification rate when pattern classes are overlapped. Since the PLNN does not assume any pattern distribution or the number of prototypes beforehand, it can be used as a general classifier in many applications. Moreover it learns faster than backpropagation network and does finer discrimination. The coarseness parameter δ can be chosen based on the tradeoff between performance and economy.

References

[1] R. O. Duda and H. Fossum, "Pattern Classification by Iteratively Determined Linear and Piecewise Linear Discriminant Functions," IEEE Trans. Electron. Comput., vol. 15 no. 2, pp. 220-232, 1966.

[2] S. Grossberg, "Adaptive Pattern Classification and Universal Recoding: I. Parallel Development and Coding of Neural Feature Detectors," Biol. Cybernetics 23, pp. 121-134, 1976.

[3] T. Kohonen, "Learning Vector Quantization," in Self-Organization and Associative Memory, pp. 199-202, Springer-Verlag, 1984.

[4] N. J. Nilson, Learning Machines: Foundations of trainable pattern-classifying systems, McGraw-Hill, New York, 1965.

[5] D. E. Rumelhart, G. E. Hinton, and R. J. Williams, "Learning internal representations by error propagation," in Parallel Distributed Processing, vol. 1, pp. 318-362, M.I.T. Press, 1986.

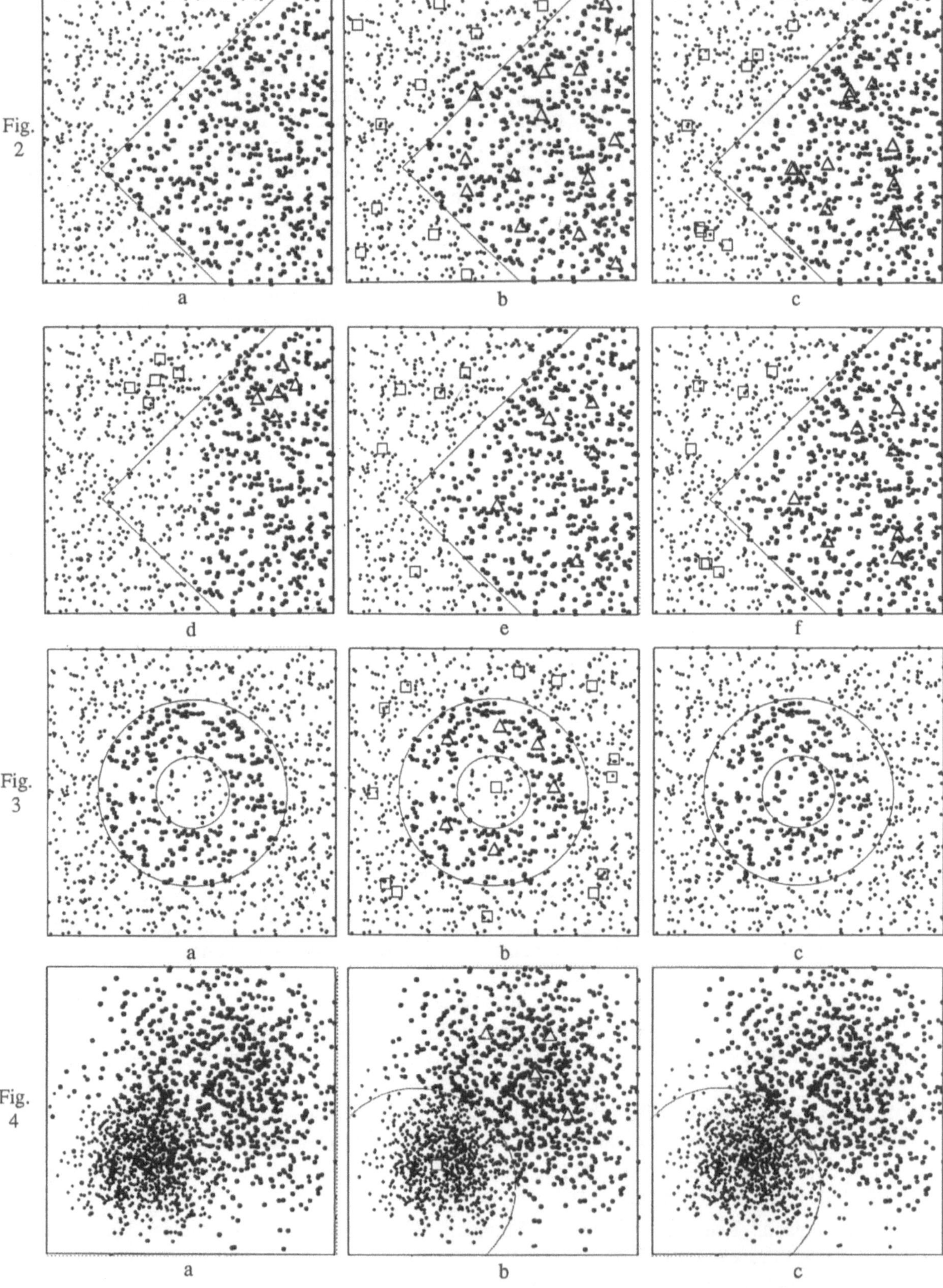
Fig. 2
a
b
c
d
e
f
Fig. 3
a
b
c
Fig. 4
a
b
c

SUPERVISED LEARNING BASED ON KOHONEN'S SELF-ORGANISING FEATURE MAPS

S. MIDENET, A. GRUMBACH,
E.N.S.T., Dpt INF., 46 rue Barrault,
75634 PARIS Cedex 13 FRANCE
e-mail:midenet@inf.enst.fr

ABSTRACT:

We present a connectionist model designed for **supervised learning** of associated patterns, which is based on Kohonen's **self-organizing feature maps.** While learning, the classification is performed on both inputs and desired associated ouputs. The learned weights are then used in exploitation phase to associate an input vector with an output one.

If we consider the supervised learning issue, our model may stand for an alternative to the *Back-Propagation* algorithm in multi-layers networks. We show experimentally that it can deal with **non linearly separable problems.**

Moreover, this model is better suited to the **emergence of regularities** in neural network dynamic and/or weights; that is the deep reason why we have investigated this type of architecture and process. This aspect however won't be developped in this communication, but will be further investigated; we focus here on **learning performances** of our model.

1 INTRODUCTION

A self-organizing feature map is an array of units on which is defined a topology, that is, a neighbourhood on each unit. Each node of the self-organizing feature map, that we call the *Kohonen layer*, receives inputs from an input layer. Thanks to an unsupervised learning algorithm, the Kohonen layer is able to classify input vectors by organizing itself in order to reflect the implicit topological relations of the input vectors given during learning [1].

A first way to handle supervised learning is to use a self-organizing feature map as a pre-processing structure on the inputs, and connect this Kohonen layer to an output layer, with a classical associator; this model can then manage a supervised learning paradigm, where the inputs are used for updating the weights W_k and the desired output for the weight W_δ (thanks to a *Delta-Rule* algorithm for example).

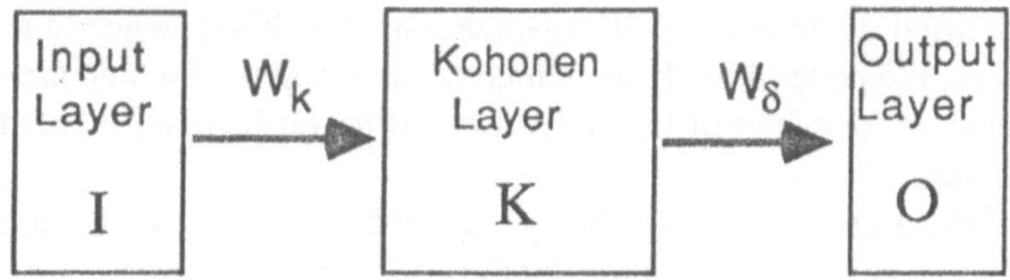

Figure 1

A few authors have studied this kind of model for simulating a sensori-motor task ([2][3][4]).

This type of architecture however, cannot be satisfactory for most supervised learning tasks. The Kohonen layer organizes itself according to the inputs only, while the relation between inputs and outputs is available (the desired output for a given input), and represents a complementary information about inputs. Besides, a treatment on the only inputs cannot allow a one-layer associator to learn every function. For instance, let's consider the problem of counting the number of adjacent "1" bits in a 8 bit vector, which we'll call the "counting problem"; this problem, studied in [5], is not linearly separable. The inputs being uniformly distributed on $\{0;1\}^8$, the Kohonen layer cannot find any topology; consequently, the *Delta-Rule* associator cannot learn "alone" the non-linearly separable problem (cf simulation results below). The relevant information in this function cannot be brought out of the inputs alone, but lies in the relation between inputs and outputs.

In order to design a general and robust pattern associator, we propose to use the Kohonen's self-organizing feature maps to reflect the implicit topological relations between input and output data. In the same way that the hidden layer of a *Back-Propagation* network receives informations from input and output layers (respectively input data and output errors), the Kohonen layer self-organizes according to both input and desired output, that is, the whole information provided for supervised learning. This is the major purpose of our approach.

2 MODEL

Architecture:

The model consists of 3 layers: an input layer, a Kohonen layer and an output one. Each Kohonen layer unit is connected to all the units of the two other layers. Connections from input units to Kohonen units (nodes from the Kohonen layer), are unidirectional; the weights are called W_{Ik}. Connections between Kohonen and output units are *bidirectional*; the weights from Kohonen layer to outputs are called W_{kO}, while W_{Ok} is used for the opposite direction.

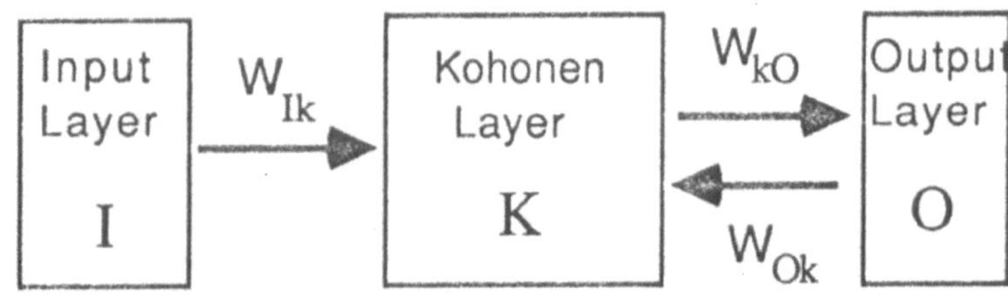

Figure 2

Principle:

The weights W_{Ik} and W_{Ok} are learned together thanks to the Kohonen's self-organization algorithm; the weights W_{kO} are set up at the end of the learning phase: $W_{kO} = W_{Ok}$.

Algorithm:

* We use the simplified Kohonen's self-organization algorithm: when an input vector is presented to the system, the node in the Kohonen layer the weights of which are nearest from the input vector, is selected as the center of the activation cluster in the map; it is called the "image-node" of the input vector.

* *Learning phase*: W_{Ik} and W_{Ok} are gradually adjusted according to Kohonen's algorithm, inputs and desired outputs being presented together at each learning step.

* W_{kO} are set up to W_{Ok}.

* *Exploitation phase*: When an input vector I_t is presented to the system, activations of the input nodes are propagated to the Kohonen layer via the weights W_{Ik}, and the image-node k_t is computed from the only inputs. Activations in the Kohonen layer are then set to 0 except in a neighbourhood of unit k_t, the width of which depends on the "representativeness" of k_t: the bigger the distance between k_t and I_t, the worse is the representativeness of k_t and the larger is the neighbourhood taken into account around k_t (in order to make up for the output estimated by

a badly representative image-node). Lastly, activations of the output layer are computed by propagation, and divided by the total amount of activation in the Kohonen layer, for normalisation; for binary outputs purpose, there is also a threshold function.

* *Remarks*:

- In some experiments, the neighbourhood around k_t has been restricted to the image-node itself whatever its representativeness ; it seems to be precise enough in a large number of applications.
- We have also investigated another kind of learning procedure for weights W_{kO}, where they are tuned individually thanks to the *Delta-Rule* algorithm. Here again, this increase in the complexity of the model doesn't seem to improve its behavior, at least in the majority of our tested examples.

3 SIMULATIONS

We have tested our model on the previously mentioned "counting problem". It consists in associating an 8 bit vector with a 3 bit one, each output node detecting respectively 3, 4 or 5 adjacent "1" bit in the input vector. In our simulations, Kohonen's algorithm is implemented according to the formalism described by Ritter and al. in [2] (the parameters used for describing the Gaussian functions are the ones they used for their first simulation). Tested with 64 (8x8) units in Kohonen layer, 20000 learning steps (one random input presented at each step), the system gives 2 errors as output, out of the 256 input vectors, that is, an accuracy of 0.8 % (against 36% for a system based on a model following Figure 1). Concerning the neighbourhood taken into account around k_t, it contains k_t and the 8 adjacent units when the square of the distance between k_t and I_t is higher than 0.6, and otherwise is reduced to k_t. The output threshold is 0.5.

The way the kohonen layer self-organizes according to the relations between input and output data, can be illustrated in our example by showing the maps of each of the 3 output nodes: a map of the (thresholded) weights W_{kO} between Kohonen layer and an output node, which can be seen as the representation of the 2 regions of the feature map where image-nodes will give respectively 1 or 0 as output.

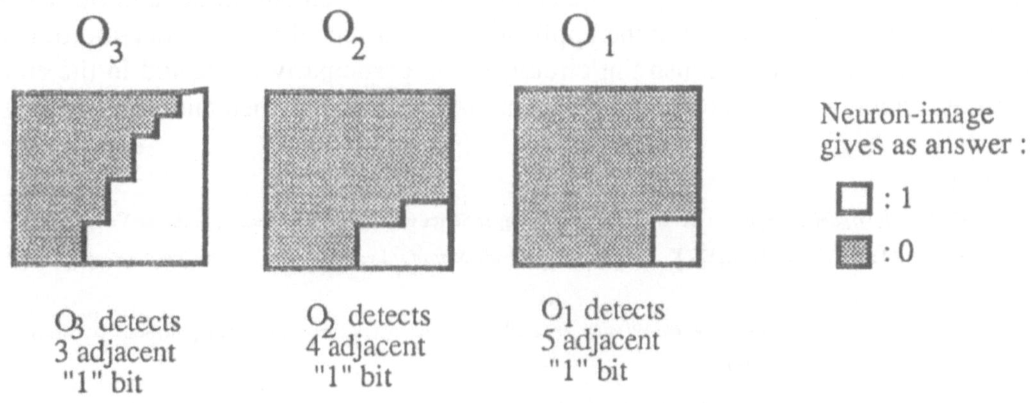
O_3
O_2
O_1
Neuron-image gives as answer :
: 1
: 0
O_3 detects 3 adjacent "1" bit
O_2 detects 4 adjacent "1" bit
O_1 detects 5 adjacent "1" bit

Figure 3

4 DISCUSSION

Comments on the respective size of the 3 layers.

The first remark deals with the number N_k of units in the Kohonen layer, which must be high enough regarding the total number of input and output nodes, to enable the self-organization of the map, and higher than the number of different possible output vectors N_{ov}. However, we have taken care of keeping it lower than the whole number of association pairs to be learned, called N_a, in order to enable generalisation in the Kohonen layer: $N_{ov} < N_k << N_a$

The second critical ratio lies in the relative numbers of input and output nodes, called N_i and N_o. During the exploitation phase, the image-node is selected as the nearest from the input, whereas the Kohonen layer has been organized while learning regarding both input and output data. This doens't seem to pose problems as soon as the self-organization of the Kohonen layer managed to reflect the topological relations between input and output data: both image-nodes, selected from the input, and from both input and output data, tend to be close in the Kohonen layer, or even to be the same unit. We think however that N_i should be at least equal to N_o: $N_i \geq N_o$.

Comparison with Back-Propagation algorithm.

Both our model and the *Back-Propagation* procedure enable the system to learn the desired function thanks to the capacity of the hidden units to generalise, that is, to detect regularities. The main difference however between both procedures stands in the way this generalisation is done. In our model, it is due to relations and cooperations between hidden units during learning, while in a *Back-Propagation* type architecture, generalisation comes from constraints on the network learning dynamic which may lead to a satisfactory equilibrium state.

This fundamental difference in learning behaviour appears again in the capacity to deal with emergence of regularities. According to us, even if this kind of phenomenon can be revealed in a *Back-Propagation* network ([5][6][7]), it greatly depends on the architecture and even on the equilibrium state reached by the network: a network state which can be interesting according to this viewpoint, is reached "by chance".

We have observed that our model is essentially much more robust regarding the initial weights, the different parameters defining the learning law (neighbourhood and so on...), and regarding its convergence, whereas the *Back-Propagation* algorithm seems to need less hidden units, to converge faster, and, in a general way, to be better suited for specific problems provided that "ad hoc" architecture and parameters have been designed.

At the moment, we are carrying on the comparison between our model and *Back-Propagation* networks, and we are investigating the application of our model in a sensory-motor task where a robot is taught to move on a constraint circuit; we're particularly interested in the emergence of regularities, which would enable to reveal some implicit rules learned and used by the system.

BIBLIOGRAPHIE:

[1]: KOHONEN T., *Self-Organization and Associative Memory*, Springer Series in Information Sciences, Springer-Verlag, 1988.

[2]: RITTER H., MARTINETZ T. SCHULTEN K., *Topology-Conserving Maps for Learning Visuo-Motor Coordination*, Neural Networks, Vol. 2, pp 159-168, 1989.

[3]: COITON Y., *Organisation sensori-motrice: modélisation et simulation d'une logique neuro-mimétique*, Memoire de D.E.A., Université de Aix-Marseille I, France, Juin 1987.

[4]: SOROUCHYARI E., *Mobile robot navigation: a neural network approach*, Journees d'electronique, E.P.F.L., Lausanne 1989.

[5]: FOGELMAN SOULIE F., GALLINARI P., LE CUN Y. THIRIA S., *Evaluation of network architectures on test learning tasks*, ICNN 1987, pp II-653 - II-660.

[6]: TOURETZSY D., POMERLEAU D., *What's Hidden in the Hidden Layers?*, BYTE August 1989, pp 227-233.

[7]: HINTON G., *Learning distributed representations of concepts*, 8th Cognitive Science Conference, Amherst, 1986.

MULTI-MODULE NEURAL NETWORKS FOR CLASSIFICATION

M. de Bollivier*, P. Gallinari*,, S. Thiria*,*****

(*)Université de Paris Sud-Orsay
Laboratoire de Recherche en Informatique
Batiment 490 - 91405 Orsay Cedex-France
e-mail: bollivi@lri.lri.fr

(**) Ecole des Hautes Etudes en Informatique
Université René Descartes-Paris V.
45 rue des Saint Pères
75006 Paris - France

(***) CEDRIC
Conservatoire National des Arts et Métiers
292 rue Saint Martin
75003 PARIS - France

ABSTRACT

In this paper we describe a new method based on the cooperation of two Neural Networks (NNs) techniques. This multi-module architecture is made up of Multi-Layer Perceptron (MLP) and Learning Vector Quantization (LVQ) nets which cooperate. This new technique combines the advantages of its two components. We will show that it is accurate for a large range of tasks and is easy to tune.

1. INTRODUCTION

Neural Networks have provided good results for many real world problems such as speech recognition and signal classification, and thus are commonly used and widely distributed. Unfortunately, given a problem to solve, there exists, until now, no accessible typology allowing the selection of the proper NN algorithm. Moreover, choosing the right parameters for a given net can be very tedious.

As a consequence, it is very interesting to develop new methods which perform, given any problem, as well as the proper available NN technique. For this purpose, we introduce multi-module architectures. To build our new methods, we will use two representative algorithms chosen among the two main types of NNs. The first one is the Multi-Layer Perceptron (MLP) [Rumelhart 86] which separates classes by computing non-linear borders. The second is the Learning Vector Quantization (LVQ) [Kohonen 88]. It is based on a Nearest Neighbor classification and uses correlation or distance measurements. Both are well suited for different data distributions and can lead to very different performances, on a common task, as will be illustrated in section 3.

We will show that the multi-module techniques we have developed combine each component quality and perform better than individual modules for a wide variety of tasks. In spite of their apparent complexity, such architectures are easy to tune.

In section 2, we introduce the data sets used throughout the paper to evaluate our methods' performances. We propose in section 3 a combination of NNs modules, whose learning phases are independent. We finally formulate in section 4 a new algorithm which allows an integrated learning.

2. DATA SETS

2.1 WAVEFORM CLASSIFICATION PROBLEM

In order to illustrate our methodologies, we choose a signal classification problem constructed by Breiman et al. [Breiman 84]. This example and a noisy version of it allows us to show the robustness of our methods.

The original version is a three-class problem which is based on the three waveforms h_1, h_2, h_3 pictured in figure 1. Each class is generated as a random convex combination of two of these waveforms. Let u be a uniform random number and ε_i normally distributed with mean zero and variance 1. A vector x in class j will be generated as follows:

$$x[i] = u\, h_m[i] + (1-u)\, h_n[i] + \varepsilon_i \text{ where m and n} \in \{1, 2, 3\}.$$

Class 1, class 2 and class 3 are respectively combinations of (h1, h2), (h1, h3) and (h2, h3). As in [Breiman 84], the learning set is made respectively of 100, 85 and 115 measurements vectors for classes 1, 2 and 3 . Five thousands test vectors have been independently generated with equal proportions for the three classes. An analytical expression can be derived for the Bayes error rate. Using this rule on a test sample of size 5000 gives a recognition rate of 86%.

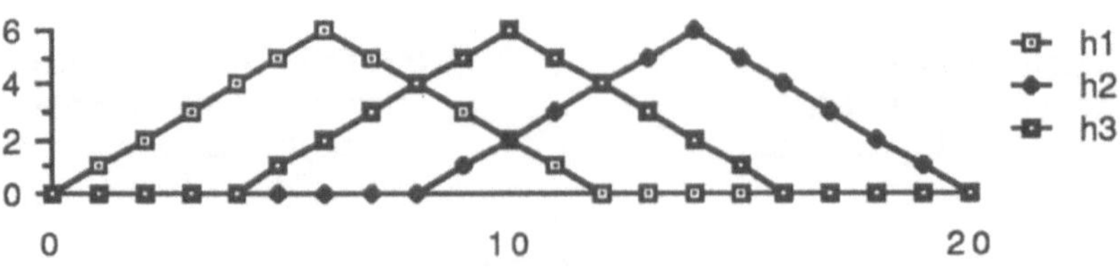

figure 1: Waveforms

2.2. Noisy waveform classification problem

Let u and ε_i be distributed as in the original problem. Each vector is 40 dimensional and generated according to:

$$x[i] = (u\, h_m[i] + (1-u)\, h_n[i])/5 + \varepsilon_i \quad \text{where m and n} \in \{1, 2, 3\} \qquad \text{if } 0 \le i \le 20$$
$$x[i] = \varepsilon_i \qquad \text{if } 21 \le i \le 40$$

We have then added a very important noisy component to the original signal. Five learning and test sets have been generated. The results below are an average of experiments on these sets.

3. MULTI-MODULE NETWORKS

3.1 INTRODUCTION

Let us introduce the two techniques we have combined that is MLP and LVQ. We have used fully connected MLP networks with hidden layers of decreasing size. Each output-layer cell is dedicated to a class: this layer activation then gives the class. Some theoretical and practical results are available. It has been proved that linear fully connected MLPs perform a Discriminant Analysis in the last hidden layer and a kind of non linear Discriminant Analysis in the non linear case [Gallinari 88]. Moreover, given a hidden layer of h cells, MLPs, due to the sigmoïd functions, tend to form clusters inside the $[-1, 1]^h$ space. Data are clustered through the successive layers in order to project each class onto a single cluster on the last hidden layer. The last layer of weights acts as a classifier. It tries to separate data on the last hidden layer using quasi-linear surfaces. This assumes of course that the projected data on the last hidden layer are nearly linearly separable, but it is not always possible as in the case of intersecting multi-modal classes distributions.

LVQ is a nearest neighbor classifier but has been designed within the framework of NNs research. As so, its formalism allows an easy combination with other NNs techniques, in the present case with MLPs. Contrary to MLPs, LVQ easily deals with multi-modal distributions.

Each NN (MLP and LVQ) was separately trained using the data sets described in section 2. Tables 1 and 2 give the optimal performances reached with an adequate tuning of the parameters. MLP have lower performances than LVQ on the first problem and the optimal recognition was reached after 250 sweeps of the training set instead of less than 20 for LVQ. On the contrary, MLP perform better on noisy data. Clearly, each data set is favourable to one technique

This suggests to build a new architecture, able to perform well in both cases, on the basis of the cooperation of the two techniques. A solution could be to use more powerful classifiers instead of the last layer of weights on the MLP in order to be able to deal with cases where the data on the last hidden layer are not linearly separable. Following the above conclusion, it could be a LVQ algorithm. By doing this, we build a system composed of two modules that cooperate in order to classify the data. The first module is a MLP performing a clustering task in its last hidden layer. These transformed data are then classified using a LVQ net (fig. 2).

Learning is performed in two steps: after a MLP clustering process, a LVQ net performs a classification task using the projection of data on the last hidden layer of the MLP. After these two learning steps, the original space is mapped to a transformed space where each class is represented by several reference vectors, contrary to MLP where only one cell is dedicated to each class.

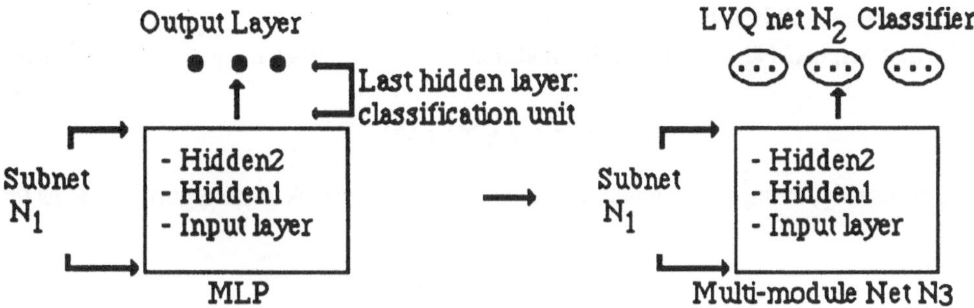

Figure 2: Networks combination. $N_3 = N_1 + N_2$.

3.2 RESULTS

For training, we have used a fixed number of reference vectors per class for LVQ. Fully connected 21-15-9-3 and 40-30-20-3 MLPs were used for non-noisy and noisy data respectively.

In the non noisy case (table 1) we compare the new architecture to Bayes rule, Discriminant Analysis, K-Means, LVQ and MLP on test sets of 5000 samples. The noisy problem was processed by K-means, LVQ and MLP (table 2).

Theoretical Bayes	Discriminant Analysis	K-means	LVQ	MLP	Multi-module Network
86%	74%	82%	84.6%	81.6%	84%

Table 1: Optimal recognition rates on test set. Non-noisy waveforms data.

LVQ	MLP	Multi-module Network
77%	80.5%	81.1%

Table 2: Performances on noisy test set of LVQ, MLP and Multi-module Network

3.3 DISCUSSION

These results show that this combination of the two techniques is very robust:

- In the two cases the new architecture performs well compared with other techniques. In this sense, the new procedure has a much larger range of applicability than MLP and LVQ.

- It converges very quickly: 20 sweeps of the learning set for each of its components is enough to reach the optimal recognition rate instead of 250 for individual MLP. Moreover, it does not require a fine tuning of its parameters.

4. INTEGRATING THE MODULES: TOWARDS NEW ALGORITHMS

In the previous architecture both modules are trained independently. Good performances were obtained but the method may be further improved. To really combine the advantages of both classifiers, we propose a new algorithm, based on the previous architecture, which will back-propagate the change in LVQ net N_2 to the MLP subnet N_1. As seen before, individual MLP project each class in a single cluster onto the last hidden layer which is not always adequate in the case of multi-modal data. The new procedure allows us to perform multi-modal clustering for each class on this layer. We thus obtain a supervised learning algorithm for the global architecture that allows multiple outputs for each class. This is expected to be easy and to improve performances.

Let us consider the architecture $N_3 = N_1 + N_2$ (see fig. 2). Let x of class class(x) be an input vector of N_3 and $N_1(x)$ the computed output of N_1. Let $\mathfrak{L}$ be the learning set and $N_1(\mathfrak{L}) = \{y \,/\, y = N_1(x),\ x \in \mathfrak{L}\}$ the learning set of N_2. Let $m_i^{class(x)}$ be the reference vector nearest to $N_1(x)$ among the reference vectors dedicated to class(x) in N_2. Let us consider $C_i(x) = \min_k \| m_k^{class(x)} - N_1(x) \|$ the distance between $N_1(x)$ and

$m_i^{class(x)}$. Classification will be achieved if the reference vector in N_2 nearest to $N_1(x)$ is one of those dedicated to class(x). For that purpose, N_3 weights will be modified so as to minimize the global cost function C defined as follows:

$$C = \Sigma_{x \in \mathcal{L}} C_i (x)$$

It can be done, as usually, by using any gradient algorithm. The main steps of the algorithm are as follows:

Initialization

1 - Let us consider an ordinary fully connected MLP in which each output cell is dedicated to a class. Train it for a few iterations to begin a clustering process. Let N_1 be its subnet obtained by subtracting its output layer. Let $\mathcal{L}$ be the learning set. For each $x \in \mathcal{L}$ define $N_1(x)$ the computed output of N_1.

2 - Define $N_1(\mathcal{L}) = \{y \,/\, y = N_1(x),\, x \in \mathcal{L}\}$. Let N_2 be a LVQ net which has been trained using $N_1(\mathcal{L})$.

Learning

3 - Build a net N_3 by adding N_2 to N_1 as in figure 2. Use the learning algorithm described below.

4 - For each x belonging to the learning set $\mathcal{L}$:
- compute the output of N_3
- select the cell i with minimum response among the cells dedicated to class(x): let C_i (x) denotes the output of this cell. $C_i (x) = \min_k \| m_k^{class(x)} - N_1(x) \|$ where $m_k^{class(x)}$ denotes a vector of N_2 dedicated to class(x).

5 - Compute the error function $C = \Sigma_{x \in \mathcal{L}} C_i (x)$.

6 - Use a gradient algorithm to modify the weights of the net so as to minimize this error function.

A few experiments were performed to test this algorithm using an adaptive version of the gradient procedure similar to Back-Propagation. A clear increase of the accuracy has been observed. For the non noisy data, the recognition rate is 85%, which compares well with the 86% of Bayes theoretical upper bound. In the noisy case, accuracy reaches 82.4%. Furthermore, this procedure is very fast: only 10 sweeps of the learning set are needed. The algorithm runs quickly and no fine tuning of the parameters or careful selection of an architecture is needed.

6. CONCLUSION

We have described new Neural Nets architectures based on the cooperation of MLP and LVQ techniques. We have shown that they are robust in the sense that they perform well on a wide range of problems which is not the case with their individual components. They combine the advantages of their components and are easy to tune. Moreover, they converge quickly. Two versions have been proposed. In the first one, each component is independently trained. The second version allows an integrated learning process. In both versions performances are improved, the integrated network being the most accurate. Further researches will be focused on the integrated learning algorithm.

ACKNOWLEDGMENT. The authors would like to acknowledge the contribution of E. Viennet to this work.

BIBLIOGRAPHY

[Breiman 84] Breiman L., Freidman J., Olshen R., and Stone C.. (1984). Classification and Regression Trees. *Wadsworth International Group.*

[Gallinari 88] Gallinari P., Fogelman Soulie F., Thiria S. (1988): Multi-Layer Perceptrons and Data Analysis. In *2nd IJCNN 88*, San Diego, vol I, 391-399.

[Kohonen 88] Kohonen T. (1988): Self Organization and Associative Memory. Springer Series in Information Science, vol. 8, *Springer Verlag*, 2nd edition.

[Rumelhart 86] Rumelhart D. E., Hinton G.E., Williams R.J. (1986): Learning Internal Representations by error propagation. In Parallel Distributed processing: explorations in the Microstructure of Cognition, *MIT Press* 318-362.

Multi-Layer Versus Single-Layer Neural Networks and an Application to Reading Hand-Stamped Characters

Yoichi Hayashi†, Masateru Sakata‡ and Stephen I. Gallant*

†Department of Information Science, Ibaraki University, Hitachi, Ibaraki 316, Japan
‡Hitachi Engineering Co. Ltd., Hitachi, Ibaraki 319-12, Japan
*College of Computer Science, Northeastern University, Boston, Massachusetts 02115, USA

Abstract

Recent advances in multi-layer learning techniques for networks have sometimes led researchers to overlook single-layer approaches that, for certain problems, give better performance. Here we examine the respective strengths and weaknesses of these two approaches for multi-class pattern recognition, and present a case study that illustrates these considerations.

The case in question—reading hand-stamped characters—is an important industrial problem of interest in its own right. Recognition rates of 99.9% and processing speeds of 86 characters per second were achieved for this very noisy application.

We conclude by recommending the following rule of thumb: *Never try a multilayer model for fitting data until you have first tried a single-layer model.*

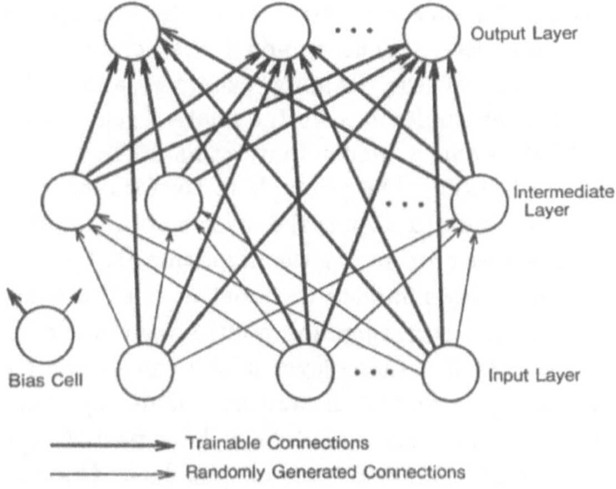

Figure 1: Distributed Method Network.

1 Introduction

The recent advances and popularization of learning algorithms for multi-layer neural networks (for example [13, 12]) has significantly increased the applicability and power of the neural network approach. However an unfortunate by-product of this success has been a tendency to overlook the speed and generalization capabilities of single-layer models. Here we examine the various trade-offs, with particular emphasis on multi-class pattern recognition. We then present a case study involving recognition of hand-stamped characters, an important problem for industrial applications, where very high recognition rates were achieved in a noisy domain.

2 Multi-Layer versus Single-Layer Models

In multi-class pattern recognition problems the inputs must be classified into one of $c \geq 2$ classes, where c is usually greater than 2. For this domain we will compare two multi-layer learning algorithms, backpropagation [13, 12] and the distributed method [3, 4], and one single-layer algorithm, the pocket algorithm [3, 5].

2.1 Model Descriptions

For backpropagation we employ the most widely used configuration consisting of c output cells, d intermediate cells ("hidden units"), and p input cells. Connections run from every input cell to every intermediate cell, and from every intermediate cell to every output cell.

For the distributed method there are the same cells and connections as with backpropagation, and in addition there are connections from input to output cells. See Figure 1. Characteristic of the distributed method is that weights for intermediate cells are not learned, but instead are generated at random. Thus the only trainable weights in this model are those for connections leading to output cells. The pocket algorithm (or pocket algorithm with ratchet) was used for training the output cells. The pocket algorithm is a modification of perceptron learning [11] that is well-behaved with non-separable or noisy data.

Finally the single-layer model has no intermediate cells at all, only input cells and output cells. (This is what we get by removing all intermediate cells and connections from Figure 1.) Again we use the pocket algorithm for training the output cells.

2.2 Comparisons

The main advantage of the multi-layer models is that they can fit training examples better than single-layer models can. This means they can give better generalization, provided there are sufficient training examples.

Conversely single-layer models give better relative generalization. By *relative generalization* we mean testing classification rate divided by training classification rate. Thus a model that correctly classifies 82% of the training examples and 81% of the test examples has better relative generalization (.81/.82) than a model that trains 100% and tests 90% (.90/1.00). The principle factor for determining relative generalization is the inverse of the number of *trainable* weights set by the algorithm [1, 6]. Therefore single-layer models have better relative generalization than corresponding multi-layer models. This means that if a single-layer model can fit the training examples almost as well as a multi-layer model then the single-layer model can be expected to give better performance on unseen test data. Conversely if a multi-layer model fits the examples significantly better, then the multi-layer model can be expected to give better performance. The deciding factor for determining how much better the multi-layer network must fit the training examples is the number of training examples. If there are few training examples then a greater difference in fit is required before a larger (ie. multi-layer) model is preferred, and conversely if there are many training examples. Note that in practice it is easy to compare generalization of single-layer and multi-layer models; we can simply test the models on unseen training examples.

Single-layer models also have faster learning speed and faster execution of the final network because they are smaller than multi-layer models.

The bottom line is: *if speed is critical or if a multi-layer model doesn't fit the training examples a lot better than a single-layer model (or a little better if a large number of training examples were used), then prefer the single-layer model.*

3 Application: Reading Hand-Stamped Characters

An important task for industrial image processing is reading characters that are stamped in aluminum (or other metal) using a hand-held or machine punch.

Figure 2: Enlarged photograph of several stamped numbers.

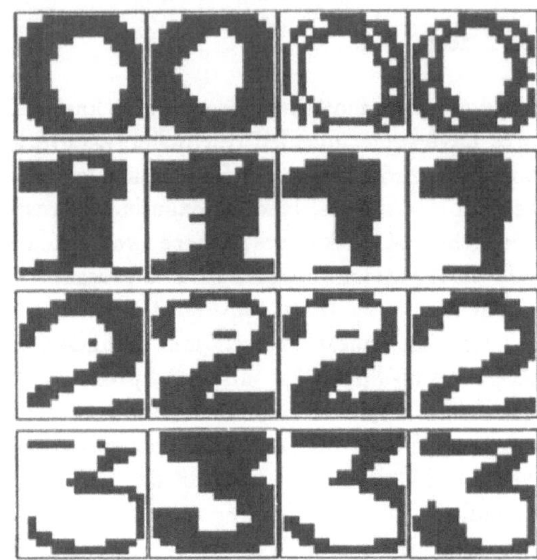

Figure 3: Some of the binarized images used for training or testing data.

Such hand-stamped characters are widely used on production lines for automobiles and other products. Typically characters are small (3 mm by 2 mm) and irregular (see Figure 2) due to the stamping process.

Digitized images are therefore very noisy and difficult to read, as can be seen in Figure 3.

Moreover industrial requirements for an automated system are typically 80–100 characters per second with 99.9% accuracy. Such a fully automatic system would have a major impact on factory automation.

Previous work in hand-stamped character recognition has been done by Nakamura and associates [8], and Nakano and associates [9] using other methods. We have also previously reported on a large-scale character recognition problem using the distributed method [7]. Space limitations preclude a discussion of this work here.

As a step toward very fast and accurate automated systems using neural networks, we gathered 1000 training images and 1000 testing images of digits 0–9 using the following procedure. First an industrial TV camera viewed fully illuminated (ring lighted) samples to produce 64-level gray scale images. We manually controlled the lighting and did a rough segmentation by hand, placing an (oversized) box around the character in question. We also manually set a threshold for binarization of the data.[1] Finally the images were precisely segmented by histogramming and normalized to fit into 16×16 grids as in Figure 3. Thus each image consisted of 256 binary values.

4 Experiments

We first used backpropagation to construct a network. Because there were 256 inputs and 10 outputs the number of trainable weights was large, so training the network was quite slow. Therefore we were only able to train on 100 of the 1000 training examples. After some initial experimentation we found that 30 intermediate cells and 2500 iterations was enough for (nearly) 100% learning.

We then used a distributed method network with 30 intermediate cells and 3000 iterations. This network also trained to 100% on the same 100 training examples used with backpropagation. Note that the number of trainable weights in the distributed method network was less than in the backpropagation network, $10 \times (256 + 30 + 1)$ versus $10 \times (30 + 1) + 30 \times (256 + 1)$. (The '+1' terms are for the biases.)

Three performance criteria were measured:

Criterion 1: A test example is correct if the correct output cell has activation +1 and all other cells have activation −1. This is the strictest criterion; all 10 cells must be independently correct. (Performance on training examples was measured using this criterion.)

Criterion 2: A test example is correct if the correct output cell has the highest weighted sum of any output cell and that weighted sum is positive.

Criterion 3: A test example is correct if the correct output cell has the highest weighted sum of any output cell. This is the most lenient criterion and is exactly the same criterion used for linear machines [10] or winner-take-all groups.

A comparison showed that the distributed method gave about the same generalization as backpropagation on the 1000 unseen test examples, but the numbers were not different enough to make a conclusive judgment from one test.[2] More importantly, backpropagation took about 750 times as long to train as did the distributed method. This was enough to convince us to continue with experiments using only the distributed method and single-layer models.

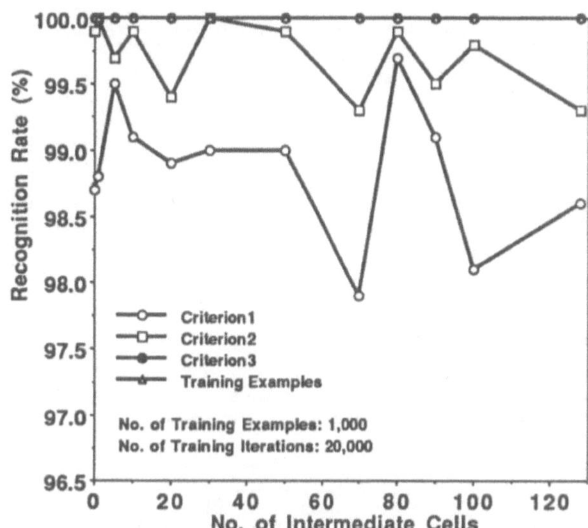

Figure 4: Generalization performance with respect to the number of intermediate cells.

Our next experiment was to compare generalization performance with various numbers of intermediate cells, including the single-layer model with 0 intermediate cells. The results are given in Figure 4.

We had expected that too many intermediate cells would cause a drop-off in performance due to overfitting the data, and this was confirmed. However we were greatly surprised that the single-layer model did as well as any of the other models, even with such noisy data. In particular we were surprised that the training examples were independently separable for every output cell, and that generalization measured by criterion 3 was 100%. In retrospect the separability seems not unreasonable, because 1000 training examples is only about twice the number for which we could expect separability by Cover's result[3] for *random* inputs [2]. Also the criterion 3 testing result seems reasonable due to the more stringent training

[1] These were the only manual operations in the image processing. In the future we will attempt to automate these steps in order to produce a totally automated system.

[2] For backpropagation testing, best results under criteria 1 and 2 were obtained by taking activations ≥ 0.5 as +1 and activations < 0.5 as −1. Generalization results were:

	Criterion #1	Criterion #2	Criterion #3
backpropagation	95.8	96.1	98.1
distributed method	91.0	97.0	99.5

[3] For a single-cell model, separability is expected if and only if the number of random training examples is less than about half the number of trainable weights.

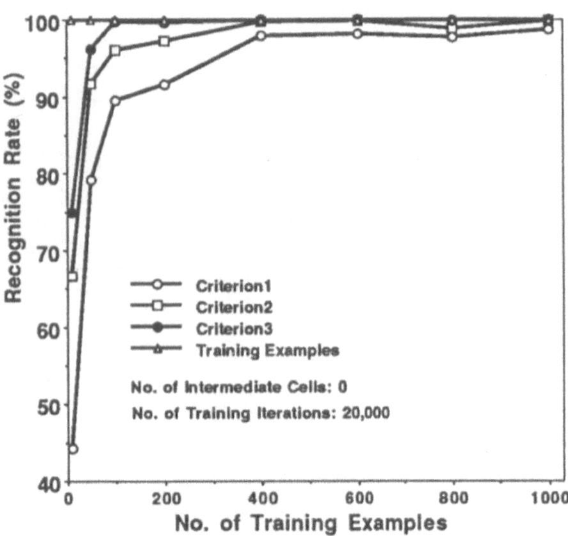

Figure 5: Generalization performance with respect to the number of training examples.

conditions where criterion 1 was used.

The final experiment we performed was to see how the number of training examples affected generalization performance. Figure 5 gives the results. As expected, the more training examples that there are the better the performance, with about 400 training examples being enough to give at least 99.9% accuracy under criterion 3 for all our tests.

5 Discussion and Conclusions

The most important conclusion we have to offer is the following rule[4]:

> **Never try a multi-layer model for fitting data until you have first tried a single-layer model.**

We would even venture to estimate that at least 10% of the multi-layer networks created with backpropagation (and used for something other than the parity problem) could be replaced by single-layer networks with improved training and execution speed and with the same or better generalization to unseen data.

The second conclusion is to note that training speed affects generalization. For example the slowness of backpropagation prohibited us from training on more data, which in turn detracted from the quality of the model that was generated.

[4] Of course there may be other considerations that motivate the use of multi-layer models, such as data compression or interest in functions computed by the intermediate layer cells.

Finally the we note that the neural network approach seems promising for hand-stamped character recognition. We achieved high accuracy with a very fast model: 10 minutes for training and 86 characters per second recognition speed on a Sun 3/50 equivalent with no special hardware. Moreover both the single-layer and distributed method models are trivial to parallelize for both training and execution. For single-layer models a speed-up by a factor of 10 can be had simply by computing the 10 output cells simultaneously.

In the future we will try to eliminate the binarization step in preparing the data and attempt to add fully automatic segmentation.

References

1. Baum, E.B. & Haussler, D. What Size Net Gives Valid Generalization? *Neural Computation*, Vol. 1, No. 1, 1989, 151-160.
2. Cover, T. M. Geometrical and Statistical Properties of Systems of Linear Inequalities with Applications in Pattern Recognition. IEEE Trans. Electronic Computers, Vol. 14, 326-334, 1965.
3. Gallant, S. I. Perceptron-Based Learning Algorithms. To appear: *IEEE Transactions on Neural Networks.*
4. Gallant, S. I., and Smith, D. Random Cells: An Idea Whose Time Has Come and Gone... And Come Again? IEEE International Conference on Neural Networks, San Diego, Ca., Vol. II, 671-678, June 1987.
5. Gallant, S. I. Optimal Linear Discriminants. Proc. Eighth International Conference on Pattern Recognition, Paris, France, Oct. 28-31, 1986. Pg. 849-852.
6. Gallant, S. I. *A Connectionist Learning Algorithm With Provable Generalization and Scaling Bounds.* To appear: *Neural Networks.*
7. Hayashi, Y., Sakata, M., Nakao, T. & Ohhashi, S. Alphanumeric Character Recognition Using a Connectionist Model with the Pocket Algorithm. Int. J. of Neural Networks: Research & Applications, Vol.1, No. 3, 175-186, 1989.
8. Nakamura, Y., Suda, M., Sakai, K., Takeda, Y. & Udaka, M. Development of a High-Performance Stamped Character Reader. IEEE Transactions on Industrial Electronics, Vol. IE-33, No. 2, 1986, 144-147.
9. Nakano, T., Takeda, N., & Yamamoto, S. Recognition Algorithm of Stamped Alphanumerals on Car Body. Trans. IEE of Japan, Vol. 109-C, No. 6, 1989, 439-445.
10. Nilsson, N. J. *Learning Machines.* (1965) McGraw-Hill, New York, NY.
11. Rosenblatt, F. *Principles of neurodynamics: Perceptrons and the theory of brain mechanisms.* Spartan Press, Washington, DC., 1961.
12. Rumelhart, D. E., Hinton, G. E., & Williams, R. J. Learning Internal Representations by Error Propagation. In Rumelhart, D. E. & McClelland, J. L. (Eds.) *Parallel Distributed Processing: Explorations in the Microstructures of Cognition, Vol. 1.* MIT Press, 1986.
13. Werbos, P. J. Beyond Regression: New Tools for Prediction and Analysis in the Behavioral Sciences. Ph.D. Thesis, Harvard University, 1974.

DELAY-INSENSITIVE LEARNING IN A FEEDFORWARD NEURAL NETWORK

Jos Nijhuis, Andreas Siggelkow and Lambert Spaanenburg
Institute for Microelectronics Stuttgart
Allmandring 30a, D-7000 Stuttgart 80
West-Germany

- Abstract - *A neural network learns from a long history of applied patterns. To create fault-tolerance, an extended learning period is required. For actual hardware such an initialization period is too long. It is investigated how under the bounded delay assumption training of the neural network can be accelerated. It is concluded that the optimal training set takes not only the smallest number of learning cycles but is also delay-insensitive and therefore allows a large speed-up using the non-normal mode of operation.*

Keywords. Fault-tolerance, Transport Delay, Artificial Neural Network, Error Back-Propagation Rule, Delay-Insensitive Training.

FAST TEACHING OF BOLTZMANN MACHINES WITH LOCAL INHIBITION

Thomas R Osborn
Computer Science Department
University of Technology, Sydney
PO Box 123 BROADWAY NSW 2007 AUSTRALIA

ABSTRACT: Local clusters of lateral inhibition are modelled *softly* by supplementing the objective function (rather than by strict competition) for the Input_to_Output Boltzmann machine.

This frustrates unwanted complexity of the induced internal representation of data.

Furthermore, incremental teaching (shaping) incorporates new data with minimal retraining of previously learned data.

Consequent learning rates are well over an order of magnitude better than the standard models, although maximum storage capacity is marginally reduced.

A Fast Connectionist Learning Paradigm

Samir I. Sayegh
Physics Department
Purdue University
Ft Wayne, IN 46805-1499
sayegh@ed.ecn.purdue.edu

ABSTRACT

A fast connectionist mapping and learning paradigm, the Optimum Path Paradigm (OPP), is introduced. It is characterized by its high speed in both feed-forward and learning phases. OPP can be thought of as a discrete version of the Principle of Least Action that enjoys universal applicability in the domain of Physics. The equivalence of OPP learning to the solution of a system of inequalities is established. The learning heuristic is deduced from this formulation and related to Hebbian learning. A number of aspects of OPP are compared to more standard paradigms such as Back Propagation.

THE EXTRACTION OF RULES FROM MULTI-LAYERED NEURAL NETWORKS

Sabrina Sestito*and Tharam Dillon
Department of Computer Science,
La Trobe University, Bundoora
Victoria, AUSTRALIA 3083

Abstract

Machine Learning is an area concerned with the automation of the process of knowledge acquisition, the bottleneck in developing knowledge base systems. Neural networks generally represent their knowledge at the lower level, while knowledge based systems use higher level knowledge representations. The methods we propose here, provide a technique which automatically allows us to extract high level knowledge representations from the lower level representations used by neural networks. Thus, we propose a method which uses neural networks as the basis for the automation of knowledge acquisition.

*Financially supported by the Aeronautical Research Laboratory, Department of Defense, Australia

QUANTUM LEARNING ALGORITHM FOR MULTILAYERED NEURAL NETWORK

Yu Shao-Bo * Hu Shou-Ren * *

Yan Jun-Yong *

* Naval Academy of Engineering, Dept. of ComputerEngineering PB. 430033, Wuhan, China

* * ChangSha Institute of Teck.Dept. of ComputerScience, PB. 410003,ChangSha, China

ABSTRACT

Back propagation rule has been shown to be an efficientlearning algorithm for multilayerd neural network. However,it is limited because it only finds local minima. Boltzmannmachine has also been shown to be an effivient learning rule. But,it is limited because it learning rate is too slow. In thispaper,we proposed and simulayed a quantum learning algorithmfor multilayerd neural network. it is shown that its learning rate is more rapid than that of Boltzmann machine, and it can tind theglobal minimun unlike back oropagation algorithm does.

Training of a Neural Network with Topology Generation for the Classification Problem

Hahn-Ming Lee and Ching-Chi Hsu
Department of Computer Science and Information Engineering
National Taiwan University
Taipei, Taiwan

Abstract

This paper is to deal with the training with the topology generation issue for the classification problem. For a separable set of training examples, which means there exists a hyperplane to discriminate the positive training examples from the negative ones, a simple iteration algorithm known as Perceptron Learning Algorithm can be applied well and there is no the topology generation issue. Real worlds problems, however, are usually non-separable due to the complex relationships between inputs and desired classifications or incorrect noisy examples. In this case, training with topology generation is important since in inappropriate network structure the known learning algorithms like back-propagation can not guarantee to find the desired results and they are time-consuming.

Here, a new neural network model called BU is presented in which the network topology does not have to be specified before training. It is named since the training algorithm starts from a single positive example, i.e. bottom up processing compared to the top down divide-and-conquer processing. This approach appears very promising due to the following features: (1) The network topology is automatically generated during training in a finite number of times, (2) We can apply the non-incremental learning or incremental learning to generate network structure, (3) All the nodes in the BU model are the same, i.e. they perform the same function, (4) The training examples are not restricted to be linearly separable, but they can define any arbitrarily complex decision regions, (5) This training method can be easily used to detect incorrect positive examples, (6) The processing time to classify an input on the generated network is constant and short, (7) It is not necessary to tune any parameters.

A MODEL OF THE NEURAL NETWORK FOR STORAGE AND RETRIEVAL OF TEMPORAL SEQUENCES

R.M. Borisyuk
Research Computing Centre, USSR Academy of Sciences
Pushchino, Moscow region, 142292 USSR

Abstract

We propose a new method for storage and retrieval of the temporal sequences of events (neural movies) using the back propagation learning procedure. During learning the weights of the connections in the network are adjusted to store several sets. During retrieval the network is able to repeat periodically one of the recorded sets.

OPTIMAL RULE INDUCTION IN NEURAL NETWORKS

G.D.Tattersall
University of East Anglia / British Telecom Research Labs
Norwich NR4 7TJ

Abstract

A central problem in artificial intelligence is the generation of rules from an incomplete set of examples of data generated by an unspecified process. The examples are normally in the form of input-output pairs of attribute vectors. If a good rule can be induced from a sparse set of examples then it can generalise and predict the correct action to be associated with a previously unseen input.

This paper shows that particular forms of artificial neural network (ANN), similar to the MLP, but which use sinusoidal non-linearities can learn optimal rules from a sparse set of examples and thereby provide the most likely generalisation when faced with previously unseen data. The paper also suggests a theoretical framework for evaluating the probability of particular generalisations from the training data by use of the multi dimensional Fourier Transform and a reformulation of the Shannon Equation.

LEARNING BY ASYMMETRIC PARALLEL BOLTZMANN MACHINES

Bruno Apolloni
Diego de Falco

Dipartimento di Scienze dell' Informazione
Università di Milano
Via Moretto da Brescia 9
I-20133 Milano, Italy

ABSTRACT

We consider the Little-Shaw-Vasudevan model as a parallel asymmetric Boltzmann machine. The resulting Hebbian learning rule draws the error signal for the updating of synaptic weights from time averages along the past history of the discrepancy between expected and actual transitions.

As we work without the hypothesis of symmetry of the weights, we can include in our analysis also feed-forward networks, for which the Hebbian learning rule turns out to be complementary to the error back-propagation rule, in that it "reinforces the correct behaviour" instead of "penalizing the wrong answers".

MULTI-CRITERIA OPTIMIZED LEARNING RULE FOR OPTICAL IMPLEMENTATION.

Ph. REFREGIER
Laboratoire Central de Recherches
Thomson-CSF 91404 ORSAY (cedex) France
TEL: 33 1 60 19 71 19, FAX: 33 1 60 19 74 16.

Abstract

We address the problem of learning rules for optical processor considered as a feature extractor for electronic or digital Neural Networks (N.N.). This approach consists in limiting the optical processing to the first layer of a N.N and yields, for the mean term, more realistic and promising solutions than 'all-optical' architectures. We propose a supervised non-iterative learning rule (i.e. with explicit formula for the weight interconnections) well suited for optical implementation. This method includes not only discrimination but also signal processing abilities. These new abilities are obtained with multi-criteria optimization which overcome the problem of over-specialization.

BACKPROPAGATION ANALOGUE MAPPING IMPROVED BY NONLINEARITIES MODIFICATION

P. Burrascano and P. Lucci
INFO-COM Dept. - Universita' di Roma "La Sapienza"
via Eudossiana, 18 - 00184 Roma - Italy

ABSTRACT

The possibility of applying Artificial Neural Networks (ANN) to real world problems such as image processing, usually implies an increase in network complexity. If the network is of the layered feedforward type and it is trained by using the backpropagation algorithm, the use of such large sized networks considerably increases the possibility of having convergence problems during learning phase.

In the present paper the generalized delta rule and its convergence characteristics are analyzed: it is evidenced in particular that important problems can rise during learning phase owing to the choice of sigmoidal functions as the non linearities of processing elements.

On the basis of the performed analysis a modification to the non linearity of the output layer processing elements is proposed. In particular the case of analogue target values is focused in order to allow the use of the modified delta rule algorithm to image processing applications. The proposed algorithm is shown to significantly better the convergence properties of the generalized delta rule in both cases of discrete and analogue targets.

SUPERVISED LEARNING USING A GENETIC ALGORITHM

Marco Muselli, Sandro Ridella

Istituto per i Circuiti Elettronici
Consiglio Nazionale delle Ricerche
via Opera Pia, 11 - 16145 Genova, Italy

Abstract

A new type of genetic algorithm for the training of neural networks is presented. The difficulty of obtaining a nearly globally optimal set of weights in a reasonable time is overcome by using a global optimization method. It maintains a set of points at every iteration and permits a parallel search of the global minimum.

The procedure for generating and choosing the set of points is essential in genetic algorithms and it is derived from the "Simulated Annealing" technique. This gives to the proposed method a better opportunity of avoiding local minima by accepting uphill moves.

Some runs have been performed on the parity and the symmetry problem, which are two classical tests in the field of neural networks. The results show an increase in reliability and convergence speed with respect to the traditional back-propagation method and a conventional Simulated Annealing algorithm.

STRUCTURED TRAINING FOR MLPs: SOME EXPERIMENTS ON NETSPEAK WITH A CHILD'S READING SCHEME

David G Bounds and Mark D Bedworth
Research Initiative in Pattern Recognition
Royal Signals and Radar Establishment
St Andrews Road
Malvern
Worcs. WR14 3PS, UK

ABSTRACT

This paper investigates the possibility of structuring training data in order to improve the performance of MLP networks. The task chosen was that originally studied by Sejnowski and Rosenberg: transcribing English text into a phonemic representation suitable for speech synthesis. The performance of networks trained on a dictionary is compared with those trained incrementally using a childs reading scheme, the Oxford Reading Tree, which is in widespread use in British schools. Both types of network were also scored against a synthesis-by-rule program. It is found that, in the limited domain of the childrens books studied, the performance of the networks trained on the reading scheme data is as good as that of the synthesis-by-rule program. However, this appears to be because of the choice of training data rather than the order in which it was presented.

ALOPEX ALGORITHM FOR SUPERVISED LEARNING IN LAYER NETWORKS

Abhijit S. Pandya
Center for Complex Systems
Dept. of Computer Eng.
Florida Atlantic Univ., Boca Raton, FL 33431

Raisa Szabo
Dept. of Computer & Info. Science
Nova University
Fort Lauderdale, FL 33314

ABSTRACT

We describe the use of ALOPEX algorithm for solving non-linear learning tasks by multi-layer feed-forward networks. ALOPEX is a stochastic parallel process which has been previously applied in a theory of perception [1,2,3]. It has also been applied to several non-linear optimization problems such as the Travelling Salesman Problem [4]. It estimates the weight changes by using only a scalar cost function which is a measure of global performance. We present the results of computer simulations for learning tasks involving the parity problem and compare them to the back-propagation method of Rumelhart et al. [5].

A LEARNING RULE IN THE CHEBYSHEV NORM FOR MULTILAYER PERCEPTRONS

P. Burrascano and P. Lucci

INFO-COM Dept., Universita' di Roma "La Sapienza"
v. Eudossiana 18, 00184 Roma - Italy

ABSTRACT

Layered feedforward topology trained with back propagation is currently the most widely applied neural network architecture. The process which is carried out by the back propagation paradigm can be interpreted as an optimization process operated in the L_2 norm. However there are situations in which the L_2 norm is not the best choice; consequently it can be useful to have available training algorithms for multilayer perceptrons which allow to operate with different norms. In the present paper an L_∞ version of the back propagation paradigm is proposed, and a comparison between the L_2 and the L_∞ paradigms is presented, taking into account computational cost and speed of convergence.

Symmetry and representability properties of feed–forward neural networks.

Edgardo A. Ferrán and Roberto P.J. Perazzo
Comisión Nacional de Energía Atómica, Departamento de Física (TANDAR).
Av. Libertador 8250, (1429) Buenos Aires, Argentina.

Abstract

We study the relation between the symmetries of feed-forward neural networks and their learning capabilities. The symmetries of the network are associated to operations on the input and output signals that can be compensated with a simultaneous transformation on the synaptic matrix. These operations consist in the permutation of any pair of input or output signals and the interchange of ones and zeros in the input. As a consequence of these symmetries the boolean functions implemented by the network are grouped into symmetry classes. We prove that all functions of the same class are represented by the same number of synaptic matrices and have the same learning curve. We also find the condition that must fulfill a boolean function in order to be represented by a network composed by neurons with odd activations functions and with all its firing thresholds equal to zero.

HIERARCHICAL ARCHITECTURES FOR OPTIMIZED TRAINING

J. Deppisch, H.-U. Bauer, and T. Geisel

Institut für Theoretische Physik and SFB Nichtlineare Dynamik
Universität Frankfurt, D-6000 Frankfurt/Main, Fed. Rep. of Germany

We have developed a procedure for optimized training of large feed-forward neural networks of arbitrary topologies which is based on hierarchical architectures combined with a backpropagation learning rule. Compared to the common backpropagation rule and more sophisticated algorithms, we have substantially reduced the huge amount of computing power needed to train large networks and also obtained a considerable improvement of the absolute learning error. We test our method by applying it to the forecasting of chaotic systems, such as the logistic map and the Rössler system. A new measure is defined for an optimum evolution of the prediction error. In both examples this limit is reached by our hierarchical procedure and the absolute predicition error is reduced considerably compared to traditional predicition methods.

UNSUPERVISED LEARNING

Chair: Stephen GROSSBERG

SELF-ORGANIZING NEURAL ARCHITECTURES FOR VISION, LEARNING, AND ROBOTIC CONTROL

Stephen Grossberg†
Center for Adaptive Systems, Boston University,
111 Cummington Street, Boston, Massachusetts 02215 USA

Abstract

Self-organizing neural architectures are described for the visual perception of static and moving forms, for autonomous real-time learning of multidimensional associative maps, and for adaptive control of variable-speed multi-joint motor trajectories. Motion filtering and segmentation are carried out by a motion Boundary Contour System. Associative learning is accomplished by a Vector Associative Map, which provides an on-line alternative to the off-line properties of Back Propagation for error-based learning. Trajectory formation is carried out by a Vector Integration to Endpoint model.

1. Self-Organizing Networks for Peception of Static and Moving Form

The first part of this lecture will analyse computational properties that clarify why parallel cortical systems exist for the perceptual processing of static visual forms and moving visual forms. These properties arise as part of a theory of preattentive vision that is called FACADE Theory. FACADE Theory clarifies how the visual cortex generates a multiplexed representation of Form-And-Color-And-DEpth, or FACADE. This process is controlled by interactions of two systems, called the Boundary Contour System (BCS) and the Feature Contour System (FCS), whose properties are computationally complementary. The BCS generates an emergent 3-D boundary segmentation, whereas the FCS discounts the illuminant and fills-in surface properties. BCS segmentations are insensitive to direction-of-contrast, so that they can detect boundary structure along contrast reversals. They therefore do not generate perceived contrast differences within the BCS. FCS properties are sensitive to direction-of-contrast, and subserve visible percepts of surface brightness, color, and depth. The BCS includes striate and prestriate projections of cortical hypercolumns. The FCS includes striate and prestriate projections of cortical blobs (Grossberg and Mingolla, 1985a, 1985b).

The BCS is further divided into parallel systems for the processing of static segmentations and moving segmentations. The static BCS generates segmentations that are insensitive to direction-of-contrast and insensitive to direction-of-motion. Segmentations within the motion BCS are also insensitive to direction-of-contrast but sensitive to direction-of-motion. The motion BCS has been used to analyse a large body of data concerning short-range and long-range motion perception, notably data about apparent motion, and to clarify why specialized motion processing areas, such as MT, exist in visual cortex. Figure 1 schematizes the Motion Oriented Contrast (MOC) Filter for visual motion preprocessing (Grossberg and Rudd, 1989). The full model also suggests how coherent global motion effects occur, such as motion capture and induced motion, thereby offering a solution of the global aperture problem (Grossberg and Mingolla, 1990).

A further analysis suggests why parallel cortical systems for analysis both of static form and of moving form exist (Grossberg, 1990), and how they generate different geometries of static and motion perception. These parallel systems compute all possible ways of symmetrically gating sustained cells with transient cells to generate output signals that are insensitive to direction-of-contrast and are organized into opponent pairs of on-cells

† Supported in part by the Air Force Office of Scientific Research (AFOSR F49620-87-C-0018), DARPA (AFOSR 90-0083), and the National Science Foundation (NSF IRI-87-16960).

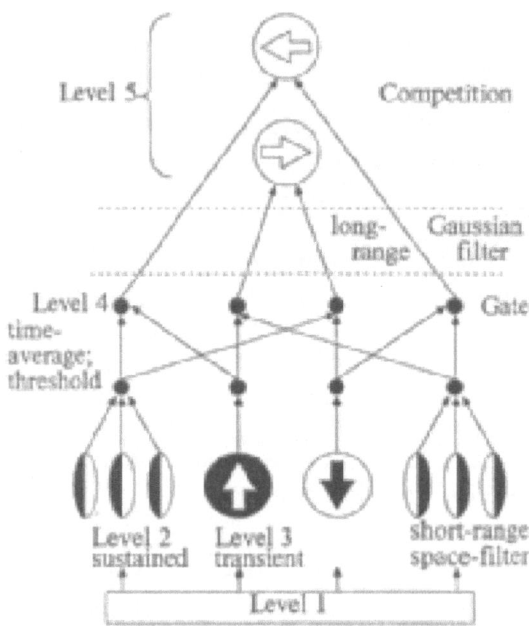

Figure 1. The MOC Filter model for visual motion preprocessing.

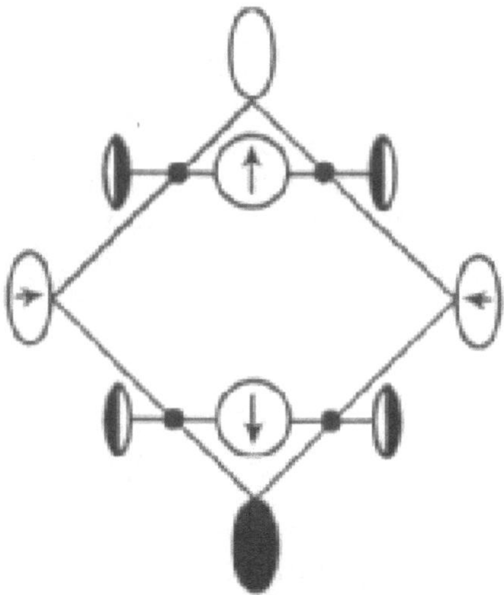

Figure 2. Four-fold symmetry of orientation cells and direction cells: Oriented sustained cells that are sensitive to direction-of-contrast are gated by transient on-cells and off-cells before being combined into opponent pairs of orientation cells and direction cells whose output signals are independent of direction-of-contrast.

and off-cells (see Figure 2). Opponent cell pairs in the static BCS define *orientations* that differ by 90° as opposites, whereas opponent cell pairs in the motion BCS define *directions* that differ by 180° as opposites. Negative afterimages, such as circular afterimages of radial images and the waterfall illusion, are clarified by antagonistic rebound properties of these opponent pairs. These antagonistic rebounds act to rapidly reset previously active emergent segmentations, which could otherwise resonate for a long time after input offset or change, thereby causing massive smearing of percepts in response to rapidly changing images.

These results collectively argue against vision theories that espouse independent processing modules. Instead, specialized subsystems interact to overcome computational uncertainties and complementary deficiencies, and to realize symmetry principles that are

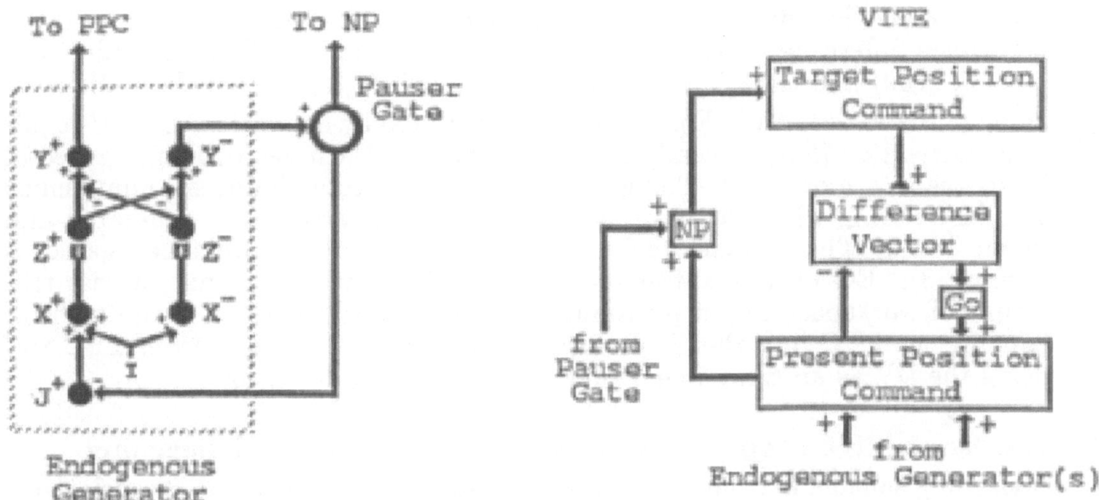

Figure 3. An AVITE circuit: The left column schematizes an endogenous generator of random training vectors composed of gated dipoles. The On channels (+ superscript) generate random unbiased training vectors, while the Off channels (-) activate the pauser gate.The right column shows how the endogenous generator influences VITE learning. The endogenous generator inputs random vectors to the Present Position Command (PPC), where they are integrated until the pauser gate is activated. Inputs to the On channels of the generator then terminate, the Now Print (NP) channel is inhibited, and the PPC is copied into the TPC. Then learning in the TPC→DV pathway zeroes the DV, and thereby adaptively calibrates TPC→DV signals to be computed in the same coordinates as PPC→DV signals. When the On channel transmitter gates recover, the cycle begins again.

predicted to govern the development of visual cortex.

2. A Self-Organizing Network for Adaptive Control of Movement Trajectories

The next part of the lecture will derive architectures for self-organization of sensory-motor coordination. These analyses clarify how a child, or infant robot, can learn to reach for objects that it sees. Piaget has provided basic insights with his concept of a *circular reaction.* As an infant makes internally generated movements of its hand, the eyes automatically follow this motion. A transformation is learned between the visual representation of hand position and the internal motor representation of hand position. Learning of this transformation eventually enables the child to accurately reach for visually detected targets. Grossberg and Kuperstein (1989) have shown how the eye movement system can make reactive movements in response to visual inputs. These movements generate visual error signals that are used to correct movement parameters via cerebellar learning. Gaudiano and Grossberg (1990a, 1990b) have shown how the arm movement system can endogenously generate movements which lead to adaptive tuning of arm control parameters. These movements also activate the target position representations that are used to learn the visuo-motor transformation that controls visually guided reaching.

These arm movement properties obtain in the AVITE model (Figure 3), an adaptive neural circuit based on the Vector Integration to Endpoint (VITE) model for arm and speech trajectory generation (Bullock and Grossberg, 1988a). In the VITE model, activation of a Target Position Command (TPC) represents the location of the desired target. The Present Position Command (PPC) encodes the present hand-arm configuration. The Difference Vector (DV) population continuously computes the difference between the PPC and the TPC. A speed-controlling GO signal multiplies DV output. The PPC integrates the (DV)·(GO) product and generates an outflow command to the arm. Integration at the PPC continues at a rate dependent on GO signal size until the DV reaches zero, indicating that the PPC equals the TPC. Several components of this circuit are in close accord with recent neurophysiological data: TPCs with parietal cortex, the GO signal with globus pallidus, and the difference vector (DV) with motor cortex. See Bullock and Grossberg (1988a, 1988b, 1989) and Grossberg and Kuperstein (1989) for further discussion.

Learning of the correct AVITE parameters is regulated by activation of a self-regulating Endogenous Generator (EG) of random training vectors. Each vector is integrated at the PPC, giving rise to a movement. The generation of each vector induces a complementary quiescent phase during which learning occurs. Then a new vector is generated and the cycle repeats itself. This biphasic behavior is controlled by a specialized gated dipole circuit (Figure 3). EG output autonomously stops in such a way that, across trials, a broad sample of workspace target positions is generated. When the EG shuts off, a gate opens, copying the PPC into the TPC. Learning of a transformation from TPC to PPC occurs using the DV as an error signal that is zeroed due to learning.

3. Vector Associative Maps: A New Model for Error-Based Learning

This learning scheme is called a Vector Associative Map, or VAM. Associative maps from other (e.g., visually activated) TPCs to the copied TPC are also learned using VAMs. The circular reaction hereby uses a random substrate to give rise to deterministic actions, enabling the AVITE circuit to learn parameters capable of generating trajectories for accurate reaching of visually-detected objects.

The VAM model is a general-purpose device for autonomous real-time error-based learning and performance of associative maps. The DV stage serves the dual functions of reading-out new TPCs during performance and reading-in new adaptive weights during learning, without a disruption of real-time operation. VAMs thus provide an on-line alternative to the off-line properties of back propagation for error-based learning.

References

Bullock, D. and Grossberg, S. (1988a). Psychological Review, **95**, 49–90.

Bullock, D. and Grossberg, S. (1988b). In J.A.S. Kelso, J.A. Mandell, and M.F. Shlesinger (Eds.), **Dynamic patterns in complex systems.** Singapore: World Scientific Publishers.

Bullock, D. and Grossberg, S. (1989). In W. Hershberger (Ed.), **Volitional action**. Amsterdam: Elsevier/North-Holland, 253–297.

Gaudiano, P. and Grossberg, S. (1990a). In M. Caudill (Ed.), **Proceedings of the international joint conference on neural networks, II**, 213–216. Hillsdale, NJ: Erlbaum.

Gaudiano, P. and Grossberg, S. (1990b). A self-organizing neural circuit for control of planned movement trajectories. In preparation.

Grossberg, S. (1990). Why do parallel cortical systems exist for the processing of static form and moving form? Submitted for publication.

Grossberg, S. and Kuperstein, M. (1989). **Neural dynamics of sensory-motor control: Expanded edition.** Elmsford, NY: Pergamon Press.

Grossberg, S. and Mingolla, E. (1985a). Psychological Review, **92**, 173–211.

Grossberg, S. and Mingolla, E. (1985b). Perception and Psychophysics, **38**, 141–171.

Grossberg, S. and Mingolla, E. (1990). In M. Caudill (Ed.), **Proceedings of the international joint conference on neural networks, I**, 11–14. Hillsdale, NJ: Erlbaum.

Grossberg, S. and Rudd, M. (1989). Neural Networks, **2**, 421–450.

ART 3: SELF-ORGANIZATION OF DISTRIBUTED PATTERN RECOGNITION CODES IN NEURAL NETWORK HIERARCHIES

Gail A. Carpenter and Stephen Grossberg
Center for Adaptive Systems, Boston University
111 Cummington Street, Boston, Massachusetts 02215 USA

ABSTRACT

Adaptive resonance architectures are neural networks that self-organize stable pattern recognition codes in real-time in response to arbitrary sequences of analog or binary input patterns. In ART architectures, top-down learned expectation and matching mechanisms are critical in self-stabilizing the code learning process. A parallel search scheme updates itself adaptively as the learning process unfolds, and realizes a form of real-time hypothesis discovery, testing, learning, and recognition. After learning self-stabilizes, the search process is automatically disengaged. Thereafter input patterns directly access their recognition codes without any search. A novel input pattern can directly access a category if it shares invariant properties with the set of familiar exemplars of that category. A parameter called the attentional vigilance parameter determines how fine the categories will be. If vigilance increases (decreases) due to environmental feedback, then the system automatically searches for and learns finer (coarser) recognition categories. Gain control parameters enable the architecture to suppress noise up to a prescribed level. This article outlines some properties of three generations of ART networks: ART 1, ART 2, and ART 3. ART 1 architectures establish recognition codes for binary inputs, ART 2 architectures for analog inputs. Learned representations are encoded in bottom-up and top-down adaptive filters whose long-term memory (LTM) traces vary slowly compared to the rapid short-term memory (STM) information processing. ART 3 architectures incorporate a third memory, on an intermediate time scale, whose dynamics may be interpreted as chemical transmitter processes. This medium-term memory (MTM) provides the extra degree of freedom needed to embed ART systems in neural network hierarchies with fast or slow learning and compressed or distributed codes. The ART 3 MTM transmitter processes can also be used in a wide variety of fully or partially connected and adaptive or non-adaptive neural networks.

Adaptive Resonance Theory (ART) emerged from an analysis of the instabilities inherent in feedforward adaptive coding structures (Grossberg, 1976a,b). More recent work has led to the development of three classes of ART neural network architectures, specified as systems of differential equations. The first class, ART 1, self-organizes recognition categories for arbitrary sequences of binary input patterns (Carpenter and Grossberg, 1987a). A second class, ART 2, does the same for either binary or analog inputs (Carpenter and Grossberg, 1987b). A third class, ART 3, solves computational problems of ART systems embedded in network hierarchies, where there can, in general, be either fast or slow learning and distributed or compressed code representations (Carpenter and Grossberg, 1990). ART 3 incorporates a search mechanism that serves at least four distinct functions: to correct erroneous category choices; to learn from reinforcement feedback or disconfirmed expectations; to respond to changing input patterns; and, when as error signal occurs, to amplify features that were previously ignored.

Acknowledgements: This research was supported in part by the Air Force Office of Scientific Research (AFOSR F49620-86-C-0037, AFOSR F49620-87-C-0018, and AFOSR 90-0128), the Army Research Office (ARO DAAL-03-88-K-0088), British Petroleum (89-A-1204), DARPA (AFOSR 90-0083), and the National Science Foundation (NSF DMS-90-00530 and IRI-87-16960). We wish to thank Diana Meyers, Cynthia Suchta, and Carol Yanakakis for their valuable assistance in the preparation of the manuscript.

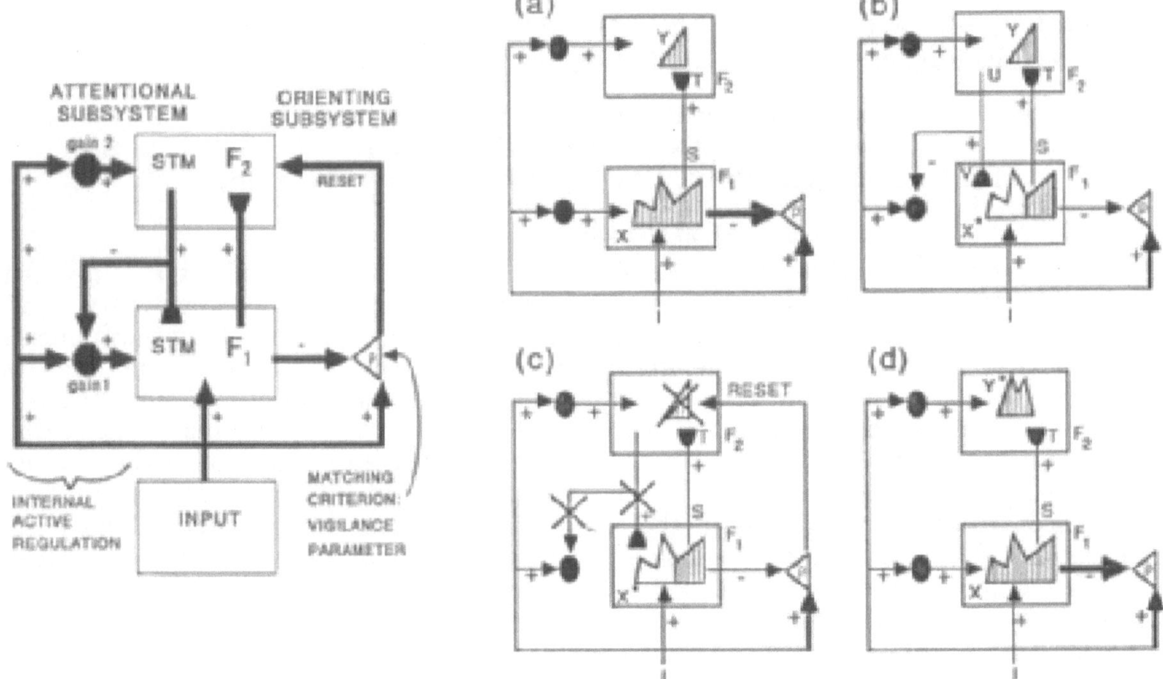

Figure 1 (left). Typical ART 1 neural network module (Carpenter and Grossberg, 1987a).
Figure 2 (right). An ART search cycle (Carpenter and Grossberg, 1987a).

The main elements of a typical ART 1 module are illustrated in Figure 1. F_1 and F_2 are fields of network nodes. An input is initially represented as a pattern of activity across the nodes, or feature detectors, of field F_1. The pattern of activity across F_2 corresponds to the category representation. The two fields, linked both bottom-up and top-down by adaptive filters, constitute the Attentional Subsystem. An auxiliary Orienting Subsystem becomes active during search.

Figure 2 illustrates a typical ART search cycle. An input pattern **I** registers itself as a pattern **X** of activity across F_1 (Figure 2a). The F_1 output signal vector **S** is then transmitted through the multiple converging and diverging weighted adaptive filter pathways emanating from F_1, sending a net input signal vector **T** to F_2. The internal competitive dynamics of F_2 contrast-enhance **T**. The F_2 activity vector **Y** therefore registers a compressed representation of the filtered $F_1 \rightarrow F_2$ input and corresponds to a category representation for the input active at F_1. Vector **Y** generates a signal vector **U** that is sent top-down through the second adaptive filter, giving rise to a net top-down signal vector **V** to F_1 (Figure 2b). F_1 now receives two input vectors, **I** and **V**. An ART system is designed to carry out a matching process whereby the original activity pattern **X** due to input pattern **I** may be modified by the *template pattern* **V** that is associated with the current active category. If **I** and **V** are not sufficiently similar according to a matching criterion established by a dimensionless *vigilance parameter* ρ, a reset signal quickly and enduringly shuts off the active category representation (Figure 2c), allowing a new category to become active. Search ensues (Figure 2d) until either an adequate match is made or a new category is established.

In ART 3, computational requirements of the ART search process can be fulfilled by formal properties of neurotransmitters, if these properties are appropriately embedded in the total architecture model. In particular, the ART 3 medium-term memory (MTM) equations incorporate the dynamics of production and release of a chemical transmitter substance; the inactivation of transmitter at postsynaptic binding sites; and the modulation of these processes via a nonspecific control signal. The net effect of these transmitter processes is to alter the ionic permeability at the postsynaptic membrane site, thus effecting excitation or inhibition of the postsynaptic cell. Specifically, the presynaptic signal, or action potential, S_i arrives at a synapse whose adaptive weight, or LTM trace, is denoted z_{ij}. The variable z_{ij} is identified with the maximum amount of available transmitter. When the transmitter at this synapse is fully accumulated, the amount of transmitter u_{ij} available for release is equal to z_{ij}. When a signal S_i arrives, transmitter is typically released. The variable v_{ij} denotes the amount of transmitter released into the extracellular space, a fraction of which is assumed to be bound at the postsynaptic cell surface and the remainder rendered ineffective in the extracellular space. Finally, x_j denotes the activity, or membrane potential, of the postsynaptic cell.

Initially the transmitted signal pattern $\mathbf{S}\cdot\mathbf{u}_j$, as well as the postsynaptic activity x_j, are proportional to the weighted signal pattern $\mathbf{S}\cdot\mathbf{z}_j$ of the linear filter. The activity pattern of the target field is then contrast-enhanced, due to the internal competitive dynamics. The primary ART 3 MTM hypothesis assumes that the transmitter release rate is greatly amplified in proportion to the level of postsynaptic activity. A subsequent reset signal may thus selectively inactivate those pathways that cause an error. Following such a reset wave, the new signal $\mathbf{S}\cdot\mathbf{u}_j$ is no longer proportional to $\mathbf{S}\cdot\mathbf{z}_j$ but is, rather, biased against the previously active representation due to transmitter depletion at those sites. A series of reset events ensue, until an adequate match or a new category is found. Learning occurs on a time scale that is long relative to that of the search process.

The ART 3 MTM serves other functions as well as implementing the ART mismatch-reset-search cycle. In particular it allows the neural network to dispense with special processes to reset STM at onset or offset of an input pattern. The representation of input patterns as a sequence, $\mathbf{I}_1, \mathbf{I}_2, \mathbf{I}_3, \ldots$, corresponds to the assumption that each input is constant for a fixed time interval. In practice, an input vector $\mathbf{I}(t)$ may vary continuously through time. The input need never be constant over an interval, and there may be no temporal marker to signal offset or onset of "an input pattern" per se.

The ART 3 MTM transmitter depletion process can also serve to enhance features that were previously ignored. For example, suppose that the input (I) signal pathways (Figure 2) contain an ART 3 MTM process, and that a reset signal is generated, say, by an internal system error or by an external teaching input. Features represented in I that were not salient in the matched STM pattern X are enhanced, due to depletion in pathways leading to those features that, for whatever reason, generated an error signal. For instance, the previously ignored color of an object may be brought forth to enhance discrimination between category exemplars.

The mechanisms described thusfar are part of the recognition learning circuit of ART 3. Recognition learning is, however, only one of several processes whereby an intelligent system can learn a correct solution to a problem. We have called Recognition, Reinforcement, and Recall the "3 R's" of neural network learning (Carpenter and Grossberg, 1988). Various types of reaction to reinforcement feedback may be useful in applications. For example, a change in vigilance alters the overall sensitivity of the system to pattern differences; a shift in attention and the reset of active features can help to overcome prior coding biases that may be maladaptive in novel contexts.

Figure 3 illustrates the results of an ART 2 simulation that demonstrates how varying vigilance during learning can lead to the stable coexistence of coarse and fine category groupings. Twenty inputs, each presented once at moderate vigilance level, establish 3 recognition categories (Figure 3a). Each of the 8 inputs in Category 2 is then presented at higher vigilance, establishing 4 new finer categories. When the original 20 inputs are then presented at low vigilance (Figure 3b), each input has direct access to a coarse category

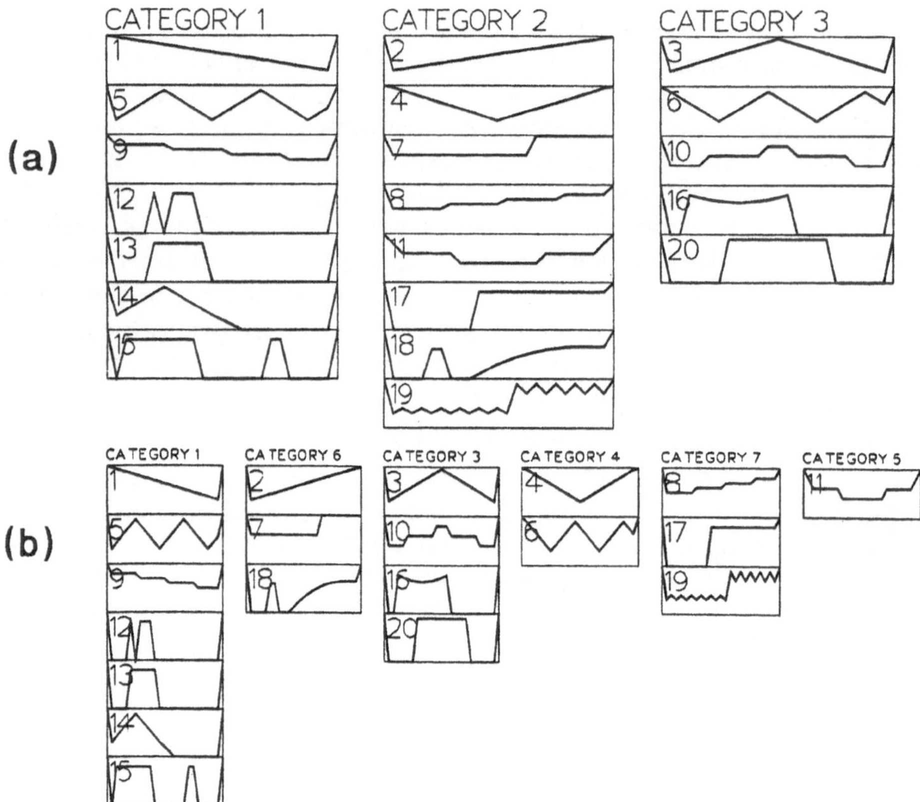

Figure 3. ART 2 simulation. (a) At moderate vigilance ($\rho = .88$), 20 inputs form 3 categories. The 8 inputs of Category 2 then form 4 new categories at higher vigilance ($\rho = .97$). (b) At low vigilance ($\rho = 0$), inputs have direct access to coarsely or finely grouped categories.

(1,3) or a fine category (4,5,6,7). The simulation demonstrates how a single ART system may establish both coarse and fine distinctions, depending on the learning history, with a non-uniform degree of discrimination across the set of inputs.

REFERENCES

Carpenter, G.A. and Grossberg, S. (1987a). A massively parallel architecture for a self-organizing neural pattern recognition machine. Computer Vision, Graphics, and Image Processing, **37**, 54–115.

Carpenter, G.A. and Grossberg, S. (1987b). ART 2: Self-organization of stable category recognition codes for analog input patterns. Applied Optics, **26**, 4919–4930.

Carpenter, G.A. and Grossberg, S. (1988). The ART of adaptive pattern recognition by a self-organizing neural network. Computer: Special issue on Artificial Neural Systems, **21**, 77–88.

Carpenter, G.A. and Grossberg, S. (1990). ART 3: Hierarchical search using chemical transmitters in self-organizing pattern recognition architectures. Neural Networks, in press.

Grossberg, S. (1976a). Adaptive pattern classification and universal recoding, I: Parallel development and coding of neural feature detectors. Biological Cybernetics, **23**, 121–134.

Grossberg, S. (1976b). Adaptive pattern classification and universal recoding, II: Feedback, expectation, olfaction, and illusions. Biological Cybernetics, **23**, 187–202.

Grossberg, S. (1982). **Studies of mind and brain: Neural principles of learning, perception, development, cognition, and motor control**. Boston: Reidel Press.

Grossberg, S. (Ed.) (1988). **Neural networks and natural intelligence**. Cambridge, MA: MIT Press.

APPLICATION OF GROSSBERG AND MINGOLLA NEURAL VISION MODEL TO SATELLITE WEATHER IMAGERY

Steve Lehar
Tim Howells
Ira Smotroff

MITRE-Bedford Neural Networks Group
The MITRE Corporation
Bedford, Massachusetts, USA 01730

ABSTRACT

Recent neural models of natural vision systems are defined in sufficiently concrete terms as to be immediately applicable to practical image processing tasks . In particular the Boundary Contour System and Feature Contour System human vision models developed by Grossberg and Mingolla, and Grossberg and Todorovicz are applied to satellite weather imagery to highlight significant features for the purpose of pattern recognition.

INTRODUCTION

Grossberg & Mingolla have made significant advances in correlating psychophysical findings with neural models in a manner consistent with neurophysiology . The Boundary Contour System (BCS) and Feature Contour System (FCS) models [3,4,5] are based primarily on psychophysical data on perceptual illusions under the rationale that a model that not only reproduces the functionality but also the failure modes of a system reflects a deeper commonality of architecture . The BCS / FCS combination explains a large body of psychophysical data, and the elements of the model correspond closely to neurophysiological data about the visual cortex,. We have developed an image processing implementation of the model and a modified implementation that has been used to segment satellite weather imagery.

BCS IMPLEMENTATION

On-center Off-surround Processing: The first stage of processing in the visual pathway is the on-center off-surround receptive field response of the ganglion cells of the retina . This we simulated using a convolving filter constructed of a difference of two Gaussian filters . Marr [6] shows that such a filter closely approximates the spatial second derivative operation when the inhibitory Gaussian has a standard deviation of 1.6 times that of the excitatory one . In the retina there are also off-center on-surround cells, but we dispensed with these by allowing our filters to produce both negative and positive responses . The product of this filtering serves as the input to both the BCS and FCS systems, which are postulated to occur in the primary visual cortex.

Oriented Edge Detection: Hubel and Wiesel have found evidence for oriented receptive fields in complex cells in the visual cortex which respond maximally to a light-to-dark edge at a particular orientation . Such responses can be accounted for by asymmetrical receptive fields from simple cells, excitatory in one direction and inhibitory in another . In our simulations we constructed convolving filters from the difference of two Gaussians of equal dimensions but with offset centers . We also made use of Gabor filters which are a product of a sinusoidal and an exponential term . Daugman [1] has shown that such filters bear a close similarity to receptive field profiles in the cortex . Grossberg notes that perceptual boundaries are insensitive to direction of contrast, so the BCS includes a stage for removing direction of contrast sensitivity . In our implementation we used six oriented filters to represent twelve orientations, taking the absolute value of the filter response for each orientation . This convolution operation produced six oriented response images which we represented in a three dimensional data structure with six planes for the orientations . Separation of the imagery into distinct orientation channels allows orientation specific processing to be performed on each channel separately before the information is then recombined into a multi-orientation representation . This is an essential aspect of the algorithm which allows for certain powerful operations which could not be performed on the image as a whole.

First Competitive Stage: An edge enhancement operation is performed on each orientation plane in the form of an on-center off-surround convolution . The effect of this step is to sharpen the edges of the oriented images, and corresponds to the first competitive stage of the BCS.

Second Competitive Stage: The next step in the model is a second competitive stage, wherein competition occurs at each spatial location between all the orientation channels . In our algorithm this was performed by an on-center off-surround convolution, this time not within the two-dimensional orientation planes, but rather in the third dimension between pixels in corresponding locations of the different orientation planes . The effect of this step is to produce a sharper tuning of the oriented filters . A vertical edge for instance will elicit some response in all of the oriented filters except the orthogonal one, the horizontal filter . The vertical filter will however produce the strongest response, so this response will be boosted at the expense of the others . Grossberg and Mingolla have shown that this operation is responsible for a large number of visual illusions, most notably orthogonal end-cuts in line terminations.

Oriented Cooperation: The next step is perhaps the most significant one in the BCS for the value that it adds to the final result . A cooperative operation is performed in each orientation channel to complete broken or incomplete edges . Data from visual illusions shows that such boundary completion does occur in human vision when two or more edges are seen to be parallel, aligned with one another, and in some proximity to each other, although such completion only occurs inward between edges, it does not extend outward beyond edges . In our image representation this is performed with a special cooperative filter operation . In each orientation plane, energy at any location represents an edge at that location, and all such points within that plane represent parallel edges . In the plane corresponding to vertical orientation for instance, any two such points which are vertically aligned represent edges on the retina which are both aligned and parallel . A two-armed vertical filter is convolved with this image plane, and a conjunction operation between the two arms produces a response in the filter only while it is straddled between such aligned points . Similar filters with corresponding orientations are applied to the other image planes, according to the equation:

$$F^{(r)}_{x,y} = \pm\, e^{-2\left(\frac{\sqrt{x^2+y^2}}{\rho}-1\right)^2} \cdot \cos\left(\left|a\tan\left(\tfrac{y}{x}\right)\right| - r\right)^P$$

where $F^{(r)}_{x\,y}$ is the filter value at x,y for orientation r, $\rho = 0.5$, and $P = 9$.

Feedback Loop: further sharpening is performed on the oriented images after oriented cooperation in the form of another on-center off-surround convolution . The resulting images are then recombined with the output of the oriented filtering to close a large feedback loop . This feedback allows the results of the oriented cooperation to contribute to the oriented competition of the second competitive stage to perhaps shift the emphasis between oriented responses where appropriate .. In our simulations all the operations described above were repeated two or three times, although the most significant changes were seen to occur in the first or second cycle.

FCS IMPLEMENTATION

The filling in function of the feature contour system begins with the same on-center off-surround ganglion cell image used by the BCS . This image encodes the intensity difference across boundaries, and is featureless in regions away from boundaries . The FCS allows the color to diffuse freely in all directions except in regions which correspond to strong boundaries in the BCS . The result is that enclosed boundaries tend to trap color, but within those boundaries the color reaches a dynamic equilibrium at a uniform value throughout the bounded region . The intensity in such regions is thus an average of the relative intensities around the entire perimeter . In this way the algorithm discounts the illuminant, since absolute illumination at the boundaries is ignored . Grossberg and Todorovicz [4] simulate the action of the FCS in reproducing visual illusions . In our simulations the FCS is implemented in two ways . The complete simulation performs a pixel-by-pixel diffusion for 30 iterations . This operation is extremely time consuming, and no color can flow farther than 30 pixels, although the result is a fairly close approximation to the FCS operation, at least for small or thin regions . In the accelerated algorithm the enclosed contours are identified instead by a standard blob segmentation algorithm . This is much faster although it does not preserve the intensity information of the bounded region . A hybrid operation is being perfected which is a combination of the two.

MODIFIED IMPLEMENTATION FOR WEATHER SATELLITE DATA

The Task: The images we are segmenting are satellite weather photographs, i.e . photographs of clouds as seen from space . These are well suited to the BCS approach because they include many subtle edges and boundaries which are clear to the eye despite the inherent indefiniteness of the image: Clouds do not provide very many clearly defined, regular boundaries.

Oriented On-center/Off-surround filters: The cooperative filters tend to blur lines as well as complete them, so after each cooperative step we applied the "difference of gaussian" filters described above to sharpen the lines up again . We ran into a problem here because the sharpening filters tend to work against the cooperative filters to some extent . The cooperative step would tend to complete boundaries, but then the following sharpening step would sometimes separate them again . Our solution was to replace the single symmetrical sharpening filter with oriented filters for each image plane. These new filters are scaled so that they sharpen strongly along the axis orthogonal to the orientation of the image plane, but not at all along the axis colinear with the orientation. These oriented filters don't interfere with the action of the cooperative filters.

Saturation Step: Initially we found that after several iterations of our feedback loop, virtually all of the energy in the images was receeding into a few points. This makes sense, considering that the filters tend to redistribute the energy, favoring areas that already have strong illumination: The strong get stronger and the weak get weaker. To an extent this kind of competition is precisely what we want - the stronger, more mutually consistent set of hypothesized lines should gather up energy at the expense of weaker hypotheses. It proved necessary, however, to prevent this competitive effect from running away. To this end we added a step to our feedback loop in which all image values are scaled according to the Grossberg shunting equation at equilibrium [2]

$$I = \frac{BI}{A+I}$$

where I is the pixel value, B is the maximum allowed value, and A is the decay constant. This has the effect of limiting the high values while boosting the lower values.

Implicit Competition : We omitted the "second competitive stage" which is intended to enhance the response of the oriented filters at or near their angle of orientation and weaken the responses to lines at other orientations. As noted above, an implicit form of competition occurs naturally during the feedback loop, and in fact it proved desirable to mitigate the "strong get stronger" effect.

RESULTS

Figure 1 shows a weather image of the east coast of the United States. Illumination is almost saturated at some points and barely exceeds detection threshold at others. The illuminant discounting properties of the algorithm serve to record the cloud patterns despite the extremes of illumination. The interesting meteorological features in this image are cloud masses and fronts. The boundaries of these features are strong in some areas and weaker in others, making it difficult for a conventional edge operator to strike a balance between missing significant weak edges and responding excessively to spurious signals. Figure 2 shows the sum total of BCS operations after three iterations of the algorithm. The bold sweeping lines illustrate how the discontinuous boundary features were completed across breaks, without introducing spurious edges. Figure 3 shows the result of featural filling in for 30 iterations. This figure illustrates how enclosed contours in the BCS image interact with direction of contrast information to trap pools of color. Note how the reconstructed image is clearer and sharper than the original, and with good contrast throughout. Figure 4 illustrates the fast alternative to the FCS processing. In this example the BCS image of figure 2 was thresholded to produce a binary contour image. A conventional blob segmentation algorithm was then applied to identify and label enclosed contours. This operation confirms the validity of using enclosed contours as a measure of the identity of image segments. This process has the additional advantage of uniquely labeling each image segment in preparation for further feature extraction.

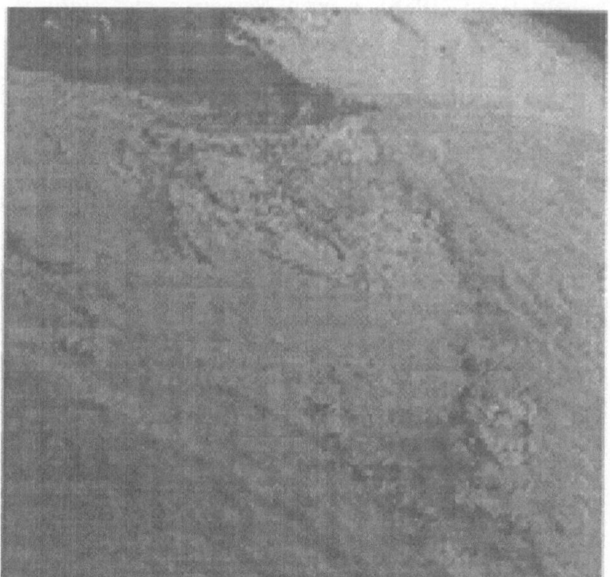

Figure 1: Original Cloud Image

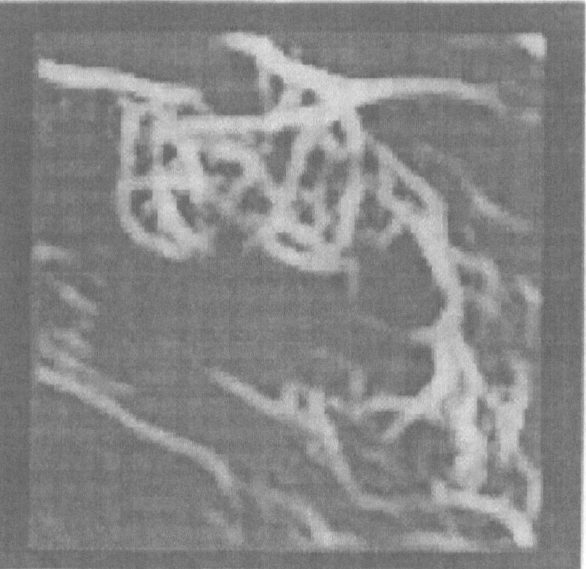

Figure 2: BCS Processed Cloud Image

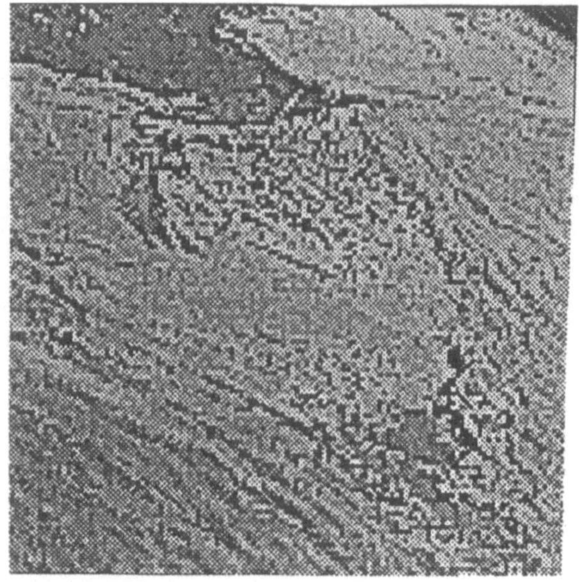

Figure 3: FCS Processed Cloud Image

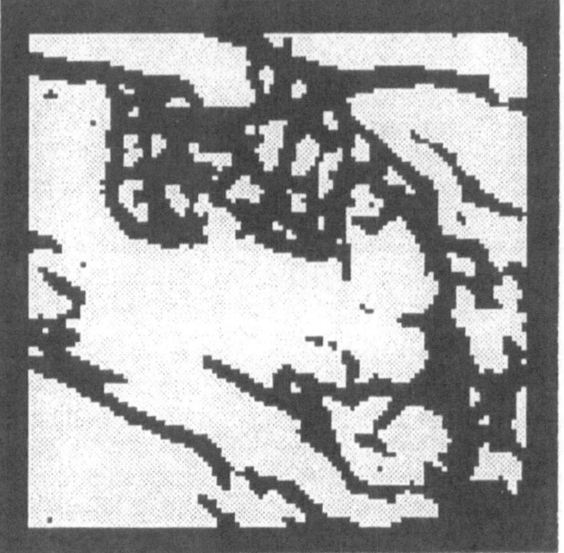

Figure 4: Blob-Segmented Cloud Image

REFERENCES

[1] Daugman, J. G. (1988). Complete Discrete 2-D Gabor Transforms by Neural Networks for Image Analysis and Compression.. *I.E.E.E. Trans. Acoustics, Speech, and Signal Processing* **36**(7), pp 1169-1179.

[2] Grossberg, S. (1968) Some Physiological and Biochemical Consequences of Psychological Postulates. In *Proceedings of the National Academy of Sciences* . Reprinted in Grossberg, S. (1982) *Studies of Mind and Brain*, Chapter 2. D. Reidel Publishing

[3] Grossberg, S. & E. Mingolla (1985). Neural Dynamics of Perceptual Grouping: Textures, Boundaries, and Emergent Segmentations. *Perception & Psychophysics* **38** (2), 141-171.

[4] Grossberg, S. & E. Mingolla (1987). Neural Dynamics of Surface Perception: Boundary Webs, Illuminants, and Shape-form-Shading. *Computer Vision, Graphics and Image Processing* **37**, 116-165.

[5] Grossberg, S. & D. Todorovic (1988). Neural Dynamics of 2-D and 1-D Brightness Perception. *Perception and Psychophysics* **43**, 241-277. Reprinted in Stephen Grossberg (Ed.) (1988), *Neural Networks and Natural Intelligence*, Chapter 3. Cambridge, MA: MIT Press

[6] Marr, D. (1982). *Vision*, Freeman & Co.

REPRESENTATION OF UNCERTAINTY IN SELF-ORGANIZING NEURAL NETWORKS

JONATHAN A. MARSHALL

Center for Research in Learning, Perception, and Cognition
and Minnesota Supercomputer Institute

205 Elliott Hall, University of Minnesota, Minneapolis, MN 55455, U.S.A.

Abstract

The ability to represent multiple hypotheses about the classification of an ambiguous input pattern is an extremely useful network property. Yet typically, self-organizing neural networks are designed to make only winner-take-all pattern classification decisions. The winner-take-all decisions may often turn out to be wrong when they are based on insufficient or ambiguous input data. In a self-organizing neural network, incorrect decisions can impair or prevent the development of efficient codes for classifying the input environment. A new method is described which allows a neural network to maintain a representation of its own uncertainty. A key benefit of the new technique is that incorrect learning is avoided and efficient representation structures can thereby self-organize. A new anti-Hebbian *inhibitory* learning rule is combined with a variant of a Hebbian excitatory learning rule. Inhibitory learning permits the superposition of *multiple* simultaneous partial neural activations, under strictly regulated circumstances. An ambiguous input pattern then partially activates a set of neurons, each of which represents a hypothesis about the pattern's likely classification. If disambiguating pattern information is subsequently added, only the neuron that codes the most likely hypothesis becomes fully active; the activations of other neurons are suppressed. Efficient pattern codes can then self-organize in neural networks exposed to input environments which, like our own perceptual world, contain a great deal of ambiguity.

Introduction: Wrong Choices Lead to Wrong Learning

Perceptual uncertainty arises from a variety of sources, both internal and external. For example, an out-of-focus or distant image of a person's face might look like one of many familiar faces; its identification or classification would thus be uncertain. In neural networks, uncertainty occurs when an incomplete, noisy, or ambiguous incoming signal pattern has more than one likely classification.

Self-organizing neural networks have often been designed to operate in a winner-take-all fashion. That is, only one neuron or a small cluster of neurons has been allowed to be active at any moment, because of heavy lateral inhibition. Such networks make a definite choice (i.e., a single winner) about the classification of an input pattern, even in the presence of uncertainty.

In complex, real-world environments, perceptual information is often initially ambiguous. Perceptual uncertainty can be resolved in such cases by the subsequent addition of disambiguating information. For instance, in vision, a monocular image is generated by the projection of a 3-D scene onto a 2-D retina. A given feature in the image could have been generated by an object at any depth in the scene. Until further information is added, the visual system cannot necessarily determine the feature's depth. Such additional information can be provided by motion parallax transformations, top-down size familiarity, or other cues. The addition of depth cues enables the visual system to resolve uncertainty about the object's depth.

If a network makes a winner-take-all choice about the classification of an ambiguous input pattern, its decision may subsequently turn out to be wrong, as disambiguating information is added. Activation of an incorrect classifier neuron may not, by itself, be bad. After all, the network could just pick a winner, and then if subsequent information warrants, change its earlier decision. However, the network's *learning* is a function of neuron activations. Because the operation of Hebbian-type excitatory learning rules (Grossberg, 1976; Hebb, 1949) is based on correlations in neuron activity, the wrong choices of active neurons could lead to wrong learning, thereby impairing or even preventing the development of stable pattern codes.

Ambiguous environments can thus seriously undermine the adaptive development of winner-take-all networks. One method by which the wrong-winner problem may be solved is to allow the network to maintain a representation of its own uncertainty.

Classification of Ambiguous Patterns

FIGURE 1 schematically illustrates the main problem. Initially, if too little inhibition were present between a set of classifying neurons, then an input pattern would activate multiple neurons (FIGURE 1A). Each active neuron would then learn the *same* input pattern (FIGURE 1B), thereby defeating the network's purpose as a self-organizing classifier. Because each neuron ought to acquire a *different* sensitivity, the inhibition strengths between the classifying neurons need to be high enough (at least initially) so that slight differences in neuron input levels would result in great differences in neuron activations (FIGURE 1C). Over a sufficient number of exposures to input patterns, each neuron would tend to acquire a different sensitivity (Grossberg, 1976; Kohonen, 1984) (FIGURE 1D).

However, if an initially ambiguous pattern (e.g., the intersection of two familiar patterns) is then presented (FIGURE 1E), the strong inhibition would force the network to make an immediate choice (which may later turn out to be wrong) of a single active neuron. As learning proceeds, the development of the proper connection strengths could be disrupted (FIGURE 1F). Furthermore, if additional disambiguating parts of the input pattern subsequently become available, then the correct classifier neuron might be unable to overcome feedback inhibition from an incorrect neuron (FIGURE 1G). Thus, pattern ambiguity can cause *instability* in the structure of Hebbian-type winner-take-all networks.

How should the network respond to ambiguous input patterns so that its behavior and structure remain stable? The problem stems from the winner-take-all nature of the network. The problem might be solved if more than one neuron were allowed to remain active in response to an ambiguous input, for instance by reducing the amount of inhibition (FIGURE 1H). Then all possible correct classifications for the input could be maintained. The simultaneous activity of multiple neurons constitutes a representation of the network's uncertainty about the correct classification of the input pattern. Subsequent disambiguating information could enhance the activity of a single correct classifying neuron, which in turn would more strongly inhibit alternate classifying neurons (FIGURE 1I).

Weakening Inhibition to Permit Coactivation

How can the nearly identical neurons in a nonspecific network become differentiated, in the presence of ambiguous input patterns? Can the conflicting requirements for both strong inhibition (to produce selectivity) and weak inhibition (to allow coactivation) be reconciled? Yes: the trick is to notice that strong inhibition is mainly needed only at the outset of the network's development, to ensure that all neurons do not respond to all input patterns. Afterward, according to the excitatory learning rule, the neurons' input pattern sensitivities become incorporated into the *excitatory* connection weights. Thereafter, less inhibition is needed, because a given input pattern would tend not to fully coactivate the neurons anyway.

A new *inhibitory* learning rule permits the network to choose appropriate intermediate levels of inhibition and thereby represent uncertainty via coactivation of multiple neurons. The inhibitory learning rule (Marshall, 1988, 1989, 1990ab) is similar to Hebbian-type excitatory rules: ***Whenever a neuron is active, its output inhibitory connections to other active neurons become stronger; its output inhibitory connections to inactive neurons become weaker.*** More formally, let the quantity z_{ji}^{-} represent the strength of the inhibitory connection from the jth neuron to the ith neuron. Then

$$\frac{d}{dt} z_{ji}^{-} = \delta\, g(x_j)\left(-z_{ji}^{-} + V q(x_i)\right),$$

where x_j represents the activity level of the jth neuron, $0 < \delta << 1$, and g and q are increasing functions. The parameter V governs the overall amount of coactivation permitted in the network. Thus, if two neurons, i and j, tend to be coactivated, then the amount of inhibition between them tends to *increase* (Easton & Gordon, 1984; Földiák, 1989; Kohonen, 1984), thereby making them less likely to become coactivated in the future. If two neurons tend *not* to be coactivated, then the amount of inhibition between them tends to decrease, so that they can become coactivated on relatively rare occasions – such as during brief periods of ambiguity. This rule is a reverse of inhibitory learning rules proposed previously (Amari & Takeuchi, 1978; Nagano & Kurata, 1981), in which coactivation results in a *decrease* of inhibition strength.

Suppose initially the reciprocal inhibitory connection strengths between two neurons are quite high. Then even if both neurons have similar excitatory input connections, only one of the neurons can, in general, become active in response to an input pattern. Thus, the two neurons are unlikely to become coactivated. Hence, according to the inhibitory learning rule, their reciprocal inhibitory connection strengths are likely to weaken gradually. Meanwhile, each of the two neurons acquires a different excitatory in-

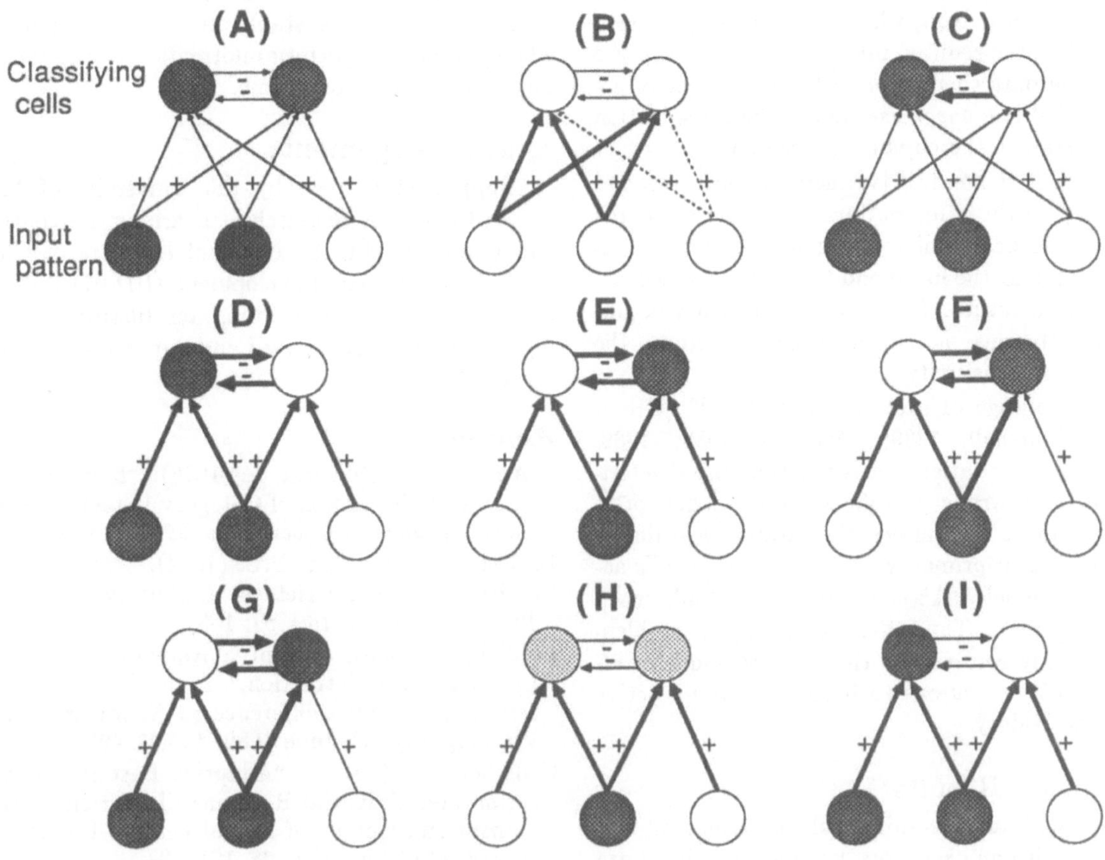

FIGURE 1. (A) An input pattern (shaded lower circles) excites a layer of classifying neurons (upper circles). The classifying neurons are initially nonspecific – i.e., they all receive the same connections from the input layer. Inhibition between classifying neurons is weak (thin horizontal arrows), so that all such neurons become active (shading) in response to the input pattern. (B) Learning would then cause the excitatory connections from the active inputs to the active neurons to strengthen (thick arrows), while the excitatory connections from the inactive inputs weaken (dotted arrows). Because the classifying neurons were both active, they both learned the same input pattern. (C) If inhibition is strong (thick horizontal arrows), then only one classifying neuron can become active at a time. (D) Each neuron can then acquire its own sensitivities after repeated exposure to input patterns. (E) But then, if an ambiguous input pattern is presented (could be in either category), only one neuron can respond. (F) This can lead to unwanted distortions of connection strengths (thin and thicker arrows). (G) Even if the input pattern is disambiguated by subsequent additional information, the correct classifier neuron may be unable to overcome hysteresis from the incorrect neuron's activation. (H) But if inhibition strength is then reduced (thinner arrows), multiple neurons could simultaneously respond to the ambiguous input pattern, albeit possibly at a lower activation level (lighter shading). (I) Then the new disambiguating information could cause the correct neuron to win, suppressing incorrect classifications.

put connection profile and begins to respond to a different pattern. Now, when an ambiguous pattern is presented, the reduced inhibition between the two neurons permits them *both* to become partially activated. Multiple *hypotheses* about the classification of a pattern can thereby be represented.

On the other hand, if two neurons both respond frequently to the same pattern, then the strengths of their reciprocal inhibitory connections tend to increase, reducing the likelihood that they can be coactivated in the future. The network's efficiency is thus promoted, because no two neurons can acquire the same pattern sensitivity.

One advantage of such a "multiplexed" (Grossberg & Marshall, 1989; Marshall, 1988, 1989, 1990ab) representation scheme is that the deleterious effects of wrong classifications on a network's learning can be minimized. If excitatory learning is allowed to occur primarily when a neuron is *fully* active, then the network's structure changes only when fully warranted. Thus, the wrong-winner problem may be easily resolved by the new technique of allowing multiple neurons to become partially active under uncertainty.

Simulation Results

The inhibitory learning rule described above, combined with an excitatory learning rule, has been implemented and applied successfully to problems in visual motion ambiguity. Simulations using the new rule allow networks to self-organize to disambiguate aspects of the *aperture problem* (Marshall, 1988, 1990a) and to disambiguate object motion and egomotion in computing visual depth from motion parallax (Marshall, 1989).

The new inhibitory learning rule has also been implemented in a network simulation in combination with both an excitatory learning rule and a neuron-growth rule (Marshall, 1990b). Besides being able to represent uncertainty, the network also self-organizes efficiently in response to overlapping input patterns and to patterns of different spatial scales. Pattern ambiguity in the simulation is again represented by simultaneous partial activation of the neurons that most closely represent the pattern.

Conclusions

The technique of adding an inhibitory learning rule to a Hebbian-type network can be applied quite generally, for self-organization in many kinds of complex environments. Using an inhibitory learning rule in conjunction with a standard excitatory rule can permit greater flexibility in representing both uncertainty and decision in pattern classification tasks. The new inhibitory rule helps the process of self-organization operate stably in realistic situations, where ambiguous pattern information becomes completed and refined over time.

Acknowledgements

Supported in part by the University of Minnesota Center for Research in Learning, Perception, and Cognition, by the National Institute of Child Health and Human Development (HD-07151), and by the Minnesota Supercomputer Institute (Visiting Research Scholar award and Supercomputer Resource Grant).

References

Amari, S. & Takeuchi, A. (1978). "Mathematical Theory on Formation of Category Detecting Nerve Cells." *Biological Cybernetics, 29*, 127–136.

Easton, P. & Gordon, P.E. (1984). "Stabilization of Hebbian Neural Nets by Inhibitory Learning." *Biological Cybernetics, 51*, 1–9.

Földiák, P. (1989). "Adaptive Network for Optimal Linear Feature Extraction." *Proceedings of the International Joint Conference on Neural Networks,* Washington, DC, June 1989, *I.*, 401–405.

Grossberg, S. (1976). "Adaptive Pattern Classification and Universal Recoding: I. Parallel Development and Coding of Neural Feature Detectors." *Biological Cybernetics, 23*, 121–134.

Grossberg, S. & Marshall, J.A. (1989). "Stereo Boundary Fusion by Cortical Complex Cells: A System of Maps, Filters, and Feedback Networks for Multiplexing Distributed Data." *Neural Networks, 2*, 29–51.

Hebb, D.O. (1949). *The Organization of Behavior.* New York: Wiley.

Kohonen, T. (1984). *Self-Organization and Associative Memory.* New York: Springer-Verlag.

Marshall, J.A. (1988). "Self-Organizing Neural Networks for Perception of Visual Motion." Technical Report 88-010, Boston University Computer Science Department.

Marshall, J.A. (1989). "Self-Organizing Neural Network Architectures for Computing Visual Depth from Motion Parallax." *Proceedings of the International Joint Conference on Neural Networks*, Washington DC, June 1989, *II.*, 227–234.

Marshall, J.A. (1990a). "Self-Organizing Neural Networks for Perception of Visual Motion." *Neural Networks*, In press.

Marshall, J.A. (1990b). "A Self-Organizing Scale-Sensitive Neural Network." UMSI Technical Report, Minnesota Supercomputer Institute.

Nagano, T. & Kurata, K. (1981). "A Self-Organizing Neural Network Model for the Development of Complex Cells." *Biological Cybernetics, 40*, 195–200.

IMPROVING THE LEARNING SPEED IN TOPOLOGICAL MAPS OF PATTERNS

Joaquim S. Rodrigues* and Luis B. Almeida*
INESC, R. Alves Redol, 9, 1000 Lisbon, Portugal

Abstract

Topological Maps of Patterns are a very powerful neural network paradigm. However, the learning method originally proposed by T. Kohonen for these maps, is a very time-consuming process. In this paper we propose a method of improving the learning speed, by starting the map with very few units and increasing that number progressively until the map reaches its final size. When the number of units increases, the locations of the new units are interpolated from the locations of the old units. The use of this method dramatically reduces the time needed for the "unfolding" phase and also yields some improvements in the asymptotic convergence phase. The improvements observed in this second phase can vary from marginal improvements for small sized networks, to very significant improvements for large networks.

Introduction

Topological Maps of Patterns were originally proposed by T. Kohonen as a means to represent complex empirical data by a self organizing network [1]. This self-organization is performed by moving all the units that belong to the topological neighbourhood of the unit that is closer (in some pre-defined sense) to the input vector, in a direction that will decrease their distance to that vector. The step size that determines the magnitude of that movement, is decreased over time to ensure asymptotic stability.

There are two phases in the formation of a map: one in which the map "unfolds" itself so as to situate the units in the correct order, and a second one in which the statistical distribution of the units in the map, will asymptotically approach the statistical distribution of the input vectors. To accelerate the "unfolding" phase it is convenient to start with a large neighbourhood that is then decreased over time. The second phase usually takes much more time than the first one.

The process of forming a map is a slow one, especially when the size of the map is not trivially small. There are two distinct factors that contribute to that slowness of convergence:

- the first one is the time required to perform each **step**, i.e., the processing of each input vector. The main part of this time is spent in the search for the unit that is closer to that vector. That search is done once for every input vector and, to be optimal, it must be an exhaustive search among all the units of the map. A method that tries to improve that search in quasi-organized maps was recently proposed by Koikkalainen et al [2].

- the second one is the number of steps needed for a map to be formed. In this paper, we will present a method of improving the learning speed, by reducing both the number of steps needed to form a map and the average time required to process each input vector.

---

* Also with the *Instituto Superior Técnico*, Lisbon, Portugal

The Improved Learning Method

The new strategy consists in starting the map with very few units and increasing that number progressively until the map reaches its final size. When the number of units increases, the locations (i.e., the internal states, s_i) of the new units are interpolated from the locations of the old units, in such a way that, if the map was already quasi-organized, it remains quasi-organized.

Fig. 1 exemplifies the new convergence method for a bidimensional map. The input data are formed by bidimensional vectors representing points uniformly distributed inside a rectangle. The map was started with 3x3 units with random locations (internal states), Fig.1a. After 1000 steps (i.e., processing of 1000 input vectors) the map was already organized (Fig. 1b). At this point the number of units of the map was increased by interpolation. The locations of the new 6x6 units are shown in Fig. 1c. This process is then repeated: 1000 presentations of input vectors (Fig. 1d); interpolation to 12x12 units (Fig. 1e); presentation of 1000 input vectors (Fig. 1f).

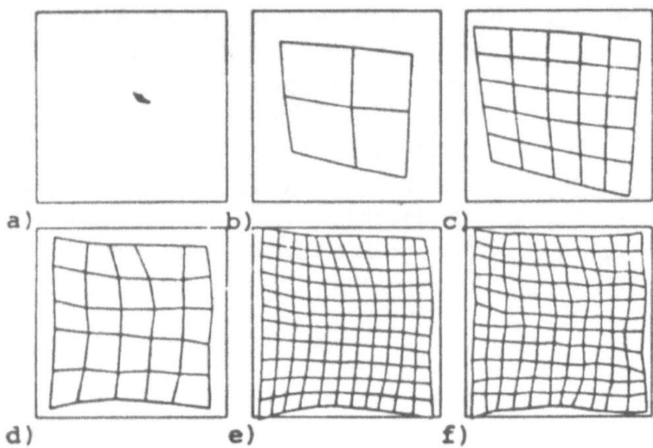

Fig. 1 - Example of the new map formation process (see text).

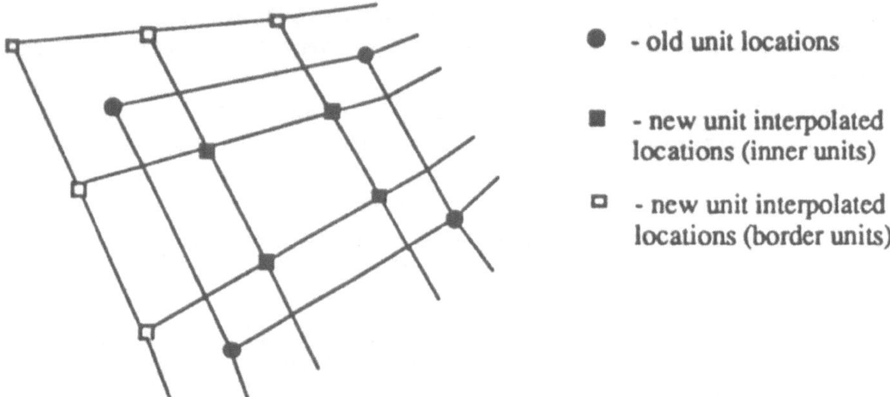

Fig. 2 - Example of the interpolation procedure for bidimensional maps.

The interpolation procedure is illustrated in Fig. 2 for a bidimensional map. Consider that we wish to interpolate from an LxL map to a 2Lx2L one. Call s_{ij} the internal state of the unit located in the position ij of the lattice before interpolation, and s'_{ij} the internal state of the unit located in the position ij of the lattice after interpolation. We use for the inner units (i.e., units that are not on the border of the map):

$$
\begin{aligned}
s'_{2i,2j} &= {}^1/_{16} \cdot (9\, s_{i,j} + 3\, s_{i,j+1} + 3\, s_{i+1,j} + s_{i+1,j+1}) \\
s'_{2i,2j+1} &= {}^1/_{16} \cdot (3\, s_{i,j} + 9\, s_{i,j+1} + s_{i+1,j} + 3\, s_{i+1,j+1}) \qquad 0<i,j<L \\
s'_{2i+1,2j} &= {}^1/_{16} \cdot (3\, s_{i,j} + s_{i,j+1} + 9\, s_{i+1,j} + 3\, s_{i+1,j+1}) \\
s'_{2i+1,2j+1} &= {}^1/_{16} \cdot (s_{i,j} + 3\, s_{i,j+1} + 3\, s_{i+1,j} + 9\, s_{i+1,j+1})
\end{aligned}
$$

The interpolation relations for the border units can be easily derived (see Fig. 2).For N-dimensional maps similar relations could be derived following the same idea.

Another difference of the new paradigm concerns the size of the topological neighbourhoods. In the classical learning paradigm the size of the topological neighbourhood is supposed to decrease according to some heuristic law in order to facilitate the "unfolding" of the map. The choice of that heuristic, although not critical in general for the asymptotic convergence phase, is essential to allow the "unfolding" to be performed in useful time. In the new method, on the contrary, we use a fixed size topological neighbourhood (a square of 3x3 units centred in the selected unit). Only near the end of the convergence process, when the map has already reached its final size, the units can be freed (the topological neighbourhood is restricted to the selected unit itself). The rationale behind this is that for small sized networks the 3x3 neighbourhood is sufficiently large for unfolding it and, when the size of the map increases, as the units get closer to their final locations, the 3x3 neighbourhood becomes sufficiently small to allow the asymptotic convergence to near the final state, ensuring at the same time the organization of the map.

In order for the improved method to achieve its best results, it is important to eliminate the border effects observed in the classical learning methods. The procedure to achieve this cannot be described here due to lack of space (see [3]).

Evaluation Tests

The evaluation of the improved learning method was performed with two distinct perspectives: finding how it behaved facing input data with different statistical distributions and compare its performance with the traditional learning method.

We will only describe one series of tests due to lack of space. More exhaustive results can be found in [3]. This series of tests was performed using a 48x48 map and input data with a distribution obtained by the superposition of 10 different gaussian distributions (Fig. 3) (this was a case in which the improvements were smallest among those tested). The training and the test sets were the same for all tests, and were composed by 10.000 input patterns each. The test set was distinct from the training one but with the same statistical distribution. The "degree of organization" was measured by averaging, for all the patterns in the test set, the distance to the closest unit in the learned map. The initial step sizes used were 0.04 (new method) and 0.05 (traditional method), which were decreased twice (when 1/3 and 2/3 of the input vectors were processed) by a factor of 1.42 (new method) and 2.0 (traditional method). These values were chosen because they were the ones that had produced the best results in preliminary tests.

In the tests performed with the new method, the maps were started with 3x3 units. The number of units was increased when 1/10, 2/10, 3/10, ..., of the total number of input patterns were processed, until the map reached its final size of 48x48 units. After that, when half (5/10) of the input patterns had been processed, the topological neighbourhood, that had been kept constant (3x3 units centred in the selected unit), was reduced to the selected unit itself.

In the tests performed with the traditional method, the net always kept the size 48x48, and the topological neighbourhood was started with 31x31 units centred in the selected unit. The size of that neighbourhood was reduced to 13x13, 7x7, 5x5, 3x3 and 1x1 when respectively 1/4, 1/3, 1/2, 2/3 and 4/5 of the total number of input patterns had been processed.

The convergence time (i.e., the CPU time in one processor of an Alliant FX/8 computer) is illustrated in Fig. 4. The total number of steps varies from 5.000 steps (both methods) to 900.000 steps (traditional method) and 1.500.000 steps (new method). Note that the tests with the larger number of steps, for both methods, used approximately the same amount of CPU time. The new method yields an improvement in the learning speed, of about one order of magnitude, for very well organized maps (average distance below 6.0). For less organized maps the improvements are even larger as can be seen from Fig.4.

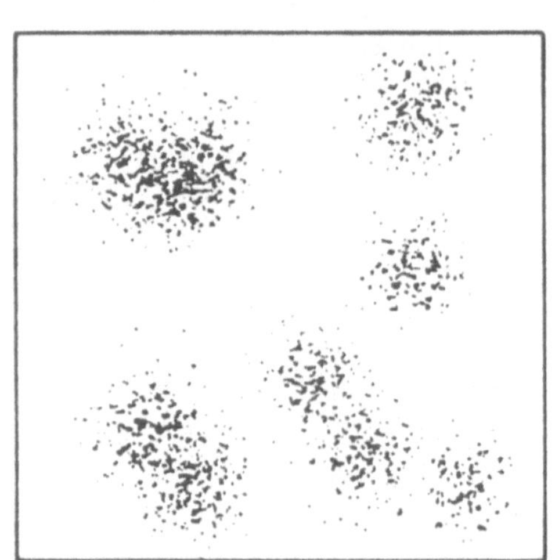

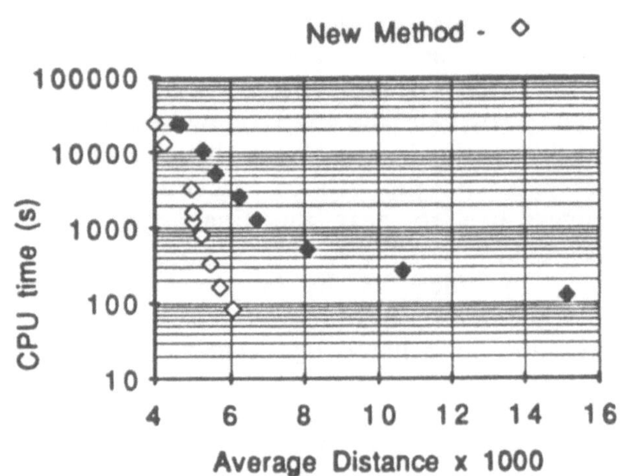

Fig. 3 - Distribution of input vectors. Fig. 4 - Convergence time versus average distance (degree of organization).

In the few tests made to evaluate the improvements in larger networks, 192x192 and 384x384, the gain in learning speed was, as expected, even bigger. Even for smaller networks, where this gain is not so big, the map attains a quasi-organized form much sooner, compared with the time it would take with the traditional training method. This fact has significant advantages, allowing for instance, the use of the technique proposed by Koikkalainen et al[2], very early in the convergence process.

Conclusion

We have proposed a new version of the learning method for Topological Maps of Patterns, that improves the learning speed more than one order of magnitude. This new method consists of starting the map with very few units and progressively increasing that number by interpolation. This technique is very efficient in unfolding the maps, and also yields significative improvements in the asymptotic convergence phase.

References

[1] Kohonen, T. "Clustering, Taxonomy, and Topological Maps of Patterns", Proc. 6th Annual Conf. on Pattern Recognition, pp 114-128, Los Angeles, 1982.
[2] Koikkalainen, P., Lampinen, J., and Oja, E. "Fast Implementations for the Kohonen Self-Organizing Learning Algorithm", Neuro-Computing: Algorithms, Architectures and Applications, F. Fogelman-Soulié (ed.), Springer-Verlag, Berlin, 1989.
[3] Rodrigues, J.S. and Almeida. L.B. "Improving the Learning Speed in Topological Maps of Patterns", INESC Internal Report, Jan. 1990.

Reinforcement Learning with Interacting Continually Running Fully Recurrent Networks

Jürgen Schmidhuber*
Institut für Informatik
Technische Universität München
Arcisstr. 21, 8000 München 2, Germany
schmidhu@tumult.informatik.tu-muenchen.de

Abstract

We describe an on-line learning algorithm for attacking the fundamental credit assignment problem in non-stationary reactive environments. Reinforcement and pain are considered as special types of input to an agent living in the environment. The agent's only goal is to maximize cumulative reinforcement and to minimize cumulative pain. This simple goal may require to produce complicated action sequences. Supervised learning techniques for recurrent networks serve to construct a differentiable model of the environmental dynamics which includes a model of future reinforcement. While this model is adapted, it is concurrently used for learning goal directed behavior. The method extends work done by Munro, Robinson and Fallside, Werbos, Widrow, and Jordan.

Introduction

Consider an agent whose movements are controlled by the output units of a neural network, called the control network, which also receives the agent's sensory perception by means of its input units. The agent potentially is able to produce actions that may change the environmental input (external feedback caused by the 'reactive' environment). By means of recurrent connections in the network the agent is also potentially able to internally represent past events (internal feedback).

The agent sometimes experiences different types of negative reinforcement or 'pain' by means of so-called *reinforcement units* or *pain units* that become activated in moments of 'pain' (e.g. the experience of bumping against an obstacle with an extremity). The agent's only goal is to minimize cumulative pain. The agent is autonomous in the sense that no intelligent external teacher is required to provide additional goals or subgoals for it.

A pain unit is treated as a special type of input unit which possesses conventional outgoing connections to other units. Unlike normal input units pain units can have desired activation values at every time. For the purpose of this paper we say that the desireable activation of a pain unit is zero for all times. In the sequel we assume a discrete time environment with 'time ticks'. At a given time the quantity to be minimized is $\sum_{t,i} y_i(t)$ where $y_i(t)$ is the activation of the ith pain unit at time t, and t ranges over all remaining time ticks still to come.

Pain corresponds to negative reinforcement. The reinforcement learning agent faces a very general spatio-temporal credit assignment task: No external teacher provides knowledge about e.g. desired outputs or 'episode boundaries'. In this paper we demonstrate how the agent can employ a combination of two recurrent *self-supervised* learning networks in order to satisfy its goal.

As Munro [3] has pointed out in the case of stationary environments and feedforward networks, one does not necessarily have to employ a 'pure' reinforcement learning algorithm for reinforcement learning. ('Pure' reinforcement learning algorithms (or *reinforcement comparison* algorithms) for temporal credit assignment in non-stationary environments have been described in [1], [11], [7] and [8].) A *supervised* learning algorithm can be applied to build a model of the relationships between environmental inputs, output actions of the

*This work was supported by a scholarship from SIEMENS AG

agent, and corresponding reinforcement. An adaptive model network representing the model can be used to propagate gradient information back into the control network in order to maximize reinforcement.

Robinson and Fallside described an extension of Munro's static approach to dynamic recurrent networks in time-varying environments [6]. (Nguyen and Widrow [4], Jordan [2], and Werbos [10] also use model networks for constructing a mapping from output actions of a control network to their effects in in 'task space' [2]. The same principle as used in Munro's work serves to provide error signals for the control network, in order to improve performance on a given control task.)

As in Munro's approach, the only aspect of the external world which is explicitly described by Robinson and Fallside's recurrent model network is the reinforcement's dependency on past inputs and outputs. There is no model for the dependency of (non-reinforcement) inputs on past outputs (or on past inputs which again may have been caused by past outputs). This makes the model for the reinforcement itself incomplete: Paths for credit assignment leading 'through the environment' can not be considered.

The system described in the next section (see also [9]) employs an adaptive model of the environmental dynamics for computing gradients of the control network's pain. Both the control network and the model network are fully recurrent.

Unlike Robinson and Fallside's approach our approach includes credit assignment passes that lead from pain units back to output units back to all input units and so on. There are also credit assignment paths that lead from input units back to the input units themselves, and from there to the output units. The latter paths are important in the common case when the environment can change even if there are no recent output actions.

The On-Line Algorithm

The discrete time algorithm below *concurrently* adjusts the fully recurrent model network and the fully recurrent control network. Williams and Zipser's on-line version [12] of Robinson and Fallside's Infinite-Input-Duration learning algorithm for fully recurrent networks [5] is used for training both networks. The algorithm is a particular instantiation of a more general form and is based on the logistic activation function for all non-input units. Notation (the reader may find it convenient to compare with [12]):

C is the set of all units of the control network, A is the set of its output units, I is the set of its 'normal' input units, P is the set of its pain units, M is the set of all units of the model network, O is the set of its output units, $O_P \subset O$ is the set of all units that predict pain, $H = M \cup C \backslash (I \cup P)$, W_M is the set of variables for the weights of the model network, W_C is the set of variables for the weights of the control network, $y_{k_{new}}$ is the variable for the updated activation of the kth unit from $M \cup C$, $y_{k_{old}}$ is the variable for the last value of $y_{k_{new}}$, w_{ij} is the variable for the weight of the directed connection from unit j to unit i, $p^k_{ij_{new}}$ is the variable which gives the current (approximated) value of $\frac{\partial y_{k_{new}}}{\partial w_{ij}}$, $p^k_{ij_{old}}$ is the variable which gives the last value of $p^k_{ij_{new}}$, α_C is a positive constant, the learning rate for the control network, α_M is a positive constant, the learning rate for the model network.

$| I \cup P | = | O |$, $| O_P | = | P |$. For each $k \in O \backslash O_P$ there is exactly one $i \in I$ such that $y_{k_{new}}$ predicts the value of $y_{i_{new}}$, which also is called $x_{k_{new}}$. For each $k \in O_P$ there is exactly one $i \in P$ such that $y_{k_{new}}$ predicts the value of $y_{i_{new}}$, which also is called $x_{k_{new}}$. Each unit from $I \cup P \cup A$ has one forward connection to each unit from H. Each unit from M is connected to each other unit from M. Each unit from $C \backslash (I \cup P)$ is connected to each other unit from this set. Each weight of a connection leading to a unit in M is said to belong to W_M. Each weight of a connection leading to a unit in $C \backslash (I \cup P)$ is said to belong to W_C. Each weight $w_{ij} \in W_M$ needs p^k_{ij}-values for all $k \in M$. Each weight $w_{ij} \in W_C$ needs p^k_{ij}-values for all $k \in H$.

First we will describe the algorithm, then some comments will be given.

INITIALIZATION:
For all $w_{ij} \in W_M \cup W_C$: begin $w_{ij} \leftarrow$ random, for all possible k: $p^k_{ij_{old}} \leftarrow 0, p^k_{ij_{new}} \leftarrow 0$ end,
for all $k \in H$: $y_{k_{old}} \leftarrow 0, y_{k_{new}} \leftarrow 0$.
For all $k \in I \cup P$: Set $y_{k_{old}}$ by environmental perception, $y_{k_{new}} \leftarrow 0$.

FOREVER REPEAT:

1. A. For all $i \in H : y_{i_{new}} \leftarrow \frac{1}{1+e^{-\sum_j w_{ij} y_{j_{old}}}}$,

for all $i \in I \cup P$*: Set* $y_{i_{new}}$ *by environmental perception.*

B. Execute all motoric actions based on activations of units in A.

2. A. For all $w_{ij} \in W_M, k \in M$*:* $p^k_{ij_{new}} \leftarrow y_{k_{new}}(1 - y_{k_{new}})(\sum_{l \in M} w_{kl} p^l_{ij_{old}} + \delta_{ik} y_{j_{old}})$

B. For all $w_{ij} \in W_M$*:* $w_{ij} \leftarrow w_{ij} + \alpha_M \sum_{k \in O}(x_{k_{new}} - y_{k_{new}}) p^k_{ij_{new}}$.

3. A.For all $k \in O$ *begin* $y_{k_{new}} \leftarrow x_{k_{new}}$*, for all* $w_{ij} \in W_M : p^k_{ij_{new}} \leftarrow 0$ *end.*

B. For all $w_{ij} \in W_C, k \in H$*:* $p^k_{ij_{new}} \leftarrow y_{k_{new}}(1 - y_{k_{new}})(\sum_{l \in H, w_{kl} \ exists} w_{kl} p^l_{ij_{old}} + \delta_{ik} y_{j_{old}})$

C. For all $w_{ij} \in W_C$*:* $w_{ij} \leftarrow w_{ij} - \alpha_C \sum_{k \in O_P} p^k_{ij_{new}}$.

4. For all $k \in M \cup C : y_{k_{old}} \leftarrow y_{k_{new}}$,

For all $w_{ij} \in W_M \cup W_C$ *and for all possible k:* $p^k_{ij_{old}} \leftarrow p^k_{ij_{new}}$.

General comments on the algorithm. 1. In step 2 the model network is updated in order to better predict the input (including pain) for the controller. Since the control network continues activation spreading based on the actual inputs instead of using the predictions of the model network, 'teacher forcing' [12] is used in the model network (step 3.A).

2. In step 3 the weights of the control network are updated in order to minimize the cumulative activations of the pain units. In the version above no teacher forcing is used for the control network. Here the philosophy is that a little pain may be informative for the agent, and may have an explicit influence on future actions.

3. The algorithm assumes that from one time tick to the next the environment changes in a fashion that is predictable by linearily separable mappings from past states. If there is a 'higher degree of environmental non-linearity' then the algorithm has to be modified in a trivial manner such that the involved networks tick at a higher frequency than the environment. In any case it suffices if there are four network ticks for each environmental tick. This is due to the fact that 4-layer-operations in principle are enough to arbitrarily approximate any desired mapping.

Comments on the on-line nature of the algorithm. Since we want an on-line learning procedure we deviate from true gradient descent in several respects:

1. Instead of accumulating contributions to weight changes over time and actually changing the weights after activation spreading, the weights are changed immediately. Immediate weight changes allow to renounce on information about 'episode boundaries' [12].

2. The weight changing mechanism of the controller acts as if the model network already was a perfect predictor (with fixed weights) which could replace the environment. However, the model may be imperfect:

2A. Jordan [2] as well as Robinson and Fallside [6] note that a model network does not need to be perfect to allow increasing performance of a control network. If the error for the control network is not given by the difference of the desired input for the control network and the model output but by the difference of the desired input and the actual input of the control network, then the minima of this difference still are fixpoints of the weight changing mechanism, as long as the model network already has reached a local minimum. The zero-points of the controller's error are fixpoints even if the model network has not yet found a local minimum. The minima of the error for the control network can be found if the inner products of the approximated gradients for the control network's weights and the exact gradients (according to a perfect model) tend to be positive.

2B. Note that the p^k_{ij}'s of the model network change independently from the p^k_{ij}'s of the control network. A situation where the control network experiences pain and where its weights are based on an inaccurate model will not remain stable, as long as not both the model network and the control network are trapped in local minima. If we assume that the model network always finds a zero-point of its error function (which means that it sooner or later always will correctly predict future inputs no matter how the controller behaves), then over time we can expect the control network to perform gradient descent in pain according to a perfect model of the visible parts of the real world. As long as the model is inaccurate the controller partly functions as a random explorer who rather uninformedly causes situations that help the model network to collect new data about the environmental dynamics, in order to 'make the relevant dynamics of the world differentiable'.

Experiments with a difficult control task. The algorithm is currently being tested on a complicated pole balancing problem (the differential equations modelling the cart-pole system described in [1] are employed). Unlike with previous pole balancing tasks no prewired decoder is used to pre-process the inputs from the environment. Additionally, unlike with previous pole balancing tasks no information is provided about temporal derivatives of the environment's state variables (pole velocity, etc.). The agent is forced to extract this kind of information by itself, by means of the recurrent connections of its model network. An additional difficulty is that no external teacher provides information about 'trial boundaries'. Thus the agent faces a complex and realistic spatio-temporal credit assignment task. The results of preliminary test runs are very encouraging, however, the experiments have not yet been completed.

Extensions to the system. In [9] it is discussed how the two recurrent networks can be used for planning future action sequences by performing gradient descent in predicted pain instead of actual pain. It is also hinted at the possibility of using probabilistic output units for the controller, thus providing the agent with explicit explorative capabilities. Furthermore it is noted that a perfect model which also predicts the controller's output can be used for 'meta-learning' ('learning how to learn').

Concluding Remarks

The weights of a network with fixed topology may be considered as its program. One of the most interesting aspects of many connectionist algorithms is that program outputs are differentiable with respect to programs. A simple program generator (the gradient descent procedure) allows to produce increasingly successful programs, if the desired outputs are known.

In typical reinforcement learning situations the environment is not *a priori* represented in a differentiable form. So the main reason for connectionist world models in the style above can be seen in 'making the world differentiable'. Thus even *program inputs* can become differentiable with respect to programs. A differentiable world model allows the program generator an informed search for better goal directed programs.

The degree of informedness of this search for suitable programs is a main difference between the very general approach presented in this paper and other reinforcement learning algorithms.

References

[1] A. G. Barto, R. S. Sutton, and C. W. Anderson. Neuronlike adaptive elements that can solve difficult learning control problems. *IEEE Transactions on Systems, Man, and Cybernetics*, SMC-13, 834-846, 1983.

[2] M.I. Jordan. Supervised learning and systems with excess degrees of freedom. Technical Report COINS TR 88-27, Massachusetts Institute of Technology, 1988.

[3] P.W. Munro. A dual back-propagation scheme for scalar reinforcement learning. *Proceedings of Ninth Annual Conference of the Cognitive Science Society, Seattle, WA*, 1987.

[4] Nguyen and B. Widrow. The truck backer-upper: An example of self learning in neural networks. In *IJCNN International Joint Conference on Neural Networks, Vol 2*, 1989.

[5] A. J. Robinson and F. Fallside. Static and dynamic error propagation networks with application to speech coding. *Proceedings of Neural Information Processing Systems, American Institute of Physics*, 1987.

[6] T. Robinson and F. Fallside. Dynamic reinforcement driven error propagation networks with application to game playing. In *Proceedings of the 11th Conference of the Cognitive Science Society, Ann Arbor*, 1989.

[7] J. H. Schmidhuber. The neural bucket brigade. In R. Pfeifer, Z. Schreter, Z. Fogelman, and L. Steels, editors, *Connectionism in Perspective*, Amsterdam: Elsevier, 1988.

[8] J. H. Schmidhuber. Temporal-Difference-Driven Learning in Recurrent Networks. In *ICNC International Conference on Parallel Processing in Neural Systems and Computers, Düsseldorf*, 1990.

[9] J. H. Schmidhuber. Making the world differentiable: On using supervised learning recurrent neural networks for dynamic reinforcement learning and planning in non-stationary environments. FKI-Report, Institut für Informatik, Technische Universität München, 1990.

[10] P. J. Werbos. Building and understanding adaptive systems: A statistical/numerical approach to factory automation and brain research. *IEEE Transactions on Systems, Man, and Cybernetics*, 17, 1987.

[11] R. J. Williams. Toward a theory of reinforcement-learning connectionist systems. Technical Report NU-CCS-88-3, College of Comp. Sci., Northeastern University, Boston, MA, 1988.

[12] R. J. Williams and D. Zipser. A learning algorithm for continually running fully recurrent networks. Technical Report ICS Report 8805, Univ. of California, San Diego, La Jolla, 1988.

HARDWARE REALISABLE LEARNING ALGORITHMS

D.Gorse
Dept. of Computer Science, University College,
Gower Street, London

J.G. Taylor
Dept. of Mathematics, King's College,
Strand, London

ABSTRACT

Learning algorithms are developed which are implementable by means of RAM hardware. In particular probabilistic RAMs are used to implement unsupervised reward and associative learning algorithms. These are considered after an introductory survey of supervised gradient descent learning in pRAM form. Such learning is quite effective for certain hard problems, but reward and associative learning are shown to be more advantageous due to their local nature. The development of topological maps in nets of pRAMs is also shown to occur.

There are many learning algorithms for artificial neural nets, but none of them has yet proven to be easily hardware implementable either by analog or digital technologies. Recent developments in learning have indicated that stochastic activity in the neural units is important for allowing a fuller exploration of the state space than can be achieved in the purely deterministic case. We have recently developed a hardware implementable stochastic model which uses probabilistic RAMs or pRAMs, in which the output from an address $\underline{u}$ (a binary n-vector) in a RAM with n input lines is 1 with a certain probability denoted $w(\underline{u})$ ([Gorse and Taylor 1989]); this generalises the three-state PLN logic node of Aleksander [1989]. It was shown that the activities of a net of pRAMs develops (in discrete time) identically to a corresponding net of synaptically noisy neurons ([Gorse and Taylor 1989]). If this identity theorom (which allows the possibility of using neurobiological insights in the context of electronic networks) is to be fully exploited it is necessary to develop learning rules for pRAMs. It is to be expected that pRAM versions of the full range of connectionist learning rules can be developed, and this has indeed become apparent ([Gorse and Taylor 1990ab]). What is emphasised in this paper is the development of rules which are hardware implementable in pRAM technology at all stages of the learning process. Before this is done it will be useful to briefly review the gradient descent learning algorithm developed in more detail in Gorse and Taylor [1990b]. The error function used is of mean square form

$$E = (1/2n_o) \sum_{j=1}^{n_o} (O_j - \langle o_j \rangle)^2 \quad (1)$$

where O_j is the desired activity on the jth output node and $\langle o_j \rangle$ is the actual mean firing rate of that node. The update rule for the memory contents $w(\underline{u},i)$ (initially assigned random values in [0,1]) is given by

$$\Delta w(\underline{u},i) = -c dE/dw(\underline{u},i)$$

$$= c/n_o \sum_{j=1}^{n} (O_j - \langle o_j \rangle)(R(\underline{u},i)/R^2) \times$$

$$\times \sum_{r=1}^{R} X(a_i(t-r), o_j(t-r)) \quad (2)$$

where a training step consists of R presentations of the input patterns and X is a correlation function between the activities (0,1) of a pair of nodes:

$$X(a_i,a_j) = (\bar{a}_i\bar{a}_j + a_i a_j - a_i\bar{a}_j - \bar{a}_i a_j)$$

with $\bar{a} = 1-a$, and $R(\underline{u},i)/R$ is the proportion of times location $\underline{u}$ in the ith pRAM is addressed. The algorithm described above can be applied to various problems. Training times for the parity problem compare extremely well with those obtained by conventional back error propagation. However it does not seem easy to implement (2) directly in pRAM hardware.

An intermediate between supervised and unsupervised learning is reward or reinforcement training ([Barto, Sutton and Brewer 1981]). That also may be implemented by pRAMs in which the update rule for $w(\underline{u},i)$ now becomes ([Gorse and Taylor 1990a]).

$$\Delta w(\underline{u},i) = c_1 r[a_i - w(\underline{u},i)] + c_2 \bar{r}[\bar{a}_i - w(\underline{u},i] \quad (3)$$

In (3) r is the reward given to the address $\underline{u}$ in the ith pRAM which will bring its output probability closer to the actual output a_i. The term proportional to c_1 gives the magnitude of the reward, that proportional to c_2 the penalty, in this situation.A non-zero value of c_2 is necessary in order to prevent the system converging on false minima. The pRAM version (3) of reinforcement learning differs from that of Barto et al [1981] in that noise is introduced at the synaptic level rather than at the threshold; the former approach is biologically more realistic. The rule (3) may be applied to various learning problems, and is more effective than the gradient descent approach in certain cases ([Gorse and Taylor 1990a]). It is also possible to implement reward learning by local rules in pRAM form. In particular the rule (3) is implementable by a pRAM with input lines carrying r, a(t) and $w(\underline{u})(t)$ and with memory contents given by

$$V(\underline{u})=(0,1,0,0,1-c_1,1-c_2,c_2,1) \quad (4)$$

It is possible to generalise rule (3) so that the reward depends on the local activity of a pRAM itself. Thus r in (3) may be generated as a stochastic process with some probability $p(\underline{u},y)$, where $\underline{u}$ is the input activity vector which generated output y. This rule for reward generation may clearly be implemented in hardware by a reward pRAM with address lines labelled by $\underline{u}$ and y and memory contents $p(\underline{u},y)$. A typical example for two input lines u_1, u_2 and one output y is p(010)=0.7, p(011)=0 (indicating the inhibitory nature of u_2) and p(100)=0,p(101)=0.9 (indicating the excitatory nature of u_1). Such learning rules lead to stable values for the $w(\underline{u},i)$ (Clarkson, Gorse and Taylor, 1990) and allow the storage of pattern sets.

Finally the generation of a topological map is also possible in a pRAM-implementable form. The Kohonen learning rule ([Kohonen 1982]) is adapted for an input pattern as a distribution of probabilities $P\underline{u}$ on the addresses $\underline{u}$ (we assume here that all inputs go to all pRAMs). The basic learning algorithm involves only the set of memory contents of the pRAM for which the 2^N vector $\underline{w}(i)= \{w(\underline{u},i)\}$ is closest to the 2^N vector $\underline{P}=\{P\underline{u}\}$ with respect to, say, the Euclidean distance:

$$\underline{w} = \{\underline{w}(i):d(P,\underline{w}(i)) \le d(P,\underline{W}(j)),\forall_j\} \quad (5)$$

This has the result that $\underline{w}(i)$ and memory vector for pRAMs geographically close in the net $\underline{w}(i)$ are rotated to $\underline{P}$. This is achieved stochastically; note that the net averaged output of a pRAM is $\underline{P}$. $\underline{w}(i)$, so the learning rule achieves maximization of this output as in the original neural case. The learning rule

$$\Delta \underline{w}(i) = \epsilon (\underline{P}-\underline{w}(i)) \quad (6)$$

can be implemented by a suitable set of pRAMs, with memory contents ($\epsilon P\underline{u}$, $\epsilon+\epsilon P\underline{u}$). The result of such training for a chain of 3 2-pRAMs is given in the figure. The integers 1,2,3 denote which pRAM was most responsive to a test input at that position in the unit square. The earlier phase of setting up the competition between the pRAMs by lateral inhibition also seems possible in pRAM-implementable form. We propose to construct a realistically-sized net of pRAMs with this self-organising property.

REFERENCES

Aleksander I. 1989, The Logic of Connectionist Systems, in Aleksander I.(ed.) Neural Computing Architectures, MIT Press.

Barto A.G., Sutton R.S. and Brouwer P.S. 1981 Biol.Cyb. 40 201

Clarkson T, Gorse D. and Taylor J.G. Local Reward Learning for pRAMs, KCL preprint

Gorse D. and Taylor J.G. 1989 Physica D 34 90

Gorse D and Taylor J.G. 1990a Training Strategies for Probabilistic RAMs, Proc ICNN Conference, Dusseldorf

Gorse D and Taylor J.G. 1990b A Gradient Descent Training Algorithm for Probabilistic Random Access Memories, KCL preprint.

FIGURE

2	2	2	2	2	2	2	2	2	2	2	2	2	3	2	2	3
2	2	2	2	2	2	2	2	2	2	2	2	2	2	2	3	3
2	2	2	2	2	2	2	2	2	2	2	2	2	2	2	3	3
2	2	2	2	2	2	2	2	2	2	2	2	2	3	2	3	3
2	2	2	2	2	2	2	2	2	2	2	3	3	3	3	3	3
2	2	2	2	2	2	2	2	2	3	2	3	3	3	3	3	3
1	2	2	2	2	2	2	2	2	2	3	3	3	3	3	3	3
1	1	1	2	1	1	1	1	2	2	2	3	3	3	3	3	3
1	1	2	1	2	1	1	1	1	3	3	3	3	3	3	3	3
1	1	1	1	1	2	1	1	1	1	2	3	3	3	3	3	3
1	1	1	1	1	1	1	1	1	1	3	3	3	3	3	3	3
1	1	1	1	1	1	1	1	1	3	3	3	3	3	3	3	3
1	1	1	1	1	1	1	1	1	3	3	3	3	3	3	3	3
1	1	1	1	1	1	1	1	1	1	3	3	3	3	3	3	3
1	1	1	1	1	1	1	1	1	1	3	3	3	3	3	3	3
1	1	1	1	1	1	1	1	1	1	1	3	3	3	3	3	3
1	1	1	1	1	1	1	1	1	1	1	3	1	3	3	3	3

Figure

Feature map generated by three 2-pRAMs trained on uniform 2D input probability distribution.

SUPERVISED AND UNSUPERVISED LEARNING IN LINEAR NETWORKS

Pierre Baldi*
Jet Propulsion Laboratory, 303-310
California Institute of Technology
Pasadena, CA 91109

Yves Chauvin†
Thomson-CSF, Inc.
630, Hansen Way, Suite 250
Palo Alto, CA 94305

Kurt Hornik
Institut für Statistik und Wahrsheinlichkeitstheorie
Technische Universität Wien
A-1040 Wien, Austria

Abstract: We give an overview of the main facts on supervised and unsupervised learning in networks of linear units and present several new results and open questions. In the case of back-propagation, the complete structure of the landscape of the error function and its connections to known statistical techniques such as linear regression, principal component analysis and discriminant analysis have been established. Here, we examine the dynamical aspects of the learning process, how in certain cases the spectral properties of a covariance matrix are learnt according to the order defined by the eigenvalues, and the effects of noise. In the low noise limit, we prove that the strategy adopted by the networks are unchanged whereas in the high noise limit, the solution adopted is one of complete redundancy. In the case of unsupervised learning, several algorithms based on various hebbian and anti-hebbian mechanisms are reviewed together with the structure of their fixed points. We show that three "symmetric" algorithms suggested in the literature (Oja, 1982; Williams, 1985; Baldi, 1988) are in fact equivalent. Results of simulations are presented.

*also with Division of Biology, California Institute of Technology
†also with Psychology Department, Stanford University

Abstract: We give an overview of the main facts on supervised and unsupervised learning in networks of linear units and present several new results and open questions. In the case of back-propagation, the complete structure of the landscape of the error function and its connections to known statistical techniques such as linear regression, principal component analysis and discriminant analysis have been established. Here, we examine the dynamical aspects of the learning process, how in certain cases the spectral properties of a covariance matrix are learnt according to the order defined by the eigenvalues, and the effects of noise. In the low noise limit, we prove that the strategy adopted by the networks are unchanged whereas in the high noise limit, the solution adopted is one of complete redundancy. In the case of unsupervised learning, several algorithms based on various hebbian and anti-hebbian mechanisms are reviewed together with the structure of their fixed points. We show that three "symmetric" algorithms suggested in the literature (Oja, 1982; Williams, 1985; Baldi, 1988) are in fact equivalent. Results of simulations are presented.

This paper addresses problems related to supervised and unsupervised learning in layered networks of linear units. In theory, such networks with m units at the input layer and n units at the output layer can always be collapsed to a single layer network of $m.n$ weights by multiplying the successive weight matrices; hence, one might expect the topic to be fairly restricted. However, it appears that a detailed mathematical analysis of these networks in noisy and noiseless conditions provides insights on the structure and organization problems arising during learning. This in turn can bring insights to the behavior and learning dynamics of complex non-linear networks.

Let us consider a linear network with a n-p-n architecture comprising one input layer, one hidden layer and one output layer with n, p and n units repectively. Let us suppose that this network is trained by minimizing the mean square error E between output and desired patterns (LMS) using the back-propagation algorithm. Let us define Σ_{XX}, as the covariance matrix of the (centered) input patterns, Σ_{YY}, as the covariance matrix of the desired outputs, and Σ_{XY} as the covariance matrix between input and desired patterns. It is then possible to show that a critical set of weights of rank p is always the product of the ordinary least squares regression matrix followed by an orthogonal projection onto the subspace spanned by p eigenvectors of $\Sigma = \Sigma_{XY}^T \Sigma_{XX}^{-1} \Sigma_{XY}$ (Baldi and Hornik, 1989). Furthermore, at the global minimum, the activities of the hidden units correspond to a rotation of the projections of the least square estimators of the input patterns to the p main principal components of Σ. All additional critical points are saddle points and can be characterized in terms of orthogonal projections onto other combinations of $q \leq p$ principal components (see also Bourlard and Kamp, 1988; Gallinari, Thiria and Fogelman Soulie, 1988).

Since the optimal solution can be expressed analytically, it can be effectively obtained using numerical analysis techniques without resorting to any descent procedure such as back-propagation. However, such solutions cannot always be obtained in non-linear cases. We have analysed the dynamical aspects of the learning process under gradient descent on LMS. In some cases, it is possible to show that the spectral properties of Σ_{XX} are learned according to the order defined by the corresponding eigenvalues. The error E can then be showed to be a sum of exponentially decreasing terms E_i linked to the principal components (λ_i, e_i) of Σ_{XX}. If noise is added to the input patterns, we have demonstrated that doing gradient descent on a subset of training patterns can result in overtraining on the complete population of patterns. Furthermore, this overtraining phenomenon is shown to depend not only on the number of training patterns but also on the initial conditions of the weight matrix and on the size of the noise level (characterized by its covariance matrix $\Sigma_{NN} = N.I$) relative to the smallest eigenvalue of Σ_{XX}. Such phenomena have been simulated with linear networks and observed with non-linear networks for a speech application. With these theoretical framework, we can now understand why learning curves can be so regular across runs but generalizations so variable, in spite of appropriate learning rates. Furthermore, we can compute when to stop training to obtain optimal generalization performance from known initial conditions (*a priori* knowledge) and known noise level.

If noise is now added to the hidden layer, it is possible to show that the representations learned by the hidden units will become redundant (Baldi and Hornik, 1990). The level of redundancy will again depend on the relative importance of the noise level and of the eigenvalues of the data covariance matrices. As N becomes very large, all the hidden units will try to do the same thing (for instance, extract the principal component associated with the largest eigenvalue in the simple autoassociative case). Similar phenomena have also been observed by Linsker (1988) in the context of a single linear hebbian cell using unsupervised learning.

Interestingly, it is possible to make other comparisons between the learning dynamics and the error landscapes of several supervised and unsupervised learning algorithms. For example, it the landscape of a *n-1-n* auto-associative back-propagation network has the same critical points (global minimum and saddle points) as a constrained linear hebbian cell (Chauvin, 1989). In the *n-p-n* autoassociative case, the two weight matrices of the network are transposed of each other at the optimum. Based on this remark, a symmetric approximation is suggested in Baldi (1988) whereby only one of the matrices is updated by gradient descent. The resulting procedure can be shown to be equivalent to an extension of a hebbian type of algorithm introduced originally for one single unit by Oja (1982). We also examine the properties of some unsupervised algorithms with lateral intralayer connections. Provided the network is started with symmetric initial conditions, we show that Williams' SEC algorithm (Williams, 1985) is completely equivalent to Baldi's (and therefore Oja's) procedures. These three algorithms can be seen as equivalent by writing the symmetric constraints directly into the error being minimized.

Acknowledgements

Thanks to Julie Holmes and Katayoun Zahedi for helpful comments.

References

[1] Baldi, P. (1988). Linear learning: Landscapes and algorithms. In D. S. Touretzky (Ed.), *Advances in neural information processing systems 1,* Morgan Kaufman: Palo Alto, CA.

[2] Baldi, P. and Hornik, K. (1989). Neural network and principal component analysis: Learning from examples without local minima. *Neural Networks, 2*, 1, 53-58.

[3] Baldi, P. and Hornik, K. (1990). Back-propagation and unsupervised learning in linear networks. In Y. Chauvin and D. E. Rumelhart (Eds.) *Back-propagation: Theory, architectures and applications.* Lawrence Erlbaum Ass. To Appear.

[4] Bourlard, H. and Kamp, Y. (1988). Auto-association by the multilayer perceptrons and singular value decomposition. *Biological Cybernetics, 51,* 291-294.

[5] Chauvin, Y. (1989). Principal component analysis by gradient descent on a constrained linear hebbian cell. *Proceedings of the 1989 IJCNN Conference, 1*, 373-380, Washington D. C.

[6] Gallinari, P., Thiria, S. and Folgelman Soulie, F. (1988). Multilayer perceptrons and data analysis. *Proceedings of the 1988 IJCNN Conference*, 391-399, San Diego, CA.

[7] Linsker, R. (1988). Self-organization in a perceptual network. *Computer*, March, 105-117.

[8] Oja, E. (1982). A simplified neuron model as a principal component analyser. *Journal of Mathematical Biology, 15*, 267-273.

[9] Williams, R. J. (1985). Feature discovery through error-correction learning. *Technical Report 8501.* Institute for Cognitive Science, UCSD, La Jolla, CA.

NEURAL MODELS FOR ORTHOGONAL AND OBLIQUE FACTOR ANALYSES : Towards dynamic data analysis of large sets of highly multidimensional objects

Alain LELU - Albert GEORGEL

INIST/CNRS - 2 Allée du Parc de Brabois
54514 Vandoeuvre-lès-Nancy CEDEX - FRANCE

Data of "pick-any" type (objects described by a few items chosen in a long list), as encountered in texts, documentary systems, machine-tasks problems, marketing research, constitute very large and highly multidimensional data sets, generally out of range of data analysis techniques ; neural models for the unsupervised learning of such data do not need sophisticated separation hypersurfaces, but do need clustering both the objects and the dimensions, with separation hyperplanes passing through the origin, or gravity center, of the data cloud ; overlapping clusters are best suited to this type of data, too. In this context a stochastic neural model implementing a hybrid representation of the data, a mix of clustering and factor Analysis, is presented ; this "Local Component Analysis" model converges to a global optimum of its objective function, contrary to the classical " K-Means" algorithm.A parameter controls the "coarseness" of the analysis. Simulations, ranging from systematic tests on a 25 X 19 table to analysis of a 3 900 X 500 real data table, and comparisons with existing algorithms are presented.

Dynamicity of the model, i.e. instantaneous adaptation to any new data vector, is a desirable feature in many applications. It is the major characteristic of Grossberg's ART2 model. We have tried to go deeper into the theoretical and operational aspects of this concept, and have built analytically two "exact" neural models : the first one implements a one-pass Principal Component Analysis ; instability problems, inherent to(constitutive of) any SVD algorithm, are adressed, and one can adjust a compromise between precision and stability. The second uses the latter as a gain control module in order to implement Linear Discrimination and Regression.

Neural networks models are generally applied to two main classes of problems : 1) Clustering or classification of numerous objects in a low dimensionality space, say 10 000 objects X 3 dimensions ; associative mapping is one of these models. 2) Clustering or classification in a few categories for sets of very highly dimensional objects (say 10 000 dimensions) ; back-propagation models for pattern-recognition are typical of this approach.Classical data analysis methods - clustering models, factor and discriminant analyses - generally synthetize medium-sized sets of objects in medium dimensional space (say 1000 objects X 100 dimensions).

None of these methods are well suited to an important class of problems which arises in practice. "Pick-any" data is the name psychologists gave to the situation where a person is asked to describe an object by a few items chosen in a very long list of elements (say 1000 to 10 000) ; in a machine environment, a similar type of problem is encountered during analysis of textual data, task allocation problems in production systems, consumer preferences... The solution lies in clustering both the objects and the dimensions. As the presence of an item does not necessarily involve the absence of others, the analysis must display homogeneous groups in which objects appear together with their features ("dimensions").A first answer is the block-seriation method (1), which can be successfully used on a relatively small-sized matrix (about 100 X 100) ; it achieves a strict non-overlapping clustering of both the lines and columns in a binary table.

However, when textual and documentary data are analysed, overlapping clusters are needed because the same word may have different meanings in different contexts. This point was already made in the 60's by English authors (2), who called these clusters overlapping "clumps". We know now that a double, symmetrical clustering of the described objects and of the describers is necessary, and that defining the membership by a stringent criterion of all or none is not enough. In fact some elements of a cluster are more "typical" than others, and it is useful to have sorted lists of these elements for an easier interpretation of the clumps. This sorting could also prove useful in building navigation tools in the (hyper-)text or the database.Our approach is related to the oblique factor analyses proposed by Thurstone (3) and his successors for retrieving the "simple structure" underlying in a data set.

One of the very few neural models which is similar to our own is Grossberg's ART2 (4). However, his main concern is with biological likeness, while the focus of our work is on the gap between data analysis and neural models. The most remarkable feature of ART2 is its ability to adapt instantaneously its internal representation of the data to any new data vector. Either this new vector is used to create a new neuron, or it serves to modify slightly the internal representation of the model. We will try in this paper to give a precise definition of what could be a dynamic data analysis, a goal which appears to be of fundamental importance in our application domain, and possibly in many others as well ! As a first step, we have derived analytically a neural model implementing one-pass Principal Component Analysis (5).

1 - LOCAL COMPONENT ANALYSIS: When clustering, it is necessary to detect areas of relatively high density in the data cloud. The "structuring function" concept (10) enables one to define in any point of the data space a scalar representative of this density. Any structuring function depends on a "coarseness" parameter varying between two bounds : at one end, one finds as many clusters as points to be clustered ; at the other end, one finds a single cluster.

This is our model for an individual neuron :

- transfer function : $\eta'_t = f_+ [\eta_t (1 - tg(m_t,x_t)/tg\Theta_0)]$

where Θ_0 is the parameter of the structuring function ($0 < \Theta_0 < 180°$),
x_t is the t-ieth data vector ; m_t is the weight vector at "time" t ; $\eta_t = \langle m_t, x_t \rangle$
$f_+ : \mathbb{R} \rightarrow \mathbb{R} \quad f_+(x) = x$ if $x>0$
$= 0$ otherwise

The derivation of η' from η is shown in Figure 1 ; it is a kind of "angular thresholding" based upon the intersection with a cone.

- learning rule : $m_{t+1} = m_t + \alpha\, \eta'_t (1 + \eta_t \cot g\Theta_0/|\underline{x}_t|)(x_t - \eta_t x_t)$

where α is a small positive constant ; $\underline{x}_t$ is the orthogonal complement of x_t on vector m :
$|\underline{x}_t| = (|x_t|^2 - \eta_t^2)^{½}$

This "gradient-climbing" rule follows from the maximization of the objective function :

$$E = \sum_{t=1,T} \eta'^2_t$$

The set of all local maxima constitutes an absolute optimum, given a value of the parameter Θ_0. When one disposes of at least as many neurons as maxima, and if one supposes that each neuron must "climb" up a different maximum, then the model provides an "axial" clustering procedure which is optimal in an absolute sense. The delicate question in this approach is not to "forget" any local maximum. The problem is not too difficult to resolve with a small data set - one has just to choose as many neurons as data vectors, and then initialize the weight vectors with the data. However the problem is more complicated with large data sets. We are currently testing different strategies. For example, starting with 50 neurons and $\Theta_0 = 90°$, we made about a hundred passes over a documentary data table of 3 900 X 500 and converged towards a partial set of a dozen half-axes, easier to interpret than in our previous tests (14) ; then we carried out another analysis with the subset of documentary elements whose η' was 0, and fusionned the new neurons with the previous ones, and so on...

We have proved that there are as many maxima as different data vectors when $\Theta_0 \rightarrow 0$, and a single one when $\Theta_0 \rightarrow 180°$ (1st principal component). We have also carried out systematical tests with a small data set presented in our paper (7) - 25 dog breeds described by 19 binary features - for different values of Θ_0. The data were transformed with the law :

$$x_{ij} := (x_{ij} - x_{i.}x_{.j}/x_{..}) \,/\, \sqrt{x_{i.}x_{.j}/x_{..}} \qquad (a)$$

where $x_{i.}$, $x_{.j}$ and $x_{..}$ are respectively the sums of columns, rows, and general sum of the data matrix X. Our Figure 2 displays the number of clusters decreasing with Θ_0, from 25 to 1 ; these clusters appear to be progressively coalescing into the two halves of the main principal component, at the limit $\Theta_0 = 180°$.

We have finally tested a well-known reference data set, the "Thurstone's boxes", that Thurstone built in order to illustrate his concept of "simple structure", obtained through oblique factor analyses. The data set was built by computing 20 generally non-linear functions of the 3 correlated dimensions (x, y, z) of 20 boxes. Our Figure 3 is based on Thurstone's graphical representation. It displays in a spherical triangle : 1) the cloud of the 20 functions, 2) the 3 "optimal" oblique axes visually adjusted by Thurstone, 3) our 6 axes ($\Theta_0 = 40°$). Our results are not as good as Thurstone's, but seem slightly better than those issued from an enhanced VARIMAX method (11) - see Table 1.

2 - "EXACT" ONE-PASS PRINCIPAL COMPONENT ANLYSIS : The design of dynamic data analysis systems, implementing representations that will automatically account for the arrival of any new

data, must be based upon a clear vision of the aim to be achieved. We think that the minimal definition of a dynamic data analysis system is <u>a system that derives the same representation after learning the t-ieth data vector as a classical method would have derived had it analyzed the whole set of t data vectors</u>. This definition is simple and operational, but the idea of a representation growing indefinitely is nonsense... Nevertheless, this definition may provide a reference point for more sophisticated systems ; for example it might be desirable : 1) to selectively forget the first learned data, according to several possible laws... 2) not to forget anything, but to increase the "coarseness" of the analysis... Moreover, one must be conscious of the fact that instability problems are inherent to dynamic data analysis : the birth of a new cluster or factor is not a continuous process, but a "catastrophic" one ! In a first step, we have derived an analytical expression for a neural model implementing Principal Component Analysis. The demonstration can be found in (5). The proposed network is displayed in Figure 4.

- <u>transfer functions</u> : $y = M\,x$

$$\hat{x}_k = \eta_k m_k + \sum_{j=1,K} \eta_j m_j \tau_j/(\tau_k - \tau_j)$$

where $y = (\eta_1, \eta_2, \dots \eta_K)$

- <u>learning rule</u> : $\delta m_k = (1/\tau_k)\,\eta_k\,(x - \hat{x}_k)$

where $\tau_k := \tau_k + \eta_k^2$

Remark - If we define t as the size of the data set, and λ_k as the eigenvalue of XX^T:
$\tau_k = t\,\lambda_k$ and : $\tau_j/(\tau_k - \tau_j) = \lambda_j/(\lambda_k - \lambda_j)$

This learning rule is of the Widrow-Hoff type ; it supplies an analytical expression for both the gain and the feed-back on internal representations of each data vector. Interestingly, each neuron has its own representation of the data vector.

<u>Implementation problems</u> : the stability/precision tradeoff. For maximum precision, we had to divide each new data vector x into n "data chunks" x/n and learn n times. But as we tried to keep n within reasonable bounds, stability problems arised when λ_k, or $\lambda_k - \lambda_j$ were very small ; we have tested several solutions, which provide the user with a "stability/precision tradeoff" parameter .

<u>Current and further developments :</u> In (5) we have shown how this "instantaneous" Principal Component analyser can be used as a gain control module for a network implementing one-pass Linear Discriminant/Regression Analysis. But we have not tested it as yet, and supervised learning is another story...

Our first model (8, 14) was based upon associative mapping ; it led to satisfactory and interpretable results in one pass, but these results were not independent from the initial conditions. All these results lead us to think that a one-pass Local Component Analyser is not necessarily an utopia.

<u>References :</u>

(1) F. Marcotorchino - "Block-seriation problems : a unified approach" - Applied stochastic models and Data Analysis, vol. 3, pp 73-91, 1987

(2) K. Spark-Jones - *"Linguistics and information science"* - Academic Press, 1973

(3) L.L. Thurstone - *"Multiple-factor analysis"* - Chicago Press - 1947

(4) G.A. Carpenter, S. Grossberg - "ART2 : self-organization of stable category recognition codes for analog input patterns" - Applied Optics, vol. 26, N°23, 1987

(5) D. Rosenblatt, A. Lelu, A. Georgel - "Learning in a single pass : A neural model for instantaneous Principal Component Analysis and Linear Regression" - proceedings of the first IEE conference on neural computing, London, Oct. 18-20 1989

(7) A. Lelu - "A neural model for highly dimensional data analysis" - Journées Analyse des Données / Apprentissage de connaissances symboliques et numériques - Antibes, September 11-14, 1989 - Published by Nova Science Publishers : *Data Analysis, Learning Symbolic and Numeric Knowledge* - E. Diday Ed., New York, 1989

(8) A. Lelu - "Browsing through Image Databases via Data Analysis and Neural Networks" - proceedings of RIAO 88 (User-Oriented Content-Based Text and Image Handling) - MIT - March 21-24 1988

(10) H. Emptoz, M. Lamure - "A systemic approach to pattern recognition" - Robotica, vol.5, pp 129-133, 1987

(11) P.M. Bentler - "Factor simplicity index and transformations" - Psychometrika, vol. 42, N°2, 1977

(14) A. Lelu - "Local Component Analysis : a neural model for information retrieval" - proceedings of International Joint Conference on Neural Networks - Washington, June 18-22 1989

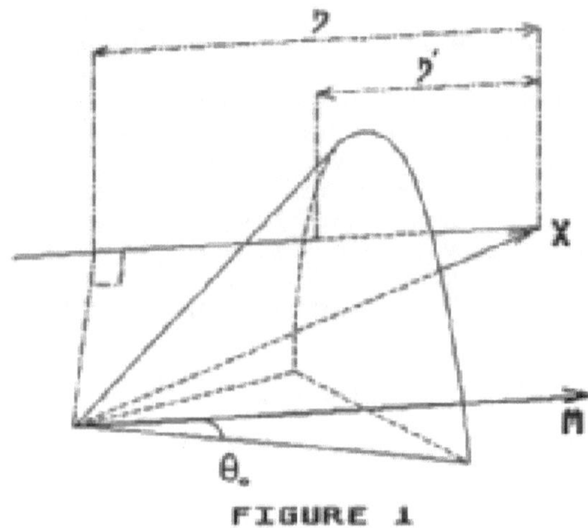

FIGURE 1

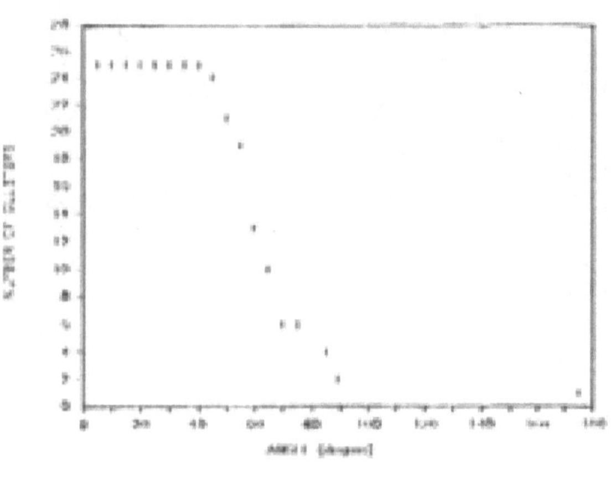

FIGURE 2

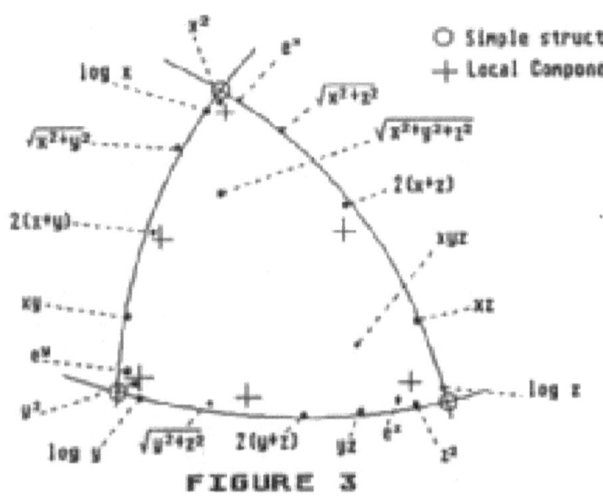

FIGURE 3

	(y,z)	(z,x)	(x,y)
Simple structure (3)	76.7	83.9	77.8
VARIMAX (11)	65	72	64
Local Component Analysis (θ_o = 40°)	64.8	73.4	68.1

"Thurstone's boxes"

Angles (°) between the 3 main oblique factors

Table 1

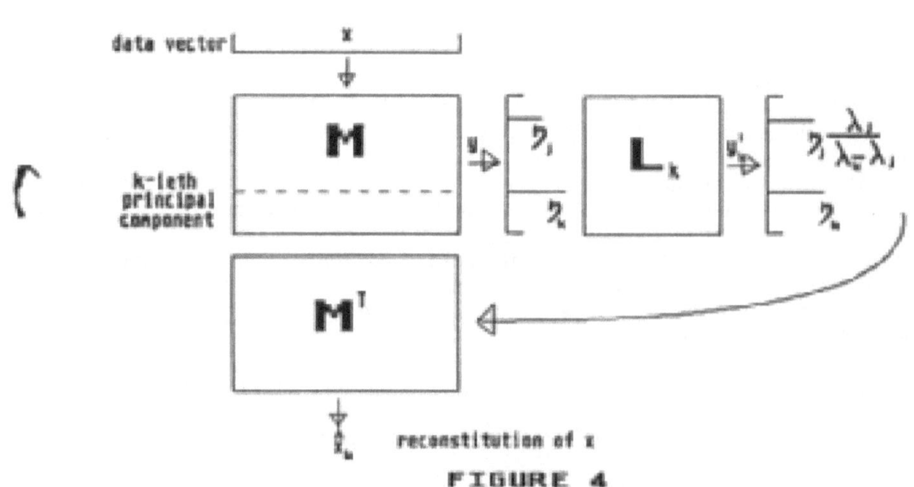

FIGURE 4

A LEARNING ALGORITHM FOR ATTRIBUTE CLASSES CONSTRUCTION *

LIJIA ZHOU
Computer Center, XIDIAN University
Xian, 710071, P.R.China

ABSTRACT

The objective of this rsearch is to seek a method for automatically constructing Attribute Class (AC) based on machine learning . The method should have high learing speed under the prerequisite that the quality of AC description will not be come down. What is called Attribute Class is a set of functions where each function has a number of inputs and only one output. AC is used to describe public and distinct features of input specimens and is a base of generating hierarchical Knowledge Base (KB) oriented to Pattern Recognition (i.e PR expert system) . this paper proposes a method using Pocket and Back-Propagation neural models to form attribute functions according to the different hierarchical requirements of KB respectively . An AC Constructing Algorithm (ACCA) is also presented here in the paper . ACCA can meet above-mentioned demands well . The further work is going on now .

Keyword : Machine Learning, Neural Network, Knowledge Acquisition, Pattern Recognition

COMPETITIVE CLUSTERING

Christine DECAESTECKER*

CADEPS (CAIRU)
Université Libre de Bruxelles (U.L.B.)
Av. F.D.Roosevelt 50, C.P. 194/7. B-1050, Brussels, Belgium.
Tel (+ 32 2) 642 2783; Email CADEPS@BBRNSF11(.BITNET)

Abstract :

This paper presents a variation of the usual winner-take-all competition model which overcomes certain of its deficiencies. The advantage of this variation is the (empirical) stability of the results, which are independent of the initial network structure (i.e. the number of output nodes, the initial connection weights). This model uses variable inhibition between nodes (which depends upon the matching qualities between input patterns and weight vectors) and reduces the amount of learning in cases of competition which are not easily decidable. The adaptation of the weighting constraints permits a direct interpretation of the weight vectors in terms of the description of clusters of input patterns.

* This work is supported by the Belgian National incentive-program for fundamental research in Artificial Intelligence. The scientific responsibility is assumed by the author.

ART: AN IMPLEMENTATION OF THE NEW DIRECT ACCESS CONDITION.

Rodriguez-Galán, Roberto[1] and Garcia-Tejedor, Alvaro[+]
[1]Knowledge Engineering Dept. ENTEL, S.A.
Castellana, 141 - 28046 Madrid (Spain) Fax: 34-1-2791074
[+]Dept. of Biochemistry; Fac. of Chemistry
Universidad Complutense; 28040 Madrid; EMAIL W055@EMDUCM11

Abstract

ART networks, developed by S.Grossberg and G.Carpenter [1], represent a relevant improvement on neural nets. They have been widely studied and theirs limitations have been also pointed out. In this paper, ART networks characteristics are introduced in a first section in order to center the topic. Afterwards, some inherent restrictions to ART systems are discussed: input patterns can be only either binary (in ART-1) or gray-scale (in ART-2), the storage method destroys the previous patterns, the network is limited to the number of nodes in F_2 and it is not possible to represent hierarchical knowledge. This last point is specially emphasized and a possible improvement is suggested. Finally, a modification of direct access condition is presented in the third section; this modification leading to an improvement on the ART network perfomance. A comparative implementation of both access was also perfomed by using Smalltalk/V 286, and the increase on pattern-matching speed upon previously learned ones is showed up.

---

[1] To whom correspondence should be addressed.

ELEMENTARY OPERATIONS IN NEURAL NETS: ADDITION AND SUBTRACTION

Salvatore Rampone
I.I.A.S.S. - Via G.Pellegrino, 19 - 84019 VIETRI SUL MARE (SA)
FACE SUD S.P.A - Via Generale Clark, 19/21 - 84100 SALERNO

Roberto Tagliaferri
Dipartimento di Informatica ed Applicazioni
Universita' di Salerno - 84081 BARONISSI (SA)

Abstract

The developement of neural net research has involved several theoretical studies, neurobiological connections, learning architectures. Nevertheless few people investigated the minimal functional requirements of a neural-like hardware. In this paper it has show that, using a learning law developed in the last years [2,3,5,6], we obtain the elementary operations of addition and subtraction [4] in a classical model of neural network. The learning law is based on Caianiello's Mnemonic Equations [1]and on the learning protocol of Valiant[7]. We obtain the net answer in a constant time (as and better than the best parallel adders). Besides, because a single neuron can perform more than one boolean operation at same time, we need a smaller number of devices.

LEP - A Neural Model learning Reliably

*Jian-Kang Wu**

L. f. Informatik 5, University of Erlangen-Nurnberg
8520 Erlangen, FRG

ABSTRACTION

Reliable learning is not only the key to keep balance between stability and adaptability of neural system, but also a tool to produce meaningful categories. A neural net system, which Learns with respect to Experiences and Perspectives (LEP), is proposed. Each learning is enabled by a successful verification of all perspectives of pattern under consideration. Amount of learning is carefully tuned by experience records and the confidence factors of both stored knowledge and input patterns. On the other hand, a so-called forgetting process is introduced to control the degree of system adaptability. After semantic association, memory reorganization, signal flows of motivation and expectation are facilitated. A meaningful preliminary results of image texture recognition with LEP have shown its prominence.

*J.K.Wu is granted by Alexander von Humboldt-Stiftung, and on leave from the University of Science and Technology of China, Hefei, China 230026

PATTERN RECOGNITION WITH A MULTILAYERED NEURAL NETWORK TRAINED WITH DYNAMIC COMPETITIVE LEARNING

János Rácz[*/**] & Tamás Klotz[**]
[*]Hungarian Academy of Sciences
[**]Computer Research and Innovation Center
H-1015 Budapest, Donáti u. 35-45.
Phone +36 1 115 1009
Hungary

Abstract

Dynamic competitive learning is an unsupervised learning technique. The only information the model gets is the pattern imposed on the input units. We don't give any information concerning the goodness of a decision as e.g. an expected answer, which it can use as an error measure to approach the proper behavior (as e.g. with the back-propagation algorithm).

The competitive learning technique classifies the input vectors, so that the vectors (samples) belonging to the same class have similar characteristics. Each class is represented by one unit. The technique is called competitive because the units within one cluster compete to win the right to be active.

In the multi-layered neural networks described here the number of clusters, their connections, and the generation of new units are determined dynamically, during learning. The training is a modified version of the competitive learning strategy. The model seems to provide a good solution for the high-level storage of complex data structures, their classification, and for restoring and filtering fragmented or noisy patterns.

WEIGHTED HEBBIAN LEARNING

Petr Božovský
Department of Computer Science, Charles University
Malostranské nám. 25, 118 00 Prague 1, Czechoslovakia

Abstract

The Hebbian learning rule is an example of simple and quick unsupervised learning. It provides easy additional learning. However, it does not make sure the pattern states learnt to be stable. Moreover, there is created a large number of spurious stable states that worsen behavior of the neural network as an associative memory.

We propose a new method of modified Hebbian learning in which the state vectors that are to be memorized are weighted. These vectors are proven to be stable including given neighborhood of them in sense of the Hamming distance. This method is polynomial. It also provides an additional learning. The problem of the weights tolerance is also solved.

SELF-ORGANIZING NEURAL NETWORK FOR NON-PARAMETRIC REGRESSION ANALYSIS

Vladimir Cherkassky and Hossein Lari-Najafi
Dept. Electrical Eng., University of Minnesota
Minneapolis, MN 55455, USA

Abstract

The idea of using Kohonen's self-organizing maps is applied to the problem of non-parametric regression analysis, i.e. evaluation (approximation) of the unknown function of N-1 variables given a number of data points (possibly corrupted by random noise) in N-dimensional input space. Simple examples show that the original Kohonen's algorithm performs poorly for regression problems of even low dimensionality, due to the fact that topologically correct ordering of units in N-dimensional space may violate the natural topological ordering of projections of those units onto (N-1)-dimensional subspace of independent variables. A modification of the original algorithm called the Constrained Topological Mapping algorithm is proposed for regression analysis applications. Given a number of data points in N-dimensional input space, the proposed algorithm performs correct topological mapping of units (as the original algorithm) and at the same time preserves topological ordering of projections of these units onto (N-1)-dimensional subspace of independent coordinates. Simulation examples illustrate good performance (i.e. accuracy, convergence) of the proposed algorithm for approximating 2- and 3- variable functions. Moreover, for multivariate problems the proposed neural approach allows to bypass "the curse of dimensionality", i.e. the size of the training set required for evaluation of the unknown function with a specified accuracy grows approximately linearly with the dimensionality of the sample space (or the number of independent variables).

Index terms: constrained topological mapping, nonparametric regression, self-organization.

A DRIVE REINFORCEMENT MODEL FOR VISUAL PERCEPTION

Omid M. Omidvar, Ph.D

Computer Science Department
University of the District of Columbia
4200 Connecticut Ave
Washington , D.C. 20008

ABSTRACT

Neural networks have been used to mimic cognitive processes which take place in animal brains. The learning capability inherent in neural networks makes them suitable candidates for adaptive tasks such as formation of visual perception. Synaptic reinforcements creates a proper condition for adaptation, which results in memorization, formation of perception, and higher order information processing activities.

In this research a model of a neural network with drive reinforcement is studied; also the operation of the network with regard to formation of visual perception and its role in recall and recognition is analyzed. Formation of perception by interconnected neurons is the mean of storing knowledge about the viewed object. As such, visual perception is a developed skill and is achieved through the learning process. A skilled perceiver utilizes a large amount of learned knowledge and many frames of reference. The human eyes as skilled perceivers do not loosely remap the light intensities of the visual image onto the sensorium. Instead, they detect pattern elements, discriminate the depth and texture of objects, and ignore irrelevant causes of variation. Furthermore, there is evidence that they give prominence to what is informationally important. The versatility of the performed functions results in the formation of visual perception.

Visual perception plays a very important role in recall and recognition. Recall is defined as retrieval of stored information where little or no matching is involved. On the other hand recognition is recall with matching; therefore it involves memorizing a piece of information with complete presentation. This research takes the generalized view of reinforcement in which all the signals are potential reinforcers. The neuronal response is considered to be the source of the reinforcement. This approach to adaptation leads to the drive reinforcement nature of the neurons as network components. In the proposed model all the synaptic strengths are reinforced in parallel while the reinforcement among the layers is done in a distributed fashion and pipeline mode.

A model of complex neurons with feedback is developed to account for inhibitory and excitatory behavior of real neurons. A drive reinforcement neural network with dynamic organization is presented to accommodate patterns of different sizes. The network creates the perception of the object by reinforcement of synaptic strengths. The visual perception is analyzed with respect to recall and recognition tasks. The performance of the model with regard to the assigned tasks is presented.

The significant departure in this system is that the network does not check for contours, edges, or connected segment to match against a set of predefined frames. The network creates the perception of the object by reinforcement of the synaptic strengths and conditioning of each neuron to its proper state. This completes the formation of the perception process. Thereafter, whenever the same set of stimuli presented to the network, the formed visual perception is activated and the recall and recognition is performed.

ASSOCIATIVE MEMORIES

Chair: James ANDERSON

BIDIRECTIONAL ASSOCIATIVE MEMORY FOR THREE PATTERNS

B. Humpert

Department of Mathematics and Computer Science
Indiana State University
TERRE HAUTE, IN 47809

ABSTRACT

The bidirectional associative memory (BAM) is extended to three- (and more-) layer nonlinear feedback networks. The energy function is defined and the most obvious generalizations to memorize triple vector associations $(\vec{A},\vec{B},\vec{C})$ are given. Most earlier insights for two-vector associations generalize in a straightforward manner. The importance of this approach for data fusion and pattern splitting is stressed.

1. INTRODUCTION

Detailed and careful analyses of the bidirectional associative memory (BAM) involving paired-data associations $(\vec{A},\vec{B})$ of binary/bipolar vectors has been carried out [1,2]. The storage capabilities of this model have been analysed and possibilities for higher storage capabilities are under intensive investigation [3]. Further, the temporal associative memory (TAM), a spring-off from content addressable memory (CAM) [4] and BAM [1], allows for interesting new analyses [5]. In this paper we extend the idea of paired data associations to triple-data associations, to be stored in the memory of a BAM-network which consists of three separate output layers, such that they eventually can be recalled in the manner familiar from CAM.

2. Homogeneous and Inhomogeneous BAM

We first summarize the main insights. Associative memories map an input-vector $\vec{A}$ [$\dim(\vec{A})=n$] to an output-vector $\vec{B}$ [$\dim(\vec{B})=m$]. The BAM-network is defined as an associative memory that uses forward and backward directional search to recall an associated vector-pair $(\vec{A}_f, \vec{B}_f)$ from an input-pair $(\vec{A},\vec{B})$. The vectors $\vec{A},\vec{B}$ are understood to have binary components: $\vec{A} \in \{0,1\}^n$, $\vec{B} \in \{0,1\}^m$, whereas $\vec{X} \in \{-1,+1\}^n$, $\vec{Y} \in \{-1,+1\}^m$ denote the analogous bipolar vectors. The homogeneous and inhomogeneous BAM distinguish themselves by the absence resp. presence of threshold terms in the units transfer function:

$$
\begin{aligned}
A_i(t+1) &= \operatorname{sgn}\{ \sum_{j=1}^{m} W_{ij} * B_j(t) - S_i \} \\
B_j(t+1) &= \operatorname{sgn}(\sum_{i=1}^{n} W_{ij} * A_i(t) - T_j \}
\end{aligned} \tag{1}
$$

The energy function of the inhomogeneous BAM reads

$$
E(\vec{A},\vec{B}) = - \vec{A} * W * \vec{B} + \vec{A} * \vec{S} + \vec{B} * \vec{T} \tag{2}
$$

with the thresholds $\vec{S}=\vec{T}=0$ in the homogeneous case. The embedding of p vector-pairs $(\vec{\xi}_A, \vec{\xi}_B)^1 \ldots (\vec{\xi}_A, \vec{\xi}_B)^p$ where $\vec{\xi}_A^{\alpha} \in \{-1,+1\}^n$ and $\vec{\xi}_B^{\alpha} \in \{-1,+1\}^m$ in the BAM-system is often based upon correlation matrix summation

$$
W_{AB} = \sum_{\alpha=1}^{p} (\vec{\xi}_A^{\alpha}) \circ (\vec{\xi}_B^{\alpha}) \tag{3}
$$

(where o stands for outerproduct summation) although other useful schemes have been proposed and are currently under study [6].

There are several basic insights:

(i) The energy-function of the (in-)homogeneous BAM decreases if and only if the state of the network changes; it has a minimum value.

(ii) The (in-)homogeneous BAM will always converge to a stable state - irrespective of the chosen form of W_{AB} and the thresholds $\vec{S}$ and $\vec{T}$. Once a stable state is reached, the BAM never changes to another state, no matter how many unit updates take place.

(iii) The BAM converges in a finite number of steps to some stable state which all occur at local minima of the energy.

(iv) The total number of stable states is finite.

(v) An inhomogeneous BAM can have stable states between 1 and $2^{min(n,m)}$ where n,m are the dimensions of the $(\vec{A},\vec{B})$-fields.

(vi) The storage capacity of the <u>homogeneous BAM</u> is limited in the embedding and (full) recall of user-selected stable states. "Even-coding" optimalizes the storage capacity. The upper bound is given by

$$p \lesssim N/(2*\ln_2 N) \quad \text{with} \quad N = n+m \,. \tag{4}$$

(vii) The storage capacity of the <u>inhomogeneous BAM</u> again depends on the number of +1 and -1 components in the embedded vectors. Assuming $\dim(\vec{A})=n=\dim(\vec{B})$ with exactly $M\equiv 4+\ln_2 n$ entries of +1's and n-M entries of -1's, then

$$p \lesssim 0.68 * \{ n/(4+\ln_2 n) \}^2 \tag{5}$$

associated vector-pairs can be embedded with approximately 98% of them being stable states.

(viii) The BAM is also subject to the unwanted feature of spurious states, and the forms of its basins of attraction have been studies [7].

3. FROM BAM_2 TO BAM_3 :

The bidirectional associative memory with two-vector fields ($\equiv BAM_2$) can be generalized to three- (or more-) vector fields ($\equiv BAM_3$) as shown in Fig.1. It is understood that any unit in one field is connected with all other units in the other field; this rule applies in both directions. The fields $(\vec{A},\vec{B},\vec{C})$ have (n,m,k) units respectively. The corresponding connection strengths are indicated by (W_{AB}, W_{BC}, W_{CA}) and the connection strengths in the opposite direction are given by $(W_{BA}, W_{CB}, W_{AC}) = (W_{AB}^T, W_{BC}^T, W_{CA}^T)$. Obviously, if the $\vec{C}$-field is removed we are back at BAM_2. Suppose the new field C is associated with field B by integrating both in the field $\vec{B}'=(\vec{B},\vec{C})$. We thus would be back at a BAM_2 however with intra-layer connections now being allowed; one thus might consider this as a generalized BAM_2. For later purposes we continue to consider it as BAM_3.

The energy function of the inhomogeneous BAM_3 is given by three terms of the form

$$E(\vec{A},\vec{B},\vec{C}) = E(\vec{A},\vec{B}) + E(\vec{B},\vec{C}) + E(\vec{C},\vec{A}) \qquad (6)$$

with E(..,..) defined in Eq.2. If the threshold vectors vanish $(\vec{S},\vec{T},\vec{U})=0$ we are back at the homogeneous case.

The embedding of vector-associations proceeds analogously. Assuming several associations of vectors from the $(\vec{A},\vec{B},\vec{C})$-fields: $(\vec{\xi}_A,\vec{\xi}_B,\vec{\xi}_C)^\alpha$ where the parameter α indicates a particular association, we have

$$\begin{aligned} W_{AB} &= \sum_{\alpha=1}^{p} (\vec{\xi}_A^\alpha) \circ (\vec{\xi}_B^\alpha) \\ W_{BC} &= \sum_{\alpha=1}^{p} (\vec{\xi}_B^\alpha) \circ (\vec{\xi}_C^\alpha) \\ W_{CA} &= \sum_{\alpha=1}^{p} (\vec{\xi}_C^\alpha) \circ (\vec{\xi}_A^\alpha) \end{aligned} \qquad (7)$$

For a homogeneous BAM_3 the maximal number of triple-vector associations p is limited by the binary-vector associations giving:

$$p \lesssim N/(2*\ln_2 N) \qquad \text{where} \qquad N = n+m+k \qquad (8)$$

For an inhomogeneous BAM_3 we can have stable states between 1 and $2^{\min(n,m,k)}$. Assuming $\dim(\vec{A})=\dim(\vec{B})=\dim(\vec{C})=n$, with exactly $M \equiv 4+\ln_2 n$ entries of +1's and n-M entries of -1's, then maximal

$$p \lesssim 0.68 * \{ n/(4+\ln_2 n) \}^2 \qquad (9)$$

associated vector-triplets can be embedded, with approximately 98% of them being stable states.

4. INCREASING THE STORAGE CAPACITY

The above definition of BAM_3 is a straightforward extension of BAM_2 with all its storage limitations. Obviously, one is looking for increased storage possibilities.

Staying in the framework of 2-node connections several methods of improving the BAM_2 storage capabilities have been suggested. We mention the outer-product scheme of Ref.[6]:

$$W_{ij} = (1/N)*(1-\delta_{ij}) \sum_{\alpha,\beta=1}^{p} \xi_i^\alpha (C^{-1})_{\alpha\beta} \xi_j^\beta \quad , \quad C^{\alpha\beta} \equiv (1/N)\sum_{i=1}^{N} \xi_i^\alpha \xi_i^\beta \qquad (10)$$

where p<N linearly independent stored patterns can be retrieved without error, which even could be correlated. There exist a local embedding rules for this model [8].

An enhanced BAM_2-model is obtained if the forward-backward adjustment process of the units in the two fields undergoes, in addition, a "clipping-type operation" leaving the vector-components: +1 or -1 or preceding value, at each iterative cycle [9].

Sofar we have assumed that all connections are between two nodes only. Early studies of the Hopfield model already revealed that the storage capabilities significantly increase if higher order node-to-node connection schemes are considered, and studies in this direction are still going on [3]. Much in the same spirit one can define

$$E = -\sum_{i=1}^{n}\sum_{j=1}^{m}\sum_{\ell=1}^{k} W_{ij\ell} * A_i * B_j * C_\ell \tag{11}$$

where

$$W_{ij\ell} = (1-\delta_{ij})\ (1-\delta_{j\ell})\ (1-\delta_{\ell i})\ \sum_{\alpha=1}^{p} (\xi_A^\alpha)_i\ (\xi_B^\alpha)_j\ (\xi_C^\alpha)_\ell \tag{12}$$

and all thresholds are assumed to vanish (homogeneous BAM_3). The updating proceeds as

$$A_i(t+1) = \operatorname{sgn}\{\sum_{j=1}^{m}\sum_{\ell=1}^{k} W_{ij\ell} * B_j(t) * C_\ell(t)\} \tag{13}$$

and analogously for the $\{\vec{B},\vec{C}\}$-fields.

Assuming "even-coding", the storage capacity of this higher-order connection model then increases significantly:

$$p \lesssim N^2/(12*\ln_2 N) \quad \text{with} \quad N = n+m+k \tag{14}$$

5. UNI- AND BIDIRECTIONAL STABILITY

The arguments for BAM_2 uni- and bidirectional stability apply to the corresponding BAM_3 in the same way. The triplet $(\vec{A},\vec{B},\vec{C})$ defines the state of BAM_3. $(\vec{A}_f,\vec{B}_f,\vec{C}_f)$ shall be a stable state which is reached after the energy has found a local maxima. The proof proceeds in complete analogy to Ref.[1]. The definition Eq.6 guarantees also commutativity among the vector-fields: $E(\vec{A},\vec{B},\vec{C})=E(\vec{B},\vec{A},\vec{C})=E(\vec{A},\vec{C},\vec{B})$ etc.. What is significantly different in BAM as compared to BAM_2 is the fact that the bidirectional updating process has to be slightly generalized. Instead of BAM_2's updating process (corresponding to a resonance in Grossberg's ART-theory [10]):

we have the requirement that two of the fields must be simultaneously active such that the units of the third field can be updated. We thus would have the sequence:

whereby the updating of the units can be asynchronously or synchronously. What is essential is that each unit in a field requires simultaneous input from the units in the other fields. One obviously could define the updating in other more selective ways assuming for instance that a unit, at a given time, receives input from one field only, thus generating a cyclic updating from the different fields. The consequences of such variations are currently under study.

6. GENERALIZATION

The above setup of BAM_3 for three vector associations can be generalized in the same way to BAM_q of any number of vector associations. The energy definition and the prescription for storing (and successfully recalling) vectors generalize in the same way. All earlier insights concerning storage capacity apply if appropriately generalized.

7. APPLICATIONS

The practical application of triple-(or more-)pattern associations in the suggested way has not occured very often although numerous examples could be listed. Limiting ourselves to image analysis: one image could be converted to two related images, or the appearance of two defined patterns would be needed for a third pattern to be generated, or, as another alternative, the completion of three patterns. One might ask whether the interaction between <u>all</u> fields is necessary instead of having a relation, say between $(\vec{A},\vec{B})$ and $(\vec{A},\vec{C})$ only. The answer depends on the interdependence of the embedded patterns, how strongly they are correlated, and how much tolerance for error there is, as well as on the rapidity of convergence to an energy minima. BAM_3-type networks have also the advantage that the relationships between different fields can be implemented such that the patterns $\vec{\xi}$ (to be stored) of two fields only are involved. Besides of the earlier sketched storage prescription, supervised learning methods [11] can also be applied.

The above presentation should also be seen in the wider perspective of data fusion where a variety of completely different types of data (image, numerical, algebraic, judgemental, etc.) have to be processed and final decisions have to be reached such as for instance in risk analysis of mortagage screening [12]. We also point to the "Jets and Sharks" example in Ref.[13] where several data-fields interact through a field of hidden units (but no other field-connections are allowed), as well as the hierarchical network architectures where competitive learning [14] is an essential ingredient.

REFERENCES

[1] B.Kosko, "Bidirectional Associative Memories", IEEE Transactions on Systems, Man, and Cybernetics, Vol.18, No.1 (Jan/Feb 1988) pp.49-60;
B.Kosko, "Constructing an Associative Memory", BYTE Vol.12, No.10 (Sep 1987) pp.137-144.

[2] K.Haines and R.Hecht-Nielsen, "A BAM with Increased Information Storage Capacity", Proc. IEEE Intl. Conference on Neural Networks, San Diego (Jul 24-27, 1988) pp.I: 181-190.

[3] C.S.Bak and M.J.Little, "Memory Capacity of Artificial Neural Networks with High Order Node Connection", Proc. IEEE Intl. Conference on Neural Networks, San Diego (Jul 24-27, 1988) pp.I:207-216;
T.-D.Chiuek and R.M.Goodman, "High-Capacity Exponential Associative Memories", Proc. IEEE Intl. Conference on Neural Networks", San Diego (Jul 24-27, 1988) pp.I:153-160.

[4] J.J.Hopfield, "Neural Networks and Physical Systems with Emergent Collective Computational Abilities", Proc.Natl. Acad.Sci.USA, Vol.79 (1982) pp.2554-2558;
B.Humpert, "Analyses of the Hopfield Associative Memory", Preprint ISU-CS/119 (Nov 1989) [submitted to: 6th Annual Academic Microcomputing Conference, Columbus OH (Apr 22-25, 1990)]

[5] D.Kleinfeld, "Sequential State Generation by Model Neural Networks", Proc.Natl.Acad.Sci.USA Vol.83 (Dec 1986) pp.9469-9473;
H.Sompolinsky and I.Kanter, "Temporal Association in Asymmetric Neural Networks", Phys.Rev.Lett. 57,22 (1986) pp.2861-2864;
B.Kosko, see Ref.[1].

[6] L.Personnaz, I.Guyon, and G.Dreyfus, "Collective Computational Properties of Neural Networks: New Learning Mechanisms", Phys. Rev. A34, 5 (1986) pp.4217-4228;
I.Kanter and H.Sompolinsky, "Associative Recall of Memory Without Errors", Phys. Rev. A35,1 (1987) pp.380-392;
S.S.Venkatesh and D.Psaltis, "Linear and Logarithmic Capacities in Associative Neural Networks", IEEE Transactions on Information Theory 35,3 (May 1989) pp.558-568.

[7] W.Krauth, M.Mezard, and J.-P.Nadel, "Basins of Attraction in Perceptron-like Neural Networks", Complex Systems 2 (1988) pp.387-408.

[8] S.Diedrich and M.Opper, "Learning of Correlated Patterns in Spin-Glass Networks by Local Learning Rules", Phys.Rev.Lett. 58, 9 (1987) pp.949-952.

[9] Y.-F.Wand, J.B.Cruz,Jr., and J.H.Mulligan,Jr., "An Enhanced Bidirectional Associative Memory", Proc. IJCNN Intl. Joint Conference on Neural Networks", Washington D.C. (June 18-22, 1983) pp.105-110.

[10] G.A.Carpenter and S.Grossberg, "Massively Parallel Architecture for a Self-Organizing Neural Pattern Recognition Machine", Comput.Vis.Graphics, Image Processing, Vol.37 (1987) pp.54-116;
G.A.Carpenter and S.Grossberg, "The ART of Adaptive Pattern Recognition by a Self-Organizing Neural Network", IEEE Computer (March 1988) pp.77-88 (and references therein).

[11] B.Kosko, "Unsupervised Learning in Noise", Proc. IJCNN Intl. Joint Conference on Neural Networks, Washington D.C. (June 18-22, 1989) pp.I:7-17.

[12] E.Collins, S.Ghosh, and S.Scofield, "An Application of a Multiple Neural Network Learning System to Emulation of Mortgage Underwriting Judgments", Proc. IEEE Intl. Joint Conference on Neural Networks, San Diego CA (July 24-27, 1988) pp.II:459-466;
B.Humpert, "Neurocomputing in Financial Services", ISU-CS/116 (Apr 1989) [submitted for publication].

[13] J.L.McClelland and D.E.Rumelhart, "Explorations in Parallel Distributed Processing", The MIT Press, Cambridge (1988).

[14] see Ref.[13];
K.Fukushima, "A Neural Network for Visual Pattern Recognition", IEEE Computer (March 1988) pp.65-73;
B.Humpert, "On a Neural Network for Visual Pattern Recognition", Proc. 5th Annual Academic Microcomputing Conference (AMC), Indianapolis (Apr 23-26, 1989) pp.113-124.

FIGURE CAPTIONS

Fig. 1 : Bidirectional associative memory with three patterns ($\equiv BAM_3$).

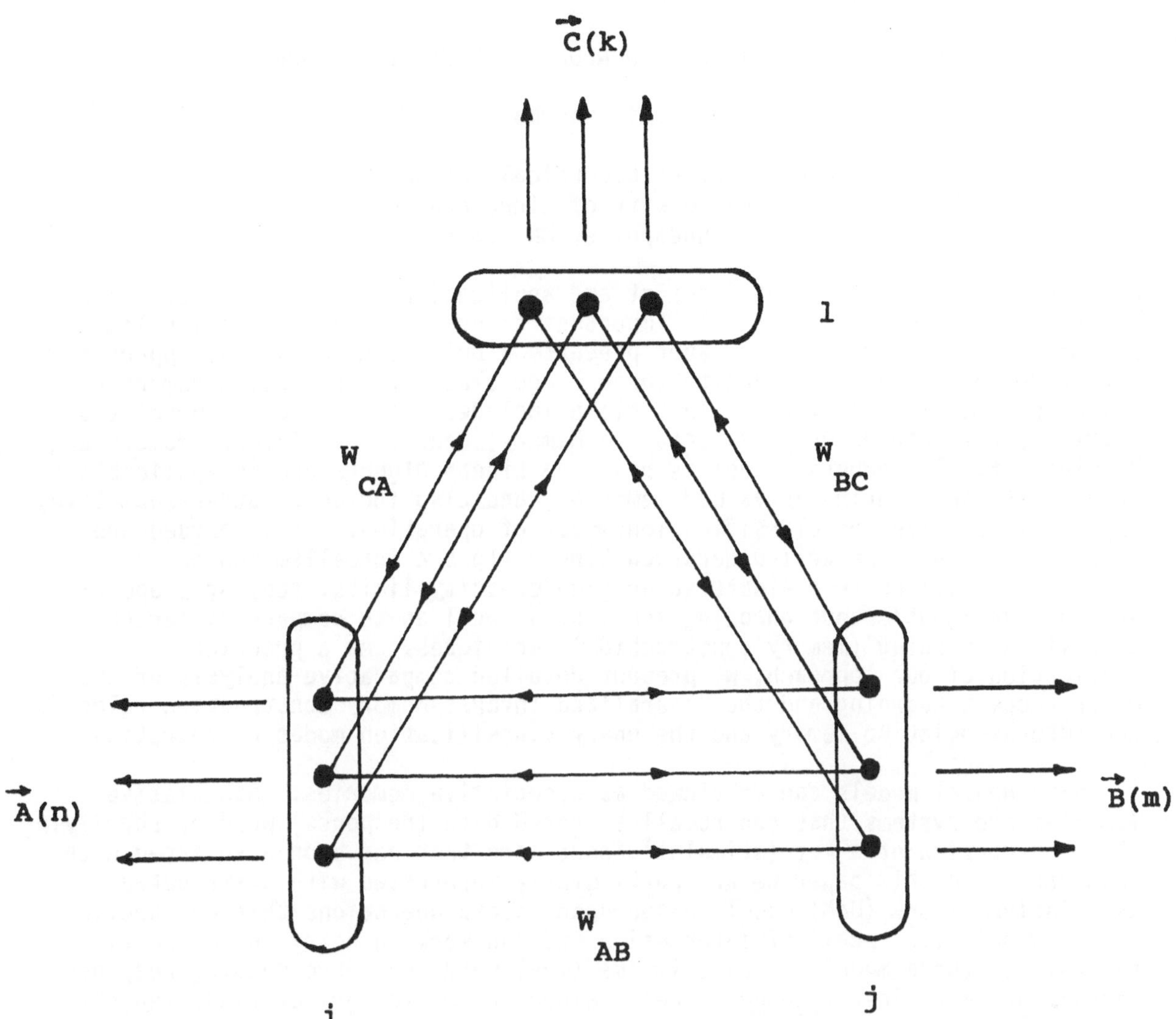

Fig. 1

MATRIX COMPUTATIONS AND NEURAL ASSOCIATIVE MEMORIES

Vladimir Cherkassky

Department of Electrical Engineering
University of Minnesota
Minneapolis, MN 55455

ABSTRACT In view of recent interest and applications of Neural Associative Memories, it becomes increasingly important to evaluate their capacity limits and saturation effects. This paper presents a unified mathematical approach to evaluation of saturation/capacity for a large class of associative memories based upon matrix operations. This class includes, among others, Correlation Matrix Memory, Higher Order Associative Memory, Generalized Inverse Memory and Hamming net. The general model is based on Linear Algebra and is applicable to both binary and continuous-valued memories, and also includes auto-associative, hetero-associative and classification modes of operation. It is argued and demonstrated that the well-understood Linear Algebra formalism can be effectively applied to evaluate saturation/capacity limits, scaling properties and various input/output encoding schemes, as well as to compare different supervised learning (memory construction) techniques. As a practical application of our approach, we present detailed comparative analysis of the Outer Product Learning and the Generalized Inverse Memory construction rules for the auto-associative memory and the unary classification modes of operation.

Many neural models can be viewed as associative memories. Associative memories are systems that can recall a stored data (response) word by specifying all or a portion of a key (stimulus) input word that has been associated with that data. In this paper we are particularly interested with Distributed Associative Memory (DAM) models based upon matrix operations that can provide rapid associative recall of information and can work on noisy and/or partial inputs. In these models, each stimulus (key) word and corresponding response (data) word are first encoded as real-valued column vectors of fixed length. The simulus and response vectors are formed column by column into stimulus and response matrices S and R, respectively. From these matrices, a memory matrix M is then generated according to some learning, or memory construction procedure During the recall phase the memory matrix is multiplied by a stimulus vector to produce the associated response vector

$$\mathbf{r}_i = M\mathbf{s}_i \quad , \quad \text{for } i = 1,2,..n \text{ (n is the number of associations)} .$$

It should be pointed out that error-correcting properties and generalization capabilities of neural networks are usually achieved at the expense of reduced storage capacity. In view of a recent interest in neural networks it seems strange that very little work has been done on saturation/capacity (with a notable exception of Hopfield nets).

Another interesting problem that we address in this paper is how data representation affect network capacity and saturation. In fact, we advocate a unified approach for evaluating distributed representations in terms of their effect on saturation/capacity for a large class of associative memories based upon matrix operations. Consider an associative memory of the form $\mathbf{r} = M\mathbf{s}$. Throughout this paper, bold lowercase letters are used to denote column-vectors. Assume that there are n associations between stimulus vectors of length m and response vectors of length ℓ. This means that the memory matrix M is of size $\ell \times m$. Also assume that the input noise $\mathbf{n}_i$ that is added to each element of a memorized stimulus vector is an i.i.d. random variable.

The recall from memory is then

$$r = r_k + n_o = M(s_k + n_i) = r_k + Mn_i$$

where n_o is the output noise vector. The input noise power is defined as the average square of input noise

$$\sigma_i^2 = (1/m)E(n_i^T n_i) \quad .$$

Since $n_o = Mn_i$, the output noise power is

$$\sigma_o^2 = (1/\ell)E(n_o^T n_o) = (1/\ell)Tr(n_i^T M^T M n_i) = (1/\ell)Tr(n_i n_i^T M^T M). \quad (1)$$

In the general case when input noise components are arbitrarily correlated and non-zero mean, there is no obvious way to simplify (1). However, when each component of the input noise vector is independent identically distributed random variable, expression (1) can be further simplified using some matrix manipulations. For example, for zero mean input noise,

$$\sigma_o^2 = (1/\ell)\sigma_i^2 \, Tr(M^T M) = (1/\ell)\sigma_i^2 \, Tr(MM^T) \; . \quad (2)$$

Notice that expressions (1) and (2) are valid for <u>any type</u> of associative memory with recall of the form $r = Ms$ and <u>any kind</u> of input/output representation. These expressions can be used to evaluate the quality of associative recall, given a specific learning rule that determines how the memory matrix M is constructed from a set of (stimulus, response) vector pairs, or stimulus and response matrices S and R, composed of stimulus and response vector-columns, respectively. Moreover, specific input/output representations are encoded in the form of stimulus and response matrices and their effects on the quality of recall an be readily evaluated.

We then apply the general model to evaluate the quality of recall for the two popular memory construction rules, e.g. the outer product training (or Correlation Matrix Memory) and the Generalized Inverse memory. For each model, we consider and analyze two commonly used associative memory paradigms, i.e. an auto-associator and a unary classifier (Hamming net).

Our analytic results are summarized in the table below that shows how the output noise power depends on the input vector dimensionality, m, and on the number of stored associations, n, for different training rules and different types of memory.

MEMORY TYPE	TRAINING (LEARNING) RULE	
	outer product	Generalized Inverse
auto-associative	$\sim n^2/m$	$\sim n/m$ when $n < m$ 1 when $n > m$
unary classifier	small constant (doesn't depend on n,m)	small when $n \ll m$ large (unstable) when $n \approx m$ small when $n \gg m$

The above result for the Generalized Inverse autoassociative memory agrees with an earlier result by Kohonen [1]. Analysis of the output noise of the Generalized Inverse Memory in hetero-associative (classification) mode of operation explains inherent instability of this model. Here instability means that the output noise becomes extremely sensitive (very large) even to small amounts of the input noise when n approaches m. This effect has been first discovered empirically in [2]. We explain this interesting non-monotonic behavior peculiar to the generalized inverse hetero-associative memory using Singular Value Decomposition techniques. Specifically, we <u>prove</u> that adding a new association monotonically increases the output noise when $n < m$, and monotonically reduces the output noise when $n > m$. The proof is based on the following

Theorem For the hetero-associative Generalized Inverse Memory $M = RS^+$, where S and R are matrices composed of stimulus and response vectors, respectively, the following statements are true:

(a) adding a new independent column-vector to S monotonically degrades its rank deficiency, i.e. decreases its smallest singular value (this occurs when the number of stored associations, n, is smaller than the input space dimensionality, m).

(b) adding a new dependent column-vector to S monotonically improves its rank deficiency, i.e. increases its smallest singular value (this happens when the number of stored associations, n, exceeds the input space dimensionality, m).

Overall, our analysis indicates that the Generalized Inverse training should be used for the auto-associative memory (in perfect agreement with [1]), but that the Correlation Matrix Memory (CMM) should be used for the unary classifier or for heteroassociative memory. Experimental results in Fig. 1 confirm above analyses for binary character recognition applications. Characters of size 8x8 are represented as binary vectors of length 64 and stored in a hetero-associative memory (classifier). Comparison of retrieval (classification) accuracy (see Fig. 1) clearly shows superiority of the CMM model over GI memory.

References

1. Kohonen, T. (1984), Self-Organization and Associative Memory, Springer.
2. Stiles, G. S. and D. L. Denq (1985), On the effect of noise in the generalized inverse associative memory, IEEE Trans. on PAMI, 7, 3, 358.

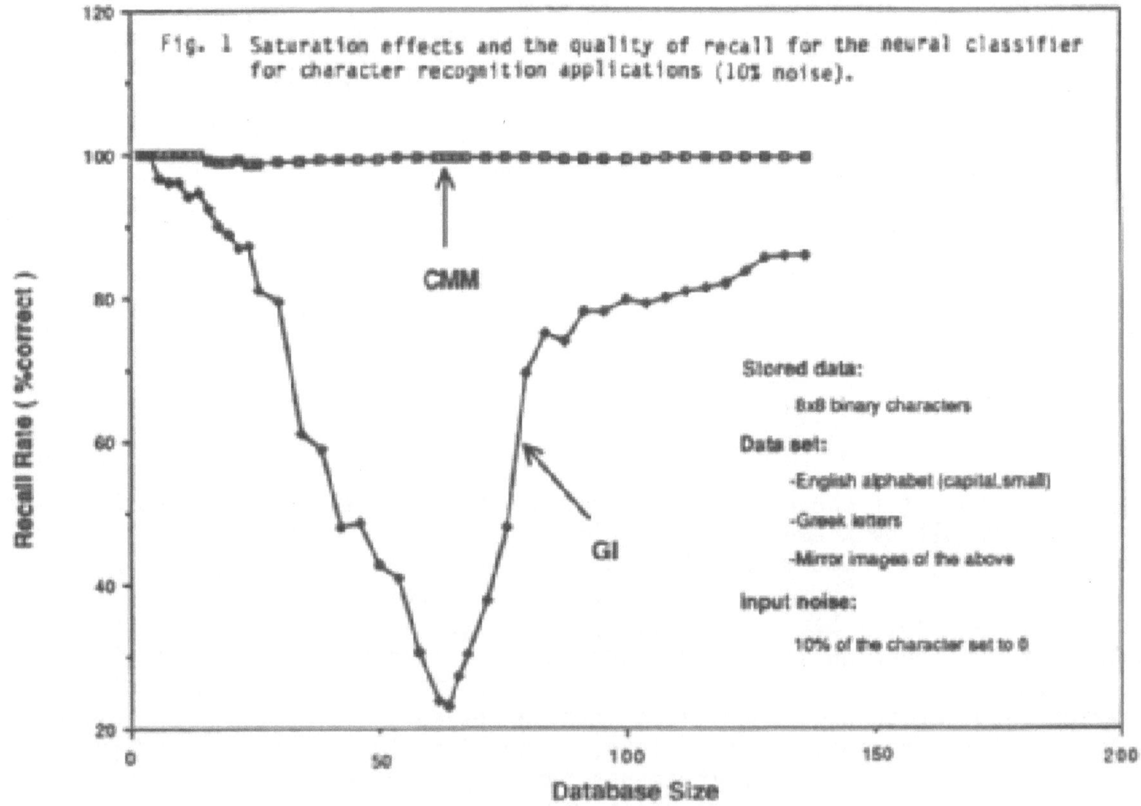

Partially Connected Models of Associative Memory

John Schotland

Bellcore
Morristown, NJ 07960
USA

We investigate the role of graph connectivity in a spin glass model of associative memory. Using replica symmetric mean field theory and spectral graph theory we characterize the storage capacity and retrieval error of partially connected neural networks. Our results have the form of topologically invariant upper and lower bounds on the storage capacity and an asymptotic expansion for the retrieval error. We find that partially connected neural networks have storage and retrieval characteristics that are comparable to fully connected networks. This is of some interest for electronic implementations of neural networks which by necessity are partially connected.

There has been considerable recent interest in spin glass models of associative memory [1,2]. As first introduced by Little [4] and Hopfield [5], these models are based on neural networks with two basic features: symmetric synaptic weights and full connectivity between neurons. However, partial connectivity is essential for the nervous system where a neuron is connected on average to only a small fraction of other neurons. Full connectivity also imposes severe constraints on electronic implementation. Here, interest in the efficient use of resources such as communications links further suggests the importance of partially connected models. In addition, only partially connected topologies can be expected to scale to large systems. Hence there is significant motivation to investigate the computational capabilities of partially connected neural networks.

In a preceding study of partially connected neural networks Derrida *et.al.*[6-8] considered an asymmetric and randomly diluted version of the Hopfield model. They showed for finite connectivity in the thermodynamic limit that the full dynamics of the model can be solved. Sompolinsky [3] has examined the properties of a randomly diluted symmetric model where a finite fraction of the synapses are absent. He found a modest reduction in storage capacity with only a small increase in retrieval error. This conclusion is further substantiated by a study of Canning and Gardner [9] who found a similar result for a dilute symmetric model on a cubic lattice.

In this work we investigate the role of connectivity in controlling the capacity and retrieval characteristics of neural networks. In contrast to randomly diluted models [3,6-8], we consider symmetric neural networks in which the neurons form the vertices of an incomplete graph. Complete graphs appear as a special case. This represents a large previously unstudied class of models for which algebraic graph theory and statistical mechanics may be used to provide an analytic description of the storage capacity and retrieval error. Our results suggest that physically realizable partially connected neural networks have storage and retrieval properties that are comparable to fully connected networks. This conclusion derives from replica symmetric mean field theory and spectral graph theory from which topologically invariant upper and lower bounds on the storage capacity and an asymptotic form for the retrieval error are obtained.

The model we consider is a neural network of Ising spins $S_i \in \{-1, 1\}$ with i a vertex of a graph G. The neural dynamics is a discrete time, zero-temperature Glauber dynamics with evolution law

$$S_i(t+1) = \mathrm{sgn}(\sum_{j\in G} J_{ij} S_j(t)) \tag{1}$$

where J_{ij} is the coupling between spins. The connectivity of G is reflected in the prescription for storing a distinguished set of spin configurations, or patterns $\{\xi_i^\mu\}$ that encode the synaptic couplings by

$$J_{ij} = \frac{1}{N} \sum_{\mu=1}^{p} C_{ij} \xi_i^\mu \xi_j^\mu \tag{2}$$

where p is the number of patterns, N is the number of vertices of G, and C_{ij} is the adjacency matrix of G, with $C_{ij} = 1$ if vertices i and j are connected and $C_{ij} = 0$ otherwise. Consequently, there is a nonzero synaptic weight between neurons i and j if there exists an edge of G connecting vertices i and j. We assume that vertex i has degree ρ_i with $0 \leq \rho_i \leq 1$ in the thermodynamic limit. Note that ρ_i is defined so that full connectivity results when $\rho_i = 1$ for all $i \in G$.

A neural network functions as an associative memory by identifying stored patterns with the fixed points of the dynamical system defined by Eq.(1). The recognition, or retrieval, of a pattern is achieved when the system evolves from an initial configuration to a state that is highly correlated with a stored pattern. The long-time asymptotics of the evolution is governed by the equilibrium statistical mechanics of the underlying spin system. Accordingly, it is possible to introduce a set of order parameters that may be regarded as the overlap of the equilibrium state of the system with each of the stored patterns. Various phases result in which some or all of the overlaps vanish. A nonergodic phase with a single nonvanishing overlap is referred to as a retrieval state. Varying the loading parameter $\alpha = p/N$ induces a phase transition from a retrieval state to an ergodic phase with vanishing overlaps. In a retrieval state the overlap is related to the retrieval error—the expected Hamming distance per spin between the equilibrium state of the system and a stored pattern. The capacity, α_c, is defined to be the largest value of α above which retrieval is not possible.

The principal results of this paper are upper and lower bounds for the capacity and an asymptotic expression for the retrieval error that is valid for small α in the thermodynamic limit. The bounds obtain from the graph spectrum and have the form

$$\alpha_c(G) \leq \alpha_0 \lambda_M^2(G)/d_m(G) \tag{3}$$

and

$$\alpha_c(G) \geq \alpha_0 \lambda_m^2(G)/d_M(G) \tag{4}$$

where α_0 is the capacity for the fully connected graph, $\lambda_M(G)$ and $\lambda_m(G)$ are the largest and smallest eigenvalues of the adjacency matrix of G, and $d_M(G)$ and $d_m(G)$ are the maximum and minimum degree of G. The spectrum of G is related to various topological invariants [10,11] through which Eq.(3) becomes

$$\alpha_c(G) \leq \alpha_0 d(G)/d_m(G) \tag{5}$$

and

$$\alpha_c(G) \leq \alpha_0 d_M^2(G)/d_m(G) \tag{6}$$

where $d(G)$ is the average degree of G. We say that G is regular if $\rho_i = \rho$ for all $i \in G$. If G is regular then the capacity takes the especially simple form $\alpha_c(G) = \alpha_0 d(G)$. Finally, for $\alpha < \alpha_c$ the retrieval error may be expressed as

$$\epsilon = (\frac{\alpha}{2\pi})^{\frac{1}{2}} \int d\nu(\rho)\rho^{-\frac{1}{2}} \exp(-\frac{\rho}{2\alpha}) \tag{7}$$

where $d\nu(\rho)$ is counting measure on $\{\rho_i\}$ with total mass one.

Several comments on these results are necessary. First, the bounds on the capacity given in Eq.'s(3) and (6) are sharp for regular graphs. Second, it should be noted that since the graph spectrum is a topological invariant it follows that the above bounds are also invariant. Third, Eq.'s (3)-(6) predict a reduction in storage capacity for partially connected neural networks. However, for regular graphs the number of bits stored per edge is a constant independent of connectivity. If $B_c(G)$ is the number of bits per edge then

$$B_c(G) \leq 2\alpha_0/d_m(G) \tag{9}$$

and if G is regular then $B_c(G) = 2\alpha_0$. Finally, if G is required to be planar then graph coloring considerations give a lower bound on the capacity of the form

$$\alpha_c(G) \geq \frac{1}{9}\alpha_0 d^2(G)/d_M(G) \tag{10}$$

In summary, we have examined the role of connectivity in a spin glass model of associative memory. We have seen that the graph spectrum determines the capacity and retrieval error of the model. These results will be described in fuller detail in an extended version of this work. In addition we will discuss the modification of zero-temperature results by replica symmetry breaking and non-Hebbian learning rules.

References

1. M. Mezard, G. Parisi, and M. Virasoro, Spin Glass Theory and Beyond (World Scientific, Singapore, 1987)
2. D.J. Amit, in Heidelberg Colloquim on Glassy Dynamics, edited by J.L. van Hemmen and I. Morgenstern (Springer Verlag, Berlin, 1987)
3. H. Sompolinsky, *ibid.*
4. W.A. Little, Math. Biosci. **19**, 101 (1974)
5. J.J. Hopfield, Proc. Nat. Acad. Sci. USA **79**, 2554 (1982)
6. B. Derrida, E. Gardner, and A. Zippelius, Europhys. Lett **4** , 167 (1987)
7. B. Derrida, J. Phys. A **20**, L721 (1987)
8. I. Kanter, Phys. Rev. Lett. **60**, 1891 (1988)
9. A. Canning and E. Gardner, J. Phys. A **21**, 3275 (1988)
10. D. Cvetkovic, M. Doob, and H. Sachs, Spectra of Graphs (Academic Press, New York, 1979)
11. N. Biggs, Algebraic Graph Theory (Cambridge University Press, London, 1974)

REINFORCEMENT LEARNING WHEN RESULTS ARE DELAYED AND INTERLEAVED IN TIME

Catherine Myers
Neural Systems Engineering
Department of Electrical Engineering
Imperial College, London SW7 2BT ENGLAND

Many real-world problems involve sequences where a automaton executes an action but there is some delay before the results of that action become apparent. A system is presented which learns to associate early stimuli with later reinforcement by buffering unfamiliar input images until that reinforcement arrives. It is shown to learn to predict the immediate results of various actions in a given state, to avoid entering negative next-states, and also to avoid entering positive next-states which lead in turn only to negative states. The system is capable of learning across indefinitely long reinforcement delays while only buffering a small number of past states locally at the nodes.

Introduction. In the physical world, most events which entail (positive or negative) reinforcement occur some time before the results actually arrive. For example, when an animal sees a food-like image and decides to approach, grasp and ingest the object, the positive reinforcement (taste) does not arrive until the end of the sequence. The original visual image of the food is supplanted by a series of intervening ones – the final ones do not even include the food image as it is out of sight inside the mouth. Yet even very simple animals learn to bridge this time gap and associate distal images with appropriate approach responses.

Solving this problem requires two abilities: first, that the memory of earlier images be available when the reinforcement arrives; second, that learning be possible even though this reinforcement is only an estimate of "goodness" rather than a full desired output as is traditionally provided for supervised learning in neural networks.

The latter issue, reinforcement learning, has received some attention, notably from Widrow [1,2], Barto and Sutton [3], Klopf [4] and Aleksander [5]. A variant of Aleksander's model, the MPLN (Multi-valued probabilistic logic node) [6] is used here.

The question of learning with delayed reinforcement is often approached in one of two ways. One solution is to maintain a buffer of all previous states, possibly each with an eligibility that decays with time, and then to update each according to its eligibility when reinforcement arrives. This quickly becomes impractical as the number of possible states grows. A second solution is to buffer only the S states immediately previous to the current; but this precludes the system from learning about images which occur S+1 time steps before their associated rewards.

The system investigated here also maintains a buffer of some N previous states, but these are not necessarily those that occurred in the immediately preceding time steps. Rather, they are the S most unfamiliar previous stimuli. A state is placed in the buffer if its outcome is more unpredictable than some item currently in storage, and it overwrites that item. Simultaneously, the longer an item has been in the buffer, the more likely it is to be ousted by a new item. If, on the other hand, the effects of the new stimulus are predictable with great certainty, it is unnecessary that the item enter the buffer – since there is nothing new to be learned about it.

In this way, the system can keep a small number of previous states available, and yet learn to associate reinforcement with states which occurred indefinitely earlier.

The Task. The problem considered here is based loosely on the idea of an automaton learning to select food. A set of M types of element (each a 64-bit pattern) exist in the world; of these some $P \subseteq M$ are positive while the remainder $N \subseteq M$ are negative. At each time cycle, the automaton is in some state $x \in M$, and has a choice of moving left, right or straight ahead; a transition matrix determines the next state: $f(x, \text{move}) \rightarrow y \in M$. If $y \in P$, a positive reinforcement is supplied immediately (as if the automaton experienced the taste of food). If the automaton enters $y \in N$, there is an immediate positive reinforcement, followed by a strong negative reinforcement delayed by some d time steps (as if there was a taste of food later followed by nausea).

Thus the task is to learn to predict the three adjoining states from the current one, and also to select the moves which result in some $y \in P$ and not $y \in N$, even though the results are delayed and contradictory signals may intervene. For example, negative reinforcement may not arrive until some time after $x \in N$ has been entered, and it may arrive just after some element of P has been entered.

The Model. The adaptive system consists of three basic parts: the Associator Module (AM) which, given the current input image and a suggested next move, predicts the resulting next state; the Judge Module (JM) which, given a predicted next state, estimates the desirability of entering that state; and the short-term store (STS) which stores recently entered states in readiness for the arrival of results. The complete system is shown in Figure 1. The move selected will be the one which results in a prediction from AM to which the JM responds most highly.

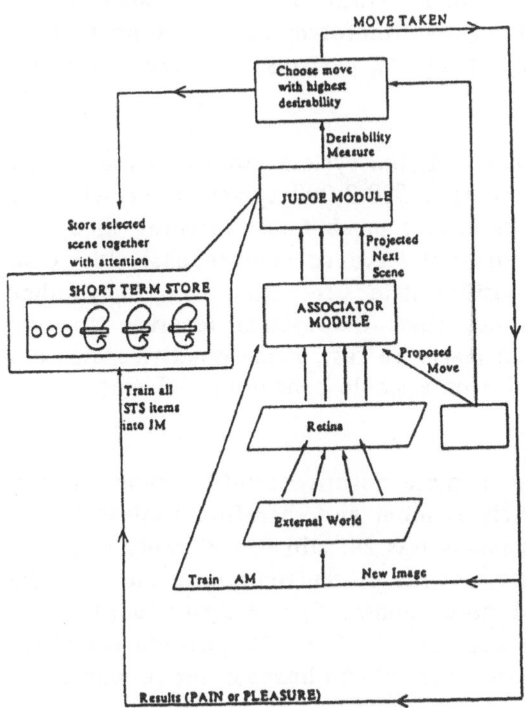

Figure 1. The system for learning with delayed reinforcement. The AM predicts next states for given moves, the JM judges desirability of next states, and move is chosen accordingly. Input scenes are stored in STS buffer until reinforcement arrives, whereupon judge is trained to adjust desirability predictions.

There is a limit, S, on the number of elements which may reside in STS at any one time. As each element enters STS, it is assigned a certain strength or attention which decays with time. When a new element is to be stored in STS, it overwrites the resident element with the weakest attention. Initial attention, in the simplest case, is a constant larger than the longest possible delay D: STS then reduces to an S-previous-element buffer of the sort discussed above.

The AM consisted of MPLN trees: each containing 8 10-input MPLNs feeding into an 8-input top node. The lower level nodes each sampled 8 input bits (randomly, but all 64 input bits were used) plus two bits encoding the move to be evaluated. 64 of these trees existed, each outputting a bit, so that the predicted next state could be reproduced. Two JMs were investigated: a multi-layer version, MLJM, of 8 MPLN trees – each similar to the AM trees, but without the move information; the total responses from all top nodes gave a score 0..8 of how desirable the projected state was judged to be; and a single-layer version, SLJM, consisting simply of 25 8-input MPLNs, sampling the input retina randomly but evenly, and outputting total response in the range 0..25.

When an action is selected, the current state is stored in STS, along with up to 4 other previous states (S=5); this is done locally at each node, and so could be synchronous. When reinforcements arrive, each STS pattern is reapplied to the JM, which is then trained by a standard MPLN learning algorithm [6] to increase or decrease its desirability measure for those patterns. The AM is updated after each cycle, by the same algorithm, to produce the received next state of the world in response to the previous state and move taken.

Results. The AM, trained alone to produce next state from current state plus move, predicted with 90% accuracy (measured in bits right) after 300 passes through the pattern set; within 900 passes it achieved 99% accuracy, and took 6,000 passes to perform perfectly.

The MLJM, also trained separately, learned within 3,000 cycles (~100 passes through its training corpus) to respond strongly to all elements of P and weakly to all elements of N. This module was then paired with the trained AM, so that temporal effects (such as "avoid positive states which lead only to negative ones (cul-de-sacs)") came into effect. Within a further 2,000 cycles, the complete system learned all the necessary associations: including avoidance of the state which led only to elements of N, but non-avoidance of states which led to an element of N but also at least one element of P. However, by the conclusion of this training, the system had experienced over 60 negative reinforcements or an average of 20 per element of N – low by neural engineering standards, perhaps, but excessive when compared with animal learning.

Training SLJMs in conjunction with the pretrained AM, the system tended to receive an average of 5.5 negative reinforcements within the first 500-2,000 cycles, after which it would never again enter a negative state. This is clearly much faster learning than was obtained with the MLJM; however, the system did not learn so comprehensively. One solution found is shown in Figure 2a. Not every positive state is re-entrant – in fact, rather less than half are. This is still a valid solution, since the system receives positive reinforcement on every cycle. Because the system learns so fast, some positive states are entered only once or twice before the system settles on a stable behaviour and may not be entered again.

Simultaneous training of AM and SLJM resulted in more negative reinforcements during training, as early output from the AM was nearly random and therefore useless to the SLJM. The average number of negative reinforcements was 24, still considerably less than with the MLJM even trained alone. Typically, most of these occurred within the first 500 cycles. Solutions found by these systems included more re-entrant states than when the AM was pre-trained; one solution, shown in Figure 2b, has 15 of the 22 positive states re-entrant, with all negative elements and the cul-de-sac transient and hence never re-entered.

The observed tradeoff is that the MLJM solves the problem perfectly, while the SLJM finds acceptable approximations to the solution within much fewer negative experiences. The advantage of using multi-layer systems is their ability to use hidden nodes to form internal representations. In this task, such representations are unnecessary, and the SLJM was perfectly adequate – particularly as, composed of MPLNs, each node could learn any boolean function of its inputs.

Conclusions. A system has been described to learn under conditions of a global scalar

Figure 2. Example solutions found by SLJM with pre-trained AM (a) and by SLJM and AM trained together (b). Filled circles = positive states, white circles = negative states, striped circle = cul-de-sac state. Box enclosed re-entrant states.

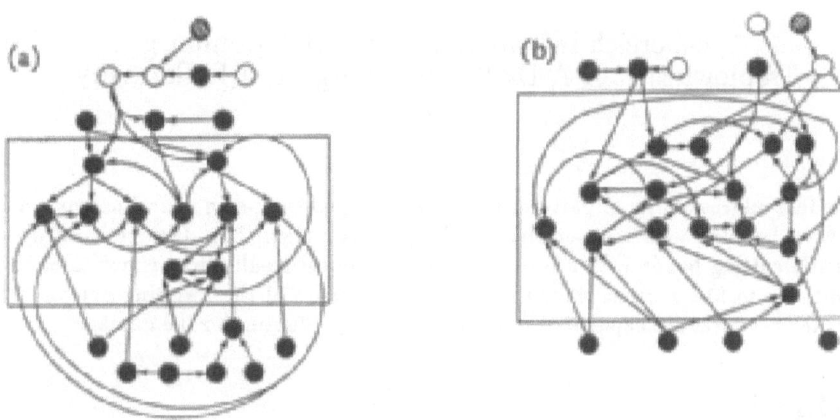

reinforcement signal which arrives with some delay; it achieves this by keeping a store of S recent inputs with unpredictable outcomes. This store may be local to the nodes, allowing parallel updates. The system has been shown capable of making second order temporal predictions, such as "avoid a state which is itself positive but which leads inevitably to a negative one." The results described here involve a buffer size which is at least as large as the maximum possible delay; a future paper will show the system capable of learning even when the maximum delay is longer.

References

[1] Widrow, B., Gupta, N. and Maitra, S. Punish/reward: learning with a critic in adaptive systems. *IEEE Trans. on Systems, Man and Cybernetics*, SMC-3(5), 455-465, 1973.

[2] Widrow, B. and Smith, F. Pattern-recognising control systems. *Computer and Information Sciences*, Eds. J. Tou and R. Wilcox. Washington, D.C.: Spartan Books, 1964, pp. 288-317.

[3] Barto, A., Sutton, R. and Anderson, C. Neuronlike adaptive elements that can learn to solve difficult learning control problems. *IEEE Trans. on Systems, Man and Cybernetics*, SMC-13(5), 834-851, 1983.

[4] Klopf, A. H. A neuronal model of classical conditioning. *Psychobiology*, 16(2), 85-125, 1988.

[5] Aleksander, I. Logical connectionist systems. *Neural Computers*, Eds. R. Eckmiller and C. von der Malsburg. Berlin: Springer-Verlag, 1988, pp. 189-197.

[6] Myers, C. Output functions for probabilistic logic nodes. *Proc. First IEE International Conf. Artificial Neural Networks*, London, October, 1989, pp. 310-314.

Fault-tolerance in Iterative Learning Neural Networks

Norman Hendrich

Fachbereich Informatik, Universität Hamburg,
Troplowitzstraße 7, D-2000 Hamburg 54, F.R.Germany

Abstract

The content-addressability of patterns stored in damaged Ising-spin neural network models is studied. After learning with the iterative Edinburgh-group algorithm, damage is introduced by independently cutting bonds with a probability p. Numerical results from simulations involving systems with up to 512 neurons show the retrieval properties of the resulting networks. Below saturation the networks are capable of re-learning, thereby adapting to the damage.

1 Introduction

Ising-spin neural networks are believed to be a massive parallel and fault-tolerant paradigm for content-addressable memory. Recent works on damaged networks [3] have concentrated on the Hopfield-model [1] [2], which is rather limited in terms of its storage capacity.

The network considered here was first proposed by Gardner [5] [6]. It consists of N neurons $S_i = \pm 1$, with couplings J_{ij} from neuron j to neuron i. Updating is parallel:

$$S_i(t+1) = \text{sgn}(h_i(t)) = \text{sgn}\Big(\sum_{j\neq i} J_{ij} S_j(t)\Big), \qquad i = 1, \ldots, N. \tag{1}$$

Content-addressable memory is realized by this dynamics, when the $P = \alpha N$ patterns (Ising-spin configurations) $\xi_i^\mu = \pm 1$, $\mu = 1, \ldots, P$, $i = 1, \ldots N$ to be stored are its fixed points, $\xi_i^\mu \cdot h_{i\mu} > 0$. To store the patterns ξ_i^μ with finite basins of attraction, positive constants $\kappa_{i\mu}$ are introduced and the couplings J_{ij} are choosen to fulfil the stronger constraints

$$\xi_i^\mu \cdot \sum_{j\neq i} J_{ij} \xi_j^\mu \geq \kappa_{i\mu} \Big[\sum_{j\neq i} J_{ij}^2\Big] > 0. \tag{2}$$

The stabilities $\kappa_{i\mu}$ may be chosen independently for each pattern and neuron, larger values of the $\kappa_{i\mu}$ implying larger basins of attraction. A network with the optimal values of the $\kappa_{i\mu}$ is called saturated. Here the case of $\kappa_{i\mu} = \kappa$ is studied.

The values of the couplings J_{ij} are calculated iteratively using the Edinburgh-group algorithm [6], [4]. In one learning step a coupling is modified if the stability of the pattern is less than the desired stability:

$$\Delta J_{ij} = \epsilon_{i\mu} \xi_i^\mu \xi_j^\mu, \tag{3}$$

$$\epsilon_{i\mu} = \Theta\Big(\kappa_{i\mu} - \xi_i^\mu \cdot h_{i\mu}\Big), \tag{4}$$

with $\Theta(x) = 1$ if $x \geq 0$ and $\Theta(x) = 0$ if $x < 0$. The process is repeated until ΔJ_{ij} is zero for all patterns and neurons. The algorithm is proven to converge towards a solution in a finite number of steps below saturation. The couplings constructed by this algorithm are not symmetric.

This model has been studied in a number of ways. The storage capacity in the replica-symmetric approximation is known to be [6]

$$\alpha_c(\kappa)^{-1} = \int_{-\kappa}^{\infty} dt \, \frac{e^{-t^2/2}}{\sqrt{2\pi}} (t+\kappa)^2, \tag{5}$$

so that $\alpha_c(0) = 2.0$. The model has finite basins of attraction for all storage ratios below $\alpha = 2.0$. Analytical results for the basins of attraction have been found for the case of saturated networks [8].

2 Performance of damaged networks

To study the fault-tolerance of the networks we introduce random damage after learning by independently cutting bonds (zeroing couplings) with a probabiliy p, while the remaining couplings were left unchanged. The result (5) for the storage capacity remains valid in the damaged network when α is defined as $\alpha = P/(1-p)N$.

The basins of attraction in the damaged networks are investigated by generating random test patterns $\xi_i^{\mu,r}$ having an initial overlap m_0 with the stored patterns ξ_i^{μ},

$$m_0 = N^{-1} \sum_i \xi_i^{\mu} \xi_i^{\mu,r} \tag{6}$$

and iterating to stability under (1).

Two measures of the basins of attraction (and hence the content-addressability) are the percentage of perfectly recalled patterns f_p and the mean final overlap m_f of the iterated test patterns with the stored patterns as a function of the initial overlap m_0.

Typical simulation-runs consisted of several hundred test patterns per m_0. The error bars, typically between 1% and 5% were dropped in the interest of clarity. The simulations were carried out on VAX-computers of the university of Hamburg and involved systems of up to $N = 512$ spins with parallel updating.

Figure 1 shows the percentage of perfectly recalled patterns for networks with $N = 256$ neurons for storage ratios $\alpha = 0.20$ and 0.40 after damaging with different probabilities p.

After damaging the percentage of perfectly recalled patterns decreases. As was to be expected, the fluctuations in $f(m_0)$ in the damaged systems are rather large. The data is well described by the scaling hypothesis

$$f(m_0)/(d(p) - f(m_0)) = C \exp(Na(m_0 - m_c)) \tag{7}$$

where $d(p)$ is a function of the concentration of damaged synapses and m_c is the critical initial overlap a test pattern must have with a stored pattern to get recalled. Best fit curves of this form are shown in figure 1. In the undamaged systems $d(p) = 1$, which confirms the finite-size scaling behaviour proposed by [4].

Figure 3 shows the fraction of perfectly stored nominal patterns after damage as a function of damage p for various values of the storage ratio α ($N = 256$). At $\alpha = 0.2$ the network stores all nominal patterns perfectly for a damage of $p \leq 0.2$, while at $\alpha = 0.5$ a damage $p \leq 0.05$ is tolerable.

The final overlap of the patterns is even less sensitive to damage, as may be seen from figure 2, which shows the final overlap m_f as a function of the initial overlap m_0 at $\alpha = 0.2$ and 0.4.

While at $\alpha = 0.2$ and a damage of $p = 0.35$ only 20 % of the patterns are recalled perfectly, the system reaches a mean final overlap of $m_f > 0.98$. The stored patterns contain some faults (wrong spins), but are highly correlated to the nominal states ξ_i^{μ}. A similar behaviour is found in the Hopfield-model at much smaller storage ratios.

3 Relearning after damage

Unlike the Hopfield-model with its Hebbian prescription for the couplings $J_{ij} = N^{-1} \sum_{\mu} \xi_i^{\mu} \xi_j^{\mu}$ which depends only on the patterns to be stored, the iterative learning algorithm uses information of the state of the net to determine the couplings.

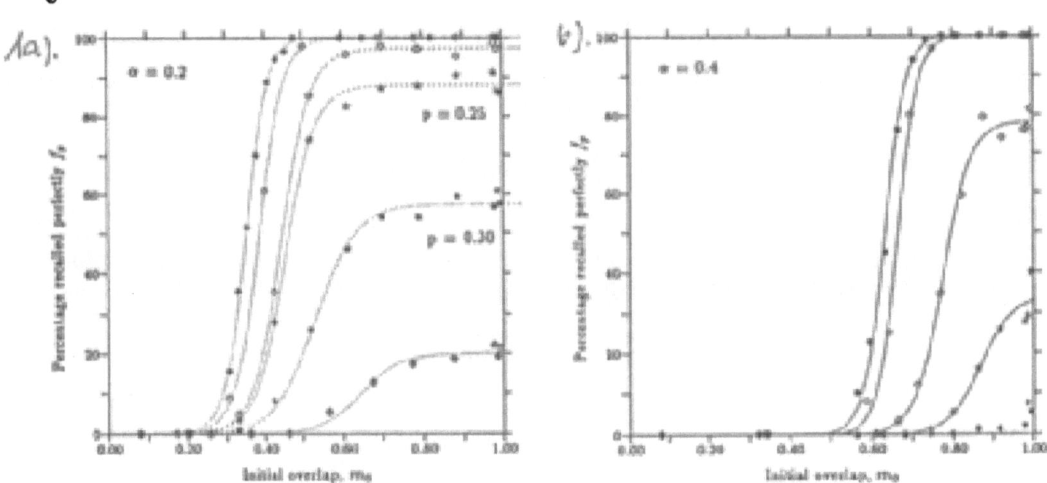

Figure 1: The fraction f_p of nominal patterns that are recalled perfectly from initial states having an initial overlap m_0 for different values of damage p. Best-fit scaling forms (7) are shown. (a) $\alpha = 0.2$, $\kappa > 1.6$, damage $p = 0$ •, 0.1 ○, 0.2 ◇, 0.25 ⋆, 0.3 •, 0.35 •. (b) $\alpha = 0.4$, $\kappa > 0.8$, damage $p = 0$ •, 0.05 ○, 0.1 ◇, 0.15 ⋆, 0.2 •.

Figure 2: Mean final overlap m_f of test patterns having initial overlap m_0 for different values of damage p. Note different scales. (a) $\alpha = 0.2$, $\kappa > 1.6$, damage $p = 0$ •, 0.2 ⋆, 0.3 •, 0.35 ○, 0.5 ◇. (b) $\alpha = 0.4$, $\kappa > 0.8$, damage $p = 0$ •, 0.05 ○, 0.1 ◇, 0.15 ⋆, 0.2 •.

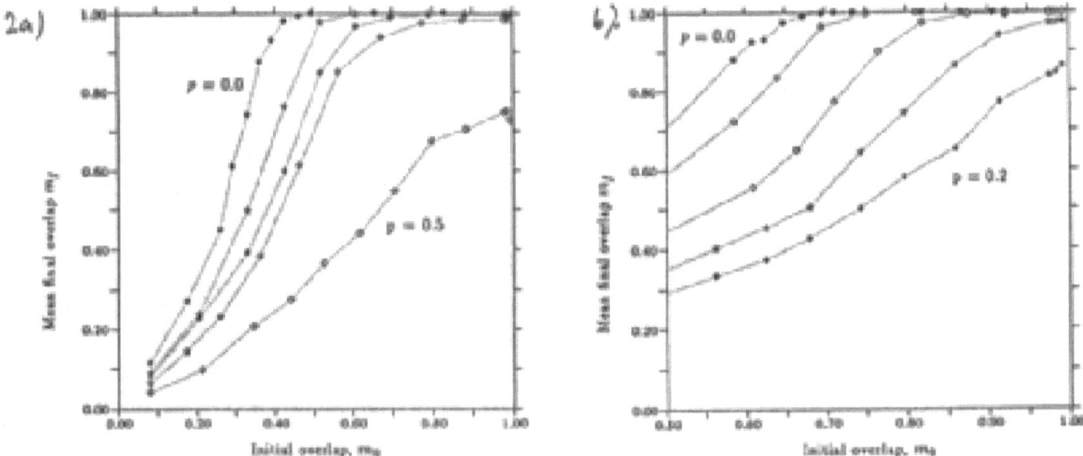

After damage most values of the $\kappa_{i\mu}$ will be lower than desired and the repetition of the learning algorithm (3) will result — as long as $\alpha < \alpha_c$ — in a new solution to (2) under the additional constraint that the damaged couplings are hold at zero.

Figure 4 shows an example of re-learning for $\alpha_i = 0.2$ and a damage of $p = 0.5$. The destruction of 50% of the couplings introduces negative values of $\kappa_{i\mu}$ and the system looses all memories. After re-learning all patterns, corresponding to $\alpha_f = 0.4$ in the damaged network, are again stored perfectly. The distribution of the stabilities $\kappa_{i\mu}$ is shown after learning, after damage and after re-learning.

4 Conclusions

Spinglass neural networks prove to be fault-tolerant under the effects of synaptic damage, especially at moderate storage ratios. However, damage results in a distribution of stabilities $\kappa_{i\mu}$ that is not optimal and degrades the content-addressability.

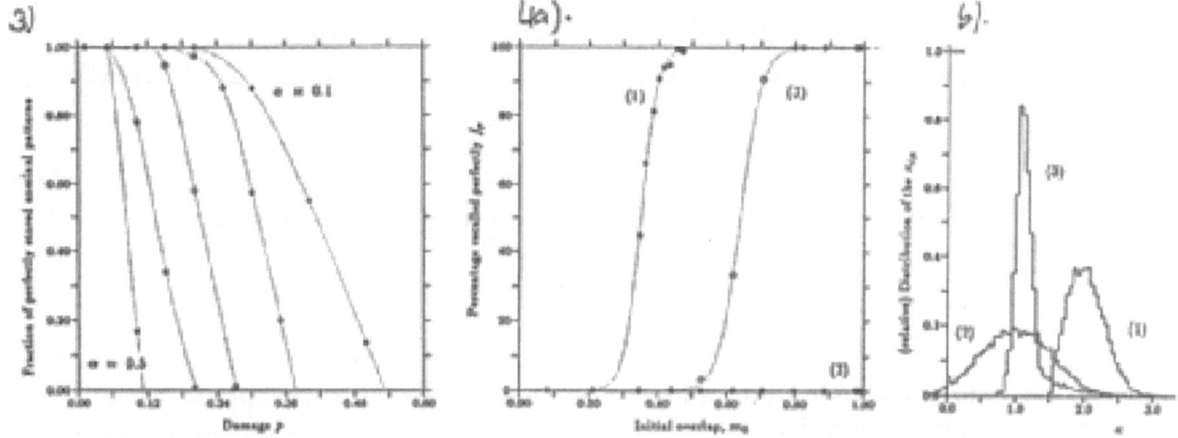

Figure 3: The fraction of perfectly stored patterns ξ_i^μ as a function of damage p for different storage ratios $\alpha = 0.1\bullet, 0.2\circ, 0.3\circ, 0.4\star, 0.5\bullet$.

Figure 4: Example of relearning, $\alpha = 0.2$, $\kappa = 1.6$. (1) after learning, no damage (2) damaged (3) after relearning.
(a) Percentage f_p of perfectly recalled patterns. (b) (relative) Distribution of stabilities $\kappa_{i\mu}$.

Iterative relearning of the connections enables the systems to adapt to damage and minimize its effects. The distribution of the stabilities $\kappa_{i\mu}$ after relearning is again optimal.

Therefore the systems may implement an efficient means of fault-tolerance and robustness in neurocomputers. Any errors introduced by defective hardware may be compensated by relearning, as long as the storage ratio α is kept below the critical value α_c (5).

We are currently designing a set of simulator-chips, which will implement iterative learning and will allow us to study much larger systems.

References

[1] J. J. Hopfield, Neural networks and physical systems with emergent collective computational abilities, *Proc. Nat. Acad. Sci.* 79, 2554–2558 (1982).

[2] Daniel J. Amit, Hanoch Gutfreund and H. Sompolinsky, Spin-glass models of neural networks, *Physical Review A* 32, 1007–1018 (1985).

[3] Eva Koscielny-Bunde, Effect of damage in neural networks, *J. Stat. Phys.*, in press (1990).

[4] B. M. Forrest, Content-addressability and learning in neural networks, *Journal of Physics* A 21, 245-255 (1988).

[5] E. Gardner, Maximum Storage Capacity in Neural Networks, *Europhysics Letters* 4, 481–485 (1987).

[6] E. Gardner, The space of interactions in neural network models, *Journal of Physics* A 21, 257–270 (1988).

[7] E. Gardner, B. Derrida, Optimal storage properties of neural network models, *Journal of Physics* A 21, 271–284 (1988).

[8] Thomas B. Kepler and L. F. Abbott, Domains of attraction in neural networks, *J. Phys. France* 49, 1657-1662 (1988).

MEMORY OF CORRELATED PATTERNS BY ASSOCIATIVE NEURAL NETWORKS WITH IMPROVED DYNAMICS

Masahiko Morita, Shuji Yoshizawa and Kaoru Nakano

Department of Mathematical Engineering and Information Physics,
Faculty of Engineering, University of Tokyo,
Hongo, Tokyo 113, Japan

Abstract: It has been considered that the autocorrelation type of associative memory does not work well unless stored patterns are mutually uncorrelated. The present paper reports that the autocorrelation model with a nonmonotonic output function can memorize the patterns which are substantially correlated; moreover, the recollection ability is even raised in some cases. The reason for this strange fact is also examined through numerical experiments.

1. Introduction

In the conventional autocorrelation associative memory [1-3], including the Hopfield model[4], patterns to be memorized in the neural network should be nearly orthogonal or uncorrelated with each other. If not so, memory capacity and recollection ability of the network are much lessened. This is an unsatisfactory condition because it strongly restricts the representation of memory: no matter how closely related two things are, they should be represented by uncorrelated patterns.

Though correlated patterns can be stored by pseudo-inverse[5] or orthogonal learning[6], these methods are complex and take much time for learning; besides, the problem of spurious memory still exists, *i.e.*, the stored patterns and other equilibrium states cannot be distinguished by only the outputs of neurons. Furthermore, as described later, they cannot make good use of correlations between the patterns.

We have found that improvement of the recalling dynamics—using a nonmonotonic output function instead of the conventional sigmoid function—greatly raises the ability of the associative neural networks[7]: both memory capacity and recollection ability are increased, and stored patterns can be retrieved without errors; moreover, spurious memory is never recalled. Then is this method effective for the case where the stored patterns are correlated?

In this paper, we will show that the autocorrelation associative memory with the improved dynamics can memorize the patterns which are mutually correlated to a considerable extent, and that the recollection ability is *raised* if the correlation is within a limit. We will also examine the process of recollection and consider why the nonmonotonic function improves the dynamical properties of the network.

2. Autocorrelation Associative Memory

Let us consider a neural network consisting of n neurons connected with each other. We describe the state of the network by an n-dimensional vector $X = (x_1, x_2, \cdots, x_n)$, $x_i = \mathrm{sgn}(u_i)$ being the state of neuron i. Here, $\mathrm{sgn}(u) = 1$ when $u > 0$ and -1 when $u < 0$, and u_i denotes the total input to neuron i (for discrete model) or the instantaneous potential (for analog model).

Let w_{ij} be the synaptic weight from neuron j to neuron i, $S^1, S^2, \cdots, S^m$ be m binary patterns to be stored, and $s_i^\mu (= \pm 1)$ be the i-th component of S^μ $(i = 1, 2, \cdots, n)$.

First we consider the case where these patterns are randomly selected out of 2^n possible patterns, *i.e.*, every component is 1 or -1 with equal probability. In this case, any two of the patterns are nearly orthogonal, or the correlation between S^μ and S^ν $(\mu \neq \nu)$ defined by

$$\mathrm{Cor}(S^\mu, S^\nu) = \frac{1}{n} \sum_{i=1}^{n} s_i^\mu s_i^\nu \tag{1}$$

is very small. Then, as is widely known, we can store these patterns in the network using the autocorrelation matrix:

$$w_{ij} = \frac{1}{n} \sum_{\mu=1}^{m} s_i^\mu s_j^\mu, \tag{2}$$

because

$$\sum_{j \neq i} w_{ij} x_j \simeq s_i^\mu \tag{3}$$

holds if X is sufficiently similar to S^μ.

Though there are some variations of neural dynamics to retrieve the stored patterns, we deal with an analog network whose dynamics is expressed by

$$\tau \frac{du_i(t)}{dt} = -u_i + \sum_{j \neq i} w_{ij} y_i, \tag{4}$$

$$y_i = f(u_i), \tag{5}$$

where τ is a time constant, y_i is the output of neuron i and $f(u)$ is an output function. Conventionally $f(u)$ is a sigmoid function (Fig. 1(a)) given by

$$f(u) = \frac{1 - \exp[-cu]}{1 + \exp[-cu]}, \tag{6}$$

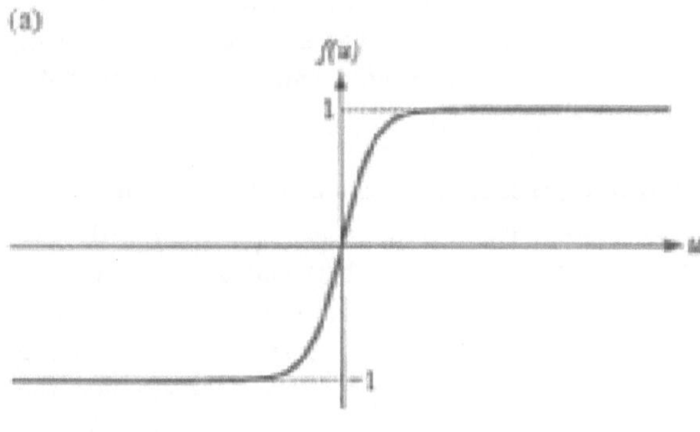

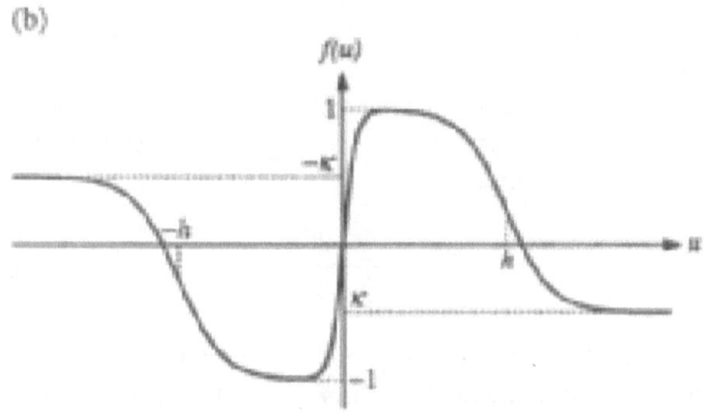

Fig. 1. Output functions of analog neurons: (a) conventional (sigmoid function); (b) improved (S-function).

c being a positive constant. Starting from some given initial state X_0 close to one of the stored pattern, say S^1, the network state X is expected to approach S^1 and settle in an equilibrium state which is identical or very close to S^1.

However, the autocorrelation memory model with the conventional dynamics has a small memory capacity. If more than about $0.15n$ patterns are stored, the network state X goes away from S^1 no matter what initial state is given, although it once comes near S^1[8]: see Fig. 2(a). It shows the time course of change in the overlap p between X and S^1 defined by

$$p = \mathrm{Cor}(S^1, X) = \frac{1}{n}\sum_{i=1}^{n} s_i^1 x_i. \tag{7}$$

Here, the initial state X_0 is randomly selected out of the states with a certain overlap with S^1, so that X_0 has little overlap with the other stored patterns.

3. Improved Recalling Dynamics

On the other hands, we can greatly improve the dynamical properties of the network using a nonmonotonic function shown in Fig. 1(b) instead of the conventional sigmoid function. This nonmonotonic function (S-function) is given by

$$\tilde{f}(u) = \frac{1-\exp[-cu]}{1+\exp[-cu]} \cdot \frac{1+\kappa\exp[c'(|u|-h)]}{1+\exp[c'(|u|-h)]}, \tag{8}$$

where c' and h are positive constants and κ is a constant (usually negative).

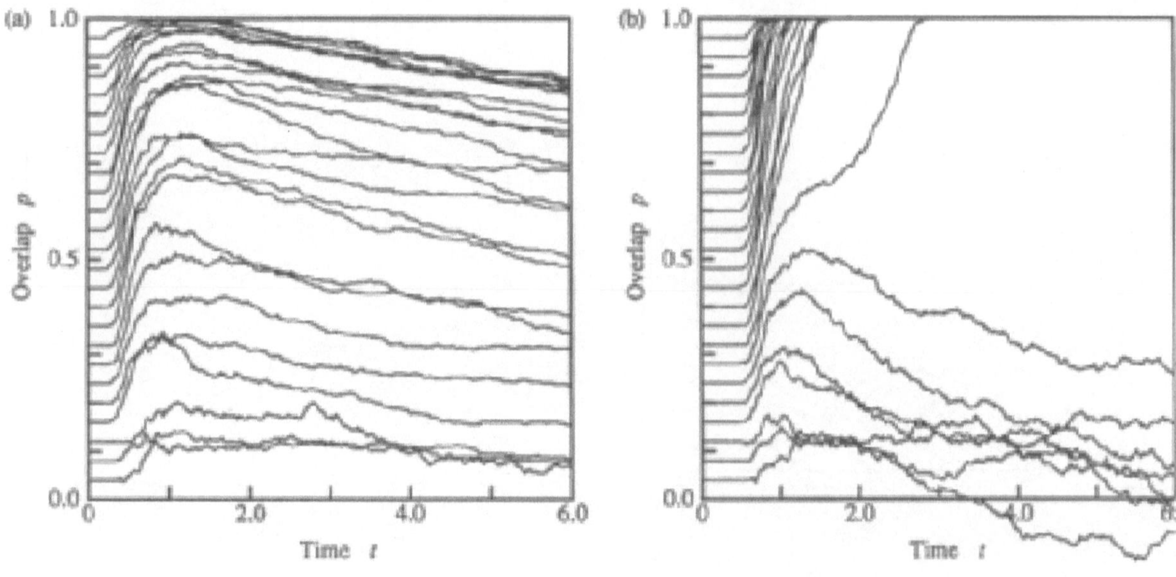

Fig. 2. Time course of change in the overlap p: (a) with conventional dynamics; (b) with improved dynamics. Numerical experiments with n = 1000, m = 200, c = 15, c' = 50, h = 0.5, $\kappa = -1$.

As shown in Fig. 2(b), if the initial overlap p_0 is larger than a critical overlap p_c, the overlap p quickly increases and becomes unity, that is, the stored pattern is recalled without errors; if $p_0 < p_c$, p continues to change chaotically and probably the network state never reach an equilibrium state. This critical overlap represents the recollection ability of the model: if p_c is smaller, the network can recall the correct pattern from a pattern with more noises.

The memory capacity of this model is more than twice as large as that of the conventional one, but it depends on the parameters and the exact capacity is not clear.

4. Memory of Correlated Patterns

In the above discussions, we assumed that all the stored patterns are uncorrelated, and thus they are distributed uniformly in the state space. Now we will consider the case where they are distributed in clusters, *i.e.*, any stored pattern is correlated to some other patterns.

Let us select k patterns $C^1, C^2, \cdots, C^k$ at random, and $k \times l$ patterns $S^{11}, S^{12}, \cdots, S^{kl}$ so that $\mathrm{Cor}(S^{\mu\nu}, C^{\mu}) = a$ for every μ and ν, where a is a positive constant. We can do so by reversing the sign of $(1-a)/2$ components of C^{μ} which is randomly selected. Then there are k clusters whose centers are C^{μ}'s, and each cluster consists of l patterns mutually correlated with a correlation of a^2.

We store these $m = kl$ patterns in the network using Eq. (2), where we put $S^1 = S^{11}, S^2 = S^{12}, \cdots, S^m = S^{kl}$.

It is considered that unless a is very small, the autocorrelation model does not work well because Eq. (3) does not hold. Indeed its memory capacity and recollection ability decrease as a increases. However, this does not necessarily hold true in the model with the improved dynamics.

5. Numerical Experiments and Discussions

Figure 3 shows the results of an experiment with $n = 1000$, $k = 50$, $l = 4$ ($m = 200$) and $a = 0.6$. In this figure, the change in p is plotted in the same manner as Fig. 2, except that the initial state X_0 has an overlap of about $a^2 p_0$ with $S^{1\nu}$ ($\nu = 1, 2, 3$). We see that the critical overlap p_c is smaller than that in Fig 2(b).

Then why does such a phenomenon occur? We do not have an exact answer, but the following examination will help us to understand it.

Let us consider the direction cosine $d_{\mu\nu}$ between $S^{\mu\nu}$ and the output vector $Y = (y_1, y_2, \cdots, y_n)$ of the neurons: we define

$$d_{\mu\nu} = \frac{\sum_i s_i^{\mu\nu} y_i}{\sum_i x_i y_i}. \tag{9}$$

Though d_{11} is similar to p and $d_{11} = 1$ means $p = 1$, they are not equivalent, because $y_i = \tilde{f}(u_i)$ is often very different from $x_i = \mathrm{sgn}(u_i)$. Note that putting the total input

$$v_i = \sum_j w_{ij} y_i, \tag{10}$$

we can write the input vector $V = (v_1, v_2, \cdots, v_n)$

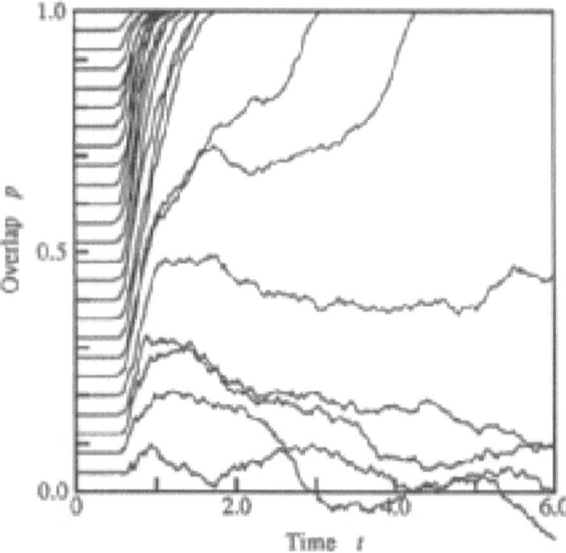

Fig. 3. Time course of change in the overlap p when the stored patterns are correlated: $n = 1000$, $k = 50$, $l = 4$, $a = 0.6$.

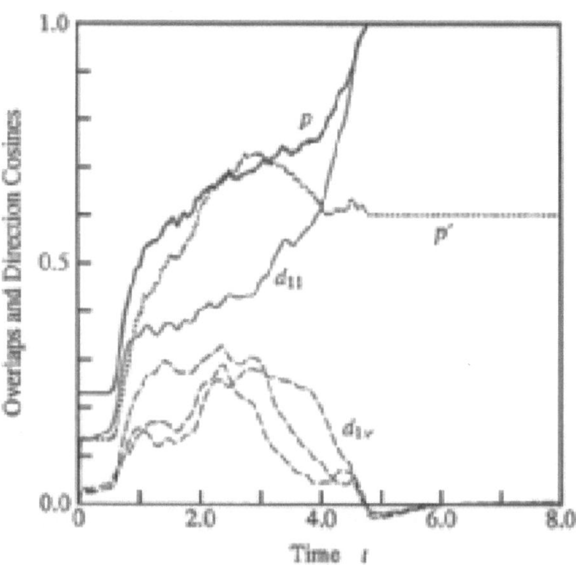

Fig. 4. Time course of change in the overlaps of X with S^{11} (thick line) and C^1 (dotted line); and the direction cosines with S^{11} (solid line), S^{12}, S^{13} and S^{14} (broken lines).

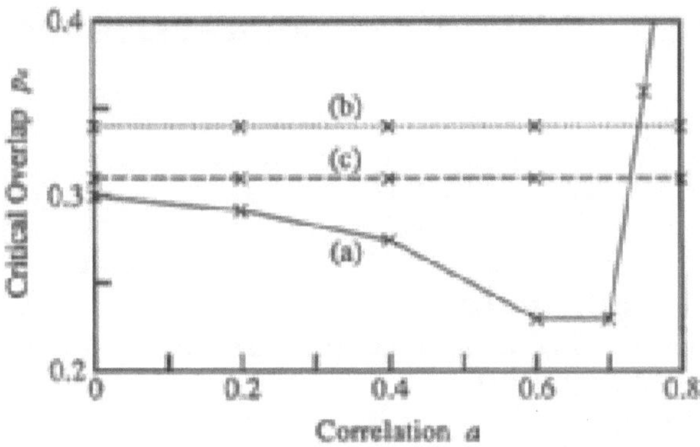

Fig. 5. Critical overlaps of (a) the autocorrelation model with the S-function, (b) the pseudo-inverse model with the conventional sigmoid function, and (c) the pseudo-inverse model with the S-function.

as

$$V = \frac{\alpha}{n} \sum_{\mu,\nu} d_{\mu\nu} S^{\mu\nu}, \tag{11}$$

where $\alpha = \sum_i x_i y_i$.

Figure 4 shows an example of changes in p (thick line), d_{11} (solid line) and $d_{1\nu}$ $(\nu = 2,3,4)$ (broken lines); the overlap p' between X and C^1 is also plotted there (dotted line). This is a case where the initial overlap is nearly equal to the critical overlap ($p_0 = 0.23 \simeq p_c$) and the network takes long time to succeed in recollection.

We see that at first p' increases more rapidly than p ($p'_0 \simeq ap_0$) and then decreases. Intuitively, this means that the whole cluster behaves like an attracter of the dynamical system when the network state X is not very close to it. We also see that as X approaches S^{11}, $d^{1\nu}$ decreases to be almost zero, although X ought to come near $S^{1\nu}$. This is because if $s_i^{11} = s_i^{1\nu}$ for many ν's, $|v_i|$ is apt to be large and thus $x_i \cdot y_i$ gets small or often negative; as a result, the outputs of the other neurons become dominant, and the network can separate S^{11} from the other patterns, that is,

$$V \simeq v_0 S^{11} \tag{12}$$

holds although Eq. (3) does not at all.

Figure 5 shows the change in the critical overlap p_c as a function of the correlation a between C^μ and $S^{\mu\nu}$. Since more than $0.15n$ patterns are stored, the critical overlap does not exist for the autocorrelation model with the conventional dynamics. We see that p_c of the improved model (solid line) decreases with an increase in a, if a is smaller than about 0.7.

If a becomes larger than that, p_c begins to increase rapidly, and finally vanishes, or the state $X = S^{\mu\nu}$ becomes unstable. Intuitively, this is because X is attracted to C^μ so strongly that the network cannot separate the individual patterns. This also causes a decrease in the memory capacity. However, these problems can probably be settled if we modify the correlation learning rule using the S-function (unpublished work).

Finally, we show the results of an experiment on the pseudo-inverse model for comparative purpose. The critical overlaps for the conventional (dotted line) and improved (broken line) dynamics are plotted in Fig. 5. Interestingly, the value of p_c does not depend on a, and is larger than the above one when $a < 0.7$. This means that the correlations between patterns do not lessen the memory capacity, but do not raise the recollection ability either.

6. Conclusion

We have shown that the autocorrelation associative memory with the improved dynamics works well even if the stored patterns are correlated with each other.

It appears strange that the recollection ability is raised by proper correlations between the patterns; but it may be reasonable if we consider the fact that total amount of information to be stored in the network decreases as the correlation increases. Anyway, the unsatisfactory restriction on memory representation is eased—we can store the patterns with some structure without transforming them or using a complex learning rule.

Our findings would also be important in understanding the dynamical properties of associative neural networks, though further study is necessary.

References

[1] Nakano, K., "Associatron—A model of associative memory", *IEEE Trans. Sys., Man, Cybern.*, SMC-2, 380–388 (1972).

[2] Anderson, J.A., "A simple neural network generating interactive memory", *Math. Biosciences*, **14**, 197–220 (1972).

[3] Kohonen, T., "Correlation matrix memories", *IEEE Trans. on Computers*, **C-21**, 353–359 (1972).

[4] Hopfield, J.J., "Neural networks and physical systems with emergent collective computational abilities", *Proc. Natl. Acad. Sci. USA*, **79**, 2554–2558 (1982).

[5] Kohonen, T., *Self-Organization and Associative Memory*, 2nd ed., Springer, New York (1988).

[6] Amari, S., "Neural theory of association and concept-formation", *Biol. Cybern.*, **26**, 175–185 (1977).

[7] Morita, M., Yoshizawa, S. and Nakano, K., "Analysis and improvement of the dynamics of autocorrelation associative memory", *Trans. IEICE Japan* (in Japanese, in press).

[8] Amari, S. and Maginu, K., "Statistical neurodynamics of associative memory", *Neural Networks*, **1**, 63–73 (1988).

HOPFIELD MODEL OF NEURAL NETWORK WITH UNRESTRICTED SELF-FEEDBACK

Arun K. Pujari & Ravindra Sharma

Artificial Intelligence Lab
School of Mathematics & Computer/Information Sciences
University of Hyderabad
HYDERABAD - 500 134
INDIA

ABSTRACT

In this work we suggest a modification to the Hopfield model of Neural Network which allows direct self-feedback for a neuron (diagonal elements of weight matrix can be non-zero) and explore the applicability of the modified Hopfield model. Stability of the modified model is discussed in detail. A method of generation of weight matrix and threshold vector is given for the modified Hopfield model. The conclusion of this work is that a higher capacity Neural Network is obtained.

A STRUCTURED DEVELOPMENT FOR THE IMPLEMENTATION OF A MULTILAYER NEURAL NETWORK (CRAM PROJECT)

Francisco J. LOPEZ ALIGUE, Isabel ACEVEDO SOTOCA, Miguel A. JARAMILLO MORAN.
Departamento de ELECTRONICA e INGENIERIA ELECTROMECANICA. Universidad de Extremadura. Av. Elvas, s/n. 06071 - BADAJOZ (SPAIN)

ABSTRACT

The main target of the CRAM (Computer Research on Associative Memory) Project is the practical implementation of a multilayer virtual Neural Network. The hardware used is a multiprocessor configuration IEEE 1014 Standard (VME Bus) with the 68020 CPU and the 68882 Floating Point Coprocessor running under the VersaDos Operating System, well suited for developing and debugging real-time programs.

The neural network has been configured as a hyper-spherical distribution of consecutive layers of planes of interconected neurons. These neurons follow the classical analogical description of the activity as a function of weighted inputs and an interactive collateral contribution coming from the rest of the neurons in the same plane, adapted for its computerized implementation. The learning algorithm was similarly redefined in order to accomplish more realistic and diverse objectives. Basically, it is a non-supervised system.

The whole system is implanted in a self-trained pattern recognition device where the neural network acts as a "descrambler" for the noise-corrupted and geometrically-deformed inputs. The output of the neural network is directed to a bidirectional associative memory to sort them, giving the final output.

PROBAM
Heteroassociative Multilayer Memory for Industrial Application

Vera Čuljak
Damir Pavuna
SP "KONČAR INSTITUT" - SAPS
41 000 ZAGREB, Baštijanova bb

Abstract: This paper deals with a kind of bidirectional associative memory (BAM) called PROBAM. The purpose of this three layer neural network is to satisfy a need for a usable associative memory for industrial application purposes like computer vision or visual inspection. Paper describes how PROBAM satisfies the two main aims: to accurately decode prelearned patterns without uncertainties as well as to be capable to recognize rather noisy patterns. Our latest work is orientated toward building a recognition prediction tool for a certain set of patterns. Example of latest PROBAM neurodynamics research is given and explained.

ARCHITECTURES

Chair: Robert HECHT-NIELSEN

CONDITIONS ON ACTIVATION FUNCTIONS OF HIDDEN UNITS FOR LEARNING BY BACKPROPAGATION

Masahiko Arai
Systems and Software Engineering Laboratory
Toshiba Corporation
70 Yanagi-cho, Saiwai-ku, Kawasaki 210, Japan

Abstract

This paper considers conditions on an activation function of hidden units for the purpose of utilizing backpropagation for three-layer-net learning. A necessary condition for the convergence of backpropagation procedures to a global minimum of a cost function is that a set of states of the hidden layer is linearly separable. A sufficient condition for the separability is that the vectors made from the states and a constant become linearly independent. This paper discusses the conditions that the vectors become linearly independent.

It is proved that when there are I-training patterns, if the (I-1)-th derivative of the activation function of hidden units exists and if it is not zero at a point, there is a set of the states which is linearly separable when there are (I-1)-hidden units.

Two examples of nets with one input unit are considered to estimate the connection weights when sets of states of hidden layers are linearly separable: the net whose activation function of the hidden units is a polynomial of degree (I-1); and the case where the connection weights between the hidden units and the input one are sufficiently small. It is shown for both nets with the (I-1)-hidden units that if all connection-weight values between the hidden units and the input unit are different, then there are separable sets of states of the hidden layers.

1 Introduction

Some problems in learning algorithms still remain in spite of many applications of feedforward nets, such as three-layer nets [1 and 2]. Backpropagation [3] is one of the most useful algorithms. However, it is well-known that learning by backpropagation is not so easy. The problem is the local minima, i.e., the learning algorithm can not distinguish global minima from local ones. Investigating the problem, backpropagation was applied to perceptrons [4, 5 and 6]. In [4], it was shown that backpropagation procedures converge to a global minimum, zero of a cost function defined by penalty functions such as the least square measure of the error if the set of input vectors is linearly separable, as in the learning of perceptrons. Of course, if a set of input vectors is not separable, the cost function does not become zero even at the global minimum since there are no perceptrons which linearly separate the set of the input vectors. That is, the separability of the input vectors is a necessary and sufficient condition for perceptrons to separate the inputs. When a three-layer net is considered, this corresponds to the condition on the states of the hidden layer, and the separability of states of the hidden layer is necessary (not sufficient) for the convergence of backpropagation to a global minimum, zero. A sufficient condition for the separability of a set of states is that the vectors made from the states and a constant become linearly independent. This paper considers conditions that the vectors become linearly independent.

There are two ways of mapping I-input vectors (linearly dependent) in the less-than-I-dimensional space to I-linearly-independent vectors in the I- dimensional space: masks [7] for binary inputs, and the other is given for continuous-valued inputs [8]. These use the threshold function. For backpropagation procedures, it is required that the activation function is differentiable. For a differentiable function, in this paper, it is proved that if it is (I-1)-times differentiable and if the (I-1)-th derivative is not zero at a point, mapping exists. In this case, (I-1)-components of the I-dimensional vector are the outputs of the hidden units and the remaining one corresponds to the input to the threshold values for the output units[8]. Two examples of the mapping are also shown.

2 Conditions on Activation Functions

First, nets with an input unit are considered. At the end of this section, obtained results are extended to nets with multiple-input units. Let V_i, $i = 1,\dots,I$, and $f(x)$ be the output value of the input unit corresponding to the i-th input and the activation function of the hidden units, respectively. Then, the output of the k-th hidden unit for the i-th input is $f(x_{ki})$, where

$$x_{ki} = w_k V_i - t_k, \tag{1}$$

for $i=1,\dots,I$, $k=1,\dots,K$. K, w_k and t_k are the number of the hidden units, the connection weight between the k-th hidden unit and the input unit and the threshold value of the k-th hidden unit, respectively. Let $J \leqq I, K$.

Lemma 1 Let $A(x_1,\dots,x_J)$ be a JxJ matrix the first row of which is defined by

$$A(x_1,\dots,x_J)_{1i} = f(x_i),$$

$$x_i = \varepsilon V_i + x_0, \qquad i = 1,\dots,J,$$

where ε is sufficiently small and x_0 is a constant, and the other row vectors be assumed to be linearly independent, i.e., there is a minor of degree (J-1)x(J-1) of **A** which is not zero. Then, if, for the j-th derivative of **f**, $f^{(j)}$,

$$f^{(j)}(x_0) \neq 0, \qquad j = 0,\dots,J-1, \tag{2}$$

where $f^{(0)}$ is **f**, there is an ε which satisfies

$$\det A(x_1,\dots,x_J) \neq 0.$$

Proof. Let Δ_{1i} be the cofactor of A with respect to the i-th component of the first row vector. Without loss of generality, it is assumed that $\Delta_{1i} \neq 0$, $i=1,\dots,J_1$, $J_1 \leqq J$. Then,

$$\det A(x_1,\dots,x_J) = \sum_{i=1}^{J_1} f(x_i)\,\Delta_{1i}. \tag{3}$$

If det A=0 in the vicinity of x_0, by taking the k-th derivative of Eq.(3) with respect to ε at $\varepsilon=0$, the following equation is obtained,

$$0 = \sum_{i=1}^{J_1} f^{(k)}(x_0)\,\Delta_{1i}(V_i)^k, \qquad k = 0,\dots,J_1-1. \tag{4}$$

From the condition (2), Eq.(4) means that the J_1-dimensional-column vectors, ${}^t(\Delta_{1i}(V_i)^0,\dots,\Delta_{1i}(V_i)^{J_1-1})$, $i=1,\dots,J_1$, are linearly dependent, where ${}^t(\dots)$ is the transposed-column vector of the row vector $(\dots)$. From $\Delta_{1i} \neq 0$, $i=1,\dots,J_1$, the column vectors ${}^t((V_i)^0,\dots,(V_i)^{J_1-1})$, $i=1,\dots,J_1$, also become linearly dependent. However, the column vectors are linearly independent. To see this, let B be a $J_1 x J_1$ matrix whose element B_{ik} is defined by

$$B_{ik} = (V_i)^{k-1}, \quad i, k = 1, \ldots, J_1.$$

Then, it is easily obtained that

$$|\det B| = \prod_{i>j\geq 1}^{J_1} |V_i - V_j| .$$

This is shown by the fact that the highest power of V is J_1-1 and the determinant becomes zero when $V_i = V_j$. V_i corresponds to the i-th input, so it differs from the others, and det B is not zero, a contradiction. ∎

Lemma 2 If $f^{(J-1)}(x_0) \neq 0$, $f^{(j)}$ is not zero for $j<J$ in the vicinity of x_0.

Proof. For any i, if $f^{(i)}(x_0) \neq 0$, there is an ε for which $f^{(i)}(x) \neq 0$, for $|x-x_0| \leqq \varepsilon$, independently of i, i.e., the intersection of the regions, $f^{(i)}(x) \neq 0$. On the other hand, for $f^{(i)}(x_0)=0$, an i exists for which $f^{(i+1)}(x) \neq 0$ for $|x-x_0| < \varepsilon$. In this case, from mean value theorem,

$$f^{(i)}(x) = f^{(i)}(x_0) + f^{(i+1)}(x')(x - x_0), \quad |x'-x_0| < |x-x_0|,$$

$f^{(i)}(x) \neq 0$, $|x-x_0| < \varepsilon$ is shown. Therefore, by repeating this procedure, $f^{(i)}(x) \neq 0$, $i=0,\ldots,J-1$, for $|x-x_0| < \varepsilon$ is obtained. ∎

Theorem 1 If $f^{(I-1)}(x)$ is not zero at a point, there is a separable set of hidden-layer states with (I-1)-hidden units for I inputs.

Proof. From Lemma 2, there is an x_0 for which $f^{(i)}(x_0) \neq 0$, $i=0,\ldots,I-1$. For an IxI matrix B whose elements are

$$B_{Ii} = 1, \quad i = 1, \ldots, I,$$

$$B_{ki} = f(x_0 + \varepsilon_k V_i), \quad k = 1, \ldots, I-1, \quad i = 1, \ldots, I, \tag{5}$$

let a JxJ matrix A(J) be defined by, for $J \leqq I$,

$$A(J)_{kj} = B_{I+k-J\ j}, \quad k, j = 1, \ldots, J.$$

From det A(1)=1, by using Lemma 1, ε_1 exists which satisfies det $A(2) \neq 0$. By repeating this procedure, it is obvious that ε_k, k = 1,...,I-1, exists for which det $A(k+1) \neq 0$. And for k=I-1, det A(I)=det $B \neq 0$. Therefore, by taking

$$w_k = \varepsilon_k, \quad t_k = -x_0,$$

k=1,...,I-1, in Eq.(1), the output values of the (I-1)-hidden units and a constant 1 make I-dimensional vectors which are linearly independent for the I-inputs, since det $B \neq 0$. In this case, a three-layer net exists which separates the I-inputs [8], i.e., the set of the hidden-layer states is separable. ∎

For multiple-input units, let the number of the input units be L. If there is an L-dimensional vector W with which all the inner products of the L-dimensional input vectors U_i, i = 1,...,I, whose components are the outputs of the input units, are different, by taking

$$V_i = WU_i, \quad i = 1, \ldots, I,$$

Theorem 1 is also obtained for the nets with multiple-input units. The existence of such a vector W has been proved in [8].

3 Examples of the Nets

Example 1 First, let $f(x)=x^{I-1}$. Then, by replacing ε_k for w_k in Eq. (5), elements of B become

$$B_{ki} = (x_0 + w_k V_i)^{I-1} = \sum_{n=0}^{I-1} a(n)(w_k V_i)^n, \quad (6)$$

for k=1,...,I-1, i=1,...,I. Let $x_0 \neq 0$, then, $a(n) \neq 0$, n=0,...,I-1. From the fact that det B becomes zero when $w_n=0$, $w_i=w_j$, and $V_i=V_j$, it is shown that

$$|\det B| = |\prod_{n=1}^{I-1} a(n) w_n \prod_{k>m\geqq 1}^{I-1} (w_k - w_m) \prod_{i>j\geqq 1}^{I} (V_i - V_j)|, \quad (7)$$

for I>2. From $V_i \neq V_j$, if $w_k \neq w_m$, det $B \neq 0$, and the vectors become linearly independent. Eq. (7) is also obtained if $a(n) \neq 0$, n=0,...,I-1, for polynomials of degree I-1 in the same way. Therefore, if all the connection weights between the hidden units and the input one are different, the net can separate the I inputs when the activation function is a polynomial of degree I-1.

Example 2 For weak connections in which $w_i \sim \varepsilon_i < \varepsilon$, if $f^{(I)}$ exists, the same result is obtained for a general activation function f by using the (I-1)-th approximation of f, as in Example 1. The complete proof will be given in the final paper.

4 Conclusion

When backpropagation is applied to nets, it must be assured that the states of the hidden layers become separable. For the conditions, it was shown that for I inputs, if the (I-1)-th derivative of the activation function of the hidden units exists and if it is not zero at a point, it is possible to make the set of the hidden-layer states separable if there are (I-1)-hidden units.

5 References

[1] R. P. Gorman and T. J. Sejnowski, "Analysis of Hidden Units in a Layered Network Trained to Classify Sonar Targets," Neural Networks Vol.1, No.1, 1988, pp.75-89.

[2] W. Y. Huang and R. P. Lippmann, "Neural Net and Traditional Classifiers," in Proc. Neural Info. Proc. Systems, Denver, 1987, pp.387-396.

[3] D. E. Rumelhalt, G. E. Hinton and R. J. Williams, "Learning Internal Representations by Error Propagations," Parallel Distributed Processing: Explorations in the Microstructure of Cognition, (D. E. Rumelhalt and J. L. MaClelland(Eds.)), Vol.1, MIT Press, Cambridge, 1986, pp.318-362.

[4] E. D. Songtag and H. J. Sussmann, "Backpropagation Separates When Perceptrons do," in Proc. IEEE Int'l Conf. on Neural Networks, Washington D. C. , July 1989, Vol.1, pp.639-642.

[5] M. Brady, R. Raghavan and J. Slawny, "Gradient Descent Fails to Separate," in Proc. IEEE Int'l Conf. on Neural Networks, San Diego, California, July 1988, Vol.1, pp.649-656.

[6] B. S. Wittner and J. S. Denker, "Strategies for Teaching Layered Networks Classification Tasks," in Proc. Conf. Neural Info. Proc. Systems, Denver, 1987, pp.387-396.

[7] M. Minsky and S. Papert, Perceptrons: an Introduction to Computational Geometry, Cambridge, MIT Press, 1986.

[8] M. Arai, "Mapping Abilities of Three-Layer Neural Networks," in Proc. IEEE Int'l Joint Conf. on Neural Networks, Washington D. C., July 1989, Vol.1, pp.419-424.

A Geometric Approach to the Structural Synthesis of Multilayer Perceptron Neural Networks

Jianbin Hao, Shaohua Tan and Joos Vandewalle
ESAT Lab., Dept. Electrical Engineering, K.U.Leuven
Kardinaal Mercierlaan 94, B-3030 Heverlee - Belgium

1 Introduction

Designing a multilayer perceptron for general purpose classification has important practical implications. Since the capacity of multilayer perceptron to realize arbitrary dichotomies (or two-class classifications) is limited, the most important step in a design procedure is the determination of the number of the layers and the amount of nodes in each layer apart from the determination of the weights and the threshold values. Unfortunately, there has been no general principle or guideline available for such a synthesis task, normal design often proceeds on an *ad hoc* and empirical basis, the methods generally lead to the structure which only deals with a particular classification problem [1] [2].

Some efforts have been made recently to understand the structural nature of the multilayer perceptron, and there have been a number of methods developed to try to access the type of classification ability and to come up with a rough estimation of the hidden units [3] [4]. These methods, however, are only valid for some restricted input patterns, and not amenable to the general case. There are also some design procedures to cope with the general case with the number of the neurons in the net being $N-1$, $N/2$, etc., where N is the number of inputs to be classified. The structures produced by these methods, however, have too many neurons, thus are considered to be impractical.

The objective of the present paper is to present a geometrical design procedure to accomplish an arbitrary dichotomy for 2^m different binary inputs. The ideas underlying our approach are based on the geometry of the n-dimensional hypercube. Our design procedure is composed of two basic steps. First, the 2^m inputs which can be alternatively considered as all the vertices of an m-dimensional hypercube are transferred to the vertices of a new n-dimensional hypercube with a special structure. The so-called n-parity configuration is especially chosen to be this special structure in our approach. This transformation is considered in section 3. Then the next step is to solve the n-parity problem with a one-hidden-layer $n+1$ neurons net structure, which is examined in the next section.

2 The n-parity problem

Let the vertices of an n-dimensional hypercube be denoted by n-vectors $\{-1,1\}^n$, and let the even (odd) vertices be defined to be those with even (odd) number of 1's. Then the n-parity problem classifies the odd and the even vertices among all the 2^n vertices into the two different classes. In this section, a one-hidden-layer network with $n+1$ nodes is constructed to solve the n-parity problem.

To begin, note that in an n-hypercube there always exist $n+1$ hyperplanes of dimension $n-1$ such that each of these $n+1$ hyperplanes includes only those vertices having the same number of 1's. Since any hyperplane of dimension $n-1$ can be represented by

$$a_1x_1 + a_2x_2 + \cdots + a_{n-1}x_{n-1} + a_nx_n = b, \tag{1}$$

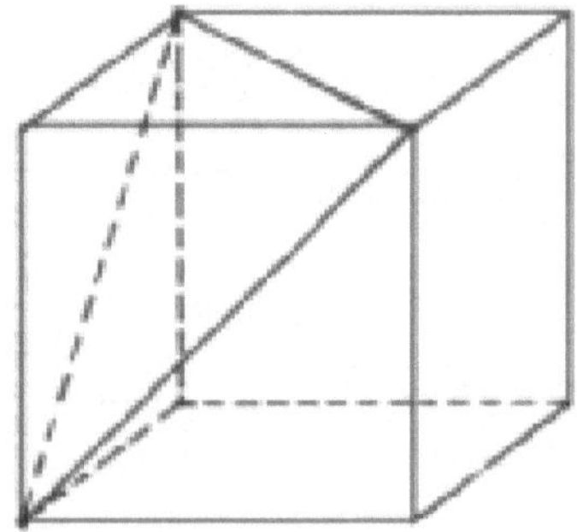
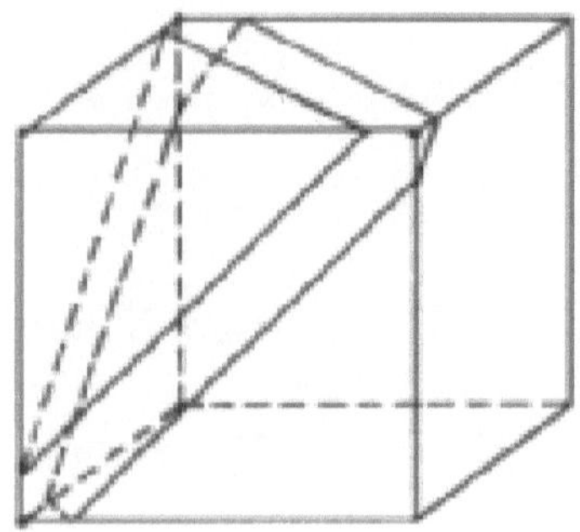

Figure 1: A simple example of the two hyperplanes in 3-dimensional space

where a_i $(i = 1, 2, \ldots, n)$ and b are some coefficients, x_i $(i = 1, 2, \ldots, n)$ are variables, the aforementioned $n+1$ hyperplanes can easily be obtained by setting $a_1 = a_2 = \cdots = a_n = 1$, $b = 2k-n$, or written explicitly

$$x_1 + x_2 + \cdots + x_n = 2k - n \tag{2}$$

where k $(= 0, 1, \ldots, n)$ is the number of 1's in the vertices. It can easily be verified that only the vertices with a number k of 1's lie on (2) and the vertices with a different number of 1's do not lie on it. With the above construction we can select p ($p = n/2$ when n is even, and $p = (n+1)/2$ when n is odd) different hyperplanes which contain, respectively, the vertices with the odd number of 1's.

With respect to each of the above p hyperplanes, we construct two parallel hyperplanes, each of which is obtained by shifting infinitesimally the original hyperplane in two opposite directions. Such construction can ensure that no even vertices lie in the slice between the two constructed hyperplanes, and each of the odd vertices is contained in one of these hyperplane slices. Clearly, the construction will result in $2p$ such hyperplanes. To illustrate, an example in 3-dimensional space is shown in Fig.1.

Recall that in the multilayer perceptron each neuron with n inputs and a single output will correspond to a hyperplane in n dimensional input space, and the converse is also true. Thus for the ith pair of the hyperplanes built in the preceeding way, we can always associate a pair of neurons, denoted by v_1^i, v_2^i so that any vertices lying in between the two hyperplanes will give rise to the output $+1$ for v_1^i, and -1 for v_2^i. Moreover, any other vertices can only exert the identical outputs for both of the neurons. By so doing, we come up with a line of $2p$ neurons. It remains to determine the weights and the threshold values for the output neuron, say y, to implement the final classification. This is easily done as follows. The weight from v_1^i to y is set to $+1$ and from v_2^i to y set to -1,and the threshold of the output y is set at 0.5. It then follows from the construction that any odd vertex which lies in between one of the hyperplane slices will give rise to the output 1 for y, and similarly, any even vertex which lies outside any of the slices can only result in the output -1.

Concerning the number of neurons used, note that $2p = n$ when n is even, $2p = n+1$ when n is odd. In the case that n is odd, it can easily be shown that $2p - 1 = n$ neurons are actually needed in this case (see Fig.2). Summarizing the above discussion, we arrive at the following theorem

Theorem 1 *A one-hidden-layer perceptron with n hidden neurons can compute the n-parity dichotomy on an n-dimensional hypercube.*

3 Mapping an arbitrary dichotomy to an n-parity problem

The aim of this section is to design a one-layer perceptron to map an arbitrary m-hypercube dichotomy to the n-parity problem. The design method to be presented constitutes one of the two

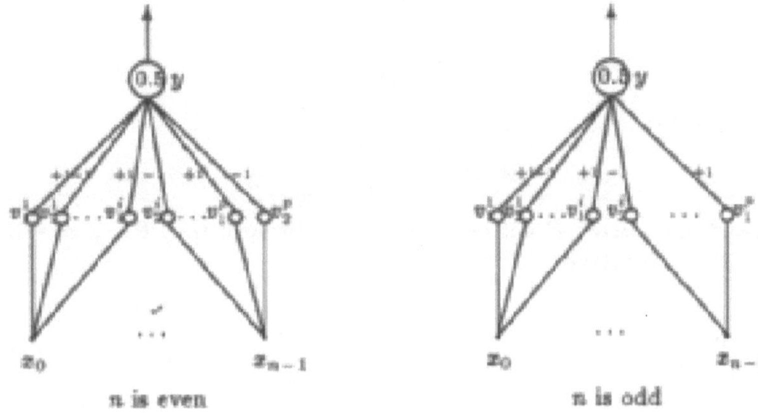

Figure 2: One-hidden-layer perceptron with n hidden neurons solving the n-parity problem

major steps in our general design approach.

Let us recall that the Hamming distance is defined to be the number of the different bits between two binary vectors. To begin the design, we first partition the given 2^m vertices into a number of sets $S(V)$ in such a way that all the vertices in one set have the Hamming distance equal to 1 with a fixed vertex V (this fixed vertex is called central vertex of this set). Note that if the central vertices are suitably chosen and their number is sufficiently large, all the 2^m vertices will at least belong to one of the sets. We can always adjust each set so that it must include m points. With such a partitioning, it follows that the vertices in one set have the Hamming distance 2 with each other. It can also be shown that the m vertices in one set are linearly independent. Furthermore, the inner product of any vertex in one set with their central vertex is $m-2$, while the inner products of any vertices in other sets with this central vertex are different from $m-2$. Another interesting property of this partitioning is given in the following theorem

Theorem 2 *All the m vertices in each set lie on an m-hyperplane, and moreover the vertices which do not belong to this set will not lie on this hyperplane.*

Proof:

Since all the m vertices in any set are linearly independent, the hyperplane on which the m vertices of a set lie must have the following form

$$a_1y_1 + a_2y_2 + \cdots + a_my_m = 1, \tag{3}$$

where a_i are the coefficients of the hyperplane, and y_i are the scalar variables ($i = 1, 2, \ldots, m$). We select an arbitrary set $S(V)$, with the central vertex V. Then the m vertices in this set $X_1, X_2, \ldots, X_m$ will satisfy the Equ.(3). Note that each of the X_i is an m-vector.

Upon defining the m-dimensional vectors $A_r = (a_1, a_2, \ldots, a_m)^T$, $E = (1, 1, \ldots, 1)^T$, and the $m \times m$ matrix

$$X = \begin{bmatrix} X_1^T \\ X_2^T \\ \vdots \\ X_m^T \end{bmatrix}$$

and noting that X_i ($i = 1, 2, \ldots, m$) will satisfy (3), we can write down the following equation

$$XA_r = E. \tag{4}$$

From the nonsingularity of X it follows that

$$A_r = X^{-1}E. \tag{5}$$

Recall that the inner product between each X_i and V is $m-2$, or equivalently

$$E = \frac{XV}{m-2}. \tag{6}$$

Substituting (6) into (5)

$$A_r = \frac{X^{-1}XV}{m-2} = \frac{V}{m-2}. \tag{7}$$

On the other hand, any vertex P not in set $S(V)$ will be such that $PV \neq m-2$. Consequently, it will not lie on the hyperplane

$$Y^T V = m - 2 \tag{8}$$

The proof is complete.

Based on this hyperplane, we now construct a new hyperplane which will divide the m vertices in set $S(V)$ into two arbitrary subsets $S(V)_1$ and $S(V)_2$ such that all the vertices in $S(V)_1$ lie above this hyperplane and those in $S(V)_2$ lie under it. When this is done, we shall obtain, say, ℓ hyperplanes (ℓ will precisely be the number of sets of the partition we have introduced) each of which corresponds to a single neuron in the only layer of the perceptron (recall that each neuron acts as a hyperplane in the input space).

Denote the entries of V which is the central vertex of set $S(V)$ by $v_1,\ v_2, \ldots,\ v_m$. Since every vertex in set $S(V)$ only has one entry different from V (recall that their Hamming distance is 1 by the definition), the vertices in this set can be expressed in the following way

$$\begin{aligned}
&X_1 = (x_{11},\ v_2,\ v_3, \ldots, v_m)^T \\
&X_2 = (v_1,\ x_{22},\ v_3, \ldots, v_m)^T \\
&\ldots \\
&X_m = (v_1,\ v_2, \ldots, v_{m-1},\ x_{mm})^T
\end{aligned}$$

where x_{ii} is the ith entry for X_i. Suppose P arbitrary vertices $X_{i_1}, X_{i_2}, \ldots, X_{i_P}$ are to be set into $S(V)_1$. To do so, we rotate slightly the hyperplane (8) so that the relative positions between it and all the vertices in the sets different from $S(V)$ will not be affected, and moreover the vertices in $S(V)_1$ are located above, the vertices in $S(V)_2$ under the rotated hyperplane. More precisely, let ϵ (> 0) be a sufficiently small number, we construct the rotated hyperplane as

$$Y^T V' = \theta \tag{9}$$

where V' can be obtained by

$$\begin{aligned}
V'^T \ = \ & [v_1 + Sgn(v_1)\epsilon,\ v_2 + Sgn(v_2)\epsilon, \ldots, \\
& v_{i_1-1} + Sgn(v_{i_1-1})\epsilon,\ v_{i_1} - Sgn(v_{i_1})\epsilon,\ v_{i_1+1} + Sgn(v_{i_1+1})\epsilon, \ldots, \\
& v_{i_2-1} + Sgn(v_{i_2-1})\epsilon,\ v_{i_2} - Sgn(v_{i_2})\epsilon,\ v_{i_2+1} + Sgn(v_{i_2+1})\epsilon, \ldots, \\
& v_{i_P-1} + Sgn(v_{i_P-1})\epsilon,\ v_{i_P} - Sgn(v_{i_P})\epsilon,\ v_{i_P+1} + Sgn(v_{i_P+1})\epsilon, \ldots, \\
& v_m + Sgn(v_m)\epsilon]
\end{aligned} \tag{10}$$

in which $Sgn(.)$ is the sign function. Substituting each X_i ($i = i_1, i_2, \ldots, i_P$) into (9) results in

$$X_i^T V' = m - 2 + [m - (P-1)]\epsilon - (P-1)\epsilon$$

and for X_i ($i \neq i_1, i_2, \ldots, i_P$)

$$X_i^T V' = m - 2 + [m - (P+1)]\epsilon - (P+1)\epsilon$$

Clearly θ should be set to $m - 2 + (m - 2P)\epsilon$. The construction of the rotated hyperplane is then completed.

If we construct this type of hyperplanes for all the sets then every vertex in the input hypercube can be located above the even or odd number of the hyperplanes, thus can be mapped into the even or odd vertex of a certain dimensional hypercube using a single layer perceptron network.

Connecting the outputs of this single layer perceptron with the inputs of the one-hidden-layer perceptron constructed in the previous section, we can obtain a two-hidden-layer perceptron networks with m inputs and 1 output. This two-hidden-layer structure, with its connection weights and thresholds determined by (9) and (2) respectively, can implement an arbitrary dichotomy among the 2^m binary inputs.

4 Discussions and conclusion

There are two major approaches to design a multilayer perceptron for the practical tasks. One is to apply some kind of learning algorithm, the other is to use direct setting strategies which are inspired by the analytical and geometrical considerations. One of the drawbacks suffered by the learning algorithm approach is that the determination of the structure is based on the trail-and-error, and there are no general guidelines available. This often results in the over or under determination of the structures. Another drawback is that it generally takes a long time for the learning algorithm to converge, or even more seriously, some learning algorithm is intrinsically not able to find the weights for certain dichotomy problems (see, for example [5]).

The direct approach can avoid the aforementioned problems, but will generally require the detailed understanding of the nature of both the problem and the net structure. The geometrical approach we have proposed belongs to this second category. Obviously, the detailed geometrical analysis has enabled us to come up with a quite systematic method for the determinations of the net structure and the setting-up of the weights. It should be noted that our net structure is aimed at arbitrary problems, and much smaller net can be obtained by exploring the nature of some specific problems. Obviously, the reduction is achieved at the expense of some lose of the arbitrariness. In this sense, our net structure provides a meaningful upper-bound (or more precisely, the maximum necessary) to the multilayer perceptron used in this purpose.

The binary assumption on the inputs is not as restrictive as might be originally thought. Since for nonbinary inputs, a preprocessing layer of neurons can always be added to map them into the vertices of a certain dimensional hypercube[4]. Moreover, this preprocessing layer also provides the basis to the so-called *generalizing ability*.

It would be interesting to know the total number of neurons used in our construction. Unfortunately, this is still an open problem as it depends on the construction of the number of sets introduced in section 3. Up till now, we are yet to find the ways to establish the minimal possible number of such sets. Nonetheless, it is known that this number greatly depends on the way the central vertices are selected and is always equal to or greater than $2^m/m$ (m is the dimension of the input hypercube).

References

[1] U.Ramacher and M.Wesseling(1989), "A Geometrical Approach to Neural Network Design," *Proc. Intern. Joint Conf. Neural Net.*, Vol.2, pp.147-153.

[2] P.Ruján(1989),"A Geometric Approach to Learning in Neural Networks," *Proc. Intern. Joint Conf. Neural Net.*, Vol.2, pp.105-109.

[3] E.B.Baum(1988), "On the Capabilities of Multilayer Perceptrons," *J. Complexity*, Vol.4, pp.193-215.

[4] G.Mirchandani and W.Cao(1989), "On Hidden Nodes for Neural Nets," *IEEE Trans. Circ. Syst.*, Vol.CAS-36, pp.661-664.

[5] M.L.Brady, R.Raghavan and J.Slawny(1989), "Back Propagation Fails to Separate Where Perceptrons Succeed," *IEEE Trans. Circ. Syst.*, Vol.CAS-36, pp.665-674.

A Rule-Based Approach to Neural Network Classifiers

Rodney M. Goodman, Chuck Higgins, John Miller

Department of Electrical Engineering (116-81)

California Institute of Technology

Pasadena, CA 91125, U.S.A.

Padhraic Smyth

Communication Systems Research Section (116-81)

Jet Propulsion Laboratory

Pasadena, CA 91109, U.S.A.

Abstract

In this paper we propose a novel classifier architecture which combines a rule based AI approach with that of the neural network paradigm. We utilize an information theoretic approach to learning a model of the domain knowledge which is explicitly encoded in the form of probabilistic conjunctive rules between attributes and the class variables. These rules are then mapped onto the weights and nodes of a feed forward neural network. The resulting classifier can be considered a hybrid between a Bayesian model and a standard rule-based expert system. When compared with conventional serial rule based expert systems, the neural network paradigm gives to the classifier architecture the advantage of high speed parallel execution; while the rule-based approach gives the neural network explicit meanings for the nodes and weights of the network. We compare the performance of the proposed classifier with a neural network backpropagation classifier. The results lead us to conclude that the rule based classifier performs roughly as well as standard neural nets over a variety of data sets in terms of prediction accuracy, but that it possesses unique advantages in terms of knowledge representation. This allows for the advantages of explaination of reasoning, and the possibility of network pruning, when compared with a traditional black-box neural network classifier.

A formal statement of the problem

We are given K discrete feature variables (attributes) comprising the set $Y = \{Y_1, \ldots, Y_K\}$. Each variable can take values in the alphabet $\{y_i^1, \ldots, y_i^{m_i}\}$, $1 \leq i \leq K$, where m_i is the cardinality of the ith attribute alphabet. For simplicity of notation we will denote y as a typical joint propositional element of these alphabets, e.g., $y = \{y_1^1, \ldots, y_K^1\}$. We define X as the 'class' variable, with a discrete alphabet $\{x_1, \ldots, x_m\}$, where m is the number of classes. Given an initial training set of N data vectors of the form $v_i = \{y_1(i), \ldots, y_K(i), x(i)\}$, $1 \leq i \leq N$, we wish to find a model or classifier C, such that given future data concerning the Y_i attributes, i.e., $\{y_1(i), \ldots, y_K(i)\}, i > N$, we can classify the unknown variable X. Hence, we are interested in the problem of supervised learning of classification models for discrete-valued variables.

Our criterion for estimating the performance of a classifier C is the standard error probability

$$\hat{p}_e(C) = 1 - \frac{1}{L}\sum_{j=1}^{L} \delta\Big(x(j) - \hat{x}(j)\Big)$$

where $\hat{x}(j)$ is our classifier's estimate of the true class $x(j)$ for the jth sample from the independent test set of size L and $\delta(x)$ is defined to be 1 if $x = 0$, else it is 0. In general, there is a lower bound on how small this error probability can be for a given pair $\{Y, X\}$, namely the optimal Bayes error rate for a given problem, p_e^B. In general, for problems of practical interest, this optimal error rate p_e^B is non-zero, i.e., there exists no perfect classifier for the problem. Another way to look at this is that there is a fundamental ambiguity in the mapping from Y to X, i.e., it is stochastic rather than deterministic. This ambiguity may arise in practice from the presence of 'hidden' causal variables (e.g., in medical domains, unobservable biological mechanisms) or inherent stochastic 'noise' on the available measurements Y. In the machine learning literature this is commonly referred to as 'learning in the presence of noise.'

In general we have that the error rate of our classifier C can never be less than the optimal error rate, i.e.,

$$\lim_{L->\infty}\left(\hat{p}_e(C)\right) \geq p_e^B$$

where p_e^B is the Bayes error rate (assuming a uniform loss function) and is defined as

$$p_e^B = \sum_{i=1}^{m}\sum_{Y}\left((1-\max_i\{p(x_i|y)\})p(y)\right)$$

Clearly we seek a procedure for finding classifiers such that the expectation

$$E[\hat{p}_e(C) - p_e^B]$$

with respect to all possible training data sets for all possible problems (i.e., sets of $\{Y, X\}$) is minimised.

This notion of learning probabilistic mappings rather than deterministic ones is important. The optimal Bayes classifier (as implicitly described in the equation above) cannot be determined in practice for the simple reason that the number of probability estimates (or decision rules) to be made grows exponentially with the cardinality of the hypothesis space Y. Hence, the 'trick' to designing good classifiers amounts to seeking good *approximations* to the joint probability distribution $p(X, Y_1, \ldots Y_K)$.

Learning classification rules from data

In previous work we have defined the notion of a probabilistic rule [1–3]. Consider the set of discrete attributes $Y = \{Y_1, \ldots, Y_K\}$ and the class variable X, as described earlier. Let $y_1, \ldots, y_K$ represent typical events from the alphabets of each of these variables. Consider a rule relating some arbitrary joint conjunction of the form $(y_1, \ldots, y_l), l \leq K$ with a particular class $x_j, 1 \leq j \leq m$, namely a probabilistic production rule of the form

If $(y_1, \ldots, y_l)$ then x_j with probability p

where p is the conditional probability $p(x_j|(y_, \ldots, y_l))$. We define the *information content* of such an "lth order" probabilistic rule as

$$\begin{aligned} J(X; y_1, \ldots, y_l) &= p(x_j, y_1, \ldots, y_l)\log\left(\frac{p(x_j|y_1, \ldots, y_l)}{p(x_j)}\right) \\ &\quad + p(\bar{x}_j, y_1, \ldots, y_l)\log\left(\frac{p(\bar{x}_j|y_1, \ldots, y_l)}{p(\bar{x}_j)}\right) \end{aligned}$$

This J-measure trades-off a simplicity term, namely $p(y_1, \ldots, y_l)$ (the probability that the left-hand side will occur), with a goodness of fit term (the cross-entropy of X and $X|y_1, \ldots, y_l$), and as such provides a useful and mathematically sound measure for induction. In more intuitive terms, it trades-off how often the rule will fire with the accuracy of the rule. The ITRULE algorithm uses the J-measure to explore the space of all possible rules relating $Y_1, \ldots, Y_K, X$, using information theoretic bounds and small sample estimators to constrain the search.

Previously we have described the use of ITRULE as a general tool for both data analysis and automated knowledge acquisition for expert systems [1, 4]. However it is also possible to run ITRULE to generate only classification rules, i.e., only search for rules with $\{x_1, \ldots, x_m\}$ on the right-hand side. In this manner we can generate the set of R most informative rules which classify X; intuitively the rules generated by ITRULE should form a good approximation in some sense. More formally, by using a probabilistic rule-based representation the type of product approximation we are using is of the form

$$p(x_j, y_1, \ldots, y_m) = C.p(x_j).\prod_{i=1}^{i=|R_j|} p(S_i|x_j)$$

where C is a normalisation constant, and the S_i represent the various left hand side combinations of y_i in the set R_j of rules. It is well known [5] that the best approximation to the *true* distribution is that which maximises the sum

$$\sum_{i=1}^{|R_j|} I(X; S_i)$$

where $I(X;S_i)$ is the average mutual information between the variables X and S_i (consider S_i, the ith rule left-hand side, to be a binary variable). Hence, the maximum likelihood solution to the problem of finding the best approximation from a given set of data, is obviously to use the most complicated rules possible, i.e., have all rules of order K. However, such an approach would suffer on generalisation to new data as it would overfit the training set. Another way to state this is that a learning algorithm which tried to maximise the above equation would only maximise goodness-of-fit — there are compelling reasons (both intuitive and theoretical) to also take account of the simplicity of the model, i.e., Occam's razor. By the expansion

$$\sum_{i=1}^{|R_j|} I(X;S_i) = \sum_{i=1}^{|R_j|}\Big(J(X;S_i) + J(X;\bar{S}_i)\Big)$$

one can interpret the role of ITRULE as that of maximising the information content of the rule-based components (the first term on the right-hand side) of the product expansion.

Building a classifier from a probabilistic rule set

Having found the most informative rules as described above, the next step is to define the actual classification algorithm. For a given input vector of evidence variables, we will have in general a set of rules R_j which "fire" (i.e., their left-hand sides are true) and which have x_j on the right-hand side. For the purposes of this paper we will assume that this set R_j has been chosen such that the left-hand sides are relatively independent of each other, *conditioned* on the particular class x_j. For example, one could subsume general versions of specialised rules, e.g., if both y_1 and y_1, y_2 are present we may prefer to keep only the more specialised rule y_1, y_2. We will refer to this problem of finding good sets of rules to use for inferencing as the "rule-pruning" problem — the details as to how this is accomplished are not relevant at this point.

Clearly we wish to estimate $p(x_j|R_j)$. We have

$$\begin{aligned} p(x_j|R_j) &= p(x_j|S_1,\ldots,S_{|R_j|}) \\ &\frac{p(x_j)\prod_{i=1}^{|R_j|} p(S_i|x_j)}{p(S_1,\ldots,S_{|R_j|})} \end{aligned}$$

where the last equation holds if we assume that the evidential propositions (the rule left-hand sides) are ***conditionally independent*** of each other given the class x_j. For a justification of this type of assumption see the excellent treatment by Pearl [6]. In particular note that no assumptions are made regarding independence conditioned on $\bar{x}_j$. Applying Bayes rule to the terms within the product we get

$$p(x_j|R_j) = \frac{\prod_{i=1}^{|R_j|} p(S_i)}{p(S_1,\ldots,S_{|R_j|})} \cdot p(x_j) \cdot \prod_{i=1}^{R_j}\left(\frac{p(x_j|S_i)}{p(x_j)}\right)$$

Let us define

$$C_j = \log\left(\frac{\prod_{i=1}^{|R_j|} p(S_i)}{p(S_1,\ldots,S_{|R_j|})}\right)$$

and

$$W_{ij} = \log\left(\frac{p(x_j|S_i)}{p(x_j)}\right)$$

then we have in general that

$$\log(p(x_j|R_j)) = C_j + \sum_{i=1}^{|R_j|} W_{ij} + \log p(x_j)$$

The classification procedure amounts to finding a set R_j of rules for each class $x_j, 1 \le j \le m$, estimating $\log(p(x_j|R_j))$ for each class according to the equations above, and then making a classification decision according to whichever estimate is largest. It is worth noting the role of each of the terms in the equation above for $\log(p(x_j|R_j))$. The C_j terms serves as a decorrelation factor which accounts for the inherent interdependence of the S_i *unconditioned* on x_j. The $\log(p(x_j))$ serves the role of a prior *bias* term which in the abscence of any rules firing ($|R_j| = 0$), results in the estimate being the prior probability for the class x_j. The remaining term, the sum of the weights W_{ij}, has the direct interpretation as evidential support provided by the rules — positive weights imply that the class is true, while negative weights imply it is false. Hence we see that while statistical in nature, this scheme possess the ability to provide direct explanations to the user in terms of the W_{ij} as to how the classification decision was arrived at. The rule-based features of this classifier make it much more likely to be accepted by a user in practice, than say an alternative 'black box' approach such as a neural network.

The implemetation of the classifier onto the neural network is now straightforward. The hidden layer neurons form the appropriate Boolean left hand side conjunctions S_i, these are linked to the output node (class variable) via the log weights W_{ij}. The output neuron then thresholds the sum of the evidence (weights) to infer the value of the class variable.

Experimental Results

We used three data sets in our experiments, each of which were sub-divided into two disjoint training and test sets: the 'LED-digits' dataset with 10% noise; the voting records of the 1984 United States Congress [7] dataset; and an artificial Boolean function of the form $X = \text{OR}(\text{XOR}(Y_1, Y_2),\ \text{AND}(Y_3, Y_4),\ \text{AND}(Y_5, Y_6))$. When compared with a three layer conjugate-gradient backpropagation classifier [8], the rule based classifier performed comparably. Specifically, the rule based system achieved 73.5%, 89.8%, 82.5%, on the respective datasets while the neural net achieved 68.4%, 93.6%, 90%. These results compare favorably with known optimal classification rates of 74%, 95%, and 90% for the three datasets.

Conclusions

The rule-based classifiers described in this paper perform well when compared with optimal classifiers, and with conventional neural network classifiers, whilst having the advantage of being based on sound statistical and information-theoretic principles, and retaining an explicit knowledge representation. Current work involves investigating how rule-based schemes can handle continuous valued attributes, improving rule-pruning techniques, and applications to areas such as speech recognition where both low-level acoustic information and higher-level contextual information can be combined using a unified rule-based representation scheme.

Acknowledgements

This work is supported in part by Pacific Bell, and in part by the Army Research Office under Contract No. DAAL03-89-K-0126. Part of this research was carried out by the Jet Propulsion Laboratory, California Institute of Technology, under a contract with the National Aeronautics and Space Administration.The authors also gratefully acknowledge the cooperation of David Aha of U.C. Irvine in providing the voting data set.

References

1. R. M. Goodman and P. Smyth, 'Information theoretic rule induction,' *Proceedings of the 1988 European Conference on Artificial Intelligence*, Pitman Publishing: London, 1988.

2. R. M. Goodman and P. Smyth, 'The induction of probabilistic rule-sets — the ITRULE algorithm,' *Proceedings of the 1989 International Workshop on Machine Learning*, Morgan Kaufmann: Palo Alto, CA, 1989.

3. P. Smyth and R. M. Goodman,'Deriving rules from databases — the ITRULE algorithm,' submitted for publication to the *IEEE Trans. on Knowledge and Data Engineering*, 1989.

4. R. M. Goodman, J. W. Miller and P. Smyth, 'An Information Theoretic Approach to Rule-Based Connectionist Expert Systems,' in *Advances in Neural Information Processing Systems I*, Morgan Kaufmann, 1989.

5. P. M. Lewis, 'Approximating probability distributions to reduce storage requirements,' *Information and Control*, 2, pp.214–225, 1959.

6. J. Pearl, *Probabilistic Reasoning in Intelligent Systems*, Morgan Kaufmann: Palo Alto, CA, 1988.

7. *Congressional Quarterly Almanac, 98th Congress*, 2nd session 1984, Congressional Quarterly Inc.: Washington D.C., 1985.

8. E. Barnard and R. Cole, 'A neural net training program based on conjugate-gradient optimization,' Oregon Graduate Centre Technical Report No. CSE 89–014, Oregon, 1989.

Choosing optimal network structure

György Barna[1,2] and Kimmo Kaski[2]

[1] *Central Research Institute for Physics of the Hungarian Academy of Sciences P.O.Box 49, H-1525 Budapest, Hungary*

[2] *Tampere University of Technology, Microelectronics Laboratory P.O.Box 527, SF-33101 Tampere, Finland*

1 Introduction

Neural network modelling is a now extensively used method both to simulate dynamic neural phenomena and to solve non-biological computational problems. One of the best known neural network models is the Hopfield model [10]. It forms an autoassociative memory which can store and retrieve certain number of patterns. Hopfield assumed that the neural networks are fully connected (i. e. their structure can be represented by perfect graphs). In the human brain 10^{11} neurons form a network by 10^{14} connections. This is rather far from the theoretically possible 10^{22} synapses: the nervous system is not a "structureless" system, but one of its characteristic properties is its remarkable precise wiring.

At a first sight cortical connection is (quasi-)random [15,6]. In mathematical terms at a zeroth approximation the structure of the nervous system can be treated as a random graph. The theory of random graphs was founded by Erdős and Rényi [7,2]. They showed that a great variety of graph properties appear rather suddenly during *graph evolution*. The *graph evolution* is a nested sequence of graphs with n nodes, and the tth member has exactly t edges, of which t–1 edges are identical with edges of the $t-1^{\text{th}}$ member of the sequence. Such the length of a sequence is $\binom{n}{2}+1$.

Within the framework of the family of Hopfield networks *diluted* Hopfield networks were investigated by Derrida [3], Derrida *et al.* [4], Evans [8], Sompolinsky [14] and others. The therm *dilution* means that some of the connection strengths are (symmetrically or nonsymmetrically) zeros; i.e. the graph structure of a *diluted* network can be considered as a member of a *graph evolution* sequence. It was found that in the thermodynamic limit ($n \to \infty$) a second order transition appears. The system is not able to recall any of the p stored patterns, if the C density parameter ($0 < C < N=\binom{n}{2}$) is below a certain threshold ($C < p/\alpha_c$).

The relationship between network structure and dynamic stability has been studied in different biomathematical contexts. There exists a general statement in the cybernetic literature that the stability of a system decreases as the connectivity exceeds a certain threshold. In other words, in order to be stable a large system should not be too strongly connected. Simulation experiments [9] and the May-Wigner theorem [12,17] support this statement. The scope and limits of this theorem has been emphasized recently [5].

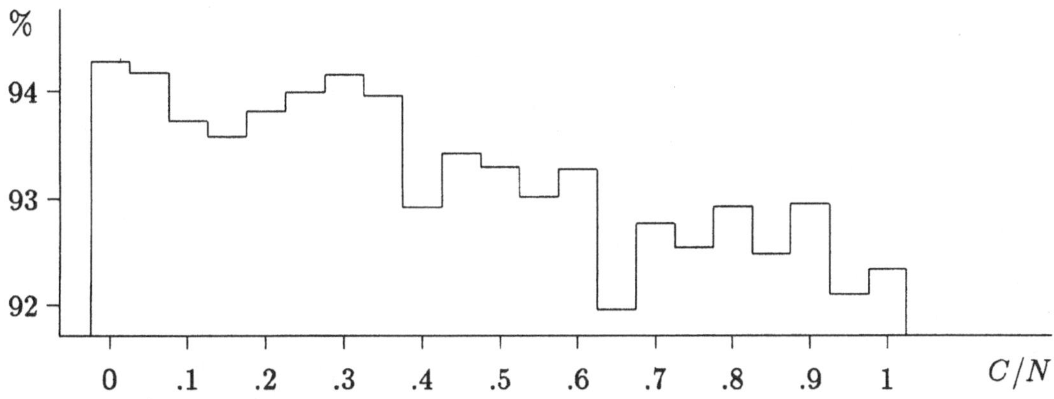

Figure 1. Numerical results for the encoder problem.

In this paper the effect of the sparse connectivity of a neural network is investigated, in a more general case. Particularly one question is raised: is there any situation in which the decrease of connection density offers some advantages over the fully connected network? The Boltzmann machine was used as the subject of the studies. There were two reasons for this selection. Firstly, it can be configured in various ways, it can form associative memory as well as input-output machine. Secondly, it does not only handle uncorrelated data, like the Hopfield network, but it is virtually capable of solving any task — provided, that the network is large enough.

2 Problems considered

There were two tasks to be considered: the encoder and the T-C problem. They are — according to Minsky and Papert [13] — first and third order problems, respectively. The exact definition of the order of problems is beyond the scope of this paper, but one could say, that the higher order the problem, the more difficult is to solve it. The natural problems for humans are higher order problems. (There are, however, higher order problems, for example the XOR problem [16], which are extremely difficult for humans.)

The encoder problem consists of two randomly selected pattern sets, and the task is to associate the elements of the two sets. In the experiments shown below the selected patterns were uncorrelated. In the T-C problem, there is a pixel matrix with torus topology. In the matrix, there is either a 'T' or a 'C' shape formed by 5 pixels. The two capital letters can be rotated or shifted in any direction. The task is to discriminate between the two letters.

3 Simulation experiments

The Boltzmann machine [1] is a probabilistic network consisting of binary units fully connected by symmetrical bidirectional links. It is based on the so called simulated annealing method [11]. To make the network sparse, the symmetry was preserved but some

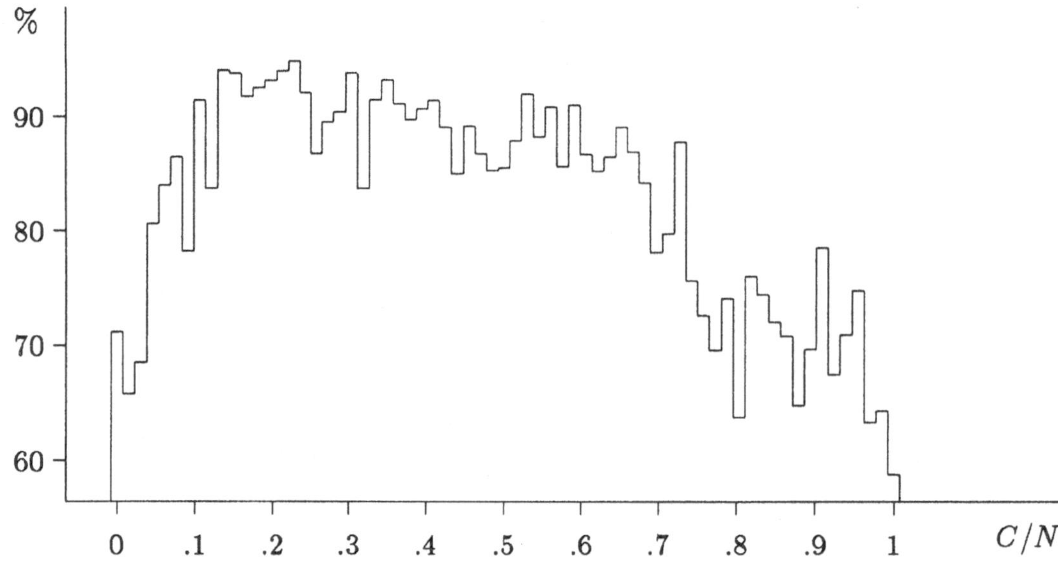

Figure 2. Achieved performance for the T-C problem.

of the connections were deleted simultaneously in both directions. In the experiments shown below the evolution of the graph of the hidden units was investigated. The input units were fully connected to the hidden units and those in turn to the output units.

The performance of the network for the encoder problem is shown in Fig. 1. The horizontal axis is the connectivity of the network, on the vertical axis the performance is shown in percentages. The connectivity is defined by the formula C/N, where C is the number of actual, and $N=\binom{n}{2}$ is the number of all the possible connections. The figure shows that adding more edges to the hidden layer decreases the performance almost linearly. The optimal network is disconnected in the hidden layer.

On Fig. 2. the results obtained from the solution of the T-C problem on a 7x7 torus are shown. There is an optimal configuration — around C/N=0.25 — where the best results can be obtained, and the extreme configurations give considerably poorer results.

It should be noted, that these results are a little "jumpy". The reason is mainly the limited time allowed for annealing the system. Given much longer time for cooling, the performance of the network could be different from those on Fig. 2. In fact, the Boltzmann machine is virtually capable to yield the best possible result. However, our aim has been to investigate a "natural", not an ideal situation.

4 Conclusions

Simulation experiments suggest that there are no universal rules, how the neural network, particularly, connections of the Boltzmann machine should be constructed in order to gain maximum performance. In contrary to the naive expectation it turned out that sometimes performance might be improved by reducing the degree of connections. The optimal structure of the network is strongly determined by the task to be solved. However, the tasks natural for neural networks seem to require sparse connectivity with density far below that of a fully connected network. This is in accordance with the kind of structure found in natural neural networks.

References

[1] D. H. Ackley, G. E. Hinton and T. J. Sejnowski, A learning algorithm for Boltzmann machines, *Journal of Cognitive Science*, **9**, 147–69 (1985).

[2] B. Bollobás, The evolution of sparse graphs, *Graph Theory and Combinatorics* (B. Bollobás ed.), Academic Press (1984).

[3] B. Derrida, Distribution of the activities in a diluted neural network, *Journal of Physics A: Mathematical and General,* **22**, 2069–80 (1989).

[4] B. Derrida, E. Gardner and A. Zippelius, An exactly solvable asymmetric neural network model, *Europhysics Letters,* **4**, 167–73 (1987).

[5] P. Érdi and J. Tóth, What is and what is not stated in the May-Wigner theorem? (submitted).

[6] P. Érdi and J. Szentágothai, Neural connectivities: between determinism and randomness, *Dynamics of Macrosystems* (J.-P. Aubin *et al.* eds.), Springer, 21–9 (1985).

[7] P. Erdős and A. Rényi, On the evolution of random graphs, *Publ. Math. Inst. Hungar. Acad. Sci,* **5**, 17–61 (1960).

[8] M. R. Evans, Random dilution in a neural network for biased patterns, *Journal of Physics A: Mathematical and General,* **22**, 2103–18 (1989).

[9] E. Gardner and W. R. Ashby, *Nature,* **228**, 784 (1970).

[10] J. J. Hopfield, Neural networks and physical systems with emergent collective computational properties, *Proc. Natl. Acad. Sci. USA,* **79**, 2554–8 (1982).

[11] S. Kirpatrick, C. D. Gelatt and M. P. Vecchi, Optimization by simulated annealing, *Science,* **220**, 671–80 (1983).

[12] R. M. May, *Nature,* **238**, 413 (1972).

[13] M. Minsky and S. Papert, *Perceptrons,* MIT Press, Cambridge (MA) (1969).

[14] H. Sompolinsky, *Phys. Rev.,* **A34**, 2571 (1986).

[15] J. Szentágothai, Specificity versus (quasi-)randomness in cortical connectivity, *Architectures in the cerebral cortex* (M. A. B. Brazier and H. Petsche eds.), Raven Press, New York(1978).

[16] S. J. Thorpe, J. K. O'Regan and A. Pouget, Humans fail on XOR pattern recognition problems, *Proceedings of the First European Conference on Neural Networks,* (1988).

[17] J. Wigner, Statistical properties of real symmetric matrices with many dimensions, *Proc. Fourth Canad. Math. Cong.* (M. S. MacPhail ed.) University of Toronto Press, Toronto, 174–84 (1959).

A GOAL SEEKING NEURON FOR BOOLEAN NEURAL NETWORKS

E.C.D.B.C.Filho, D.L.Bisset, and M.C.Fairhurst
Electronic Engineering Laboratories,
University of Kent,
Canterbury,
Kent, CT2 7NT, U.K.
e-mail: dlb@ukc.ac.uk

Abstract

This paper proposes a novel Boolean neural model known as the *Goal Seeking Neuron (GSN)*. The operation of the neuron and its associated learning algorithm are discussed in detail. The GSN model has been generated in response to a number of observed weaknesses in the *Probabilistic Logic Node (PLN)* proposed by Kan[1]. The paper identifies these problems and shows how the goal-seeking nature of the GSN overcomes them. The GSN is designed to make efficient use of its memory space by compacting its internal representation, and allowing new patterns to be learned without corrupting existing memories. This is achieved without losing the potential for direct hardware implementation, or its local processing characteristics. By a simple modification to the learning mechanism it is also possible to achieve acceptable performance with only a single pass of the training data.

Introduction

It is possible to identify a class of neural networks that use programmable Boolean logic elements as processing nodes, rather than computing the more familiar sum-of-products functions. These neural networks are often described as *RAM-based* networks because the elements are similar to random access memories, their main advantage being the ease with which they can be implemented in hardware. This derives from their essentially logical rather than continuous nature. However there are a number of disadvantages with current Boolean architectures which become apparent when they are applied to pattern recognition problems. The *Goal Seeking Neuron (GSN)* proposed in this paper overcomes some of these limitations and as a result is much better suited to pattern recognition problems, while at the same time is able to perform other types of tasks such as pattern association, or logic problems.

The Probabilistic Logic Node (PLN)

The PLN was first proposed by Kan[1] and subsequent papers have explored its properties and extended its description[2,3]. The PLN is used primarily in a pyramid architecture which maps the input image to a single neuron at the top of the pyramid. Many pyramids may be needed for a particular recognition task. The PLN has a number of disadvantages when it is applied to pattern recognition problems where there are a significant number of classes to be recognised, and these may be categorised as follows:

a) Saturation of storage space

After a small number of training patterns the available storage space in the cells becomes too small for learning to take place. The PLN learning algorithm then erases paths through the pyramid, creating space for new learning. Unfortunately these paths are quickly blocked and little overall gain is made in recognition ability. In this state the network is said to be saturated. This presents a severe problem when many classes must be learned, because it becomes difficult to expose the network to sufficient training data for each of the classes to be learned.[4]

b) Corruption of previous learning

Nearly all neural networks suffer from this problem to some extent in the short term as the training data is cycled. Learning mechanisms that employ some kind of global error measure, taken over the total result set, are able to minimise this corruption in the long term. Boolean networks are particularly prone to this problem because of the all or nothing nature of their storage, and the PLN algorithm tries to overcome this problem by repeatedly cycling through the training set.

c) Non-deterministic response

The internal storage in the PLN element allows an undefined state to be stored. When the input signals address such a value a random decision is made by the element as to what value is output, either **0** or **1**. This means that during the operation of the network different responses can be obtained for the same input pattern. In order to determine the correct response a number of presentations of the input pattern must be made and a vote taken to decide the most probable output, or to reject the pattern. This mechanism is inefficient because of the need to repeat the input pattern presentation.

The Goal Seeking Neuron (GSN)

It is important in any neural network, and particularly in Boolean networks where memory size is a limiting factor, to make the best use of the storage capacity available, while at the same time maximising the likelihood of a correct result. It is important that these optimisation properties are either contained within the design of the neural element or are imposed through the training mechanism. This should be achieved without sacrificing simplicity or the local processing requirement of neural networks. The GSN has been designed to try and overcome some of the limitations described, while satisfying these basic design criteria.

In common with the PLN the GSN is based on a random access memory cell, where the inputs address storage cells which can contain values from the set **{0,1,u}**, where **u** is the undefined value. In contrast with the PLN the GSN is able to output all three values, rather than having to make a choice when an undefined cell is addressed.

Because the inputs to the neuron can contain undefined values the operation of the neuron is not the same as a simple random access memory. When a single input is undefined this is taken to mean that two memory cells inside the neuron are being addressed. These are the cells that would be addressed by setting the undefined input to either the **1** or **0** value. Therefore for each combination of the input lines, *input state*, there are a collection of addressed cells called the *addressable set*, and the contents of these cells are called the *addressable contents*. The cells addressed can thus range from a single cell, when the input state contains no undefined values, to the entire contents of the cell when all of the inputs are undefined. The interpretation of this scheme depends upon the state of the neuron.

a) The *Seeking State*

In the seeking state the neuron responds in the following manner:

> The output is a **1** if all the addressable contents are **1**.
> The output is a **0** if all the addressable contents are **0**.
> The output is a *u* for all other values of the addressable contents.

Thus when the addressable contents contain an undefined value, or when there is a mixture of **0** and **1** values then the output is undefined. This mode of operation helps to propagate undefined paths through networks of elements, and to seek out unused or conflicting values.

b) The *Learning State*

In the learning state the neuron tries to associate the desired output with an existing cell in the addressable set. If it fails to do this it chooses an undefined cell and sets its value to the desired output. If there are a number of possible choices a random decision is made. It is necessary to choose one particular cell to represent the output value because the address of this cell is passed back down the input connections to become the desired output values for the previous layer.

c)The *Recall State*

In the recall state the neuron produces outputs according to the following scheme:

> If the number of ones in the addressable contents is greater than the number of zeros, then the neuron outputs a **1** value.
> If the number of **0** values is greater then the neuron will output a **0** value.
> If the numbers of ones and zeros is equal then the neuron outputs an undefined value.

Thus, even if the addressed contents only contains a single **1** value, and the remainder are undefined, the neuron will output a **1** value. The reason for adopting this scheme is to minimise the propagation of undefined values.

Networks of GSNs

In this paper the GSN elements will be used in a pyramid structure as has been used for PLN elements[2], although alternative configurations are possible. The connectivity is kept as low as possible to minimise both the memory requirements, and the size of the pyramids. The numbers of neurons in each pyramid and the number of pyramids is problem dependent. Each pyramid can be viewed as an independent processing unit since there is no cross connection between pyramids. The first layer of the pyramid usually covers the input space, and the final layer consists of a single neuron. To train the pyramid the set of input patterns are clamped to the inputs and the desired outputs made available at the final output unit of the pyramid, and the training then proceeds as follows:

Step 1

with all elements in the *Seek State*, propagate the input pattern through the pyramid, until a final output is obtained.

Step 2

If the output of the final element is either undefined, or the same as the desired pyramid value for this input pattern,

then (a)

set the elements into the *Learning State* and start the reverse propagation of the desired outputs from the final element to the input element.

(b) else

this indicates a conflict of learning. This can be handled in a number of different ways, the simplest of which is to reset the addressed cells for that pattern, throughout the pyramid, to the undefined state.

Step 3

Repeat for each pattern in the training set.

In the light of this learning algorithm the functions of the seeking and learning states can be analysed in more detail:

The *seeking* phase establishes the possibility of learning the desired output without corrupting any of the previously stored information, while at the same time trying to maximise the representation capacity. The pyramid is working in a forward direction, and the output from the final element indicates what final value can be learned for this input pattern. Ideally this will either be set to the undefined value, or have the same value as the desired output for the pyramid.

The *learning* phase chooses the best possible adaptation for each element, trying to maximise memory capacity. The pyramid is working backwards in this phase updating the cell values, and forming desired values for all of the elements in the previous layer.

It is possible to define a number of alternative learning strategies based on the action taken when a learning conflict occurs. If the paths in the pyramid are erased then data is lost, and the presentation of the pattern set must be repeated. This not only lengthens the training time but also increases the probability of saturation, which will create the need for more path erasure. There is one particular alternative which does not require the erasure of information in the network, and thus allows learning to take place with a single pass of the training data. This scheme works as follows:

a) Each pattern class is assigned a code vector which specifies the desired output for each pyramid.

b) The pyramids are trained on these code vectors, but when a conflict occurs a note is kept of which class and which pyramid was involved.

c) After the training data has been presented the code vectors are modified to represent the most probable outputs for each class, based on the response to the training data.

d) During the recalling phase the modified code vector closest to the actual output vector is taken as the class of the input pattern.

Although this involves a more complex recalling process, it is preferable to a repeated cycling of the training data.

Results

Figure 1 shows the relative saturation levels of a PLN pyramid and a GSN pyramid for the same training data. It can clearly be seen that the GSN is able to absorb more training data while maintaining significantly more undefined cells within its neurons. It is also possible to measure the number of training patterns that do not cause a conflict as a percentage of the total amount of training data. As can be seen from Figure 2 the GSN is able to learn a larger number of patterns before conflicts start to occur. These graphs represent average results for many runs of different data sets taken from the same problem, and represent the typical advantage that can be obtained by using GSN. Preliminary results on a realistic pattern recognition problem indicate that GSN performs some *10-15%* better than PLN, while providing a significant speed advantage in training due to the single pass over the training data.

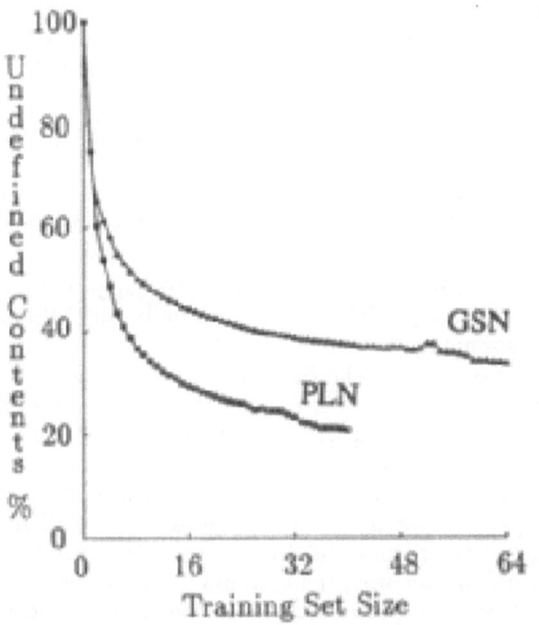

Figure 1: Saturation vs training set size for PLN and GSN.

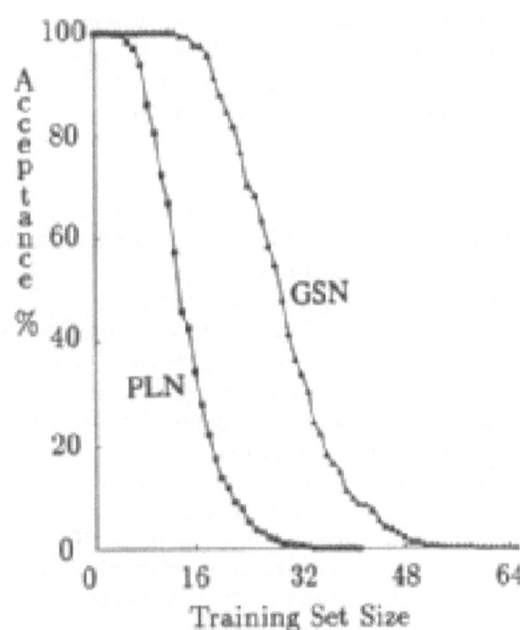

Figure 2: Percentage of patterns not causing learning conflicts vs training set size.

Conclusions

This paper has discussed a number of problems with the application of the PLN model to pattern recognition tasks. It has presented a novel neuron and associated learning algorithm for Boolean neural networks that is less prone to the problems associated with the PLN network model. Preliminary results also indicate that there is an expected improvement in recognition performance. The GSN has been designed with final implementation in hardware in mind, thus rendering it a viable candidate for high speed implementation.

References

[1] *A probabilistic logic neuron network for associative learning*, W.Kan, and I.Aleksander, in Proceedings of IEEE International Conference on Neural Networks, San-Diego, **1987**.

[2] *Learning algorithm for probabilistic neural networks*, C.Myers, and I.Aleksander, in Proceedings of First INNS Annual Meeting, Boston, **1988**.

[3] K.Y.M. Wong, and D.Sherrington, **Europhys. Lett.**, *7* 197, 1988.

[4] *A comparative study of neural network structures for practical application in a pattern recognition environment*, D.L.Bisset, E.Filho, and M.C.Fairhurst, in Proceedings of the First IEE International Conference on Artificial Neural Networks, London, **1989**.

A Spatial Approach to Feature Linking

Helge Ritter
Beckman Institute and Department of Physics
University of Illinois at Urbana-Champaign
Urbana IL 61801, USA
E-Mail: 11074@ncsavmsa.bitnet

Abstract: A spatial mechanism for the dynamical linking of features is proposed. Linking is achieved by a set of layers of feature-selective cells with competitive, topographic interactions between layers, and a suitable lateral interaction within layers. Analytical and simulation results are given for the simple case of an "on-center-off"-surround interaction.

1. Introduction

One major problem of brain theory is the *binding problem*: what is the neural mechanism that is responsible for dynamically linking object features appropriately together, so that the proper conjunctions between features are preserved?

One candidate mechanism have been temporal correlations among different neurons [1-3,5,6]. Linking is achieved by temporal synchrony of neurons coding for related features. This is a conceptually very elegant scheme, which recently has also received some experimental support [1,2].

However, at the present stage of our understanding, and in view of the fundamental relevance of the issue, it seems important to explore also alternative mechanisms for feature binding. In this contribution we suggest an approach which uses *the spatial domain* for feature binding and which, therefore, can be viewed as complementary to the temporal scheme.

The approach uses several identical layers of feature-selective cells. For each particular feature cell there is an identically tuned replica per layer. Therefore, each feature that is present in the input can elicit a response of any of these replicas. Suitable competitive interactions among the replicas force the system to choose one replica as the solely responding cell for each feature. In this way, the system can break the set of input features in a meaningful way into as many subsets as there are layers, each subset being indicated by the active feature cells of one layer.

In its present simple form, the model appears less flexible than a temporal binding scheme: the number of possible groups is limited and fixed beforehand by the number of available layers. However, finite temporal resolution for the synchronization of independent spike trains will also impose a limit for the temporal scheme, and the fixed number of groupings could be overcome by a dynamical scheme for "blocking" and "releasing" entire layers.

In its current form, the model is meant to explore an interesting alternative mechanism for dynamical feature binding. A possible role for this spatial mechanism could be to complement and assist the hypothesized temporal feature binding mechanism in the brain.

2. The Model

The model employs a fixed number L of identical layers $\alpha = 1, 2 \ldots L$ of feature-selective cells. For each spatial position $\mathbf{r}$ there is one feature-selective cell per layer, whose activity shall be denoted by $x_{\mathbf{r}\alpha} \geq 0$. Cells belonging to different layers α, but to the same location $\mathbf{r}$ are assumed to respond to the same feature and share a common input line $h_\mathbf{r}$, which can be either active ($h_\mathbf{r} = 1$) or silent ($h_\mathbf{r} = -c_0$, $c_0 > 0$).

In a simple setting, we may think of $h_\mathbf{r}$ encoding the presence or absence of a "dot" at a location $\mathbf{r}$ in some imaginary "retina". We would then want the system to bind those "dots" together that form salient subgroups ("Gestalts") of the input pattern. Of course, any actual grouping in the visual field requires in addition to retinal location many further features, such as e.g. line orientation and velocity.

The model can process more abstract features, if the input $h_\mathbf{r}$ arises from some suitable, spatially organized topographic "feature-map" instead of the "retina". In such a map each location is associated with some more abstract feature of the input. Various such maps have been found in the brain, and Kohonen has given an elegant model for their formation through a self-organizing process in artificial networks [see e.g. 7-9]. Here we will not pursue such modifications. Instead we shall present the model in its most basic form, so that a mathematical analysis of some important aspects remains feasible and its operation can be visualized using simple two-dimensional patterns of dots.

The activity of each cell $x_{\mathbf{r}\alpha}$ is subject to the following dynamics:

$$\dot{x}_{\mathbf{r}\alpha} = J_1 \Big(h_\mathbf{r} - \sum_{\beta=1}^{L} x_{\mathbf{r}\beta} \Big)^{2p-1} + \sum_{\mathbf{r}'} f_{\mathbf{r}\mathbf{r}'} x_{\mathbf{r}'\alpha}, \tag{1}$$

but constrained by the condition $x_{\mathbf{r}\alpha} \geq 0$. The first term is a competitive interaction between cells of different layers β, which obeys a topographic structure by coupling only cells with identical location $\mathbf{r}$ in each layer†. The coupling strength of the cells of a common location $\mathbf{r}$ to the external input $h_{\mathbf{r}}$ is determined by the parameter $J_1 > 0$ and the positive integer p. The second sum is a lateral interaction between cells of the same layer α, but different locations $\mathbf{r}$, $\mathbf{r}'$. The interaction coefficients $f_{\mathbf{r}\mathbf{r}'}$ determine which feature configurations are preferably linked together by the model. We will discuss the case

$$f_{\mathbf{r}\mathbf{r}'} = \begin{cases} 1 & \text{if } |\mathbf{r}-\mathbf{r}'| < R_0 \\ -J_2 & \text{else.} \end{cases} \tag{2}$$

This corresponds to an excitatory interaction over distances smaller than R_0 and an inhibitory (assuming $J_2 > 0$) interaction over distances greater than R_0. We shall see that this very simple type of lateral interaction leads to a grouping of dots which are closer than some distance of the order of R_0. This kind of grouping tendency is somewhat similar to the "Law of Proximity", formulated by the "Gestaltists" for visual perception (see e.g. [4]).

Eq. (1) admits the energy function

$$E = \frac{1}{2p} J_1 \sum_{\mathbf{r}} \Big(\sum_{\alpha=1}^{L} x_{\mathbf{r}\alpha} - h_{\mathbf{r}} \Big)^{2p} - \frac{1}{2} \sum_{\alpha=1}^{L} \sum_{\mathbf{r}\mathbf{r}'} f_{\mathbf{r}\mathbf{r}'} x_{\mathbf{r}\alpha} x_{\mathbf{r}'\alpha}. \tag{3}$$

For the system to be stable E must be bounded from below. This requires that the coupling of the cells to the external inputs must be sufficiently strong compared to the self-excitation of the network (determined by the radius R_0). This is automatically guaranteed if $J_1 > 0$ and $p > 1$. For the case $p = 1$ a sufficient condition can be shown to be $J_1 > \sum_{|\mathbf{r}|<R_0} 1$ ([10]). An interesting generalization occurs, if higher order lateral interactions such as $f_{\mathbf{r}\mathbf{r}'\mathbf{r}''}$ are included, by which the system can be made sensitive to more subtle feature correlations, such a e.g. collinearity ([10]). Then the value $2p$ must at least equal the order of the highest order lateral interaction term.

Under these conditions, the system settles into a local minimum of E for any input $h_{\mathbf{r}}$. For $p = 1$, E is a quadratic function of the neuron activities $x_{\mathbf{r}\alpha}$ and has a single, unique minimum when the values of all $x_{\mathbf{r}\alpha}$ are unrestricted, in which case the minimum can be easily calculated. In the present model, this "linear" situation does not hold even for $p = 1$, because the constraints $x_{\mathbf{r}\alpha} \geq 0$ on the range of the cell activities introduce a nonlinearity which is essential to the operation of the model. It is then possible to show for the stationary states of eq. (1) the following three theorems ([10]):

Theorem 1: Choose $0 < J_2 < 1 < J_1$ and p such, that E is bounded from below. Then all stationary states of (1) are of the form

$$x_{\mathbf{r}\alpha} = \begin{cases} z_{\mathbf{r}} & \text{for precisely one layer } \alpha = \alpha_{\mathbf{r}} \\ 0 & \text{in all other layers } \alpha \neq \alpha_{\mathbf{r}} \end{cases} \tag{4}$$

For each site $\mathbf{r}$ the integer $\alpha_{\mathbf{r}}$ denotes one of the layers, and $z_{\mathbf{r}} \geq 0$ are real numbers (which for sufficiently large J_1 approach zero, if $h_{\mathbf{r}} < 0$, and $h_{\mathbf{r}}$, if $h_{\mathbf{r}} > 0$).

Theorem 1 states that the stationary state admits no simultaneous activity of two cells in different layers, but with same location $\mathbf{r}$. This forces the system "to decide" and to assign each input $h_{\mathbf{r}} = 1$ unambiguously to one of the layers.

The next theorem states that a sufficiently localized cluster C of activated cell positions must always arise from active cells within the same layer: clusters of inputs become grouped into a common layer and are not divided and scattered over several layers. This constitutes the feature binding property of the model.

Theorem 2: Let for one of the stationary states of Theorem 1 the set C denote a subset of all active (i.e. $z_{\mathbf{r}} > 0$) cell locations $\mathbf{r}$ with the property

$$\begin{aligned} \mathbf{r}, \mathbf{r}' \in C &\Rightarrow f_{\mathbf{r}\mathbf{r}'} = 1 \\ \mathbf{r} \in C, \mathbf{r}' \notin C &\Rightarrow f_{\mathbf{r}\mathbf{r}'} = -J_2 \end{aligned} \tag{5}$$

Then $\alpha_{\mathbf{r}} = const.$ for all $\mathbf{r} \in C$.

Theorem 2 remains valid, if the sets where $f_{\mathbf{r}\mathbf{r}'} = 1$ are of arbitrary, but translationally invariant shape. For the special choice (2) C can be taken to include all active cells within an area of some radius $R \leq R_0$.

† We note that the feature maps in different layers could be distorted with respect to each other. In this case it is understood that the labelling of the cells follows this distortion in each layer such that $\mathbf{r}$ can still be used to refer to the same feature in each layer.

If then all other active cells are farther away than R_0 from the *boundary* of this area, then the active cells within the circle must lie in the same layer.

It would be conceivable, that solutions can exist for which one or several layers end up unused, i.e. without any activity. From Theorem 1 we can conclude that such solutions must exist, if the number of active inputs is less than the number of layers, or if the number of simultaneous clusters satisfying Theorem 2 is less than the number of layers. Theorem 3 states that in all cases in which sufficiently many inputs are active, all layers will develop responses:

Theorem 3: Let $S_\alpha = \sum_{\mathbf{r}} x_{\mathbf{r}\alpha}$ be the total activity of layer α, and let $S_{\mathbf{r}\alpha} = \sum_{\mathbf{r}',|\mathbf{r}'-\mathbf{r}|<R_0} x_{\mathbf{r}'\alpha}$ be the fraction of S_α contributed from cells closer to $\mathbf{r}$ than a distance R_0. Then, each stable stationary state that contains at least one active cell $x_{\mathbf{r}\alpha} > 0$ such that

$$S_{\mathbf{r}\alpha} < \frac{J_2}{1+J_2} S_\alpha, \tag{6}$$

has active cells *in all layers.*

This shows that inactive layers can only exist if within each active layer most (i.e. at least a fraction $J_2/(1+J_2)$) of its activity is clustered within a single region of diameter R_0, which is only possible, if the input pattern contains sufficiently few and tight clusters.

Taken together, the above three theorems show that a "stack" of laterally inhibited layers with suitable topographic interactions between layers can use its layers to link subsets of input features together. In the next section we shall give a few illustrations of this capability.

3. Simulation Results

For a computer simulation of the system we used parameter values $p = 1$, $J_1 = 10$, $J_2 = 0.5$ and $R_0 = 0.2$. Each layer was assumed to be a planar array of cells occupying lattice positions within a unit square. As mentioned earlier, we restrict ourselves in the following to the use of simple "dot" features as inputs. The presence of a "dot" at location $\mathbf{r}$ means a value of $h_{\mathbf{r}} = 1$ and external excitatory input to the cell at position $\mathbf{r}$ in each layer. Those cells whose locations do not coincide with any of the "dot" positions of the input pattern receive inhibitory external input $h_{\mathbf{r}} = -c_0$. In the limit of very large c_0, the activity of these cells is effectively "tied" to zero, and the dynamics of the system changes only very little if they are then omitted altogether. This approximation reduces the computational cost of simulating the system (1) considerably and was, therefore, adopted for the examples given below.

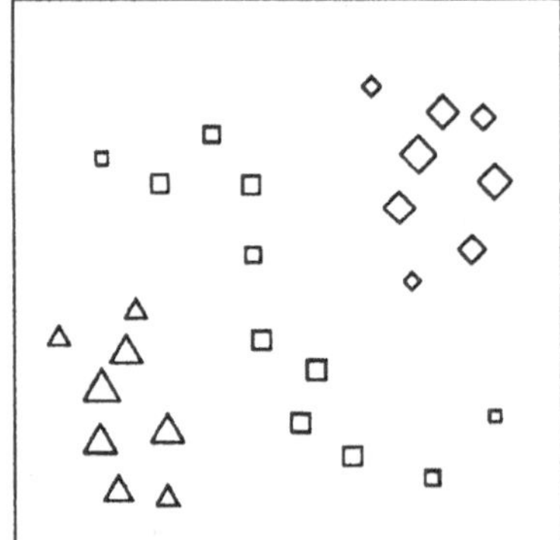

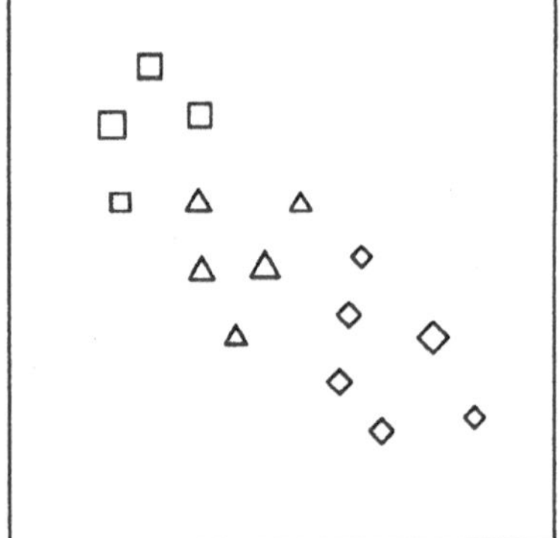

Fig.1a-c Responses of a three-layer system to various "dot" patterns. From left to right: (a) Successful clustering. (b) If the number of feature clusters exceeds the number of layers, some clusters are not resolved. (c) In the absence of any clusters, the system enforces a clustering.

Figs.1a-c show the response of a three-layer system to various input patterns. Each diagram can be viewed as a perpendicular projection of the cell layers onto a common plane, and the activities $x_{\mathbf{r}\alpha}$, $\alpha = 1,2,3$ of cells of different layers, but identical $\mathbf{r}$ are visualized by the size of different symbols superimposed at the location $\mathbf{r}$ in the diagram. In all cases, the cell responses also coincide with the "dots" of the input pattern, i.e. with those locations, where $h_{\mathbf{r}} = 1$.

In Fig.1a, the system has linked the "dot" features into three clusters. The detection of the elongated cluster in the middle shows that already the very simple choice (2) for the lateral interaction $f_{\mathbf{r}\mathbf{r}'}$ provides the system with some degree of sensitivity to global properties of the input pattern.

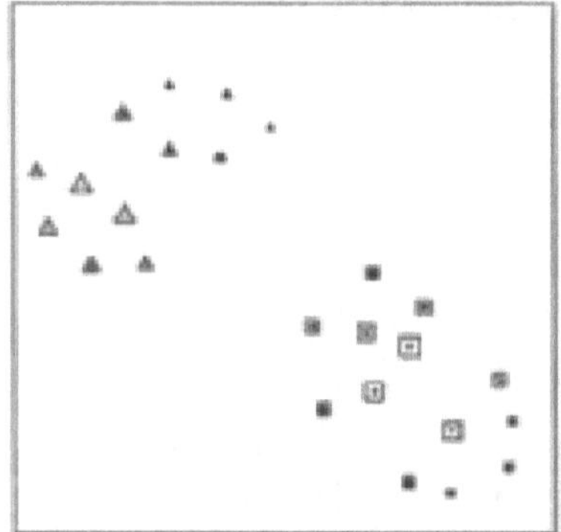
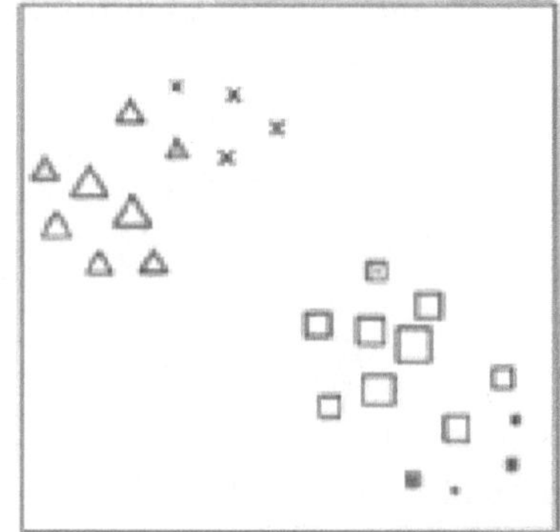
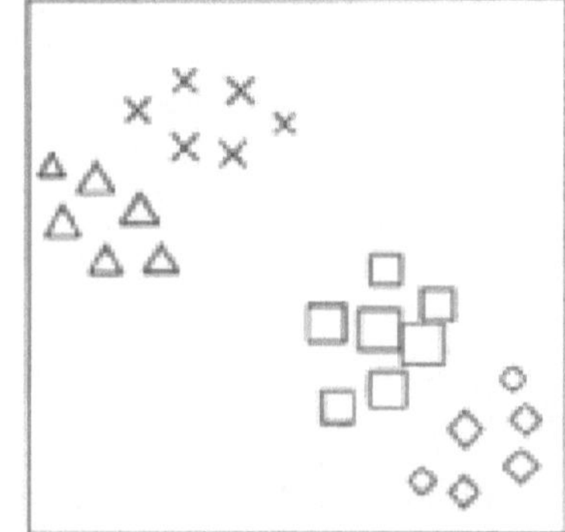

Fig.2a-c Temporal course of feature linking for a four-layer system. From left to right: (a) shortly after input presentation two clusters are differentiated. (b) Later, these are further differentiated until finally the correct features are linked together (c).

Figs.1b-c illustrate situations when the number of layers does not match the number of input clusters. In Fig.1b, the system succeeded to link together the features of each of the two larger clusters separately. With only a third layer available, it then could not resolve the remaining part of the input into its three constituents. Fig.1c shows the response of the same system to a pattern admitting no natural clustering. In this case, the system enforces a subdivision of the input into three compact and roughly equally sized portions.

On its way to equilibrium the system may temporarily link features together which are separate in the final state. Figs.2a-c illustrate this phenomenon with a four-layer system. Shortly after presentation of the input the system has formed two large groups ("triangles" and "squares"), but the simultaneous activation at the same site of cells from different layers indicates the transient nature of this state (Fig.2a). After a while, a further subdivision of the two large groups is clearly on the way (Fig.2b), until finally all features have been linked together according to their cluster structure (Fig.2c). Comparison with Fig.2b shows, that several previous assignments have become revised.

4. Conclusion

A remarkably simple, layered system is capable of some interesting kind of feature linking. In contrast to the correlation approach to dynamical feature linking, the present system is based on a *spatial mechanism* and uses a set of layers to link features together. The linking mechanism requires two different kinds of neural interactions: a *competitive, topographically structured interaction between layers*, and a *lateral interaction within layers*. The latter determines the linking properties. The case of an "on-center-off"-surround interaction leads to the capability of spatial clustering of simple dot patterns. For this case, analytical and simulation results are reported. The main limitation of the model in its current form is the requirement that the number of layers must match the number of desired feature clusters.

References:

[1] Eckhorn R., Reitboeck H.J., Arndt M., Dicke P. (1989) Feature Linking via Stimulus-Evoked Oscillations: Experimental Results from Cat Visual Cortex and Functional Implications from a Network Model. Proceedings of the IJCNN89, Vol.I, pp.723-730.

[2] Gray C.M., König P., Engel A.K., Singer W. (1989) Oscillatory responses in cat visual cortex exhibit inter-columnar synchronization which reflects global stimulus properties. Nature 338, pp. 334-337

[3] Hartmann G., Drüe S. (1990) Feature Linking by Synchronization in a Two Dimensional Network. Proc. of the IJCNN90, Vol.I, pp. 247-250, Washington D.C.

[4] Köhler W. (1929) Gestalt Psychology. New York, Liveright

[5] v.d. Malsburg C. (1981) The Correlation Theory of Brain Function. Internal Report 81-2, Max Planck-Institute for Biophysical Chemistry, Göttingen, FRG

[6] v.d. Malsburg C., Singer W. (1988) In: Racic P., Singer W. (Eds.) Neurobiology of Neocortex (Proceedings of the Dahlem Conference) pp. 69-99, Wiley, Chicester

[7] Kohonen T. (1982) Self-organized Formation of Topologically Correct Feature Maps. Biol. Cybern. 43, pp. 59-69.

[8] Kohonen T. (1984) Self-Organization and Associative Memory. Springer Series in Information Sciences 8, Heidelberg.

[9] Ritter H., Kohonen T. (1989) Self-Organizing Semantic Maps. Biol. Cybern. 61, pp. 241-254

[10] Ritter H. (1990) (*in preparation*)

The Discrete Neuronal Model and The Probabilistic Discrete Neuronal Model

Ron Sun
Brandeis University
Computer Science Dept.
Waltham, MA 02254

ABSTRACT

In this paper, we present and analyze a class of neural network models previously proposed by the author: the discrete neuronal model and the probabilistic discrete neuronal models (DN and PDN). The models have been applied to model some biological neural networks (e.g. the stomatogastric ganglion in lobsters) and domenstrated their generality and usefulness in simulating the information processing capabilities of real neural networks given their exact or inexact connectivity patterns and endogenous firing patterns. Another application of this model is in building rule based inference systems. The advantages of this type of models is that it can handle variable bindings easily. We also incorporate certainty factors propagation mechanisms into the system. Systems for sequential processing based on this models have been worked on too.

Because of these interesing properties of the models, we need to perform some rigerous analysis to better understand the models. But, because of the generality and complexity of the models, it is very difficult to analyze them in the same way as conventional PDP models are analyzed. This paper is an initial step towards a better unbderstnding.

DN can be described as:

```
          W=<N, M>

where
          N={<S, A, I, T, C>}

          S= the set of all the possible
             states of a neuron,
          A= the set of all the actions
             to be taken by the neuron,
          I= inputs,
          T= State transition function:
             S x I --> S,
          C= action function:
             S x I --> 2^A,

          and M is the connectivity among
              neurons in the set N.
```

PDN is an extension to DN by introducing probabilistic state transition and action functions.

We show that there are some striking similarity between this model and the following computational models: hidden markov models, learning automata, team decision theory, finite state machine etc. Based on the things we know about these models, we derived some useful learning algorithms for various purposes in this type of networks. The difference and advantages of these algorithms, compared with existing algorithms for similar kinds of network models, are discussed and analyzed. These algorithms can be used for various purposes in different domains.

1 Introduction

In this paper, we present and analyze a class of neural network models previously proposed by the author: the discrete neuronal model and the probabilistic discrete neuronal models (DN and PDN). The models have been applied to model some biological neural networks (e.g. the stomatogastric ganglion in lobsters) and domenstrated their generality and usefulness in simulating the information processing capabilities of real neural networks given their exact or inexact connectivity patterns and endogenous firing patterns. For details, see Sun et al (1989). Another application of this model is in building rule based inference systems. The advantages of this type of models is that it can handle variable bindings easily. We can also incorporate certainty factors propagation mechanisms into the system. The detailed description of the system and also some description of the neuronal models are contained in Sun (1989). Systems for sequential processing based on this models have been proposed too.

Because of these interesing properties of the models, we need to perform some rigerous analysis to better understand the models. But, because of the generality and complexity of the models, it is very difficult to analyze them in the same way as conventional PDP models are analyzed. This paper is an initial step towards a better unbderstnding.

Neural network models did not start with weighted sum models, and did not end with them. Although the most frequently seen model is the weighted sum (thresholding) model, its biological validity and computational sufficiency are questioable. The popularity of this model mainly stems from its simplicity. Many alternative neuronal models may provide better computational properties and biological properties.

To a large extent, finding alternative neuronal models is an area yet to be explored. Devising more complicated models seems to be an inevitable step toward useful, effecient and implementable neural networks. In this endeaver we could look into some past efforts in this and lessons to be learned.

2 DN and PDN

A automaton-theoretic description follows. Basically, a discrete neuronal model (DN) is a 2-tuple

```
          W=<N, M>

where
          N={<S, A, I, T, C>}

          S= the set of all the possible
             states of a neuron,
          A= the set of all the actions
             to be taken by the neuron,
          I= inputs,
          T= State transition function:
             S x I --> S,
          C= action function:
             S x I --> 2^A,

          and M is the connectivity among
              neurons in the set N.
```

In this model a set of discrete states is explicitly specified instead of a continuous activation function. By introducing the state variable we hope to be able to model more aspects of the working of real neurons.

The DN units can be implemented with the simple PDP neuron types. A linear neural network model can be expressed as

$$\frac{d}{dt}X = AX + BI$$

where X is the activation vector, and I is the input vector. This is exactly the description of a linear dynamic system. A decides the contribution of current X in deciding the next X, and B that of input I. In case of semilinear systems, we can easily replace X in the right hand side of the equation with a function h(X). I want to show that it is functionally equivalent to a Finite State Automaton, and therefore equivalent to the Discrete Neuron model. The above equation can be rewritten as

$$\frac{d}{dt}X = f(X, I)$$

Discretizing it, we have

$$\Delta X = f(X, I)\Delta t$$

or

$$X_{t+1} = g(X_t, I_t)$$

which is exacly a FSA! We can apply standard system theory to study if a particular FSA is implementable or not.

PDN (Probabilistic Discrete Neuronal Model) is a simple extension of this. It assumes that each entry (of the table specifying the state transition function and the action function) is probabilistic instead of deterministic. If a state transition or action is not specified by the table, then it is completely random, based on a uniform distribution. If it is specified with a probability less than 1, then the other choices might be specified by other entries, or in case there is no other entry, by a uniform distribution over all other choices with the rest of the total probability.

Biophysiological justifications can usually be argued but will not be conclusive at all, considering the state of the art of neural network research. We simply do not know enough at this time to fully describe the operation of the neuronal cells. But we can get some hint from biophysiological research to further our modeling efforts. The probabilistic nature of the model can be attributed to the unreliability and noise in the cell, stattistical nature of membrane operations (channel openings and closings), and random environmental influence etc.

3 Learning Algorithms

A learning algorithm for a network of nodes of PDN type can be easily devised based on the idea of reinforcement learning.

Scheme 1 *If a unit fires correctly (acocording to external teachers or internal critic, error signal or reinforcement signal), then increase the firing probability of the corresponding entry, in case that entry exists, or create an entry for that action, if the corresponding entry does not exist.*

It is based on the same idea as in William (1985).

The above algorithm addresses the issue of training a single neuron, with input, output, and a reinforcement signal. In case of a large, complicatedly connected network, we have a simple algorithm which is a simple extension of the above scheme and can tune the network into performing the required input/output mapping.

Scheme 2 (persimistic) *Clamp the input nodes and look at the output nodes. If all the output nodes are correct or all are wrong, then apply the above scheme to each node, assuming that every node in the network is correct (or wrong). Otherwise, apply the above scheme to the correct output nodes for positive reinforcement, and then, assuming the rest of the nodes are all wrong, apply the above scheme for negative reinforcement.*

An alternative is to reduce the number of negatively reinforced nodes as much as possible, and also at the same time, to try to preserve the part of the network that preduced the correct results (if we negatively reinforce all of the network units except those that produce correct results, we may inadvertantly reverse the correct subnetworks. So we arrived at this:

Scheme 3 (optimistic) *Clamp the input nodes and look at the output nodes. If all the output nodes are correct or all are wrong, then apply the above scheme to each node, assuming that every node in the network is correct (or wrong). Otherwise, apply the above scheme to the incorrect output nodes for negative reinforcement, and then, assuming the rest of the nodes are all correct, apply the above scheme for positive reinforcement.*

This algorithm has a different problem, namely if there are too few negatively reinforced units, we may never get the optimal or near optimal solution, because we lose many degrees of freedom due to our decision to give positive reinforcement to the rest of the network.

The rate of change made in each node has a profound impact on the perfomance of the networks. It can decide if it will converge, how fast it will converge and if it will converge to a correct value, etc. An expedient scheme is the linear reward-penalty algorithm. i.e.

$$f_j(p) = ap_j$$

$$g_j(p) = bp_j + b/r - 1$$

where f is the amount of increment to the probability of an action if positively reinforced and g is the amount of decrement if negatively reinforced. There are a lot of other schemes that have various natures and performance.

As indicated by Barto (1988), such a network can be viewed as a confederation of units that face difficulties in optimize their own performance because of the lack of knowledge of the behavior of the other units due to distributivity, limited communication, and inability to access centralized control information. Each of them wants to maiximize a evaluation function to its own internal goal (see also Klopf 1982). A learning algorithm is a way that each of them can learn to cooperate with one another and adapt itself so that its own decisions fit into the overall situation, including not only environment factors but also the decision-making of its peers. This is a new kind of problem, namely distributed decision-making (Ho 1980). The problem is ubiquitous: it exists in economical systems, ecological systems, and in societies in general. It is not "societies of mind". Rather it is mind of societies.

The type of algorithms we devised above are often called Run-and-Twiddle Strategy. If things are going well, then keep doing the same thing or even do more of it; otherwise, stop doing the same thing and go find an alternative. The problem is that, if not all units are observable (or there may be no reinforcement signal for some of them), then how we decide if they performed correctly or not, or in other words, how we assign credits and blames. The above two algorithms try to tackle the problem, and we need some mathematical analysis and empirical study on this in order to furthur our understanding of this type of situations.

The above algorithms are stateless, namely the state variables in the generic definition of DN and PDN are not used at all. On one hand it greatly simplifies the problem of learning in PDN networrks, but on the other hand it loses some degrees of freedom, computational power or generality. To fully appreciate the difficulty of developing a general learning algorithm, we can compare these neuronal models with some known computational models. namely, HMM, learning automata, team decision theory, PLN (Aleksander 1989), finite state automata, reinforcement learning, etc. We can use the knowledge and insight we have of other models for DN and PDN. These learning algorithms can integrate the useful aspects of each related field, taking the advantage of the known algorithms for more restricted domains or subproblems.

Generally, learning algorithms for PDN can be classified into three catagories:

1. learning only action functions – The algorithms presented above falls into this catagory. There are two possible ways of dealing with states in the model: (1) ignore them, i.e. asuming there

is only one state and it never changes, or (2) set up a fixed sequence of states to go through for each node, useful for dealing with temporal sequences which are not time invariant.

2. learning only state transitions – With fixed action probability, the model is grossly equivalent to a hidden Markov model. This is an interesting class of problems and will be discussed below in detail.

3. learning both action functions and state transition functions – This is a combination of the first two cases. there are however some extra complications which will be discussed later.

In fixed (as in (1)) or adjustable (as in (2) and (3)) state transition process, we can analogize the sequential transition with a trajectory through a state space, each dimension of which is the state of an individual node. An important thing that can be used to determine the optimal trajectory is the energy landscape, which is over the entire state space. While each node wants to maximize its own gain (" hedonistic neurons"), the overall system has to reach its global maximum or the bottom of the deepest valley. This leads to the team decision theory, which studies the ways that the strategy of each node to achieve its own manximum gain can help to guide the whole system into a global maximum. The global cooperativity is achieved through the choice of local individual strategies. By choosing an appropriate strategy for each node (usually the same for all), the system will be able to follow an appropriate trajectory to settle into an optimal state in the state space.

By allowing communication through synaptic links, each node knows the previous actions of a subset of nodes that have synapses with it. Then what it has to do is to try to predict what that subset of nodes will do next and then choose an action (or increase the chance that action will be choosen) to maximize the energy. I call it Predictive Actions algorithm.

Scheme 4 (Predictive Actions (learning action functions only)) *each node should choose an action that achieve*

$$max(\sum_i P_i E_{ij})$$

where P_i is the probability of being in the ith combination of inputs next, and E_{ij} is the local energy of choosing jth action in the face of the ith input combination, assuming global energy is an monotone increasing function of local energy values.

By introducing multiple states into the model, we have a more powerful model (more powerful in expressiveness mainly), and thus need more powerful learning algorithms to be used in the model. From analogy to HMM, we can extend well-known Viterbi algorithm to learn an optimal state space trajectory, optimal in the sense that a state is most likely individually at a particular time for each of the time period. After calculating this trajectory, we can use the linear reward-penalty algorithm to modify the system. This algorithm is supervised learning, because a sequence of desired outputs are needed. With this algorithm, oscilation is possible, because of mutual dependency among connected nodes.

Scheme 5 (Viterbi/LRP) *Use Viterbi algorithm to calculate a most likely state at each moment, given a squence of desired outputs. Run the network and use the linear reward-penalty algorithm to modify parameters.*

When we want to learn both state transition and action functions at the same time, we need to consider the interaction between these two functions. For example, we could be punishing a state transition and an action at the same time. While we are gradually reducing the probability of that particular state transition to near zero, we are messing up the action function in the target state at the same time unnecessarily. With this in mind, we can blend the Predictive algorithm for action fuctions and Viterbi/LRP algorithm for state transitions together in the following way:

Scheme 6 (Blending Viterbi/LRP and Predictive Algorithm) *When both deserve rewards, give both rewards. When both deserve penalty, give penalty only to the state trnasition. When the state transition deserves reward but the action deserves penalty, give them correspondingly. When the state transition deserves penalty but the action deserves reward, give only penalty to the state transition.*

It will also work if we replace the predictive algorithm with straight comparisons of the desired output and the actual output. It is also possible to devise an algorithm without using Viterbi algorithm so that there is only one reinforcement signal for both state transitions and actions. It can work this way:

Scheme 7 (Learning both state transition and action functions) *If the node deserve reward, give it to both state transitions and actions. If the node deserve penalty, give it to the action only. The reinforcement signal can be generated by the predictive algorithm or by straight comparisons etc.*

Another interesting algorithm which is more or less retrospective instead of predictive is Baum-Welch algorithm from HMM theory.

Scheme 8 (Baum-Welch) *see Rabiner and Juang (1986)*

This algorithm is absolutely expedient. It learns both state transition and action probabilities.

Compared with PLN learning algorithm (Aleksander 1989), these algorithms have the following advantage:

- After learning, it still has the versitility due to the probabilistic nature, and the variability of behavior due to further learning.
- During learning, new information does not wipe out old one. Rather, the change is gradual and cumulative.
- It allows contradictary information, and thus works in ill-structured domains, because it is based on statistical sampling.

4 Concluding Remarks

We lifted the requirement that a unit has to be very simple in terms of computational resouces. So we get into a brand new type of networks, which has many very interesting properties that are not present in conventional PDP models. There is a strong possibility that this type of network may solve many difficult problems for PDP models and gives a new vitality to this area of study. For example, it might be used to solve decision problems, variable bindings, symbol manipulations, and sequence generation and recognition, etc.

REFERENCES

I. Aleksander, The Logic of connectionist system, in: Neural Computing Architectures, MIT Press, 1989

Y. Ho, Decision Theory and Information Structures, Proc. IEEE, V.68, 1980

A. Klopf, The Hedonistic Neuron, Hemisphere, 1982

K. Narendra and M. Thathachar, Learning Automata, IEEE Trans. SMC, 13, 1974

R. Sun, E. Marder and D. Waltz, Model local neural networks in the lobster stomatogastric ganglion, IJCNN, 1989

R. Sun, A discrete neural network model for conceptual representation and reasoning, 11th Cognitive Science Society Conference, 1989

R. Williams, Reinforcement learning in connectionist networks, TR 8605, ICS, University of California, San Diego, 1986

COMPUTING WITH ARRAYS OF COUPLED OSCILLATORS

Pierre Baldi *
Jet Propulsion Laboratory, 303-310
California Institute of Technology
Pasadena, CA 91109

Joachim Buhmann
Department of Computer Science
University of Southern California
Los Angeles, CA 90089

Ronny Meir
Division of Chemistry
California Institute of Technology
Pasadena, CA 91125

Abstract: Over the past two years, several experimental results have shed new light on the question of cortical oscillations and their possible roles in neural information processing. In the labeling hypothesis, temporal characteristics such as the phases and/or frequencies of pools of oscillating neurons are used to transiently encode information, in particular to label various features of an object by synchronous activity of the corresponding feature extracting neurons. We propose that two dimensional arrays of weakly and locally coupled oscillators be considered as a new class of architectures, particularly well suited for low level sensory processing such as early vision. Using the coupled limit cycle approach, we investigate several computational properties of such architectures, how they intrinsically differ, for instance, from discrete spin systems, and the importance of two dimensional effects. We show that such arrays are characterized by a very enhanced sensitivity to external inputs and the ability to organize themselves into flexible "patchy" structures over extremely brief transients. Relevance for the neurophysiological data is discussed. Possible applications to early vision, motion analysis, figure ground separation, pattern recognition are demonstrated by coupling oscillator architectures with arrays of filters or other networks.

In a series of recent experiments (see Gray et al. (1989) and Eckhorn et al. (1988)), the existence of stimulus specific oscillatory responses in the 40-60 Hz range in the visual cortex of both anesthesized and alert cats has been demonstrated. In addition, inter-columnar synchronizations over distances of several millimeters reflecting global stimulus properties have also been observed. One possible interpretation of these results has been advanced in the form of the so called labeling hypothesis. In the labeling hypothesis, temporal characteristics such as the phases (and/or frequencies) of pools of oscillating neurons are used to encode information, in particular to label various features of an object

* also with: Division of Biology, California Institute of Technology.

by synchronous activity of the corresponding feature extracting neurons (von der Malsburg (1981)). Phaselockings then serve to link associated features in different parts of the visual field and, in particular to represent the coherency of an object. For instance, in the experiments, elongated or colinear moving light bars of specific orientation elicit zero phaselocked periodic responses in separated columns with non overlapping receptive fields. In contrast, uncorrelated oscillations are observed in the case of similar but non-colinear stimuli. On the basis of these results and of some preliminary theoretical investigations using, among others, the coupled limit cycle approach, it is suggested in Baldi et al. (1989) that large arrays of locally and weakly coupled oscillators be considered as a new class of architectures, particularly well suited for low level sensory information processing such as early vision.

The coupled limit cycle approach seeks generality by omitting most of the details of the single nonlinear oscillators and observing that, once relaxed to its limit cycle, a single oscillator can be described by one parameter: its phase θ_i along the cycle. The behavior of a population of n interacting oscillators can then be approximated by the system

$$\frac{d\theta_i}{dt} = \omega_i(t) + f_i(\theta_1, ..., \theta_n) \tag{1}$$

where $\omega_i(t)$ allows to represent the internal frequencies and/or the external driving inputs, when their action is independent of the current phases. The functions f_i take into account the coupling amongst the oscillators assuming that such effects depend only on the phases. The oscillators are located at the vertices of a graph of interactions and, typically, the coupling functions f_i are symmetric of the form $\sum_{j \in V_i} f(\theta_j - \theta_i)$, where V_i is the set of vertices j adjacent to i and f is an odd periodic function such as $f(\theta) = \sin(\theta)$. To capture some of the essential characteristics of cortical sheets, the graph considered here are two dimensional lattices with local couplings. The variables θ_i represent the phase of a large fraction of synchronized neurons within a cortical column. In the absence of external driving inputs, consider first the very simple system defined by

$$\frac{d\theta_i}{dt} = K \sum_{j \in V_i} \sin(\theta_j - \theta_i). \tag{2}$$

The dynamics can be described in terms of an energy function H for $d\theta_i/dt = -\partial H/\partial\theta_i$ with

$$H(\theta_1, ..., \theta_n) = -\frac{K}{2} \sum_{i,j} \cos(\theta_i - \theta_j). \tag{3}$$

This is in fact the XY model of statistical mechanics (Kosterlitz et al. (1973)) and important differences dependent on the dimension have been described in the literature. In one dimension, there is no phase transition and the correlation among two spins at sites i and j decays exponentially fast with the distance $d(i,j)$ at any temperature. In two dimensions, there is a phase transition and below a critical t_c the correlation decays only as $d(i,j)^{-\alpha}$. However, most of the known results on the XY model, even in the case of Monte Carlo simulations, are mainly concerned with equilibrium properties and not with the transient dynamics suggested by the neurophysiological data (an exception can be found, for instance, in Loft et al. (1987)).

We have simulated the XY model (and other related models) on two dimensional lattices of size up to 128×128, focusing on the short term dynamical evolution. In all cases, we observe the following remarkable behavior. Starting from an initial random configuration of phases:

(i) after a very brief transient, the spins θ_i tend to cluster into a system of patches of constant phase;

(ii) the system of patches continues to evolve but on an extremely slow time scale.

These previous phenomena occur in the absence of external driving stimuli, i.e. with $\omega_i(t) = 0$. To simulate the experiments of Gray et al. (1989), we have considered arrays of size 64×64, where a constant external input w is applied to two rectangular regions of size 4×20 and spatially separated by up to 8 lattice intervals. The effect of ω is to increase the frequency of the corresponding oscillators which, in turn, is to be interpreted as an increase in activity and firing frequency in the corresponding columns. Random small inputs ω_i are applied to the background and all the initial phases are randomly assigned. After a brief transient:

-*(i)* the region covering the two bars and the space in between oscillates at one frequency;

-*(ii)* the two bars are zero phaselocked (while the oscillators occupying the space in between have slightly different phases). As in the experiments of Gray et al., presentation of one single elongated bar greatly enhances the synchronization effects.

In summary, at least in principle, transient spatial synchronizations of groups of oscillators within a two dimensional array are possible over spatiotemporal scales which are roughly consistent with the experimental findings. The establishment of such correlations is robust and requires only local couplings but no global feedback. The models used are extremely simplistic and are not meant to closely fit what is presently known of cortical neuroanatomy or neurophysiology. Rather, a minimal set of assumptions is introduced to qualitatively produce certain behaviors. Plenty of room exists for successively incorporating more realistic details and manipulating more complex dynamics. One obvious extension is to include a more sophisticated set of local connections and coupling functions. The short term dynamics can be expected to be more sensitive to such changes than the more "universal" equilibrium properties. One simple prediction from our simulations is that, in the two colinear bars experiment, the oscillators "in between" become entrained at the same frequency. However, a lot of caution is necessary since, for example, what is seen as a cluster in the simulations could represent a set of columns with similar orientation and therefore have a scattered appearance on the cortical surface.

If arrays of locally weakly coupled oscillators are to be used in neural computations, whether natural or synthetic, one may wonder whether conceptually they provide any advantages over other forms of processing such as, for instance, discrete spin systems. Spin systems have been used to model associative memory and classification tasks. In these examples, information is stored in the connections which are not weak in general and many different input patterns can be mapped on a same output. This however may not be desirable in the initial stages of sensory information processing where the diversity in the inputs needs to be preserved. A network with a dynamics which is strongly dominated by the interactions between the different individual components is not able to process weak stimuli and tends to be biased in its interpretation of incoming information. The architectures considered here are translation invariant and not well suited for classification. Yet in contrast, the sensitivity of a two dimensional weakly coupled oscillator network is very high. It could be measured by the change in the average phase as a function of the input strength. In the context of magnetic systems, this quantity is called the susceptibility. For oscillators, the phase can vary continuously and even a very small external input will always produce a macrocopically visible effect after a certain time whereas this is not the case for discrete spin systems which have only a finite susceptibility (nor is it the case for oscillator networks which are one dimensional, fully interconnected or with diffuse feedback).

Finally, to demonstrate how useful computations can be carried through the transient establishment of spatiotemporal wave patterns across sheets of coupled oscillators, we shall report our results on several problems in vision (figure/ground segregation, motion analysis, pattern recognition). In all cases, oscillators architectures need to be coupled with other networks, for instance with a preprocessing array of Gabor filters with different scales of resolution.

Acknowledgement

This work is supported by NSF grant DMS-8914302 and ONR contract NAS7-100/918 to P. B., AFOSR grant 89-0274 to J. B. (C. von der Malsburg, P. I.), ONR grant N00014-87-K-0377 (J. J. Hopfield, P. I.) and a Weizmann Fellowship to R. M.

Some References

(1) Baldi, P. and Atiya, A. (1989) Oscillations and synchronizations in neural networks. An exploration of the labeling hypothesis. International Journal of Neural Systems. In Press.

(2) Baldi, P., Buhmann, J. and Meir, R. (1989) Computing with arrays of coupled oscillators. Two dimensional effects. Submitted for publication.

(3) Eckhorn, R., Bauer, R., Jordan, W., Brosch, M., Kruse, W., Munk, M. and Reitboek, H. J. (1988) Coherent oscillations: a mechanism of feature linking in the visual cortex? Biological Cybernetics **60**, 121-130.

(4) Gray, C. M. and Singer, W. (1989) Stimulus specific neuronal oscillations in orientation columns of cat visual cortex. PNAS USA, **86**, 1698-1702.

(5) Gray, C. M., König, P., Engel, A. K. and Singer, W. (1989) Oscillatory responses in cat visual cortex exhibit inter-columnar synchronization which reflects global stimulus properties. Nature **338**, 334-337.

(6) Kosterlitz, J. M. and Thouless, D. J. (1973) Ordering, metastability and phase transitions in two-dimensional systems. Journal of Physics C, Vol. **6**, 1181-1203

(7) Loft, R. and DeGrand, T. A. (1987) Numerical simulations of dynamics in the XY model. Physical Review B, **35**, 16, 8528-8541.

(8) von der Malsburg, C. (1981) The correlation theory of brain function. Internal Report 81-2, Department of Neurobiology, Max Planck Institute for Biophysical Chemistry

NOVELTY DEPENDENT CATEGORIZATION AND LEARNING IN CALM MODULES

Jacob M.J. Murre, R. Hans Phaf[1]*, and Gezinus Wolters*

Leiden University
Unit of Experimental and Theoretical Psychology
P.O.Box 9555, 2300 RB Leiden
The Netherlands

Abstract CALM (Categorizing And Learning Module) forms a basic unit for the construction of multi-modular networks suited for supervised and unsupervised learning. Its design is guided by considerations from a practical, neurobiological, and psychological nature. In CALM networks, modules may or may not be interconnected. If modules are linked, interconnections exist which are modifiable according to a modified Hebb rule. Bidirectional (nonsymmetric) connections are possible. Categorization and learning speed in a CALM module are dependent on the novelty of a local activation pattern. Single module simulations indicate that module convergence and pattern discrimination of new inputs are learned relatively fast. A CALM module can both discriminate and generalize, depending on module size and structure of the pattern set. CALM networks have, among other things, been applied to pattern (character) recognition tasks, and to the modeling of psychological experiments on the dissociation of explicit and implicit memory. For the construction and evaluation of CALM networks a high level tool has been developed. CALM is currently also being implemented in a 400 processor parallel machine.

Basic principles: modularity and activation/elaboration learning

CALM (Categorizing And Learning Module) has been developed as a basic unit for the construction of large multi-modular networks which can learn either with or without supervision (see also Murre, Phaf, and Wolters, submitted). The structure of the CALM module is motivated primarily by practical (engineering), biological, and psychological considerations. Two basic principles are particularly important: modularity and activation/elaboration learning.

Modularity. From a practical point of view modularity may be seen as a universal engineering principle. It allows for incremental development, re-use of modules, and easy maintenance and repair of independent parts. From a biological perspective the central nervous system may also be considered modular. On a microscopic level we find the mini-columns in the cortex (Mountcastle, 1978; Szentágothai, 1975). On a more macroscopic level the various cortical and sub-cortical structures indicate a modular structure as well. In experimental psychology, models for learning, memory, attention, and motor control often employ more or less modular architectures to constrain task interference in accordance with experimental data.

Elaboration/activation learning. Many learning neural networks suffer from what has been referred to as the stability-plasticity dilemma (Carpenter and Grossberg, 1988), which entails that networks on the one hand have to remain plastic to encode new stimuli, on the other hand they must be sufficiently stable to prevent overwriting of already learned stimuli. In CALM this problem has been remedied by implementing two different forms of learning: activation and elaboration learning. The transition from one type of learning to the other is dependent on the novelty of the activation pattern to a module. The distinction was originally introduced by Graf and Mandler

---

[1] This research was partially supported by the Dutch Organization for Scientific Research (NWO), grant no. 560-259-027.

(1984) to explain the dissociation between implicit and explicit memory found in a large number of studies (Schacter, 1987). In CALM, it enables a module to react plastic to novel activation patterns (elaboration learning), yet retain stability in response to familiar patterns (activation learning).

Evidence from neurophysiology and neurology suggests that elaboration type learning may be mediated by certain sub-cortical structures, such as the hippocampus and amygdala (Mishkin, 1978; Warrington and Weiskrantz, 1970). These have often been attributed some kind of arousal function, which may be accompanied by an orientation reaction (Näätänen, 1986; Sokolov, 1975). New or unexpected stimuli cause arousal, which is hypothesized to stimulate learning (Hebb, 1955; Luria, 1973). Such influences of arousal on learning have also been incorporated in CALM. In our approach new input patterns result in a high level of local competition which is associated with amount of 'arousal'. A high level of 'arousal' gives rise to fast elaboration learning, a low level results in much slower activation learning. Other findings in neurophysiology, not discussed here, further help to constrain the structure of CALM, for instance, the distribution and function of long-range and short-range connections and Dale's law.

Structure of CALM

Internal wiring in a CALM module (see Figure) is fixed (i.e., all intramodular connections are non-modifiable). All nodes in a module are similar (i.e., they all use the same sigmoid activation rule with continuous activations), but it is useful to distinguish three groups on the basis of the connections they give off. (1) A row of V(eto)-nodes. A V-node has inhibitory connections to all other nodes in a module. In particular, V-nodes have strong mutually inhibitory connections, resulting in a winner-take-all com- petition. A winning V-node further inhibits all other nodes in a CALM module. (2) A row of R(epresentation)-nodes, giving off only excitatory connections. Every R-node excites one V-node, thus feeding the competition. Only R-nodes have inter-modular connections. Connecting two modules A and B means connecting each R-node in A to all R-nodes in B. These are always learning connections, which are nonsymmetric in case of bidirectional connections. (3) A single A(rousal)-node. Because of its wiring, an A-node functions as a competition detector: its activation value is a measure of the ongoing competition. The A-node receives excitation from the R-nodes, and inhibition from the V-nodes. With strong competition going on, R-nodes will have higher net activations than V-nodes, because of their strong mutual inhibitory connections. The weights to the A-node are chosen in such a way, that after the competition has been resolved (with only one R-V-pair active) the net input to the A-node will be below zero. Competition will mainly occur with new patterns resulting in a highly activated A-node during the prolonged competition phase. For a familiar pattern the competition will be low since relevant connections to one particular R-node have been strengthened in previous learning. Therefore, such a pattern will give one node a headstart, which will quickly resolve the competition, if any.

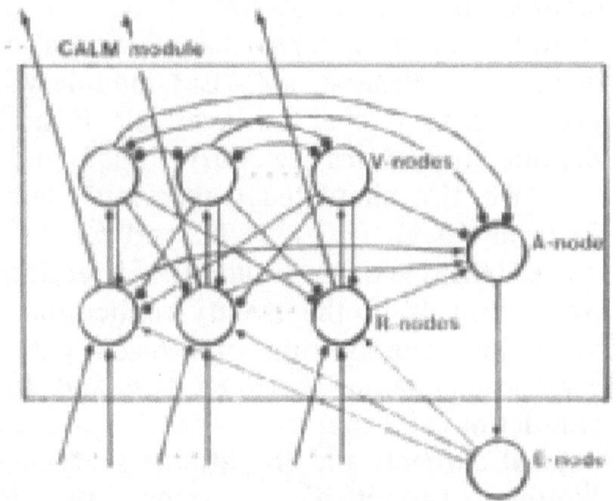

There is still another node, the E(xternal)-node. It integrates the A-node activation. The E-node has two important functions. First, it distributes noise over R-nodes with an amplitude that is proportional to the E-node activation. This noise enhances the search process for a new representation. It may also help to escape from local minima. Second, it boosts the learning rate. The learning rule used, is a slightly modified version of a learning rule introduced by Grossberg (1976), which in turn is an extended Hebb rule. The learning parameter in this rule is also proportional to the E-node activation. This causes fast learning of novel patterns. If the E-node is not activated learning proceeds at a fixed, slow rate.

In summary, novel patterns tend to cause higher and longer activation of the E-node, which in turn causes noise activation to be distributed and learning to be boosted. We call this the elaboration mode of learning. A familiar pattern gives rise to only a weakly activated E-node, with little noise and base-rate learning. We call this activation learning.

Simulations and applications

A large number of single module simulations were carried out to assess the properties of the CALM module. In CALM, number of iterations required for categorization of a single pattern decreases with pattern repetition. Simulations with modules of increasing sizes indicate that categorization time at first pattern presentation increases only very slowly with the size of the CALM module. After repeated presentation of a pattern categorization times usuallly drop below 15 iterations. In another set of simulations, number of pattern presentations until near perfect (95%) discrimination was measured. Modules of sizes N = 10, 15, 20, .., 95 (number of R-nodes) were presented with pattern sets of N orthogonal patterns. A multiplicative regression yielded the following relation for the scaling performance of CALM: discrimination time = .2 * N^1.7. In another simulation we prepared eight (non-orthogonal) stimuli with relative (Hamming) distances chosen in such a way that two principal clusters were formed. This set was offered to modules of sizes 2, 3, .., 8. The results showed that relatively small CALM modules (more patterns than R-nodes) will generalize over patterns, based on shortest distance. In general, a CALM module will try to spread as much as possible all patterns presented to it over the available R-nodes. For a more extensive description of single module simulations see Murre, Phaf, and Wolters (submitted).

We also simulated (supervised) learning of the XOR in a multimodular network. Patterns (.1,1), (1,.1), (1,1), and (.1,.1) were offered to a network with one three R-node CALM module (middle) bidirectionally connected to a two R-node CALM module (top). In half of the twenty replications the network correctly categorized the patterns after just one training run (each pattern offered once for 45 iterations), the other replications taking between two and four runs. CALM has also been applied to the simulation of experiments on the dissociation of implicit and explicit memory, both in normal subjects and in patients suffering from anterograde amnesia (Phaf, Postma, and Wolters, submitted). A more practical application, i.e., handwritten character recognition, is described in Happel, Phaf, and Murre (1990).

In order to quickly set up simulations and evaluate their results an extensive simulation environment has been developed, that allows for high level specification of CALM models without any prior programming knowledge. This tool is currently being used by students in our department for the simulation of more experimental results. Furthermore, because of their modular structure CALM networks may be suited for implementation in parallel hardware. To explore this possibility, CALM has been implemented in several types of hardware (e.g., Hoekstra, Heemskerk, Hudson, and Klaassen, 1990).

Much of the structure of CALM has been motivated by biological, and psychological considerations. Among its functional characteristics are relatively fast convergence and discrimination speed, a feasible remedy of the plasticity-stability dilemma, the ability to realize both discrimination and generalization in a single approach, as well as suitability for hardware implementation. We feel that attempts to integrate the fields of neurophysiology and psychology with the help of models formulated in the language of connectionism, may not only be fruitful to these fields, but may also contribute towards solutions of practical problems encountered in the broader field of neural networks.

References

Carpenter, G.A., and S. Grossberg (1988) The ART of adaptive pattern recognition by a self-organizing neural network. *Computer*, 21, 77-88.

Graf, P., and G. Mandler (1984) Activation makes words more accessible, but not necessarily more retrievable. *Journal of Verbal Learning and Verbal Behavior*, 23, 553-568.

Grossberg, S. (1976) Adaptive pattern classification and universal recoding, II: Feedback, expectation, olfaction, and illusions. *Biological Cybernetics*, 23, 187-202.

Happel, B.L.M, R.H. Phaf, and J.M.J. Murre (1990) Categorization in multi-module CALM networks: recognition of handwritten digits. *INNC 90 Paris*, july 9-13.

Hoekstra, J, J.N.H. Heemskerk, A.J. Klaassen, R.H. Phaf, P. Knoppers, and P.T.W. Hudsoln (1990) Hardware design concepts for a CALM neural network using 400 simple processors: architecture and implementation. *INNC 90 Paris*, july 9-13.

Hebb, D.O. (1955) Drives and the conceptual nervous system. *Psychological Review*, 62, 243-254.

Luria, A.R. (1973) *The working brain: an introduction to neuropsychology*. New York: Basic Books.

McClelland, J.L., and D.E. Rumelhart (1981) An interactive activation model of context effects in letter perception. Part I: an account of basic findings. *Psychological Review*, 5, 375-407.

Mishkin, M. (1978) Memory in monkeys severely impaired by combined but by separate removal of amygdala and hippocampus. *Nature*, 273, 297-298.

Mountcastle, V.B. (1978) An organizing principle for cerebral function: the unit module and the distributed system. In G.M. Edelman and V.B. Mountcastle (eds.) *The mindful brain*. Cambridge, MA: MIT Press.

Murre, J.M.J., R.H. Phaf, and G. Wolters (submitted) CALM: Categorizing And Learning Module.

Näätänen, R. (1986) The orienting response: a combination of informational and energetical aspects of brain function. In: G.R.J. Hockey, A.W.K. Gaillard, and M.G.H. Coles (eds.) *Energetics and human information processing*. Dordrecht: Martinus Nijhoff, 91-111.

Phaf, R.H., E.O. Postma, and G. Wolters (submitted) ELAN-1: a connectionist model for implicit and explicit memory tasks.

Schacter, D.L. (1987) Implicit memory: history and current status. *Journal of Experimental Psychology: Learning, Memory, and Cognition*. 13, 501-518.

Sokolov, E.N., (1975) The neuronal mechanisms of the orienting reflex. In: E.N. Sokolov, and O.S. Vinogradova (eds.) *Neuronal mechanisms of the orienting reflex*. Hillsdale, NJ: Erlbaum.

Szentágothai, J., (1975) The 'module-concept' in cerebral cortex architecture. *Brain Research*, 95, 475-496.

Warrington, E.K., and L. Weiskrantz (1970) Amnesic syndrome: consolidation or retrieval? *Nature*, 228, 628-630.

EQUIVALENT TLU- AND ΣΠ-NETWORKS FOR INVARIANT PATTERN RECOGNITION

Helmut Glünder
Institut für Medizinische Psychologie, Ludwig-Maximilians-Universität, Gœthestraße 31, D-8000 München 2, FRG

Abstract

Two universal types of networks for the invariant recognition of pictorial patterns are compared with respect to function, structure and costs. The main stage of both networks serves for the extraction of features that are invariant under certain types of *unrestricted* geometric transformations, e.g. rigid translations. Both approaches are conceptualized for *unequivocal* class definitions and thus for the feasibility of *perfect* pattern reconstructions. Although the networks are structurally different, they are to a high degree functionally equivalent. The costs, i.e., the number of weights per class that must be adjusted in order to obtain ideal and invariant classification, turn out to be almost the same for both approaches as well as for the reference network (list classifier). In practice, however, the ΣΠ-network is superior to the TLU-network; it is more robust and even single invariant features are unequivocally defined. The investigations reported here do not concern any aspects of learning.

Introduction

Most network approaches to geometrically invariant recognition of pictorial patterns are based on invariant classification, i.e., explicit representations of invariant features do not occur. The most powerful approach of this kind that can be adapted to meet arbitrary demands for invariance and pattern reconstruction, is the so-called list classifier. This straightforward method uses many holistic template-matching (sub)classifiers for each invariant class in parallel – one for every transformed version of the class prototype. A class-specific decision $\mathbf{z}_{List,j}$ is obtained by the summation of the corresponding subclassifier outputs. Each of the holistic subclassifiers is a threshold logic unit (TLU) that computes the inner product of the normalized signal vector $\mathbf{x}$ and the normalized weighting vector $\mathbf{w}_{j\nu}$ and finally thresholds the resulting cross correlation coefficient $y_{j\nu}$ according to

$$u_{j\nu} = \begin{cases} 1 & \text{if } y_{j\nu} = 1 \\ 0 & \text{else} \end{cases} \qquad \text{with} \quad y_{j\nu} = (\mathbf{x}^T/|\mathbf{x}|)(\mathbf{w}_{j\nu}/|\mathbf{w}_{j\nu}|) \,. \tag{1}$$

Obviously, invariant list classification allows for unequivocal pattern characterization and thus reconstruction which, of course, does not include the description parameters under which the classification is invariant. The advantages of this approach are opposed by the total lack of generalization which can only be attained at unreasonably high costs.

In this paper, however, two network approaches to geometrically invariant pattern recognition are presented and compared which essentially deal with the generation of representations of invariant features that may subsequently be linearly classified. This decision making second stage consists of TLUs – one for every class j –, whose thresholds can be replaced by a single maximum detector if generalization is demanded. Unlike most of the existing investigations that deal with 'learning' and recognition rates, the main goal of this contribution is to determine the necessary and sufficient conditions (structure and costs) under which the following demands for a classifier can be met:

1. invariance under certain types of *unrestricted* geometric transformations
2. *unequivocal* class definitions and thus *perfect* reconstructability also of nonbinary patterns
3. no *a priori* knowledge about the kind and number of the pattern classes

Owing to these demands, more economic but less universal standard approaches, such as linear reduction of signal dimensions, as well as subclass techniques, must be discarded.

The investigations reported here are exemplified for unrestricted 2D pattern translations on a toroidal image array of n pixels. Consequently, the maximum number of translation invariant classes that exist for signals of m levels, is $k_{max} \geq m^n/n$. – Comparable considerations hold for unrestricted rotations and scale changes which can be imagined as converted into 2D translations by log-polar mapping (Brousil and Smith, 1967; Casasent and Psaltis, 1976).

Direct Invariant Classification

A list classifier that is invariant under unrestricted translations of signals on a toroidal image array of n pixels typically requires n TLUs per class – one for every shift position. Because the TLUs comprise holistic templates, i.e., nD vectors $\mathbf{w}_{ji}$, the costs of ideal invariant class definitions, expressed by the number of weighting coefficients per class, are $N \approx n^2$ (see Fig.1a, roman inscriptions).

Direct invariant classification by a linear classifier, i.e., by cross correlations and subsequent maximum detection, is generally impossible. Thus, nonlinearities, at least of the polynomial degree p=2, must be applied to the signal components before linear classification becomes possible. Owing to the involved thresholds θ, list classification relies on nonlinearities of degree $p \rightarrow \infty$. Hence p=2 is a necessary and $p \rightarrow \infty$ a sufficient condition for ideal and invariant class definitions in such networks.

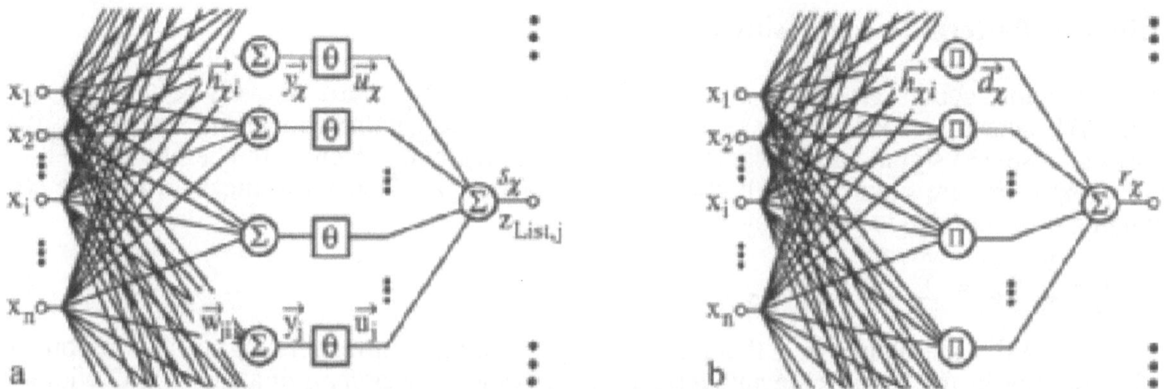

Figure 1a, b. Network structures for list classification (roman inscriptions) and for polynomial pattern decomposition (PPD) (italic inscriptions) **(a)**, as well as for multilinear pattern decomposition (MPD) **(b)**

Polynomial Pattern Decomposition

Instead by holistic comparisons, one can try to describe patterns in terms of their composition from pattern elements. For this purpose the weighting vectors consist of only q<<n nonzero components and shall be called mask vectors ${}^q\mathbf{h}_\chi$. As with list classification, each type of mask (denoted by χ, with $1 \leq \chi \leq \kappa$) must be applied at every position if shift invariance is demanded, and the desired invariant features must be determined from the outputs of each of the thereby defined κ sets that consist of typically n TLUs. Commonly the number of suprathreshold signals, i.e., their sum s_χ, is used for this purpose which leads to a network structur that is known from the list classifier (see Fig.1a, italic inscriptions).

Unfortunately separate normalization of signal and mask vectors according to Eq.(1) cannot be applied for the unequivocal detection of pattern elements. Because normalization is only feasible at the enormous costs of κn individual operations which are difficult to realize within network structures, normalization is sacrificed. Consequently, the threshold of the TLUs must be replaced by a nonlinear and strictly monotonic transfer characteristic of non-negative derivative which leads to so-called generalized TLUs. If the nonlinearity is expressed by a polynomial of degree p, then every invariant feature is a sum of typically n polynomials that are computed from a linear discriminant function $y_{\chi i}$ (cf. Eq. (1)). Therefore, this kind of extraction of invariant features shall be called a polynomial approach to pattern decomposition (PPD).

$$ {}_p s_\chi = \sum_{i=1}^{n} {}_p u_{\chi i} = \sum_{i=1}^{n} \sum_{\rho=0}^{p} \alpha_\rho \cdot (y_{\chi i})^\rho \qquad (2)$$

In conjunction with a final linear classifier stage the PPD network represents a *perceptron* network with one hidden layer (Lippman, 1987; Rosenblatt, 1962).

Obviously, non-binary masks do not make much sense without normalization. Therefore, and in order to prevent the combinatorial explosion of mask types, all further investigations are confined to 0/1-valued masks – which turns out to be an appropriate strategy. Irrespective of this restriction, the fundamental question, how many masks of what type are necessary and sufficient for perfect translation invariant pattern descriptions, is not yet answered. Empirical investigations reveal that they are not achieved if masks with $q<3$ are applied, or if the nonlinearity is of polynomial degree $p<3$. Therefore, a necessary condition and, with the fact that a subsequent template-matching classifier accesses κD features, the costs for translation invariant classification with ideal class definitions are

$$(p = 3) \wedge (q = 3) \Rightarrow (N = {}^{3}\kappa \approx n^2/6) \qquad \text{for} \quad n >> 3 \; . \tag{3}$$

Although more specific than the statements that were given at the end of the previous section, it does not yet clarify whether $p=3$ and $q=3$ are sufficient for unequivocal pattern reconstructions.

Multilinear Pattern Decomposition

From the previous conditions it must be concluded that only a single term of every polynomial ${}^{3}_{3}u_{\chi i}$ (cf. Eq.(2)) is actually responsible for perfect pattern descriptions, namely the trilinear term ${}^{3}d_{\chi i}$ that comprises those $q=3$ signal values which are selected by mask ${}^{3}\mathbf{h}_{\chi i}$. Hence, the hereby defined invariant features ${}^{3}r_{\chi}$ are sums of n well-chosen trilinear terms, i.e., special trilinear forms. The corresponding q-linear form is defined by

$${}^{q}r_{\chi} = \sum_{i=1}^{n} {}^{q}d_{\chi i} = \sum_{i=1}^{n} \prod_{\sigma=1}^{q} h_{\chi i \sigma} \cdot x_{\chi i \sigma} \; , \tag{4}$$

with the selected signal values $x_{\chi i \sigma}$ that may be weighted by the corresponding mask components $h_{\chi i \sigma}$. For 0/1-masks the features are autocorrelation coefficients of order q of a signal $\mathbf{x}$. Obviously, pattern elements can be defined through the multiplication schemes that are given by 0/1-masks, although the element detection relies on mechanisms that fundamentally differ from mask matching. Owing to the multilinear forms, this kind of extraction of invariant features shall be called a multilinear approach to pattern decomposition (MPD). Figure 1b shows a network for the computation of a single feature. The generalized TLUs of Fig.1a are replaced by so-called product units (Durbin and Rumelhart, 1989; Giles and Maxwell, 1987; Glünder, 1986). According to Eq.(4), this network belongs to the category of the ΣΠ-networks (Rumelhart *et al.*, 1986).

Merely the minimal order q that guarantees perfect invariant pattern descriptions remains to be determined. Fortunately it has been proven by several authors – firstly, by McLaughlin and Raviv (1968), for binary pattern signals by Minsky and Papert (1969), and for complex-valued signals by Lohmann and Wirnitzer (1984) –, that any pattern of finite extent is perfectly described by its complete triple-autocorrelation function, except for its translatory position (translation invariance). Actually, all possible ${}^{3}\kappa$ and ${}^{2}\kappa$ trilinear and bilinear autocorrelation coefficients, in conjunction with a subsequent template-matching stage, permit ideal translation invariant classifications. Therefore, the necessary and sufficient condition and the costs for ideal MPD classification are

$$(q = 2) \wedge (q = 3) \Rightarrow (N = {}^{2}\kappa + {}^{3}\kappa \approx n^2/6) \qquad \text{for} \quad n >> 3 \, . \tag{5}$$

This result expresses the primary role polynomial order q plays in ideal invariant classification and it demonstrates that the polynomial degree p is relevant only insofar, as multilinear terms are to be generated through polynomials, e.g. by generalized TLUs. Features that are invariant under other geometric transformations can be evaluated via generalized autocorrelation (Glünder, 1987).

Conclusions

Somewhat surprisingly, the costs for list, PPD, and MPD classification turn out to be almost the same, namely $N \approx n^2$ weighting coefficients per class that must be adjusted or 'learned' in order to obtain unrestricted translation invariant and unequivocal class definitions. While list classifiers suffer from the mentioned shortcoming, the PPD and MPD approaches appear functionally equivalent, at least theoretically. In practice, however, PPD networks must cope with a high signal dynamic that is caused by the large number of polynomial terms that constitute each feature. On the other hand, the processing must be highly accurate in order to represent the crucial trilinear terms which otherwise are lost in the polynomials. In this regard, MPD networks are more economic and robust.

All three approaches permit a proportional decrease in the number of templates, masks, or multiplication schemes if the variance is restricted. This property is a consequence of the fact that invariance is exclusively due to the averaging of nonlinearly weighted correlation coefficients, or multilinear terms which are extracted at different positions, according to the extent of the geometric variance, as it was already pointed out by Pitts and McCulloch (1947).

Because every single PPD or MPD feature is invariant, a reduction of the number of features does not affect the invariance but the precision of the pattern descriptions. In this sense, good pattern descriptions even require certain types of variance, namely all those geometric transformations that relate the pixels of a pattern. That is why 2D translations, as well as expansions in conjunction with rotations, are well suited for ideal pattern characterizations. – For example, a reduction of the number of mask types by a factor n, to all ${}^2\kappa \approx n/2$ possible features, implies that patterns with identical second order autocorrelation functions become indistinguishable.

Unfortunately, the decrease in descriptive power that is caused by fewer features happens more rapidly for PPD than for the MPD features, at least if non-negative signals are considered. This is due to the principal fact that PPD features cannot unambiguously signal the presence or absence of form elements in patterns which in turn is a consequence of the normalization problem. Owing to the coincidence-detection character of multiplication, MPD features are ideally suited for such decisions, especially if they are computed from sparsely coded signal representations, i.e., those that contain only a few nonzero values. The MPD approach even permits unequivocal reconstructions of binary pattern versions from binarized multilinear terms. Consequently, the replacement of the multiplication nodes by real coincidence detectors drastically improves the noise immunity of the feature extraction process. PPD systems lack this valuable property because pattern information is essentially coded in the amplitudes of polynomials.

References

Brousil, J.K. and Smith, D.R. (1967) A threshold logic network for shape invariance. *IEEE Trans. Electronic Computers*, **EC-16**, 818-828

Casasent, D. and Psaltis, D. (1976) Position, rotation and scale invariant optical correlation. *Applied Optics*, **15**, 1795-1799

Durbin, R. and Rumelhart, D.E. (1989) Product units: a computationally powerful and biologically plausible extension to backpropagation networks. *Neural Computation*, **1**, 133-142

Giles, C.L. and Maxwell, T. (1987) Learning, invariances, and generalization in high-order neural networks. *Applied Optics*, **26**, 4972-4978

Glünder, H. (1986) Neural computation of inner geometric pattern relations. *Biol. Cybern.*, **55**, 239-251

Glünder, H. (1987) Invariant description of pictorial patterns via generalized autocorrelation functions. In Meyer-Ebrecht, D. (ed.), *ASST '87*, Springer Verlag, Berlin, 84-87

Lippmann, R.P. (1987) An introduction to computing with neural nets. *IEEE ASSP Magazine*, **4**, 4-22

Lohmann, A.W. and Wirnitzer, B. (1984) Triple correlations. *Proc. IEEE*, **72**, 889-901

McLaughlin, J.A. and Raviv, J. (1968) Nth-order autocorrelations in pattern recognition. *Information and Control*, **12**, 121-142

Pitts, W. and McCulloch W.S. (1947) How do we know universals. The perception of auditory and visual forms. *Bull. Math. Biophys.*, **9**, 127-147

Rosenblatt, F. (1962) *Principles of Neurodynamics – Perceptrons and the Theory of Brain Mechanisms*, Spartan Books, Washington/DC

Rumelhart, D.E.; Hinton, G.E. and McClelland, J.L. (1986) A general framework for parallel distributed processing. In Rumelhart, D.E. and McClelland, J.L. (eds.), *Parallel Distributed Processing 1*, The MIT Press, Cambridge/MA, 45-76

The Concept of Learning Equilibrium Dynamics in the Development of Solid Networks

R. C. Berkan, L. Tsoukalas, B. R. Upadhyaya and R. E. Uhrig

Department of Nuclear Engineering
The University of Tennessee
Knoxville, Tn 37996-2300

Abstract

A new approach is developed and tested to improve standard performance measures in neural networks. The concept consists of using connection weights as equilibria establishing functions in network dynamics and identifying the total state change in weights as the classifying information. The equilibrium concept is associated with "internal" learning where the connection weights are assumed to learn the dynamics of their environment prior to any training activity in the network. The method facilitates the identification of nonlinear decision boundaries and offers several desirable features including small memory size, no convergence problems and deterministic network architecture. Preliminary performance evaluations compared with the standard back-propagation network showed significant improvement in reducing the memory usage and training time for exclusive-or type problems.

1 Introduction

The existing artificial scenarios of neuronic events in the biological brain have led successful simulations of human-like learning and recalling in mathematical terms. Among the most significant ones, the learning schemes like Hebbian, Grossberg's instar-outstar, or back-propagation techniques [1] have proven adequate with certain limitations. The limitations, in general, arise from the diversified options in selecting certain network or rule parameters which are known to be uninterpretable to a degree (number of hidden layers or learning coefficients). The most widely encountered trouble in neural networks (NN) is the convergence problem during training.

Excitatory, inhibitory and competitive types of neuronic behavior have constituted the main source of inspiration in the development of some of the existing learning rules. In the scope of this work, we investigate the merits of another property, the sense of "equilibrium" in weights' behavior. We assume that the synaptic weights learn the dynamics of their environment through an external stimulus before performing any standard learning activity. After a period of maturation, the weights can establish equilibrium around the desired output in their connected neuron. Knowing the network nonlinearity (sigmoid thresholding), we can formulate the mature weight behavior.

We present a sample network structure that uses the knowledge of equilibrium dynamics. A measure is produced in terms of weight dynamics in order to identify the nonlinear decision boundaries. The network consists of two vertical sections. The lower layer network (LLN) requires no random network parameters (solid network). The weights are designed to follow their equilibrium routes to establish equilibrium in processing elements (**PEs**) at all times during the iteration dynamics. The output of the LLN is the integral weight change for a given input. The upper layer network (ULN) is optional and it can be one of the standard NN structures or a simple computational routine. Its primary function is to perform output classification. The overall network is called the equilibrium route network (**ERN**).

2 Equilibrium Route

The equilibrium route is the final functional form which describes weight dynamics as a function of the dynamics of its environment. The environment of a weight includes the upper layer PE it is connected to (master PE), and all downward connections of the master PE at one layer depth.

Consider a standard processing element shown in Fig. 1 which is widely used in neural networks.

The output of the j_{th} master PE in layer p is given by

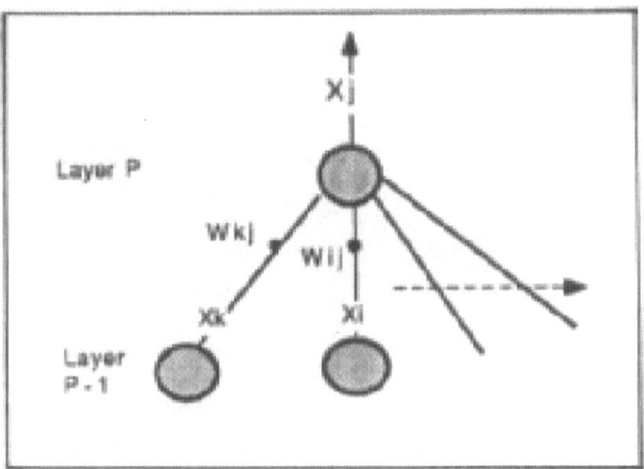

Figure 1: A master PE and its downward connections

$$x_j^p = F(\sum_i w_{ij}^p x_i^{p-1}) \quad (1)$$

where the thresholding function F is assumed to have the sigmoidal form.

$$F(x) = \frac{1}{1+e^{-\beta x}} \quad ,\beta > 0 \quad (2)$$

The rate of change in x_j^p from one iteration to another is expressed by the following equation.

$$\frac{dx_j^p}{dt} = F(\sum_i w_{ij}^p x_i^{p-1}) - x_j^p \quad (3)$$

Note that the steady-state solution of Eq.(3) yields the original network equation (1). Assume that w_{kj} is one of the weights connected to the master PE of x_j^p. Then the desired equilibrium route [2] is given by

$$w_{kj}^* = -\frac{1}{n\beta x_k^{p-1}} \ln[\frac{1}{x_{rj}} - 1] - \sum_{i \neq k} \frac{w_{ij} x_i^{p-1}}{x_k^{p-1}} \quad (4)$$

$$0 < x_{rj} < 1$$

where
n = type index
x_{rj} = reference state

The index n determines the type of cross-talk among the weights. For $n = 1$, the collective weight behavior becomes entirely competitive in searching for equilibrium. When n equals the number of connections to a particular PE, then the weight behavior is partially competitive. It can easily be shown that if the kth weight w_{kj} dynamically follows its desired route ($w_{kj} = w_{kj}^*$) then the substitution of w_{kj} into Eq. (3) with $n = 1$ would yield

$$\frac{dx_j^p}{dt} = x_{rj} - x_j^p \quad (5)$$

which gives an exponential solution for x_j^p reaching the reference state x_{rj} in a few iterations. Thus, the weight absorbs all the information embedded in the master PE's dynamics. The choice of reference and initial states determines the magnitude of the target state transition. In a solid structure, x_{rj} and initial conditions are functions of inputs. Thus the magnitude of the target state transition changes with different inputs introduced to the network. In this work, the reference state is simply the initial state of PEs.

The weight w_{kj} following its desired route is given by

$$\frac{dw_{kj}}{dt} = K(w_{kj}^* - w_{kj}) \quad (6)$$

where K is a control parameter and equal to 1 in most cases. The solid network dynamics for one weight and its master PE is represented by Eqs.(3), (4) and (6). A learning rule may also be derived using Eq.(4) and substituting the desired output as the reference state. However, each weight must be accompanied by a coupling weight and the coupling should produce a stationary condition. This is because the concept of learning requires one set of weights to describe input-output dynamics.

Consider a network with all the connection weights designed to follow their equilibrium routes. When an input pattern is applied to the network, all the weights work for a target state transition in the network. The integral weight change gives us a new measure. For the ith output PE, the integral weight change is given by

$$E_i = \int_{t_0}^{t_f} \sum_{i,f} w_{if}(t)dt \quad (7)$$

where
t_0 = initial time,
t_f = final time.

The subscript f in Eq.(7) indicates the connections used in the summation for the ith output. According to the given problem the selection of weights is determined by the number of output PEs. In the examples presented in this paper, f includes all the connections.

In atomic physics, an atom can be characterized by its energy state. It jumps to an excited state when externally disturbed, and stays there if the state is

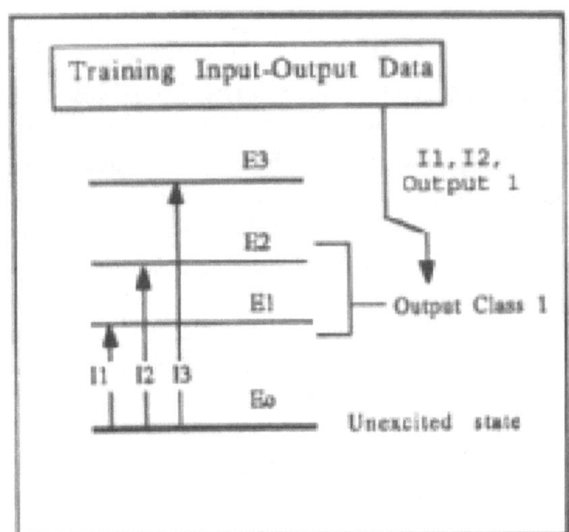

Figure 2: Energy grouping of input patterns 1 and 2 by the output pattern 1

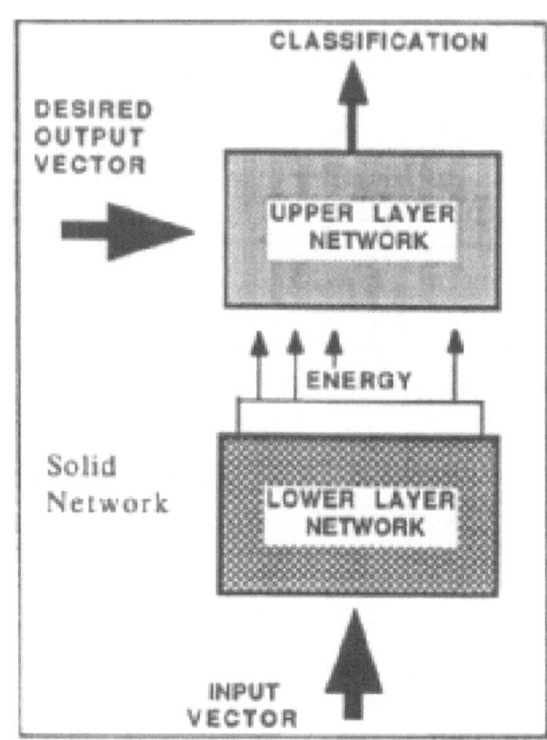

Figure 3: ERN architecture

stable. Its state transition depends on the conservative atomic forces and the external force. The solid ERN functions in an analogous way. It undergoes a state transition and reaches a stable state for each input pattern. The integral weight change indicates the magnitude of the energy gap between the initial and the excited states. The equilibrium routes are the conservative forces and the input pattern is the external force. Thus, the term *energy space* is employed to remind of this analogy. The different energy levels are grouped into energy clusters by the desired output vector. This process is the task of the upper layer network. The energy grouping is illustrated in Fig. 2.

3 ERN Architecture

The solid LLN consists of three layers. The development for a given problem starts from the top layer. The number of output PEs coincides with the desired output vector. Every output PE is connected to two downward PEs in the middle layer. The middle layer PEs are fully connected to the input buffer. The thresholding nonlinearity is a sigmoid function. Every connection weight recalls the equilibrium dynamics and the network equations are generated using Eqs. (3), (4), (6) and (7).

Input vector is introduced to the network in a ramp fashion. Therefore there is an input development period which takes few iterations. The network dynamics is realized during this period as the PEs undergo a target state transition. The energy clusters are identified at the end of the input development period. The initial conditions of weights are equal to the raw input they receive. In solid structure, the reference state is equal to the initial value of the corresponding PE. Initial states for PEs are obtained by steady-state calculations. The ULN structure is optional. The ERN may contain any type of computational routine as the upper layer network. Figure 3 shows the block diagram of ERN.

4 Applications in Binary Space

A typical challenge to a neural network is the nonlinear decision boundaries in exclusive-or/nor problem. The applications presented in this paper includes two input-one output ex-or problem and three input-two output, coupled ex-or/ex-nor problem. The network performance is compared with the performance of a standard back-propagation network (**BPN**) [3].

4.1 Ex-or problem using ERN

An ERN is designed for the ex-or problem. The LLN structure is developed with one output PE, two middle layer PEs and two PEs for the input buffer. The

Table 1: ERN-LLN energies for ex-or problem.

Input	Energy	Output
.1 .9	-2.554	.1
.9 .1	-2.554	.1
.1 .1	1.410	.9
.9 .9	5.267	.9

Table 2: Heuristic test inputs classified by ERN

Input	Energy	Output
0.7 0.2	-1.59	0.1
0.6 0.6	2.945	0.9
0.8 0.5	-1.24	0.1
0.1 0.5	-1.106	0.1

LLN consists of six connection weights all designed to follow their equilibrium routes. The network is represented by nine state equations and six algebraic equations. The input and desired output vectors include 0.1 and 0.9 representing binary numbers 0 and 1. The inputs are developed in two iterations. Energies are calculated at the sixth iteration for each input. The number of total iterations is 24 since there are four input pairs in ex-or problem. Table 1 shows the calculated energies for each input pair and their desired outputs.

It can easily be seen from Table 1 that a single number between -2.554 and 1.410 will identify the output classes. Therefore, the number of necessary memory storage is only a single energy boundary value. The ULN is an expression of the form

$$y = 0.9\,,\ E > -0.57$$

$$y = 0.1\,,\ E < -0.57$$

where, y is the ULN's output and E is the LLN's energy output. Several heuristic inputs are classified by the network as shown in Table 2. Note that heuristic inputs are only applied to the network as the initial conditions of weights where the main input stream is kept over the range of 0.1-0.9. Because the ULN is represented by a simple expression, the network only performs classification in the 0.1-0.9 range. A heuristic answer to a heuristic test input can be made possible by chosing an appropriate ULN or by simply calculating the eucledian-distance in the energy space.

Table 3: ERN energy clusters for exnor-exor problem

Output 1	E1	Output 2	E2
0.1	5.267	0.9	5.267
0.1	1.410	0.9	1.410
0.9	-2.55	0.1	-2.55

4.2 Exnor-exor problem using ERN

Consider the network shown in Fig.4a. Assume that the new training patterns are generated by applying ex-nor and ex-or logic to left and right input PE pairs. This case can be handled by the ERN pair shown in Fig. 4b. The number of training iterations is 48 since there are 8 input patterns. Table 3 shows the corresponding energy clusters.

The energy clusters are distributed with the same boundary value between the output classes as in the single ex-or case. The trained network recalls only this boundary energy value during the test stage. However the output classification requires two expressions for the two output PEs of the ULN. If a NN was used as the ULN, the input training pattern would be the same as singe ex-or problem. Note that the energy clusters would have formed differently if the output generation logic was selected different than the ex-nor/ex-or combination. Also note that the ERN-LLN of Fig. 4b is an invariant structure.

4.3 Comparison with BPN performance

A back-propagation network [4] is trained for the two problems stated above. The network training is performed in a batch environment where the training is terminated if the steady-state errors are unreasonably high. After each termination, a new set of network parameters (learning rate, momentum term and shape factor) is selected and the training is repeated. An over-night batch process yielded two best networks for each of the problems. Table 4 shows a comparative summary on the BPN and ERN performance.

As it can be seen from the Table 4, the training iterations increase in BPN as the nonlinearity of decision boundaries gets higher. The training iterations only increase as a function of the number of training patterns in ERN lower-layer network. Table 4 also indicates several other desirable features with ERN method such as no random network parameters, and small memory size.

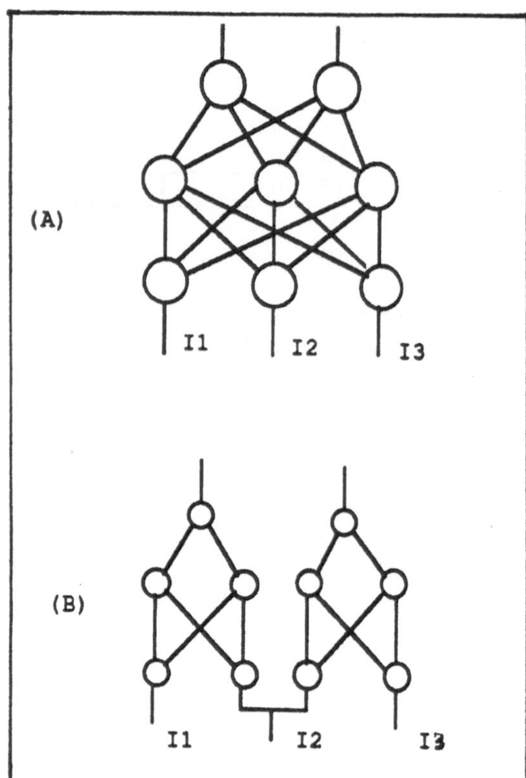

Figure 4: (a) BPN structure for exnor-exor problem, (b) ERN-LLN structure for the same problem

Table 4: BPN and ERN networks for exor and exnor-exor problems

EXOR	BPN	ERN
Training iter.	1266	24
Training CPU	27 sec	2.1 sec
Training time	batch	2.1 sec
Memory size	9 with bias	1
Connections	6	6
PEs	5	5
Random parameters	12	0
1 Pattern Test iter.	1	6
1 Pattern Test CPU	0.012 sec	0.25 sec
EXNOR-EXOR	BPN	ERN
Training iter.	2740	48
Training CPU	55 sec	4.2 sec
Training time	batch	4.2 sec
Memory size	20 with bias	1
Connections	15	12
PEs	8	10
Random parameters	23	0
1 Pattern test iter.	1	6
1 Pattern test CPU	0.017 sec	0.25 sec

5 Conclusions

The equilibrium route network (ERN) development and a preliminary performance evaluation with comparisons to back-propagation network (BPN) are presented. The results of the applications indicate the validity of the approach for ex-or type nonlinear decision boundaries. For this type of problems, ERN compares very favorably with BPN. A detailed comparison is tabulated in Table 4. In particular, the lower-layer ERN structure performed a space conversion in which the nonlinear decision boundaries were identified easier than their original form. This outcome is encouraging because it leads to a drastic memory reduction in neural network applications. The solid network structure and the concept of internal learning are potential areas for future investigation.

References

[1] S. Grossberg, "The Adaptive Brain I: Cognition ,Learning, Reinforcement and Rhythm; The Adaptive Brain II: Vision, Speech, Language, and Motor Control, Elsevier / North-Holland, Amsterdam, 1986.

[2] R. C. Berkan, B. R. Upadhyaya, R. B. Perez and R. A. Kisner," A New Nonlinear "Reconstructive" Control Approach Applied to the Axial Xenon Oscillation Problem in PWRs," *Proc. Seventh Power Plant Dynamics, Control and Testing Symposium*, Knoxville ,TN, Vol.2, pp 77.01-77.18, 1989.

[3] D. E. Rumelhart and J. L. McClelland, Parallel Distributed Processing, Vol 1: Foundations, The MIT Press, Cambridge, Massachusetts, 1987.

[4] G. Mathai, "Automated Methods for Signal Validation and Anomaly Detection", MS thesis, University of Tennessee, Knoxville, 1989.

M. Reuter & R.H. Kluwe
Inst. für Kognitionsforschung, Universität Bw Hamburg
D-2000 Hamburg 70, Holstenhofweg 85, Tel. 040/6541 2876
Telefax 040/6530413

FD-spectrums and their simulation by neural networks

F - AI,connectionist,networks

Introduction

The research reported here is directed at the elaboration of an efficient system for the classification and identification of acoustical signals. In this context,a new frequency representation of acoustical signatures has been developed which is invariant under fade out of frequency bands or under masking of single signature shares. This "metaform" of the signatures results from the projection of all frequencies that are parts of the single oscillations on their fundamental frequencies.

The spectral presentation in which one can find the metaforms is called FD-spectrum and can be generated by rather simple neural networks. Preliminary experiments with these networks suggest that neural modules are of central importance for the computation of these spectrums. This led to the assumption that specific neural substructures rather than single neurons are the fundamental components of an efficient network in pattern recognition.

The modification of the operators that generate socalled eFD- and mFD-spectrums allow the construction of frequency filters and of higher order presentations which can be chosen to select individual qualities for the classification and identification of acoustical sources.

RAPID LEARNING OF PATTERN SEQUENCES: A NOVEL NETWORK MODEL

Valeriy I. Nenov
Artificial Intelligence Lab, CSD, UCLA, Los Angeles CA 90024

Abstract

This paper describes a neural architecture -- the KATAMIC memory, designed specifically to store sequences of binary patterns. The model integrates continuous learning with concurrent recall based on a step by step sequence predictions which allows memorized sequences to be recalled in response to cues -- short sub-sequences. A novel neuron-like computing element -- the PREDICTRON is introduced which learns to generate at each time step a prediction of the next input pattern based on the interaction of the memories of previously learned sequences (long term memory) and the recent states of the network (short term memory). A complete mathematical description of the KATAMIC algorithm is presented. The memory was implemented on the CM-2 Connection Machine and in *Lisp and results of some basic simulation studies are discussed.

MULTI CLASS PATTERN ASSOCIATION USING DIGITAL N-TUPLE NETWORKS

J.M.Bishop, P.R.Minchinton, R.J.Mitchell.
The Neural Network Research Group,
Cybernetics Department, Reading University, Whiteknights,
READING, Berkshire, Great Britain. RG6 2AL.

The n-tuple Pattern Separation Network (PSN) has a significant practical advantage over conventional Pattern Associator systems, as it exploits the parallelism inherent when addressing memory in the standard Von Neumann architecture. The PSN has been demonstrated to be effective at two class pattern association problems [1]. This paper examines Pattern Separation behaviour in more detail on multi class pattern association problems.

A NOTION OF SEQUENTIAL COMPOSITION FOR NEURAL NETS

Robert Zimmer
Department of Electrical Engineering
Brunel University
Uxbridge, Middx. UB8 3PH
England

Abstract:

This paper describes a new notion of sequential composition for neural nets, a composition that mirrors the more readily defined notion of composition of the state spaces of the nets. The composition is defined for any two nets that have the same number of nodes and compute the same sort of functions. In this paper we give the definition, which we motivate by considering in some detail the case of boolean nets, and give some preliminary remarks on the algebraic properties of the space of boolean neural nets under the composition.

CLASSIFYING ARTIFICIAL NEURAL NETWORK ARCHITECTURE

ABSTRACT

Most Artificial Neural Network architecture appears to be heuristic in conception; devised by the skill and experience of the creator rather than derived from first principles. In contrast, however, we feel that it would be preferable to develop a formal taxonomy for Neural Networks in order that network size and designs may be matched more closely to the underlying problem. To this end, a theoretic and algebraic classification of Neural Network architecture has been developed.

Dr J P Evans
Sir George Cayley Institute,
School of Computing Science and Information Systems Engineering
The Polytechnic of Central London
115 New Cavendish Street
LONDON
W1M 8JS
United Kingdom.

Most Neural Networks Compute in Steps

Armand de Callataÿ

IBM A.K. Watson IEC, La Hulpe, Belgium (callatay at bbribm11 on EARN)

Summary. *Periodic processing in neural networks has been suggested as an organizing principle for robot control (Callataÿ, 1969). The present paper studies which neural networks can or must have a continuous or rhythmic processing for real-time control of non-stop systems. The properties of each processing mode are reviewed in the following neural networks: linear, with threshold, with classifiers, with latches, self-timed, clock-timed, multi-layered, circular, and recursive. Learning is studied for 3 modifiable types: self-tuned, adaptive and recording. The conclusion is that most neural networks must perform discrete computations.*

A MACRO ARCHITECTURE FOR DYNAMIC PROCESSING OF TEMPORAL STRUCTURES

Christian Balkenius

Department of Cognitive Science,
Lund University,
Kungshuset, Lundagård, S-223 50 LUND, Sweden

Abstract

This paper describes the macro architecture of a neural model for dynamic processing of temporal structures. It consists of a number of modules which have the ability to self-organize in order to handle constraint satisfaction, representation, recognition and production of sequentially ordered activity patterns. The system incorporates mechanisms for short term memory (STM), long term memory (LTM) and intermediate memory (IM). IM is used to implement a set of virtual slots and avoids problems associated with models in which a fixed number of slots are used. It is able to construct abstract classes of patterns and sequences in order to generalize learned sequential structures to novel inputs and has been used in computer simulation of language acquisition where it proved to be able to learn some aspects of the syntax of natural language.

A MODULAR AND EXPANDABLE ANALOG INTEGRATED NEURAL NETWORK

O. ROSSETTO, C. JUTTEN

L.TIRF/INPG 46, Avenue Félix Viallet 38031 Grenoble

For high speed application using a neuromimetic algorithm, classical computers are not adapted. Only special architecture, highly parallel, can provide the computing potential required to efficiently implement neuromimetics algorithms. In such a case, it is necessary to use a dedicated hardware. Since 1986, a great interest for designing neuro-computers and neural VLSI has emerged [see in 1].

We propose a modular analog neural network based on basic integrated cells used as building blocks. A various kinds of neural networks can be implemented simply by changing the connections between the different circuits. In the paper, we present briefly the architecture of each functional blocks with their possibilities and their limits.

MODEL OF HYPOTHETICAL VISUAL PERCEPTIVE SYSTEM

Babic Ranko
Ramiz Sadika 12
38000 Pristina
Yugoslavia

ABSTRACT - The description of a model, including the definition of input-pattern structure, is based on its central part - the structural memory. Each input pattern can activate the respective contents of the structural memory, which is than incorporated in the pattern itself, making it meaningful for further conceptual processing. In terms of this idea we attempted to explain two phenomena: the existence of mental field of vision and the visual system's capability to evoke mental images.

ANALYSIS OF NETWORK DYNAMICS I

Chair: David WALLACE

NEURAL NETWORK UNIT DENSITY : A CRITICAL BIOLOGICAL PARAMETER.

N. AZMY, J-F. VIBERT
Université Pierre et Marie CURIE-CNRS UA 1162
Laboratoire de Physiologie, CHU Saint Antoine
27 rue Chaligny F-75571 Paris Cedex 12 (France)
e-mail: vibert@frsim51.bitnet

Abstract:

This paper deals with the problem of the implementation of biological properties or characteristics in theoretical formal neurons. Some of these properties are rarely considered, such as the neural network unit density. We developped a computer model implementing neurons with a high biological plausibility, in which it was possible to modify the unit density of the network i.e *the length of the axons. The influence of the neural network unit density on dynamic behavior was observed as a nonlinear function. Large nets that don't display spontaneous cyclic mode become cycling when their unit density decreases and reaches a threshold value. Quasi instantaneous transition from chaotic to stable cyclic activity was found for a critical density value specific to a given network. It is concluded that the process by which a cyclic mode emerges in a neural network is unit density dependent. This phenomenon is reversible and may offer means of studying dynamic transition in quasirandom networks of threshold neurons.*

INTRODUCTION: Computational neurobiology plays a particularly critical role in the study of neural networks. The subject of the present study is a more general question than those developped to give specific solutions to specific problems, *e.g.*, learning or data recognition. Our interest here, is to point out some clues as that biological parameters as unit density of neural networks has some influence in computational processing. Anatomical organization of the brain shows a great diversity of architecture of clusters of neurons. The most streaking caracteristic of this organisation is the metric euclidean distance that separates neurons. Our connectionist model enables one to simulate interactions between different structures such as nucleed, stratified or reticulated ones varying the length of the axons. This modulation of axons length led to consider the notion of **neural unit density** in a network, and to study its effects on dynamical behavior in computational processing.

METHODS: This work was done using a computer model implementing formal neurons with a high biological plausibility.

The model: `Neuro_clusters`[1] deals basically with the problem of implementing biological parameters (Vibert, 1990). We believe that knowledge of connectivity and synaptic weight alone are not sufficient to account for the operation and capabilities of neural networks nor is relevent to implement a model with properties or algorithms that has no biological foundement, *e.g.*, retro-

[1]Readers interested in further information on the simulator should send requests by e-mail to vibert@frsim51 via EARN/bitnet.

propagation. Thus, we took into consideration in our approach, fundamentally reductionist, some of cellular, synaptic, and connectivity biological parameters such as threshold, post-burst hyperpolarisation, absolute refractory period, connection sign, synaptic strength and the length of axon so that individual model neurons exhibit some useful resemblance to actual biological systems, *e.g.*, an all-or-none character of the action potential, a spatio-temporal summation of postsynaptic potentials, and both absolute and relative refractory period. As described somewhere else (Vibert et Azmy, 1989) neurons in `neuro_cluster` maybe distributed in a set of clusters each one may have its own parameters, and located at a distance that represents the spatial constraint observed in biological neural networks.

In the present study, we examine the case of one cluster of N neurons that are linked by a network of axons and dendrites that synapse onto one another at a fixed distance. Input from the external world produces local depolarizations that are integrated in the cell body to produce a discharge. We assume that inputs are of two origins. On the one hand, incoming signals from local interactions such as inter-netlets fead-back loops, on the other hand, a more global input that acts as a noise feeding the totality of the network. Output is a non-linear function of local depolarisations. Each neuron is assigned a binary state variable x_i, which takes on the value $+1$ when neurone i is *on* (firing an action potential) and the value 0 when i is *off* (inactive or not firing). The synaptic organization of the network obtained by a random-number generator is summarized in a connexion matrix $[W_{ij}]$ of coupling strengths. In general, $W_{ij} \neq W_{ji}$.

Signal transmission delay: Say τ, the minimal universal delay time for signal transmission from one neuron to another. Signal sent at time t from neuron i reaches neuron j whithin a variable time interval $(t+k\tau)$, where k is an integer positive number. Thus, we represent a spatial constraint materializing the length of the axons by its temporal effect on data transmission.

Unit density of the network: The parameter k may vary from the minimal value $k = 1$ to a given integer positive number L as a maximal value that is compatible with a biological scale of sizes. When neurons are distant the ones from the others by $(t+\tau)$, then we say that the network is in its highest density D_l. Density of the network decreases to a minimal density D_L as k increases. In this study the matrix of the organization of axons length $[L_{ij}]$ is symmetrical.

Thus, the state of the net at time $(t+k\tau)$ depends on the state vector at time t, the connection matrix, and the threshold vector formed from s_i, $i = 1, ..., N$, which can symbolically be written as,

$$X_i(t+k\tau) = \theta[\textstyle\sum_j v_{ij} x_j(t) + \text{accumulated signal} + \text{external input} + \text{noise} - s_i],$$

where $\theta[y]$ is the step function, s_i is the threshold of the neuron i and $x_i(t)$, $i = 1, ..., N$, forms the state vector.

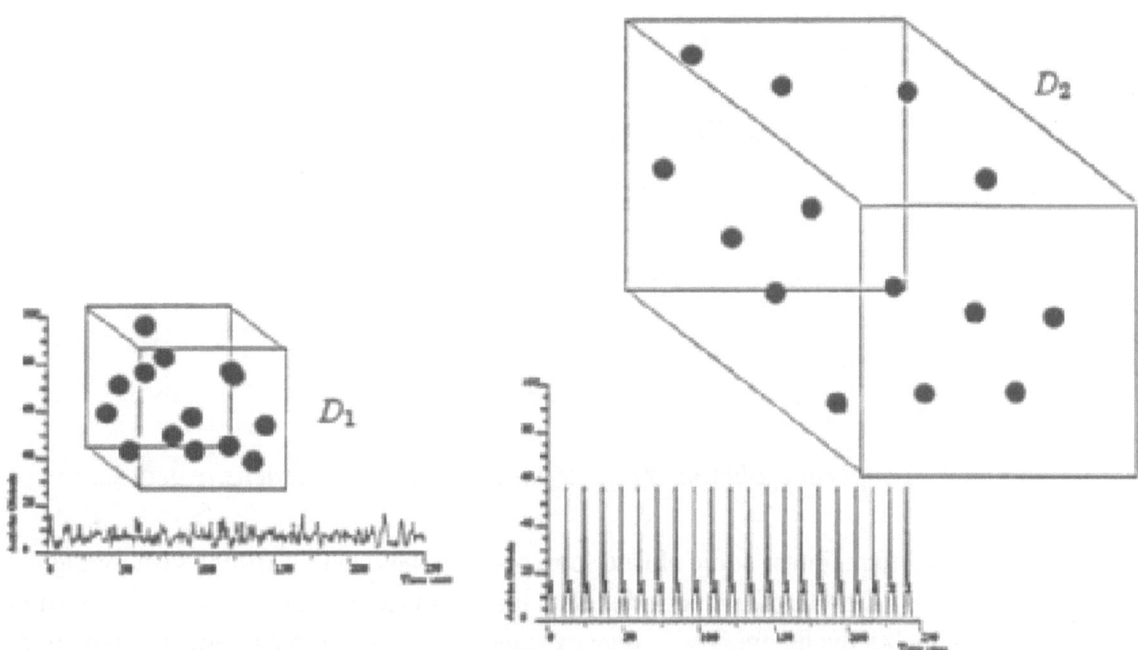

Figure 1 : *Effect of changing the neural network unit density*

RESULTS and DISCUSSION: A large number of computer simulation of large N-networks ($N = 100 - 1000$) in which the inter-unit connections were randomly distributed both in weight and sign (inhibitory, excitory and no connection) were performed using `neuro_clusters`. Generally, such networks presented a spontaneous cyclic activity usually stabilized within 500 time units (*i.e.*, ms). It can be estimated that, time from initialisation of a net's activity to the entrance into a cycle (the transient period) is short in most cases. Sometimes, such networks did not displayed such a regular cyclic activity. We empirically searched, by running several computer simulations, for such chaotic networks (Fig. 1 A). When this occured, their dynamic evolution, underlied by their connection matrix, did not led to any cyclic mode within at least 1000 time units when $k = 1$. Self-organization may not occur, and dynamic evolution may keep on a quasi-random activity (Kohonen, 1988). When this was observed, we increased largely the axons length (thus decreasing the unit density), all the other parameters remaining constant. The neural network output became cyclic and stable (Fig. 1-B). Then it appears to exist a structural relationship between output pattern and unit density. The way the pattern changed was then studied. When density decreased, the network output remained unchanged until a threshold value was reached, where the output abruptly switched to a cyclic stable pattern (Fig 2). Above this value the net kept on cycling, even if the unit density continue to decrease. This phenomenon was reversible in most cases, and when reincreasing the neural density, it failed back in its previous chaotic mode for the same axon length. Thus rapid convergence toward stable, ordered behavior for a given critical value of axons length was observed. Furthermore, we established that whenever the network reaches a stable activity, its neurons organise themselves spontaneously into three or four netlets beating at a common period, that constitute toghether a stable organized loop of activity (Clark *et al.*, 1988). Their individual firings repeat with a period equal to the number of netlets involved.

Large nets, *e.g.*, $N = 100 - 1000$, are generaly spontaneously and rapidly divided into netlets

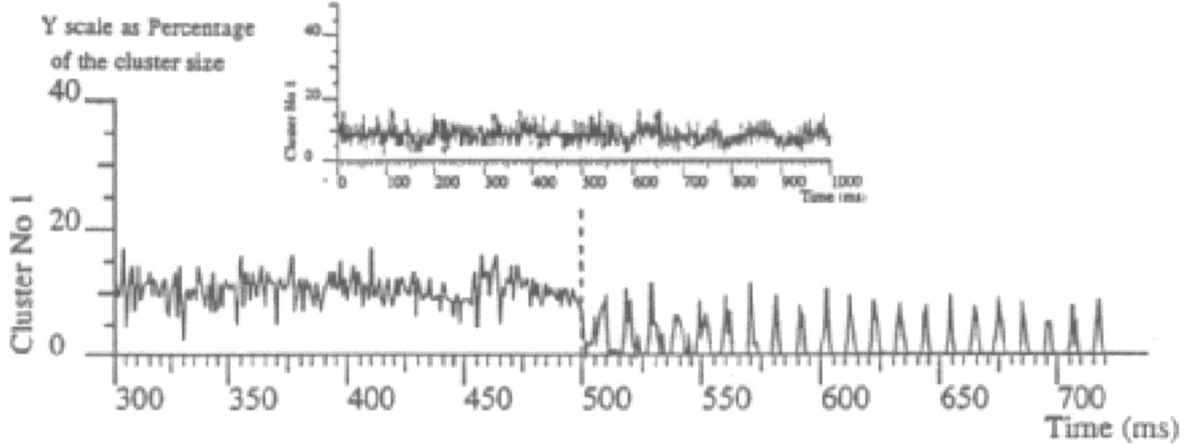

Figure 2 : *Temporal evolution of output pattern when the unit density decreases.*

which are supposed to maintain some closed loop of successive neural firings visualizing a cyclic mode, with relatively short transient periode. Nets that don't reach spontaneous stability show that the process by which a cycle emerges is density dependent. Quasi instantaneous transition from chaotic to stable cyclic activity were found for a critical density value that seems specific to a given pattern of synaptic connectivity. This phenomenon is reversible, and may offer means of studying the transition from a stable point attractor to a chaotic attractor *et vice versa* (Bergé *et al.*, 1988). Quasi-random activity can be turned on and off virtually instantaneously, as with a switch through a bifurcation point that is, in this study, function of the critical value of unit density. This result emphisizes the importance of the network architecture parameters on its dynamic mode activity, and suggests that a form of *a priori* data processing already exists in biological neural networks according to a morphogenetic plan of the brain.

References:

P.Bergé, Y. Pomeau, Ch. Vidal. *L'ORDRE dans le CHAOS*, (1988), Hermann, Paris.

J.W. Clark, K.E. Kürten and J. Rafelski. Topology, structure and distance in quasirandom neural networks. Chapter 7 in Rodney M.J. Cotterill, eds. (1988), *Computer simulation in brain science.* Cambridge University Press, Cambridge.

T. Kohonen. *Self-Organization and Associative Memory*, (1988) Springer-Verlag, Berlin.

J-F. Vibert et N. Azmy. Simulation de la genèse d'un rythme biologique par des réseaux interconnectés, in proceedings & exhibition catalog EC2, eds. (1989), *Neural networks & their applications*, Neuronimes'89., Paris.

J-F. Vibert. Neuro_clusters: Un simulateur de réseaux de neurones en interaction. In Symposium DECUS-France, Paris(1990).

EMERGENCE OF COMPLEXITY IN THE DYNAMIC OF A DILUTED NEURAL NETWORK.

by Michel BENAIM* and Manuel SAMUELIDES* **

* Office National d'Etudes et de Recherches Aérospatiale ,TOULOUSE, FRANCE
** Ecole Nationale Supérieure de l'Aéronautique et de l'Espace,
BP 4032, TOULOUSE 31055 Cedex .

Abstract **:**
Sensitive dependance on initial conditons is studied for a diluted and non symmetric model of neural networks . Two parameters appear to be relevant: the connectivity which allows local difference to spread all over the network and the noise temperature which smooths these difference. Their influence is shown by theoretical estimations in the thermodynamic limit and by simulations which give exactly the same results for medium size networks: when the connectivity reaches a critical threshold, transition from a "sensitive" dynamic to a single attractor dynamic is exhibited for a critical temperature which can be computed. This result is proved first ior constant connectivity models then for randomly diluted models.

1.Introduction.

The study of generic properties of neural networks consists in computing probability of dynamic features in the frame of quenched randomness: namely, the architecture of the network is randomly selected once for all. Then it is frozen while the network is processing according to its dynamic. This method is successfully practised for various kinds of non-linear dynamic systems with a large number of freedom degrees as popultation dynamics, boolean automata, spin glasses and neural networks. Since exact results for this kind of random models are hard, the annealed approximation was proposed in [1] for large systems: in this approximation the architecture of the network is randomly reconfigurated at each timestep. This approximation is shown to be correct in the thermodynamic limit in [2]. Numerical simulations [1],[3],[4],[6] show this approximation is quite good and works for systems of medium size which is far better than one can expect from theoretical estimations. Similar results were obtained in [5] from simulation of analogic deterministic neural network . In [6] the noise is replaced by a common threshold for all the neurons. All these papers are showing the evidence of a phase transition from a "chaotic" phase to a "frozen" one which is interesting to interpret with regard to information theory: it is clear that **when connectivity increases, the local variations between two configurations will spread all over the network. On the contrary when the temperature is growing, the thermal noise will smooth and cut down the significative variations inherited from two different initial states.**

Analog results are proved here for diluted asymmetrical neural networks. Moreover, gaussian modelisation of the variability of the connexion weights and of thermal noise allow to compute the critical noise temperature of the phase transition towards a noise-driven dynamic. In section 2, a constant connectivity model is studied first in the annealed approximation (2.1) then in the actual operating model by using simulations in section (2.2). The results of the simulations were predicted by the previous estimations of (2.1). In section 3, another random model of diluted network is studied where each possible connection is randomly realized independantly one from another. This allows to take the average connectivity as a continuous parameter. The results for this new model are qualtatively analog to the first one and the critical value of the connectivity agrees numerically with the results from the first model. These results tend to prove that the observed phase transition is rather robust with repect to the modelization of the variability of the configuration of the network by a probability distribution.

In section 4, we briefly sketch out the points which we are presently investigating and the general interest of such research for real or artificial neural networks.

2. Study of a constant connectivity model.

2.1. Estimation of the annealed approximation.

Let us consider a set of N neurons with binary activation states belonging to {-1,1}. Each neuron has κ inputs. At time t=0 the κ input neurons connected to each neuron are chosen randomly among the set of all neurons and the weights of each connection are chosen according to a gaussian centered distribution . The configuration of the network is therefore defined as a realisation of a random valuated directed graph of order κ. This configuration actually carries the information which is given to the system during the learning phase whatever it is.

The dynamic of the network is discrete-time and defined by the synchronous updating of the neurons . It follows the classical rule:

$$x_i(t+1)=\mathrm{Sgn}(\sum[J_{ji}.x_j(t)+\sqrt{T}.B_i(t)]) \qquad (1)$$

where T is the noise temperature which is the inverse of the power ratio signal upon noise and B_i is the normalized discrete time white noise.

To study the dependance on intial condition, we select two activation states of the network x and x' and we shall study the evolution of the normalized Hamming distance d(x,x') between the two states evolving with time.

The input of neuron i at time t=0 has three components: the weighted sum IN^+_i of inputs from neurons which holds the same activation in states x and x', the weighted sum IN^-_i from neurons which holds different activation in states x and x' and the noise. As the noise is supposed to be the same for the two processes in order to compare its action on the variations between x and x', $x_i(1)$ is different from $x'_i(1)$ if and only if the real quantities $IN^+_i + IN^-_i + \sqrt{T}.B_i$ and $IN^+_i - IN^-_i + \sqrt{T}.B_i$ have got different signs. If K_i is the number of input neurons which hold different activation states, one gets

$$P[x_i(1)\neq x'_i(1)\,|\,K_i=k] = \frac{2}{\pi}\,\mathrm{Atan}\sqrt{\frac{k}{\kappa-k+T}} \qquad (2)$$

Following the annealed approximation one can reproduce this computation for the evolution from time t to time t+1 . Then the evolution of the average distance d(t) ,between two states is given by the recurrence relation $d(t+1)=f_{\kappa,T}(d(t))$ where $f_{\kappa,T}(d)$ is the polynomial mapping of the interval [0;1] into itself defined by :

$$f_{\kappa,T}(d) = \frac{2}{\pi}\sum_{k=0}^{\kappa} C_\kappa^k\,\mathrm{Atan}\sqrt{\frac{k}{\kappa-k+T}}\; d^k\,(1-d)^{\kappa-k} \qquad (3)$$

When studying the sign of $f'_{\kappa,T}(0)-1$; one observes two behaviours :

For $\kappa \le 2$ and $T > 0$,0 is always a stable fixed point of (3).

For $\kappa > 2$, 0 is unstable for $T < T_\kappa$ and stable for $T > T_\kappa$, where $T_\kappa = \dfrac{1}{\sin^2(\frac{\pi}{2.\kappa})} - \kappa$.

is given by the equation $f'_{\kappa,T}(0) = 1$. Furthermore, the local bifurcation of the map (4) at the point $(0,T_\kappa)$ is a *transcritical* bifurcation. More precisely, one has :

$$d^* = 0 \text{ for } T > T_\kappa \text{ and } d^* = \alpha.(T-T_{c,\kappa}) + o\,(T-T_\kappa) \text{ for } T < T_\kappa \qquad (4)$$

where d^* is the stable fixed point of (3) and α is computed from explicit partial derivation. For instance, if $\kappa=3$, then $T_\kappa = 1$ and $\alpha = -\frac{1}{6\pi}$.

2.2 Results of simulation.

The reported simulations have been performed for a net of N neurons (N = 1000, 5000,..) a given temperature T and a number n of iterations. In the figure 1, $d^{*}= d(x(n),x'(n))$ is plotted as a function of the temperature T for different values of κ, N and n . The full curve represents the theorical bifurcation diagram obtained from the study of the map (3).

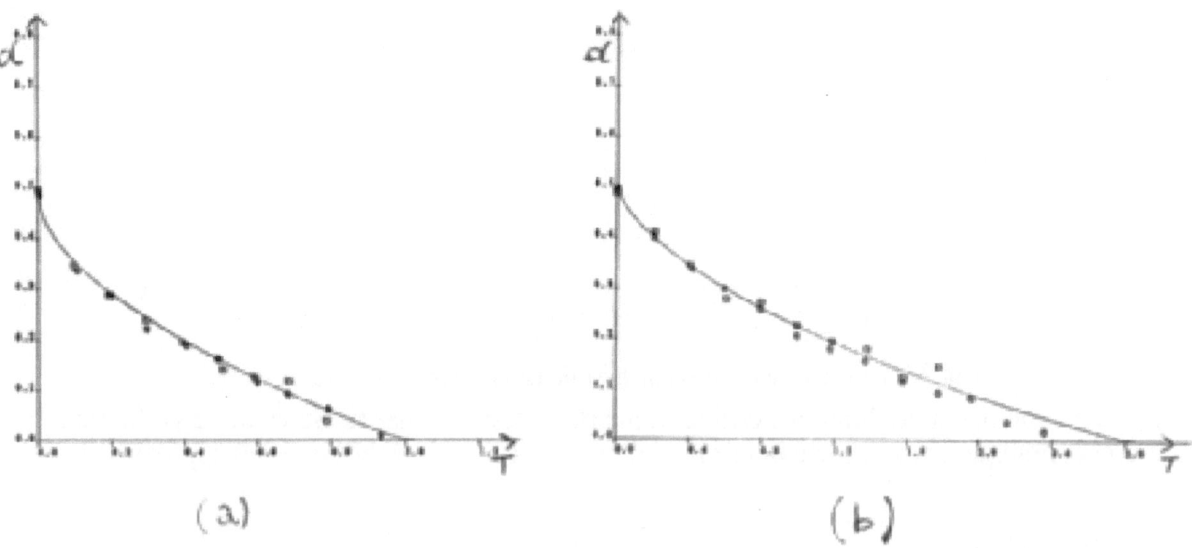

Fig 1. Distance at time t = n between 2 randomly chosen states with an initial distance $d_0 = 0.01$ in a random net of N neurons of connectivity κ.
(o) N = 5000, n = 1000 ; (□) : N = 1000 , n = 10000 . (a) : $\kappa = 3$; (b) $\kappa = 4$.

The figure shows that , even in the case where n is large against $\mathrm{Log}(\sqrt{N})$ (N=1000, n= 10000) and where the distance computed is not an average distance but the distance at time n between **two** individual states of **one** randomly given net , the result performed by computer simulation agree remarkably with the theorical estimation obtained for the mean distance in the thermodynamic limit.

3. Study of the randomly connected model.

In order to give a better description of the behaviour of a diluted neural network we are now going to study the sensitive dependence on initial condition in terms of a continous connectivity parameter α. Let us suppose now that for each ordered couple (i,j) of neurons there is a connection line from i to j with probability $\frac{\alpha}{N}$. Following the main lines of computation of section 2, and according to the convergence of binomial distribution to Poisson distributions the following recurrence is shown to predict the evolution of d(t) :

$$d(t+1) = \frac{2}{\pi} . e^{-\alpha} . \sum_{k \geq 0}^{\infty} f_{k,T}(d(t)) . \frac{\alpha^k}{k!} \qquad (5)$$

We are now able to describe the set of critical parameters (α_c , T_c) which is the bifurcation set of the map (5) . This set , is a smooth curve which cuts the (α,T) space into two distinct regions (I and II) which correspond to the "frozen" and "chaotic" phases.

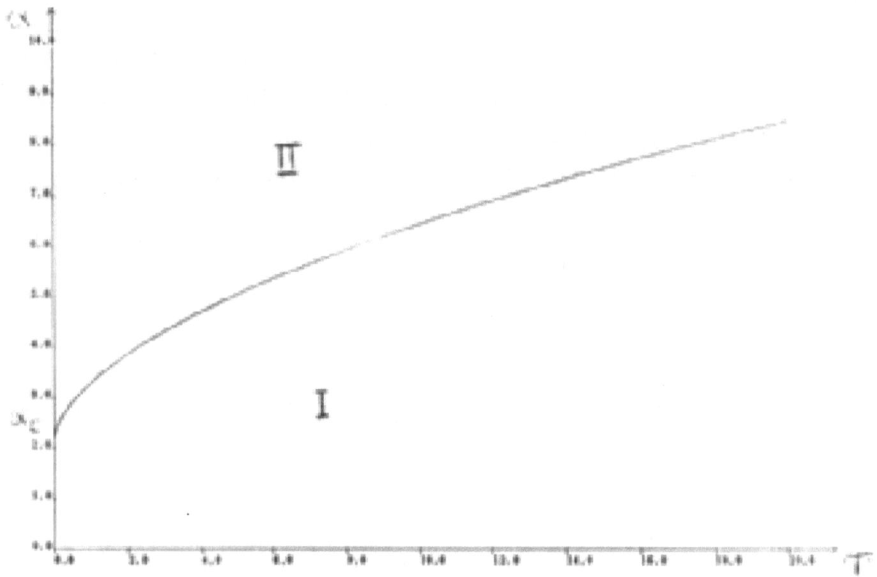

Fig 3. Phase diagram illustrating the differents regimes in the (α,T) plane.

In the zero temperature limit the critical value α_c of α converges to the value: $\alpha_0 = 2.206...$ Further simulation have shown in this model too the accuracy of the annealed approximation.

4. Discussion.

In this communication, we have considered the sensitive dependance on initial conditions in a diluted neural network in presence of thermal noise. We have checked that in the thermodynamic limit, there occurs a transition from a point attractor dynamic to a "chaotic" dynamic when crossing a critical curve in the parameter space. The genericity of such dynamic supports the conjecture that "chaotic behaviour serves as the essential ground state for the neural perceptual apparatus" [8]. Its occurence in simple models provides encouraging perspectives for introducing it in artificial signal processing software.

The accuracy of the "annealed" approximation is confirmed far from the thermodynamic limit and deserves further investigation. Besides, the fact that in the thermodynamic limit there is no loop in the tree of ancestors of one given neuron in the second model is not a straight consequence of the same result for the first model. Its proof uses generating functions and techniques which were recently introduced in computer science to estimate complexity of software. These techniques may allow us to derive further statistical properties of the attractors of diluted networks. By doing so, one can hope to get a better understanding of the chaotic phase in order to use its sensitivity to code temporal information.

References:

[1] DERRIDA , POMMEAU. Random Networks of Automata: A simple Annealed Approximation. *Europhys.Lett, 1, (2), 44-49 (1987)*

[2] DERRIDA, WEISBUCH. Evolution of Overlaps between configurations in Random Boolean Networks. *J.Phys (Paris) , 47 (1986) 1297.*

[3] DERRIDA, WEISBUCH. *Europhys.Lett, 4 ,(1987).*

[4] DERRIDA, GARDNER, ZIPPELIUS. An exactly Solvable Asymmetric Neural Network Model. *Europhys.Lett, 4,(2),(1987)*

[5] BAUER, MARTIENSTEN. Quasy_Periodicity Route to chaos in Neural Networks. *Europhys.Lett, 10,(5), (1989)*

[6] K.E.KURTEN. Phase transitions in quasirandom neural networks,*IJCNN San Diego 1988*

[7] PERETTO, NIEZ. Stochastic dynamic of Neural Networks. *IEEE Trans Man Cybern 16. 1 (1986)*

[8] C.A.SKARDA, W.J.FREEMAN.How brains make chaos in order to make sense of the world. *Behavioral and Brain Sciences (10),161-195* (1987)

INFORMATIONAL VERSUS BIFURCATIVE USE OF CHAOTIC DYNAMICS IN NEURAL NETWORKS

Gianfranco Basti, Antonio Perrone
Pontifical Gregorian University, Piazza della Pilotta 4, 00187 Rome, Italy
Valerio Cimagalli, Massimiliano Giona
Faculty of Engineering, University of Rome «La Sapienza», Via Eudossiana 18, 00184 Rome, Italy
Eros Pasero
Department of Electr. Engineering, University of Rome «Tor Vergata», Via O. Raimondo, 00173 Rome, Italy.
Giovanna Morgavi
Institute for Electonic Circuits, National Research Council, Via Opera Pia 11, 16145 Genoa, Italy.

ABSTRACT

We discuss some basic problem in the mathematical theory of neural networks and suggest a new *informational* approach on the modeling of a chaotic architecture. Some example of this approach are given in connection to the building of coincidence detectors and 2D asymmetric spin glass systems that stabilize themselves in a non bifurcative way.

1. THE CONVERGENT PARADIGM IN SIMULATED NEURAL NETWORKS

In a recent essay devoted to a mathematical analysis of the state of art in neural networks [1], M.Hirsch affirmed that among the three different behaviors of dynamical systems in modeling neural networks:

1) The *convergent* one: every trajectory converges asymptotically to some equilibrium;
2) The *oscillatory* one: every trajectory is asymptotic to a periodic (perhaps stationary) orbit;
3) The *chaotic* one: «most» trajectories do not tend to periodic orbits;

though the biological systems practically never display a convergent dynamics, nevertheless the great majority of neural net models are convergent because of the exceeding difficulty in studying the opposite cases.

So, in the classic *convergent paradigm* in continuous neural nets, in view of granting an analytical tractability of the dynamic system [1-2, 12], there is a strong distinction between:

1) the *Code-Defining Function* (CDF) of the pattern learning phase (= LTM or weight dynamics);
2) the *Coding Function* (CF) of the pattern recognition phase (= STM or activation dynamics).

After the learning phase (i.e., with *fixed* weights W_{ij}), during the recognition phase, the network (the activation dynamics x_i) may be considered as a dynamic system governed by a set of differential equations:

$$\dot{x}_i = F_i\,(x_1,..., x_n)\,; \quad i = 1,..., n \qquad (1)$$

The CF of the activation dynamics consists thus in establishing the following biunivocal association:

$$y_k <\!\!-\!\!-\!\!> x_{p_k}$$

where y_k is the set of input pattern of the pattern space P, and x_{y_k} is the corresponding attractor (i.e., a *code*) of the net phase space X (i.e., the code-space). Such a biunivocal association defines thus the notion of *Content Addressable Memory* (CAM), as typical of this kind of dynamic associative memories [3]. So, the activation dynamics may be considered as the evolution in the phase space of the temporal flow associated with the system of Eq.1:

$$\lim_{t\to\infty} \Phi_t\ (x_i\,(0), W_{ij}) = x_{y_k} \qquad (2)$$

Two are the parameters of this dynamics:

1) the fixed input pattern y_k^l ($l=1,...,m$), described by the static initial condition x_i (0) in which all the input information is stored;
2) the weights W_{ij} parametrizing biunivocally the correspondence between the attractor x_{y_k} and the pattern y_k. The CDF consists thus in the biunivocal definition of this correspondence.

All this implies a *dichotomy* in the convergent paradigm:

1) this absolute separation between CDF and CF makes these nets deficient in reckoning with *non-static environments* (i.e. time-varying environments);
2) if we treat simultaneously the activation and the weight dynamics we obtain a *non-linear problem* of tremendous difficulty, for which no general approach exists.

2. OSCILLATORY AND CHAOTIC BEHAVIOR IN THE MAMMALIAN BRAIN: THE BIFURCATIVE PARADIGM

There is an experimental evidence about the presence of an *oscillatory* behavior on a *chaotic* basis in the mammalian (rabbit and cat) sensory cortex [4-6]. In both cases, the recognition process seems to be related with a dynamic transition from a chaotic to a regular behavior of the cortical oscillators.

Particularly, the evidence of a *synchronization in phase* (after a *chaotic* state during the resting phase for the resetting of the STM) of the oscillatory behavior of the cells within the same column of feature detectors or among homologous columns in the primary visual cortex of the cat, might be a way to establish relations between features in different parts of the visual field [6-9].

On the contrary, different synaptic mechanism, related with the modification of the dendritic spine geometry and the mediator releasing [10], let the neurons be considered as *coincidence detectors* in the time domain. Such a net can display a fast cell-assembling called *synaptic pattern* [11], founded on good timing relationships among the cells, depending on such a fast variability of the weights. Globally, a net of coincidence detectors can show both an aperiodic or a periodic oscillatory behavior.

Nevertheless, the problem of the theoretical explication of the role of chaos in the dynamics of the cognitive process remains to be solved. C.A.Skarda and W.J.Freeman offered a first interpretation of such a role [5]. Two are the main points of their interpretation: 1) Each limit cycle to which the chaotic dynamics *bifurcates* corresponds to a learned pattern (i.e., an odor); 2) The role of chaos is simply to make instantaneous and unbiased for the dynamics the access to the whole set of the limit cycles (i.e., of the learned patterns). In other terms the chaos is intended as the simplest dynamic device to avoid the local minima problem in neural nets.

More recently M.W.Hirsch [1] summarized the criticism to Freeman's approach to chaos in the following way: 1) The role of chaos is only in the background for granting the bifurcative behavior of the global dynamics: only the limit cycles have an informational role in this approach. 2) It is not specified how the chaotic system learns to bifurcate upon presentation of learned inputs.

On the other hand, if we want to try an *informational* (see Sect.3) and not purely dynamic use of chaos in neural nets, but we remain conceptually in the context of the convergent paradigm, a lot of problems arise. They are summarized by M.Hirsch in the following way:

> The limit set of a chaotic orbit is generally some sort of fractal; in what sense can it represent useful information? How do we retrieve information from the fractals? How can we use it as input to another net? How can we train the weights? In what sense can a chaotic net be stable?

As we see, the dichotomy remembered at the end of Sect.1 results to be confirmed. On the other hand, it is not easy to answer the questions pointed out by Hirsch about the informational use of chaos in neural nets: the necessity of an adequate study of chaos is urgent. Nevertheless, all the discussion points out a necessary conclusion. The essential problem bound to the bifurcative use of chaos depends on the persistence of the classical paradigm also in treating chaotic neural nets. I.e., it depends on the insistence in considering *the input as a parameter* of the dynamic system (i.e., of the net). In this way the chaos has effectively only disadvantages.

3. INVARIANT EXTRACTION AND THE INFORMATIONAL USE OF CHAOS

The alternative use of chaos consists in considering the input as a dynamic system (at last chaotic) from which the cognitive system *extracts (interesting) invariants* to construct an inner dynamic representation of the input [13]. Indeed, the classical characterization of chaos (Ljapunov exponents, entropies, limit set dimensions) are essentialy descriptive. On the contrary, the *fine structure of chaos* is related with its *dynamic invariants*. The set of them coincides with the set *Per (f)* of all the critical elements (fixed points, j-cycles. $j = 1,..., n,...$) of a given trasformation f [14-16]. We are studying two possibilities related with **two types of invariants**: 1) The **set *Per (f)*** constituting the unstable skeleton of an aperiodic bounded dynamics; 2) The **hierarchical system of fundamental frequencies**, i.e., of the unstable periodic orbits organized with respect to the **subharmonic order *n***. In our approach the notions of STM and LTM of the convergent approach are to be reinterpreted in the following way (they depend no longer on the distinction activation *versus* weight dynamics):

1) **STM** = dynamic processing of a complex (chaotic) input in view of determining its invariants;

2) **LTM** = dynamic storing of the invariants in (chaotic) distributed memories.

4. TWO FIRST DEVELOPMENTS OF THE APPROACH

4.1 CRF and coincidence detectors net

A first way for extracting dynamic invariants is the **Chaotic-Recursive-Filtering** (**CRF**: [17]). Let us consider the dynamical system generated by *the discrete map* $f: I \rightarrow I$ (I=interval of the euclidean line) $x_{n+1} = f(x_n)$, the CRF corresponds to a dynamic correlation of the actual evolution with the outputs of a finite number (N, relatively small) of preceding states:

$$x_{n+1} = f(H_N(x_n)) = f\left(\sum_{k=0}^{N} p_k^{(n)} x_{n-k}\right)$$

where the weights $p_k^{(n)}$ vary periodically with period equal to the order j of subharmonics to be extracted. The extraction of a generic element of $Per(f)$ does not depends critically on the values of $p_k^{(n)}$ but on the dimension N of the memory.

With respect to convergent architectures making an **extensive** use of the phase space we can make a **selective** use of it. I.e., with respect to a convergent net that can display only a competitive behavior in STM we have a hierarchical structure of **code-defining STM**: 1) *Selective level*: among the cycle orders. 2) *Competitive level*: among different cycles of the same order. 3) *Superselective level*: among different functional classes correspondent to different «thesauri» (i.e., maps).

To sum up, to the ordering: *COMPETITION* $\rightarrow$ *SELECTION* $\rightarrow$ *SUPERSELECTION* corresponds roughly the ordering: *ORBIT* $\rightarrow$ *FUNCTION* $\rightarrow$ *FUNCTIONAL CLASS*.

Using CRF we have developed a **first modeling of the synchronization behavior** of the sensory cortex intended as a **net of coincidence detectors** by a net of filtered chaotic oscillators (=maps). Let us consider a generic filtered map $x_{n+1} = f(s, H_N(x_n))$ depending on an input variable s. By the coupling between filtered maps such that the j-th map processes and filters the i-th map dynamic evolution, we have the following equations:

$$x_{n+1}^{ij} = f(s_i, H_N(x_n^j))\,; \quad i,j = 1,2$$

The mismatch $\Delta_{ij} = |x^{ij} - x^{ji}|$ between the limit values of the cross-related maps ($x^{ij}, x^{ji}; i \neq j$) is greater than a threshold ε if their respective *inputs* differ more than a second threshold related to the topology of the maps. So, equal inputs on *spatially* separated oscillators (maps) are *temporally* connected. To display these **locking properties** we have plotted in Fig.1 the temporal behavior of the filtered map $y_{n+1} = f(g(\Delta_{ij}), H_N(y_n))$ where $f(g, x) = 4gx(1-x)$, $g \,\varepsilon\, [0, 1]$ (sigmoidal) with a variable input $s = s(t)$ on the oscillators for different values of the dimension N of the CRF memory. The locking properties displayed by such a net seems to demonstrate that it is powerful in dealing with non-steady inputs.

4.2 CRF and a 2D asymmetrical spin-glass model of dynamic network

We consider a modified Hopfield 2D spin glass of which state dynamics is given by $S_{ij}^{(n+1)} = T_{ij}[H^{Nmem}(S_{ij}^{(n)})]$ where T_{ij} is the evolution operator and H^{Nmem} is the **self-correlation degree** relative to the spin state $S_{ij}^{(n)}$ according to the following formula:

$$H^{Nmem}(S_{ij}^{(n)}) = \sum_{k=0}^{Nmem-1} S_{ij}^{(n-k)}$$

H is thus the filter operating the **time-correlation** with dimension $Nmem$. The dynamics of the weights $J_{ij\,ml}^{(n)}$ connecting the spins S_{ij} and S_{ml} at the time (n) follows an asymmetrical Hebbian evolution owing to the presence of the term $H^{Nmem}(S_{ij}^{(n)})$. This model has been studied in [18]. It shows a stabilization from a chaotic to a regular (cyclic) behavior without a *bifurcative* mechanism according to our *informational* use of chaos.

REFERENCES

[1] M.W. Hirsch, "Convergent Activation Dynamics in Continuous Time Networks", *Neural Networks*, vol. 2, pp.331-349,1989.

[2] S.Grossberg, "Nonlinear neural networks: principles, mechanisms and architectures", *Neural Networks*, vol. 1, pp.17-61, 1988.

[3] T. Kohonen, *Self-Organization and Associative Memory. Second Edition.*Berlin-Heidelberg-New York, 1988.

[4] W.J. Freeman, "Simulation of Chaotic EEG Patterns with a Dynamic Model of the Olfactory System",*Biol. Cybern.*, vol 56, pp. 139-150, 1987.

[5] Ch. Skarda & W.J. Freeman, "How Brains Make Chaos in Order to Make Sense of the World", *Behavioral and Brain Sciences*, vol. 10, pp. 161-195, 1987.

[6] W.Singer, "Self-organization in cognitive systems". In Eccles J.C. & Creutzfeldt O. (Eds.). *The Principles of Design and Operation of the Brain. Proceedings of the Study Week Organized by the Pontifical Academy of Sciences, Vatican City, October 19-24, 1988*, Vatican City and Berlin-Heidelberg, New York, 1989. In Press.

[7] C.M.Gray, P.Koenig, A.K.Engel & W.Singer, "Oscillatory responses in cat visual cortex exhibit intercolumnar synchronization which reflects global stimulus propereties",*Nature*, Vol. 338-6213, pp.334-337, 1989.

[8] C.M.Gray & W.Singer, "Stimulus-specific neuronal oscillations in orientation columns of cat visual cortex". In *Proceedings of the National Accademy of Sciences USA*, vol. 86, pp. 1698-1702, 1989.

[9] M.Livingstone & D.Hubel, "Segregation of form, color, movement and depth: anatomy, physiology and perception", *Science*, Vol. 240, pp. 740-749, 1988.

[10] W.Singer, "The role of acetylcholine in use-dependent plasticity of the visual cortex". In M.Steriade & D.Biesold (Eds.), *Brain Cholinergic Systems*, Oxford-New York, 1989. In Press.

[11] C.Von der Malsburg & E.Bienenstock, "Statistical coding and short-term synaptic plasticity: a scheme for knowledge representation in the brain". In E.Bienenstock (Ed.), *Disordered System and Biological Organization*, NATO ASI Series, Vol. F20, Berlin-Heidelberg-New York, 1986, pp.247-271.

[12] G.A. Carpenter & S. Grossberg, "A Massively Parallel Architecture for a Self-Organizing Neural Pattern Recognition Machine", *Computer Vision, Graphics and Image Processing*, vol. 37, pp. 54-115, 1987.

[13] G. Basti & A.Perrone, "Time-dependent short-term memories in neural networks". In *Proceedings of Second Italian Workshop on Parallel Architectures and Neural Networks*, Vietri sul Mare, April 26-28, 1989, London, 1989. In Press

[14] I. Procaccia, "The Organization of Chaos by Periodic Orbits: Topological Universality of Complex Systems". In R. Jullien et Al. (Eds.), *Universalities in Condensed Matter*, Berlin-Heidelberg, 1988, pp. 213-215.

[15] Ikegami T. & Tsuda I., submitted to *Complex Systems*; Tsuda I., in: *Proceedings of the International Workshop «Neurocomputers and Attention», USSR Academy of Sciences*, Moscow, 1989. In Press

[16] G. Basti, V. Cimagalli, M. Giona, E. Pasero & A. Perrone, in: *Proceedings of the International Workshop «Neurocomputers and Attention»*, USSR Academy of Sciences, Moscow, 1989. In Press

[17] M. Giona, "Recursive Filtering of Chaotic Maps", submitted to *Signal Processing*.

[18] V. Cimagalli, M. Giona, G. Basti, A. Perrone & E. Pasero, " An Asymmetric Spin-Glass Model of Long Term Memory in a Dynamic Network Architecture". In: *IJCNN-90-WASH-DC: International Joint Conference on Neural Networks*, Washington D.C., January 15-19, 1990.

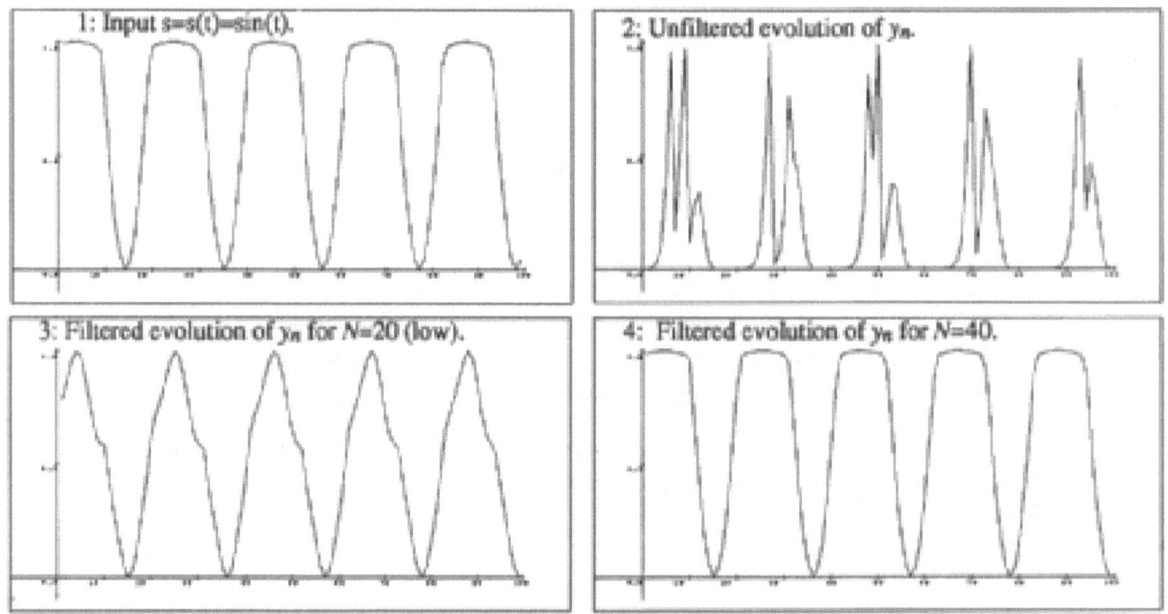

Figure 1.

ON DYNAMICS OF HIGHER ORDER NEURAL NETWORKS: EXISTENCES OF OSCILLATIONS AND CHAOS

Lipo Wang
Department of Electrical Engineering, University College, University of New South Wales
Australian Defence Force Academy, Campbell ACT 2600 Australia

John Ross
Department of Chemistry, Stanford University, Stanford, CA 94305 USA

Abstract We present a discussion on the dynamics of a higher order network of McColluch-Pitts neurons connected via Hebbian-type rules. Both first and second order synaptic connections are randomly cut off to model observed incomplete connectivity in real neurophysiological systems and thereby we obtain an exact solution of the network dynamics. We find a variety of dynamical behavior such as stable retrievals, oscillations and chaos in the neural network. We show that the rescaled noise level which represents the combined effects of the random synaptic dilution, intereference between stored patterns, and additional background noise, acts as the bifurcation parameter in the present system.

Introduction

Oscillatory activities are abundant in chemical (1), physical (2), as well as living system (3-8). Locomotion, respiration, heart beat are just a few examples of rhythmic excitations of the corresponding neural systems. Oscillations and chaos in neural networks have received some attention (3,4,6,8). The are two ways that oscillations can occur in a neural system: (i) one or more single neurons in a network are capable of generating oscillations (see,e.g.,ref.6), and (ii) the synaptic connectivity between neurons produces oscillations (see,e.g.,refs. 3,4,8).

We study oscillations and chaos caused by Hebb-type connectivity by proposing an exactly solvable model of neural network that consists of simple McColluch-Pitts neurons (9) and yet exhibits a rich spectrum of stable retrieving, oscillatory, and chaotic behavior.

Formulation and Results

The system considered here consists of N McColluch-Pitts neurons (9) that have two states, i.e., $S_i = \pm 1$ and are connected by both the first order and the second order Hebbian rule. The total input for the i-th neuron is (10,11)

$$h_i(t) = \gamma_1 \sum_{j=1}^{N} T_{ij} S_j(t) + \gamma_2 \sum_{j,k=1}^{N} T_{ijk} S_j(t) S_k(t) + \eta_i , \qquad (1)$$

where $S_j(t)$ represents the state of the j-neuron at time t, $\vec{S}^\mu$ is the $\mu - th$ stored pattern, and p is the number of patterns stored. γ_1 and γ_2 measure the relative strengths of the first order and second order interactions and

$$T_{ij} = C_{ij} \sum_{\mu=1}^{p} S_i^\mu S_j^\mu , \quad T_{ijk} = C_{ijk} \sum_{\mu=1}^{p} S_i^\mu S_j^\mu S_k^\mu \qquad (2)$$

are the synaptic efficacies according to the Hebbian rule (10-13). Random asymmetric dilution is introduced in the efficacies T_{ij} and T_{ijk} through random variables C_{ij} and C_{ijk} with the following distributions:

$$\rho(C_{ij}) = (\frac{C}{N})\delta(C_{ij} - \frac{1}{N}) + (1 - \frac{C}{N})\delta(C_{ij}) \qquad (3)$$

and

$$\rho(C_{ijk}) = (\frac{C}{N}) \quad \delta(C_{ijk} - \frac{1}{N^2}) + [1 - (\frac{C}{N}) \quad]\delta(C_{ijk}) . \qquad (4)$$

Dilution is essential in assuring an *exact* solution. We also include in eq.(1) a random Gaussian noise η_i with a standard deviation σ_o in order to take into account the presence of noise.

We consider parallel dynamics where all neurons are updated simultaneously. Suppose that the initial state of the network is set in the neighborhood of pattern $\vec{S}^1$. The overlap between the state of the system at time t and pattern $\vec{S}^1$, $m(t) = (1/N)\vec{S}^1 \cdot \vec{S}(t)$ obeys the following dynamical equation (details in derivation are omitted here):

$$< m(t+1) > = 1 - 2\,\psi\{\gamma_1 m(t) + \gamma_2 [m(t)]^2\} \equiv F\,[m(t), \sigma]\,. \tag{5}$$

with $\psi(y) = \frac{1}{\sqrt{2\pi}} \int_{y/\sigma}^{+\infty} e^{-x^2/2} dx$, and a rescaled noise deviation:

$$\sigma \equiv \sqrt{(\gamma_1^2 + \gamma_2^2/N\,)(\alpha/b) + (\sigma_o/b)^2}. \tag{6}$$

In eq.(6), $\alpha \equiv p/N$, and $b \equiv C/N$.

Eqs.(5) and (6) are generalizations of the formulation of Derrida, Gardner, and Zippelius (14), who derived an exact solution of a diluted neural network with first order interaction only. Also, they used Little's definition of temperature (15), instead of the Gaussian noise used in the present work.

Eq.(5) has been first derived by Keeler, Pichler, and Ross (11) for a fully connected network via the approximation that the interference is a random Gaussian noise in the absence of dilution. They have found that compared to the case with only first order interaction ($\gamma_1 = 1$, $\gamma_2 = 0$), the final retrieval ability is enhanced by letting $\gamma_1 = 1$, $\gamma_2 > 0$. We show that the formulation of Keeler, Pichler, and Ross (11) becomes exact in the diluted limit.

In the present paper, we discuss the case where $\gamma_1 > 0$, $\gamma_2 < 0$: the system exhibits a variety of dynamical phenomena, such as oscillations, period doubling bifurcations, chaos, as well as stable retrievals. The fixed points of the system can be obtained by iterating eq.(5) for many time steps. A special case where $\gamma_1 = 1$, $\gamma_2 = -1$ is calculated explicitly and the solutions of $< m(\infty) >$ as a function of the noise level σ are presented in Figs.1 and 2.

When $\sigma > \sqrt{2/\pi} = 0.799$, the only non-negative fixed points are zeros. For $\sqrt{2/\pi} = 0.799 > \sigma > 0.193$, the system converges to stable positive fixed points with any positive initial overlap. As the noise level decreases below $\sigma_1 = 0.193$, oscillations starts to appear, as shown in Fig.1. There is a complete period-doubling sequence between $\sigma_1 = 0.193$ and $\sigma_\infty = 0.1234$, which is the saturation point of this period-doubling sequence. Space-filling chaotic structures and periodic windows can be seen beyond this saturation point. A period-3 window is presented in Fig.2. "Period-3 implies chaos"(16). Around the periodic window there are the typical chaotic "explosions": space-filling bands appear abruptly at certain point of bifurcation.

Acknowledgement

This work was supported in part by the National Science Foundation.

Reference

1. Field, R.J. and Berger, M. (1984) *Oscillations and Traveling Waves in Chemical Systems* (John Wiley and Sons, New York).
2. Behringer, R.P. (1985) *Rev. Mod. Phys.* **57**, 657 - 687.
3. Grossberg, S. (1970) *Stud. Appl. Math.* **44**, 135.
4. Amari, S.-I. (1971) *Proc. IEEE* **59**, 35 - 46.
5. Ross, J. and Schell, M. (1987) *Ann. Rev. Biophys. Chem.* **16**, 401 - 422.
6. Guevara, M.R., Glass, L., Mackey, M.C., and Shrier, A. (1983) *IEEE Trans.* **SMC-13**, 790 - 798.
7. Skarda, C.A. and Freeman, W.J. (1987) *Behavioral Brain Sci.* **10**, 161 - 195.
8. Sompolinsky, H., Crisanti, A., and Sommers, H.J. (1988) *Phys. Rev. Lett.* **61**, 259 - 262.
9. McCulloch, W.S. and Pitts, W.(1943) *Bull. Math. Biophys* **5**, 115 - 133.
10. Peretto, P. and Niez, J.J. (1986) *Biol. Cybern.* **54**, 53 - 63.
11. Keeler, J.D., Pichler, E.E., and Ross, J. (1989) *Proc. Natl. Acad. Sci. USA* **86**, 1712 - 1716.
12. Hebb, D.O. (1949) *The Organization of Behavior* (John Wiley, New York) p.44
13. Shaw, G.L., Harth, E., and Scheibel, A.B. (1982) *Exp. Neurol.* **77**, 324 - 358.
14. Derrida, B., Gardner, E., and Zippelius (1987) *Europhys. Lett.* **4**, 167 - 173.
15. Little, W.A. (1974) *Math. Biosci.* **19**, 101 -120.
16. Li, T.Y. and Yorke, J.A. (1975) *Am. Math. Monthly* **82**, 985 - 992.

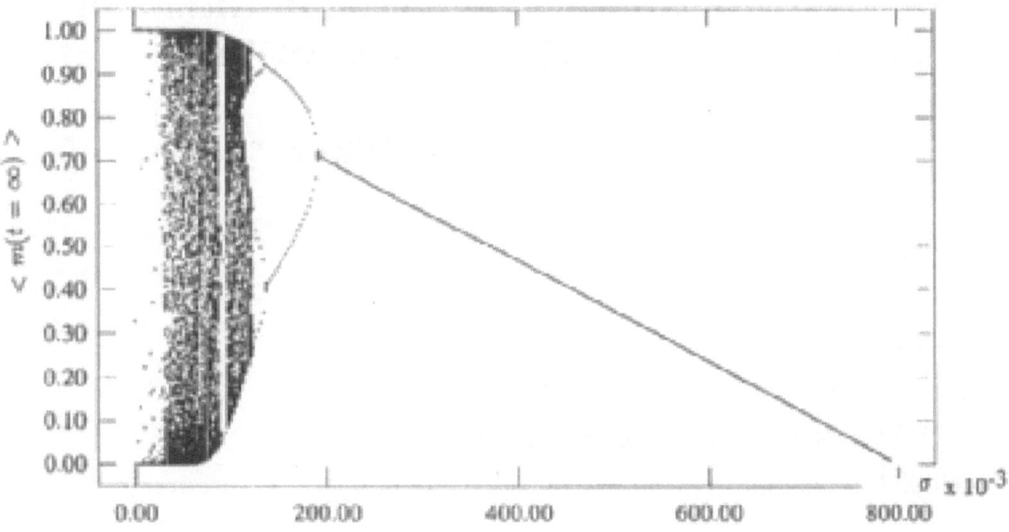

1. Plot of fixed points $< m(t = \infty) >$ given by eq.(5) vs. standard deviation of the Gaussian noise σ given by eq.(6).

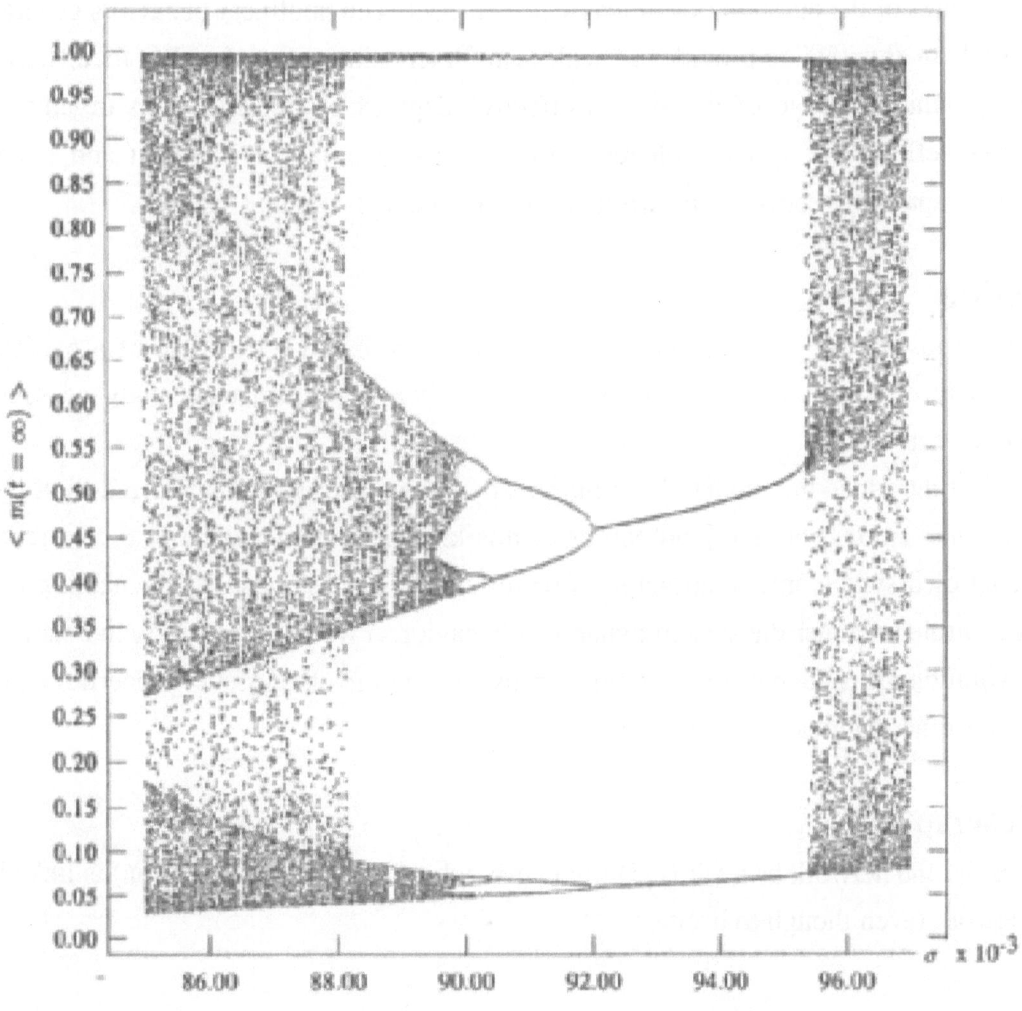

2. A period-3 window in the oscillatory region shown in Fig.1.

DYNAMICAL CHANGE OF EFFECTIVE DEGREES OF FREEDOM IN FRACTAL CHAOS MODEL

J.M. BERTILLE (*,) and J.C. PEREZ (*)**

(*) IBM Montpellier, Advanced Technics Group
B.P. 1021, 34006 Montpellier Cedex, France
Phone : 67-34-69-19
(**) Montpellier Sciences University (USTL)
CRIM Laboratory, Place Eugène Bataillon
34000 Montpellier

ABSTRACT

We are concerned here with the theoritical study of a globally coupled network using the features of the autonomous Fractal Chaos model. The nonlinear behaviors generated by such an architecture provides us with many interesting properties. We focus here on the dynamical change of the system effective degrees of freedom. This organisation change reflects the system coherence towards its external environment and history. Such a capability seems crucial while trying to build self-learning systems.

Introduction

In previous papers, we presented experimental work based on the use of a particular neural model called Fractal Chaos [1-6]. We describe here the theoritical study of such a globally coupled network.

We first introduce the model, describing its dynamics and the tools realized to analyse its behaviors. Then among the many properties of this kind of system, (dynamical hierarchical clustering, controled switch among attractors, structural contraint establishments enabling us to modify in a controled manner the system dynamical behavior...) we choose to look in details the dynamical evolution of its degrees of freedom. Discussions on the other model properties should be available in further publications.

Model description

Consider the network as a Dimx×Dimy matrix where each element (neuron) is linked to the whole network (even though to itself).

The system dynamical evolution is described by the parallel iteration of N equations with the following form :

$$x_{n+1}[i,j] = (1-c)f(x_n[i,j]) + c/N \sum_{k=1}^{Dimx} \sum_{l=1}^{Dimy} f(x_n[k,l])$$

where n,n+1 are the indexes relative to a discrete time scale

x[i,j] is the neuron state within row i and column j

c is the coupling strength of the neuron, $c \in [0,1]$

and N represents the system size, that is N = Dimx×Dimy.

The function f stays for the logistic map given by :

$$f(x) = 1 - ax^2$$

where x is the function state, $x \in [-1,1]$

and a represents its nonlinearity parameter, $a \in [0,2]$.

To get more informations on the physical discipline aspects of this kind of coupling, the reader should look at the work done by Kaneko [7]. The study of the logistic map f [8] shows that it exhibits an infinite number of distinct behaviors (attractors) depending from the value of its a parameter. Some of them periodic, while others chaotic. Given the diversity and complexity of a single entity, the reader may easily see that for the moment it is foolish to fully describe in details all of the possible behaviors of any network performing a structural coupling of such units. That is the reason why we restrict ourselves to only study a uniform global coupling (the network parameters c and a are all identical). The system considered here has chaotic behaviors, so it is impossible to exactly predict its long-term evolution (strange attractors sensibility dependance on initial conditions). In such situations, we rather look for determining qualitative criteria (order parameters) which enable us to classify the different phases of system behaviors. To establish such average results we realize the tests on large randomly generated initial conditions sets.

Dynamical change of the system effective degrees of freedom

To bring up this property we introduce a cluster notion as follows : Two neurons x[i,j] and x[k,l] belong to the same cluster C at time n if we have :

$$x_n[i,j] = x_n[k,l]$$

Although arbitrary, this notion enables us to reduce the amount of information provided by the spatio-temporal evolution of the network state ($x_n[i,j]$). That is, while considering clusters, we get rid of the temporal character of this evolution. In fact, this kind of reduction is only valid when we consider a stable system. Then we can assert the following property :

> If two neurons belong to the same cluster at time t_1, then they will still belong to the same cluster at time t_2, with $t_2>t_1$, in the absence of any external disturbances.

This property is self-explainable if we consider the complete connexion structure of the network and the exclusive use of deterministic functions within it. We will insist on the network stability necessity to observe this property. In effect, if a cluster is unstable, then it can explode under the communication process stress of the network and due to the fact of the chaotic behaviors of its parts.

When the whole network is stabilized, it may be reduced to its constituent clusters and so whatever their own dynamics. The whole system state ($x_n[i,j]$) can be described using a code of the form :

$$[k;(N_1,N_2,...,N_k)] \text{ with } N_1 \geq N_2 \geq ... \geq N_k$$

where k is the number of clusters in the network
and N_i represents the number of elements in cluster C_i.

Using this kind of coding, we get rid of the spatial character of the information. That is the reason why this coding is not bijective, because many network configurations share the same code.

Once the system fallen in a stable k-clusters attractor, it will stay within it indefinitly in absence of external disturbances. That means that the system effective degrees of freedom are not any longer N but k. Then the N-equations initial system can now be described using only k equations of the form:

$$X^u_{n+1} = (1 - \sum_{i=1,i\neq u}^{k} ce_i) f(X^u_n) + \sum_{i=1,i\neq u}^{k} ce_i f(X^i_n)$$

where $X^u_n = x_n[i,j]$ with $x_n[i,j] \in C_u$ is the state of the uth cluster
and $ce_i = c \times N_i/N$ represents the effective coupling of the cluster C_i.

The effective degrees of freedom of such a globally coupled network evolve dynamically until the system reaches a perfect stabilization configuration. The complete study of this phenomenon enables us to give the following remarks :

> The final network organisation (stable code) is influenced by the choice of the c and a parameters, but is independant from the initial network state.

> The codes temporal sequence driving the system to this final organisation reacts to the initial conditions.

If we submit the system to disturbances, this impacts its codes temporal sequence and quite often its final code.

The submission order and the kind of disturbances applied also affects the network behavior.

Conclusion

We briefly described here a system which organisation corresponds to a kind of coherence in regard to its environment and history. We realized some tools able to capt this organisation evolution, that is to bring up this coherence. We think that such a study may be usefull when we investigate self-learning and autonomy properties, when knowledge spontaneously spreads out of the network structural organisation without any previous use of acquisition processes.

References

[1] Bertille, Perez : "Fractal Chaos : a new neural holographic model", INNS 88 Boston Conference, Neural Networks Pergammon Journals

[2] Bertille, Perez : "The neural holographic fractal chaos model : Theoric foundations and industrial applications", Neuro-Nimes Conference 88, EC2 Paris (french)

[3] Bertille, Perez : "A new neural net family : the holographic fractal chaos model and its applications", Convention IA Conference 89, Hermès Paris (french)

[4] Bertille, Perez : "A spatio-temporal novelty detector using fractal chaos model", IJCNN 90 San Diego Conference

[5] Perez : "New ways towards Artificial Intelligence : Pluri-disciplinarity, self-organisation, and neural networks", Masson 88, Paris (french)

[6] Perez : "The neural networks revolution", Hermès 90, Paris (french)

[7] Kaneko : " Clustering, Coding, Swithching, Hierarchical Ordering and Control in Network of Chaotic Elements", to be published

[8] May : Nature London 261, 459 (1976)

TEMPORAL PATTERNS AND LEAKY INTEGRATOR NEURONS

J.G. Taylor
Department of Mathematics, King's College,
Strand, London, U.K.

ABSTRACT

The leaky integrator model of neurons leads to time summation of inputs, with time-delaying weights which in general may be independent of each other. A Hebbian-type learning rule is given for these weights, which allow for the storage of temporal patterns of activity in a net. Storage of pattern sequences is analysed analytically under simplifying assumptions. The relation of the model to real neurons and hardware realisations is briefly discussed

The storage of sequences of patterns by a neural net is of great importance to understand and to implement artificially. Episodic memory in humans is known to be very effective in such storage, but the algorithms by which such storage is achieved are not yet known. Attempts have been made to construct nets of artificial neurons which can achieve storage of sequences of patterns. In the spin glass approach a form of the synaptic weight matrix has been suggested [Sompolinsky and Kanter 1986] which does produce a pattern sequence if initiated in one of the states of the sequence. The somewhat artificial form of this matrix was avoided by Coolen and Gielen [1988] where axonal time delays were introduced to achieve sequence storage. The approach used here to address the problem is to look more directly at the temporal activity of the living neuron.

All detailed temporal modelling of real neurons uses the leaky integrator model ([Hodgkin and Huxley, 1952]). Account needs also to be taken of cell geometry, various ionic channels, possibly active patches of dendritic membrane and the quantal (stochastic) nature of information transmission at synapses. The latter has already been discussed elsewhere at this conference ([Bressloff and Taylor 1990a]) and elsewhere ([Bressloff and Taylor 1990b]). Cell geometry may be handled by the compartmental model ([Rall 1964]), while ionic currents and active membrane introduce terms depending non-linearly on the membrane potential. We will simplify by disregarding the details of these latter-features, but attempts to lump them together in a suitable manner.

The leaky integrator model equations, neglecting compartments, etc, have been shown ([Bressloff and Taylor

1990a]) to reduce, in discrete time, to time-summating Caianello equations in the activities $a_i(t)$ of the ith neuron at time t ($a_i(t)$ is one if the neuron is active, zero if not)

$$a_i(t)= Y \left(\sum_{j=1}^{N} \sum_{r=1}^{t+1} W_{ijr} a_j(t-r) - S_i\right) \tag{1}$$

where Y(x) is the unit step function (Y(x)=1 if x>o, Y(x)=o if x<o), S_i is a threshold for the ith neuron and the weights W_{ijr} may in principle, for different r, be independent of each other. In the case that synaptic noise is present the W_{ijr} involve different weights of the vesicle distribution function.

An important question to be resolved for (1) is as to the learning rules for W_{ijr}. At any given times, these latter coefficients measure the correlation between activity on the ith and j^{th} neurons, separated by a time difference r. A Hebbian learning rule may thus be naturally formulated as

$$\Delta W_{ijr} \propto a_i(s) a_j(s-r) e^{-r/d} \tag{2}$$

The exponential on the r.h.s. of (2) is present in order to take account of the temporal decay of activity impinging on the ith neuron over the time period r; d is the membrane time constant (in dimensionless units). For cortical neurons, using the natural time unit as 1 msec, d may be of the order of 60-100 ([K. Stratford et al, 1989]). A non-Hebbian conjunctive learning rule may also be constructed, with

$$\Delta W_{ijr} \propto \sum_{k} a_k(s-r) a_j(s) e^{-r/d} \tag{3}$$

where k is a set of neurons with axons synapsing onto regions near the jth synapse. Only (2) will be considered henceforth, whose result is that we may take

$$W_{ijr} = (1/NT) \int_0^T a_i(s) a_j(s-r) e^{-r/d} \tag{4}$$

where T is the learning period and N the number of neurons. The value of (4) may be considered as a combination of effects on the modifiable pre-and post-synaptic structures, such as the speed of opening of ionic gates, second messenger releases and spine apparatus.

The arguments of Coolen and Gielen [1988] may be followed to show that (1) and (4) give effective storage of pattern sequences. This will be shown using neuronal spin instead of activity, and the threshold set to zero. Thus (1) becomes

$$S_i(t) = \mathrm{sgn}\,\Big(\sum_{j=1}^{N}\sum_{r=1}^{t+1} W_{ijr} S_j(t-r)\Big) \tag{5}$$

with

$$W_{ijr} = \sum_m 1/NT \int_0^T S_i^{(m)}(s) S_j^{(m)}(s-r) e^{-r/d}\, ds \tag{6}$$

It is assumed that there is a learning stage, when a set of patterns $\{S_i^{(m)}(s)\}$ are learnt. The index s denotes that state in the pattern m, the index in differentiating patterns. Independence of patterns leads us to drop the latter index, although the usual bounds on the number of such patterns will be required to avoid interference. Then it is straightforward to show that after learning has occurred the dynamics (5) is satisfied by the sequence of spins $S_i^{(1)}(S)$ at an earlier time, since

$$(1/NT \sum_{ijr} \int_0^T S_i^{(1)}(s) S_j^{(i)}(s-r) e^{-r/d} S_j^{(1)}(t-t_o-r)$$

$$= (1/T)\Big(\sum_r e^{-r/d}\Big).\int_0^T S_i^{(1)}(s)\,\delta^*(s-t+t_o)\,ds \tag{7}$$

where $\delta^*(x) = 1$ if $x=o$, $\delta^*(x)=o$ otherwise, and the lack of correlation implies that

$$(1/N) \sum_j S_j^{(1)}(s)\, S_j^{(1)}(s^1) = \delta^*(s-s^1) \tag{8}$$

is used. Then the time development (5) is satsfied at a later time t_o by the pattern $\{S_i^{(1)}(s)\}$; pattern storage is achieved.

There are numerous questions which this approach produces:

(a) how effective is recall under simulation?
(b) how more subtle are real neurons when plastic channels and spine structure are properly taken into account?
(c) what is the effect on pattern storage of cell geometry?
(d) is the conjunctive learning rule (3) as effective as (2) (if not more so)?
(e) can this approach be implemented in hardware, as have other non-temporal storage properties of neurons ([Gorse and Taylor 1990])?

REFERENCES

Bressloff P.C.and Taylor J.G. 1990a (paper at this conference)

Bressloff P.C. and Taylor J.G. 1990b Neurons with Synaptic Noise, GEC preprint

Coolen A.C.C. and Gielen C.C.A.M. 1988 Europhys. Letts 7 281

Hodgkin A.L. and Huxley A.F. 1952 J.Physiol (Lond) 116 500

Gorse D.G. and Taylor J.G. 1990 (paper at this conferenc)

Rall W. 1964, Neural Theory and Modelling, ed R. Reiss, Stanford Univ. Press

Sompolinsky H. and Kanter I. 1986 Phys.Rev.Lett. A 57 28 2861

Stratford K, Larkman A, Mason A. Major G, and Jack J, J.Physiol. 1989 (in press)

AN ANALYSIS OF THE DELTA RULE

S.W.Ellacott
Information Technology Research Institute
Brighton Polytechnic
Lewes Road
Brighton
BN2 4AT
United Kingdom

Abstract

The delta rule and its generalisations is one of the most practically useful learning algorithms for connectionist systems. (See [5] and, e.g. [1]). It is often stated that the method is gradient descent for the least squares error. However, neither part of this statement is true for finitely large learning rates. The purpose of this paper is to employ certain mathematical tools, largely borrowed from numerical analysis, to provide a rigourous framework for discussing the behaviour of learning algorithms. To this end, a more or less complete analysis of the original linear version of the delta rule is presented. It is shown that when the rule is applied by repetitively cycling through a fixed epoch of patterns, updating the weights after each pattern, the algorithm generates a limit cycle. The least squares error of this limit cycle appoaches that of the true minimum quadratically as the learning rate tends to zero. The algorithm is convergent and numerically stable subject only to a simple normalisation condition. By contrast, if the weights are updated after the complete epoch of patterns has been presented, the iteration has a fixed point which is the true least squares minimum. However the algorithm may have very bad numerical stability and convergence properties, even for problems which are "good" from the point of view of learning. This simple linear case is of limited practical use, but heuristic and numerical evidence suggests that the analysis does give insight into the behaviour of the method for more useful cases such as back propagation networks. Current work is directed to a rigorous justification of this. It is the author's belief that the methods can be extended to other learning paradigms.

1. Basic Iteration Properties of the Delta Rule

Consider a linear single layer network. Denote the training vectors (generically) by $\underline{x}$ and desired output vectors by $\underline{y}$. Assume initially that the weights are updated after each training pattern. In the linear case there is in fact no coupling by the iteration between the rows of the weight matrix. Thus _without loss of generality_ the analysis may be restricted to weight matrices with only one row, which we denote generically by a (column) vector $\underline{w}$. (See [2] p2 for a detailed discussion of this. This decoupling property is not, of course, shared by nonlinear networks such as multilayer perceptrons with back propagation.) The basic iteration to be analysed ([5] p.322) becomes in matrix form

$$\underline{w}_{k+1} = (I - \eta \underline{x}\underline{x}^T)\underline{w}_k + \eta \underline{y}\underline{x}^T,$$

but, of course, the pattern used will be different for eack k.

Let $$B = (I - \eta \underline{x}\underline{x}^T).$$

Lemma 1.1

B has only two distinct eigenvalues: $(1 - \eta \|\underline{x}\|_2^2)$ corresponding to the eigenvector $\underline{x}$ and 1 corresponding to the subspace of vectors orthogonal to $\underline{x}$.

Proof

A direct verification ■

Lemma 1.2

Provided $0 \leq \eta \leq 2/\|\underline{x}\|_2^2$, we have $\|B\|_2 = \rho(B) = 1$, where ρ denotes the spectral radius and the norm is the matrix norm corresponding to the 2 vector norm. (See [3], ch.1. for information on matrix norms)

Proof

The 2 norm and spectral radius of a symmetric matrix are always the same [3]. That the spectral radius is 1 follows from Lemma 1.1. ■

Now consider t training patterns $\underline{x}_p$, $p = 1, \ldots t$ with corresponding matrices B_p. Consider also the cyclic iteration

$$(1.1) \qquad \underline{w}_{k+t} = \Omega \underline{w}_k + \eta \underline{h} \qquad \text{where} \quad \Omega = B_t B_{t-1} \ldots B_1$$

and $$\underline{h} = y_1(B_t B_{t-1} \ldots B_2)\underline{x}_1 + \ldots y_{t-1} B_t \underline{x}_{t-1} + y_t \underline{x}_t .$$

Lemma 1.3

If the condition of Lemma 1.2 holds for each training pattern, and if the $\underline{x}_p$ span the space of training pattern vectors, $\|\Omega\|_2 < 1$.

Proof

By definition, $\exists\ \underline{v}$ such that $\|\Omega\|_2 = \|\Omega \underline{v}\|_2$ and $\|\underline{v}\|_2 = 1$. Thus $\|\Omega\|_2 = \|B_t B_{t-1} \ldots B_1 \underline{v}\|_2 \leq \|B_t B_{t-1} \ldots B_2\|_2 \|B_1 \underline{v}\|_2$. Now $\|B_t B_{t-1} \ldots B_2\|_2 \leq \|B_t\|_2 \|B_{t-1}\|_2 \ldots \|B_2\|_2 = 1$. If $\underline{v}^T \underline{x}_1 \neq 0$, $\|B_1 \underline{v}\|_2 < 1$, since the component of $\underline{v}$ in the direction of $\underline{x}$ is reduced (compare the proof of Lemma 1.1)

Conversely if $\underline{v}^T \underline{x}_1 = 0$, then $B_1 \underline{v} = \underline{v}$ (Lemma 1.1). Hence $\|\Omega\|_2 =$

$\|B_t B_{t-1} \ldots B_2 \underline{v}\|_2$ and B's may be removed until the previous case applies. ■

Corollary 1.4

Under the conditions of Lemma 1.2, the iteration (1.1) has a fixed point.

Proof

The mapping $F(\underline{w}) = \Omega\underline{w}+\eta\underline{h}$ satisfies $\|F(\underline{w})-F(\underline{v})\|_2 \leq \|\Omega\|_2\|\underline{w}-\underline{v}\|_2$. i.e. it is contractive with contraction paramenter $\|\Omega\|_2$. The result follows from the contraction mapping theorem ([4] p.267) ■

It is easy to extend this result to the case when the $\underline{x}_p$ do not span: [2] p10.

The fixed point of (1.1) is not in general a fixed point for each individual pattern: however, since it is returned to after each epoch of patterns, it may be seen that in the limit the algorithm will simply cycle through a fixed set of values. Neither is the fixed point a minimum of the least squares error over the patterns. Denoting the matrix whose columns are the patterns by X, it will be found that

$$(XX^T+H(\eta))\underline{w}(\eta) = \underline{h}(\eta),$$

where $\underline{w}(\eta)$ is the fixed point of (1.1) for a given η, $\underline{h}(\eta)$ is a vector such that as $\eta \to 0$, $\underline{h}(\eta) = y_1\underline{x}_1 + \ldots + y_t\underline{x}_t + O(\eta)$, and $H(\eta)$ is a matrix which is $O(\eta)$. It follows that as $\eta \to 0$, the limit cycle approaches the true minimum at the same rate. However, since at the true minimum of a least squares problem the error is orthogonal to the approximating space, it can be shown that the error of the limit cycle approaches that of the true minimum at a rate proportional to the square of η. Full details are given in [2], p15.

2. Choice of η and Numerical Stability

In view of the previous remarks, it is not necessary to choose a very small value of η so that the algorithm behaves like gradient descent. A value which minimises the contraction parameter may well also give good separation of the patterns: a very limited amount of numerical experimentation has been carried out which supports this contention. The fact that the algorithm gnerates a limit cycle rather than a single limit does not really matter, although it is well to be aware of it. Note that to get convergence, we only need the conditions of Lemma 1.2 to hold for each pattern: this is a simple normalisation condition. Moreover, Lemma 1.1 means that the algorithm is numerically stable: the importance of this may not be appreciated by non numerical analysts, but in fact this is an essential property of any usable algorithm.

It has been suggested that the weights be updated only after a complete epoch of patterns [5]. However, this method can be numerically unstable, except for very small η. The equivalent to (1.1) in this case is

$$\underline{w}_{k+t} = \Lambda\underline{w}_k + \eta\underline{h} \text{ where } \Lambda = (I - \eta XX^T) . \tag{2.1}$$

Let $\lambda_1 \ldots \lambda_n$ be the eigenvalues of XX^T, with $0 < \lambda_1 < \ldots < \lambda_n$. Consider for example the case when the patterns cluster around two mutually orthogonal directions: an ideal example from the point of view of machine learning. Clearly X is close to a matrix of rank 2, so λ_1 may be arbitrarily small. On the other hand, it is possible [2] to show that for any k, k=1, ... p, $\lambda_n \geq \|x_k\|_2$. Since the eigenvalues of Λ are $1-\lambda_j$, j=1, ... n, an arbitrarily small η may be required to make $\rho(\Omega) < 1$, for convergence and stability.

3. Non Linear Networks

The real purpose of this analysis is to understand the delta rule for more useful networks such as semilinear feedforward nets with back propagation. Assuming differentiable activation functions, it is plausible to suppose that the system will be linearisable near a minimum of the error, and that the asymptotic behaviour will therefore be similar to the linear case. M. Evans (private communication) has indeed observed numerical instability in this case for the analogue of (2.1) applied to a vision problem. However, further work is required for a rigorous discussion of the nonlinear case.

References

[1] M.R.Devos and G.A.Orban: "Self-adapting back propagation", in the proceedings of Neuro-Nimes '88, Nimes, France, November 1988, from EC2, 269-287 rue de la Garenne, 92000, Nanterre, France.

[2] S.Ellacott: "Some working papers on the delta rule", ITRI technical report no.79, 1989. (Address as above.)

[3] E.Isaacson and H.B.Keller: "Analysis of numerical methods", Wiley, 1966.

[4] R.D.Milne: "Applied functional analysis: an introductory treatment", Pitman, 1980. (ISBN 0-273-08404-6)

[5] D.E.Rumelhart and J.L.McClelland: "Parallel and distributed processing: explorations in the microstructure of cognition", vol.1, MIT, 1986. (ISBN 0-262-181270-7)

Classification capability of the two-layer perceptron

Peter Butovitsch
Dept. of Telecommunication Theory
S-100 44 Stockholm, Sweden
Telephone +46 8 790 84 37

P.O. Lindberg
Dept. of Mathematics
S-100 44 Stockholm, Sweden
Telephone +46 8 790 73 14

Abstract

This paper treats the classification capability of a two-layer perceptron [1]. The discussion is confined to the problem to separate two classes of objects. The aim of the study has been to investigate what shapes the decision regions of a two-layer perceptron can have. Two theorems are established for this purpose. One which can be used to decide whether a given partition of the input space is realizable or not and one that states that a general property of a non-realizable partition of the input space leads to a number of inconsistent inequalities. Three corollaries treating a few special geometrical shapes are also presented.

INTRODUCTION

The prime interest in these studies has been to find out what topological and geometrical shapes that can be realized by a two-layer perceptron. Therefore no attention has been paid to either what kind of problems that gives rise to the specific kinds of decision regions nor how a specific net actually approaches the solution to a problem.

The input space to a two-layer perceptron is partitioned into a number of linearly bounded convex cells [2] in the first layer of the network. The work has basically been based on two problem formulations. The first problem is to determine whether, for a given linear partition of the input space, two given output sets S_1 and S_2 of cells representing the two classes can be realized or not, when using a perceptron with only one layer of hidden nodes. The two sets S_1 and S_2 satisfy the relations $S_1 \cap S_2 = \emptyset$ and $S_1 \cup S_2 =$(all cells of the input space partition). So $S_1 \cup S_2$ is a partition of the set of cells. The second problem is to find a general property of the non-realizable decision regions. This property is not given in topological terms, but as a relation between the weights of each of a few cells, since there does not seem to be a simple relationship between the topological structure and the realizability of the decision regions.

Three corollaries treating a few special classes of decision regions are also presented. The first one states that any linearly bounded convex region can be realized. This is also the only general structure that can be realized. The second corollary states that any linearly bounded hypershell with both convex outer and inner boundaries can be constructed. The third corollary states that an arbitrary number of the hypershells described above, for each two one is outside the other, can be constructed.

ANALYSIS

To be able to make calculations on the problem the Heavyside function, H, is chosen as the nonlinear function in the nodes. Let $\mathbf{x}$ be the input vector of dimension m, the matrix $\mathbf{W}$ of dimension $m \times n$ be the weight from the inputs to the layer of hidden nodes, the vector $\mathbf{v}$ of dimension n be the weight vector from the layer of hidden nodes to the output node (as the problem is formulated only one output node is needed), the vector $\mathbf{\Theta}$ of dimension n be the vector of thresholds in the layer of hidden nodes and θ_o be the threshold in the output node. Then the output will be: $H(\mathbf{v}^T H(\mathbf{Wx} - \mathbf{\Theta}) - \theta_o)$.

Let a partition of the input space and the sets S_1 and S_2 be given. Let the elements in one set e.g. S_2 be supposed to generate 0 from the output node and the elements in the other set S_1 be supposed to generate 1. Then the sum of the weights for each cell in S_2 must be greater than the sum of weights for each cell in S_1.

Problem 1 Decide whether the given sets S_1 and S_2 can be realized or not.

If for a given placement of the weights $\mathbf{v}$ any of the weights v_i are required to be negative due to the inequalities generated between the elements of the two sets S_1 and S_2 these can be moved to the opposite halfspace and thus be positive. If there appear to be contradictory constraints of the type $v_i > 0$ and $v_i < 0$ the partition of the cells into the sets S_1 and S_2 is definitely not realizable otherwise the weights can be assumed to be greater than zero and realizability will be given by theorem 1.

Define a matrix $\mathbf{A}$ of dimension $r \times n$ in which the rows represents the cells and the columns represents the weights. Let the element a_{ij}=1 if weight j is found in cell i and a_{ij}=0 otherwise. Number the regions so that the regions in S_2 have the lowest indices. Partition the matrix as $\mathbf{A}= \begin{bmatrix} \mathbf{A}_1 \\ \mathbf{A}_2 \end{bmatrix}$, where the matrices $\mathbf{A}_1$ and $\mathbf{A}_2$ corresponds to the cells in S_1 and S_2 respectively. Define a vector $\mathbf{t}_k$ as $\mathbf{t}_k = [1,1,..,1]^T$, where $[1,1,..,1]^T$ is a vector of dimension k. Let n_1 be the number of cells in S_1. Then the problem can be reformulated as: Find v and θ_o so that $\begin{cases} \mathbf{A}_1\mathbf{v} - \theta_o\mathbf{t}_{n_1} & > 0 \\ \mathbf{A}_2\mathbf{v} - \theta_o\mathbf{t}_{(n-n_1)} & < 0 \end{cases}$ Now augment the matrix $\mathbf{A}$ with the vector $\mathbf{t}_n$ and the vector $\mathbf{v}$ with θ_o. The problem can then be reformulated as: Find the solution to $\begin{cases} \mathbf{A}'\mathbf{v}' > 0 \\ \mathbf{v}' \geq 0 \end{cases}$,where $\mathbf{A}' = \begin{bmatrix} \mathbf{A}_1 & ,-\mathbf{t}_{n_1} \\ -(\mathbf{A}_2 & ,-\mathbf{t}_{(n-n_1)}) \end{bmatrix}$ and $\mathbf{v}' = \begin{bmatrix} \mathbf{v} \\ \theta_o \end{bmatrix}$.

Theorem 1 Exactly one of the systems I) and II) has a solution.

$$\text{I)}\begin{cases} \mathbf{A}'\mathbf{v}' > 0 \\ \mathbf{v}' \geq 0 \end{cases} \quad \text{II)}\begin{cases} \lambda^T\mathbf{A}' \leq 0 \\ \lambda \geq 0 \end{cases} \quad \lambda \neq 0$$

A given partition of the cells into the sets S_1 and S_2, where the cells in S_1 and S_2 are supposed to generate 1 and 0 respectively from the output node, can be realized if system I) has a solution.□

Proof of theorem 1: This theorem is proved in [3].□

Through ordinary linear programming it is possible to decide if system I) has a solution and if it has a solution this can also explicitly be found.

Problem 2 To find a general property of non-realizable partitions of the cells into the sets S_1 and S_2.

Since the sum of weights in each of the cells in S_1 must be strictly greater than the sum of weights in each of the cells in S_2, inequalities between the sums of weights in every two cells, one from S_1 and one from S_2, respectively can be formulated. These will be of the type $\mathbf{b}^T\mathbf{v} > \mathbf{d}^T\mathbf{v}$. Also consider the inequalities $v_j > 0$. The following theorem can then be formulated.

Theorem 2 For a given partition of the input space a given partition of the cells into the sets S_1 and S_2 can <u>not</u> be realized if and only if it is possible to construct a cyclic "chain" of inequalities, originating from the inequalities above, of the type

$$\mathbf{b}^T\mathbf{v} > \mathbf{d}^T\mathbf{v} > \mathbf{f}^T\mathbf{v} > .. > \mathbf{b}^T\mathbf{v}$$

where

$$b_j, d_j, f_j, .. \in \{0,1\}, \text{ ,j=1,..,n;} \quad v_j > 0, \text{ j=1,..,n.}\square$$

Theorem 2 is proved in [4].

In the two theorems the sum of weights in each of the cells are related to the sum of weights in the other cells. Thus, because each of the weights appear in every cell in a halfspace and not in any cell in the complementary halfspace, the geometrical placement of the cells in relation to each other is important. Some classes of decision regions with a common structure can however be proven to be realizable independantly of how the specific cells are arranged. A few examples of these classes are here presented.

Corollary 1 Every linearly bounded convex region can be realized.□

Proof of corollary 1: Suppose that we have a number, n, of hyperplanes which bound the convex region exactly. Let the the weights corresponding to each hyperplane be in the same halfspace as the convex region. The convex region would then be constituted of only one cell, which also is the only cell containing all weights and thus any threshold:

$$\theta = \sum_{i=1}^{n} v_i - \Delta; \qquad \Delta < \min_{i \in \{1,..,n\}} v_i$$

, would solve the problem.□

Corollary 2 Every linearly bounded convex region with a linearly bounded convex hyperhole can be realized.□

Proof of corollary 2: Let a convex region with the shape of the desired hyperhole be realized. Let $v_i \rightarrow -v_i$ and $\theta \rightarrow -\theta$ to switch the output of the elements of the sets S_1 and S_2. Now form the outer boundary of the region with a number of hyperplanes and put the weights corresponding to these hyperplanes in the halfspace outside the desired region and let its value be negative and sufficiently large to reduce the sum of the weights in all cells outside the hyperplane to a value beneath the threshold. These cells would be added to the cells of the interior of the inner boundary S_1. Since the region inside of the outer boundary is not affected by the new hyperplanes the old values of the weights and the threshold connected to the boundary of the hyperhole, the problem will be solved with these values kept as they are.□

Corollary 3 An arbitrary number of hyperholes with linearly bounded convex outer and inner boundaries, for each two one is outside the other, can be constructed.□

Proof of corollary 3: The corollary is proved by induction. The first step is already done in corollary 2. Suppose that we have constructed N numbers of hypershells. Now the $N+1$:st hypershell is constructed by using corollary 2 two times in succession. The first time treating everything inside the outer boundary of the N:th shell as a hyperhole. This would change the output of the elements in the sets S_1 and S_2. This is put right when using corollary 2 the second time when everything inside the inner boundary of the $N+1$:st hypershell is treated as a hyperhole. This treatment can be done, since no changes are made to the interior of an added hypershell. □

SUMMARY

Two theorems have been established. One of them concerning the realizability of a specific decision region. It states that it is possible to decide whether a given linear partition of the cells that the input space are partitioned in is realizable or not. This can e.g. be done by using linear programming methods. The other theorem concerns the property of non-realizable decision regions. It states that the demand that the sum of weights in each cell, which is supposed to generate a one, shall be greater than the sum of weights in each cell, which is supposed to generate a zero, gives rise to a number of inconsistent inequalities. Since these theorems only concern the cells independantly of the topological shape the decision regions form, the geometrical placement of the cells in relation to each other rather than the topology seems to be important for the realizability of a given problem.

Three corollaries concerning decision regions of certain shapes are presented. Because of the constraints put on the problem the proofs are valid for the whole classes independently of the placement of the specific cells.

References:

[1] Rumelhart and McLelland "Parallel distributed processing" vol. 2, 1986.

[2] J. Makhoul, R. Schwartz, A. El-Jaroudi "Classification capabilities of two-layer neural nets" IEEE ICASSP 89.

[3] D. Gale, "The theory of linear economic models", McGraw-Hill, NewYork, 1960.

[4] P. Butovitsch "Fundamental Study of the Perceptron", TRITA–TTT–8916, Dept. of Telecomm. Theory, Royal Institute of Technology, Stockholm, Sweden.

QUALITATIVE ANALYSIS OF EQUILIBRIUM CONFINEMENT AND EXPONENTIAL STABILITY OF A CLASS OF DYNAMICAL NEURAL NETWORKS

S. I. Sudharsanan and M. K. Sundareshan
Department of Electrical and Computer Engineering
University of Arizona
Tucson, AZ 85721

ABSTRACT

Fundamental to a systematic synthesis of neural networks for optimization, associative memory and continuous mapping applications is a qualitative analysis of the network equilibrium conditions and the convergence of network trajectories to these equilibria. Specific conditions using the theory of $\mathcal{P}$-matrices are developed in this paper for characterizing the equilibrium conditions and the stability properties for an important class of neural networks that has received wide attention. Network structures for ensuring equilibrium confinement are identified by determining conditions for uniqueness of stable equilibrium points in specified quadrants of the state space. Conditions on network parameters for ensuring exponential stability with an arbitrarily specified degree within the basins of attraction of the equilibrium points are also presented. These results serve as valuable guidelines in the synthesis of neural networks for any specific application.

1. Introduction

A neural network structure that has received wide attention for its numerous computational capabilities is a nonlinear dynamical system described by

$$\dot{\mathbf{u}} = -\mathbf{u} + T\mathbf{g}(\mathbf{u}) + \mathbf{b} \tag{1}$$

where $\mathbf{u} \in \Re^n$, $T \in \Re^{n\times n}$, $\mathbf{b} \in \Re^n$, and $\mathbf{g} : \Re^n \rightarrow \Re^n$ is a vector valued function with sigmoidal elements. This network and its variations have been proposed to function as an optimizer, an associative memory and a nonlinear input-output mapper [1-4]. For any of these applications, a characterization of the network equilibrium conditions and a rigorous analysis of the qualitative properties of the network are of fundamental importance in order to serve as guidelines for developing systematic synthesis procedures and also to identify appropriate policies for training the neural network. Several earlier works have addressed these questions, particularly the evaluation of stability properties based on the consideration of energy functions, and a number of important results have been reported [1,5,6]. Although such results are not application specific, there exist some fundamental differences in the types of questions one would like to pose depending on the eventual use of the neural network for associative memory applications or for nonlinear mapping and optimization applications.

In optimization or nonlinear mapping problems one would like to have the network with a single stable equilibrium point, whereas the associative memory requires that there exist many stable equilibria. These issues have been discussed in the context of synthesizing a neural network for optimization applications in [7], where conditions for the existence of a unique equilibrium in the entire state space [8] are given. For a neural network to function as an associative memory, however, one should be interested in a different set of conditions to meet the requirement that the network should store the arbitrarily specified memory vectors as stable equilibria confined to specified quadrants of the state space. Moreover, there are several additional complexities that one should contend with. When the vectors are stored, it is not guaranteed to have only the stored vectors as the stable equilibria and many other vectors may also be introduced as stable equilibrium points. This will result in smaller regions of attractions of the stored vectors and also lead to spurious stable outputs. Since this issue is closely related to the memory capacity of the network, a number of researchers have paid attention to this problem for the discrete-time Hopfield network (i.e., the recognition of a binary memory with respect to its Hamming distance) [9,10].

For continuous-time systems described by (1), the maximum number of stable equilibria, i.e, the capacity, was first shown to be 2^n by Li *et al* [6]; however, this evaluation was disproved by counter examples through numerical simulations by Salam [11]. The inaccuracy in the computation in [6] was traced to erroneous application of the mean value theorem to show that there can exist only one equilibrium point in a quadrant of the state space [9]. In this paper, we shall present conditions for the uniqueness of the stable equilibrium in one quadrant and further show that these conditions vividly explain the discrepancy observed in [11]. We shall also present another important stability result which is considerably less restrictive compared to the previous results. This condition ensures the system trajectories to exponentially converge to the stable equilibria at a predetermined rate.

2. Exponential Stability

In this section, we shall present the stability results in the form of theorems. The basic principles underlying this development come from the theory of large scale systems as may be found in [12]. The proofs will not be included due to page limitations and will appear elsewhere. First, the assumptions on the sigmoidal nonlinearity and a few definitions are stated below. We are interested in the stability of the system

$$\frac{du_i}{dt} = -u_i + \sum_{j=1}^{n} t_{ij} g_j(u_j) + b_i; \quad i = 1, 2, \cdots, n \tag{2}$$

where u_i, b_i and t_{ij} are elements of the vectors $\mathbf{u}, \mathbf{b}$ and the matrix T.

The equilibrium points of system (2) are given by

$$u_i^* = \sum_{j=1}^{N} t_{ij} g_j(u_j^*) + b_i \quad i = 1, 2, \cdots, n\,. \tag{3}$$

With a coordinate transformation $x_i = u_i - u_i^*$, we can rewrite (2) as

$$\dot{x}_i = -x_i + \sum_{j=1}^{n} t_{ij} \tilde{g}_j(x_j)\;; \quad i = 1, 2, \cdots, n \tag{4}$$

where $\tilde{g}_j(x_j) = g_j(x_j + u_j^*) - g_j(u_j^*)$. Now let us state some properties of the nonlinearity $g_j(.)$.

Assumption 1: The nonlinear functions $g_i(.)$, $i = 1, 2, \cdots, N$, which are the elements of $\mathbf{g}$, satisfy the following:

1) $u_i g_i(u_i) > 0; \quad u_i \in \Re,$
2) $\lim_{|u_i| \to \infty} g_i(u_i) = \mathrm{Sgn}(u_i),$
3) $g_i(u_i)/u_i \geq g_i(v_i)/v_i \quad \forall |u_i| \leq |v_i|$
4) $g_i'(u_i) = dg_i(u_i)/du_i > 0; \quad u_i \in \Re$

Remark 1: It is easily seen that the above conditions are satisfied for the generally used sigmoidal nonlinearities $g(u_i) = \tan^{-1}(\pi \lambda u_i/2)2/\pi$ and $g(u_i) = \tanh(\lambda u_i)$ [1,4,6].

Remark 2: There exists a real number $r_i > 0$, such that $\forall \; |x_i| < r_i, \;, \tilde{g}_i(x_i)/x_i \leq \gamma_i$, where $\gamma_i > g'(u_i)$.

Definition 1: A square matrix A is said to be a $\mathcal{P}$ -matrix if all the principal minors of A are positive or every real eigenvalue of A is positive [13].

Let us also define a region $B \subset \Re^n$ specified by the positive constants $r_i, i = 1, 2, \cdots, n$, such that $B = \{\mathbf{x};\; |x_i| \leq r_i,\; i = 1, 2, \cdots, n\}$. For the next definition, let us assume that the equilibrium point $\mathbf{x}^*$ of the dynamical system

$$\dot{\mathbf{x}}(t) = \mathbf{f}(\mathbf{x}(t)),\; \mathbf{f}(\mathbf{x}^*) = 0 \tag{5}$$

where $\mathbf{x}(.) : \Re \to \Re^n$ and $\mathbf{f}(.) : \Re^n \to \Re^n$ is an isolated equilibrium contained in the region B (i.e, no other equilibria of (5) are inside B).

Definition 2: The isolated equilibrium point $\mathbf{x}^*$ of the system described (5) is *locally exponentially stable* in $B \subset \Re^n$ *with degree* η if every trajectory starting at any feasible initial state $\mathbf{x}(t_0) = \mathbf{x}_0 \in B$ satisfies the condition

$$||\mathbf{x}(t) - \mathbf{x}^*|| \leq \pi ||\mathbf{x}_0 - \mathbf{x}^*|| \exp(\eta(t - t_0)) \;\forall\, t \geq t_0,\; \forall \mathbf{x} \in B, \tag{6}$$

where π and η are positive constants independent of the initial conditions (t_0, x_0) and $\|.\|$ is the $\mathcal{L}_2$-norm. We then have the following result.

Theorem 1: The equilibrium point $\mathbf{u}^*$ of the neural network (2) is locally exponentially stable in $B \subset \Re^n$, with degree η, if the vector $\mathbf{h}(.) : \Re^n \rightarrow \Re^n$ defined by

$$\mathbf{h}(\mathbf{x}) = [h_1(\mathbf{x}), h_2(\mathbf{x}), \cdots, h_n(\mathbf{x})]^T$$

$$\mathbf{x} = [x_1, x_2, \cdots, x_n]^T, \quad h_i(\mathbf{x}) = \sum_{j=1}^{n} t_{ij}\tilde{g}_j(x_j) \tag{7}$$

can be factored in the form,

$$h(x) = [U(x) - S(x)]Px$$

where $P \in \Re^{n \times n} \ni P = \mathrm{Diag}\{p_{11}, p_{22}, \cdots, p_{NN}\}$, with $p_{ii} = 0.5(1-\eta)$, $\forall i = 1, 2, \cdots, n$, $U : \Re^n \rightarrow \Re^{n \times n}$ is an arbitrary skew symmetric matrix and S is an arbitrary symmetric matrix that satisfies the inequality

$$\mathbf{x}^T[I + 2PS(\mathbf{x})P]\mathbf{x} \geq 0, \quad \forall \mathbf{x} \in B \ , \tag{8}$$

I being the $n \times n$ identity matrix.

For the synthesis of a neural network whose equilibrium points are exponentially stable, one can translate the conditions of Theorem 1 into corresponding conditions on the parameters of the neural network.
Theorem 2: If $W = I - TG \in \Re^{n \times n}$ is a $\mathcal{P}$-matrix, then the equilibrium point $\mathbf{u}^*$, described by (3) is locally exponentially stable with degree $\eta = \lambda_{\min}(I - TD^{-1})$, $\forall \ |x_i| = |u_i - u_i^*| \leq r_i$, where, $G = \mathrm{Diag}\{g'(u_1^*), g'(u_2^*), \cdots, g'(u_N^*)\}$ and $D^{-1} = \mathrm{Diag}\{\tilde{g}_1(x_1)/x_1, \tilde{g}_2(x_2)/x_2, \cdots, \tilde{g}_N(x_N)/x_N\}$.

The proof of Theorem 2 consists of a number of steps involving the construction of matrices U and S, and to show that such a selection satisfies the inequality (8) under the condition that W is a $\mathcal{P}$-matrix. Note that the usual assumption that T is a symmetric matrix is relaxed and this will permit one to store patterns which would not be possible if T were assumed to be a symmetric matrix. Equipped with a stability result for a network with a nonsymmetric interconnection matrix, we are now prepared to characterize the stable equilibrium points of the network.

3. Equilibrium Confinement

Definition 3: A quadrant in the state space, $\Re^n$, is defined as

$$\Gamma(\xi, \mathbf{z}) = \{\xi; \ \xi_i z_i > 0, \ i = 1, 2, \cdots, n\}$$

where $\mathbf{z}$ is a fixed point in $\Re^n$.

Theorem 3: If there exists an equilibrium point, $\mathbf{u}^*$, such that $Q = I - TB$, where $B = \mathrm{Diag}\{g(u_1^*)/u_1^*, g(u_2^*)/u_2^*, \cdots, g(u_n^*)/u_n^*\}$, is a $\mathcal{P}$-matrix, then there exist no other equilibrium points in $\Gamma(\xi, \mathbf{u}^*)$.

It should be noted that the condition in Theorem 3 is a stronger condition than the one that ensures the stability of the equilibrium point $\mathbf{u}^*$. Therefore, whenever an equilibrium point exists such that the corresponding Q is a $\mathcal{P}$ matrix, that equilibrium point is also exponentially stable and under this condition, there exist no other equilibrium points in the same quadrant. This clearly explains the reason for the discrepancy demonstrated in the example shown in [11]. The above two results can be very effectively used to obtain a systematic synthesis procedure for associative memories which will be given elsewhere. Although the above stated result is only a sufficient condition for the uniqueness of the equilibrium point in a particular quadrant, it can be employed in design procedures to minimize the spurious stable vectors. The following numerical simulation gives an illustration of the conditions of the theorem.

Example: A set of dynamical equations as in (1) were simulated using fourth order Runge-Kutta methods with the following design parameters:

$$T = \begin{bmatrix} 0.0 & -1.0 & -1.0 \\ -1.0 & 0.0 & -1.0 \\ -1.0 & -1.0 & 0.0 \end{bmatrix},$$

$g_i(u_i) = 2\tan^{-1}(0.7\pi u_i)/\pi$. When $\mathbf{b}$ was selected as $\mathbf{b} = [0.001, 0.000, 0.001]^T$, there were three stable equilibrium points, $\mathbf{u}^{1*} = [.0342, -0.6128, 0.5501]^T$, $\mathbf{u}^{2*} = [0.5501, -0.6128, 0.0342]^T$ and $\mathbf{u}^{3*} = 0.2859, -0.7222, 0.2859]^T$, in the quadrant $\Gamma(\xi, \mathbf{z})$, where $\mathbf{z} = [1.0, -1.0, 1.0]^T$. The corresponding eigenvalues of Q_i matrices were evaluated for each each of the stable equilibrium points, $\mathbf{u}^{i*}$, and found out to be $\lambda(Q_1) = \lambda(Q_2) = \{3.256, 0.0009, -0.2568\}$, and $\lambda(Q_3) = \{3.2619, 0.001, -0.2629\}$. This clearly shows that the Q_i matrices are not $\mathcal{P}$-matrices and there may be more than one equilibrium point in $\Gamma(\xi, \mathbf{z})$, and there are indeed three of them. Then the $\mathbf{b}$-vector was changed to $[0.5, 0.0, 0.5]^T$ and it was found out that there existed only one equilibrium point, $\mathbf{u}^* = [0.6592, -1.2390, 0.6592]^T$, and the corresponding Q matrix had the eigenvalues, $\lambda(Q) = \{2.6539, 0.2858, 0.06032\}$. In this case, the Q matrix is a $\mathcal{P}$-matrix which implies that there can only be one equilibrium point in $\Gamma(\xi, \mathbf{z})$ and it is indeed the case.

REFERENCES

[1] J.J. Hopfield, "Neurons with graded response have collective computational properties like those of two-state neurons," *Proc. Natl. Acad. Sci.* Vol. 81, pp 3088, 1984.

[2] D. W. Tank and J. J. Hopfield, "Simple 'neural' optimization networks: an A/D converter, signal decision circuit, and a linear programming circuit," *IEEE Trans. Circuits Syst.*, vol. CAS-33, pp. 533-541, 1986.

[3] F. J. Pineda, "Dynamics and architecture for neural computation," *Journal of Complexity*, vol.4, pp 216-245, 1988.

[4] B. Kosko, "Adaptive bidirectional associative memories," *Applied Optics*, vol. 26, No.23, pp. 4947-4960, 1987.

[5] M. A. Cohen and S. Grossberg, "Absolute stability of global pattern formation and parallel memory storage by competitive neural networks," *IEEE Trans. Syst. Man, Cybern.*, vol. SMC-13, pp. 815-826, 1983.

[6] J.-H. Li, A. N. Michel and W. Porod, "Qualitative analysis and synthesis of a class of neural networks," *IEEE Trans. Circuits Syst..*, vol. CAS-35, pp 976-986, 1988.

[7] S. I. Sudharsanan and M. K. Sundareshan, "Neural network computational algorithms for least squares estimation problems," Presented at the 1989 Int. Joint. Conf. on Neural Networks, Washington D.C., June 1989.

[8] A. Atiya, "Learning on a general network," *Proc. IEEE Conf. on Neural Information Processing Systems*, Denver, Co,1987.

[9] J.-H. Li, A. N. Michel and W. Porod, "Analysis and Synthesis of a class of neural networks: variable structure systems with infinite gain," *IEEE Trans. Circuits Syst.*, vol.36, no.5, pp. 713-731, 1989.

[10] Y. S. Abu-Mostafa and J.-M. St. Jacques, "Information capacity of the Hopfield model," *IEEE Trans. on Info. Theory*, vol. IT-31, pp. 461-464, 1985.

[12] M.K. Sundareshan, "Exponential stabilization of large scale systems: decentralized and multi-level schemes," *IEEE Trans. on Systems, Man and Cybernetics*, vol. SMC -7, pp 478-484, 1977.

[13] M. Fiedler, *Special Matrices and Their Applications in Numerical Mathematics*, Prague, Czechoslavakia: Martinus Nijhoff Publishers, 1986.

A GLOBAL STATE DRIVEN COOLING SCHEMA FOR SIMULATED ANNEALING USED WITH BOLTZMANN MACHINE

A. BOUJU, P. BOURRET, C. GASPIN
ONERA - CERT/GIA
2, avenue Edouard Belin - B.P. 4025 - 31055 TOULOUSE Cédex - FRANCE
Tel. 61.55.71.92
Fax (33) 61.55.71.72

Abstract - In this paper we present a general purpose cooling schema for using simulated annealing in a Boltzmann machine. Because this cooling schema is deduced from the behaviour of each unit, we can expect that it is not problem dependent and therefore can be used for any kind of problem.

I - Introduction

The simulated annealing process [KIRKPATRICK 83] is now widely used especially in order to get the global optimum of an energy function of a Boltzmann machine in the area of neural networks [HÉRAULT 89] [MATSUBA 89]. This process amounts to give a non null probability for a variable to change of state even if the result is worst for the value of the energy function. This probability is a function of a parameter, so called temperature, which decreases slightly during the process. This probability is null when the temperature is equal to zero. The main difficulty of this process is to find a good "cooling schema" which describes the decreasing of the temperature. A lot of cooling schemas have been proposed [JOHNSON 87] [NAHAR 85]. As a matter of fact quite each application uses its own cooling schema. Our purpose is to exhibit a general purpose "cooling schema" which can be used for any kind of problem with respect to the following constraints : to converge towards the global optimum with a convergence speed not slower than the particular cooling schema. In this paper we firstly give an intuitive idea of our cooling schema, then we give a formal presentation of it and we prove that it converges in an infinite time towards the global optimum. At last we give a few results of simulation which show that our second constraint is met.

II - Intuitive presentation

The simulated annealing process is often represented by a curve which represents the "energy landscape" and a ball falling in the wholes of the energy landscape with a given probability to climb a hill in order to escape from local optima. Let us assume that we can divide this curve in 2^n parts ($n = 2$ see figure 1) each part representing the energy lanscape for given values of n variables. In a given part of the curve, these variables may be seen like constants and we can say that these variables are "frozen" (i.e. their values may not change).

Therefore, for a given set of values this is a temporary global optimum which can be achieved or not with respect to the temperature value. In any case the system converges, with this set of values for the frozen variables, towards a given state of the system. Let us call "cycle" a try to change the value of a variable according to the Metropolis algorithm. Let us assume that a variable is frozen iff its value has not changed since kN tries of change where N is a parameter . Thus in a part of the curve the system converges towards constant values of the not frozen variables. After kN cycles at least one variable becomes frozen. When this condition occurs, one variable is chosen among the already frozen variables and the Metropolis strategy is applied ; if the value of this variable is changed thus we have changed the part of the curve on which we move and the variable is no more frozen thus the number of frozen variables remains constant. If the value of the variable is not changed the variable remains frozen and the number of frozen variables is increased by one.

Now let us assume that the temperature decreases when the number of frozen variables increases. When the temperature is low the number of areas of the energy landscape becomes large but the probability of change from one area of the energy landscape to another decreases. But, more or less the probability to visit each part of the area remains constant. When every variable is frozen the temperature is equal to zero and the process is finished. Thus the decay of the temperature is linked to an attribute of the global state which is the number of tries since each not frozen variable has kept the same value. In other words, the temperature decreases if an area of the energy landscape has been sufficiently explored in such a way that we can expect that half of the energy landscape can be supposed to be well explored. (In fact we have also a given probability to create a new area which must be explored)

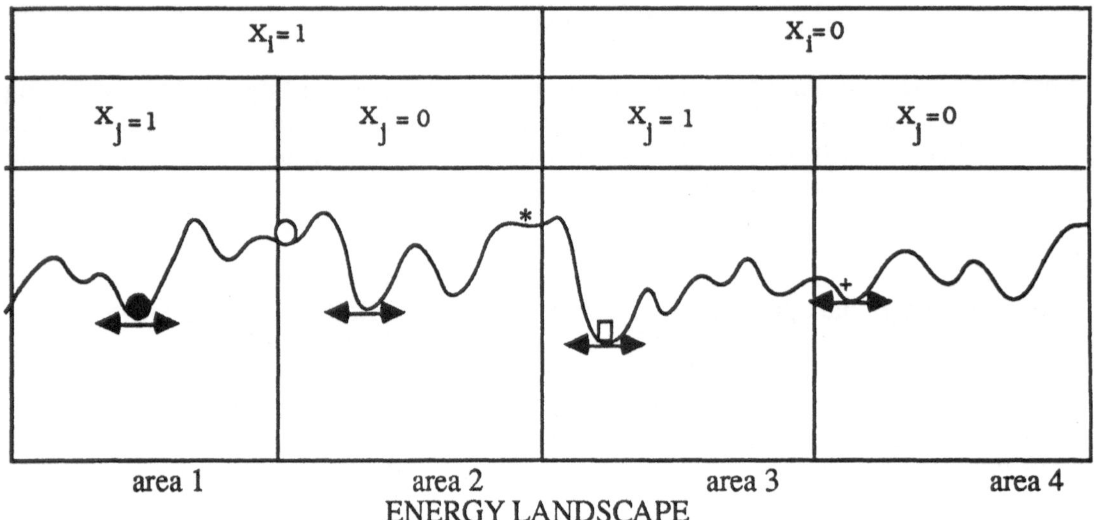

Global state	Probability of area change given that X_i or X_j is no more frozen	
•	1 to 2 : low	
O	1 to 2 : very high	2 to 1 very high
*	2 to 3 : low	2 to 1 very low
□	3 to 2 very low	3 to 4 very low
+	4 to 3 high	

Figure 1

III - Formal presentation

III.1 - Notations`

Let $S_1, S_2, \dots S_p$ be the possible states of the system
($p=2^N$ where N stands for the number of units)

Let T be the parameter "temperature"
Let $G_{S_k}(\frac{T}{S_l})$ be the probability of trying the state S_l if the system is in the state S_k

Let $f(S_i)$ be the energy function of the system
Let $A(S_i\ S_j)$ be the probability that the system moves from state S_i to the state S_j, given that S_j is tried
$x_j(1) = 1$ if $i=j$
$= 0$ otherwise

Let F_n be the set of frozen units at the cycle n (i.e. units which have kept the same value after k tries to change their values)
Let $\Sigma_i\ (F_n)$ be the set of states which differ from S_i by the value of only one not frozen unit.

III.2 - Theorem [AARTS 89]

If 1°) $\forall T\ \forall i,j \in \{1,...p\}\ \exists r \geq 1\ \exists S_0, S_1,...S_r : S_0=S_i, S_r=S_j$
$G_{S_k S_{k+1}}(T) > 0 \quad k=0,...r-1$

2°) $\forall T>0 \quad G_{S_i S_j} = G_{S_j S_i}$

3°) $\forall T>0 \quad \forall S_i, S_j \quad A_{S_i S_j}(T) = 1$ if $f(S_i) \leq f(S_j)$
$A_{S_i S_j} \in [0,1]$ if $f(S_i) > f(S_j)$

4°) $\forall S_i, S_j, S_k \ \forall T>0 \quad f(S_i) \geq f(S_j) \geq f(S_k) \Rightarrow A_{S_i S_k}(T) = A_{S_i S_j}(T) \times A_{S_j S_k}(T)$

Then there is a stationary distribution $q_i(T) = \dfrac{A_{i_{opt},i}(T)}{\sum_j A_{i_{opt},j}(T)}$

and limit $T \rightarrow 0 \quad q_i(T) = \dfrac{1}{|S_{opt}|} \chi_{S_{opt},i}$

Let $V(S_i)$ be a neigbourood of S_i the elements of which differ from S_i by only one unit value, say U_I.

$G_{S_i S_j} = 0$ if $S_j \notin V(S_i)$

$G_{S_i S_j} = \dfrac{1}{N-|F_n|}$ if $S_j \in V(S_i)$ and $U_I \notin F_n$. and $U_I \notin PF_n$

(where PFn stands for the parts of the units which have been tried k-1 times whithout changing their value)

$G_{S_i S_j} = \dfrac{1}{|F_n|} A(S_i, S_j)$ if $S_j \in V(S_i)$ and $U_I \notin PF_n$

$G_{S_i S_j} = 0$ if $S_j \in V(S_i)$ and $U_I \in P_n$

$G_{S_i S_j} = 0$ if $S_j \notin V(S_i)$

Let us assume that $T = To\ (1- \dfrac{|F_n|}{N})$

$A(S_i,S_j) = 1$ if $f(S_j) \geq f(S_i)$

$A(S_i,S_j) = 1 - \dfrac{1}{1 = e^{\frac{f(S_j)- f(S_i)}{T}}}$ otherwise

It is easy to show that the conditions of application of the previous theorem are true for

$f = \sum_{i \leq j} w_{ij}\, U_{ij} \qquad i,j \notin 1,...N$

Therefore to apply the previously defined dynamic yields to the global optimum in an unbounded time. But we can easily know the finite time at which this global optimum is reached. It is when every unit is frozen because in this case $G_{S_i S_j} = 0\ \forall\ S_j$.

IV - Experimental results

In this section we present the solution get for two assignment problems. These assignment problems consist of, respectively, 0 tasks with 20 resources and 20 tasks with 30 resources. We give the results get with the following methods :

A) Hopfield method
B) Hopfield method with sensitization/habituation units (see [BOURRET 88a]
C) Boltzmann machine with a NAHAR cooling schema
D) Boltzmann machine with our cooling schema ($T_0 = 0.007$ N=9)
E) Boltzmann machine with our cooling schema ($T_0 = 0.0005$ N=8)

	20 x 20 Assignment Problem					20 x 30 Assignment Problem				
Method	A	B	C	D	E	A	B	C	D	E
Number of needed cycles	120 [234]	90 [210]	4420	1272 [1346]	1711 [1943]	79 [147]	80 [141]	1820	378 [383]	339 [343]
Number of assigned tasks	20 [19]	20 [18.5]	20 [19.2]	20 [20]	20 [18,6]	20 [20]	20 [19,9]	20 [20]	20 [19.6]	20 [19.8]

N.B. For A,B,C,D,E the mean value is between brakets,on the other is the best value.

V - Conclusion

The various simulations (around one hundred) which have been done so far show that in every checked case our cooling schema converges faster than the other one. But we are unable to prove that this property is always true. The number of needed cycles is related to the two parameters T0 and N which should verify a relationship that we have not yet exhibited. We are trying to mathematically study this schema and to improve it with a more sophisticated definition of the relationship between the temperature and the number of frozen units.

REFERENCES

[AARTS 89] E. AARTS and J. KORST, Simulated annealing and Boltzmann machines : a stochastic approach to combinatorial optimization and neural computing - John Wiley and Sons, Ed.89

[BOURRET 88a] P.BOURRET, C.GASPIN, M.SAMUELIDES, Affectation dynamique des ressources d'un satellite en opération par une machine de Boltzmann à sensibilisation, Proceedings of Neuro-Nimes 88, Nimes, France, Novembrer 15-17 - 1988.

[BOURRET 88b] P.BOURRET, C.GASPIN, Scheduling Space Operations by a Boltzmann machine Neural Networks from Models to Applications - L.Personnaz, G.Dreyfus Ed. IDSET Paris 1988

[HERAULT 89] L.HERAULT, J.J.NIEZ, How neural networks can solve hard graph problems : a performance study on the graph K-partitioning, Proc.Neuro Nîmes 89, EC2 Ed. Nanterre (France)

[HUANG 86] M.D.HUANG, F.ROMEO, A.L.SANGIOVANNI-VINCENTELLI, An efficient general cooling schedule for simulated annealing - Proc. IEEE Int.Conf. on CAD Santa Clara - Nov. 1986

[JOHNSON 87] D.S.JOHNSON, C.R.ARAGON, L.A.McGEOCH, C.SCHEVON, Optimization by simulated anneling, A.T. & T Bell Laboratories. Internal Report 1987

[KIRKPATRICK 83] S.KIRKPATRICK, C.D.GELATT, M.P.VECCHI, Optimization by simulated annealing , Science, 220,(1983) pp. 671-680

[MATSUBA 89] I.MATSUBA, Optimal Simulated annealing method and its application to combinatiorial problems - Proc. IJCNN Washington, June 1989

[NAHAR 85] S.NAHAR, S.SHANI and E, SHRAGOWITZ, Experiments with simulated annealing Proc. 22nd Des. Automation Conf. Las Vegas, June 1985.

INDEPENDENCE FROM TEMPERATURE OF DOMAINS OF ATTRACTION IN NEURAL NETWORKS

G. NARDULLI (&) and G. PASQUARIELLO ($)

(&)Dipartimento di Fisica-Universita' di Bari and I.N.F.N. Bari, Italy
($)I.E.S.I-C.N.R.,Via Amendola 173, 70126 Bari, Italy

ABSTRACT

We consider finite temperature neural networks learned by the Edimburgh's algorithm in the symmetric version. We prove that domains of attraction of stored patterns are independent from thermal noise.

In this paper we study the effect of thermal noise on the stability of domains of attraction of stored patterns in neural networks. The basin of attraction of the patterns ξ_i^μ (i=1,...N; μ=1,...αN; N= numbers of neurons; αN= number of stored patterns; $\xi_i^\mu = \pm 1$) is basically determined by the parameter

$$K = \min(\gamma_i^\mu) = \min\left(\frac{\xi_i^\mu}{|J_i|} \sum_{j=1}^{N} J_{ij}\xi_j^\mu\right) \tag{1}$$

where J_{ij} is the coupling strength between neurons at sites i,j (J_{ii}=0) and $|J_i| = (\sum_j J_{ij}^2)^{1/2}$. For ξ^μ to be a fixed point of the dynamics one needs K > 0; the (positive) value of K actually fixes the domain of attraction [1]; the latter can be defined as follows:

$$d_c = 1 - m_c \tag{2}$$

where m_c is such that any initial configuration $S_i(0)$ having overlap m_0 with the pattern ξ^μ:

$$m_0 = 1/N \sum_i S_i(0)\xi_i^\mu, \tag{3}$$

will be driven by the dynamics towards the fixed point ξ^μ provided that $m_0 > m_c$.

The introduction of a random noise, described by a "temperature" parameter $T=1/\beta$, amounts to consider a probabilistic dynamics [2] instead of a deterministic one; the probability of the configuration $\{S_i'\}$ at time t+1, given the configuration $\{S_i\}$ at time t, is:

$$P(\{S_i'\}|\{S_j\}) = \prod_i P(S_i'|\{S_j\}) = \prod_i \frac{\exp(\beta S_i' H_i)}{\exp(\beta S_i' H_i) + \exp(-\beta S_i' H_i)} \tag{4}$$

where $H_i=(\alpha/n)^{1/2}\sum_j J_{ij}S_j$. Deterministic dynamics is recovered in the limit $T \longrightarrow 0$.

We have considered parallel dynamics (all sites are simultaneusly updated according to the rule (4)) for a neural network learned by a symmetric version of the so-called Edimburgh's algorithm, which implements the condition $\gamma_i^\mu \geqslant K$ [3],[4]. and $J_{ik} = J_{ki}$; couplings are normalized according to $\sum_{l,m} J_{lm}^2 = N^2$, which in the thermodynamical limit reduces to $|J_i| = N^{1/2}$. Moreover we have worked with networks near the saturation, i.e. with α related to K by the equation

$$\alpha = \alpha_s = [\int_{-K}^{\infty} dt\ (t + K)^2 \exp(-t^2/2)]^{-1} \tag{5}$$

which can be proven for both the asymmetric [4] and the symmetric [5] case.

The domain of attraction is obtained as follows. The overlap after one time step:

$$m_1 = 1/N \sum_i S_i(1)\xi_i^\mu, \tag{6}$$

is a sensitive parameter which determines whether the initial configuration $S_i(0)$ will be attracted by the fixed point ξ^μ or not; as a matter of fact we numerically find that $P(S\text{-->}\xi) = 1$ if and only if

$$\frac{m_1 - m_0}{1 - m_0} \geqslant \lambda(T) = a - bT \tag{7}$$

where a = 0.5 and b depends on K (for istance b = 0.80 for K = 4.8,

$b = 0.35$ for $K = 2.3$ and $b=0.1$ for $K = 1.02$). Our results have been obtained by a numerical simulation with $N = 200$, 8,000 initial inputs and several values of K; they coincide for $T = 0$ with the findings of Ref.[1]. Full retrieval of stored patterns was obtained for $T < T_s$, with $T_s = 0.40$, 0.37 and 0.25 for $K = 4.8$, 2.3 and 1.02 respectively.

By a generalization of methods of Ref.[1], one can obtain a formula connecting m_1 and m_0, which is given by

$$m_1 = m_1(m_0,T) = \int d\gamma\rho(\gamma) \, [\, \mathrm{erf}(u) - \\ - \sum_{n=1}^{\infty} (-1)^n \exp(-u^2)[\mathrm{erfc}(z_+) \exp(z_+^2) - \mathrm{erfc}(z_-) \exp(z_-^2)] \tag{8}$$

with $z_{\pm} = n[2\alpha(1-m_0^2)]^{1/2}/T + u$; $u = m_0\gamma/[2(1-m_0^2)]^{1/2}$ and $\rho(\gamma) = \exp(-\gamma^2/2)/(2\pi)^{1/2} \, \vartheta(\gamma - K) + \{1+\mathrm{erf}[K/(2)^{1/2}\}/2 \, \delta(\gamma - K)$. In Fig. 1 we compare the results of our numerical simulations and Eq.(8) for different values of K and different temperatures.

The domain of attraction is obtained by solving the equation

$$m_1(m_c,T) - m_c = (1 - m_c) \, \lambda(T) \tag{9}$$

for m_c and substituting in Eq. (2). The result is displayed in Fig. 2 which shows no dependence of the domain of attaction d_c on the temperature. In other words, domains of attraction are uniquely determined by the parameter K: the presence of a random noise slows down the convergence, as shown by positive values of b in Eq. (7), but does not alter the minimum initial overlap m_c that is needed in order to obtain 100% retrieval of stored patterns.

Acknowledgements. We thank A. Maritan, G. Cicuta, P. Colangelo and N. Cufaro-Petroni for useful discussions.

REFERENCES

1. T. B. Kepler and L. F. Abbott, J. Phys. France 49 (1988) 1657.

2. W. A. Little, Math. Biosci. 19 (1974) 101.

3. A. D. Bruce, E. Gardner and D. J. Wallace, J. Phys. A20 (1987) 2909.

4. E. Gardner, J. Phys. A21 (1988) 257.

5. G. Nardulli and G. Pasquariello, in preparation.

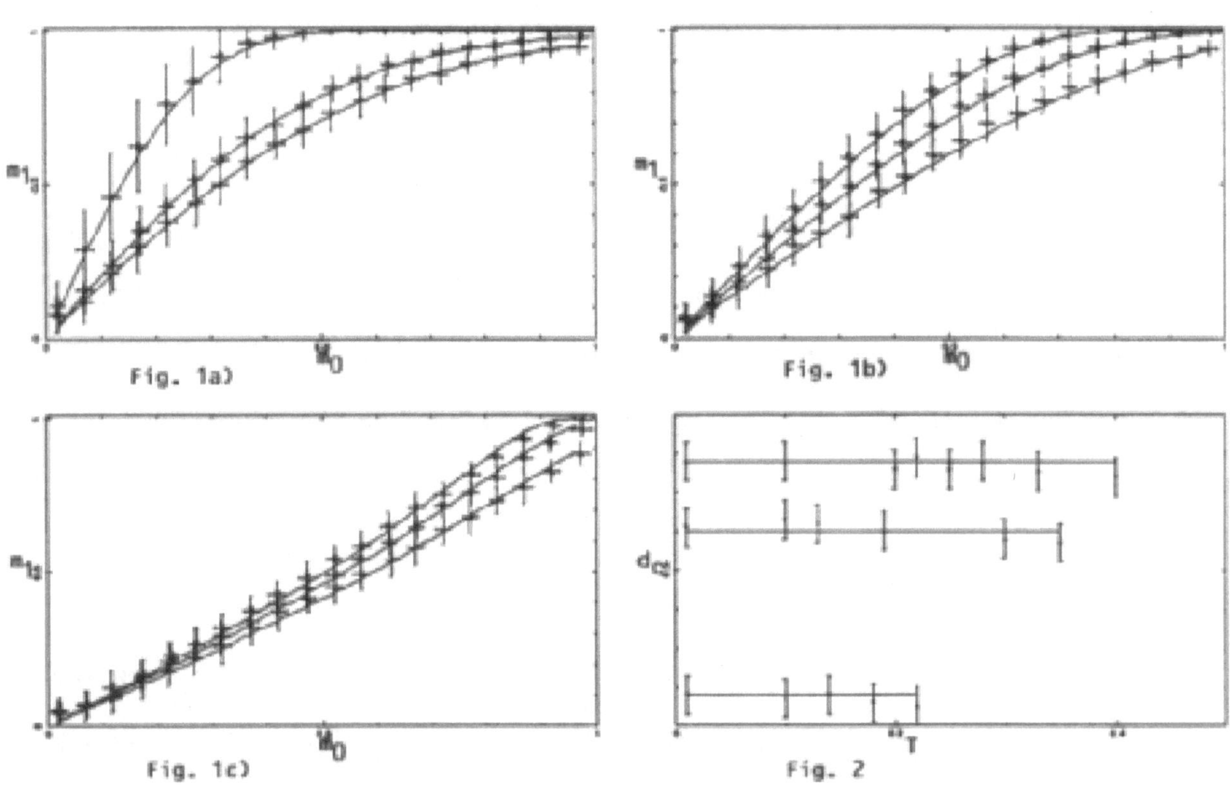

FIGURE CAPTIONS

1. First overlap m_1 vs. m_0 for different temperatures T = 0, 0.3, 0.5 (going from the top curve to the bottom); a) corresponds to K = 4.8, b) to K = 2.3 and c) to K = 1.02.

2. Domain of attraction vs. temperature for three values of K: from the top curve to the bottom, K = 4.8, K = 2.3 and K = 1.02.

SELF-INDUCED NOISE IN NEURAL NETWORKS

V.I.Makarenko and A.B.Kirillov
Research Computing Center of the USSR Academy of Sciences
Pushchino, Moscow Region 142292 USSR

Abstract Presently there exists a large amount of models simulating neural networks. It has become a common practice to pay no attention to noise component included in many of the models, do not consider seriously its physiological substantiation and its effect on the behavior of the network. Its relatively large contribution traditionally is regarded as the inevitable consequence of the great parameter and process variability typical of biological systems. The majority of the modelers treat this component as a natural noise somewhat weakening the desired effect and only few of them approach it as a critical condition of the observing phenomena to exist.

This paper is an attempt of a more detailed analysis of the origin of "noise" in the functioning neural network. Here we have assumed that noise is a sum of several independent processes acting on different structural levels (synapse, neuron, neural ensemble, etc). In this way noise is assumed to be distributed upon structural levels, as well as upon temporal and spatial components of activity.

The main result of our simulations of this network is that it exhibits self-supported metastability mode. The initial state (a localized excited subset) survived for unusually long time - at least by 4-5 orders more than neuron characteristic time. Moreover, excited subset drifted over the network and slowly oscillated. The observed effects could be of great importance in further modeling of long lasting neurophysiological phenomena, e.g. long term potentiation, as well as of short-term memory itself.

This paper is an attempt of detailed analysis of the origin of "noise" in the functioning neural network. Here we have assumed that

- noise is a sum of several independent processes acting on different structural levels (synapse, neuron, neural ensemble, etc) ;
- relative contribution of these processes to the total noise depends on the state of the network, being defined ultimately by the interaction of a single unit with the whole network;
- noise depends on the architecture of neural connections.

In this way noise is assumed to be distributed upon structural levels, as well as upon temporal and spatial components of activity.

We considered a 50*50 rectangular lattice with model neurons (of the integrate and fire type [1]) at the nodes. Each neuron was bilaterally connected to its four nearest neighbors. This model exhibits metastability phenomena [2,3]. The distinctive feature of our model (compared to the model of [2]) was a way how noise input to the neuron was formed. Every neuron had 100 randomly chosen "distant neighbors".

It is evident that only positive or only negative influence of distant neighbors is unrealistic, leading to a very primitive behavior of the network when neurons either all fire (upper steady state) or all silent (lower steady state) and oscillations between these two states. Sometimes we observed oscillations between geometrically ordered structures.

We tried to avoid this unilateral influence of distant neighbors by comparing their activity for given neuron with its mean value over the network. The difference is a physiological characteristic of the contribution of such a subset to the collective behavior of the network.

The main result of our simulations of this network is that it exhibits self-supported metastability mode. The initial state (a localized excited subset) survived for unusually long time - at least by 4-5 orders more than neuron characteristic time. Moreover, excited subset drifted over the network and slowly oscillated.

It should be noted that the behavior of the network was independent of the size of the initial excited spot, if this was within the limits 0.02-0.7 N*N - the behavior was invariant to the initial level of excitation and could keep it almost unchanged in a very wide range.

This effect was used to simulate a neuronal oscillator a model of which was based on the properties of a neural net with Gaussian noise [2,4]. When we introduced self-induced noise, the maximal period of oscillations largely increased (by more than one order).

The observed effects are the phenomena of self-organized criticality type which are well known in physics. Due to their features they could be of great importance in further modeling of long lasting neurophysiological phenomena, e.g. long term potentiation, as well as of short-term memory itself.

References

[1] Kryukov, V.I. (1978) Markov interaction processes and neuronal activity. *Lect. Notes In Math.*, **653**, pp. 122-139.

[2] Kovalenko, E. I., Borisyuk, G. N., Borisyuk, R. M., Kirillov, A. B., Kryukov, V. I. (1984), Short-term memory as a metastable state. II. Simulation model, in *Cybernetics and Systems Research*, vol. 2, R. Trappl (ed.), Elsevier, pp. 266-270.

[3] Kirillov, A. B., Makarenko, V. I. (1989), Phase transitions in simulated 2D neural nets and statistical methods of Markov random fields, in *Neurocomputers and attention,* Extended Abstracts of international workshop, Pushchino , USSR, pp.126-127

[4] Kirillov, A. B., Borisyuk, G. N., Borisyuk, R. M., Kovalenko, E. I., Kryukov, V. I., Makarenko, V. I., Chulaevsky, V. A. (1989), A unified submodule for a model of attention, in *Adv. in Neural information processing systems I*, D.S.Touretzky (ed.), Morgan Kauffman, San Mateo, CA, pp. 560-567.

SOME REMARKS ON NONLINEARITIES IN NEURAL NETWORKS

A. M. Vepsäläinen
Technical Research Centre of Finland

Abstract - *To achieve superior classification properties the neural network modellers have very often used nonlinear mappings, for example the sigmoid function. For multilevel networks or networks with feedback these nonlinearities can cause unpredictable and chaotic behaviour. At first sight, one might keep these chaotic properties undesired to which only statistical approaches are applicable. However, real nonlinear dynamical systems also possess chaotic properties, and these properties can only be described with a network having same properties. In this paper the bifurcation route to chaos for some simple neural networks is examined.*

ON PATTERN RECOGNITION FROM THE EDGE-IMAGES WITH A NONLINEAR, DYNAMICAL NETWORK

Ari Vepsäläinen
Technical Research Centre of Finland
SF-00340 Helsinki
Finland

Abstract - *An approach to model the shape of boundaries in an image with a network is presented. In the neural network the connections are described with nonlinear second order differential equations. With using a variety of adaptation rates an excellent consistency of the estimated model and the real shape can be achieved.*

LINEARLY INDEPENDENT TRANSFORMATION BY HIDDEN UNITS WITH ANALYTIC ASYMPTOTIC ACTIVATION FUNCTION

Xingren Ying

Institute of Automation
National Laboratory of Pattern Recognition
Chinese Academy of Sciences
P.O. Box 2728, Beijing, People's Republic of China

Abstract

The asymptotic function is a very general class of functions which includes the unit step functions, exponential functions, sigmoid functions, and any non-constant function which is convergent in one of the direction. The Generated Hidden Unit Vector (GHUV) is the "state" of the input in the hidden layer, which consists of the output values of hidden units and constant 1. The k hidden units with the asymptotic activation function are able to transfer any given $k+1$ different inputs to linearly independent GHUV's by properly setting weights and thresholds. For the hidden units with Analytic Asymptotic Activation Function (AAAF) and given inputs, this Linearly Independent Transformation (LIT) is with probability 1 ability when setting weights and thresholds randomly. It is a "generic" ability for the weights and thresholds, i.e. the set of weights and thresholds can implement this LIT for the given inputs is open and dense. And it is a "generic" and with probability 1 property for any $k+1$ inputs if the weights and thresholds setting has the LIT ability for some $k+1$ inputs. Therefor, the k hidden units with random setting transfer almost any $k+1$ different inputs to the linear independent GHUV's using AAAF. That is not true that any nonlinear function could be the activation function, and these nets have LIT ability for any number of inputs. The number of hidden units with this LIT ability for polynomial activation function is limited by the order of polynomials.

For three-layer nets with k hidden units, in which the activation function is asymptotic, and the output layer is without activation function, they are sufficient to precisely record $k+1$ arbitrary real sample pairs. And it is with probability 0 to record $k+2$ random real sample pairs if the activation function is a unit step function. So is true for the sigmoid function in the case of associative memory.

Mathematical theorems are provided for above conclusion. A scheme for understanding the associative memory in three-layer net is provided in the last part of this paper.

Reducing the Weight Space of a Net With Hidden Units to a Minimum Cone.

Alexander Shustorovich

Image Electronics Center,
Eastman Kodak Company,
Rochester, New York 14653-5719

ABSTRACT

In his recent talk [1] on the theory of Back-propagation (BP) at IJCNN-89, Dr. Hecht-Nielsen made an important observation that any single meaningful combination of weights can be represented in the net in a huge number of variants due to the permutations of hidden units. He remarked that if it were possible to find a cone in the weight space such that the whole space is produced from this cone by permutations of axes corresponding to the permutations of the hidden units, it would greatly reduce the volume of space in which we have to organize the search for the solutions.

In this paper such a cone is built. Besides the obvious benefits mentioned above, the same procedure enables the direct comparison of different solutions and trajectories in the weight space, that is, the analysis and comparison of functions performed by individual hidden units.

The Kink Representation for Exclusive–OR

Christopher J. Thornton

Dept. of Artificial Intelligence,
80, South Bridge,
University of Edinburgh,
Edinburgh,
EH1 1HN,
UK
Chris_Thornton@ed.ac.uk

Abstract

This paper considers the reprepresentation for the XOR problem which is constructed by backpropagation in a regular 2–>2–>1 network; i.e. a network consisting of 2 input units, 2 hidden units and 1 output unit, with total connectivity between layers and no connectivity within layers. Since there are no more than two units in any one layer, it is possible to visualise the activation and weight vectors for network as 1 or 2–dimensional vectors. This means that the development of the representation can be traced using simple diagrams, and justified using geometric arguments. Although this particular representation has been discussed frequently in the literature, a full geometric trace has not yet been provided.

1. Introduction

The paper considers the case in which a feed–forward network consisting of 2 input units, 2 hidden units and 1 output unit with total connectivity between layers and no connectivity within layers (cf. Figure 1) is trained to implement the exclusive–or mapping. The training set for this mapping is just the truth table for exclusive–or:

```
[0 0] -> [0]
[1 0] -> [1]
[1 1] -> [0]
[0 1] -> [1]
```

It has been noted (1, p. 331) that, using this network, the learning algorithm can fall into a local minimum. However, in most cases, backpropagation succeeds in producing a set of weights which enables the network to perform correctly.

EVALUATION OF THE FUNCTIONAL CAPACITIES OF MULTI-LAYERED LOGICAL NEURAL NETWORKS

R. Al-Alawi and T. J. Stonham

Department of Electrical Engineering and Electronics, Brunel University, Uxbridge, Middx ,UB8 3PH, U.K.

Abstract: **The ability of neural networks to solve real data processing tasks is dependent on the network topology being able to support the desired functionality required by the problem under consideration. An analysis of popular network topologies reveals that they have very restricted functionality and no strategy for devicing a suitable topology is given. This paper presents a method for calculating the functional capacity of a multilayer neural network with analogue and digital neurons.**

THEORY OF INFORMATION PROPAGATION

Tetsuya Takahashi

Laboratory for Neural Networks
The Frontier Research Program
The Institute of Physical and Chemical Research

Hirosawa 2-1, Wako-shi, Saitama, 351-01 Japan

Abstract

Since information is processed in a parallel fashion in the brain, one must understand how information is transmitted before discussing the computational aspects of the brain. We propose an entirely new mathematical model to explain how information flow propagates in biological neural networks. Under assumptions on a structured neural network, it is showed information propagation satisfies a rather simple equation. Finally, it is showed that some result of the well-known Hubel-Wiesel experiment can be reproduced theoretically in our scheme.

Monte-Carlo Learning : the LMS Rule and Random Weights can Form Non-Convex Decision Surfaces and Generalize

Ronnie Lau Hing
Department of Computer Science
University of the Witwatersrand
P O Wits, South Africa, 2050
122ron.witsvma@f4.n494.z5.fidonet.org

Abstract

Multi-layer feed-forward networks that use intermediate layers with fixed random weights have proved successful in learning a number of classical problems difficult for perceptron based approaches. These networks can be trained using any single-layer learning algorithm since the intermediate units with fixed random weights do not need to be adapted. This method of training multi-layer networks has less computational overhead than training networks where all the weights are modified. We have used this approach to train two-layer networks using the LMS rule to adapt the weights of the output layer. The LMS rule is used because of its gradient descent approach to learning, and its minimization of the mean squared error. We have termed this combination of using intermediate layers with fixed random weights and using the LMS rule to train the weights of the output layer, Monte-Carlo Learning (MCL).

This paper describes how two-layer networks using MCL can form complex non-convex decision surfaces. Results show that to correctly classify all p training examples the network is dependant on $(-r,+r)$, the range of random weights generated for the intermediate layers, and s the random weight generator seed. Moreover, for a two-layer network at least p intermediate units should be used for a high probability of classifying all the training examples. This is a large number of intermediate units and is cause for concern since too many intermediate units allows grandmothering or rote memorization of the training examples, which makes the network fail to extract the general features of the training examples. However, results indicate that two-layer networks with greater than p intermediate units have the ability to generalize the features of a set of analog training examples. The generalization performance of these networks compares favourably to networks trained using the Back-Propagation algorithm which is currently the most widely used learning algorithm for multi-layer networks.

USING LOCAL MINIMA AS SEQUENCES MEMORIES

Remis Balaniuk and Philippe O. Navaux
Curso de Pos-Graduação em Ciência da Computação - II - UFRGS
Av. Osvaldo Aranha 99, cx. postal
Porto Alegre RS cep 90000 - BRASIL
Phone: 0512/218499 Telex: 051 2680 CCUF
Fax: 0512/244164 E-Mail: NAVAUX@SBU.UFRGS.ANRS.BR

Abstract

This paper presents a Neural Network model to sequence memorization based on the Hopfield's model. The model uses neuron continuous activation to accept and retain the entries, composing an internal context. Local minima are created by frequent contexts learning. This local minima are the sequence memories.

THE PHYSICAL CORRELATES OF LOCAL MINIMA

L.F.A.Wessels E.Barnard
E van Rooyen
Institute for Micro Electronics /
Department of Electronic and Computer Engineering
University of Pretoria
Pretoria 0001, South Africa

Abstract

The training of various supervised neural-net classifiers is accomplished by optimization of a criterion function. For reasons of efficiency a local optimizer is usually employed for this purpose. Thus, the training of neural-net classifiers is often hampered by the occurrence of local minima, which results in the attainment of inferior classification performance. We study the following problem: which physical states of the classifier tend to correspond with local minima in the criterion function? Such an understanding of the physical correlates of local minima is important, since it may be utilized to choose the weights from which training is initiated in a more sensible manner. Specifically, it may be possible to decrease the probability of arriving at a local minimum. For the particular case of backpropagation classifiers, we show that the occurrence of a local minimum in the criterion function can often be related to specific patterns of defects in the classifier. In particular, three main causes for local minima are identified:

- the straying of hidden nodes so that they are either strongly active or very inactive for all training samples;
- duplication of function by pairs of hidden nodes; and
- arrangements of hidden neurons so that they are all highly inactive in certain regions of feature space.

These are the most common causes of local minima, but certain other types of local minima are also shown to exist.

Dynamics and Associative Mapping in Additive Systems[1]

M. Ceccarelli, A. Petrosino
Centro di Studio sui Calcolatori Ibridi, CNR
Via Claudio, 21
I-80125 Napoli, ITALY

R. Tagliaferri
Dipartimento di Informatica ed Applicazioni
Universita' degli Studi di Salerno
I-84081 Baronissi (Salerno), ITALY

Abstract

In this paper we present a model of Additive Automata, which is particularly useful to accomplish some specific associative tasks. For such systems and, generally, for neural nets a functional requirement is a non-ergodic evolution of net states, therefore it is useful to know some relationships between the structure of the net state space and the evolution law of the single processing element. In this way the aim of this paper is to analyze the dynamics of such systems (structure of attraction basins, cycle lenghts, number of transient states, etc....) and present a tool (well known constants of motion) to factorize the space of the net states into subspaces. Finally we give two simple laws which enable us to memorize either patterns or associations among them.

[1]This work was supported by CNR, Progetto Finalizzato "Sistemi Informatici e Calcolo Parallelo", by MPI 40% and by IIASS

ANALYSIS OF NETWORK DYNAMICS II

Chair: Shun-Ichi AMARI

Learning Filter Systems

Reiner Lenz, Mats Österberg
Linköping University, S-58183 Linköping, Sweden

Abstract

This paper studies properties of so-called basic units. First we investigate an eigenvalue problem that turns up in the study of the stable states of such units. It is well known that basic units using Hebb-type learning rules converge to stable states which are eigenfunctions of an integral equation whose kernel is given by the covariance function of the input process. In the first part we investigate one basic unit and we assume that the set of input patterns of this basic unit is regular in the sense that all patterns can be derived from a single prototype pattern by a group theoretically defined transformation. We will then show that the stable states of the unit are uniquely determined by the symmetry of the input set.

In the second part of this paper we investigate systems consisting of several such basic units and we describe a learning procedure for such a system. We demonstrate that the system stabilizes in a state in which the different basic units are characterized by the group theoretically derived filter functions. We train the system with an input set consisting of rotated edge and line patterns and we show that the stable states of the system are characterized by pure line and pure edge detectors. This does not only demonstrate that the system stabilizes in the group theoretically defined states but it shows also that the system learned to discriminate between two different classes of patterns. Finally we demonstrate how the system can be used in texture segmentation.

1 Introduction

In the study of neural networks one often investigates so-called basic units that act as linear filters (see [1]). As an example consider a cell in the visual system that processes input from a limited region of the retina. The input of the cell describes the gray value distribution in this region, the receptive field of the cell. As a result the cell responds to this input by sending a signal to cells in the next layer of the network. Although the outgoing signal is in general a nonlinear function of the input it is interesting to approximate the response by a linear combination of the input signals. This simplifies the analysis and such processes may also by useful in the design and analysis of artificial networks.

The behaviour of such basic units using different learning rules was intensively studied in the past but many of these investigations do not seem to take into account the fact that the environment in which the network exists is not just a set of randomly selected signals but that it usually possesses a large amount of regularity. In the first part of this paper we will summarize some of our results from the study of the behaviour of a basic unit in the case where its input consists of set of functions which possesses regularities that can be described by group theoretical methods. We will concentrate on the case where the input to the basic units consists of rotated versions of a 2-D gray value distribution. In this case we show that the resulting basic unit implements a filtering of the incoming signal with a rotation-invariant operator.

In the second part of this paper we consider systems that consist of several of these units. At a given point in time all the basic units of the system receive the same input signal. From this signal they compute their output signals in the same way as the single basic units do. Finally they update their current internal state where the new state of each unit is a function of its own internal state, the input signal, the length of the learning period and the output signals of the other units. The vector consisting of all the output signal of the different units will be called a feature vector. We describe a learning rule that ensures that the set of feature vectors has a maximum variation and that the components in the feature vectors are maximally concentrated.

In one experiment we trained the system with a set of rotated edge- and line patterns and we demonstrated that the stable states of the units correspond to (pure, lowest order) edge- and line filters constructed by the group theoretical method (see [2] and and [3]). In a second experiment we trained the system with two different types of textures. Later we used the filter functions obtained to segment an image consisting of these two textures.

2 Basic units and regular signal spaces

In the following we define a basic unit as a device that correlates an incoming signal $s_t(X)$ with a weight function $w_t(X)$ and transmits the results $o_t = s_t(X) \star w_t(X)$ via its output connection to other basic units in the network. The function $w_t(X)$ describes the state of the basic unit. Sometimes we will call $w_t(X)$ a filter function since the basic unit filters the incoming signal with $w_t(X)$. In this section we denote the time dependency by the subscript t. Note also that the unit sends the raw correlation result to the other units, not a thresholded result as is usual in the theory of neural networks.

The unit can change its behaviour, as mentioned before, and in first part of this paper we will mainly assume that the unit is updated with a rule of the form:

$$w_t'(X) = w_{t+1}(X) - w_t(X) = \gamma_1(t)w_t(X) + \gamma_2(t)o_t s_t(X)$$

We will say that $w_t(X)$ is a stable state of the unit if $w_{t+1}(X) = w_t(X)$. In this case we write $w_t(X) = w(X)$.

In [1] Kohonen showed that stable states are solutions of an eigenvalue problem. In his solution he did not pose any conditions on the type of input signals of the unit. In reality we have however often more information about the possible input signals. Often we know that the space of possible input signals is highly structured. As an example consider the case where s is a fixed signal, the so called prototype pattern. Related to this signal is a whole class of similar patterns: the rotated patterns $s_R(X) = s(R(X))$. A unit that receives only inputs signals of this form and that is trained according to the update rule mentioned above will recognize patterns that are similar to the signals $s_R(X)$.

In the cases where the class of similar patterns is produced from a prototype signal by rotation it can be shown that the stable states of the unit are all of the form:

$$w(X) = w(r, \varphi) = h_n(r) \cos n\phi \qquad \text{or} \qquad w(X) = w(r, \varphi) = h_n(r) \sin n\phi$$

where r, φ are polar coordinates in the plane and where $h_n(r)$ is a radial weight function that depends on the prototype pattern s. We see thus that such a unit (if it reaches a stable state) implements a filter process with rotation invariant filter functions (see [2] and [3]).

The fact that all stable states of a unit have the same form can be generalized as follows: We select again a fixed prototype pattern $s(X)$. Now we assume that there is a group G and a rule that maps the pair (s, g) consisting of the prototype signal s and a group element g into a new input signal $s_g(X)$. If this mapping $s \mapsto s_g$ and the group G itself satisfies certain, mild conditions then it can be shown that the stable states of the units are all connected to so-called representations of the group G. In the case where G is the 3-D rotation group and s is a gray value distribution in a sphere we find as filter functions the 3-D rotation-invariant operators: $w(X) = w(r, \varphi, \psi) = h_n^m(r) Y_n^m(\varphi, \psi)$ where Y_n^m are the surface harmonics and (r, φ, ψ) are the polar coordinates in 3-D.

Note that the update rule described above specifies only how one unit must be updated, it doesn't tell us how to construct a whole set of such filter functions. In the next section we will describe some of our experiments that demonstrate that it is also possible to find a trainings procedure that leads to a system consisting of several independent filter functions.

3 Systems of basic units

We now turn our attention to coupled systems of such simple basic units. We will assume that such a system has the following properties:

1. The primary input signal $s(t)$ is the same for all basic units in the system, i.e. all units see the same thing.
2. Each unit knows the output signals of all units, but not the internal states of the other units.

At each time step the system performes two operations:

1. Every unit convolves the input signal with its own status vector, the result is the output signal of this unit.

2. Using the input and the output signal of all units the unit updates its own status vector.

The system now acts as a system of linear filters with a time-varying filter kernels.

We denote the number of a basic unit by a subscript and we get the following relations:

1. $o_k(t) = w_k(t) * s(t)$ is the computed output value of unit k at time t and

2. the update rule is of the form: $w_k(t+1) = F(w_k(t), o_1(t), ..., o_N(t), s(t))$ where N is the number of units in the system.

From the form of the update rule one sees at once that gradient based learning rules produce update vectors that depend on the input signal, the output values of all units and on the internal state of the updating unit itself.

The selection of our quality function is based on two principles: The maximum variation and the maximum concentration principle.

By the maximum variation principle we mean the requirement that the feature vectors extracted from the trainings set should have a maximum variation. We could also say that the extracted feature vectors should contain a maximum of information about the trainings set. We measure this variation with the function:

$$Q_V(w_1(t), ..., w_N(t)) = \det(C(t))$$

where $C(t)$ is the mean covariance matrix of the output values at time t. The ij-th element c_{ij} of $C(t)$ has thus the value: $c_{ij}(t) = \text{mean}_{\tau \leq t} o_i(\tau) \cdot o_j(\tau)$.

From the definition of the quality function one sees immediately that a system $S(t) = (w_1(t), ..., w_N(t))$ and the transformed system $S_R(t) = (R \cdot (w_1(t), ..., \cdot w_N(t)))$ (where R is an orthogonal transformation) have the same quality measure $Q_V(S) = Q_V(S_R)$. The result of this trainings procedure can thus not be unique.

This ambiguity can be removed if we introduce other requirements on our filter system. In the following we assume that the set of input patterns often consists of several classes. In this case it seems natural to require that the filter system reflects in some way these different classes. In this case a better system of filter functions would be a system in which certain filter functions respond only to pattern from specific classes but ignore the patterns from the other classes.

One possible choice of a quality function that measures this behavior is the following entropy-like function:

$$Q_E(w_1(t), ..., w_N(t)) = \text{mean}_{\tau \leq t} \sum_{n=1}^{N} o_n(\tau)^2 \log(o_n(\tau)^2).$$

Finally we combine these two quality functions to new quality function that leads to a maximum variation and a minimum entropy filter system:

$$Q(w_1(t), ..., w_N(t)) = \frac{Q_V(w_1(t), ..., w_N(t))}{Q_E(w_1(t), ..., w_N(t))}$$

Using this quality function we trained the filter systems using a gradient based optimization technique.

4 Experiments

In the following series of figures we demonstrate how this system works when it is confronted with simple one-dimensional patterns. In the first figure 1 we show two typical input signals, a step edge and a line-like signal. We can think of them as the angular variation of two-dimensional edge and line patterns. The set of all trainings patterns consists of all shifted versions of these two patterns, corresponding to the rotated edges and lines in 2-D. In the next figure 2 we see the status vectors of the units after 5000 iterations. These filter functions are pure sine and cosine functions and they respond thus either to edge or to line patterns, but not to both. The experiment described above shows that there are procedures that lead to the same filter functions as predicted by the theory on group-invariant filter systems.

In a second experiment we trained a filter system consisting of four filter functions with 10000 randomly selected 16×16 neighborhoods from the image consisting of two textures. Then we applied the filter functions obtained in the trainings procedure to the image shown in figure 3. Figure 4 shows the magnitude of the resulting filtered image when we applied filter number three. A comparison shows that the filter functions one and two where adapted to the background pattern whereas the filters three and four learned the foreground texture. These preliminary results show that the filter design strategy may also be useful for more complicated image processing tasks than edge and line detection.

References

[1] Teuvo Kohonen. *Self organization and associative memory*. Springer Verlag, Heidelberg, Berlin, New York, 1984.

[2] Reiner Lenz. A group theoretical model of feature extraction. *Journal of the Optical Society of America A*, 6(6):827–834, 1989.

[3] Reiner Lenz. *Group Theoretical Methods in Image Processing*. Lecture Notes in Computer Science. Springer Verlag, Heidelberg, Berlin, New York, in print.

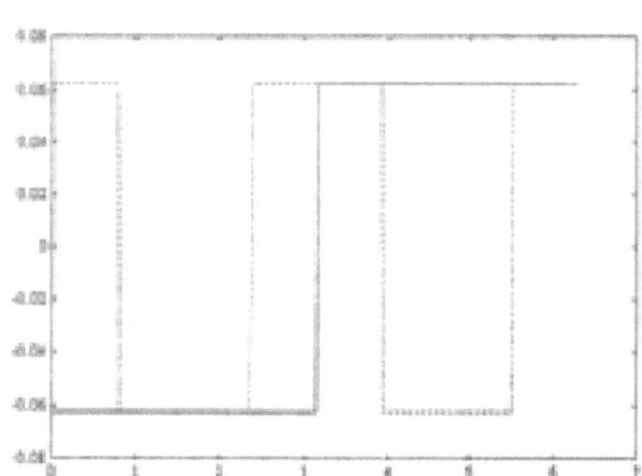

Figure 1: Two Edge and Line Patterns.

Figure 3: Test Image with two different Textures

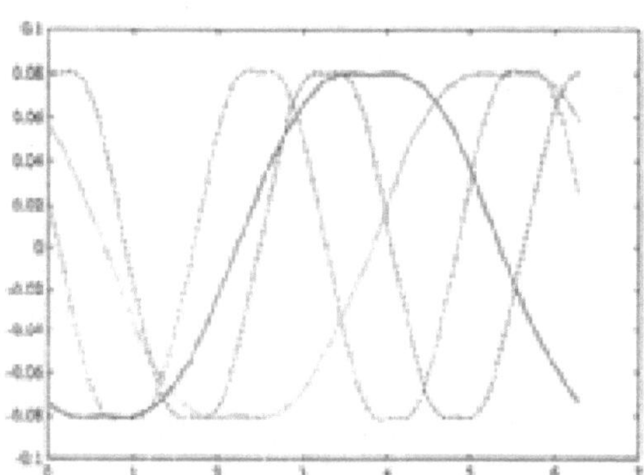

Figure 2: Filter Functions using Variation/Entropy after 5000 Iterations

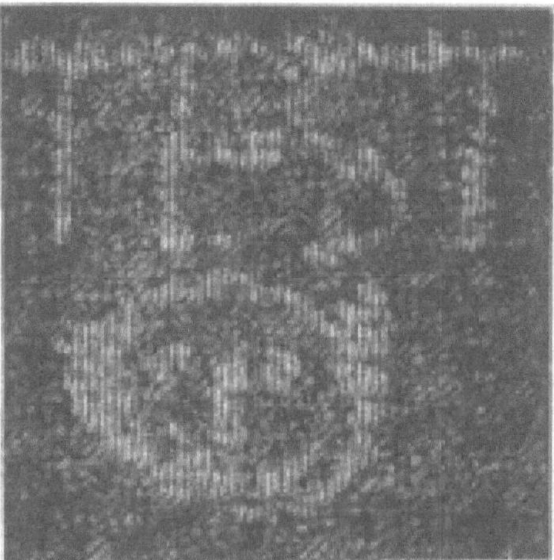

Figure 4: Result Image with Filter function 3

Fokker-Planck Description of Learning in Backpropagation Networks

G. Radons, H.G. Schuster and D. Werner
Institut für Theoretische Physik, Universität Kiel, D-2300 Kiel, FRG

Stochastic pattern presentation induces fluctuations in the weigths of Backpropagation networks, which enable the system to escape from local minima in parameter space. For small learning rates we find that learning is governed by a Fokker-Planck equation. The parameter dependence of the resulting diffusion tensor suggests how for perfectly trainable networks parameters can be made to converge to globally optimal values corresponding to an errorfree implementation of the desired input–output relations. For cases where perfect learning is impossible we demonstrate the usefulness of a simulated annealing-like procedure to reach the minimal error state. We also propose a new activation function which can drastically improve learning as is demonstrated for the parity problem.

Backpropagation (BP) [1] is currently the most prominent neural network and has proven to be able to e.g. convert text to phonems [2], to predict chaotic time series [3], to predict the secondary structure of proteins [4] or to learn how to drive a car [5].

BP is an algorithm which iteratively adjusts internal parameters $\vec{w} = (w_1, w_2, \cdots, w_N)$, the "synaptic strengths" and thresholds of a multilayer network, to implement relations between input patterns $\vec{\xi}^\mu = (\xi_1^\mu, \xi_2^\mu, \cdots, \xi_M^\mu), \mu = 1, 2, \cdots$ and L-dimensional target outputs $\vec{f}(\vec{\xi}^\mu)$. The number of patterns can be finite or infinite. If an input pattern $\vec{\xi}^\mu$ is fed into the network, it produces an output $\vec{g}(\vec{\xi}^\mu; \vec{w})$ which is in general different from the desired output $\vec{f}(\vec{\xi}^\mu)$. The task consists in finding network parameters $\vec{w}^*$ which minimize the ***averaged error***

$$E(\vec{w}) = \frac{1}{2} \sum_\mu \left[\vec{f}(\vec{\xi}^\mu) - \vec{g}(\vec{\xi}^\mu; \vec{w})\right]^2 \tag{1}$$

between actually produced outputs $\vec{g}(\vec{\xi}^\mu; \vec{w})$ and target outputs $\vec{f}(\vec{\xi}^\mu)$. Whithin BP this problem is tackled by performing gradient descent steps on the surfaces of ***individual errors*** $E^\mu(\vec{w}) = \frac{1}{2}[\vec{f}(\vec{\xi}^\mu) - \vec{g}(\vec{\xi}^\mu; \vec{w})]^2$, i.e. iteratively changing $\vec{w}$ according to

$$\vec{w}(t+1) = \vec{w}(t) - \eta\, \vec{\nabla} E^\mu(\vec{w}(t)) \tag{2}$$

where η is the learning rate. Often an additional term the so called momentum term is included in (2) to improve the performance of Backpropagation [1]. In the following this term is omitted for simplicity but can easily taken into account. In Backpropagation networks with e.g. 3 layers the nonlinear function $\vec{g}(\vec{\xi}^\mu; \vec{w})$ can always be written as

$$\vec{g}(\vec{\xi}^\mu; \vec{w}) = \vec{\sigma}\left(W^{(2)}\, \vec{\sigma}(W^{(1)} \vec{\xi}^\mu) \;+\; W^{(0)} \vec{\xi}^\mu\right) \tag{3}$$

where the elements of the synaptic matrices $W^{(i)}$ constitute the vector $\vec{w}$ and $\vec{\sigma}(\vec{x}) \equiv (\sigma(x_1), \sigma(x_2), \cdots)$, with $\sigma(x)$ denoting the monotonically increasing activation function. The following discussion and conclusions, however, are largely independent of the form of $\vec{g}$ and therefore apply also to more general nonlinear optimization problems.

Note that a trajectory $\vec{w}(t), t = 0, 1, 2, \cdots$ generated by system (2) depends in an essential way on how the patterns $\vec{\xi}^\mu$ are presented. For instance if the $\vec{\xi}^\mu$ are presented periodically, i.e. $\vec{\xi}(t) = \vec{\xi}(t+p)$ with p being the number of patterns, the force $\vec{F}^\mu(\vec{w}) \equiv -\vec{\nabla} E^\mu(\vec{w}) = -\vec{\nabla}_w\, E(\vec{\xi}^\mu; \vec{w})$ is explicitly time dependent with $\vec{F}^\mu(\vec{w}) = \vec{F}(\vec{\xi}(t); \vec{w}) = \vec{F}(\vec{\xi}(t+p); \vec{w})$. Thus in this case Eq. (2) describes a discrete periodically driven nonlinear dynamical system. Such systems are known to exhibit chaotic behaviour and one wonders if chaos may help the system to escape from local minima in the error landscape $E(\vec{w})$. The problem of getting stuck in local minima is well known for gradient descent algorithms but can indeed be avoided via chaotic diffusion as a result of periodic pattern presentation [6]. In the following, however, we will consider only stochastic pattern presentation, where escape from local minima is possible, even if trapping occurs for the periodic case. For stochastic pattern presentation the update steps in Eq. (2) are performed by changing the individual error functions E^μ in every step in a random fashion. This turns the corresponding force $\vec{F}(\vec{\xi}(t); \vec{w}) = -\vec{\nabla}_w\, E(\vec{\xi}(t); \vec{w})$ into a stochastic force. Since now every realization of the random sequence $\vec{\xi}(t), t = 0, 1, 2, \cdots$ gives rise to a different solution $w(t)$ of Eq. (1), only probabilistic statements can be made.

The central quantity corresponding to the stochastic process $\vec{w}(t+1) = \vec{w}(t) + \eta\vec{F}(\vec{\xi}(t)\,\vec{w}(t))$ is the probability density $P(\vec{w},t)$ of finding the system at time t in state $\vec{w}$ of the parameter space. It is straight forward to show that the time evolution of $P(\vec{w},t)$ is governed by a Chapman-Kolmogorov equation of the form

$$P(\vec{w},t+1) = \int T(\vec{w},\vec{w}')\,P(\vec{w}',t)\,d\vec{w}' \tag{4}$$

where the transition matrix is given by

$$T(\vec{w},\vec{w}') = \left\langle \delta(\vec{w} - \vec{w}' - \eta\,\vec{F}(\vec{\xi};\vec{w}'))\right\rangle \tag{5}$$

and $<\cdots>$ denotes the average over all patterns $\vec{\xi}^{\mu}$ i.e. $<f(\vec{\xi})> = \int d\vec{\xi} f(\vec{\xi})\rho(\vec{\xi})$ with the distribution function $\rho(\vec{\xi})$ which may be continous or discrete and is normalized to one. To proceed we write the r.h.s. of (4) formally as a power series in the learning rate η and use that $P(\vec{w},t)$ vanishes at infinity. For simplicity we give the result in one dimension

$$P(w,t+1) - P(w,t) = \sum_{l=1}^{\infty} \frac{(-\eta)^l}{l!}\frac{\partial^{(l)}}{\partial w^{(l)}}\left[\langle F^l(\xi;w)\rangle\,P(w,t)\right] \tag{6}$$

Eq. (6) is a Kramers-Moyal like expansion [7] which is fully equivalent to (4) but equally intractable. For small learning rates η, however, it is justified to truncate the r.h.s. of Eq. (6) after the second term and to replace the difference on the l.h.s. by a differential. We obtain for the general case a multidimensional Fokker-Planck Equation (FPE).

$$\frac{\partial P(\vec{w},t)}{\partial t} = -\eta\sum_l \frac{\partial}{\partial w_l}\left[F_l(\vec{w})\,P(\vec{w},t)\right] + \frac{\eta^2}{2}\sum_{kl}\frac{\partial^2}{\partial w_k \partial w_l}\left[D_{kl}(\vec{w})\,P(\vec{w},t)\right] \tag{7}$$

with drift coefficients $\eta F_l(\vec{w})$ which are simply the components of the average force $\vec{F}(\vec{w}) = -\vec{\nabla}E(\vec{w})$ resulting from the averaged error surface Eq. (1) as expected and a diffusion tensor $\eta^2 D_{kl}(\vec{w})$ with $D_{kl}(\vec{w}) = <F_k(\vec{\xi};\vec{w})\,F_l(\vec{\xi};\vec{w})>$. We are mainly interested in stationary solutions $P^*(\vec{w},t+1) = P^*(\vec{w},t) = P^*(\vec{w})$ of Eq. (4) i.e. the asymptotic distribution of network parameters after long times. It is known that stationary solutions of (4) can be nowhere differentiable e.g. fractal measures, whereas stationary states obtained from approximation (7) are differentiable. In the following we demonstrate the usefulness of (7) for practical purposes. In one dimension the stationary solution $P^*(\vec{w})$ of Eq. (7) is easily obtained by integration [7]

$$P^*(w) = \frac{\mathcal{N}}{D(w)}\exp\left(\frac{2}{\eta}\int^w \frac{F(w')}{D(w')}dw'\right) \tag{8}$$

where $\mathcal{N}$ is a normalization constant and the parameter dependent diffusion tensor is reduced to $D(w) = <F^2(\xi,w)>$. This solution is compared in Fig. 1a with results from numerical simulations of Eq. (4) for a simple example which nevertheless exhibits the typical features of the full many-dimensional problem. In this example we consider only one "neuron" with one-dimensional input ξ and target output $f(\xi)$ and the nonlinear output function g coincides with the transfer function of a single neural unit $g(\xi;\vec{w}) = \sigma(w_1\xi + w_2)$ where σ is the typical sigmoid function $\sigma(x) = \frac{1}{2}(1+\tanh x)$ used in Backpropagation networks. The task of the "network" consist in learning of only two patterns ξ^1 and ξ^2 but with only one free parameter w since we fix w_1 at a constant value $w_1 = a$ and try to optimize the "threshold" parameter " $w_2 = w$. The resulting one-dimensional error landscape is given by

$$E(w) = \sum_{\mu=1,2} E^{\mu}(w) = \sum_{\mu=1,2}\frac{1}{2}[f(\xi^{\mu}) - \sigma(a\xi^{\mu} + w)]^2 \tag{9}$$

and is depicted in Fig. 1a. Here the errors E^{μ} could be made zero individually but not the total or averaged error E. This, of course, stems from the fact that both conditions $f(\xi^{\mu}) = \sigma(a\xi^{\mu} + w), \mu = 1,2$ cannot in general be fulfilled simultaneously by only one parameter w. This competition between individual patterns leads to local minima in $E(w)$.

It is observed that the numerically obtained distribution $P(w,t)$ indeed evolves towards the stationary solution (8) of the FPE (7), showing that there is a high probability to find the system near the globally optimal value w^*.

There is an interesting aspect of Eq. (8) with respect to perfectly trainable networks. These networks are characterized by the fact that there exist paramters $\vec{w}^*$ such that *all* individual errors E^{μ} are minimal at w^* namely zero. Since the E^{μ} are differentiable functions of $\vec{w}$ all individual forces $\vec{F}(\vec{\xi}^{\mu};\vec{w})$ must also vanish at $\vec{w}^*$. This means that $\vec{F}(\vec{w}) = <\vec{F}(\vec{\xi}^{\mu};\vec{w})>$ as well as $D_{kl} = \left\langle F_k(\vec{\xi};\vec{w})\,F_l(\vec{\xi};\vec{w})\right\rangle$ is zero at $\vec{w} = \vec{w}^*$. In (8)

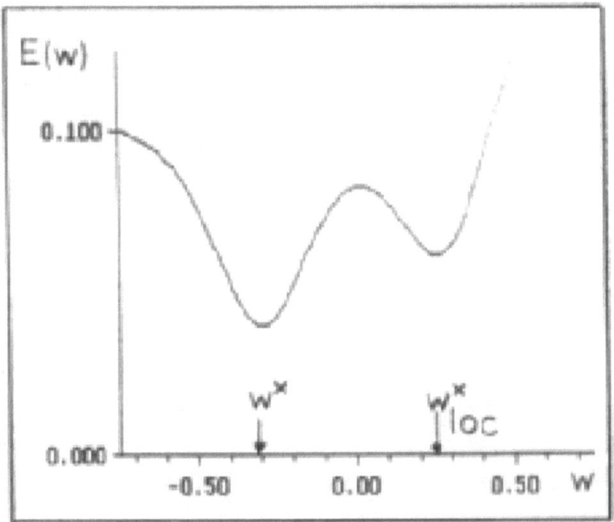

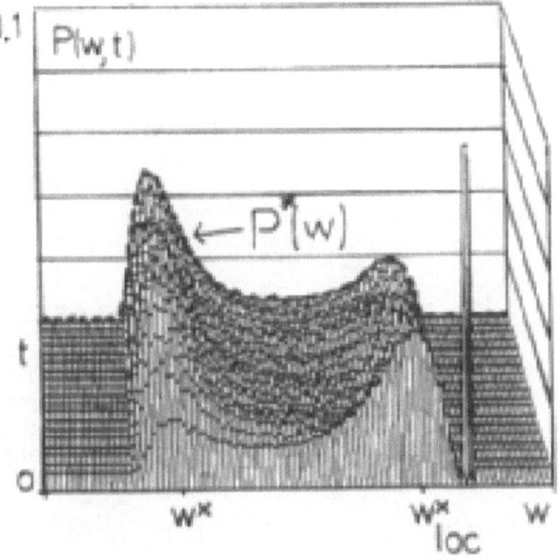

Figure 1: a) Error landscape $E(w)$ of Eq. (9). b) Corresponding Evolution of the probability distribution $P(w,t)$ to its equilibrium $P^*(w)$ for fixed $\eta = 0.5$. The dashed line is approximation $P^*(w)$ of Eq. (8).

this leads to a logarithmic divergence of the integral at $\vec{w}^*$ because $\frac{D(w)}{F(w)} \to 0$ for $w \to w^*$. As a consequence $P^*(w)$ diverges at $w = w^*$ i.e. becomes a δ-distribution. Although multidimensional generalizations of solution (8) are difficult to obtain because the potential conditions[7] are not fulfilled (no detailed balance), one can see directly from Eqs. (4), (5) that also in many dimensions $\delta(\vec{w} - \vec{w}^*)$ is a solution in the perfect learning case.

From this we cannot conclude, however, that the total probability is concentrated in the δ-distribution. If one could show that there are no local minima other than possibly degenerate absolute minima, then the drift term in Eq. (8) would drive all probability into absolute minima of $E(\vec{w})$ and with probability one parameters would take their optimal values. Though this often happens in perfectly trainable networks, we cannot exclude the occurence of local minima even in this case.

To drive the system from these local minima into absolute minima is, however, possible by choosing high learning rates η. This causes strong fluctuations of $\vec{w}$ in local minima but not at absolute minima where in perfectly trainable networks the fluctuations are zero. Thus trajectories $\vec{w}(t)$ are "heated up" everywhere in parameter space except at absolute minima where they "freeze" into their optimal values.

This picture is confirmed directly by calculating the induced parameter fluctuations from Eq. (2) near a minimum $\vec{w}^*$ of $E(\vec{w})$ in the quadratic approximation for $\eta \ll 1$

$$\overline{\vec{w}^2} - \overline{\vec{w}}^2 = \frac{\eta}{2} \langle \vec{F}(\vec{\xi};\vec{w}^*)^T < \vec{\nabla} \otimes \vec{F}(\vec{\xi};\vec{w}^*) >^{-1} \vec{F}(\vec{\xi};\vec{w}^*) \rangle \tag{10}$$

Comparing this in the one-dimensional case with the coordinate fluctuation of an overdamped particle in a quadratic potential in a heat bath of temperature T shows that $\frac{\eta}{2}\langle F^2 \rangle$ takes the role of the temperature, which vanishes at the absolute minima in perfectly trainable networks. This analogy suggests that in networks where perfect learning is impossible a simulated annealing-like procedure [8] is appropriate to find global minima of the error surface. For BP nets this simply means decreasing η in the course of time. In Fig. 2b we demonstrate the usefulness of this procedure again for our simple example Eq. (9) with η decreasing exponentially in time. We see that all samples $w(t)$ finally relax into the global minimum at w^* although there was initially a high probability of finding them in the local minimum at w^*_{loc} . From the above general considerations and from numerical simulations it follows that the same picture holds also for higher dimensional problems i.e. full networks with many parameters providing us with an optimized strategy of finding optimal parameters in Backpropagation nets. Another interesting consequence of Eqs. (7), (10) is that other choices of the activation function σ of Eq. (3) may lead to an accelerated convergence of learning parameters. E.g. choosing $\sigma(x) \sim \sinh^{-1}(ax)$ instead of $\tanh(x)$ avoids flat regions in $E(\vec{w})$ and increases the escape probability from local minima according to (10), thus resulting in an improved performance of Backpropagation. This is shown in Fig.2b for the parity problem[6] with 3 input and 6

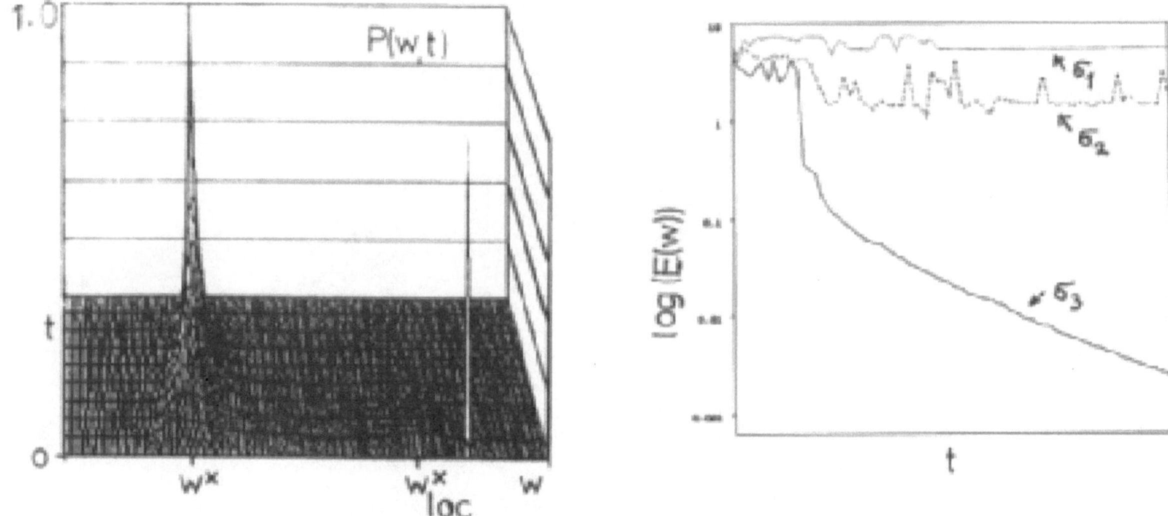

Figure 2: a)Evolution of the probability $P(w,t)$ to its equilibrium $P^*(w)$ as Fig. 1b, but for exponentially decreasing $\eta(t)$: All samples converge to the global minimum $\vec{w^*}$ (Note: Same initial condition as in a), different P-scale).
b) Comparison of learning dynamics for 3 different activation functions *sigma* for the parity problem (see text).

hidden units. In Fig. 2b we compare the decrease of the error Eq.(1) during learning ($\eta = 0.75$) for $\sigma_1 = \frac{1}{2}(1 + \tanh x)$,$\sigma_2 = \frac{2}{\pi}(\arctan \pi x/2)$, $\sigma_3 = \frac{1}{a}(\sinh^{-1}(ax))$. The results of Fig. 2a) and b) suggest that combining σ_3 with decreasing $\eta(t)$ should lead to optimal convergence properties.

One of us (D.W.) acknowledges support from the SFB 185 "Nichtlineare Dynamik"

References

[1] D. Rumelhart and J. McClelland. *Parallel Distributed Processing.* Volume 1, MIT Press, Cambridge, MA, 1986.

[2] C.R. Rosenberg and T.J. Sejnowski. *Complex Systems*, 1:145, 1987.

[3] A.S. Lapedes and R. Farber. Technical Report LA-UR-87, Los Alamos National Laboratory, 1987. submitted to Proc. IEEE.

[4] N. Qian and T.J. Sejnowski. *J.Mol.Biol.*, 202:865, 1988.

[5] D.A. Pomerleau. In D.S. Touretzky, editor, *Advances in Neural Information Processing Systems 1*, Morgan Kaufmann Publishers, San Mateo, CA, 1989.

[6] G. Radons, H.G. Schuster, and D.Werner. In R. Eckmiller, editor, *Proc.Int.Conf. on Parallel Processing in Neural Systems & Computers ICNC '90*, Elsevier Science Publ., Amsterdam, 1990.

[7] C.W. Gardiner. *Handbook of Stochastic Methods.* Springer Verlag, Berlin, 1983.

[8] S. Kirkpatrick, C.P. Gelatt.Jr., and M.P. Vecci. *Science*, 220:671, 1983.

FUNDAMENTAL STRUCTURE/BEHAVIOUR RELATIONSHIPS IN SYNCHRONOUS BOOLEAN NEURAL NETWORKS

Derek K Milligan and Manissa J Dobrée Wilson

Department of Electrical Engineering & Electronics
Brunel University
UXBRIDGE
Middx. UB8 3PH
UK.

ABSTRACT

In order to successfully synthesise Boolean Neural Networks with specific capabilities, it is important to understand the relationships between network behaviour and network structure. The essential obstacle is the lack of an analytic algebraic mapping between transformations within the hyper-state space and transformations within the function space of the network . This paper suggests ways of identifying structure within both the behaviour and functional structure of classes of simple Boolean Neural Networks.

I. Introduction

The majority of past and present work in Neural Networks is concerned with the identification of algorithms, functions and structures which enable the attainment of behaviours exhibited by natural neural systems. Some of the results have a formal theoretic basis, e.g., [Amari],[Hinton],[Kohonen]. The theory is useful in explaining the results of simulations, but unfortunately is not so helpful in an engineering sense in providing neural systems designers with synthesis tools.

In the Boolean Neural Net sphere, several analytic theories have been developed [Cull],[Caianiello],[De Luca]. Work on the synthesis of function from behaviour is less apparent, apart from, of course, the considerable body of knowledge in the engineering area of combinational/sequential logic synthesis and Automata theory. The latter field has produced useful results in the context of the design of general logic systems but not neural networks. Currently, training algorithms, architectures and theoretical models of networks of probabilistic Boolean nodes are being studied [Aleksander],[Myers]. In order to assess the requirements for probabilism, however, we believe that a better understanding of deterministic nets is necessary.

As the use of Neural Networks increases, designers will require tools and techniques for the synthesis of Neural Networks. In the Boolean Neural Networks sphere much work remains to be done in the analysis of behaviour/structure relationships before the problem of synthesis can be addressed. Indeed, at present, most results depend on a large amount of simulation of empirically-derived networks.

II. Mapping Net Structure and Behaviour

The essential obstacle to the full understanding of sequential Boolean Neural Networks is the lack of an analytic algebraic mapping between transformations within the function space of the network and transformations within the hyper-state space. That is, there is, at present, no known direct link between changing the function or connectivity of a network and the resulting change to the state structure of the network, and vice versa. The need for an analytic mapping as described can be easily justified by considering the requirements of any Boolean Neural Network design aid. The properties of any designed network which are of interest are firstly and primarily the behaviour of the network which. at the lowest level, is represented by the state structure and secondly the economics of the implemented network such as the degree of connectivity.

Neural Networks can be classified as having behaviours which are either adaptive or fixed. In the case of the latter, it is evident that synthesis tools will be needed to produce network structures from given behavioural requirements. An additional problem is the lack of a succinct language for the description of network behaviour at an abstraction level above that of state structures. In the case of adaptive networks, the design process would aim to achieve a given range of learnable behaviours for the given application constraints. Learning algorithms can be viewed as simply 'walks' through the hyper-state space of the network and therefore predicting the capabilities of networks with given constraints for particular training algorithms requires the previously described mapping between function space and hyper-state space transformations. In both cases, the establishment of synthetic and analytic tools and techniques requires a deep understanding of the lowest level relationships between network structure and behaviour.

Figure 1 illustrates the relationships within and between the set of network structures (including functions) and the corresponding set of state structures, where N1, N2 are network structures and S1, S2 state structures. Mappings a and b relate a given network to its state structure and have been investigated by several workers, including Cull and Caianiello, who have established linearised matrix representations for them. It is not clear whether these representations will be useful in completely investigating the structures of the mappings of Figure 1. Mappings x and y are the mappings between the network function space and the hyper-state space which are the goal of the present work. In order to find a structure for x and y, it is necessary to investigate the structure of mappings n and s with given constraints (such as the maximum number of connections to a network element).

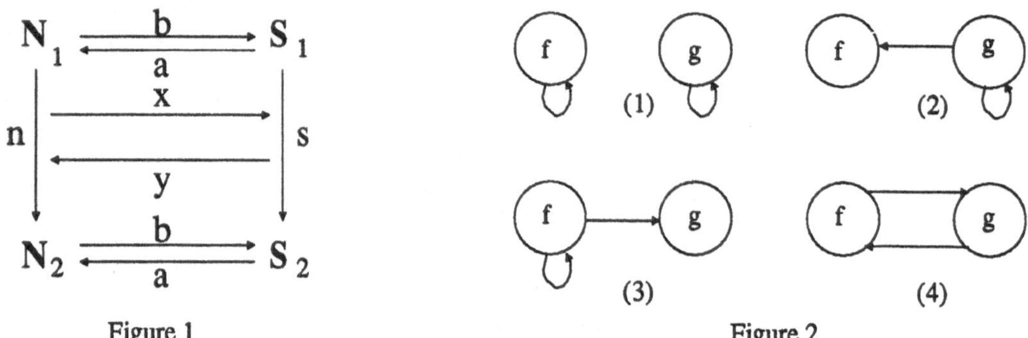

Figure 1

Figure 2

In order to study these structures, our approach is to investigate very small networks and to then attempt to expand the results to cover more complex networks. The rest of this paper describes some results for very simple networks.

III. Composition of Classes of Nets

The networks studied are autonomous, synchronous nets composed of r Boolean functions with n inputs each. For example, Figure 2 shows the four possible connection structures for nets of two elements having one input each, performing Boolean functions f and g. Each net structure can achieve a set of 16 state structures, each of which can be represented as a string of 4 numbers such that the position of a number in the string indexes the present state, whilst the value gives the next state. Thus, for example, the string 0213 represents the state structure having the following transitions: 0->0, 1->2, 2->1, 3->3. This representation of state structures is compact (for small nets) and allows a straightforward method for their composition. The total number of state structures belonging to the complete set of nets of Figure 2 is given by

$$(2 + 2\,{}^{2}C_1)^{2} = 36.$$

There are, of course, 256 possible state structures for a completely connected two-element net. This set of 36 state structures can be partitioned into subsets according to the number of distinct next states in a state structure which can, in this case, be only 1,2 or 4.

It can be shown that the set of the 36 state structures is closed under composition. The composition defined here is as follows. For state structures S1, S2 considered as mappings from the set of states to the set of states then (S1 * S2) is the natural left composition of the two mappings. For example, if S1 = 3120 and S2 = 0213 then (S1 * S2) = 3210.

Each such composition has a corresponding composition of the network structure/function. For the two-element net, the structure/function is defined by a pair of recursive functions:

$$f_t = f(a_{t-1})$$
$$g_t = g(b_{t-1}) \qquad (a, b \in \{ f, g \}).$$

The connection structure is defined by the values of a and b, so that a=g and b=g defines the connection structure of Figure 2(2). The composition of two nets, defined by

$$(f_t, g_t) \cdot (f_t', g_t') \quad \text{where} \quad \begin{array}{l} f_t' = f'(a'_{t-1}) \\ g_t' = g'(b'_{t-1}) \end{array} \quad (a', b' \in \{ f', g' \})$$

is
$$\begin{array}{l} f_t'' = f'c(x_{t-1}) \\ g_t'' = g'd(y_{t-1}) \end{array}$$

where
$$c = \begin{cases} f & \text{if } a' = f' \\ g & \text{if } a' = g' \end{cases} \quad d = \begin{cases} f & \text{if } b' = f' \\ g & \text{if } b' = g' \end{cases} \quad x = \begin{cases} a & \text{if } c = f \\ b & \text{if } c = g \end{cases} \quad y = \begin{cases} a & \text{if } d = f \\ b & \text{if } d = g. \end{cases}$$

For the example above, the net structure N1 equivalent to S1 is given by
$$\begin{array}{l} f_t = f(g_{t-1}) \\ g_t = g(f_{t-1}) \end{array}$$

and the net structure N2 equivalent to S2 is given by
$$\begin{array}{l} f_t' = f'(g'_{t-1}) \\ g_t' = g'(f'_{t-1}). \end{array}$$

Thus, both N1 and N2 have the form of Figure 2(4). Using the described composition, we achieve the net structure:
$$\begin{array}{l} f_t'' = f'g(f_{t-1}) \\ g_t'' = g'f(g_{t-1}) \end{array}$$
which has the form of Figure 2(1).

Clearly, this composition is expandable to all sequential nets with r elements having n inputs each. In addition, it is clear from the form of the composition that every class of nets of r elements with one input each is closed under the composition. Therefore the class of state structures corresponding to each class of such nets is also closed under state structure composition. The representation of the net structure/function as described has the useful property that it represents two levels of description; by not elaborating the functions we represent simply the connection structure, whilst elaboration of the functions gives a complete net description.

The set of 36 state structures corresponding to the set of nets of Figure 2 is given in Figure 3, where the set is partitioned into 3 equivalence classes depending on the number of distinct next states in each state structure. For this example, each state structure can be viewed as compositions of reflections, foldings or counter-foldings of the present state into the next state on the state space hyper-cube, exampled in Figure 4. The number of connections to a unit restricts the possible axes allowed for these transformations. As the composition of state structures can now be viewed as applying only the transformations in Figure 4, it is clear that a natural partial ordering exists in the hyper-state structure for this set of 36 state structures. Thus, set A in Figure 3 is the set of reflections and, in fact, forms a group. Sets B and C are the sets of foldings and counter-foldings, respectively, applied to set A, whilst set D corresponds to the composition of 2 (counter)foldings. It should be noted that foldings and counter-foldings have no inverses.

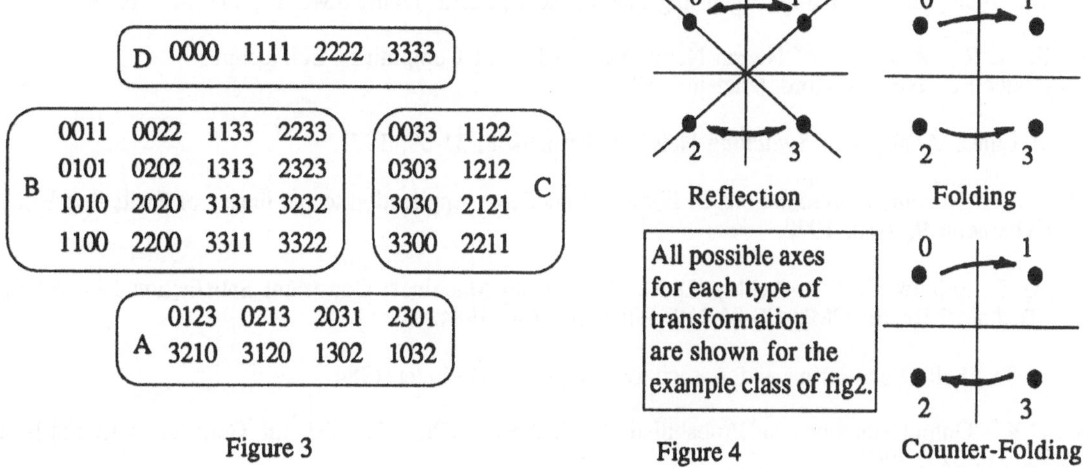

Figure 3 Figure 4

A minimum generator set for the complete set of 36 state structures contains 5 single transformations with 3 reflections, 1 folding and 1 counter-folding, for example

{1032, 2031, 0213, 0022, 0303}.

It is useful to calculate the size N_j of any subset of state structures having j distinct next states for the class of nets of r elements having n inputs each. This calculation is by no means trivial. For the case of n = 1, however, N_j is given by

$$N_j = 2^r m!\,{}^rC_m \sum_{i=m}^{rH(m)} {}^rC_{r-i}\; S(i,m)$$

where $m = \log_2 j$,
$H(m) = 1$ if $m > 0$
$= 0$ otherwise
and S(i,m) is the Stirling Number of the second kind.

For the case of n = r,

$$N_j = {}^{2^r}C_j\; j!\; S(2^r,j)$$ where S is as defined above.

For intermediate values of n, the computation is much more involved and is not yet fully understood.

IV. Conclusion

We have argued the need for a fuller understanding of the fundamental relationships between the structure/function and behaviour of Boolean Neural Networks and have outlined the areas that need to be addressed. Compositions in the hyper-state space and the function space of classes of networks have been described which indicate that useful structure can be established which will develop the required understanding. Although the nets studied to date have been very small, the results obtained can be extended to cover classes of more complex networks.

V. Acknowledgements

The authors are grateful to Dr. R M Zimmer for his suggestions on a formalism for network composition.

VI. References

Aleksander, I., "Canonical Neural Nets Based on Logic Nodes", Proc. 1st. IEE Int. Conf. on Artificial Neural Networks, 110-114, 1989.

Amari, S., "Field Theory of Self-Organising Neural Nets", IEEE Trans. SMC-13, 741-748, 1983.

Caianiello, E R., "A Theory of Neural Networks" in Neural Computing Architectures, I. Aleksander Ed., North Oxford Academic 1989.

Cull, P., "Linear Analysis of Switching Nets", Kybernetik 8, 31-39, 1971.

De Luca, A., "On some representations of Boolean Functions. Application to the theory of Switching Element Nets", Kybernetik 9, 1-10, 1970.

Hinton, G F., Sejnowski, T J., Ackley, D H., "Boltzmann Machines: Constraint Satisfaction Networks that Learn", Technical Report CMU-CS-84-119. Pittsburgh PA, 1984.

Kohonen, T., "Self-Organisation and Associative Memory", Springer 1984.

Myers, C E., "Output Functions for Probabilistic Logic Nodes", Proc. 1st. IEE Int. Conf. on Artificial Neural Networks, 310-314, 1989.

Dynamics of Lateral Interaction Networks

John Moody
Yale Computer Science, PO Box 2158 Yale Station
New Haven, CT 06520, USA; Email: moody@cs.yale.edu

Abstract

Recurrent lateral connectivity in a layer of processing units gives rise to a rich variety of nonlinear response properties such as overall gain control, emergent periodic response on a preferred spatial scale (*collective excitations*), and distributed *winner-take-all* (WTA) response. This diversity of response properties is observed in several different classes of simple network architectures including the additive linear network, the additive sigmoidal network, and the nonlinear shunting network. When Hebbian learning is coupled with network dynamics, these models have been shown to support the development of modular connectivity structures analogous to cortical columns.

1 Laterally Interconnected Network Architectures

Layered network architectures with recurrent lateral connection patterns are ubiquitous in vertebrate nervous systems and are likely to become increasingly important in neural network models for real-world information processing. It is important, therefore, to understand the full range of possible dynamic behaviors which can be displayed by such systems. As we shall see, these behaviors are surprisingly rich.

This short paper focuses on a few models with *localized*, rather than global, lateral interconnections. A more complete treatment of lateral interaction networks is given in Moody (1990). A study of how laterally interconnected network dynamics can drive the development of modular structures in cortex through Hebbian learning is presented in Chernjavsky and Moody (1990).

To demonstrate some of the possible behaviors, we consider two general classes of models (see figure 1): additive models and shunting inhibition models. Both classes possess a single layer of receptor cells which provide input to an internal layer of laterally-interconnected cells. The additive models contain a single population of internal cells which make both lateral excitatory and inhibitory connections. Both connection types are additive. The shunting inhibition models have two populations of cells in the internal layer: excitatory cells which make additive synaptic axonal contact with other cells and inhibitory cells which shunt the activities of excitatory cells. The shunting models are more realistic biologically.

The additive models are further subdivided into models with linear internal units and models with nonlinear (particularly sigmoidal) internal units. The shunting inhibition models have linear excitatory

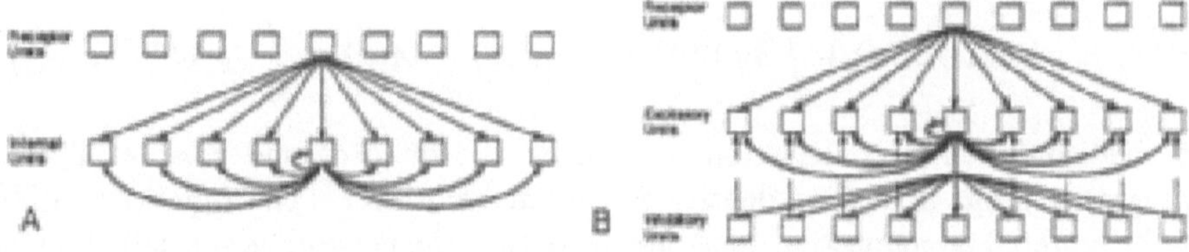

Figure 1: Network Models. A: Additive Model. B: Shunting Inhibition Model.

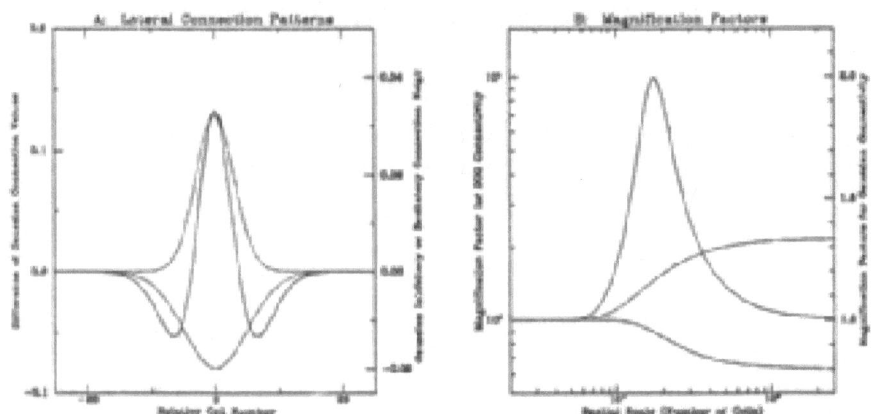

Figure 2: A: Excitatory, Inhibitory, and Difference of Gaussian Lateral Connection Patterns. B: Magnification Functions for the Linear Additive Model.

units and sigmoidal inhibitory units. We have considered two variants of the shunting models, those with and without lateral excitatory connections.

2 Nonlinear Dynamics in the Linear Additive Model

Of the various models mentioned above, only the linear additive model is exactly soluble. Furthermore, although the internal units are linear, the lateral recurrent connections give rise to overall nonlinear response properties. The characteristics of the response depend upon the pattern of lateral connections and on the gain of the internal units. The lateral connections can amplify or inhibit the overall response or produce *collective excitations* (amplified response on specific spatial scales). [Note: Kohonen (1988) has presented a model exhibiting collective excitations (which he called "bubbles"), but did not provide an analysis of the model.]

The network relaxation equations for the linear additive model are:

$$\tau_d \frac{d}{dt} V_i = -V_i + \sum_j W_{ij}^{aff} R_j + \sum_j W_{ij}^{lat} E_j \tag{1}$$

where R_j and E_j are the activities (firing rates) of the jth receptor and internal cells respectively, V_i is the somatic potential of the ith internal cell, W_{ij}^{aff} and W_{ij}^{lat} are the afferent and lateral connections respectively, and τ_d is the dynamical relaxation time. The somatic potentials and firing rates of the internal units are linearly related by $E_i = (V_i - \theta)/\epsilon$ where θ is an offset or threshold and ϵ^{-1} is the gain.

The steady state solutions of the network equations can be solved exactly by reformulating the problem in the continuum limit ($i \mapsto x$):

$$\tau_d \frac{d}{dt} V(x) = -V(x) + A(x) + \int dy \, W^{lat}(x-y) E(y) \quad , \quad A(x) \equiv \int dy \, W^{aff}(x-y) R(y) \tag{2}$$

The functions $R(y)$ and $E(y)$ are activation densities in the receptor and internal layers respectively. $A(x)$ is the integrated input activation density to the internal layer. The functions $W^{aff}(x-y)$ and $W^{lat}(x-y)$ are interpreted as connection densities. Note that the network is spatially homogeneous since the connection densities depend only on the relative separation of post-synaptic and pre-synaptic cells $(x-y)$. Examples of lateral connectivity patterns $W^{lat}(x-y)$ are shown in figure 2A. These include local gaussian excitation, intermediate range gaussian inhibition, and a scaled difference of gaussians (DOG).

The exact stationary solution $\frac{d}{dt}V(x) = 0$ of the continuum dynamics of equation 2 can be computed by fourier transforming the equations to the spatial frequency domain. The solution thereby obtained

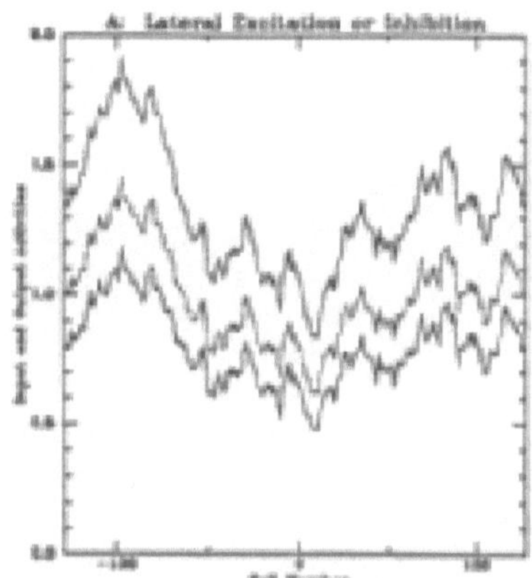

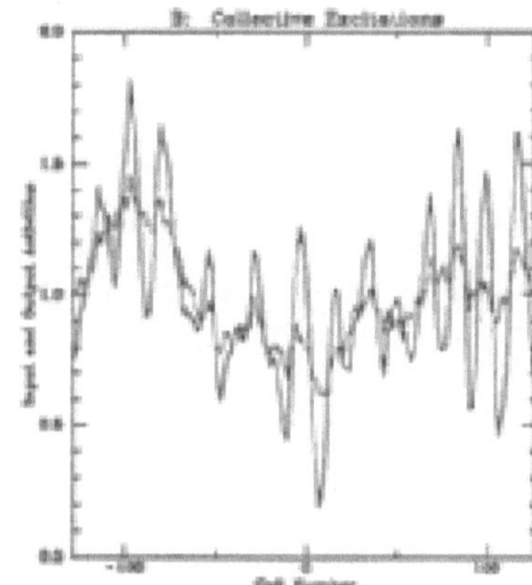

Figure 3: Response of a Linear Network to Random Input. A: Response of neutral (dashed), lateral excitatory (upper solid), and lateral inhibitory (lower solid) networks. B: Collective excitations (solid) as response to random input (dashed) in network with DOG lateral connectivity.

(for $\theta = 0$) is $E(k) = M(k)A(k)$, where the variable k is the spatial frequency and $M(k)$ is the network *magnification function*:

$$M(k) \equiv \frac{1}{\epsilon - W^{lat}(k)}. \tag{3}$$

Positive magnification factors correspond to stable modes. When the magnification function is large and positive, the network magnifies afferent activity structure on specific spatial scales. This occurs when the inverse gain ϵ is sufficiently small and/or the fourier transform of the pattern of lateral connectivity $W^{lat}(k)$ has a peak at a non-zero frequency.

Figure 2B shows magnification functions (plotted as a function of spatial scale $2\pi/k$) corresponding to the lateral connectivity patterns shown in figure 2A for a network with $\epsilon = 1$. Note that the gaussian excitatory and gaussian inhibitory connection patterns (which have total integrated weight ± 0.25) magnify structure at large spatial scales by factors of 1.33 and 0.80 respectively. The scale DOG connectivity pattern (which has total weight 0) gives rise to no large scale or small scale magnification, but rather magnifies structure on an intermediate spatial scale of 17 cells.

We illustrate the response of linear networks with unit gain $\epsilon = 1$ and different lateral connectivity patterns in figure 3. The networks correspond to connectivities and magnification functions shown in figure 2. Part A, shows the response $E(x)$ of neutral, gaussian excitatory, and gaussian inhibitory networks to net afferent input $A(x)$ generated from a random $1/f^2$ noise distribution. The neutral network (no lateral connections) yields the identity response to random input; the networks with the excitatory and inhibitory lateral connection patterns exhibit boosted and reduced response respectively. Part B shows the emergence of collective excitations (solid) for the scaled DOG lateral connectivity. The resulting collective excitations have a typical period of about 17 cells, corresponding to the peak in the magnification function shown in figure 2. Note that the positions of peaks and troughs of the collective excitations correspond approximately to local extrema in the random input (dashed).

3 Dynamics in the Nonlinear Models

The nonlinear models, including the sigmoidal additive model and the shunting models, exhibit the collective excitation phenomenon as well, although in more interesting and varied ways. Of particular interest is the possibility of distributed *winner-take-all* (WTA) response.

Figures 4 and 5 show sample responses of the sigmoidal and shunting networks. See Moody (1990) for a detailed description of the model formulations and their analyses.

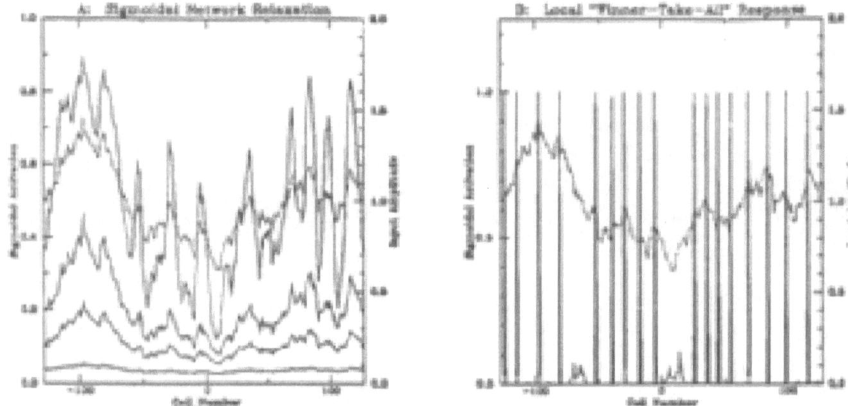

Figure 4: A: Time Development (solid curves) of Collective Excitations in the Sigmoidal Model. B: Distributed Winner-Take-All Response Limit (solid curve) of the Sigmoidal Model. The WTA limit has high gain and an excess of intermediate range inhibition over short range lateral excitation. The input for both A and B (dotted curves) is the same as that of figure 3 for comparison.

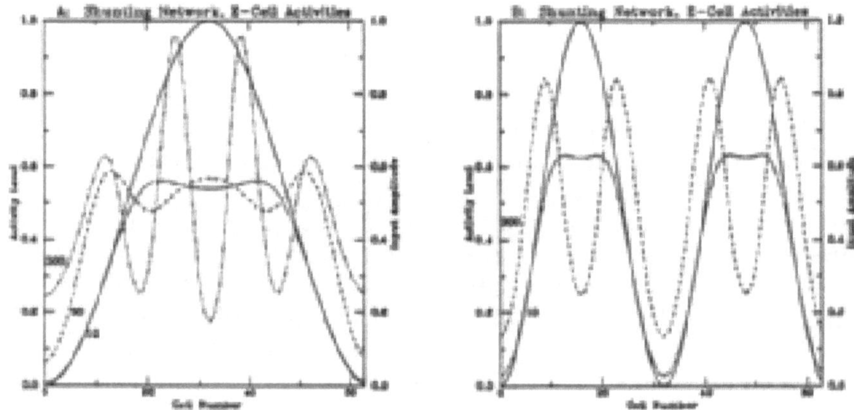

Figure 5: Time Development (dotted and dashed curves) of Collective Excitations in the Shunting Model. Responses to a single cycle (A) and two cycles (B) of spatially-sinusoidal input (solid curves). In both A and B, the natural final state of the network (iteration 300) has four spatial cycles. Initially, the networks are inactive. When the non-zero inputs are presented at iteration 1, the network responses goes though a series of bifurcations before reaching their stable states.

Acknowledgements

The author wishes to thank Alex Chernjavsky for assistance in preparing the figures. The work was supported by ONR Grant N00014-89-J-1228 and AFOSR Grant 89-0478.

References

Alex Chernjavsky and John Moody. (1990) Spontaneous development of modularity in simple cortical models. Submitted to *Neural Computation.*

T. Kohonen. (1988) *Self-organization and Associative Memory, 2nd ed.* Springer, New York.

John Moody. (1990) Dynamics of lateral interaction networks. Technical report, Yale University. (In Preparation.)

A SPARSELY CONNECTED ASYMMETRIC NEURAL NETWORK AND ITS POSSIBLE APPLICATION TO THE PROCESSING OF TRANSIENT SPATIO-TEMPORAL SIGNALS

W. Banzhaf, T. Ishii, S. Nara, T. Nakayama
Central Research Laboratory
Mitsubishi Electric Corporation
1-1, Tsukaguchi Honmachi 8-chome
Amagasaki, Hyogo, 661, Japan

We consider randomly connected neural networks with sparse asymmetric connections. Low connectivity is achieved by chosing a certain number of input and output connections for every cell. We show that rudimentary layered structures of cells automatically emerge. Then we examine the autonomous behaviour of this network and some modified models, both from a non-linear dynamical and from a functional point of view. We not only study the occurring attractors of limit-cycle type by statistical means but also report on the observation of chaos-like attractors. We demonstrate the ability of the network to react to transient spatio-temporal patterns in real-time by showing that attractor transitions due to time correlations of the transient signals happen.

1. Introduction

Artificial neural networks for associative memory are usually plagued by a serious problem: In fully connected networks, a reasonably small number of neurons (in terms of real applications) of the order of 10^5 would require 10^{10} connections which is much larger than any number of elements we can hope to position on a single chip in the foreseeable future.

As is well known from the human brain, in contrast, the number of neurons is at least 10^{10}, whereas the number of synaptic connections is of the order of 10^{14} [1]. The average connectivity in the brain is therefore 10^{-6}, a very small number, but the brain is able to solve complicated tasks. The natural conclusion is that the brain follows a very clever strategy to perform well. This is based on a modular and hierarchically structured architecture which enables a filtering and recycling of relevant information [2].

Given the mere number of synapses, there cannot be at all enough genetic information to contain detailed instructions for all the synapses to form [3]. Rather, after birth the brain looks - at least in parts - as if neurons were randomly connected. A major part of brain's organization is therefore provided by the environment it is exposed to.

Is such a strategy applicable(beneficial) for technical systems, too? Usually, this question is related to the method with which neurons adapt to their environment, in other words to learning. Another aspect of the question, however, may be elucidated by studying various types of random networks, as e.g. Amari [4] and Kauffman [5] have done.

In this paper we shall study a class of randomly connected neural networks, and one of the question we want to address is: How can we generate as much structure as possible from as few information as necessary?

The general method to generate random networks we shall apply here is this: Take N neurons and fix only the number M_I of input connections and M_O of output connections which every cell tries to form by choosing randomly from the entire set of N neurons. (Here $M_I, M_O \ll N$.) What will be the result? We get sparse connections between neurons which will have a direction and thus are asymmetric. Moreover, we will see the emergence of structure in the form of rudimentary layers, that is to say layers with inner cross-connections as well as with feedback connections. Functionally, we have got a network which spontaneously correlates spatio-temporal patterns. As can be seen from Figure 1 any neuron delays its input by a time τ.

2. The model

Our standard model consists of N neurons, $10 \leq N \leq 1000$, N_I input neurons, $N_I = 2$, N_O output neurons, $N_O = 2$. This results in $N_H = N - N_I - N_O$ hidden neurons. We have M_I input connections,

$M_I = 2$ and M_O output connections, $M_O = 2$ which every cell forms individually by chosing randomly from the set of neurons.

The network dynamics is described by

$\vec{x}(t) = \{x_1(t), ..., x_N(t)\} \in [-1, +1]$,

the activity of cells at time step t,

$y_i(t+1) = \sum_j W_{ij} x_j(t)$,

the weighted sum of its inputs (activities at time t)

$\vec{x}(t+1) = (\frac{1}{1+e^{-\beta[\vec{y}(t+1)-\vec{\Theta}-\vartheta]}} - a) \cdot b$,

the nonlinearity imposed on cells.

Here W_{ij} are the connections cell j receives from other cells i with time independent strength from a random Gaussian distribution $G(0, \sqrt{N})$

β is the steepness of the nonlinearity

Θ_i are the individual thresholds

ϑ is a global threshold.

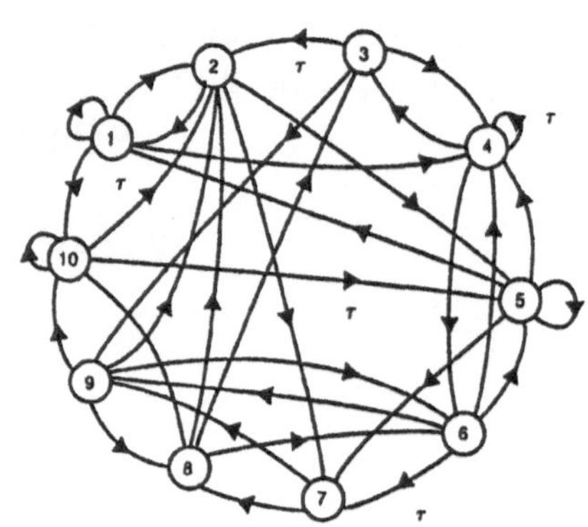

Figure 1: $N = 10$ random network. All connections (indicated by arrows) delay the signal by τ.

The motivation for choosing this combination as standard model is:

1. The activity in the range between -1 and 1 results in a richer overall behaviour of the network.
2. Starting with a fully connected net and loading more and more *heteroassociations* [6] into the synaptic matrix, one finally arrives at a Gaussian distribution of weights, the difference to here being merely the sparsity of connections. Compared to disordered spin-glas networks [7] we use asymmetric connections.
3. Since the activity of cells is not bound to the positive region, the thresholds do not have a unique influence on the network. This is due to the fact that the quiet state of a cell, $x_i = -1$, does equally contribute to the weighted sum as the full activity does. Therefore we have chosen to set the thresholds to 0.

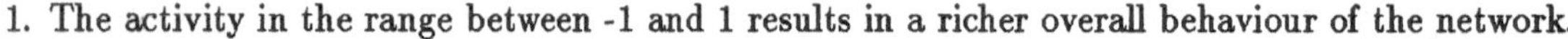

Table 1 shows the variations of the standard scheme we have also considered. Later on, we shall refer to the variations by their respective model numbers. The variation in model 6 in the distribution of weights gives every cell equal rights and will allow us to detect chaos-like attractors.

We now want to turn to a discussion of the structural aspects of the networks generated by the method described.

3. Static features

Let us first consider the distribution of synaptic connections among neurons. Due to the strategy mentioned in the previous section, how will the connections be distributed finally?

Since the strategy is local in the sense that locally every cell performs the same operations without regarding the synapse-forming activity of the other cells, the outcome will definitely be not a uniform distribution. Indeed, it is a binomial distribution, as will become clear in a moment.

Every neuron will have at least M_I inputs. What is the probability that a neuron i has more than M_I inputs? Suppose that another neuron, j, has chosen i for its output synapse. This happens with a probability $p = M_O/N$ and results in an additional input to neuron i. (We allow for multiple connections.) Thus, the probability $P_N^I(M_I + k)$ that any neuron i gets M_{I+k} inputs in total is

$$P_N^I(M_I + k) = \binom{N}{k} p^k (1-p)^{N-k} \tag{1}$$

A similar arguments holds for the symmetrical case. Therefore, the probability $P_N^O(M_O + l)$ that a neuron i gets $M_O + l$ output connections is given by

$$P_N^O(M_O + l) = \binom{N}{l} q^l (1-q)^{N-l} \tag{2}$$

where $q = M_I/N$ is the probability that neuron i is chosen by another neuron as one of its input cells.

A comparison of actual results with the predictions derived from equ. (1) and (2) shows a good agreement. For small p or q, the probability distribution approaches the Poisson distribution. In either case, the average numbers of connections are given by

$$\langle M_I \rangle = M_I + p \cdot N = M_I + M_O \equiv M = M_O + q \cdot N = \langle M_O \rangle \tag{3}$$

In summary, we have seen that the local strategy of every cell to seek to form connections with other cells leads to a distribution of the number of connections. Considered on the level of averages, the fan-in and fan-out of all neurons is the same.

Let us now turn to the second point: The emergence of structure. We analyze the structure by giving input at time step $t = 0$ to an arbitrarily chosen cell i and subsequently observing the time delay of the other cells' responses. To this end we prepare the entire network in the quiet state, i.e. $x_j = 0, \forall j \in N$. We then measure the time step t at which a cell firstly shows a deviation from activity 0. We shall say that a cell j is in 'rudimentary layer' K relative to cell i, with K identical to the number of discrete time steps t.

Certainly, there are feedback connections between layers and cross-connections in each layer, cf. Fig. 1. These connections interfere with the layered structure of the network, therefore we only call it 'rudimentary layered'. One can, however, imagine learning rules which weaken the interference so as to strengthen the structured aspect of the network in the course of a possible development.

Given a certain number of cells, N, and a number of input/output connections, $M = M_I + M_O$, how many layers n_L of cells can we expect to emerge, relative to an arbitrary cell i? It turns out that the minimal number of layers is given by $n_L \geq \lceil log_M N \rceil + 1$, where $\lceil x \rceil$ means the next higher integer to x. This is well confirmed by our data of networks between $N = 40$ and $N = 1000$

If we provide for input on a couple of cells we can see a change in the structure of layers due to the individual contribution of every cell. The number of layers tends to become smaller and smaller. Figure 2 shows the average number of layers if one varies the number of input cells for the three networks above.

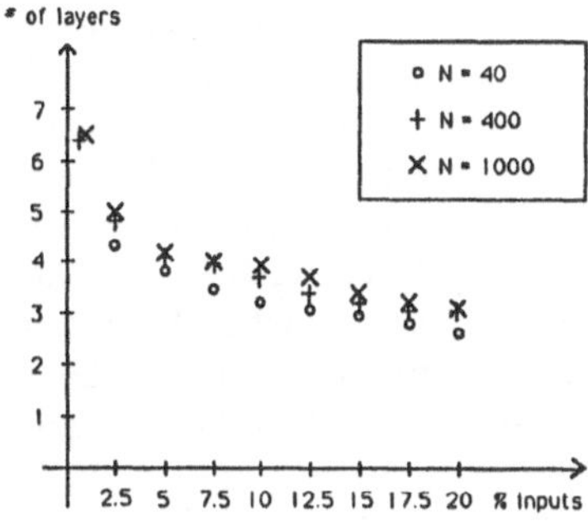

Figure 2: Number of layers formed

4. Dynamical features: Spontaneous behaviour

This section is devoted to report on the spontaneous dynamical behaviour of the network of size $N = 40$. In the standard model, starting from 1000 sample initial conditions we have observed a fluctuating number of limit-cycle attractors, depending on the particular architecture and weight distribution in the network. Usually, the number of attractors is not very large, ranging from 1 to 10. The cycle length is fluctuating heavily, the largest discovered being of length 506 iterations, the smallest being of length 1.

Model 2 looks not very much different, the different randon generator seems not to have a major impact on the behaviour. Model 3, on the other hand, looks different: We encounter much fewer limit cycles, ususally of shorter length. Quite often we observed that there was only one limit cycle of intermediate length.

It seems to us that the concept of forcing structures introduced in the framework of random boolean networks [5] can be used to explain this result. Forcing structures are interacting subsets of elements (neurons) capable of keeping similar behaviour under different circumstances and initial conditions. These structures may be considered as strengthened by the addition of (not activity dependent) individual thresholds. Restricting the activity to the positive range results in a somewhat intermediate pattern of behaviour. Sometimes, however, stable networks with constant cell activities arise (model 4 and 5).

Model 6 provides an exeption from the previous series because it shows chaos-like attractors. To see this we consider the development of trajectories $\vec{x}_1(t), \vec{x}_2(t)$ in phase space the distance of which is originally very small:

$$\vec{x}_1(0) = \vec{x}, \qquad \vec{x}_2(0) = \vec{x} + \vec{\delta x}(0), \qquad \vec{\delta x}(0) \ll 1.$$

If we made $\vec{\delta x}(0)$ very small, stable attractors would show up as a distance approaching 0

$$d(\vec{x}_1, \vec{x}2) = \mid \vec{x}_1 - \vec{x}_2 \mid \rightarrow 0$$

as $t \rightarrow \infty$. An exponentially growing distance $d(\vec{x}_1, \vec{x}2)$ as time goes on, on the other hand side, indicates the existence of chaos-like attractors. We actually observed that small flutuations grew rapidly and we could not discover any periodicity within our simulation time.

The synaptic weights in model 6 are generated by imposing on every cell j an additional normalization condition $\sum_i W_{ij}^2 = N$ which deformes and rescales the original gaussian distribution.

Table 1: The standard model and its variations

Model	Activity	Weights	Thresholds	Attractors	Period
1 (Standard)	$[-1,1]$	$G(0,\sqrt{N})$	$\Theta_i = 0$	$+/-$ [a]	$+/-$ [b]
2	$[-1,1]$	$R(0,N)$	$\Theta_i = 0$	$+/-$	$-$
3	$[-1,1]$	$G(0,\sqrt{N})$	$G(0,\sqrt{N})$	$-$	$-$
4	$[0,1]$	$G(0,\sqrt{N})$	$\Theta_i = 0$	$-$	$-$
5	$[0,1]$	$G(0,\sqrt{N})$	$G(0,\sqrt{N})$	$-$	$-$
6	$[-1,1]$	$\sqrt{N} * G(0,1), normalized$	$\Theta_i = 0$	chaos-like	∞

[a] +: large number, −: small number
[b] +: long cycles , −: short cycles

Random networks of models 1 - 5 are able to perform attractor transitions as a reaction to transient spatio-temporal input patterns. In our experiments, the spatial dimension is provided by 2 different cells each receiving input restricted to values +1 and -1. Thus, there are 2^2 different input patterns and we examine the reaction of the network after presenting them as transient signals of four time steps in 4! different orders. An eventual transition is generated since the incoming time-dependent input signals are automatically correlated with the ongoing self-sustained activity of a network. For a clean demonstration we have chosen a network from model 3 with 2 attractors of nearly equal basin size. The network is first allowed to go into a limit cycle and is then confronted with the input on 2 input cells with special weights of the order $(\sqrt{N})$ to the external world. Table 2 summarizes the results of one run.

Remains in attractor 1	Transites to attractor 2
(1,2,4,3) (1,4,2,3) (2,1,3,4)	(1,2,3,4) (1,3,2,4) (1,3,4,2)
(2,3,1,4) (2,4,1,3) (2,4,3,1)	(1,4,3,2) (2,1,4,3) (2,3,4,1)
(3,2,4,1) (3,4,2,1) (4,1,3,2)	(3,1,2,4) (3,1,4,2) (3,2,1,4)
(4,2,1,3) (4,2,3,1) (4,3,1,2)	(3,4,1,2) (4,1,2,3) (4,3,2,1)

Table 2: Attractor transitions in a $N = 40$ network with 2 input units.
1: (-1,-1) 2: (-1,+1)
3: (+1,-1) 4: (+1,+1)

The spontaneous transition to another attractor is certainly only the very beginning. Our future work will be aimed mainly at obtaining some mechanisms to manipulate the behaviour of the network and to apply it to the recognition or classifiction of transient signals. In this context, we regard a temporal development of synapses, i.e. learning, as the necessary feature which is to be introduced into the network. In order to achieve this, however, our understanding of the network itself must be deepened by more statistical examinations.

References

[1] Colonnier, M. in: Schmidt, F.O. et. al. (Eds.), The organization of the cerebral cortex, MIT Press, Cambridge, Mass., 1981

[2] Braitenberg, V., J. theor. Biol. **46** (1974) 421

[3] Gierer, A., Biol. Cyb. **59** (1988) 13

[4] Amari, S., IEEE Transact. **SMC-2** (1972) 643

[5] Kauffman, S., J. theor. Biol. **44** (1974) 167

[6] Mori, Y., Davis, P. Nara, S., J.Phys. A **22** (1989) L525

[7] Bienenstock, E., et.al. (Eds.), Disordered Systems and Biological organization, NATA ASI Series F, Springer, Berlin, 1986

DISCRETE TIME MODELS OF NOISY NETWORKS.

P. C. Bressloff

General Electric Company, Hirst Research Centre, East Lane, Wembley, Middlesex.

J. G. Taylor

Department of Mathematics, King's College, The Strand, London.

ABSTRACT

The leaky-integrator model of a neural network with synaptic noise is studied in a discrete time approximation. The resulting dynamical equations describe a network of time-summating binary neurons with thresholding activity which can be non-linear in weights. The weights and thresholds are random variables independently updated at every time step from the same time-independent probability distributions. In a certain limit the theory reduces to a single time step model whose dynamics may be formulated as a random map on $\{0,1\}^N$; functional techniques may then be used to study the pattern storage of such networks. This is an extension of the spin-glass approach to the Little model, which is a special case of the above. The dynamics of a time-summating network are given by a random map on the continuous space R^N and for linear weights correspond to a stochastic extension of a model of chaotic networks.

Consider a network of N neurons and assume for simplicity that it is fully-connected. Let $V_i(t)$ be the membrane potential of the ith neuron at time t and denote the synapse linking the axon of neuron j to a dendrite of neuron i by (ij). In the single-compartment leaky-integrator model ([Hodgkin and Huxley, 1952]), V_i obeys the differential equation

$$C_i \frac{dV_i}{dt} = -\frac{V_i(t)}{R_i} + \sum_j \Delta g_{ij}(t)[S_{ij} - V_i(t)] \tag{1}$$

Here C_i, R_i are the membrane leakage capacitance and resistance and Δg_{ij} is the change in synaptic conductance at synapse (ij), with membrane reversal potential S_{ij}, due to the release of chemical transmitters into the synaptic cleft. Let $N_i(t)$ be the number of times that the ith neuron has fired since some initial time t_0. Then a simple choice for Δg_{ij}, neglecting spontaneous emission, is

$$\Delta g_{ij}(t + t_d) = \sigma_{ij} \sum_{n\geq 1} \delta(t - T_n^j) q_{ij}(T_n^j), \quad t \geq t_0 \tag{2}$$

In equation (2) σ_{ij} is a constant related to synaptic efficiency and $q_{ij}(T_n^j)$ is the amount of transmitter chemicals released into synapse (ij) due to the arrival of an action pulse from the jth neuron at time T_n^j where T_n^j is the time at which the jth neuron fires for the nth occasion. That is, $T_n^j = inf\{t \geq t_0 | N_j(t) \geq n\}$. Note that a synaptic time-delay has been included in equation (2); the specification of $\Delta g_{ij}(t)$ over the interval $[t_0 - t_d, t_0]$ then corresponds to a boundary condition of the dynamics. The firing times are determined by the iterative threshold condition

$$T_n^j = inf\{t \mid V_j(t) \geq \kappa_j;\ t \geq T_{n-1}^j + t_R\} \tag{3}$$

where t_R is the refractory period and κ_i the threshold. In other words, supposing that the jth neuron last fired at time T_{n-1}^j, then either (i) after one refractory period the potential is above threshold and the neuron fires irrespective of incoming activity, $T_{n+1}^j = T_n^j + t_R$ or (ii) after one refractory period the potential is below threshold and the neuron fires a time t_d after the arrival of an action pulse from some other neuron k at time T_m^k, where $T_{n+1}^j = T_m^k + t_d$ and $T_m^K + t_d > T_n^j + t_R$. We conclude that the firing times T_n^j, $j = 1, ..., N$, $n \geq 1$ lie on a lattice generated by the synaptic delay time, the refractory period and the first firing times T_1^j.

It is known that the predominant source of noise in neurons arises from the quantal and stochastic nature of neuro-transmitter release at the synapses ([Katz, 1969]). This may be incorporated into the above model by taking each $q_{ij}(T_n^j)$, $n \geq 1$, to be an independent random variable generated from the same time time-independent probability distribution ([Taylor, 1972]) $\rho_{ij}(q) = \sum_{n\geq 0} \delta(q_0 n - q)p_{ij}(n)$ where $p_{ij}(n)$ is the probability that n packets of transmitter are released. The size q_0 of each packet is assumed to be fixed. The distribution function $p_{ij}(n)$ is usually taken to be either a Poisson distribution or a Binomial distribution. Threshold noise may also be included by taking κ_i to be a random variable updated at every time step from some distribution $\rho_i(\kappa)$; however, there is no evidence that such noise is present to any significant degree. In the presence of noise both $V_i(t)$ and T_n^j become random variables.

It is not in general possible to solve the closed set of equations (1) and (3) without further approximations. As an initial simplification we shall neglect the term $-V_i(t)\sum_j \Delta g_{ij}(t)$ in equation (1). Then we integrate (1) (with t_0 set to zero) by splitting the integral over the interval $[0, t]$ into two parts with the second integral being determined by the boundary condition on the interval $[-t_d, 0]$. We shall denote this boundary term by $Y_i(t)$. Setting $V_i(0) = 0$ we find

$$V_i(t) = \sum_j \sum_{n\geq 1} \omega_{ij}(T_n^j)\theta(t - t_d - T_n^j)\, e^{-(t - t_d - T_n^j)/\tau_i} + Y_i(t) \tag{4}$$

where $\omega_{ij}(t) = \sigma_{ij}S_{ij}q_{ij}(t)/C_i \equiv \varepsilon_{ij}q_{ij}(t)$ and $\tau_i = R_iC_i$ is the time constant for potential decay. To proceed further we discretise time in terms of the delay t_d such that $t = mt_d$ for integer m (for simplicity we set $t_d = 1$). We can interpret this discretisation as a special choice of boundary conditions in which the first firing times T_1^j and refractory period become multiples of t_d. In terms of the neuronal activity $a_i(t) \in \{0, 1\}$, which indicates whether or not the ith neuron fires at time t, we have for any function f that

$$\sum_{n\geq 1} f(T_n^j) = \sum_{m=0}^{\infty} f(m)a_j(m) \tag{5}$$

The state of the network at time t is effectively characterised by the vector $\mathbf{a}(t)$. Moreover, the boundary term becomes $Y_i(t) = \sum_j \omega_{ij}(-1)a_j(-1)e^{-t/\tau_i}$ and together with equations (4) and (5) implies that

$$V_i(t) = \sum_j \sum_{r=1}^{t+1} \omega_{ij}(t - r)a_j(t - r)e^{-(r-1)/\tau_i} \tag{6}$$

The boundary term $Y_i(t)$ corresponds to the case $r = t + 1$. Equation (6) must be combined with the threshold condition (3) which in discrete time becomes

$$a_i(t) = \theta(V_i(t) - \kappa_i) \prod_{r=1}^{R} \hat{a}_i(t-r) \tag{7}$$

where the product enforces the refractory period constraint with $t_R = Rt_d$. Hence we obtain from a leaky-integrator model in a discrete time approximation, the stochastic equations

$$a_i(t) = \theta(\sum_j \sum_{r=1}^{t+1} \omega_{ij}(t-r)a_j(t-r)e^{-(r-1)/\tau} - \kappa_i) \prod_{r=1}^{R} \hat{a}_i(t-r) \tag{8}$$

These describe the dynamics of a time-summating, binary neural network.

As it stands equation (8) expresses the state of the network at time t as a function of the complete past history of the network. However, we may reformulate equation (8) as a random iterative map ([Kifer, 1986]). For simplicity we shall set the refractory period to be unity (R =1). First note that in the limit $\tau_i \to 0$ we obtain a random iterative map on the discrete space $\{0, 1\}^N$ based on the Caianello equations ([Caianello, 1961])

$$a_i(t+1) = \theta(\sum_j \omega_{ij}(t)a_j(t) - \kappa_i(t)) \tag{9}$$

The random weights $\omega_{ij}(t) \equiv \varepsilon_{ij}q_{ij}(t)$ and thresholds $\kappa_i(t)$ are independently updated at every time-step according to the probability distributions $\rho_{ij}(q)$ and $\rho_i(\kappa)$. This should be contrasted with the random nerve net approach of Amari et al [1977] in which quenching occurs at the beginning of the dynamics, i.e. the weights and thresholds are chosen at random and then held fixed during the evolution of the net. Associated with the random iterative map (9) is a homogeneous Markov chain in which the time-independent transition probability of going from state $\mathbf{a}$ to state $\mathbf{b}$ in one time-step is

$$Q_{\mathbf{ba}} = \prod_{j=1}^{N} [p(j \mid \mathbf{a})b_j + (1 - p(j \mid \mathbf{a}))(1 - b_j)] \tag{10}$$

where $p(i \mid \mathbf{a})$ is the conditional probability that the ith neuron fires given that the state of the network at the previous time step is $\mathbf{a}$,

$$p(i \mid \mathbf{a}(t)) = \int d\kappa_i \; \rho_i(\kappa_i) \int \prod_j dq_{ij}\rho_{ij}(q_{ij})\theta(\sum_j q_{ij}\varepsilon_{ij}a_j(t) - \kappa_i) \tag{11}$$

Functional techniques have been developed to study the statistical dynamics of such networks in terms of the generating functional

$$Z[\mathbf{l}] = \int D\mathbf{W} D\mathbf{u} D\bar{\mathbf{u}} \exp\{i\sum_{t,i}\bar{u}_i(t)[u_i(t+1) - \theta(\sum_j \omega_{ij}(t)u_j(t) - \kappa_i(t))]\}\exp\{i \sum_{t,i} l_i u_i\} \tag{12}$$

where $D\mathbf{W} = \prod_{t,j}[d\kappa_i \; \rho_i(\kappa_i(t))\prod_j dq_{ij}\rho_{ij}(q_{ij}(t))]$ and $D\mathbf{u}D\bar{\mathbf{u}} = \prod_{t,i} du_i(t)\, d\bar{u}_i(t)$. The functional (12) contains complete statistical information about the dynamics, e.g. the mean activity is $\langle u_i(t)\rangle = \int \delta/\delta l_i(t) Z[\mathbf{l}] \mid_{\mathbf{l}=0} \equiv \int D\mathbf{W}D\mathbf{u}D\bar{\mathbf{u}}e^L$ where the action L is the argument of the first exponential of (12). In particular, $Z[\mathbf{l}]$ can be used to study the pattern storage of such networks in the large N limit ([Bressloff, 1989], [Bressloff and Taylor, 1990a]). This generalises the spin-glass approach ([Amit, 1989] and references therein) to the Little model ([Little, 1974]) which can be derived from (11) by taking the distribution of the weights to be a delta function about some mean and the distribution of the threshold to be given by $\rho_i(\kappa) = d[1+e^{(-\kappa/T)}]^{-1}/d\kappa$, where T is a temperature parameter. We also note that there exists a hardware realization of noisy networks described by equation (9) ([Gorse and Taylor, 1989]).

For finite τ equation (8) may be rewritten as

$$V_i(t) = kV_i(t-1) + \sum_j \omega_{ij}(t-1)\theta(V_j(t-1) - \kappa_j) \tag{13}$$

where $k = e^{-1/\tau_i}$. Equation (13) is a random iterative map on the continuous space $\{V_i : i = 1, ..., N\} \equiv R^N$. The neuronal activities a_i are then determined by equation (7) which decouples from (13). Equation (13) is a stochastic extension of a model of chaotic networks introduced by Aihara et al [1990] which incorporates synaptic processing in a more realistic manner allowing an analysis of the effects of synaptic noise on the dynamics. Moreover, parameters in the the theory correspond more directly to biological features such as post-synaptic efficiency ε_{ij}, distribution functions of vesicular release etc. and thus more realistic learning rules may be considered. ([Bressloff and Taylor, 1990b]). Note that similar expressions to (13) occur if the refractory period satisfies $R > 1$, the main difference being that equation (13) becomes a higher order difference equation. Moreover, the analysis is easily extended to the case of the full leaky-integrator model in which the term $-V_i(t)\sum_j \Delta g_{ij}(t)$ in equation (1) is no longer neglected. Then the theta function in (8) becomes a non-linear function of the weights ([Bressloff and Taylor, 1990b]),

$$a_i(t) = \theta(\sum_k \sum_{r=1}^{t+1} \omega_{ik}(t-r) a_k(t-r) k^{r-1} \exp[-\sum_j \sum_{s=1}^{r} \omega_{ij}(t-s) a_j(t-s)/S_{ij}] - \kappa_i) \tag{14}$$

In the limit $S_{ij} \to \infty$, for fixed weights, we recover the linear approximation (8). Furthermore, as in the linear case, we may formulate the dynamics as a random iterative map which for unit refractory period is given by

$$V_i(t) = [kV_i(t-1) + \sum_k \omega_{ik}(t-1)\theta(V_k(t-1) - \kappa_k)] \exp[-\sum_j \omega_{ij}(t-1)\theta(V_j(t-1) - \kappa_j)] \tag{15}$$

It is at present unclear how the non-linearities in equation (15) effect the dynamics of time-summating nets, either with or without noise. This is currently under investigation, as are various extensions of the model such as leaky-integrator compartmentalised models of the neuron ([Rall, 1964]).

REFERENCES

Aihara K, Takabe T and Toyoda M 1990 *Chaotic Neural Networks,* (to appear in Phys. Lett. A)

Amari S, Yoshida K and Kanatani K 1977 *SIAM J. Appl. Math.* **33**

Amit D J 1989 *Modelling Brain Function* (Cambridge University Press, Cambridge)

Bressloff P C 1989 in *New Developments in Neural Computing* ed. Taylor J G and Mannion C L T (Adam Hilger, Bristol)

Bressloff P C and Taylor J G 1990a *Phys Rev A* **41**

Bressloff P C and Taylor J G 1990b (GEC Preprint)

Caianello E R 1961 *J. Theor. Biol.* **1** 209

Gorse D and Taylor J G 1989 (King's College Preprint)

Hodgkin A L and Huxley A F 1952 *J Physiol (Lond)* **117** 500

Katz B 1969 *The Release of Neural Transmitter Substance* (Thomas, Springfield)

Kifer Y 1986 *Ergodic Theory of Random Transformations* (Birkhauser)

Little W A 1974 *Math. Biosci.* **19** 101

Rall W 1964 in *Neural Theory and Modelling* ed. Reiss R F (Stanford University Press, Stanford, Calif.)

Taylor J G 1972 *J. Theor. Biol.* **36** 513

ON THE COMPUTATIONAL POWER OF NEURAL NETWORKS AND NEURAL AUTOMATA

MOSHE LESHNO
FACULTY OF MANAGEMENT
TEL AVIV UNIVERSITY
TEL AVIV, 69978
ISRAEL
Phone: 972-3-5450368

ABSTRACT

In the following paper we discuss mainly the problem of computability i.e. which functions can be computed by a neural network?
The answers to this question determine the capabilities and limitation of a neural network as "general purpose" computers.
A computation process is defined by a dynamic motion of input states to an output states. Regarding the problem above we shall compare several types of deterministic neural networks models with the classical computational model of Turing Machine (TM), and determine there computational capabilities.

Stable Neurodynamics and Symbolic Computation

Bernd Schürmann

ZFE IS INF 2, Siemens AG, 8000 München 83, W. Germany

Dongming Wang

RISC-LINZ, Johannes Kepler University, A-4040 Linz, Austria

Abstract

An approach is presented which handles, analyzes and constructs neurodynamics and its related objects by using and extending methods and software systems of symbolic computation.

EIGENVALUE METHODS AS A TOOL FOR GLOBAL ANALYSIS

Ken-ichi MAEDA
AI Applications Institute, University of Edinburgh
80 South Bridge, Edinburgh EH1 1HN, U.K.
TOSHIBA R & D Center
1, Komukai Toshiba-cho, Saiwai-ku, Kawasaki 210, Japan

Abstract

Eigenvalue methods are a good tool for analyzing the global properties of a neural network. Instead of non-linear equation methods in which non-linear differential equations are used for describing the characteristics of a neuron or a connection, I propose to use eigenvalue methods specifically for analyzing global properties of the network, discarding time varying factors.

I present examples of a non-linear equation method and several eigenvalue methods in order to show the similarity between the two methods and the potential of the latter. Some eigenvalue methods are able to give the same result as some non-linear equations at less computing cost.

COARTICULATION INVARIANCE IN FEEDBACK MULTILAYER PERCEPTRONS

Hans-Ulrich Bauer* and Theo Geisel

Institut für Theoretische Physik and SFB Nichtlineare Dynamik
Universität Frankfurt, D-6000 Frankfurt/Main 11, Fed. Rep. of Germany

Neural networks for speech recognition must cope with several invariance requirements. These include invariance with respect to presentation speed fluctuations as well as tolerance towards coarticulation of patterns. We show that multilayer perceptrons with feedback from the output to the input layer meet these requirements. They can be trained simply, their feedback states can be made stable, and they are automatically robust with respect to presentation speed fluctuations. The coarticulation performance is expressed by the detection probability for gradual transitions between patterns. If this performance is not satisfactory anyhow, it can be improved by extending the learning set. These properties make feedback multilayer perceptrons promising candidates for speech recognition.

* Supported by the Deutsche Forschungsgemeinschaft (Grant Ge 385/7-1).

ANALYSIS OF PARALLEL AND SEQUENTIAL BOLTZMANN MACHINES

Alberto BERTONI *, Paola CAMPADELLI •*

* Dipartimento di Scienze dell'Informazione
Università degli Studi di Milano
20133 Milano, Italy
Tel.: (02) 7575213

• Istituto di fisiologia dei centri nervosi
CNR
20133, Milano, Italy

Abstract: It is well-known that sequential Boltzmann machines converge to the uniform distribution on global minima of some proper state function E (asymptotic behaviour). They are therefore considered an interesting tool for solving combinatorial optimization problems. Trying to speed up the convergence process, parallel Boltzmann machines have been introduced.

It has been proved that asymptotic behaviour of the parallel and the sequential Boltzmann machine with the same structure <W,T> (i.e. the connection matrix W and the thresholds vector T) may be quite different and that the difference can be stated taking into account properties of the deterministic symmetric networks of binary threshold neurons with sequential or parallel mode.

This paper analyzes the computational complexity of the following decision problem: given a network structure <W,T>, are the asymptotic behaviours of the parallel and the sequential Boltzmann machine different ? The main results are:

1) the problem is "very difficult" to solve; it is in fact both NP-hard and coNP-hard (with respect to many-one reduction)

2) the problem is proved to be in P^{NP} (i.e. the class of problems solvable in polynomial time using oracles in NP)

3) a non trivial subclass of networks for which the problem is solvable in P is shown.

Keywords: Boltzmann Machines, Computational Complexity, Lyapunov Functions, Sequential Updating, Parallel Updating.

STATISTICS AND PROBABILITIES

Chair : Shun-Ichi AMARI

STOCHASTIC DYNAMICS AND INPUT DIMENSIONALITY IN A TWO-LAYER NEURONAL NETWORK FOR MODELLING MULTISTABLE PERCEPTION

Massimo Riani, Francesco Masulli, and Enrico Simonotto

Department of Physics - University of Genoa
Via Dodecaneso 33 - 16146 Genova - Italy
Bitnet addresses: masulli@genova.infn.it ; riani@genova.infn.it

Abstract

A two-layer neuronal network model of the perceptual alternation of multistable figures is presented. Results of computer simulations have shown that the model makes it possible to obtain the stochastic Gamma distributions of the experimental perceptual durations of the alternating interpretations, as well as some other characteristics of the perceptual alternation phenomenon such as the dependency of reversal times on the complexities of pattern interpretations.

1. Introduction

The problem of "ambiguity" often arises from modelling the biological process of coding the external environment into its "inner" representation. In visual perception, an example of this kind of situations is the "perceptual alternation phenomenon" related to some visual patterns, called "ambiguous" or "multistable" figures. The visual input associated with an ambiguous figure can elicit two different interpretations, thus giving rise to a cyclic perceptual alternation of the two competitive interpretations. In Figure 1, three examples of this kind of figures are shown. The first is the Mach pyramid, which can be seen as a room or as a truncated pyramidal structure, (e.g. a roof); the second is the "Necker" cube, allowing two identical, though perspectively different, interpretations; and the third is a figure called "Rubin vase", which can be perceived as two faces on a white background or as a white vase on a black background.

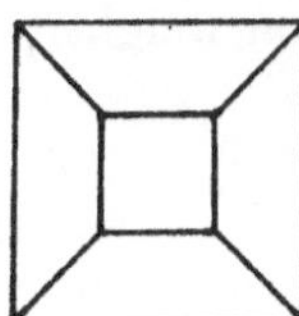
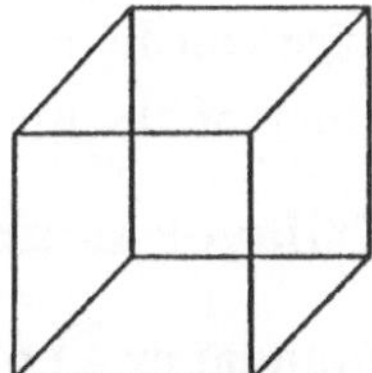

Fig. 1: Three Multistable Figures.

It should be stressed that if the observer is not aware of the existence of two different interpretations of the pattern, frequently the alternation process does not start, and the perceived configuration does not change [8]. Moreover, during a prolonged observation of an ambiguous drawing, a stationary phase is reached, in which both percepts appear with some regularity, and the perceptual durations of the competitive interpretations are well represented by a Gamma distribution, as pointed out by the extensive experimental work of our group [2),3),5)].

Different models of visual perception alternation that are based on artificial neuronal networks have been proposed in the last few years [6),9),11)]. In this paper, we propose a multilayer neuronal model that takes into account the main characteristics of the perceptual alternation phenomenon and, in particular, the stochastic distribution required.

Our model is based on a multilayer network (MLN) of redundant structures made up of identical single-layer neuronal nets (SLNs), working in parallel and independently of one another. The SLN was extensively discussed in [12)]. Here we briefly recall that the SLN is a simple recognizer, based on the properties of the "Brain State in a Box" [1)], in which the state vector $\vec{f}$ can be split into two parts: $\vec{f}_A$, associated with the first interpretation, A, of an ambiguous figure (e.g. the "faces" of the Rubin pattern) and $\vec{f}_B$, associated with the alternative interpretation, B, (the "vase" in the Rubin pattern); $\vec{f}_A$ and $\vec{f}_B$ consist, respectively, of n_A and n_B elements, which are "neurons" linked to the features of the two interpretations.

The connection matrix, C, is obtained by the Hebbian learning of both interpretations during different trials, and contains two square blocks, E_{AA} and E_{BB}, representing the positive autoconnections of each subvector ($\vec{f}_A$ or $\vec{f}_B$) to itself, and two rectangular blocks, I_{AB} and I_{BA}, which are the inhibitory cross-connections of the subvectors:

$$C = \begin{pmatrix} E_{AA} & -I_{AB} \\ -I_{BA} & E_{BB} \end{pmatrix}. \qquad (1)$$

The oscillating dynamics of the SLN is described by the following equations:

$$f_i(t+\tau) = LIMIT\Big[\Big(\sum_{j=1}^{n_A+n_B} C_{ij} f_j(t) + G_i\Big)\Big(1-\sigma_i(t)\Big)\Big]; \quad i = 1, ..., n_A + n_B, \qquad (2)$$

where $LIMIT$ is a function restricting the values of the state vector components into the range $[0,1]$, to avoid non "physiological" levels of neuron activity; C_{ij} is the element of the connection matrix; G_i is the i-th component of the stimulus; and $\sigma_i(t)$ stands for the habituation process. The effect of this process is to lower the input sensitivity of the neurons when they are firing at their maximum rate: $\sigma_i(t)$ is usually zero but when the $LIMIT$ function becomes active, it assumes the value $\sigma = \sigma_o$ ($\sigma_o \in (0,1)$) over a fixed time interval (usually, .6-.7 for a period T equal to 30 τ). A biologically plausible noise can be added to the connection weight (i.e., the elements of C), without affecting the stability of the limit cycle.

2. Stochastic dynamics of the multilayer neural network

The MLN includes the probable redundancy [7)] of the neural assemblies acting as "recognizers" in the brain, and is composed of elementary blocks made up by single SLNs. The network consists of two layers: the lower one is made up of r redundant SLNs which work in parallel and independently of one another.

We assume that the visual input stimulus, $\vec{G}$, is shifted among such parallel recognizers, for instance, by eye movements. The input $\vec{G}$ is mapped on an SLN of the lower layer, with such a probability that the stimulus will be present, on average, in only one or two SLN(s) of the lower layer over each time interval.

The upper layer is constituted by a single SLN. The input to a neuron of the upper layer is equal to the sum of the activities of the corresponding neurons in the lower layer:

$$G_i^U(t) = \sum_{k=1}^{r} f_i^k(t) \tag{3}$$

After a short transient period, the activity of the upper layer of the MLN reaches a stationary phase in which the two subvectors, $\vec{f}_A$ and $\vec{f}_B$, become alternatively dominant, and in which the durations of the two percepts, t_A and t_B are stochastically distributed around their respective mean values. We found a range of parameter values for which the simulated distributions were well fitted by a Gamma distribution, with a good χ^2 [11].

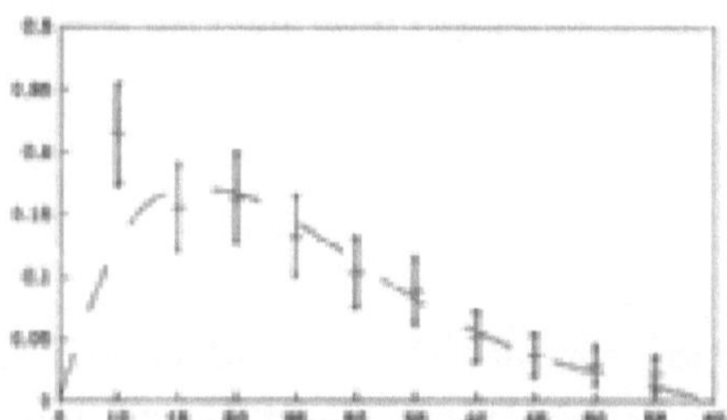

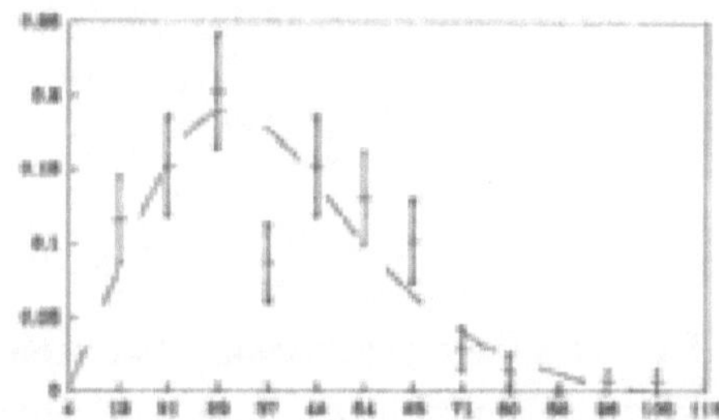

Fig. 2: Comparison between the theoretical Gamma distributions and computer simulations of the reversal times of percepts A and B.

Figure 2 shows a comparison between the stochastic distributions of the reversal times of the two percepts t_A and t_B (obtained via a computer simulation) and the corresponding theoretical Gamma distributions (dashed lines). It is worth noting that we use continuous values of the neurons' activation; this can be done by integrating the firing rate of the neurons over a suitable time interval, τ, covering the duration of some tens of spikes. A plausible value of the time τ, which can be regarded as the unit time for the MLN, is about a tenth of a second. This choice allows both the mean duration times (of the order of few seconds) and the values of the Gamma parameters to be very close to experimental ones.

3. Dependence of dynamics on the interpretations' complexities

An important phenomenological aspect of the perception of multistable figures is the dependence of perceptual alternation on the complexities of the two alternative interpretations. Following Structural Information Theory [4], the complexity of a pattern interpretation is linked to the minimum number of rules required to generate that interpretation.

Experiments on ambiguous patterns with interpretations of identical complexity, such as the Necker Cube [2),3)], gave nearly equal mean duration times for the two interpretations; by contrast, when the two interpretations of an ambiguous pattern are different in complexity, the simplest interpretation becomes dominant. Such results were obtained, for instance, by studying a series of eight patterns based on the Mach pyramid [10]. The series started with the "classic" Mach pyramid (shown in Figure 1), where the two interpretations (roof and room) differed by 2 information units, and finished with an unbalance of 20 information units in the last drawing.

In Figure 3a, the mean duration times, $\bar{t}_A$ and $\bar{t}_B$ of the two interpretations versus the interpretation complexities, I_A and I_B, are shown. These data are reported in [10]. As one can see, when the difference in complexity between the interpretations A and B becomes greater, the interpretation B becomes more dominant.

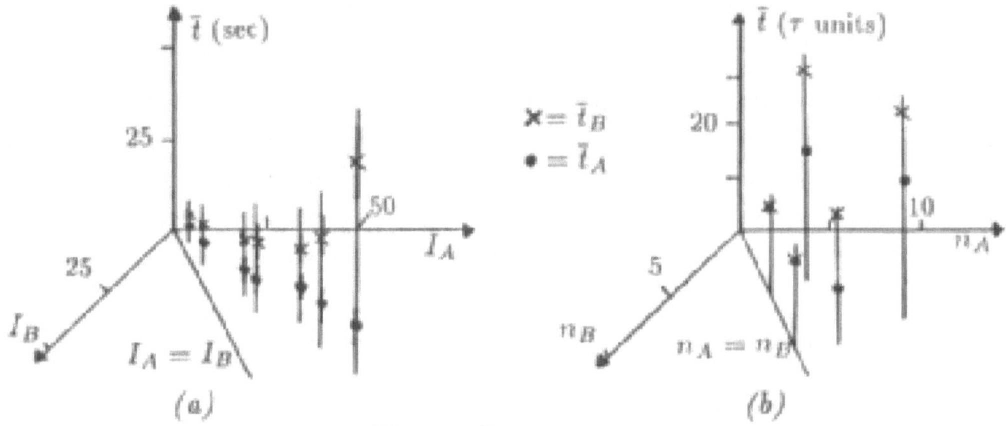

Fig. 3: See text.

If we assume a relationship between the interpretation complexities I_A and I_B and the number of SLN neurons, n_A and n_B, linked to the two interpretations A and B, the model shows a qualitative agreement between the results of computer simulations of the MLN (Figure 3b) and experimental results. In fact, the simulations give almost identical $\bar{t}_A$ and $\bar{t}_B$ values for n_A equal to n_B, whereas, when n_B is less than n_A, $\bar{t}_B$ becomes longer than $\bar{t}_A$.

References

1. Anderson, J.A., Silverstein, J.W., Ritz, S.A. and Jones, R.S., Psychol. Rev., 84, 413-451 (1977).
2. Borsellino, A., De Marco, A., A. Allazetta, A., Rinesi, S., Kybernetik, 10, 139-144 (1972).
3. Borsellino, A., Carlini, F., Riani, M., Tuccio, M.T., De Marco, A., Penengo, P., Trabucco, A., Perception, 11, 263-273 (1982).
4. Buffart, H. and Leeuwenberg, E., in "Modern Issues in Perception", Geisler, H.G., Buffart, H., Leeuwenberg, E. & Sarris, V., Eds., Amsterdam: North-Holland, 48-72 (1983).
5. De Marco, A., Penengo, P., Trabucco, A., Borsellino, A., Carlini, F., Riani, M. and Tuccio, M.T., Perception, 6, 645-656 (1977).
6. Ditzinger, T. and Haken, H., Biol. Cybern., 45, 279-287 (1989).
7. Edelman, G.M., in "The Organization of the Cerebral Cortex", Schmitt, F.O., Worden, F.G., Adelman, G., and Dennis, S.G., Eds., The MIT Press (1981).
8. Girgus, J.J., Rock, I. and Egatz, R., Percept. & Psychoph., 22, 550-556 (1977).
9. Kawamoto, A.H. and Anderson, J.A., Acta Psychol., 59, 35-65 (1985).
10. Masulli, F. and Riani, M., Percept. & Psychoph., 45, 501-513 (1989).
11. Masulli, F., Riani, M., Simonotto, E., in "International Joint Conference on Neural Networks", Caudill, M., Ed., vol. 1, 185-188, L.E.A. Publishers (1990).
12. Riani, M. and Masulli, F., in "Second Italian Workshop on Parallel Architectures and Neural Networks", Caianiello E. R., Ed., World Scientific (1989).

THE EFFECT OF NON-LINEAR SYNAPSES ON HIERARCHICAL ASSOCIATIVE MEMORY

M. Herrmann[1], M. V. Tsodyks[2]

[1] Karl-Marx-Universität, Dept. of Informatics, Leipzig, GDR

[2] Inst. of Higher Nervous Activity and Neurophysiology, USSR Academy of Scieces, Moscow, USSR

Extended Abstract

A Hopfield-like network of neurons assuming values 0, 1 is used to store hierarchically correlated patterns at a low level of activity [4, 5]. The interconnections [2]

$$J_{ij}=\sum_{l=1}^{L} C_l \sum_{\alpha^l} (X^{\alpha_l}_i - \langle X^{\alpha_l}_i\rangle_{\alpha_l}) (X^{\alpha_l}_j - \langle X^{\alpha_l}_j\rangle_{\alpha_l}), \qquad (1)$$

where $\alpha^l=(\alpha_1, \ldots, \alpha_l)$ is a multiindex and $\langle\ \rangle_{\alpha_l}$ denotes the average over the l-th level of the hierachical tree, are modified [1]

$$J_{ij}^{\Phi}=\sqrt{\langle J_{ij}^2\rangle}\ \Phi(J_{ij}/\sqrt{\langle J_{ij}^2\rangle})/N \qquad (2)$$

by the nonlinear w-step function

$$\Phi_w(x)=\begin{cases} b_k \text{ if } x\in(a_{k-1}, a_k) \\ -b_k \text{ if } x\in(-a_k, -a_{k-1}) \end{cases}, \quad 0<k<[(w+1)/2], \quad \begin{cases} a_0=0, \\ a_{[(w+1)/2]}=\infty, \\ b_{[(w+1)/2]}=1, \end{cases} \qquad (3)$$

such that Φ_w depends on w-2 parameters. The learning rule (1) with suitable chosen constants C_l allows for w-ary synapses to retrieve patterns of the hierarchical tree at levels higher than w with higher quality than lower levels. This is already for one step dynamics (perceptron) in parallel to the effect of synapses destruction investigated in [3]. We consider the case where the number of patterns is proportional to the inverse activity and the case of non-vanishing activity in the thermodynanic limit. By application of the mean field theory these results were sharpened.

References

[1] Englisch, H.; Herrmann, M.: Studia Biophysica 132 (1988) 1224.
[2] Parga, N.; Virasoro, M. A.: J. Phys. (Paris) 47 (1986) 1857.
[3] Virasoro, M. A.: Europhys. Lett. 7 (1988) 293-298.
[4] Tsodyks, M. V.: To be published.
[5] Tsodyks, M. V.; Feigelman, M. V.: Europhys. Lett. 6 (1988) 101.

A STRUCTURE FOR NEURAL NETWORK PATTERN CLASSIFIERS

Terrence L. Fine and Thomas W. Parks
School of Electrical Engineering
Cornell University
Ithaca, NY 14853, USA

Abstract

We indicate an approach to pattern classification using neural networks that clarifies and simplifies the roles of the various layers in the networks. The first layer can always be considered as quantizing, and thereby smoothing, the d-dimensional feature vector that is its input. We indicate how to determine the width w_Q of this layer in terms of the number N of training vectors and a single confidence parameter τ. The next two layers provide us with a clear method for implementing any Boolean function of the binary-valued outputs of this quantization layer. The first of these two layers has universal connections that can be fixed in advance or even randomly selected. The final layer is simply a Perceptron and is trained using the familiar Perceptron Training Algorithm to provide the desired binary classification output.

Our design approach has been applied to the construction of neural network pattern classifiers for both artificially generated pattern sources and acoustic transients embedded in ocean noise. Performance results will be reported.

I. Setup

We assume that we wish to classify a d-dimensional feature vector, $\underline{x} = \{x_i\}$, generated by a pattern source that can select one of two pattern types, $\theta \in \{0, 1\}$. The source selects category θ with probability $\pi(\theta)$ that is unknown to us. The source then selects a feature vector $\underline{x}$ with conditional density $f(\underline{x}|\theta)$ that is also unknown to us. At the outset, for each θ, we are provided with N_θ independent and identically distributed as $f(\underline{x}|\theta)$ feature vectors known to be from class θ.

The restriction to binary classification is not essential, and we can easily generalize to classify K pattern classes through parallel operation of about $\log_2 K$ binary classifiers of the type designed below. Our goal is to implement the pattern classifier as a feedforward neural network. We expect a neural network to be an appropriate choice for a pattern classifier when we have a high-dimensional feature vector, $d >> 1$, and do not anticipate that any small subset of components contains all of the relevant information for classification of the feature vector. By analogy with the uniform asymptotic negligibility conditions for central limit theorems, we expect that almost all feature vector components make a negligible contribution towards correct classification. For example, in image representation through a large grid of pixels, (e.g., the zip code reader using a 20-by-20 array of pixels to read a single handwritten digit described in Le Cun [1989]) it would be unusual if any individual pixel were critical to the success of the classifier. We also agree with the recent observation by Carver Mead [1990] that sensory perception mechanisms in successful biological organisms may be based upon utilization of a large number of crude features, perhaps ones capable of distinguishing only a few response levels.

We construct a feedforward neural network in terms of three layers. The feature vector discussed above provides the input to the neural network. The d nodes for the input vector are connected to a first processing layer through weights $\{q_{ij}\}$ linking the jth input x_j to the ith nonlinear device σ_i. We assume, with little loss of generality, that all nonlinear devices used in our network are ideal linear threshold elements;

$$\sigma(z) = U(z - \tau),$$

where U denotes the unit step function. The first layer will be referred to as the *quantization layer* or *Q-layer*, and it will have width (number of nonlinear devices) w_Q. The outputs $\{\kappa_i\} = \underline{\kappa}$ of the Q-layer are connected to the second layer, called the *expansion* or *E-layer*, through weights $\{e_{ij}\}$. The weight e_{ij} links the output of the jth nonlinear device in the Q-layer to the ith nonlinear device in the E-layer. The width of the E-layer is w_E. The outputs $\{\eta_i\} = \underline{\eta}$ of the E-layer are then connected to the third layer, called the *classification* or *C-layer* through weights $\{c_i\}$ connecting the output η_i to the single output nonlinearity in the C-layer. The binary response θ of the C-layer is the desired classification provided by the net.

II. Design of the Q-Layer

The Q-layer operates to quantize the d-dimensional input space into a finite number M of convex polyhedral cells generated by the intersection of w_Q half-spaces. The Schafli formula (Cover [1965]) gives the number of such cells as

$$M(w_Q, d) = \sum_{k=0}^{d} \binom{w_Q}{k}.$$

If, as is often the case, $d \geq w_Q$, then simply

$$M = 2^{w_Q}.$$

The larger the width w_Q, the more finely we can resolve inputs and potentially finely discriminate between pattern classes. However, the larger we make w_Q, the less data is available from the training set to tell us about the correct category to assign to feature vectors lying in each of the large number M of cells we have created. For example, if $M >> N_0 + N_1$ then many cells will perforce be empty even though they may possess a significant probability of occupancy by feature vectors from both classes, and we will not have an informed basis upon which to assign categories to these empty cells. Our approach will restrict the size of w_Q so as to ensure that we can reliably estimate the net parameters by taking into account the size of the training set. In this fashion we hope to maintain reasonable resolving power and avoid the trap of overtraining to the point of significantly impaired ability to generalize.

Our neural net architecture has sufficient flexibility to implement the assignment of any category in $\{0, 1\}$ to any binary vector $\underline{\kappa}$ that is an output of the Q-layer. However, we can only train meaningfully the remaining layers of the network to make this assignment if the data set provides statistically reliable information as to which category to assign to each possible vector. If w_Q is too large, then we cannot expect to reliably assign categories to each possible $\underline{\kappa}$ and hence we have no way to reliably select the network parameters that are to implement this assignment. If w_Q is small enough based upon the available training set, then we can accurately infer the correct category to assign to each Q-layer response $\underline{\kappa}$ and therefore have a clear basis for our selection of net parameters to implement this assignment.

The preceding considerations lead us to select w_Q small enough such that for each of the M cells generated by the Q-layer, the observed relative frequencies of cell occupancy have a high probability of being close to the true probabilities of cell occupancy. More precisely, we would like to achieve a confidence or probability of at least $1-\tau$ that every cell R induced by the Q-layer contains a fraction $\nu_\theta(R)$ of training set feature vectors of category θ lying within some prescribed δ of the true, but unknown, probability $P(R|\theta)$ of a feature vector $\underline{x}$ for a sample of class θ lying in the cell R. Equivalently,

$$\max_\theta P(\max_R |\nu_\theta(R) - P(R|\theta)| > \delta|\theta) < \tau.$$

We can select M conservatively, and without knowledge of the true underlying probability model, by employing Vapnik-Chervonenkis (VC) theory (Vapnik [1982]). If we select the weights $\{q_{ij}\}$ independently of the training set (e.g., either randomly or in accordance with some analytical design to optimize coverage of the input space) then VC theory yields

$$\tau > 2M\epsilon^{-2N\delta^2},$$

where $N = min(N_0, N_1)$.

If we attempt to adapt these weights so as to implement any of a number of methods of clustering the data from the two classes, then VC theory yields a different bound in which M is replaced by d^{w_Q}, that is only useful for low-dimensional feature vectors. However, even with randomly selected weights a pattern classifier such as the one we are designing can be proven to have performance converging to the Bayes optimum probability of correct decision (the performance achievable with known statistics) as the sample size grows.

We may proceed further by relating M and δ. If the precision δ, with which we are measuring the approximation to the cell probabilities by relative frequencies, exceeds the scale-size of the probabilities themselves, then we will be unable to resolve typical probabilities– the measurement error will be of the order of the quantities being measured. It is trivially true that the average probability of a cell must be $1/M$ when there are exactly M cells. This suggests that the scale-size of the probabilities of the cells is $1/M$. Hence $\delta \leq 1/M$, and we suggest taking δ as approximately $1/M$. We can now relate M to the minimum sample size N and the confidence level τ, when we take $\delta = 1/M$ and adopt the basic design wherein the connections from the input to the Q-layer are determined independently of the training set, through

$$M \approx 2\sqrt{N/\log(16N/\tau^2)}.$$

We see that M grows slowly with sample size N and even for large sample sizes we will have only to construct layers of moderate width. For purposes of orientation, we note that if $N = 1000$ and we desire a confidence of 0.9, then $M = 16, w_Q = 4, w_E = 16$, and we have a network of moderate size.

III. Design of the Remaining Layers

The output $\underline{\kappa}$ of the Q-layer is now a w_Q-dimensional binary vector whose category we can reliably estimate. The binary classification problem then requires us to provide a network architecture capable of implementing any Boolean function of w_Q variables. In the approach to neural network design that we are developing we carry out this task by dividing it into two stages. The final stage of classification through the C-layer can only be a Perceptron operating on the output $\underline{\eta}$ of the E-layer. For this Perceptron to be successful, the function of $\underline{\eta}$ we should be evaluating must be linearly separable. This observation determines our design of the E-layer.

We must select the E-layer so that it expands the dimension of its input space to the point where all pairs of disjoint subsets of input vectors yield linearly separable subsets of output vectors. If this requirement is not met then we may not be able to train the C-layer to properly identify inputs. Fortunately, this can be accomplished by making the E-layer wide enough. Discussions of the capacity of Perceptrons (e.g., Cover [1965]) show that in the limit of a large number of points selected in i.i.d. fashion the point set will be linearly separable if it is in a space of dimension at least half its size. Furthermore, we know that the VC dimension for a half-space equals the dimension of its space. Hence, with high probability, choosing $w_E > M/2$ will suffice for linear separability of outputs, and linear separability can always be achieved when $w_E = M$. Indeed, with high probability, linear separability will be achievable even with random connections between the Q-layer and the E-layer, provided that $w_E \geq M$. Clearly, so long as M is a moderate number (and recall from the preceding that it grows more slowly than $\sqrt{N}$), our approach will yield an easily implemented net.

The only layer that must be trained is the C-layer, and for it we can rely upon the simple Perceptron Training Algorithm. Furthermore, this algorithm is guaranteed to converge when we are dealing with linearly separable sets, as is the case for an E-layer chosen as we have described.

In sum, the structure we are urging uses the easily implementable fundamental linear threshold units for all nodes, need only be trained by using the familiar Perceptron Training Algorithm, and with probability of at least $1 - \tau$ will classify the quantized input feature vectors in agreement with the statistically optimal Bayes pattern classifier whenever the true cell occupancy probabilities under the two hypotheses differ significantly (that is by at least 2δ.) While we may have given up some ability to discriminate between neighboring feature vectors, this becomes provably negligible as the size N of the training set increases.

References

Cover, T. [1965], "Geometrical and Statistical Properties of Systems of Linear Inequalities with Applications in Pattern Recognition," *IEEE Trans. Electron. Comput.*, Vol EC-14, pp. 326-334.

Le Cun, Y., et.al. [1989], "Handwritten Digit Recognition: Applications of Neural Network Chips and Automatic Learning," *IEEE Comm. Mag.*, Vol. 27, no. 11, pp.41-46.

Mead, C. [1990], "Time, the Essential Dimension," *Trans. IJCNN* , Vol. 2, pp. 25.

Vapnik, V. [1982], *Estimation of Dependences Based on Empirical Data*, Springer-Verlag, New York.

Convergence of the Vectors in Kohonen's Learning Vector Quantization

John S. Baras and Anthony LaVigna
Systems Research Center, University of Maryland
College Park, Maryland 20742

Abstract

Kohonen's Learning Vector Quantization is a nonparametric classification scheme which classifies observations by comparing them to k templates called Voronoi vectors. The locations of these vectors are determined from past labeled data through a learning algorithm. When learning is complete, the class of a new observation is the same as the class of the closest Voronoi vector. Hence LVQ is similar to nearest neighbors, except that instead of all of the past observations being searched only the k Voronoi vectors are searched.

In this paper, we show that the LVQ learning algorithm converges to asymptotically stable zeros of an ordinary differential equation. It is shown that the learning algorithm performs stochastic approximation. Convergence of the vectors is guaranteed under the appropriate conditions on the underlying statistics of the classification problem. We also present a modification to the learning algorithm which results in more robust convergence.

1.1 Learning Vector Quantization

The LVQ algorithm is now described. Let $\{(x_i, d_{x_i})\}_{i=1}^N$ be the training data or past observation set. This means that x_i is observed when pattern d_{x_i} is in effect. Let θ_j be a Voronoi vector and let $\Theta = \{\theta_1, \ldots, \theta_k\}$. We assume that there are many more observations than Voronoi vectors (Duda & Hart [1973]). Once the Voronoi vectors are initialized, training proceeds by taking a sample (x_j, d_{x_j}) from the training set, finding the closest Voronoi vector and adjusting its value according to equations (1) and (2). After several passes through the data, the Voronoi vectors converge and training is complete.

Suppose θ_c is the closest vector. Adjust θ_c as follows:

$$\theta_c(n+1) = \theta_c(n) + \alpha_n \left(x_j - \theta_c(n)\right) \tag{1}$$

if $d_{\theta_c} = d_{x_j}$ and

$$\theta_c(n+1) = \theta_c(n) - \alpha_n \left(x_j - \theta_c(n)\right) \tag{2}$$

if $d_{\theta_c} \neq d_{x_j}$. The other Voronoi vectors are not modified.

This update has the effect that if x_j and θ_c have the same decision then θ_c is moved closer to x_j, however if they have different decisions then θ_c is moved away from x_j. The constants $\{\alpha_n\}$ are positive and decreasing, e.g., $\alpha_n = 1/n$.

1.2 Convergence of the Learning Algorithm

The LVQ algorithm has the general form

$$\theta_i(n+1) = \theta_i(n) + \alpha_n\, \gamma(d_{x_n}, d_{\theta_i(n)}, x_n, \Theta_n)\ (x_n - \theta_i(n)) \tag{3}$$

where x_n is the currently chosen past observation. The function γ determines whether there is an update and what its sign should be. It is given by

$$\gamma(d_{x_n}, d_{\theta_i}, x_n, \Theta_n) = 1_{\{x_n \in V_{\theta_i}\}}(1_{\{d_{x_n}=d_{\theta_i}\}} - 1_{\{d_{x_n}\neq d_{\theta_i}\}}). \tag{4}$$

Here $1_{\{\}}$ represents the indicator function and V_{θ_j} represents the set of points closest to θ_j.

The update in (3) is a stochastic approximation algorithm (Benveniste, Metivier & Priouret [1987]). It has the form

$$\Theta_{n+1} = \Theta_n + \alpha_n\ H(\Theta_n, z_n) \tag{5}$$

where Θ is the vector with components θ_i; $H(\Theta, z)$ is the vector with components defined in the obvious manner from (3) and z_n is the random pair consisting of the observation and the associated *true* pattern number. If the appropriate conditions are satisfied by α_n, H, and z_n, then Θ_n approaches the solution of

$$\frac{d}{dt}\bar{\Theta}(t) = h(\bar{\Theta}(t)) \tag{6}$$

for the appropriate choice of $h(\Theta)$.

Let $p_i(x)$ represent the pattern density for pattern i and let π_i represent its prior. Suppose there are ℓ patterns. It can be shown (Kohonen [1986]) that

$$h_i(\Theta) = \int_{V_{\theta_i}} (x - \Theta_i)\, p_i(x)\, \pi_i\, dx - \sum_{\substack{j=1\\ j\neq d_{\theta_i}}}^{\ell} \int_{V_{\theta_j}} (x - \Theta_j)\, p_j(x)\, \pi_j\, dx \tag{7}$$

The following hypotheses are assumed:

[H.1] $\{\alpha_n\}$ is a nonincreasing sequence of positive reals such that $\sum_n \alpha_n = \infty$, $\sum_n \alpha_n^{\lambda} < \infty$.

[H.2] Given d_{x_n}, x_n are independent and distributed according to $p_{d_{x_n}}(x)$.

[H.3] The pattern densities, $p_i(x)$, are continuous.

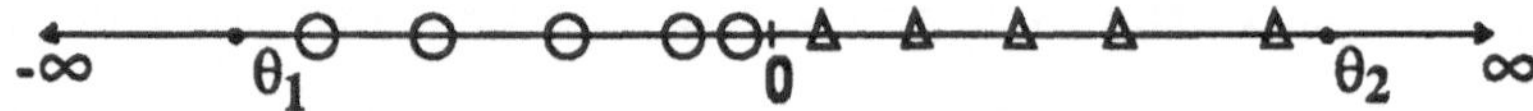

Figure 1: *A possible distribution of observations and two Voronoi vectors.*

With these assumptions it is possible, using techniques from (Benveniste, Metivier & Priouret [1987]) or (Kushner & Clark [1978]), to prove the following theorem.

Theorem 1 *Assume that [H.1]–[H.3] hold. Let $\bar{\Theta}^*$ be a locally asymptotic stable equilibrium point of (6) with domain of attraction D^*. Let Q be a compact subset of D^*. If $\Theta_n \in Q$ for infinitely many n then*

$$\lim_{n\to\infty} \Theta_n = \bar{\Theta}^* \tag{8}$$

Proof: (see (LaVigna [1989]))

Hence if the initial locations and decisions of the Voronoi vectors are close to a locally asymptotic stable equilibrium of (6) and if they do not move too much then the vectors converge.

1.3 Modified LVQ Algorithm

The convergence results above require that the initial conditions are close to the stable points of (6) in order for the algorithm to converge. In this section we present a modification to the LVQ algorithm which increases the number of stable equilibrium for equation (6) and hence increases the chances of convergence. First we present a simple example which emphasizes a defect of LVQ and suggests an appropriate modification to the algorithm.

Let ○ represent an observation from pattern 2 and let △ represent an observation from pattern 1. We assume that the observations are scalar. Figure 1 shows a possible distribution of observations. Suppose there are two Voronoi vectors θ_1 and θ_2 with decisions 1 and 2, respectively, initialized as shown in Figure 1. At each update of the LVQ algorithm, a point is picked at random from the observation set and the closest Voronoi vector is modified. We see that during this update, $\theta_2(n)$ is pushed towards ∞ and $\theta_1(n)$ is pushed towards $-\infty$, hence the Voronoi vectors do not converge.

This divergence happens because the decisions of the Voronoi vectors do not agree with the majority vote of the observations closest to each vector. As a result, the Voronoi vectors are pushed away from the origin. This phenomena occurs even though the observation data is bounded. The point here is that, if the decision associated with a Voronoi vector does not agree with the majority vote of the observations closest to that vector then it is possible for the vector to diverge. A simple solution to this problem is to correct the decisions of all the

Voronoi vectors after every adjustment so that their decisions correspond to the majority vote. In practice this correction would only be done during the beginning iterations of the learning algorithm since that is when α_n is large and the Voronoi vectors are moving around significantly. With this modification it is possible to show convergence to the Bayes optimal classifier (LaVigna [1989]) as the number of Voronoi vectors become large.

1.4 Conclusions

We have shown convergence of the Voronoi vectors in the LVQ algorithm. We have also presented the majority vote modification of the LVQ algorithm. This modification prevents divergence of the Voronoi vectors and results in convergence for a larger set of initial conditions. In addition, with this modification it is possible to show that as the appropriate parameters go to infinity the decision regions associated with the modified LVQ algorithm approach the Bayesian optimal (LaVigna [1989]).

1.5 Acknowledgements

This work was supported by the National Science Foundation through grant CDR-8803012, Texas Instruments through a TI/SRC Fellowship and the Office of Naval Research through an ONR Fellowship.

1.6 References

A. Benveniste, M. Metivier & P. Priouret [1987], *Algorithmes Adaptatifs et Approximations Stochastiques*, Mason, Paris.

R. O. Duda & P. E. Hart [1973], *Pattern Classification and Scene Analysis*, John Wiley & Sons, New York, NY.

T. Kohonen [1986], "Learning Vector Quantization for Pattern Recognition," Technical Report TKK-F-A601, Helsinki University of Technology.

H. J. Kushner & D. S. Clark [1978], *Stochastic Approximation Methods for Constrained and Unconstrained Systems* , Springer-Verlag, New York–Heidelberg–Berlin.

A. LaVigna [1989], "Nonparametric Classification using Learning Vector Quantization," Ph.D. Dissertation, Department of Electrical Engineering, University of Maryland.

Recurrent Networks for Learning Stochastic Sequences

Neil McCulloch
Research Initiative in Pattern Recognition
Royal Signals and Radar Establishment
St. Andrews Rd.
Malvern
WORCS WR14 3PS
U.K.

Abstract

This paper describes some experiments exploring the ability of networks to learn the underlying statistics of artificially generated temporal data. In the first experiment, data generated by two simple Markov chains was fed into a multi-layer perceptron (MLP). The desired output was an indication of whether a transition out of one of the models had been made. The network produced a close approximation to the probability that a transition had just been made. In the second experiment hidden Markov models were used to generate the data. This made the determination of whether a transition had occurred much more difficult, and the network produced a much poorer approximation to the correct probability.

Introduction

Many researchers have noted the usefulness of structuring neural networks to be suitable for the task at hand [DM89, Bed87]. Consequently, the structured back-propagation network shown in figure 1 was used in all the experiments. The input was produced by the Markov models shown in figure 2. The desired outputs correspond to (a) having just made a transition out of model A (b) having just made a transition out of model B (c) not having made a transition out of either model. The usual back-propagation with momentum algorithm was used [RM86]. Since the number of no transitions greatly exceeded the number of transitions, the error was inversely weighted in proportion to the number of targets of each category.

Experiment 1 - Deterministic Markov Models

Data

One hundred 'words' (i.e. 100 runs through A or B) were generated from the deterministic distributions shown in figure 2. These were used to train the network. A different 100 words were generated from the same model to test the trained networks. Ten runs were performed with different random weight starts.

Form of the Results

The results are displayed in two forms. Firstly, a confusion matrix averaged over all the runs shows the number of transition and no transition targets correctly recognised. An output was considered to be correct if it was closer in Euclidean distance to the correct target than to any other. The confusion matrices for the training and test sets are shown in figure 3.

Secondly, because the statistics of the data were known, the probability of having just made a transition out of either of the models could be calculated. This is the best output that the network can produce. In each run the Euclidean distance between these probabilities and the network outputs was calculated and normalised with respect to the number of outputs and the number of training vectors. The distances, averaged over all runs, are shown in columns 2 and 3 of table 1.

Discussion

The confusion matrices show that the network is correctly recognising all the transitions. This was expected since they are unambiguously indicated by a 1→0 transition on the input. Precisely correct values for the probabilities were not obtained, although the values chosen were better than random and also better than the simple guessing strategy of setting the transition outputs to 0 and the non-transition output to 1 as shown in table 1.

Experiment 2 - Hidden Markov Models

Data

How does the network perform on a more complicated model? Two new 100 word data sets were generated, this time using a Gaussian p.d.f. at each state. The means of these 1-D Gaussians were 0, 1, 0, and 1 respectively. The variances were the same for each state - for the first dataset they were 0.25; in the second dataset they were 1.0. Ten runs were performed with the 0.25 variance set and eight with the 1.0 set. It is now impossible to unambiguously determine when a transition has occurred, even from the true probabilities.

Results

Figure 4 shows the averaged confusion matrices for the training and test sets for the 0.25 variance runs. The fourth and fifth columns of table 1 show the averaged Euclidean distances. Similarly figure 5 shows the averaged confusion matrices for the 1.0 variance runs and the last two columns of table 1 show the averaged Euclidean distances.

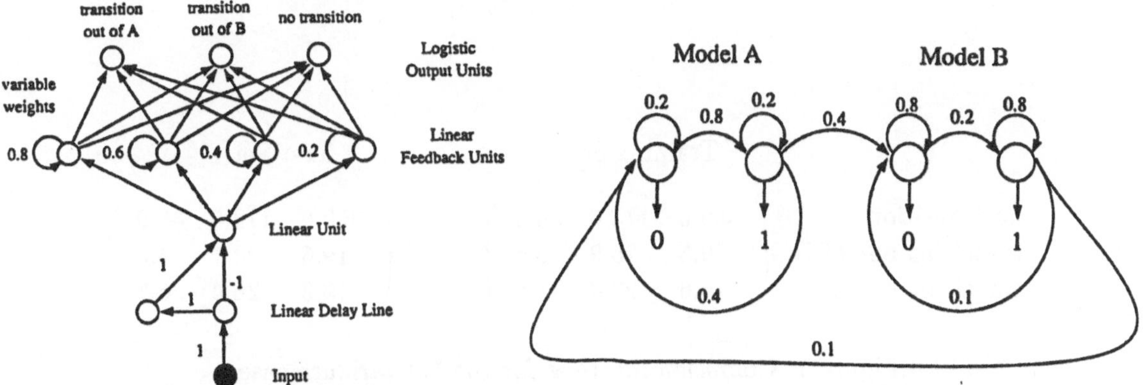

Figure 1: A structured back-propagation network

Figure 2: Two simple Markov models

	Training Set			Test Set		
no transition	594.0	0.0	0.0	522.0	0.0	0.0
transition out of B	0.0	36.5	23.5	0.0	33.1	14.9
transition out of A	0.0	3.6	36.4	0.0	8.2	44.4

Figure 3: Confusion matrices for the zero variance case

Sources	Variance					
	0		0.25		1.0	
	Training Set	Test Set	Training Set	Test Set	Training Set	Test Set
network outputs	0.017	0.018	0.077	0.073	0.097	0.094
random outputs	0.324	0.202	0.366	0.146	0.142	0.125
continuous no transition guess	0.089	0.100	0.044	0.046	0.021	0.021

Table 1: Normalised Euclidean distance between the true probabilities of the zero, 0.25 and 1.0 variance streams and other output sources

	Training Set			Test Set		
no transition	408.5	31.2	87.3	418.9	29.8	73.3
transition out of B	9.8	18.1	20.1	12.9	16.7	17.4
transition out of A	11.2	20.6	19.2	10.1	21.7	20.2

Figure 4: Confusion matrices for the 0.25 variance case

	Training Set			Test Set		
no transition	359.0	113.6	54.4	369.6	112.5	39.9
transition out of B	19.5	25.9	3.6	19.6	23.1	5.3
transition out of A	16.6	23.4	11.0	19.3	26.0	6.7

Figure 5: Confusion matrices for the 1.0 variance case

Discussion

As the uncertainty about the time of transition is increased, the performance decreases. In the 0.25 case the time of transition was still reasonably clear and so the number of transition errors was less than in the 1.0 case.

When the variance was increased to 1.0 a large number of transition timing errors were made and the ability to distinguish A and B transitions was lost completely. In both cases the Euclidean distance scores were better than random, but in both the 0.25 and 1.0 variance experiments a better fit with the true probability can be obtained by guessing 'no transition' all of the time! This is because the true probability of each output is usually close to zero or one, whereas the logistic outputs of the network cannot produce zero or one unless the magnitude of their inputs is infinitely large.

Conclusions and Future Work

The results of the first experiment indicate that it is possible to learn the underlying statistics of a Markov model using a recurrent back-propagation network. The difficulties encountered when hidden Markov models were used may be solved by the use of a more appropriate network structure. The automatic determination of appropriate network structures for different problems is an area of current research [Dod90]. An alternative approach is to directly implement the hidden Markov model in a recurrent network structure [Bri89]. We are currently investigating this approach.

References

[Bed87] Mark D. Bedworth. *Using The Error Back Propagation Algorithm: A Few Alternatives To Logistic Networks.* R.S.R.E. Memorandum, Speech Research Unit, Royal Signals and Radar Establishment, Malvern, UK, December 1987.

[Bri89] J.B. Bridle. Alpha-nets: a recurrent 'neural' network archirecture with a hidden markov model interpretation. *Speech Communication*, 1989.

[DM89] N. Dodd and N.A. McCulloch. Structured neural networks for markovian processes. In *Proceedings of the First IEE International Conference on Artificial Neural Networks*, pages 319–323, 1989.

[Dod90] Nigel Dodd. Optimisation of network structure using genetic techniques. In *Proceedings of the International Neural Network Conference, Paris*, July 1990.

[RM86] David E. Rumelhart and James L. Mcclelland. Learning internal representations by error propagation. In *Parallel Distributed Processing: Explorations in the Microstructures of Cognition*, pages 318–362, MIT press, 1986.

BRAIN BUILDING WITH GenNets

Hugo de Garis

CADEPS Artificial Intelligence Research Unit,
Universite Libre de Bruxelles (U.L.B.),
Ave F.D. Roosevelt 50, C.P. 194/7, B-1050, Brussels, Belgium.
tel: + 32 2 642 2783, email: CADEPS@BBRNSF11(.BITNET)

Abstract :

This paper shows how the Genetic Algorithm can be used to evolve time-dependent neural network modules. These functional and control modules can be assembled into increasingly complex structures, thus allowing the possibility of "brain building", using a new methodology known as Genetic Programming. The major advantage of GenNets (Genetically Programmed Neural Nets) as compared with traditional neural net learning techniques is that they need only a time-independent scalar performance measure (of the dynamical process they are controlling) to steer their evolution. GenNet modules can be evolved to perform incredible varieties of dynamic and control behaviours. Future technologies will allow Genetic Programming to be performed directly in hardware in real time, thus introducing the concept of the Darwin Machine. To illustrate the power of Genetic Programming, this paper shows how a 12 neuron, fully (self) connected, time dependent GenNet was evolved which taught a pair of stick legs to walk.

Keywords :

Genetic Algorithm, Time Dependent Neural Network Modules, GenNets, Genetic Programming, Evolvability, Brain Building, Darwin Machines.

Introduction :

Earlier experience [de GARIS, 1989, 1990] has convinced the author that a fruitful marriage is to be made between the Genetic Algorithm [GOLDBERG 1989] and Neural Networks [RUMELHART et al 1986]. This conviction has led to the development of the concept of Genetic Programming, which is the employment of the Genetic Algorithm to evolve the signs (excitory or inhibitory) and the weights of fully (self) connected neural networks (called GenNets), such that their dynamical performance is "optimalised". These GenNet modules can then be combined to form more complex structures capable of more complex behaviours. Control signals can be sent from control GenNets to functional GenNets, where the strength and timing of these control signals is itself evolved using the Genetic Algorithm. Genetic Programming thus opens the way towards "brain building". It will now be possible to build simple "brains" in this hierarchical GenNet fashion.

Genetic Programming is a conceptually simple approach to the design of time dependent systems using a single number (the quality measure of the dynamical performance) to guide the evolution. Normally, recurrent neural networks (i.e. those with feedback links between neurons) require a certain "settling time" for the output neurons to stabilize their output signal values, given fixed (clamped) input values to the input neurons. However with time-dependent GenNets, it is usual that the inputs change faster than the settling time, so that the neural network never settles. Under such circumstances, it is not obvious how the usual neural network learning algorithms (such as Backprop) can be applied to teach such a network. The Genetic Algorithm (GA) however does not really care how the neural network performs its task, so long as it performs it, and that the quality of the performance can be measured.

Section 1 of this paper introduces the Genetic Algorithm. Section 2 introduces the concept of Genetic Programming. Sections 3 describes the GenNets used to teach the stick legs to walk. Section 4 presents the results of the experiments.

1. The Genetic Algorithm :

The Genetic Algorithm is a form of simulated evolution to solve optimization problems in a Darwinian "survival of the fittest" approach. Solutions to problems are coded onto (usually binary) strings called "chromosomes", (e.g. parameter values in a control problem), which compete with each other to reproduce the next generation. A quality value for the encoded solution of each chromosome is determined, and the probability of reproduction of each chromosome into the next generation is proportional to this value. The number of chromosomes per generation remains fixed. Genetic operators can be applied to chromosomes, such as mutation (bit flipping), crossover (cutting two chromosomes at the same position and swapping portions), inversion (inverting a section of a chromosome). Occasionally, the application of these operators to the offspring causes them to have higher quality values than their parents. Hence they will reproduce with higher probability, and squeeze out inferior chromosomes. Over time, the average quality of the population will increase. The GA can be seen as a form of hill climbing where there may be many hills in the configuration space. For an excellent introduction to the principles of Genetic Algorithms see [Goldberg 1989].

2. Genetic Programming :

Genetic Programming is a new programming methodology which uses the Genetic Algorithm to design neural network modules. The programmer specifies the behavioural characteristics that the neural network should possess, the number of neurons in the network (which will be fully (self) connected), identifies the input and the output neurons, the quality criterion for the performance of the network, the number of binary places after the binary point of the numbers which represent the size of the weights connecting neurons, the initial input signal values, the number of cycles, etc. The user also provides a list of parameters necessary to control the GA, such as the number of generations, the size of the population, the probability of mutation, the size of the scaling factor to avoid premature convergence, etc.

The GA is then used to find both the signs and the values of the weights of the network which provides the functionality desired. Once these weights are found, they are frozen (fixed) and the GenNet module thus evolved can be used as a component in a more complex structure. Usually a set of GenNet modules consists of a subset of low level modules which execute some simple functions. These low level modules are usually managed by control modules which are themselves GenNets, i.e. they too are evolved with the GA. The outputs of the control modules are the inputs to the low level modules (without any intervening weights). Once the control and low level modules (considered now as a unit) are functioning together as desired, the weights of the control modules are frozen and the unit can then be considered as a module or component for an even more complex structure.

3. The Walker GenNets:

As mentioned in the abstract, the primary aim of this paper is to show that GenNets are flexible enough to provide highly time dependent control. The vehicle used to test this possibility is a simple pair of stick legs, which is to be taught to walk. FIG. 1 shows the basic setup. Initially, in planning the experiment, the output of the GenNet was considered to be control signals to "muscles" which would move the legs. This approach was soon abandoned as being too complex, and a simpler interpretation was chosen, namely that the output values of the GenNet are the angular accelerations of the four components of the legs. Knowing the values of the angular accelerations (assumed constant over one cycle - where a cycle is the time period over which the neurons calculate (synchronously) their outputs from their inputs), and knowing the values of the angles and the angular velocities at the beginning of the cycle, one can calculate the values of the angles and the angular velocities at the end of the cycle. As input to the GenNet (control module) were chosen the angles and the angular velocities. FIG.2 shows how this feedback works. Knowing the angles, one can readily calculate the positions of the two "feet" of the stick legs. Whenever one of the feet becomes lower than the other, that foot is said to be "on the ground", and the distance (whether positive or negative) between the positions of the newly grounded foot and the previously grounded foot is calculated. The aim of the exercise is to evolve GenNets which make the stick legs move/walk as far as possible to the right in the user specified number of cycles, and cycle time.

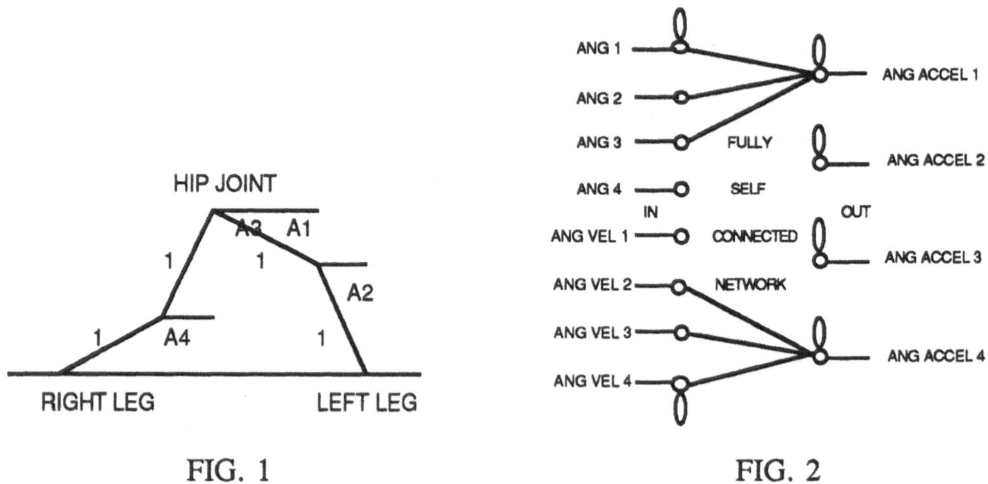

FIG. 1 FIG. 2

The GenNet used here consists of 12 neurons; 8 input neurons and 4 output neurons (no hidden neurons). The 8 input neurons have as inputs the values of the 4 angles and the 4 angular velocities. The input angles range from -1 to +1, where +1 means one half turn (i.e. 180 degrees). The initial (start) angles are chosen randomly, ranging between 0 and 1. The initial angular velocities are chosen randomly between -1 to +1, and are given in half turns per second. The activity of a neuron is calculated in the usual way, namely the sum of the products of its inputs and its weights, where weight values range from -1 to +1. The output of a neuron is calculated from this sum, using the symmetrical sigmoid function (-1 + (2/(1 + exp(-sum)))), which ranges from -1 to +1. The outputs of neurons were restricted to have absolute values of less than 1 so as to avoid the risk of explosive positive feedback.

The chromosomes used to evolve the weights and their signs in the GA are simple binary strings. The user specifies the number P of binary places after the binary point of the numbers representing the values of the weights (where weights have an absolute value less than 1). Imagine this is 6. One bit is used per weight to specify the sign of the weight (0 is positive, i.e. an excitory "synapse", 1 is negative, i.e. an inhibitory "synapse"). Thus, for a GenNet of N neurons, a chromosome will be N*N*(P + 1) bits long, since there are N*N weights in a fully (self) connected network.

The initial population of chromosomes is generated randomly with each bit equally likely to be a 0 or a 1. The ith group of (P + 1) bits corresponds to the sign bit followed by the P bits giving the weight value of the ith weight, e.g. 100101 represents a weight of -0.15625. For our experiments, no crossover was used. GenNets are so highly interdependent that crossing over chromosome portions is detrimental. GenNet GAs typically function with no sex (i.e. no crossover) and no inversion, using only mutation and selection. The selection technique used was standard "roulette wheel", (see [GOLDBERG 1989]). The quality criterion used for selecting the next generation was usually the total distance covered by the stick legs in the total time T, where (T = C*cycletime) for the user specified number C of cycles and cycletime. The quality is thus the velocity of the stick legs moving to the right. Right distances are non negative. Stick legs which moved to the left scored zero and were eliminated after the first generation.

4. Results :

A series of experiments was undertaken. In the first experiment, no constraint was imposed on the motion of the stick legs (except for a selection to move right across the screen). The resulting motion was most un-lifelike. It consisted of a curious mixture of windmilling of the legs and strange contortions of the hip and knee joints. However it certainly moved well to the right, starting at random angles and angular velocities. As the distance covered increased, the speed of the motion increased as well and became more "efficient", e.g. windmilling was squashed to a "swimmers stroke".

In the second experiment, the stick legs had to move such that the hip joint remained above the floor (a line drawn on the screen). During the evolution, if the hip joint did hit the floor, evolution ceased, the total distance covered was frozen, and no further cycles were executed. After every cycle, the coordinates of the two feet were calculated and a check made to see if the hip joint did not lie below both feet. This time the evolution was slower, presumably because it was harder to find new weights which led to a motion satisfying the constraints. The resulting motion was almost as un-lifelike as in the first experiment, again with windmilling and contortions, but at least the hip joint remained above the floor.

In the third experiment, a full set of constraints was imposed to ensure a lifelike walking motion. The result was that the stick legs moved so as to take as long a single step as possible, and "did the splits" with the two legs as extended as possible and the hip joint just above the floor. From this position, it was impossible to move any further. Evolution ceased. This was a valuable lesson and focussed attention upon the important concept of "evolvability", i.e. the capacity for further evolution. A change in approach was needed. A form of multi-step (sequential) evolution was undertaken. A GenNet was evolved over a small number of cycles, which took the stick legs from a left foot forward, right foot back position to the reverse, with starting and finishing angular velocities all at zero. The same GenNet was then used over a larger number of cycles (thus causing a stepping motion on the screen) and evolved to produce many short steppings (by having the quality measure being the product of the distance covered and the number of net positive steps taken). The fact that this sequential evolution is possible shows that GenNets have a dynamical "memory" in a certain sense. This same GenNet in turn was evolved to give maximum length covered in a given number of cycles. The result was a definite walking action with large strides. A video of the results of the above experiments has been made. To obtain a real feel for the evolution of the motion of the stick legs, one really needs to see it.

Finally, it is likely that Genetic Programmers will have to abandon the hope of having a full understanding of the dynamical behaviour of their GenNets. The internal behaviour of GenNets is simply too complex. Nevertheless, using Genetic Programming, functional neural nets can still be built and to user specified requirements. It is also likely that Genetic Programming will have implications for both robotics and future "brain building". One can imagine future robots being (self) Genetically Programmed, initially by software, and later by real time Darwin Machine components in their robotic "brains".

References :

[de GARIS 1989] "COMPO : Conceptual Clustering with Connectionist Competitive Learning", H. de Garis. Proceedings IEE International Conference on Artificial Neural Networks, London, October 1989.

[de GARIS 1990] "Genetic Programming : Modular Neural Evolution for Darwin Machines", H. de Garis, Proceedings IJCNN90 WASH DC, International Joint Conference on Neural Networks, January 1990, Washington DC.

[GOLDBERG 1989] "Genetic Algorithms in Search, Optimization, and Machine Learning", D.E. Goldberg, Addison-Wesley, 1989.

[RUMELHART et al 1986] "Parallel Distributed Processing", Rumelhart D.E., McClelland J.L., Vols 1 & 2, MIT Press, 1986.

TWO LOWER BOUNDS FOR PROBABILISTIC ALGORITHMS

Farid Ablayev
Department of Theoretical Cybernetics
Kazan University, Kazan 420008, USSR

ABSTRACT

The lower bound O(n log log n) has been proved for the time for recognizing a non-regular languages by one-tape off-line probabilistic machine with bound error probability. This lower bound proves the correctness of Freivald's long-standing hypothesis, first announced more than ten years ago.

Secondly, the lower bound $2^{\log D(L) - t h(1/2+e)}$ has been proved for the number of states of finite probabilistic automata recognizing an arbitrary regular language L with 1/2 - e error probability, where D(L) is the number of states of deterministic automata recognizing regular language L and $h(\alpha)=-\alpha \log \alpha - (1-\alpha) \log(1-\alpha)$. Finally we have presented two examples of languages which compare our lower bound with Rabin's well known lower bound.

Thermal Comfort Sensor based on Probabilistic Energy Neural Network

Toshikazu Takemori
Osaka Gas Co.,Ltd. Fundamental Research Labs..
6-19-9,Torishima,Konohana,Osaka 554 JAPAN

Shozo Hirose
OGIS Research Center

abstract

This paper describes a new type of neural network for pattern recognition, which we call, ***Probabilistic Energy Neural Network(PENN)***, and ***Thermal Comfort Sensor(TCS)*** using PENN. The idea of PENN is basically Bayes rule and the learning mechanism of it is motivated by the conventional neural networks such as Restricted Coulomb Energy(RCE)[1]. PENN is a supervised three-layered feedforward network. It can be regarded as a network that outputs ***a posteriori probability*** after learning ***a priori probability*** and ***state conditional probability density distribution.*** The wonderful features of PENN are 1) real time learning capability, 2) pattern classification ability on non-linearly separable data, 3) probabilistic nature of the decision rule.

TCS developed here only is a computer simulation system and deals with the most important variables that influence the condition of thermal comfort according to ***Fanger's idea***[2]. Those variables are activity level(heat production in the body), thermal resistance of the clothing(clo-value), air temperature, mean radiant temperature, relative air velocity, and water vapour pressure in ambient air. And in fact, PENN that is used for TCS has a specially attached unit to the top level of it, ***thermal comfort energy unit.*** This unit indicates the degree of thermal comfort of room conditions. This unit is very useful for making a total decision from the values of multiple output units, and also it enables PENN to have a time dependent characteristic of thermal comfort.

And the actual experiments under 247 thermal comfort conditions were done with 13 subjects in order to know how the thermal comfort conditions look like. These data are divided and used for recognition test of PENN. The correct performance of TCS is 93% (including 1-class-difference) and the learning time is less than 1 second on a Sun-3 workstation when the number of training data is 30. This high correct performance and surprising learning speed show that PENN could be a promising tool for building TCS or may even be useful for "***Comfort Sensor***" in the future.

ON THE BACK-PROPAGATION TRAINING OF NEURAL NETWORKS WITH NOISY DATA

Petri A. Jokinen
NESTE Technology, P.O. Box 310
SF-06101 Porvoo, Finland

SUMMARY:

In this paper the performance of back-propagation algorithm is studied in the presence of noise in the training data. A general method of using robust error estimators with back-propagation algorithm is explained and a family of so called robust M-estimates is investigated in detail.

The back-propagation algorithm for minimization of the M-estimates is derived and simulation results with these estimators are presented. Simulation results with different types of robust output error estimates have shown that these estimators are more tolerant to the noise in the training data and can provide significantly better solutions than the original back-propagation algorithm with mean square error function. As a result, the least squares error estimate should be replaced by some robust error estimate, while training of the network must be done with small amount of noisy data. This is typical in many practical measurement signal analysis and system identification applications.

EEC AND GOVERNMENT SUBSIDIZED PROJECTS

Chair: Jean-Jacques LAUTURE

NEURAL-COMPUTING WITHIN ESPRIT

J.J. Lauture, CEC, ESPRIT, DGXIII/A/4
BRE 9/191, Rue de la Loi 200, B1049 Brussels

ABSTRACT

In this document, I intend to justify why and how we should support Neuro Computing activities within ESPRIT. There is no major contradiction with the policy adopted up to date. The initial actions, covering support environment & SW-HW tools, are quite good means of strengthening the European Industry in NC. However, initial actions such as the projects ANNIE or PYGMALION need to be pursued, enlarged and completed with VLSI design, through the second ESPRIT call and the new Framework Programme.

I. INTRODUCTION

Our aim is to build computing systems which are more complex and more efficient. This implies the urgency for new architectures, which are better structured and more understandable, for which solutions are not necessarily determined in a combinatorial way (e.g. travelling salesman for optimization problems), but in a quicker and more natural way.

In comparison to the classical sequential programming for complex systems engineering, it appears that in order to allow better performances and a higher freedom degree during system development, the integration of a more and more implicit parallelism is necessary. This is actually underway with prolog and other 5GLs, supported by chips alike transputers.

The assessment of MINSKY and PAPERTS (MIN69), which shows the limitations of nets twenty years ago, has reinforced research on symbolic programming. However, on its turn symbolic programming showing its limits (e.g. memory access and data processing, learning, adaptivity) invite to go back toward nets.

That is the reason why new research now focuses on the lack of performance for:

(i) associative memories,
(ii) learning,
(ii) adaptive and self organizing architectures.

The nets models have now been improved (KOH82, HOP84, FUK83, HIN85 AND86, RUM86, CAR86, GROS86), and are now mature enough to provide basic elements to partly answer the engineering queries mentioned above.
At the same time, VLSI technology, is now providing opportunities for chipment of neural nets and biocomponents (MEA89,SIV87).

To appraise the role of neural computing and the support which needs to be given in Europe, I will now draft an overview of R&D, Markets and Applications.

II THE WORLD SCENE: R&D & MARKET

II.1 R&D PROGRAMME IN US, JAPAN AND EUROPE

A minimum of 300 universities and companies are involved in NC in the US, implying a minimum of 1000 man-years spent (IRD89, MIL89, DUR87). We most likely have an equivalent effort in JAPAN, equally characterised by large teams (about 20 people) in large companies like MITSUBISHI, NEC, NHK, SANYO, etc ...

In our list we have about 200 European organisations, but until now we do not have as large a European action as DARPA or any action in the Japanese 5th generation programme.

This gives an estimate of the total budget spent for R&D, which should be between 250 M$ and 400 M$ in 1990.

II.2 CEC R&D FUNDING

Following industry's request, the budget allocated by the CEC is globally increasing but is, as I said above, far behind the budget devoted by the US DARPA programme (400 M$, with 30 M$ for next year).

87 BRAIN	2 MECUs
88 BRA-ESPRIT II	3 MECUs
89 IPS-ESPRIT II 1st call	5 MECUs
90 IPS+CIM+BR, ESPRIT II 2nd call	15-20 MECUs

II.3 MARKET FEATURES

About 300 companies are currently, on a world scale, active in Neuro Computing (130 in the US, 100 in Japan , 40 in Europe, 30 elsewhere). The estimate market to be shared in 1990 by these companies is 113 MEcus (IRD89). This may be a weak ratio (with an average of a market of 0.3 MEcus per company and per year), but this market has only appeared since 1987, with an important growth rate, therefore it is necessary to analyse this market to understand which priority is to be given in R&D support.

a) Market progression:

The IRD report (IRD89) foresees a 1,3 BEcus market for 1998. This may be inferior to reality, for instance, if we simply estimate the costs saved for wear monitoring using nets (1 B/year saved only for ECs) (ANN87). These provisions may be exceeded if we consider that many NC-market prospectives have been underway three years already now , and are more and more optimistic, and that many new products, such as mentioned below (STE88) are now under development in the US and Japan:

- Digital filters for EKG processing;
- Process-control solving complicated non-linear functions (e.g for a broom mounted on a servo-controlled cart);
- Reading hand-printer numbers;
- Quality-control tools;

- Recognition of facial images;
- Consumer-loan credit screening;
- Mortgage Processing;
- Insurance claims processing;
- Reading text (saving a considerable processing time);
- etc.

This list (which is only a small subset of all new commercial applications which we may expect) is, in fact, more than creating new AIP or SP market segments which substitute old ones. In any case, the progression rate seems high. For instance, from IRD89 it appeared that, in comparison to 1989, the market is foreseen to multiply by ten before 1995 and thirty before 1998.

b) comparison between countries:
A typical feature of the NC scene is to compare the three actors US,JAPAN, and EUROPE, it appears that Japan, which has the smallest internal market (half the European or the US market (IRD89)), is going to invest more - and at longer term - (e.g. bio-components, bio-chips, 2nd generation networks) than any other country. In conclusion, Europe has a potential market equivalent to the US and should not invest less than its two competitors.

c) comparison with the overall AI and SP-based product markets:
AI : If we compare the IRD forecast (IRD89) of the neural-network market to the SRI87 estimation of the global AI market, we can deduct that nets market is in size at least 10% of the AIP market until 1998 (20B in 98, SRI87).

SP: Looking at the SP-market provisions from Frost & Sullivan (FROS88), it appears that until 1994, the nets represent about one third of the SP market (1.4 B in 1994, FROS88)

d) comparison between products:
As we said previously, market segments have already been identified (IRD89) with the following priorities for the coming years:

	for '95: position	%	for '98: position	%
Dedicated cards (ASIC'S)	2	28	1	35
Neurocomputers	5	10	2	22
Optical components	3	14	3	21
Software Simulators	1	29	4	12
Accelerators-VLSIcards	4	13	5	7
"Turn key systems"	6	6	6	3

The market progression is going from software tools to dedicated hardware, components and neurocomputers. This corresponds with the necessary time for the maturing of R&D, and performance optimization. Therefore, it is logical that components and systems markets are progressing quicker than the software-tools market, and it is clear that dedicated hardware, optical components and neurocomputers will have a major part of the market in a few years (in 1997 from IRD89).

II.4 THE EUROPEAN SCENE

The NC scene in Europe is actually characterised by a major interest and involvement from the academic side (about 200 universities and research centres), but only minor from the industrial side (20-30 identified sites).

The ESPRIT programme might help to strengthen the current labs, contributing the exchange from the theoretical side to the application side. In this sense the AIP or SP application experts would see new opportunities to solve specific problems by NC, or to create new services. Therefore considering that the current expertise from both sides (NC and application) is mature, support projects for cooperation are necessary.

II.5 CONCLUSION ON THE WORLD SCENE

Despite some inertia to industrialise NC in Europe (versus the US or Japan), the major players in Europe (Thomson, Philips, Siemens, British Telecom) now have teams between 15 to 30 people each. This shows that the challenge is taken seriously. Some small enterprises are now set up for services, consultansies, tool deliveries, all pushing us (CEC R&D programmes) to support the movement, which may hopefully be followed by concrete actions.

III APPLICATIONS

What makes seeing the industrial potential of Neural Networks confusing, is that we have - at this moment - two overlapping classifications without any links; the theoretical and the industrial. The theoretical corresponds to the taxonomy of existing models (LIP87). The industrial corresponds to the experience of companies, which could be limited to one or two categories of models, and to some application products (MIL89).
Indeed, the potential applications of neural-network technology are manifold, and this may explain the non-structured progression of the area; some products will be seen almost immediately, some are more distant. To understand the range and the chronology of NC applicability, we first have to bear in mind the categories of problem solving, which are the following:

(i) Optimization;
(ii) Pattern recognition;
(iii) Non destructive testing;
(iv) Control;
(v) Any application of the new generic abilities for information processing (e.g continous mapping from n to m dimensional space based upon a set of examples, or instantaneous nearest neighbour classification...)

and secondly, main categories of application for the above problems:

(i) Sensor Processing (speech, vision, others);
(ii) Combined sensor motor Control (robotic);
(iii) Knowledge processing;
(iv) Administrative and business Appl.;
(v) Quality Control;
etc.

As we have seen above (II.3), the estimation of the neural computing market may be around 10% of AI or 30% of SP markets. This is realistic, when we look at the new services which might be provided.

For instance I will present lists of services (not exhaustive) which are currently under development in the US and Japan.

III.1 SENSOR PROCESSING (SPEECH, VISION, OTHERS)
- Low-level processing, filters (HEI87, KAD87);
- Reading hand printer numbers (RIC87);
- Recognition of facial images;
- Reading text (saving considerable processing time) (BUR87,GAR89);
- Machines which read to the blind (JOG87);
- Icon recognition (GUL87);
- Voice typewriters;
- etc.

III.2 COMBINED SENSOR MOTOR CONTROL (ROBOTIC)
- Autonomous robot navigation (JOR87, GOG89);
- Broom-balancing problem (with a servo controlled cart) (nets works particularly well in complicated non-linear situations-STE88);
- Specialised robots for dangerous industrial tasks (mining and smelting); - Household robots (to be available on the market within ten to fifteen years, at the same cost as a car);
- etc.

III.3 KNOWLEDGE PROCESSING
- Autonomous (no programming required) extraction of knowledge from large databases;
- Connectionist expert systems (MOZ87, FOZ89, ROD89);
- Medical Applications (eg: filter for EKG processing);
- Financial Applications,
  . equities trading,
  . consumer loan credit screening (eg: in the US a neural network running on a HNC ANZA neurocomputer has been developed for AVCO-financial that outperforms their current credit-scoring system);
- Inventory control .(eg: training of a net, on examples simulated by an expert system, for inventory management) (with a large database and a large number of control parameters);
- etc ...

III.4 ADMINISTRATIVE AND BUSINESS APPLICATIONS
- Crew scheduling problem (POL87);
- Airline tactician (HUT87);
- etc ...

III.5 QUALITY CONTROL
- Application of nets for wear monitoring (NDT);
- Application of nets to automated piece-by-piece inspection for industry for 100% quality (pattern recognition);
- Information system security (adaptive reference monitor, KEL87);
- etc.

IV ESPRIT CURRENT ACTIONS AND FUTURE

IV.1 CURRENT NEURAL NETWORK ACTIVITIES

There are two industrial precompetitive ESPRIT Projects (in the IPS area) that have the acronyms PYGMALION (P2059) and ANNIE (P2092), respectively.

The objectives of PYGMALION are:
- to demonstrate the potential of a neural network approach in two main industrial application areas, viz. image and speech processing;
- to develop a portable neural network specification language and a testing environment for neural network applications supported by Transputer based Hardware.

PYGMALION has a duration of two years.

1st Year results of PYGMALION

Environment

The high and intermediate languages (N and nC) have been specified. They are close to the C++ language. The development of compilers is under way. All the well-known neural models are already available in the library.

Hardware

A first cascadable VLSI demonstrator has been produced and tested. This chip offers full connectivity which allows us to programme the most commonly used neural networks models (BP, Hopfield, MLP).

Low level imaging

Results of applications of neural models (BP constrained) have proved robustness and good restoration for image compression. Applications for realtime HDTV compression are under study. Further work on image segmentation and texture is underway..

High level imaging - Speech recognition and acoustic signal classification

For these three areas, fairly positive results have been obtained and demonstrated, with some potential short-term commercial applications.

The objective of ANNIE is to develop demonstrations of neural network software technology in selected industrial areas such as optimisation and adaptive control of robot movement, pattern recognition in ultrasonic inspection of welds, and fusion of multi-channel acoustic data in the continuous monitoring of machine faults. Results will be generalised to related classes of applications and used to assess hardware-dependent aspects of silicon and optical implementations.
The ANNIE Project has a duration of three years.

The results are as follows:

A survey on the systematics and capabilities of neural networks, their architecture, applications and publication of a taxonomy.

A survey of the software and hardware available for neural network simulation has been carried out. Areas of applications where neural network might complete or outperform conventional techniques have been chosen.

(1) For image processing: two problem areas related to non-destructive testing have been chosen.

(i) visual inspection of solder joints,

(ii) ultrasonic testing of pressure vessels.

An investigation on the large supply of training data is to be made.

(2) For robotics: a prototype of an autonomous roving vehicle was demonstrated at the last ESPRIT Technical Week. This work has been found very encouraging with regard to the problem of collision avoidance.

(3) For optimisation, application for airline end-user and crew scheduling problems: A first prototype is to be finished. It is planned that the follow-up work will refine and extend the prototype.

ANNIE and PYGMALION have to cooperate to further improve the quality of the testing environment developed in PYGMALION.
During the first half of 1989, the following ESPRIT Basic Research Actions in neural networks have been launched (BERN'89), all having a duration of 2,5 years:

Innovative architectures and VLSI implementations for Neurocomputing (NERVES A3049)

This action has three aims:

1. Connectionist algorithms and architectures for adaptive information processing, pattern recognition and learning.

2. Design of a neurocomputing machine suitable for distributed algorithms with high speed and parallelism abilities. Various technologies will be studied, based either on specific arithmetic accelerators inside a reconfigurable architecture, or on custom pulse stream modulation circuits.

3. ASINCs (Application-Specific Integrated NeuroCircuits) dedicated to particular applications: both analog and digital basic cells will be designed:

Expected results of the action include: machine architectures suitable for simulation of general connectionist algorithms, and integrated circuits for machines and for specific applications.

Self-Organisation and Analogical Modelling using Subsymbolic Computing (A3234).
This action explores a subsymbolic approach to artificial intelligence using feature maps and the theory of complex dynamical systems. The "subsymbolic" mechanisms to be studied and demonstrated in problems of vision and motor control include:

1) analogical representations that keep part of what is represented implicit in the representation, and

2) analogous dynamics that exploits the implicit properties of the analogical representation to derive new information.

Two actions deal with neural network applications in speech recognition, namely:

High Resolution Speech Recognition (A3207).
This highly interdiscipline action aims at transcending the limitations of current speech recognition systems by combining modular high-resolution auditory preprocessing of speech input with connectionist speech recognition modules for feature extraction, feature-phonology conversion, and phonology-word conversion.

Speech Processing and Recognition using Integrated Neurocomputing Techniques (A3228).
This action seeks to improve the capability of automatic speech recognition systems to handle speaker adaption and noise reduction. Neural networks will be designed to perform transformations between the following levels of representation of speech input: signal to parametric, parametric to parametric, parametric to phonetic, phonetic to sublexical, parametrical to lexical.

IV.2 ESPRIT POTENTIAL ROLE AND CONTRIBUTION

Taking into account the previous characteristics, we can assess priorities and set up a frame for European NC activities. This would supposedly take place in ESPRIT, so as to follow the directions presented below:

1. Prepare tools, environments and methods, to support Neural Network technology (software tools and simulators) (This is what is done in Pygmalion, but with a small budget, and will be finished in 1990).

2. Prepare tools, environments and methods to support dedicated hardware, design, optical components, neuro computers.

3 Support of and experiments on neural networks application to promising and innovating areas (SP, HCI, PDP, and KE).

4 Foundations, implementation, and testing of hybrid technologies (Computing-science side and application side).

5 Basic research on new nets architectures, models, algorithms, associative memories, learning, etc...

6 Components technology.

7 Neural Networks solutions applied to CIM.

8 Technology and information transfer.

9 International Cooperation in Basic Research.
Major US and Japanese companies are giving high priority to NC within their R&D programmes (IBM, Fujitsu, ...). The main NC labs (Grossberg, Fukushima, Hopfield, ...) are not European. This implies that ESPRIT and national programmes should support European research activities and international cooperation with the US and Japan through international congresses, or through the HFSP programme, as far as basic research on NC is concerned.

10 The industrial infrastructure (as any other area) should be reinforced, inspite of some of the big 12 are already involved in NC (Thomson, Siemens, Philips). It is essential that the European industry can associate, and define for itself a common strategy for technology transfer and market share.

SMI's also have a key role to play in providing services, and quickly industrialising , nets on precise market segments.

V CONCLUSION

The 5th ESPRIT call and future calls and actions must ensure a good coverage of the following complementary aspects:

1. genericity and uniformity of neurocomputing;
2. applications;
3. foundations of nets.

This may enable us, to boost industry by selecting and diffusing the most promising actions in the field, in order to prepare the necessary background (methods, tools, material) for this new market segment.

ACKNOWLEDGEMENTS

I would like to thank Janis Folkmanis, Neill Edwards, Charlotte Brandenburg and Phyllis Cooke for their excellent help in the finalization of this document.

REFERENCES

SYN89: For further information, project synopses can be obtained at the CEC information office, DGXIII A2, rue de la Loi 200, B1049 Brussels.

As there is a bibliography of four pages and limit place, the full document may be requested at the same address.

PYGMALION - Neurocomputing

ESPRIT II Project 2059

Bernard Angéniol, Gaël de La Croix Vaubois
Thomson-CSF
DSE/DILS/SIA
BP 150
F-92223 Bagneux Cedex

Abstract

The PYGMALION project aims to promote the application of neural networks by European industry, and to develop European "standard" computational tools for programming and simulation of neural networks.

Our major objective is to produce of a complete environment for developing neural network algorithms and applications. The software aspects are largely developed in this project. To prepare the future production of neural-oriented hardware, a preliminary study of silicon integration is undertaken.

The other part of PYGMALION is the development of a number of applications, mainly in image and speech processing, to demonstrate the neural networks' capabilities expected from their properties of massive parallelism, fault tolerance, adaptivity and learning capabilities...

PYGMALION started in January 1989, for a duration of two years.

SOFTWARE ENVIRONMENT

One objective of PYGMALION is to ensure the widest usage of the neural programming environment, by making it as flexible as possible. The software environment we develop comprises 5 major parts :

The High Level Language called N is an object-oriented language for defining neural network algorithms and applications. This is the language a programmer will use, in conjunction with the algorithm library, to describe the topology and dynamics of a neural network.

The Algorithm Library is mainly a library of common neural networks, written in the high-level language. It provides the user with a number of validated modules for constructing new applications. The user will also have the ability to store his own algorithms as well as applications he has developed by his own, at any stage of their learning or relaxation phase.

The Intermediate Level Language called nC is the low level machine-independent network specification language for representing the partially or totally trained neural network applications.

The Graphic Monitor is the software environment for controlling the execution and monitoring of a neural network application simulation. This includes a simulation command language for setting up a simulation, monitoring its execution, interactively changing values, and saving a trained network.

Compilers are developed to the target UNIX-based workstations and parallel Transputer-based machines.

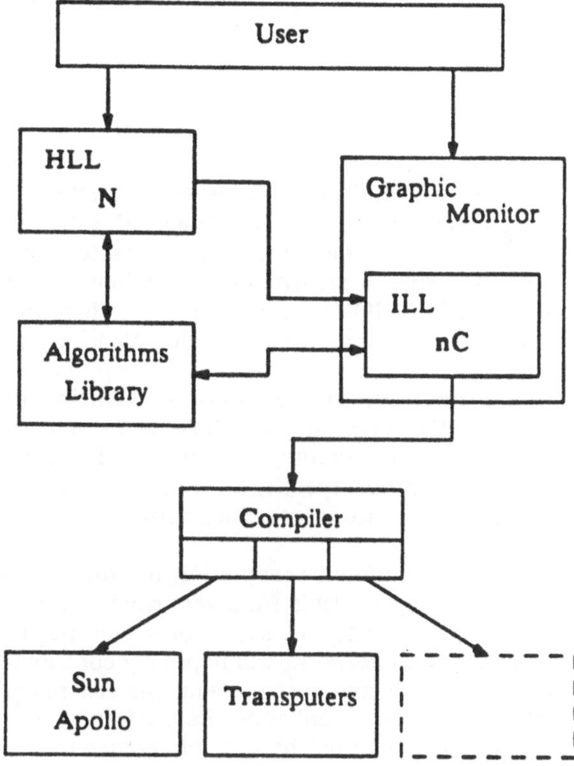

PYGMALION Neural Programming Environment

HARDWARE STUDIES

This small activity addresses the implementation of neural networks in WSI (Wafer Scale Integration). Considering the high level of connectivity in neural networks, the purpose is to have a large number of "neurons" on one chip. Fault tolerance is a characteristic of neural networks, which makes them a good candidate for WSI. PYGMALION is studying this feature and produces a VLSI demonstrator, to demonstrate the ability to make neural chips for a future neurocomputer or some application dedicated chips.

IMAGE PROCESSING

PYGMALION is investigating the application of neural networks to two important fields in image processing. The objective is to demonstrate neural network capabilities for specific applications, compared to more classical techniques.

The first one, **remote-sensing**, deals with texture and contour analysis on Spot images of the Earth's surface. Mainly low-level processing, like data compression and image segmentation, are developed to enable the extraction of elements such as roads, fields and various kinds of grounds. As good results have already been obtained in supervised learning, efforts are now being done without supervision. One approach we explore is a collaboration between contour extraction and texture classification.

The second one, **factory inspection**, covers the recognition and classification of workpieces in a factory automation context. Problems relating to position, overlap and orientation of workpieces in different lighting conditions are handled both in 2 and 3 dimensions. Two main approaches are used to apply neural networks. One feeds the neural network with statistical or syntactical features extracted from the image. This imposes to be able to find out which features are sufficiently significant to do the recognition. In the other one, neural networks' characteristics are used to deal with the image as direct input with no other preprocessing.

SPEECH PROCESSING

Many heuristic and even sophisticated methods have been tried for automatic speech recognition, but research investments into this "natural" task have, however, not yet yielded adequate results. Unlike traditional systems, neural network characteristics lead us to deal simultaneously with all the main current problems in speech recognition : noise interference, speaker insensitivity and vocabulary size. The objective of PYGMALION is to develop and to investigate a variety of artificial neural network architectures, in accordance with appropriate training algorithms, for some individual tasks.

Speaker–adaptive isolated word recognition for a medium–sized vocabulary in an office environment, with the recognition of whole words, aims at the design of products for the office and telecommunication areas. The learning capabilities of neural networks are exploited with the objective to adapt a recognizer to a new speaker or noisy environment during only a brief enrollment phase. This study explores the ability to differentiate between about hundred words with good noise tolerance.

Speaker–independent recognition of isolated words in a telecom environment is industrially appealing, even if only available for a very small vocabulary. Many useful voice–based services could be offered to the telephone network subscribers. Although good results have already been obtained with classical methods, neural networks will hopefully contribute to reach commercial application needs in the typically uncontrolled telecom environment. **Speech preprocessing**, with feature extraction and noise reduction, is also addressed to enhance the performance. The goal is to extract relevant features like spectral information, timing of phonetic phenomena...

Studies devoted to **vowel–consonant discrimination, coarticulation and subword units** are fundamental and critical issues for multi–speaker and connected word recognition. The aim of this task is to demonstrate that a non–arbitrary structure of discrete (subword) units exists within words, which a self organizing neural network structure is able to detect.

ACOUSTIC SIGNALS

The last field of application we are exploring is the classification of underwater natural sounds. A relatively small study will try to assess the abilities of neural networks compared to many kinds of algorithms already developed and optimized. Neurocomputing is applied in two ways. One uses a processed version of the signal, obtained through well–known and efficient preprocessing algorithms. The other one directly applies neural classification to the raw signal.

This approach also illustrates how efficient neural networks may be in terms of the amount of time and effort needed to develop an application. Provided the basic knowledge of the type of signal to be treated is taken into account in the global structure of the network, automatic neural adaptivity is quicker to implement than theoretical studies and more efficient than empirical research for significant characteristics to be extracted from the signal before the recognition process. This field is expected to give rise to operational applications in a very close future.

ROLE OF PARTNERS

CSELT has in charge the speaker–independent recognition of isolated words in a telecom environment. **CTI** is involved in the algorithm library and in the image pattern recognition in 3D. **ENS** is involved in pattern recognition in 2D images. **INESC** participates in the algorithm library definition. **IRIAC** handles the implementation of the algorithm library, and does the speech preprocessing. **Philips** is involved in image segmentation through texture analysis, and handles the pattern recognition in 2D. **SEL** is in charge of the speaker–adaptive isolated word recognition for a medium–sized vocabulary in an office environment. **Thomson–CSF** is developing the N language, and does the hardware integration. The company is also involved in image processing for compression, segmentation and pattern recognition, and has in charge the acoustic signal classification. **UCL** develops the nC language, the graphic monitor, and the compilers to UNIX–based machines and Transputer–based machines. **UPM** has undertaken the vowel–consonant discrimination, coarticulation and subword units.

CONTRACTORS

Thomson–CSF (Prime contractor)
B. Angeniol
DSE/DILS
BP 150
F–92223 Bagneux Cedex
Tel: 33–1–40844239
Fax: 33–1–40844696
Email: Gael@eurokom.ie

CSELT
G. Giandonato
Via G. Reiss Romoli 274
I–10148 Torino
Tel: 39–112169367
Fax: 39–112169520
Email: Giancarlo_Pirani_CSELT@eurokom.ie

CTI
T.–S. Papatheodorou
P.O. Box 1122
GR–26110 Patras
Tel: 30–61225073–61273496
Fax: 30–61222086
Email: ptheodor@grpatvx1.bitnet

ENS
R. Azencott
Cellule Applications
d'informatique et Mathematiques
Groupe de recherche DIAM
45, rue d'Ulm
F–75231 Paris Cedex 05
Tel: 33–1–43291225 33–1–69416019
Fax: 33–1–46340531

INESC
L. Borges de Almeida
Rua Alves Redol, 9
P–1000 Lisboa
Tel: 351–1–544607
Fax: 351–1–525843
Email: lba@inesc.uucp

IRIAC
F. Fogelman Soulie
10, rue Vandrezanne
F–75013 Paris
Tel: 33–1–42862230 33–1–69416369
Fax: 33–1–64461992
Email: ff@lri.lri.fr

Philips
J.–B. Theeten
LEP
3, avenue Descartes
BP 15
F–94451 Limeil Brevannes Cedex
Tel: 33–1–45699610
Fax: 33–1–45694088

SEL
H. Hackbarth
Lorenzstrasse 10.
P.O. Box 400749
D–7000 Stuttgart 40
Tel: 49–711–8692191
Fax: 49–711–8692185

UCL
P. Treleaven
Department of Computer Science
Gower Street
London WC1E 6BT
UK
Tel: 44–13807288
Fax: 44–1–3871397
Email: treleaven@cs.ucl.AC.UK

UPM
F. Aldana
Universidad Polytecnica de Madrid
Computer Science Faculty
Campus Montegancedo
E–28660 Madrid
Tel: 34–1–7157412

THE RESEARCH INITIATIVE IN PATTERN RECOGNITION: A UK COLLABORATIVE PROJECT IN NEURAL NETWORKS

David G Bounds
Deputy Director
Research Initiative in Pattern Recognition
Royal Signals and Radar Establishment
St Andrews Road
Malvern
Worcs. WR14 3PS, UK

ABSTRACT

The Research Initiative in Pattern Recognition (RIPR) was one of the first projects to be set up, in December 1986, under the Department of Trade and Industry's National Electronics Research Initiative. The aim of this scheme is to use the UK research resources to better effect by providing an efficient means of technology transfer of research findings to industry. This scheme also fulfills a second Government aim: to provide a more effective spin-off to industry of advanced research from the Ministry of Defence.

Eight companies are collaborating in the project: British Telecom, British Aerospace, Plessey, Pilkington, The Royal Signals and Radar Establishment (RSRE), Smiths Industries, STC and Thorn EMI. To date, the project has generated 65 technical reports, many of which appear in the open literature, and has been highly successful in helping to establish the use of neural networks as a practical application tool within the participating companies.

INTRODUCTION

In order to develop a research area to the point where it becomes useful for practical commercial applications, it is necessary to have a research team that exceeds some critical mass. For all but largest organisations, this is difficult to justify when the research is "blue sky" and commercial prospects are distant. Thus, although many companies have active research teams in the field of neural networks, now that the commercial possibilites are obvious, very few research teams existed in 1985 when the RIPR was conceived.

The idea of a research initiative in the area of pattern recognition grew out of meeting of the research directors of some of the major UK electronics and defence companies, together with the Ministry of Defence and Department of Trade and Industry, in the Autumn of 1985. The goal was to link more closely the research activities of the MOD and Industry, with a potential gain for both in faster and more effective exploitation of new technology.

Of the possible projects discussed, two gained widespread support from a number of potential collaborators. The Research Initiative in Silicon Hybrids (RISH) is developing multi-chip modules with high speed interconnect on a silicon substrate. The Research Initiative in Pattern Recognition is carrying out research in the area of machine intelligence and pattern recognition, primarily in neural networks. Both initiatives were established in late 1986 at the Royal Signals and Radar Establishment (RSRE), Malvern to take advantage of their existing expertise and facilities. In both areas RSRE had strong research programmes that could provide significant benefit to UK industry if exploited properly. Although the initial focus of the programmes was built on RSRE's expertise, the collaborating companies now play a leading role as the research moves out into their laboratories. Although there are a number of overlaps in both the collaborators and the organisational structure between the two research initiatives, the remainder of this paper will deal solely with the RIPR.

PROJECT OBJECTIVES

The project mission is to carry out research in pattern processing, control of intergrated pattern processing systems and the application of neural networks to pattern recognition tasks; and to transfer the knowledge gained to the participating companies.

STRUCTURE OF THE RESEARCH INITIATIVE

Like ESPRIT projects, or their forerunners in the UK - ALVEY projects, a research initiative is a collaborative exercise involving a number of organisations who work together on research and then share the results. The programmes are design to be pre-competitive, undertaking basic research in new technologies. This allows organisations who may be competitors in the market place to work together before the stage of individual product development. The RIPR project currently has 8 collaborators: British Telecom, British Aerospace, Plessey, Pilkington, RSRE, Smiths Industries, STC and Thorn EMI.

There is one feature which makes research initiatives unique: the research is all located at one site. The host organisation, in our case RSRE, provides space, infrastructure and support services, all of which are paid for out of the project funds. The collaborators second members of staff to work at the host site, usually for the duration of the project. Thus, instead of scientists working independently in their company laboratories and communicating by meetings, electronic mail and the telephone, they are in the same laboratory so that they are in constant daily contact. This organisation has resulted in a number of benefits for the research programme.

First, it has helped to create a critical intellectual mass, which has been particularly important in the area of neural networks where there was little or no expertise within the collaborating industrial organisations prior to the start of the project. This has been enhanced by establishing ties with university based research groups: Edinburgh University on Neural Networks, and the Information Technology Group at Oxford University and the Artificial Intelligence Vision Research Unit at Sheffield University for the Vision topics. Subject to the requirements to safeguard intellectual property rights and commercial exploitation, open publication of the research results and interaction with the international research community has been encouraged.

Second, the initiative has been able to establish a common development environment for software, which greatly improves the productivity of the staff involved. This environment is based on a network of SUN workstations together with the standard UNIX operating system to enable the development of common software modules and tools used by researchers across the topics. Since the major output of the project is information, a considerable effort has been extended in establishing a common text and document processing system. This is based around LaTeX with both graphics support for diagrams and an online library data base. Further computational resource for the project is provided by a MEIKO computing surface with 64 transputers. This is used both as a processing farm for multiple simulations and as tool to investigate the problems associated with parallel software.

During the first two and half years of the project, the research programme was split into two linked topics: one on image understanding systems, control and hardware and the second on neural networks. The two linked topics formed a single project with free information flow and intellectual property rights across the combined project to all collaborators, even though individual companies may have seconded a researcher to one particular topic. The project has now evolved into a more homogeneous research programme, primarily in neural networks with vision as the major application area.

A single site collaborative project does however have some potential disadvantages. Setting up a new research programme from scratch is a time consuming exercise, which is significant in a project with an initial phase of only 3 years. The time taken to establish the infrastructure was approximately 3 months for the neural network topic and over 9 months for the vision topic. The latter time reflects the need to integrate a wide range of technologies before useful work can be undertaken. Our experience shows that a start up phase for such a project, with skeleton staff to organise facilities and equipment purchase would be a more effective way to use the project resources.

There could also be a problem in finding staff who are prepared to re-locate for the duration of the project. In practice this has not occured and we have been fortunate in building a team of young highly motivated researchers. Care is taken to ensure that they remain in regular with their own organisations to ensure that the research results are disseminated as fast as possible, and that the project remains aware of the changing need of its market place.

Finance for the project is provided jointly by the collaborators and the DTI. Each collaborator provides for the cost associated with the member of staff seconded to the project. This includes travelling, overheads and the cost of half a person within the home organisation to maintain contact with the secondee and assist in the exploitation of the research findings. A total of 9 people are seconded from Industry with two from the MOD. The costs born by the collaborators are matched by the DTI to provide a cash budget which pays for capital equipment, host costs, the hiring of additional support staff, and consumables. The total budget for the first 3 years of the project is approximately £3.6 million.

The project is managed on a day to day basis by two topic leaders who are responsible for the technical direction and management of the research. A Project Director is responsible for the administration and organisation of the initiative. Overall control of the initiative rests with the Project Management Committee comprising of representatives from each of the collaborators, the host organisation and the DTI.

ACHIEVEMENTS

The RIPR has produced a vast amount of information for the companies, embodied both in the 65 technical reports to date (15 of which are now in the open literature), in the software produced, and in the less formal contacts between project scientists and scientists and managers back in the participating companies.

Areas of research include theoretical and simulation work on network algorithms and architectures, and applications to image processing, speech synthesis and recognition, sonar recognition, medical diagnosis, texture recognition and robot control.

Efficient software modules have been produced, both for a range of models for which commercial software packages have become available subsequently, and for capabilities which are not provided for by commercial software packages. This software is in use in every collaborating company where it has saved software development time and enabled companies to explore their own private applications. Some of these applications are already saving or making money for the companies, but they are outside the scope of this talk!

Many of the companies have set up their own in-house research project and have drawn heavily on experience gained in the RIPR to found these project. Perhaps the largest project is British Telecom Connex project which is described elsewhere in this conference.

The best sign of the success of this project is the participants desire to continue. The project was initially funded for a 3 year term, and would have expired in December 1989. The participants were unanimous in their desire that the project should continue for a further 2 years, and the project is continuing with reduced Government funding, to December 1991.

CONCLUSIONS

The ultimate test of the project will be the amount of revenue generated by companies involved in the initiative using the research done there. Inevitably, this will be very difficult to judge. I believe that one of the reasons that science is so seriously under-funded is the difficulty of making a connection between company profits and the research which led to them, which may have been done 10 years before. However, a number of conclusions can be drawn already.

First, from a standing start the project has made a significant contribution to UK research in pattern recognition using neural networks and has made some impact on the world scene in neural computing.

Second, the project has provided a focus for the collaborators to view their own in-house programmes and to build upon the preliminary results obtained. The collaborators own in-house programmes would have started later, started more slowly, or not started at all had the initiative not existed. The initiative now makes a very useful contribution in bringing in-house staff up to speed in the new technology.

Third, the project gives secondees invaluable experience in working together in an inter-company group. As potential research and general managers of the future, this exposure to other companies cultures could be vital in adapting to competition in an increasingly global market place.

In conclusion, the success of the current project provides a good model for future collaborative projects in areas which are difficult to fund initially because they appear to be "blue skies" research.

GALATEA: A C-LIBRARY FOR CONNECTIONIST APPLICATIONS

C. Mejia, L. Bottou, F. Fogelman Soulié

Laboratoire de Recherche en Informatique, bât. 490
Université de Paris Sud- 91 405 ORSAY Cedex - FRANCE

1 Introduction

Neural Networks are techniques which are increasingly used for real world applications. However, developing such applications does not simply mean applying academic ideas or algorithms: efficient programming tools and environments are required if one wants to efficiently design networks with optimal performances.
The ESPRIT-Pygmalion project has been set up to achieve a complete programming environment for connectionist applications. This environment will include both a Hich Level and Intermediate Level languages, a graphic monitor, and 2 Libraries: one in C and the other in the HL Language. Partners in the project are Thomson (HLL), UCL (ILL and graphic), IRIAC-LRI (Libraries) and INESC (C-Library). We present here the C-Library, Galatea, which has been designed so as to allow the partners in Pygmalion to easily develop and test their applications.

2 Galatea: the Algorithms C-Library in the Pygmalion Project

The C-algorithms library Galatea was intended to provide to partners in the ESPRIT-Pygmalion project a tool including efficient versions of the algorithms most commonly used for the applications developed within the project, i.e. image and speech processing. When developing the Library, we tried to enforce a unified description of the algorithms to ease the designing process of algorithms modules and further developments in the High Level Language of the Pygmalion environment.

2.1 User paradigm

A critical decision consists in defining the kind of objects the user will deal with. We might have defined a very general data structure, but it would have been too complex for most algorithms, and too restrictive for some others. Unless we use an ad-hoc compiler, such a choice would thus lead to inefficient programs for most algorithms and almost useless programs for other algorithms!
We describe a network as a black box, which has an input vector and an output vector. We also provide some ways to access internal variables as vectors. The user will first *define* a variable for its network, he will also *initialize* this network, with a specific network topology. He can then access the input and output vectors, the internal states, the weights; and set parameters. He can also *operate* the network: i.e. train the network, or retrieve results with it.

2.2 Vector-oriented data structures

We do not enforce a high level data structure, but a rather low level data structure: vectors of floating point numbers. The states of a single layer are grouped into a single vector, as well as the weights, and more generally all the internal data. This philosophy has a drawback: neural computations can poorly be deduced from the vector structure, except in some very simple cases (fully connected networks, for instance). However, in many applications, data naturally come as sets of vectors.

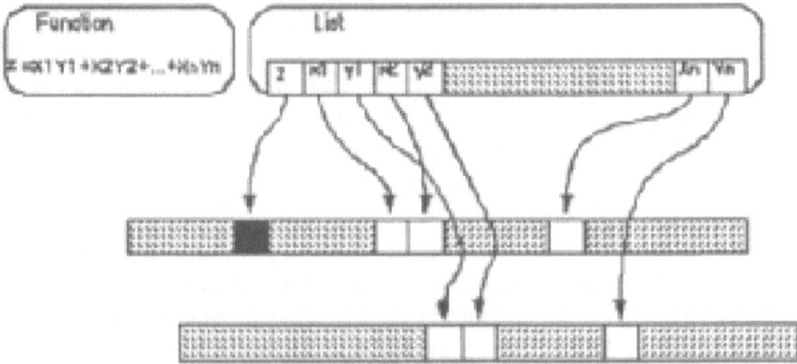

Figure 1: A recomputation rule example. The execution of this rule will compute the dot product x.y into z.

2.3 Computation rules

We also provide support routines for specifically describing neural computations. These routines are based on the idea of *recomputation rules*. A recomputation rule (fig.1) closely looks like a formula in a spreadsheet. They are basically composed of a function, and of a list of pointers to vectors elements. When a rule is executed, the list of pointers is passed to the function, which then performs a given computation.

Rule management functions include rules creation, extension, deletion, and execution. The rule system avoids the definition of complex data structures for designing the network architecture. They will simply describe the computations to be done rather than the network structure. These functions will be more precisely described below.

3 Galatea organization

3.1 Components

We describe now more precisely the Galatea components. Galatea has five main parts (fig.2):

- The *Algorithm Independent Part* (AIP) contains the computation support routines. It includes functions performing on vectors, functions for dealing with recomputation rules, and some basic routines for memory and error management.
- The*Tools Library* contains routines for managing standardized data files, network architecture files, and functions for implementing the most common classification criteria.
- The *Algorithm Modules*, contain the programs for the connectionist algorithms implemented.
- *Algorithm Evaluation Programs* are text oriented front-ends for programmer-designed modules. These front-ends allow the users to rapidly test various algorithms for their applications.
- These programs will be written using a standard *Environment Library* that implements most of the text oriented front-ends. Writing an evaluation program merely amounts in defining the command names, the associated C functions, and calling a MainLoop function. The Environment Library provides functionalities for managing help files, command files, and calling the system functions.

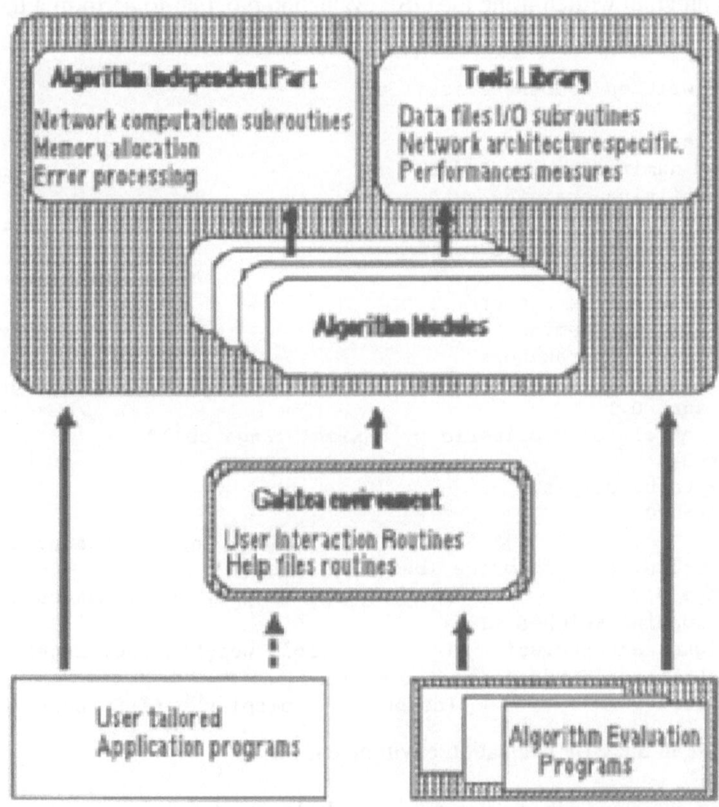

Figure 2: Organization of Galatea

3.2 Algorithms modules in Galatea

Galatea includes at present a large range of algorithms (fig.3). More will be added in the near future.

LAM	Linear Associative memories (Pseudo-Inverse Methods)
Hop	Hopfield nets
Kan	Kanerva associative memory
BAM	Bi-Directional Associative Memory
GBP	Gradient Back Propagation algorithm
GBPF	Gradient Back Propagation with Feed-back
TMap	Kohonen Topological maps
SimAnn	Simulated Annealing
BM	Boltzmann Machines
ART1	Adaptive Resonance Theory
LVQ	Learning Vector Quantization

Figure 3: Modules implemented in Galatea

4 How to use Galatea?

4.1 Using the C library.

The easiest way to use Galatea is through the Algorithm Evaluation Programs. These small programs are text-oriented interfaces to the Galatea functions. They are able to work with one algorithm and one network only. They share a common environment, common file formats, and similar commands.
Many users, however, need more powerful or more specific network programming. They should then write a C language program, that directly calls the Galatea routines. So, they gain the ability to deal with many networks, using different algorithms. They also can compile and link a complete C program.

4.2 An example

We show in figure 4 a program written using the GBP evaluation program to perform a task of hand-written digit recognition [Bottou 89].

```
#
# a hand written numbers classifier
#
echo "--loading the data ..."
data load numin.dat numout.dat                              { load the data files}
echo "--defining training set"
data training 0 319                      {use 320 first patterns for training}
echo "--defining test set"
data test 320 479                        {use 160 following patterns for test}
echo "--loading the network ..."
network load handnum.net                           {load network topology file}
echo "--setting parameters"                 {this is a MLP with shared weights}
forget inverse 2.4                                 {set weights initial values}
epsilon sqrt 0.1                                           {set iteration step}
echo "--set classify criteria by maximal range cell"
classify max
echo "--status display"
show width 80
status                                  {display the status of network and data}
echo "--training performance (before learning)"
perf range 0 9              {compute and display performances for 10 first digits}
echo "--loading weights ..."
load-weights handnum.wei              {load weights from another learning session}
echo "--training performance (after learning)"
perf range 0 9              {compute and display performances for 10 first digits}
```

Figure 4: Command file in the GBP evaluation environment

As can be seen from this command file, programming a multi-layer network, even with a complicated architecture (here with shared weights [Lang, 88, Bottou 90]), is very easy because of the use of the predefined functions provided by Galatea in the Tools Library.
The previous command file produces the following script file (fig. 5):

```
Pygmalion C library - GBP evaluation program
(C) IRIAC 1989
--loading the data ...
--defining training set
--defining test set
--loading the network ...
--setting parameters
--set classify criteria by maximal range cell
--status display
Network      :  loaded from <handnum.net>
                aged <0> sweeps, epsilon sqrt <0.1000>
                momentum not allowed, decay not allowed
Data         :  <480> patterns (input+output) loaded
Patterns     :  a <480 x 256> matrix
Desired      :  a <480 x 10> matrix
Training set :  from <0> to <319>
Test set     :  from <320> to <479>
Classify     :  <max>
Function     :  <standard sigmoid (-1.7 to 1.7)>
--training performance (before learning)
Set {0,9}: sweep 0      , Error= 0.29643, Performance= 20.00
--loading weights ...
--training performance (after learning)
Set {0,9}: sweep 0      , Error= 0.06378, Performance=100.00
gbp> forget sqrt 1.5              {here Galatea lets you type-in new commands}
gbp> epsilon sqrt 0.1
gbp> run 1 4
---------------------                                           {performances on}
Set {0,319}: sweep 1      , Error= 0.16296, Performance= 55.00    {learning set}
Set {320,479}: sweep 1      , Error= 0.17202, Performance= 38.75      {test set}
---------------------
Set {0,319}: sweep 2      , Error= 0.13862, Performance= 77.81
Set {320,479}: sweep 2      , Error= 0.15280, Performance= 67.50
---------------------
Set {0,319}: sweep 3      , Error= 0.12355, Performance= 89.38
Set {320,479}: sweep 3      , Error= 0.13975, Performance= 79.38
---------------------
Set {0,319}: sweep 4      , Error= 0.11456, Performance= 93.75
Set {320,479}: sweep 4      , Error= 0.13253, Performance= 85.63
```

Figure 5: script file in Galatea

5 Conclusion

We have presented the Galatea Library which has been developed for the ESPRIT-Pygmalion project. Galatea intends to offer a fast access to the most commonly used algorithms at the present time. In Galatea, the user can easily incorporate the library functions into his own program implementation for his specific application. In addition to the inherent ability to switch from one algorithm to another, when testing an application, through its friendly text-environment, Galatea invites us to standardize the programming style. Galatea has been widely distributed both within and out of the Pygmalion consortium. Comments about its use for developing real size applications are expected from the users.

Galatea has been developed by the IRIAC team, at LRI (Université de Paris-Sud). INESC (Portugal) contributed some of the modules. The Library is available upon request from the authors.

References

L. Bottou, F. Fogelman Soulié, P. Blanchet, J.S. Blanchard: Speaker independent isolated digit recognition: multi layer perceptrons vs dynamic time warping. Neural Networks. To appear.

K. Lang, G.E. Hinton: The development of TDNN architectures for speech recognition. Tech. Report CMU-CS-88-152, 1988.

C. Mejia, L. Bottou: C Library preliminary specifications. Document 1, ESPRIT Pygmalion project n°2059, 1989.

L. Bottou, X. Driancourt, C. Mejia, E. Viennet: Evaluation of the C-Library. R140-3 Report M12. ESPRIT Pygmalion project n°2059, 1989.

AN OVERVIEW OF THE FIRST YEAR OF ESPRIT PROJECT ANNIE

J C Collingwood

AEA Technology, Harwell Laboratory
Didcot, Oxon OX11 0RA, UK

Abstract

Project ANNIE is a all European collaboration of nine industrial and academic partners lasting three years and costing 5 million ECU. The project aims to evaluate Artificial Neural Networks and their applicability to real industrial problems. The industrial partners were selected for their expertise in solving problems in Control, Optimisation and Pattern Recognition, all of which were seen as potential application areas for Neural Networks. After one year the project has completed its evaluation phase and now moves into a demonstration phase. This paper describes the results of the theoretical study and the decisions made concerning demonstrators.

1. Introduction

This paper provides an overview of technical work carried out during the first of the three years of project ANNIE (Applications of Neural Networks for Industry in Europe). The project, number 2092 of the CEC's ESPRIT 2 programme, brings together nine industrial enterprises and research institutions in an attempt to understand where neural network approaches are most useful to industry, and to pioneer their application through software simulation. Figure 1 shows the interrelationships between the partners in the project and their principal application interests.

2. Aims and objectives of ANNIE

A central aim of the ANNIE project is to define those areas of industrial problem solving where neural networks perform well relative to conventional methods. Criteria of performance are required to make an assessment of this type. These criteria may be of a mathematical nature, such as accuracy; they may also include speed and other less quantifiable measures of user satisfaction. Over the ANNIE project as a whole, it is hoped to establish performance against criteria of all three types. Assessments of accuracy and the suitability of different algorithms have already been undertaken in the areas of pattern recognition, control and optimisation. The question of speed depends on both the hardware and the software environment that is chosen and on the suitability of the algorithm for the particular system on which it is implemented. Necessarily, the scope for using special hardware is limited, but pointers to efficient hardware environments have been made in a survey report. The final question of user satisfaction can only be answered in the context of concrete applications, for which prototype demonstrators are planned during 1990/91. The scaling of performance with problem size is an important issue which will be addressed in more detail in the context of particular applications and hardware systems.

3. Some technical achievements

3.1 Theoretical support

A continuing level of theoretical support is being maintained within the project with the aim of providing authoritative information on the mathematical capabilities of neural net algorithms and advice on which networks to use for a given problem type. A review of neural network architectures has been produced and is due to appear in the open literature. In addition, investigations of the properties of learning by error back-propagation have pointed up problems concerning convergence. Methods to deal with some of these problems are being developed.

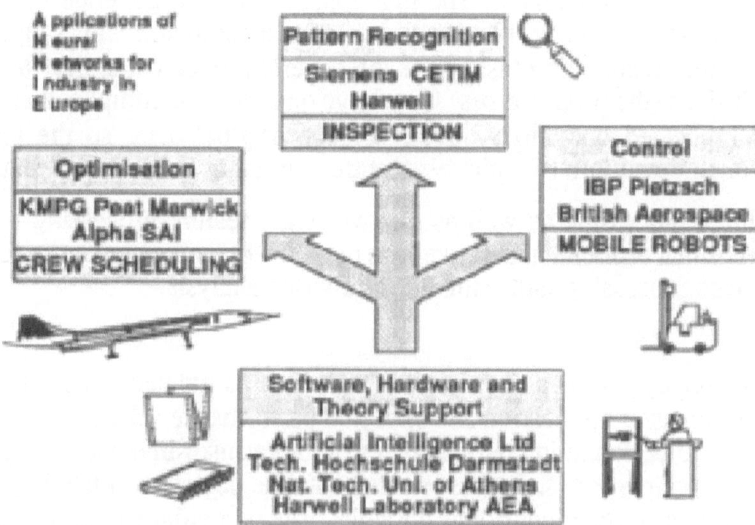

Figure 1: the interests and interactions of the ANNIE partners

3.2 Problem analysis

During 1989, the first full year of the project, work has been carried out to compare neural and conventional methods of problem analysis. Three subgroups have worked on pattern recognition, control and optimisation, with comparisons at present being restricted to matters which depend primarily on the algorithm, not on its implementation.

A further goal for the problem analysis team has been to identify applications for which prototype demonstrators will be produced during 1990. This has been done in the light of knowledge of potential applications within user companies, and the general conclusions emerging on which types of problem are best suited to a neural network approach. One year of the project represents the end of the comparative phase and provides the starting point for development of the three prototypes. Those recommended for further development within ANNIE are:

(i) **Pattern recognition.** Automated material inspection in which defects are distinguished on the basis of their shape. Both ultrasonic and optical (scanning laser) inspection are included.

(ii) **Control.** Collision avoidance and path planning for autonomous vehicles.

(iii) **Optimisation.** Crew scheduling for airlines, an example of a difficult planning problem.

Pattern recognition

Many papers in the neural network literature compare neural network approaches to problems with conventional approaches. A problem with many of these comparisons is that researchers tend to use simple conventional methods for data analysis. The applications group set out to compare neural techniques with the best available conventional techniques.

The ANNIE partners have considerable experience in solving specific pattern recognition problems for real industrial problems. Application areas of concern to the partners involved were selected and used as test problems for both conventional and neural techniques. The problems chosen were: defect detection in welds via non-destructive ultrasonic scanning; defect detection in printed circuit board solder joints using optical scanning; and composite material quality analysis using acoustic emission monitoring.

Two approaches to comparison were taken. The first approach was to generate synthetic data which approximated real data from the above problems and was equivalent in size and complexity; the second was to use real data. Pattern recognition tasks have the generic property that they are concerned with extracting simple conclusions (the pattern types) from large quantities of complex data. It is not usually feasible to perform complicated operations directly on the original data, so the process of pattern recognition is often broken down into two phases: feature extraction followed by feature processing.

The team addressed direct processing as well as the two-phase feature-processing route. One of the conclusions of the work is that artificial neural methods have an edge over conventional methods if they are applied to direct image processing rather than feature space analysis.

Control

The control subgroup started by getting an overview of the use of neural networks in the field of control, and determining possible areas of application and expected performance. Based on this knowledge and on the technical problems to be found in control, such as motor/sensor interaction, signal classification and sensor fusion, a single prototype system was selected with the objective of tackling a problem which contained as many different aspects of control as possible. The particular system chosen consisted of a simulator for automated guided vehicles moving in a fixed environment, initially using manually derived ultrasonic sensor input data. Later a full simulation of both the vehicle and the sensor system has provided a most effective tool for the assessment of neural network performance. The simulator was demonstrated successfully at the ESPRIT week meeting in Brussels, November 1989, and has since been extended to cope with multiple vehicles in the same space.

Comparison of the performance of the sensor/activator systems using neural network techniques combined with conventional AI methods has shown their superiority in the areas of response time, flexibility and reduced programming effort. The work has been summarised to provide the basis for a set of guidelines for the application of neural networks to control problems. The control team will continue to test further application simulations based on industrial applications and on several neural network paradigms with variations of parameter sets. The work will include multiple vehicle simulations and hierarchies of networks interacting to accomplish different navigation tasks. Investigations of implementations with special purpose hardware will be made using binary associative networks.

Optimisation

Optimisation problems fall into two categories. The first is characterised by the existence of very efficient conventional algorithms delivering the optimum solution, while in the second such algorithms do not exist. The second category contains problems such as the travelling salesman problem (TSP), integer programming and set covering. The main field of application for neural network in optimisation is this category of hard problems.

Historically speaking, the Travelling Salesman Problem was the first to be tackled by Hopfield. By now there is a large body of literature which reports attempts by researchers to apply neural networks to a variety of optimisation problems. It is important to note that neural networks (as they are applied in optimisation) should be regarded as inherently approximate algorithms. In which case, they should be compared with conventional heuristics. In order to deliver reasonably "good" solutions they rely heavily on the structure of the problem. The task of comparing neural networks and heuristics for a set of representative hard problems would be a formidable task on its own, and would prohibit the execution of the application task. So, instead of performing the comparison work on the generic level, it was decided to do it on the application level; this group will compare conventional and neural network methods for the airline crew scheduling problem.

3.3 Hardware for neural network simulation

Within project ANNIE, the suitability of computer systems for neural network simulation have been

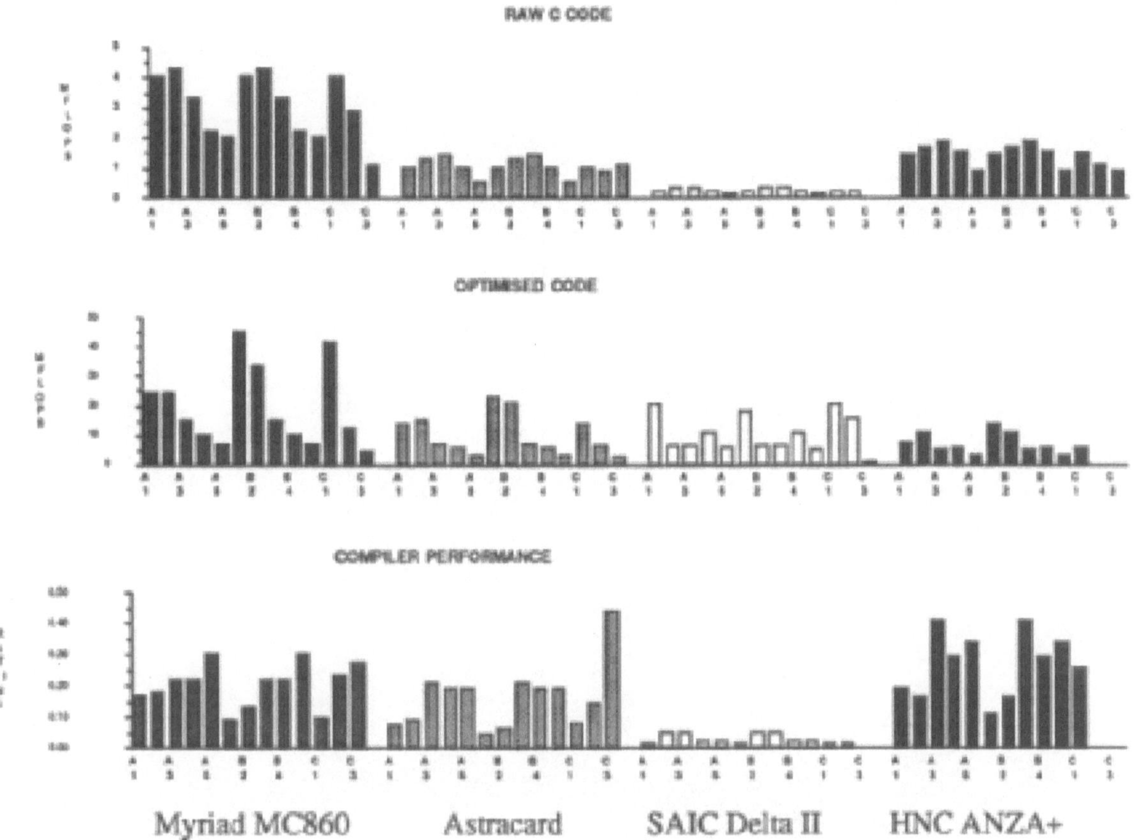

Figure 2: Performance of Accelerator Cards

evaluated with respect to a seven-point checklist: software availability; compatibility with other systems; status; support; price; architecture; and benchmark performance. Here we will describe briefly the methodology behind the benchmarking tests. The full results of this survey are to be published in a separate report.

Network simulation involves a large number of floating point calculations being performed on a large dataset. A suitable hardware system should therefore have fast floating point performance, sufficient memory capacity to store the network state, and a high rate of data flow through the floating point units. Benchmarks were therefore developed to reveal the cases where these features are lacking. Three sets of array processing kernels, which are time-critical components of network simulators, were chosen in order to give both a general impression of hardware performance, and an understanding of the effect on execution rate resulting from small changes in code and coding approach. A variety of systems were tested, from co-processors for a PC up to an i860-based accelerator card, and the benchmarks were executed at three different levels of code optimisation: unchanged source code; modified source code; and hand coded routines using assembly language. Figure 2 shows the results of comparing the performance of four commercially available accelerator cards.

4. Conclusions

Work on problem analysis has provided detailed results in the areas of pattern recognition, control and optimisation. In addition, theoretical support work and implementation studies continue to contribute towards future applications, with advances being achieved in the understanding of networks and in their effective implementation in the next phase of the project: production of application demonstrators.

COMPANY STRATEGY

Chair: Wolfram BUTTNER

Neural Networks Activities at Thomson-CSF

F. VALLET
Laboratoire Central de Recherches
Thomson-CSF 91404 ORSAY (cedex)
France

February 14, 1990

1 INTRODUCTION

Thomson-CSF is strongly involved in the domain of neural networks [1,2,3,4]. The reason of this commitment is the potential capabilities of neural networks for discrimination and classification tasks, numeric-symbolic interfaces, signal and image processing, optimisation and data fusion. These capabilities are important for several equipments developed in the company (radar, sonar, telecommunication, IR/visible/radar image processors, simulators, video equipments), as well as for systems (air traffic control, weapon systems, telecommunication networks, battlefield management).

In these domains of application, the main features of interest are the following:

- Learning from examples. This is important when no clear model of the problem is available.
- Neural architectures are intrinsically massively parallel.
- Easyness of use. Experience shows that from a software point of view, neural nets are quick and easy to implement and give generally good results.
- Failure robustness. In large networks, tolerance to failures seems to be a common feature of neural nets. This can be particularly important in military applications.

2 APPLICATIONS

2.1 Discrimination/classification

Discrimination and classification represent the domain for which neural nets are the most used in the company. They are usually compared with standard numerical methods. Neural networks are fed in with data coming from a signal processing stage, and are therefore considered as data analyser. The addressed applications are the following:

- passive [5] and active [6] sonar recognition,
- passive sonar for transient noise analysis (biological or mechanical) [7,8],

- aerial acoustic identification,
- identification and emission-mode recognition of transmitters,
- target and sources recognition, echos filtering for RADAR,
- identification in IR images,
- identification of flight phases (helicopters and aircrafts),
- identification of industrial pieces.

2.2 Numeric-symbolic interfaces

It can happen that between a low-level signal processing stage (numeric data) and a high level decision making stage (symbolic data) some parts are lacking. In such cases neural method are used:

- symbolic extraction from time-frequency sonar data,
- shape description for sea mines,
- interpretation of aerial scenes.

2.3 Signal and image processing

Neural nets are used as low level processors, dealing directly with row signal:

- image processing: compression (video equipments), contour extraction and noise reduction in satellite images [9],
- temporal regression analysis in financial market,
- demultiplexing in reconstruction of radar images.

2.4 Other applications

- Optimisation: processor allocation in sonar systems [10],
- Control: last landing phase of aircrafts,
- Data fusion: multi-localisation for radar sources,
- Associative memories: under-sea sources localisation,
- Pattern matching: cartographic correlation.

3 FUNDAMENTAL STUDIES

To support all the addressed applications, as well as to evaluate soundly the neural technology, fundamental studies are done (physicists, mathematicians, computer scientists):

- theoretical investigation on mono- and multi-layer perceptrons, especially on learning and generalisation performances, overfitting, architecture and algorithms [11,12,13,14,15,16,17],
- statistical and probabilistic aspects of neural nets for discrimination tasks,
- combinatorial optimisation methods [18,19],
- control of dynamical systems,
- signal processing [20,21] and image processing (compression [22,23,24], 2-D wavelets).

Preliminary results show that neural nets are a new way to tackle old problems, but do not contain intrinsically new fundamental mathematical aspects. They integrate themselves quite well in standard mathematical models. Their main interest is mainly to enlighten old problems from a new point of view, and to reactive research for standard approaches.

4 SOFTWARE

Language A neural network user programming object-oriented language for defining neural network algorithms, architecture and dynamics is developed [25,26]. A target hardware description sub-language will then be used to generate the appropriate code for each specific hardware base, after compilation (see PYGMALION project).

NeuroClass ^{TM}. The *Laboratoire Central de Recherches* is developing a interactive toolkit devoted to discrimination tasks: *NeuroClass*^{TM}. It contains a library of standard and neural discrimination algorithms, pre- and post-processing routines and a supervisor for the experiments.

5 HARDWARE IMPLEMENTATION

Transputers. The back-propagation and Kohonen self-organising nets haves been successfully implemented on transputers [27].

Optical processing. An optical architecture for the first stage of a pattern recogniser on video images is studied on the basis of static and dynamic holographic technology. Thomson is involved in the related ESPRIT P2288 NAOPIA project. At long term this could provide high level performances ($\sim 10^{12}$ Ops) [28,29,30] .

6 PYGMALION: Neurocomputing ESPRIT II Project 2059

Thomson-CSF is prime for the european PYGMALION project, which aims to promote the application of neural networks by European industry, and to develop European "standard" computational tools for programming and simulation of neural networks.

The major objective is produce of a complete environment for developing neural network algorithms and applications. The software aspects are largely developed in this project. Although it is also planed to produce neural-oriented hardware, only a small preliminary study of silicon integration is undertaken.

The other part of PYGMALION is the development of applications in image and speech processing.

PYGMALION started in January 1989, for a duration of two years.

The consortium is the following: Thomson-CSF prime (France), CSELT (Italy), CTI (Greece), IRIAC (France), ENS (France), INESC (Portugal), Philips (France), SEL (Germany), UCL (UK) and UPM (Spain).

The current status is that the Back propagation algorithm implemented after adequate preprocessing, appears to be a good classifier in most of the cases. Further work will concern implementation of unsupervised classification algorithms and more precise performance evaluation tests. The whole software environment is fully specified and will be finished by the end of 1990.

7 Conclusions

A significant effort is being made in Thomson to develop an expertise in the field of neural nets and to assess their performances in various application areas. This is important because of the potentially very wide class of domains this new technology can influence. The teams involved in this work do master quite well the fundamental aspects as well as the practical use of neural nets for applications. Neural nets show at the present state of progress promising capabilities which have still to be deeply studied and validated on real data.

References

[1] B. Angéniol. Les réseaux neuro-mimétiques: de la recherche à l'industrie . *Proceedings of the international workshop Neuro-NIMES 88 (Nimes, France, nov.88)*, p. 17, 1988.

[2] B. Angéniol. Thomson-CSF strategy in Neural Networks. *Conférence: Neural Computing, Commercial Prospects (Londre 16-17/2/89).*

[3] D. Potier and B. Angéniol. Activités de Thomson-CSF en Réseaux de Neurones. *Bulletin de Liaison de la Recherche en Automatique et Informatique, Inria*, 1989.

[4] B. Angeniol F.vallet. Thomson-CSF: activities in Neural Networks. *Conférence: Neural Computing, Commercial Prospects (Londre 15-16/2/90).*

[5] M. De Bollivier, A. Lemer, and J. Tanguy. Reconnaissance de bruits sous-marins par réseaux multicouches. *Proceedings of the international workshop Neuro-NIMES 88 (Nimes, France, nov.88)*, pp. 423–430, 1988.

[6] Y. Ammirati and D. Neveu. Classification de signaux sonar en mode actif. *Proceeding of GRETSI 89 (12ème colloque, Juan-les-Pins, 12-16/6/89)*, pp. 391–394.

[7] A. Lemer, JM. Nicolas, and P. Giancone. Identification automatique de bruits impulsifs sous-marins. *Proceedings of GRETSI 1989*, pp. 403–406, 1989.

[8] JM. Nicolas, A. Lemer, and D. Legitimus. identification automatique de bruits impulsifs en acoustique sous-marine par réseaux multicouche. *Proceedings of the international workshop Neuro-NIMES 89 (Nimes, France, 13-16/11/89)*, pp. 269–278, 1989.

[9] J. Loncelle. Détection de contours par rétro-propagation de gradient. *Proceedings of the international workshop Neuro-NIMES 89 (Nimes, France, nov.89)*, pp. 373–393, 1989.

[10] F. Grizard and M. Revol. Application d'une méthode de recuit simulé à l'implantation de traitements sonar. *To appear in proceedings of GRETSI 89.*

[11] J.-G. Cailton, B. Angéniol, and E. Marcadé. Constrained Back-propagation. *Abstracts of the First Annual INNS Meeting*, p. 539, 1988.

[12] Ph. Réfrégier and J-M. Vignolle. An improved version of the pseudo-inverse solution for classification and neurl networks. *Europhysics Letters*, 10 (4):pp. 387–392, 1989.

[13] F Vallet. The Hebb rule for learning linearly separable functions: learning and generalisation. *Europhysics Letters*, 8 (8):pp. 747–751, 1989.

[14] F. Vallet, J-G. Cailton, and Ph. Réfrégier. Linear and nonlinear extension of the pseudo-inverse solution for learning boolean functions. *Europhysics Letters*, 9 (4):pp. 315–320, 1989.

[15] F. Vallet and J-G. Cailton. Recognition rates of the Hebb rule for learning boolean functions. 1990. To appear in *Physical Review A*.

[16] F. Vallet, J-G. Cailton, and Ph. Réfrégier. Solving the problem of overfitting of the pseudo-inverse solution for classification learning. 1989. International joint conference on neural networks, Washington D.C., June 18-22 (1989), vol.II, p. 443-450.

[17] F. Vallet. Optimization of the number of hidden cells in a mutilayer perceptron. Validation in the linear case . *Proceedings of the workshop NATO ARW (february 89, Les Arcs)*, 1989.

[18] B. Angéniol, G. de La Croix Vaubois, and J.Y. Le Texier. Self-Organizing feature maps and the travelling salesman problem. *Neural networks*, vol 1:pp. 289–293, 1988.

[19] P. Goyoneix. Self-organization and an optimization problem . *Proceedings of the international workshop Neuro-NIMES 88 (Nimes, France, nov.88)*, pp. 149.

[20] P. Comon. Séparation de mélanges de signaux. *To appear in proceedings of GRETSI 89 (12ème colloque, Juan-les-Pins, 12-16/6/89)*, 1989.

[21] P. Comon. Séparation of stochastic processes whose a linear mixture is observed. *ONRNSF-IEEE, Workshop on higher order spectral analysis, Vail (28-30/6/89).*

[22] B. Angéniol. Extraction of prototypes. *Proceedings of the workshop NATO ARW (février 89, Les Arcs).*

[23] M. Mougeot, R. Azencott, and B. Angéniol. Compression d'images, réduction de données et rétropropagation. *Communication in euro-image 88 (octobre 1988).*

[24] M. Mougeot, R. Azencott, and B. Angéniol. A study of image compression with backpropagation. *Proceedings of the workshop NATO ARW (february 89, Les Arcs).*

[25] B. Derot, Ph. Escande, and C. Moulinoux. NACRE: A neuron-oriented programming environment. *Proceedings of the international workshop Neuro-NIMES 89 (Nimes, France, 13-16/11/89)*, pp. 183–200, 1989.

[26] C. Moulinoux, B. Derot, and G. de La Croix Vaubois. A high level language for neural networks specification. *Proceedings of the workshop NATO ARW (february 89, Les Arcs).*

[27] C. Ernoult. Performance of backpropagation on a parallel transputer-based machine . *Proceedings of the international workshop Neuro-NIMES 88 (Nimes, France, nov.88)*, p. 311.

[28] Ph. Réfrégier. Phase optimisation of sdf filters. *International symposium: Optics in computing*, 1:p. 15, 1989.

[29] Ph. Réfrégier and JP. Huignard. Multi criteria optimization approach for sdf filters. *To appear in proceedings of the International topical meeting on Optical computing (Japan 1990).*

[30] B. Loiseaux, G. Illiaquer, and JP. Huignard. Dynamic optical cross-correlator using a liquid crystal light valve and a bso crystal in the fourier plane. *Optical engineering*, 24(1), 1985.

THE BRITISH TELECOM CONNECTIONISM PROJECT

David J Myers and Charley Nightingale
British Telecom Research Laboratories,
Martlesham Heath,Ipswich IP5 7RE U.K.

ABSTRACT

The British Telecom Connectionist Project is a multi-disciplinary project that has been running since 1988. This paper covers the work being undertaken in the areas of speech, natural language, and image processing, describing some current results and indicating the direction of future work. Implementation issues are briefly mentioned.

1. INTRODUCTION

The British Telecom Connectionism Project was set up to investigate and exploit the potential of neural net techniques in areas such as Speech, Vision and Natural Language. At the start of the project a few specific applications areas which are of particular interest were identified, and used to form the basis of test problems. Test data for these problems were collected, enabling comparisons to be made between different neural techniques, and between neural and 'conventional' techniques on the same data. The origins and nature of the test problems and data have been described elsewhere[1]. This paper reviews some of the successes that have been obtained in the past year in applying neural networks to these test problems, and briefly describes other work taking place within the project on the implementation of neural net systems.

2 SPEECH TELECOMMUNICATIONS

The main focus of work in the speech area has been on applying neural nets to speech recognition. The customer base of the Telecom network represents a large, diverse range of speakers. The design of speech recognition systems which will cover such a large user base is challenging, and will allow the provision of sophisticated voice activated services. Neural network techniques offer the prospect of recognition systems with better performance than those currently available.

Two of the test problems that have been defined in this area are speaker independent recognition of i) letters of the alphabet, and ii) "Yes/No" utterances over the public telephone network. Work on these test problems has yielded promising results.

Using the Multi-layer Perceptron (MLP), trained using the back propagation (BP) algorithm, a recognition score of over 89.5% on the test data has been achieved for the alphabet recognition problem. This is a marginally better performance than the conventional Hidden Markov Model (HMM) and Dynamic Time Warping (DTW) techniques that the MLP based solution has been benchmarked against. However, the MLP has considerably lower computational requirements. The MLP based alphabet recogniser has been incorporated into a proof-of-concept voice-interrogation Directory Enquiry system demonstrator.

For the Yes/No recognition problem, a number of neural net techniques have been tried. The use of Grossberg's biologically inspired Adaptive Resonance Theory (ART) nets has given disappointing results so far, with a recognition score of around 80%. Kohonen's Self Organising Network gave a result of 92% on the test data, which was comparable to the performance of the conventional Nearest Neighbour classifier[2], whilst the MLP/BP yielded a best result of over 95% [2,3].

The success achieved in isolated word recognition have led to investigations of hybrid neural/conventional systems, and to the extension of neural techniques to cope with varying dimensional patterns or sequences of patterns. This will allow the application of neural networks to the problem of connected speech recognition.

3 NATURAL LANGUAGE

The recognition of natural language will allow more natural communication with machines, and the development of facilities such as automatic translation.

Two test problems that have been defined in this area are (i) the recognition of grammatically valid sentences, and (ii) the production of a phrase to phrase English to French translator in a limited domain (hotel complaints). Neural network techniques offer the prospect of more robust systems than are obtained using conventional rule-based approaches.

For tackling the natural language problems, a unique neural net has been developed. Known initially as DTN (Dynamic Topology Net)[4], it has been developed to produce HODYNE (Higher Order Dynamic Topology Net)[5]. In the DTN, an input is associated with each item in the input string (eg a word). In HODYNE this is extended so that a sub-string or 'tuple' is associated with each input to provide context information. This allows the net to discriminate between eg "The chicken ate the man" and "The man ate the chicken". Both nets are dynamic - they can add input nodes to accommodate inputs that have not been encountered previously. In the case of HODYNE this dynamic growth takes place probabilistically, with probabilities falling as the net "ages" in order to constrain its growth.

HODYNE has been remarkably successful on the grammaticality problem, scoring over 99% on the test set. On the limited domain translation problem, quantitative assessments of performance are difficult, but HODYNE has shown good subjective results when subject to fairly free input.

4 VISUAL TELECOMMUNICATIONS

In visual communication, image recognition can be used to produce low bit-rate video codecs for the transmission of images over low bandwidth channels. For restricted images such as head and shoulders (eg for videophone applications) one possible approach is to produce an animated model of the speaker at the far end, and then to transmit only the data required to animate it [6]. The problem here is to map the facial features of the subject on to the animated model initially, and to track those features subsequently to provide animation data. Neural networks may be used to perform facial feature location and tracking, and this is the basis of the test problem defined in this area.

Initial work has concentrated on the location of eyes in human portrait images. Initial experiments centred around the use of an MLP to detect eye locations in a very low resolution image (16 x 16 pixels)[7]. This gave encouraging results, although it did tend to locate spurious 'eyes' in the chin and elsewhere in the image. Further work on much higher resolution images[8] has compared the use of Kohonen Self Organising Nets, MLPs, and conventional template matching methods. In the case of Kohonen nets, the output was labelled using an MLP. For unpreprocessed facial images of 128x128 pixels both the Kohonen net and the MLP achieved a success rate of 88.6% on the test data, better than the best of the template matching methods which achieved a much lower score. The MLP score was later improved to 93%, and averaging the outputs of nets operating on differently preprocessed images has resulted in overall scores of 98% [9].

For the higher resolution work, the 16x16 or 32x16 inputs of the MLP were windowed or scanned across the image. It is therefore very computationally expensive to scan the whole image at this resolution, so the search area was artificially restricted to the vicinity of the eye for this work. Future work will be based on using low resolution eye detection to constrain the search area for the more accurate high resolution detectors. It is also planned to improve performance on moving image sequences by exploiting inter-frame correlation to improve location scores.

5 HARDWARE IMPLEMENTATIONS

In parallel with the work going on in the application of neural network techniques, the problem of how ultimately to implement Neural Net systems is being addressed. Activity in this area divides into two approaches: the use of available technologies, and the exploitation of new technologies.

In the first of these approaches, the design of a digital systolic array based neurocomputer is being undertaken [10,11]. This is intended to be optimised for the efficient implementation of MLPs, with on-chip learning using the backpropagation algorithm.

In the area of new technologies, investigations centre on optical and electro-optical devices, and the use of amorphous silicon. An optical neurochip demonstrator, which implements the MLP preprogrammed to perform the EXOR function is due to be produced in the second quarter of 1990.

REFERENCES

1. NIGHTINGALE, C.:"The British Telecom Connectionist Project - Applications" in The Second European Seminar on Neural Computing, London, 16-17 February 1989.
2. WOODLAND, P.C. and SMYTH, S.G.:"An Experimental Comparison of Connectionist and Conventional Classification Systems on Natural Data" to be published in the Neurospeech issue of Speech Communication 1990.
3. WOODLAND, P.C.:"Weight Limiting, Weight Quantisation and Generalisation in Multi-layer Perceptrons" Proc 1st IEE Int. Conf. on Artificial Neural Networks, London 16-18 October 1989.
4. NIGHTINGALE, C.:"A Neural Net Whose Topology Varies Dynamically During Training" submitted to Connection Science (Ed N Sharkey) Carfax.
5. WYARD, P., NIGHTINGALE, C. and MARSH, R.:"A Higher Order Dynamic Topology Neural Net and its Application to Natural Language Problems" submitted to Connection Science (Ed N Sharkey) Carfax.
6. WELSH, W. J.:"Model Based Image Coding" in Image Processing in Telecommunications and Information Systems, D Pearson (Ed) McGraw-Hill 1989.
7. HINES, E. L. and HUTCHINSON, R. A.: "Application of Multi-layer Perceptrons to Facial Feature Location" Proc 3rd IEE Int. Conf. on Image Processing and its Applications, Warwick 18-20 July 1989.
8. HUTCHINSON, R. A. and WELSH, W. J.: "Comparison of Neural Networks and Conventional Techniques for Feature Location in Facial Images" Proc 1st IEE Int. Conf. on Artificial Neural Networks, London 16-18 October 1989.
9. HUTCHINSON, R. A.:"Development of an MLP Feature Location Technique using Preprocessed Images" submitted to INNC Paris July 9-13 1990.
10. MYERS, D. J. and BREBNER, G.:"The Implementation of Hardware Neural Net Systems" Proc 1st IEE Int. Conf. on Artificial Neural Networks, London 16-18 October 1989.
11. MYERS, D. J. and HUTCHINSON, R. A.: "Efficient Implementation of Piecewise Linear Activation Function for Digital VLSI Neural Networks" Electronics Letters Vol 24 No 24 November 1989.

THE IMPACT OF NEURAL NETWORKS ON THE AMERICAN ECONOMY: 1990 - 2010

Fletcher Crowe, Ph.D
Equifax, Inc.
5775 Peachtree-Dunwoody Rd.
Building G, Suite 300
Atlanta, GA 30394 USA

ABSTRACT: In this paper, projections are made for the impact of neural networks on American economy, 1990 - 2010. The study forecasts the sales and penetration of neural network technology resulting from anticipated gains in the power and capability of connectionist technology during this time period. Standard projections of employment and productivity in the American economy are examined, and then are contrasted with projections made by the author based on the forecasted penetration of neural network technology.

The study concludes that the power of neural networks will increase exponentially during the next twenty years, and predicts that by 2010 connectionist technology will have become the predominant computer architecture. Given this projected enormous increase in power and capability, and the anticipated high degree of market penetration, the study forecasts profound impacts of neural network technology on employment, labor, training, and GNP.

Because neural network technology will enable machines to perform a wide range of tasks heretofore reserved for humans, the study predicts that the widespread adoption of neural network technology will produce a far greater degree of economic dislocation than resulted from the introduction of conventional computers in previous decades. Widespread penetration of advanced neural network technology will have pronounced impact on key economic factors such as: balance of payments, length of the work week, age of retirement, immigration policy and age at which formal education is completed.

This study forecasts the economic impact of neural network technology on the American economy, 1990 -- 2010. The paper consists of four major sections:

1. Projections of the growth in sales and power of neural network technology, 1990 -- 2010, including forecasts of predicted market penetration and commercial applications.

2. Baseline forecasts of the GNP, employment levels unemployment, productivity, and levels of training in American economy, 1990 -- 2010, using standard baseline forecasts and assuming conventional rates of productivity.

3. Projections of the productivity gains anticipated from commercial applications of neural network technology, such as voice recognition systems, industrial control systems, image understanding, medical analysis applications, and character recognition systems.

4. Projections of employment, unemployment, GNP, and training for the American economy, assuming the projected penetration rate of neural network technology; these projections will be contrasted with standard, conventional baseline forecasts described in Section 2 which do not assume powerful impacts of neural network technology.

1. Market Penetration of Neural Network Technology

Using standard forecasts of the future growth of computing power in areas other than neural networks -- including forecasts prepared by DARPA, the Office of Technology Assessment (OTA), Intel, the National Research Council (NRC), the National Academy of Engineering (NAE), the American Electronics Association (AEA), James Martin, Tom Schwartz, and Abraham Peled -- the study predicts that by the year 2000, commercial neural network designs will routinely offer speeds of 10^{12} interconnections per second (IPS). Coupled with dramatic advances in massively parallel processing, optical computing, and applications of superconductivity, by 2010 commercial products based on neural network speeds in excess of 10^{13} IPS should be widely available.

By the turn of the century, speeds this great, linked with parallel advances in software engineering, are projected to make possible commercial speech comprehension, speech generation, advanced robotics, and limited image understanding. Later, by 2010, with increasing power and sophistication of neural network technology, widespread commercial penetration of these technologies is forecast in the study.

Applying the Fisher-Pry model of technological market penetration, as well as historical technology penetration rates found in studies conducted by the Stanford Research Institute (SRI) and the Rand Corp., the study predicts that neural network technology will become the "Sixth Generation" of computing. By the year 2010, the model forecasts that neural network technology, linked closely with concurrent developments in parallel processing, will have substantially succeeded in displacing Fourth Generation technology (technology of relational databases and conventional office automation software); and will have begun to supplant Fifth Generation technology (based on symbolic logic and symbolic processing). Nevertheless, these earlier technologies will live on in hybrid form, integrated into overall Sixth Generation architectures.

Using software and hardware sales projections developed by OTA and the NRC, the model predicts the dollar value of neural network sales as Sixth Generation technology inexorably penetrates the market for computer-based technology during the twenty-year period examined by the study.

2. Baseline Economic Predictions

To be able to predict the economic impacts of neural network technology, the study reviews standard baseline economic and demographic projections of employment, unemployment, productivity, population, and training. A baseline model of the U.S. economy, 1990 -- 2010 is developed using *Workforce 2000*, prepared by the Hudson Institute for the U.S. Department of Commerce, as well as studies of the Bureau of Labor Statistics, Paul Osterman, Vasily Leontief and Faye Duchin, the Census Bureau, and the Rand Corp.

Among other studies, the paper establishes baseline economic projections using reports such as: *Labor Force Projections: 1986 to 2000* by BLS (1987); *Economic Projections to the Year 2000* by BLS; *The U.S. Workforce, 1988 -- 2015* by Jeffery Hallett (1988); and *Projections of the Population of the United States, 1988 -- 2080* by the Bureau of the Census (1989). These studies reflect, in part, both the DRI model of the U. S. economy and the Wharton model. It is against these projected baseline reference statistics that the effects of neural network technology are gauged in the paper.

3. Productivity Gains from Neural Network Technology

The study examines historical gains in productivity in the U. S. economy resulting from computerization and other factors. Productivity studies published by Martin Neil Baily of the Brookings Institution, Alan Fechter of the National Research Council, and by the National Academy of Sciences Panel of Technology and Employment edited by Richard Cyert and David Mowery are reviewed.

Using these studies, base-line projections of employment and productivity in ten sectors are developed: durables, non-durables, services, retail trade, wholesale trade, FIRE (finance/ insurance, real estate), construction, transportation, govern-ment, communications, mining, and utilities. Education and training levels for each of these sectors developed by the Hudson Institute for the Dept. of Commerce are also examined. The study *Trends in Manufacturing Productivity and Labor Costs in the U.S. and Abroad* by BLS (1987) is used to project baseline productivity levels.

For each of these sectors, estimates are made for the penetration of neural network technology. Additionally, estimates of productivity gains that should be achieved with neural network technology in each of these areas are developed.

4. Econometric Impacts of Neural Networks

Finally, given the projected penetration rate of Sixth Generation technology in the American economy and the productivity gains anticipated, overall macroeconomic impacts are forecasted.

Although this is a preliminary study, not meant to provide a definitive answer to the question of economic impacts of technology, the study suggests that profound economic impacts from the penetration of neural network technology can be anticipated. A number of studies -- such as those of the National Research Council and the National Academy of Sciences -- have found that, to date, the macroeconomic impact of conventional computer technology has had a decidedly positive impact on the American economy. By contrast, this study concludes that because neural network technology allows the computer to learn on its own, its impact will be fundamentally different than that of conventional computer technology. Conventional serial computers increased the efficiency by which tasks were performed but did not generate a measurable net displacement of workers. Neural network technology will enable machines to *learn*, and thereby will be able to perform far more economically many tasks now reserved for humans.

Once neural network technology becomes exceedingly powerful and has attained a considerable degree of commercial penetration, substantial and widespread economic consequences are predicted. Sharply increased levels of unemployment for lesser-trained workers, significant economic dislocation, and reduced costs of production are predicted. The advent of widespread penetration of neural network technology will mean earlier retirement ages, reduced immigration, and later ages for entering the workforce. The technology will increase the trend to a shorter work week, and will significantly increase labor productivity.

By allowing manufacturers to use computer-based processes where formerly relatively highly-paid laborers were required, neural network technology will bolster domestic manufacturing production and reduce balance of payments deficits. The paper concludes that the widespread penetration of sophisticated Sixth Generation technology is expected to result in the need for increased levels of formal training for workers, earlier retirement ages, increased pressure to restrict immigration, and shorter work weeks.

TEXT-DEPENDENT SPEAKER IDENTIFICATION USING LEARNING VECTOR QUANTIZATION

Younès BENNANI [*], *Françoise FOGELMAN SOULIE* [*] *and Patrick GALLINARI* [+,*]

[*] Université de Paris Sud, Centre d'Orsay
Laboratoire de Recherche en Informatique
CNRS UA 410, Bâtiment 490
91405 Orsay FRANCE

[+] Ecole des Hautes Etudes en Informatique
Laboratoire d'Intelligence Artificielle
45 rue des Saints-Pères
75006 Paris FRANCE

Abstract

This paper presents a connectionist approach to automatic speaker identification, based for the first time on the LVQ (Learning Vector Quantization) algorithm. For each "subscriber" to the identification system, a number of references is fixed. The algorithm is based on a nearest neighbor principle, with adaptation through learning. The identification is realized by comparing to a given threshold the distance of the unknown utterance to the nearest reference. Preliminary tests run on a 10 speakers set show an identication rate of 97% for MFC coefficients. We present the identification system and data base used, and indicate the results obtained for different combinations of parameters. We further evaluate our system, by comparing its performances with a Bayesian system.

1. INTRODUCTION

We introduce a complete system for Automatic Speaker Identification (ASI) and Automatic Speaker Verification (ASV). Connectionist models have recently shown good performances for classification tasks [11], [14], [15], [16], [18]. We are thus investigating the possibility to best integrate connectionist methods into such a system so as to get enhanced performances.The main problem in integrating connectionist techniques is to determine, at each step, the best combination of classical and connectionist processings. We have started working in this area for the ASI problem.

We present in this paper preliminary results which aim at demonstrating the viability of such an approach in ASI: we have built a simple system (shown in fig 1) investigating the optimal use of a few techniques (different coefficients: LPC and MFCC, different sentences models, and different classifiers: Bayesian and LVQ). LVQ is a nearest neighbor classifier recently proposed in [14] which has produced good performances in classification tasks [7].

We are currently working at building a larger and more complex system which will combine more processings and seek their optimal combinations. In particular, other connectionist techniques are being investigated: multi-layer networks and TDNN, topological maps...

The paper is organized as follows. In section 2, we present the architecture of our system, the speech data base, and its pre-processing, the sentences modelization and the LVQ2 algorithm ; in section 3, our experiments and results. Conclusions are drawn in section 4.

2. THE AUTOMATIC SPEAKER IDENTIFICATION SYSTEM

2.1 System architecture

A speaker identification system (fig.1) includes various successive steps: a parameterization step (to produce, from the microphone signal, a population of vectors in RP), a modelization step (to build a model of the speaker's voice). The next step depends on the use of the system: in learning mode, for each speaker, models for training and for test are chosen to serve as references and -in the case of LVQ- further adapted during training. In recognition mode, a classifier is implemented.

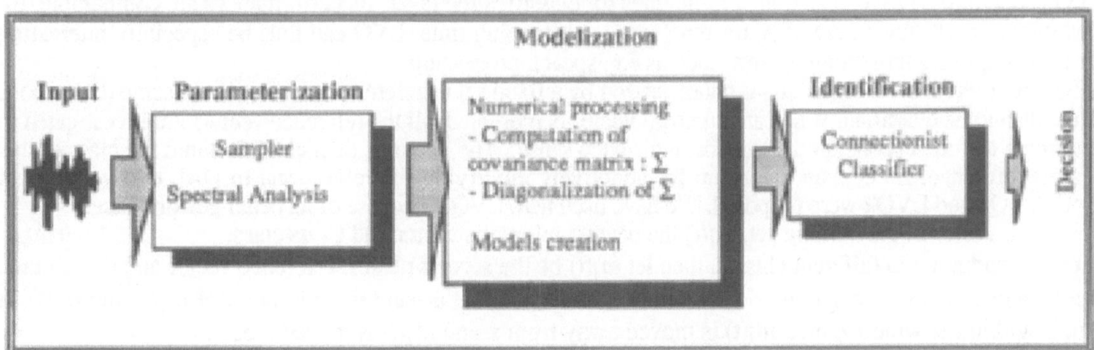

Figure 1 : System architecture

2.2 The Data base

We have tested our ASI system on a population of 10 speakers, half male and female. The data base contains ten sentences, in french, phonetically balanced [5]. Each sentence is very short, lasting from 1,5 to 3 s. Each of the 10 speakers has pronounced each sentence 10 times, in a unique session. The total number of sentences is thus 10x10x10= 1000. In the preliminary results presented here, we have used the first sentence only.

We have already tested an ASI system on isolated words [1]. We now use sentences instead of words, because they allow to make better use of a speaker's identity. Balanced sentences provide short signals with optimal variability distribution .

The recordings have been realized on a Memorex equipment, in our office, where the background noise is relatively high. Noise sources originate from: conversations, steps, doors opening and closing, exterior noises (telephone...). The energy of these different sources is concentrated in certain frequency bands and does not correspond to a white noise.

2.3 Preprocessing of the analogic signal

The analogic recordings have been first digitized at 10 KHz on 16 bits with an OROS card, after low-pass filtering (0-4000 Hz). The samples are then pre-emphasized by a first order digital filter with transfer function $1 - 0.95.z^{-1}$.

2.4 Digital signal analysis

We have tested two different parameterization methods:

LPC : a 12th order autocorrelation analysis is carried out, every 10 ms, using 25,6 ms overlapping Hamming windows. Each frame is then converted into a 12th order LPC vector. Each of the 1000 sentences is thus converted into a 12xN array, where N is the number of frames in the sentence. N is thus different for the various sentences.

This parameterization technique has been widely used for automatic speaker recognition [2], [12] ...

MFCC : the MFCC (Mel Frequency Cepstral Coefficient) parameterization comes from a Fourier analysis of the signal. The Fourier spectrum is computed from a 25,6 ms frame (256 points at 10 KHz) obtained through Hamming windowing. 24 triangular filters are then passed to obtain the Mel scale. For each window, we obtain an 8-dimension MFCC vector.

2.5 Sentences modelization

Since the signal has been parameterized, a sentence is now a pxN array or a set of N points in R^p, where p = 12 (for LPC) or 8 (for MFCC) and N is the number of frames. For each speaker, and each utterance of a sentence, we will model this cloud of points through a combination of its mean and the two first eigen vectors of the covariance matrix, in Principal Component Analysis (PCA). The mean gives the position of the cloud and the two eigen vectors its "shape".

A voice model may thus be characterized by 3 vectors in R^p: one for the mean and two eigen vectors. In [12], it has been shown that such a model sufficiently captures the speaker characteristics so as to allow for good identification. In this paper we have chosen to use the mean and first eigen vector only.

Models can then be compared by using the euclidean distance.

2.6 The learning algorithm: LVQ2

LVQ is one of the best connectionist techniques for classification tasks. Its performances are comparable, for example, to multi layer networks, for much reduced learning time. LVQ can thus be especially interesting for tasks requiring large training sets, such as e.g. speech processing.

LVQ works as follows: each class is characterized by a fixed set of reference vectors, of the same dimensions as the data to be classified. When an unknown vector is presented, all the reference vectors are investigated to determine the nearest one, in the Euclidean distance sense. The vector is then classified into the class of this nearest reference. LVQ is an algorithm for adaptively modifying the references. In [14], two versions of LVQ, LVQ1 and LVQ2 were proposed. We have used here LVQ2, because of its better performances.

Let $\mathbf{x}$ be a vector in the training set, $\mathbf{m}_i(t)$ the nearest reference vector and C_i its class.

- If $\mathbf{m}_i(t)$ and $\mathbf{x}$ are in different classes, then let $\mathbf{m}_j(t)$ be the second nearest reference vector and C_j its class. If $\mathbf{x}$ is in class C_j, then a symmetrical window, of size w, is set around the mid-point of $\mathbf{m}_i(t)$ and $\mathbf{m}_j(t)$. If $\mathbf{x}$ falls within the window, then $\mathbf{m}_i(t)$ is moved away from $\mathbf{x}$ and $\mathbf{m}_j(t)$ is moved closer.
- In all other cases, nothing changes.

More precisely, the adaptation rule writes:

$m_i(t+1) = m_i(t) - \alpha(t)\,(x - m_i(t))$

$m_j(t+1) = m_j(t) + \alpha(t)\,(x - m_j(t))$

We have taken here :

- $\alpha(t) = 0.1*(1 - t / nmax)$ where nmax is the maximum number of iterations allowed.
- $w = | m_i(t) - m_j(t) |$

The computations required for LVQ2 are thus very simple. On top of that, LVQ can be made to converge very fast, by a careful initialization: for example, reference vectors for each class can be initialized by a K-means technique on the examples. This initial choice is thus already approximately correct, LVQ just has to refine it by making use of the class identification information.

3. EXPERIMENTS AND RESULTS

3.1 Experiments

The speaker identification system has been first simulated on a population of 10 speakers. The system must work independently of the sentence ; as a first step, we have started here working on the first sentence only: *"Il se garantira du froid avec ce bon capuchon"*. This sentence is very short, lasting between 2,4 s and 3 s. We thus have 100 sentences (10 pronunciations x 10 speakers) and their 100 models. We have run different experiments, for both the LPC and MFC coefficients, by using vectors with the mean m and the first eigen vector V_1. m behaves like a long term spectrum. Other possibilities have also been tested, but led to poor results.

For each pair (vector, coefficients), the system has been tested, on the same data base, with the cross validation technique "leave-one-out" described in [13] and [6]. The LVQ algorithm always converged in less than 50 sweeps through the data base, for each combination 9 utterances * 10 speakers. The result is thus, in our case, an average on the 10 possible utterances "left out" (for each speaker). LVQ was initialized with a K-means with K=2 or 3, for each class.

Moreover, the results of the LVQ technique have been compared to a Bayesian system [1],[3], [4].

3.2 Results

The results are given in table 1.

	LVQ2	LVQ3	BS
LPC	76	78	74
MFC	95	97	89

Table 1: Identification rate (in%) for the LVQ-based technique (with K=2 and K=3 references: LVQ-2 and LVQ-3) and the Bayesian system (BS)

Those results show that, for both the Bayesian (BS) and the LVQ systems, the MFC coefficients do significantly better than the LPCs; but their computation time is twice larger than for LPCs. However, the huge difference obtained for the two parameter sets might also be due to some numerical instability in our computation of the LPCs (through the auto-correlation method).

These results indeed show that the information contained in the mean m and first eigen vector V_1 by itself is sufficient to identify the different speakers.The classification errors can be better understood by looking at the confusion matrix, which is here an average on the different passes.

The confusion matrices are given in Table 2.

	a	b	c	d	e	f	g	h	i	j	a	b	c	d	e	f	g	h	i	j
a	10	0	0	0	0	0	0	0	0	0	9	0	0	0	1	0	0	0	0	0
b	0	10	0	0	0	0	0	0	0	0	0	8	0	2	0	0	0	0	0	0
c	0	0	10	0	0	0	0	0	0	0	0	0	9	1	0	0	0	0	0	0
d	0	0	0	10	0	0	0	0	0	0	0	1	0	9	0	0	0	0	0	0
e	0	0	0	0	10	0	0	0	0	0	2	0	0	0	8	0	0	0	0	0
f	0	0	0	0	0	10	0	0	0	0	0	0	0	0	0	7	0	0	1	2
g	0	0	0	0	0	0	10	0	0	0	0	0	0	0	0	0	8	1	1	0
h	0	0	0	0	0	0	1	9	0	0	0	0	0	0	0	0	2	6	2	0
i	0	0	0	0	0	0	1	0	8	1	0	0	0	0	0	0	1	1	6	2
j	0	0	0	0	0	0	0	0	0	10	0	0	0	0	0	0	0	1	1	8

Table 2: confusion matrices for LPCs (right) and MFCCs (left). Speaker in line i is identified as speaker in column j. (a .. e) : males and (f .. j) : females

In this table, it is interesting to see that the two methods are not complementary: the errors with MFCCs are the same as those for LPCs. As expected, confusions arise for females more than for males, females being always mistaken for other females and males for males.
These results are only preliminary. With an increased number of speakers, we intend to use two different data bases for males and females and different models.

4. CONCLUSION

The experimental system that we have realized has the following properties:
- simple algorithm
- fast learning
- fast identification
- maximal error rate of 3%
- short signal (3s): classical systems, with similar performances, use much longer durations (20 to 30s).

We think that these preliminary results demonstrate the interest of using connectionist models for speaker identification tasks. By mapping the speech problem to a pattern recognition problem, we have been able to provide a connectionist classifier leading to an error rate less than 3%, with a model based only on the statistical properties of the frames cloud.
Of course, many problems have yet to be solved: increase of the number of speakers, addition/suppression of a speaker, automatic control of the learning process, update of the references (because of the speaker adaptation).
We are presently working on a larger system, with an increased number of speakers, combining various connectionist algorithms.

ACKNOWLEDGEMENTS

This work was sponsored by the I.R.I.A.C Institut de Recherche en Intelligence Artificielle et Connexionnisme.

5. REFERENCES

[1] **E.Daudin, Y.Bennani, G.Chollet** *"Simulation de Technique de Vérification Automatique du Locuteur"* Centre International de Rencontres Mathématiques, Luminy 20-21 juin 1989, Variabilité et spécifité des locuteurs.
[2] **B.S.Atal.** *"Effectiveness of LPC characteristics of the speech wave for A.S.I and A.S.V"*, JASA-Vol.55.1974.
[3] **J. B.Attili.** *"On the development of a real-time text-independent speaker verification system."* Ph.D Rensselaer Polytechnic Institute, 1987.
[4] **J. B.Attili.** *"A TMS320C20 based real time, text-independent automatic speaker verification system."* ICASSP 1988.
[5] **P.Combescure.** *"20 listes de dix phrases phonétiquement équilibrées. "* Revue d'acoustique, n° 56, 34-38, 1981.
[6] **Efron.** *"Estimating the Error Rate in a Predictive Rule: Improvement on Cross-Validation"*, 316-331, JASA,78. 1983.
[7] **E.McDermott et S. Katagiri.** *"Shift-invariant, Multi-Category Phoneme Recognition Using Kohonen's LVQ2"* proc. of ICASSP, S3.1, 1989.
[8] **F. Fogelman-Soulié** *"Méthodes connexionnistes pour l'apprentissage"*. 2èmes Journées du PRC-GRECO Intelligence Artificielle, Toulouse, Teknea, 275-293, 1988.
[9] **F. Fogelman-Soulié, P. Gallinari, Y. Le Cun, S. Thiria.** *"Network learning"* In "Machine Learning", Y. Kodratoff, R. Michalski Eds. **Morgan Kaufmann**, to appear.
[10] **S. Furui.** *"Cepstral Analysis technique for automatic speaker verification."* IEEE Trans. on ASSP, vol 29 N°2, april 1981.
[11] **P. Gallinari, S. Thiria, F. Fogelman-Soulié.** *"Multilayer Perceptrons and Data Analysis"*. Second annual International Conference on Neural Networks, San Diego, I-391-401, 1988.
[12] **Y.Grenier.** *"Identification du locuteur et Adaptation au locuteur d'un système de Reconnaissance Phonémique"*. thèse de Docteur-Ingénieur, ENST-E-77005, Paris 1977.
[13] **Lachenbruch & Mickey.** *"Estimation of Error Rates in Discriminant Analysis"*.Technometics.1-11. 1968.
[14] **T.Kohonen.***"Self-Organization and Associative Memory "* (2nd Ed.), Springer, Berlin-Heidelberg-New York-Tokyo, 1988.
[15] **T. Kohonen.***"The neural phonetic typewriter"*. IEEE Computer, 11-22, march 1988.
[16] **T. Kohonen, G. Barna et R. Chrisley.** *"Statistical Pattern Recognition with neural networks: Benchmarking Studies "*. Second annual International Conference on Neural Networks, San Diego,IEEE, proc. of ICNN. vol. I 61-68, July 1988.
[17] **F.K.Soong, A. E. Rosenberg.** *"On the use of instantaneous and transitional spectral information in speaker recognition."* . IEEE Trans. on ASSP, vol.36 N°6, june 1988.
[18] **A. Waibel, T.Hanazawa, G. Hinton, K. Shikano et K. Lang** *"Phoneme Recognition: Neural Networks vs. Hidden Markov Models"*. Proc. of ICASSP. S3.3. 107-110. avril 1988.

AUTHOR INDEX